W9-BNY-923

THE HANDBOOK OF INFORMATION AND COMPUTER ETHICS

THE HANDBOOK OF INFORMATION AND COMPUTER ETHICS

Edited by

Kenneth Einar Himma and Herman T. Tavani

Published by John Wiley & Sons, Inc., Hoboken, New Jersey
Published simultaneously in Canada

Library of Congress Cataloging-in-Publication Data:

The handbook of information and computer ethics / edited by Kenneth Einar Himma and Herman T. Tavani.

p. cm.
Includes index.
ISBN 978-0-471-79959-7 (cloth)
1. Electronic data processing–Moral and ethical aspects. I. Himma, Kenneth Einar. II. Tavani, Herman T.
QA76.9.M65H36 2008
004.01'9–dc22

2007044568

Printed in the United States of America
10 9 8 7 6 5 4 3 2 1

For my wife, Maria Elias Sotirhos, and my nieces, Angela and Maria Katinas
KEH

In memory of my mother-in-law, Mary Abate
HTT

CONTENTS

FOREWORD

The publication of *The Handbook of Information and Computer Ethics* signifies a milestone in the field of computer ethics. The field began to emerge as a scholarly field in the late 1970s and early 1980s. Joseph Weizenbaum's *Computer Power and Human Reason* (1976) was the first extended work to draw attention to the potentially deep social implications of the new technology. During this period, privacy had been subject to a number of major studies, including Alan Westin and Michael Baker's *Data Banks in a Free Society* (1972). The first works by philosophers began to appear in the 1980s, and in 1985 Terrell Bynum published a special issue of *Metaphilosophy* pulling together these first works and making them more available to the philosophical community. That year, 1985, was also the year in which my own *Computer Ethics* was first published.

Perhaps it is an understatement to say that in the twenty-plus years since the appearance of these first works, the field of computer ethics has flourished enormously. Of course, the development of the field has gone hand-in-hand with the development of computer and information technology. In one of the seminal articles in the field, Jim Moor identifies malleability as a key feature of computers; that malleability has meant that computer and information technology has permeated almost every domain of human activity. And, of course, wherever the technology goes, ethical issues can be found. While the flourishing of the field of computer ethics is to be celebrated, growth inevitably means pressure to split the whole into parts. The topics that need to be addressed continue to expand, and perspectives from a wide range of disciplines are relevant. Thus, there is pressure for the field to become splintered into subfields (for example, with a distinction between computer ethics and information ethics); for scholars to become specialists in one subfield (for example, to choose to become an expert in privacy or intellectual property or professional ethics); or to have subfields merged into already existing fields such as media studies, business ethics, information sciences, etc.

In this context, the publication of *The Handbook of Information and Computer Ethics* is particularly important because it aims to keep the field whole. It is intended to provide an overview of the issues and controversies in a field that has become increasingly unwieldy. As a handbook, the volume defines the field as a whole; it identifies foundational issues, provides theoretical perspectives, and includes analyses of a range of applied and practical issues. The volume does this through chapters by individuals who have been working in the field from the beginning, as well as works by scholars who have come to the field more recently. For this reason, I applaud the efforts of Kenneth Himma and Herman Tavani and welcome the publication of *The Handbook of Information and Computer Ethics*.

DEBORAH G. JOHNSON

PREFACE

In the last 10 years, information and computer ethics has emerged as an important area of philosophical and social theorizing, combining conceptual, metaethical, normative, and applied elements. Interest in the area has increased dramatically in computer science departments, philosophy departments, communications departments, business schools, information and library schools, and law schools. Information ethics has become one of the most important areas of applied philosophy in terms of professional, student, and popular interest. Many of the most pressing new ethical issues we face have arisen in connection with the use and development of new information technologies. For example, debates about the ethics of online music file sharing have led academics and ordinary citizens to reconsider the arguments for the legitimacy of intellectual property protection. New developments in information technology threaten privacy in ways that could not have been imagined 50 years ago, raising new ethical issues about the rights to privacy and anonymity. The growing dependence of large-scale economies on the Internet creates new vulnerabilities that can be exploited by hackers, cybercriminals, and terrorists, raising novel ethical issues about computer intrusions and security.

The Handbook of Information and Computer Ethics responds to this growing professional interest in information ethics with 27 chapters that address both traditional and current issues in information and computer ethics research. Each chapter, written by one or more of the most influential information ethicists, explains and evaluates the most important positions and arguments on the respective issues. As a result, the *Handbook* reader will be able to come away from each chapter with an understanding of the major positions and arguments, their strengths and weaknesses, and the author's original take on the issue. In addition, each chapter not only contains useful summaries of the most important research on the topic but also makes an important new contribution to the literature, and ends with a bibliography that identifies the most important books and articles on the topic.

Because a number of very good anthologies on information and computer ethics already exist, one might ask: Why another book of readings on ethical aspects of information and computer technology? One justification for the book is that, as noted above, each chapter in the present volume is written in a style that conforms to the objectives of a handbook and thus provides the conceptual background that is often not found in papers comprising other volumes. Consider that many papers included in those volumes are compiled from disparate sources and, thus, can reflect various styles and diverse objectives. With one exception, every chapter in this volume is an original piece that was written specifically for the *Handbook*. As such, each paper provides an accessible but sophisticated overview of the most important positions and supporting arguments and objections, along with the author's state-of-the-art take on these positions, arguments, and objections.

Another justification for this book is that existing anthologies tend to be narrower in scope than *The Handbook of Information and Computer Ethics*. For example, many anthologies cover only a limited set of topics that affect one or more subfields of information ethics; as a result, these works often exclude some of the controversies and issues that arise in information ethics as a broader field of inquiry. Consider that some anthologies have focused on Internet- or cyber-specific issues involving information ethics,[1] while others have centered mainly on professional ethics issues affecting responsibility.[2] Other volumes are dedicated to information ethics concerns affecting specific topical areas such as privacy, security, and property.[3] Still other anthologies have focused on ethical aspects of information technology that converge with ethics-related concerns affecting medicine and genetics/genomics research.[4] And other anthologies are dedicated to the examination of ethical issues in information technology that intersect either with disciplines, such as philosophy, or with new or emerging fields, such as nanotechnology.[5] So even though there is no shortage of anthologies that examine ethical issues centering on these, and related, ethical aspects of information technology, none addresses the breadth of topics covered in the present handbook.

The *Handbook* is organized into six main parts, which cover a wide range of topics—i.e., from foundational concepts and methodological approaches in information ethics (at the theoretical level) to specific problem areas involving applied or practical ethical issues.

At the theoretical level, conceptual frameworks underlying topical areas such as intellectual property, privacy, and security are examined. These frameworks provide *Handbook* readers with some conceptual tools needed to analyze more systematically the kinds of issues examined in the chapters comprising the remaining sections of the book. At the practical level, a number of contemporary controversies ranging from professional-ethical issues to issues of responsibility, regulation, and access are examined. For example, these chapters examine controversies affecting open-source software, medical informatics and genetic research, cyber-conflict, risk assessment, the digital divide, information overload, e-mail spam, online file sharing, plagiarism, censorship and free speech, and so forth. Thus, *Handbook* readers will gain an

---

[1]See Langford, D. (Ed.). *Internet Ethics*. Macmillan, 2000; Baird, R., Reagan, R., and Ramsower, S., (Eds.) *Cyberethics*. Prometheus, 2000; Spinello, R. and Tavani, H. (Eds.). *Readings in CyberEthics*, 2nd ed. Jones and Bartlett, 2004.

[2]See Bynum, T. and Rogerson, S. (Eds.). *Computer Ethics and Professional Responsibility.* Blackwell, 2004.

[3]See Moore, A. (Ed.). *Information Ethics: Privacy, Property, and Power.* University of Washington Press, 2005; Himma, K. (Ed.). *Internet Security: Hacking, Counter Hacking, and Society.* Jones and Bartlett, 2007.

[4]See Goodman, K. (Ed.). *Ethics, Computing, and Medicine*. Cambridge, 1998; Tavani, H. (Ed.). *Ethics, Computing, and Genomics.* Jones and Bartlett, 2006.

[5]See Moor, J. and Bynum, T. (Eds.). *Cyberphilosophy.* Blackwell, 2002; Allhoff, F., Lin, P., Moor, J. and Weckert, J. (Eds.). *Nanoethics.* Wiley, 2007.

understanding of both the general frameworks and specific issues that define the fields of information and computer ethics.

ACKNOWLEDGMENTS

The editors would like to acknowledge the support they received from their respective institutions while composing the *Handbook*. Kenneth Himma's work on this project was partially supported by Seattle Pacific University through a Faculty Research Grant, which released him from teaching duties and made it possible for him to devote most of his efforts to this project. Herman Tavani also received institutional support in the form of a Summer Faculty Research and Professional Development Grant from Rivier College, during the summer of 2006, to work on the *Handbook*. The editors are very grateful to Seattle Pacific University and Rivier College, respectively, for the institutional support they received.

The editors are especially grateful to the contributing authors, without whom this volume would not exist. The contributors' willingness to revise drafts of their papers to comply with the specific objectives of this handbook is greatly appreciated. We also appreciate the extraordinary patience, as well as the ongoing cooperation and support, the contributors displayed throughout the long, and sometimes tedious, process required to complete this book.

We are also grateful to the editorial staff at Wiley, especially to Paul Petralia, Anastasia Wasko, and Whitney Lesch, for managing to keep the book on a reasonable schedule, despite some of the obstacles that we encountered along the way.

Finally, we wish to thank our spouses, and our families, for their unwavering support throughout this project. To them, we dedicate this book.

K. E. H.
H. T. T.
August, 2007

Contributors

Alison Adam, PhD, is Professor of Information Systems and Director of the Informatics Research Institute at the University of Salford, UK. Her recent publications include *Gender, Ethics and Information Technology* (Palgrave Macmillan, 2005).

Yeslam Al-Saggaf, PhD, is a Senior Lecturer in Information Technology at Charles Sturt University and a Research Fellow at the Centre for Applied Philosophy and Public Ethics. His research interests lie in the areas of online communities (both social and political) and the online political public sphere in the Arab world.

Alan Borning, PhD, is Professor in the Department of Computer Science and Engineering at the University of Washington, adjunct professor in the Information School, and Co-Director of the Center for Urban Simulation and Policy Analysis. His current research interests are in human–computer interaction and designing for human values, particularly as applied to land use, transportation, and environmental modeling.

Maria Canellopoulou-Bottis, PhD, is a Lecturer at the Information Science Department of the Ionian University, Greece. Her recent publications include *The Legal Protection of Databases* (2004) and *Information Law* (2004), in Greek, and numerous articles in Greek and foreign journals.

Philip Brey, PhD, is Professor of Philosophy and Director of the Center for Philosophy of Technology and Engineering Science at the University of Twente, The Netherlands. He is a member of the board of the International Society for Ethics and Information Technology and the author of numerous articles in philosophy of technology and computer and information ethics.

Elizabeth A. Buchanan, PhD, is Associate Professor and Director, Center for Information Policy Research, School of Information Studies, University of Wisconsin-Milwaukee. She is Co-Director of the International Society for Ethics and Information Technology (INSEIT), Chair, Ethics Working Group, Association of Internet Researchers (AoIR), and Chair, Intellectual Freedom Round Table, Wisconsin Library Association.

Terrell Ward Bynum, PhD, is Professor of Philosophy at Southern Connecticut State University and Director of the Research Center for Computing and Society. His recent publications include *Computer Ethics and Professional Responsibility* (Blackwell, 2004), coedited with Simon Rogerson, and *Cyberphilosophy: The*

Intersection of Philosophy and Computing (Blackwell, 2002), coedited with James Moor.

Rafael Capurro, PhD, is Professor of Information Science and Information Ethics at Stuttgart Media University (Germany) and Director of the International Center for Information Ethics (ICIE). His recent publications include *Localizing the Internet: Ethical Aspects in Intercultural Perspective* (Fink Munich, 2007), coedited with J. Fruehbauer and T. Hausmanninger, as well as numerous book chapters and journal articles.

Tony Clear, is Associate Head of School in the School of Computing and Mathematical Sciences at Auckland University of Technology, New Zealand. He edits a regular column in the *ACM SIGCSE Bulletin*, is on the editorial board of *Computer Science Education*, and has research interests in software risk assessment.

Dorothy E. Denning, PhD, is Professor of Defense Analysis at the Naval Postgraduate School. She is the author of *Information Warfare and Security* (Addison Wesley, 1999) and of numerous articles and book chapters relating to conflict in cyberspace and information security.

Charles Ess, PhD, is Distinguished Research Professor and Professor of Philosophy and Religion at Drury University. His recent publications include *Information Technology Ethics: Cultural Perspectives*, coedited with Soraj Hongladarom (Idea Group, 2007), and special issues of *Ethics and Information Technology* devoted to cross-cultural approaches to privacy and to Kantian approaches to topics in information ethics.

Don Fallis, PhD, is Associate Professor of Information Resources and Adjunct Associate Professor of Philosophy at the University of Arizona. His articles have appeared in the *Journal of Philosophy*, *Philosophical Studies*, *Library Quarterly*, and the *Journal of the American Society for Information Science and Technology*.

Luciano Floridi, PhD, holds the research chair in philosophy of information at the Department of Philosophy of the University of Hertfordshire and is Fellow of St. Cross College, Oxford University. His books include the *Blackwell Guide to the Philosophy of Computing and Information* (2004) and *The Philosophy of Information* (Oxford University Press, forthcoming).

Batya Friedman, PhD, is Professor in the Information School, an Adjunct Professor in the Department of Computer Science & Engineering, and Co-Director of the Value Sensitive Design Research Lab at the University of Washington. Her recent publications include the development of an open source privacy addendum (Ubicomp, 2006), the value-sensitive design of a corporation's groupware systems (GROUP, 2007), and numerous journal articles and book chapters.

Kenneth W. Goodman, PhD, is Associate Professor of Medicine and Philosophy at the University of Miami, where he directs the Bioethics Program. He has written extensively about ethics and health informatics, including hospital, public health, and genetics applications.

Don Gotterbarn, PhD, Director of the Software Engineering Ethics Research Institute at East Tennessee State University, has been active in computer ethics for more than 20 years. Most recently, his work has focused on ethical decision support methodologies.

Frances S. Grodzinsky, PhD, is Professor of Computer Science and Information Technology at Sacred Heart University and Co-Director of the Hersher Institute of Ethics. Her recent publications include numerous book chapters and journal articles.

Kenneth Einar Himma, PhD, JD, is Associate Professor of Philosophy at Seattle Pacific University and formerly taught at the University of Washington in the Information School, Law School, and Philosophy Department. He has published more than a hundred journal articles and is on the editorial boards of several journals on information technology and ethics.

Deborah G. Johnson, PhD, is the Anne Shirley Carter Olsson Professor of Applied Ethics and Chair of the Department of Science, Technology, and Society at the University of Virginia. Her anthology, *Technology & Society: Engineering our SocioTechnical Future* (coedited with J. Wetmore), is forthcoming from MIT Press in 2008, and she is currently working on the fourth edition of *Computer Ethics* (forthcoming from Prentice Hall).

Peter H. Kahn, Jr., PhD, is Associate Professor of Psychology at the University of Washington. His books include *The Human Relationship with Nature: Development and Culture* (MIT Press, 1999) and *Children and Nature: Psychological, Sociocultural, and Evolutionary Investigations* (MIT Press, 2002), and his research publications have appeared in such journals as *Child Development*, *Developmental Psychology*, *Human-Computer Interaction*, and *Journal of Systems Software*.

Choon-Tuck Kwan is a Lecturer at the Auckland University of Technology (New Zealand) and Manager of the New-Zealand-based Software Engineering Practice Improvement Alliance. He was formerly a senior IT manager in a large governmental Statutory Board in Singapore.

David M. Levy, PhD, is Professor in the Information School of the University of Washington. He is the author of *Scrolling Forward: Making Sense of Documents in the Digital Age* (Arcade, 2001).

Antonio Marturano, PhD, an Adjunct Professor of Business Ethics at the Faculty of Economics, Sacred Heart Catholic University of Rome, has held several academic posts at universities in the United States, United Kingdom, and Italy. His main research area is applied ethics, with a special focus on ethical and legal problems spanning genetics and information technology.

Kay Mathiesen, PhD, teaches courses on information ethics and policy at the School of Information Resources and Library Science, University of Arizona. Her articles have appeared in journals such as *Library Quarterly*, *Computers and Society*, the *Annual Review of Law and Ethics*, and *Business Ethics Quarterly*.

Keith W. Miller, PhD, is Professor of Computer Science at the University of Illinois at Springfield and Editor of *IEEE Technology and Society Magazine*. His research interests include computer ethics, software testing, and computer science education.

James H. Moor, PhD, is Professor of Philosophy at Dartmouth College, an Adjunct Professor with the Centre for Applied Philosophy and Public Ethics at The Australian National University, and President of the International Society for Ethics and Information Technology. His ethical writings are on computer ethics, artificial intelligence, and nanotechnology.

Adam D. Moore, PhD, is Associate Professor in the Philosophy Department and the Information School at the University of Washington. He is the author of *Intellectual Property and Information Control* (Transaction Pub. University, hardback 2001, paperback 2004), and editor of *Intellectual Property: Moral, Legal, and International Dilemmas* (Rowman and Littlefield, 1997) and *Information Ethics: Privacy, Property, and Power* (The University of Washington Press, 2005).

John Snapper, PhD, is Associate Professor of Philosophy and an Associate of the Center for the Study of Ethics in the Professions. He is coeditor of *Owning Scientific and Technical Information* (Rutgers Press, 1989) and *Ethical Issues in the Use of Computers* (Wadsworth, 1985) and author of a numerous journal articles on related subjects.

Richard A. Spinello, PhD, is Associate Research Professor in the Carroll School of Management at Boston College. He has written and edited seven books on ethics and public policy, including *CyberEthics: Morality and Law in Cyberspace* (Jones and Bartlett, third edition, 2006) and *Intellectual Property Rights in a Networked World* (Idea Group, 2005).

Bernd Carsten Stahl, PhD, is Reader in Critical Research in Technology in the Centre for Computing and Social Responsibility at De Montfort University, Leicester, UK. As the Editor-in-Chief of the *International Journal of Technology and Human Interaction*, his interests cover philosophical issues arising from the intersections of business, technology, and information.

Herman T. Tavani, PhD, is Professor of Philosophy at Rivier College, a Lecturer in the Carroll School of Management at Boston College, and a visiting scholar/ethicist in the Department of Environmental Health at the Harvard School of Public Health. His recent books include *Ethics and Technology* (Wiley, second edition 2007) and *Ethics, Computing, and Genomics* (Jones and Bartlett, 2006).

Jeroen van den Hoven, PhD, is Professor of Moral Philosophy at Delft University of Technology (the Netherlands). He is the Editor-in-Chief of *Ethics and Information Technology* (Springer), Scientific Director of the Centre of Excellence for Ethics and Technology of the three technical universities in the Netherlands, and a member of the High Level Advisory Group on ICT (ISTAG) of the European Commission in Brussels.

Anton Vedder, PhD, is Associate Professor of Ethics and Law at the Tilburg Institute for Law, Technology, and Society of Tilburg University (the Netherlands). His recent publications include *NGO Involvement in International Governance and Policy: Sources of Legitimacy* (Martinus Nijhoff, 2007) and several book chapters and journal articles on privacy, reliability of information, accountability, and legitimacy of newly emerging governance regimes.

Kathleen A. Wallace, PhD, is Professor of Philosophy at Hofstra University. Her recent publications include "Educating for Autonomy: Identity and Intersectional Selves," in *Education for a Democratic Society* (Rodopi Press, 2006), "Moral Reform, Moral Disagreement and Abortion" (*Metaphilosophy*, 2007), and "Morality and the Capacity for Symbolic Cognition," *Moral Psychology* (MIT Press, 2007).

John Weckert, PhD, is Professor of Computer Ethics and Professorial Fellow at the Centre for Applied Philosophy and Public Ethics, Charles Sturt University, Australia. He is founding editor-in-chief of the journal *Nanoethics: Ethics for Technologies that Converge at the Nanoscale,* and author of numerous book chapters and journal articles.

Marty J. Wolf, PhD, is Professor of Computer Science and the Computer Science Program Coordinator at Bemidji State University in Minnesota. He has over 15 years experience using and administering Linux and has published numerous book chapters and journal articles in areas ranging from graph theory to computer ethics.

INTRODUCTION

KENNETH EINAR HIMMA and HERMAN T. TAVANI

As noted in the Preface to this volume, *The Handbook of Information and Computer Ethics* covers a wide range of topics and issues. The 27 chapters that comprise this work are organized into six main parts: I. Foundational Issues and Methodological Frameworks; II. Theoretical Issues Affecting Property, Privacy, Anonymity, and Security; III. Professional Issues and the Information-Related Professions; IV. Responsibility Issues and Risk Assessment; V. Regulatory Issues and Challenges; and VI. Access and Equity Issues.

I FOUNDATIONAL ISSUES AND METHODOLOGICAL FRAMEWORKS

Part I, comprising four chapters, opens with Luciano Floridi's examination of some key foundational concepts in information ethics. Floridi points out that the expression "information ethics," introduced in the 1980s, was originally used as a general label to discuss issues regarding information (or data) confidentiality, reliability, quality, and usage. He also notes that "information ethics" has since come to mean different things to different researchers working in a variety of disciplines, including computer ethics, business ethics, medical ethics, computer science, the philosophy of information, and library and information science. Floridi is perhaps best known among computer ethicists for his influential methodological (and metaethical) framework, which he calls *Information Ethics* or *IE*. He contrasts his framework with traditional views that have tended to view IE as either an "ethics of informational resources," an "ethics of informational products," or an "ethics of the informational environment." Floridi argues that his alternative view of IE, as a "macroethics," is superior to the various microethical analyses of IE that have been suggested.

Floridi's discussion of foundational issues in IE is followed by Terrell Ward Bynum's chapter, "Milestones in the History of Information Ethics." Bynum is generally considered to be one of the "pioneers" in computer ethics, helping to establish the field as an independent area of applied ethics in the 1980s. In Chapter 2, Bynum argues that the origin of computer and information ethics can be traced to the work of philosopher/scientist Norbert Wiener, who, during World War II, worked with a group of scientists and engineers on the invention of digital computers and radar. His chapter begins with a discussion of Wiener's "powerful foundation" for information and computer ethics, and then it describes a number of additional "milestones" in the history of what Bynum describes as a "new and vital branch of ethics."

Next, Jeroen van den Hoven examines some methodological issues in his chapter, "Moral Methodology and Information Technology." One question that has been considered by some theoreticians in the fields of information and computer ethics is whether a new and distinct methodology is needed to handle the kinds of ethical issues that have been generated. Van den Hoven suggests that we need a methodology that is "different from what we have seen thus far in applied ethics," but which does not call for "cataclysmic re-conceptualizations." He begins with an overview of some of the main methodological positions in applied ethics that are relevant for computer ethics, before sketching out his proposed method that aims at making moral values a part of technological design in the early stages of its development. This method assumes, as van den Hoven notes, that "human values, norms, moral considerations can be imparted to the things we make and use (technical artefacts, policy, laws and regulation, institutions, incentive structures, plans)."

Part I closes with Batya Friedman, Peter Kahn, and Alan Borning's chapter, "Value Sensitive Design and Information Systems." The authors note that *value sensitive design* (VSD) is a theoretically grounded approach to the design of technology that accounts for human values in a "principled and comprehensive manner throughout the design process." It also includes a tripartite methodology, consisting of conceptual, empirical, and technical investigations. In explicating VSD, Friedman, Kahn, and Borning consider three case studies: one concerning information and control of web browser cookies (implicating the value of informed consent); a second study concerning using high-definition plasma displays in an office environment to provide a "window to the outside world" (implicating the values of physical and psychological well-being and privacy in public spaces); and a third study concerning an integrated land use, transportation, and environmental simulation system to support public deliberation and debate on major land use and transportation decisions (implicating the values of fairness, accountability, and support for the democratic process). In the concluding section of their chapter, the authors offer some practical suggestions for how to engage in VSD.

II THEORETICAL ISSUES AFFECTING PROPERTY, PRIVACY, ANONYMITY, AND SECURITY

Part II comprises four chapters that examine conceptual and theoretical frameworks in information ethics. Unlike the chapters in Part I, however, they examine some topic- or theme-specific frameworks that underlie many of the practical issues considered in the remaining parts of the Handbook. Specifically, the chapters in Part II examine theoretical and conceptual aspects of intellectual property, informational privacy, online anonymity, and cyber security. In the opening chapter, Adam Moore discusses three different kinds of justifications for intellectual property (IP), also noting that we need to be careful not to confuse moral claims involving IP with legal ones. His chapter begins with a brief sketch of Anglo-American and Continental systems of IP that focuses on legal conceptions and rights. Moore then examines arguments for the personality-based, utilitarian, and Lockean views of property. He concludes that there are justified moral claims to intellectual works, that is, "claims that are strong enough to warrant legal protection."

Moore's analysis of IP is followed by Herman Tavani's examination of some key concepts, theories, and controversies affecting informational privacy. Beginning with an overview of the concept of privacy in general, Tavani distinguishes among four distinct kinds of privacy: physical, decisional, psychological, and informational privacy. He then evaluates some classic and contemporary theories of informational privacy before considering the impact that some specific information technologies (such as cookies, data mining, and RFID techologies) have had on four subcategories of informational privacy: consumer privacy, medical privacy, employee privacy, and location privacy. His chapter closes with a brief examination of some recent proposals for framing a comprehensive informational-privacy policy.

Next, Kathleen Wallace examines the concept of anonymity in her chapter, "Online Anonymity." Wallace points out that anonymity and privacy are closely related, with anonymity "being one means of ensuring privacy." She also notes that anonymity can be brought about in a variety of ways and that there are many purposes, both positive and negative, that anonymity could serve. For example, on the positive side, it can promote free expression and exchange of ideas, and it can protect someone from undesirable publicity. On the negative side, however, anonymity can facilitate hate speech with no accountability, as well as fraud or other criminal activity. Wallace believes that there are two thoughts regarding anonymity as a "byproduct" that are worth distinguishing; it could be the "byproduct of sheer size as when one is among a throng of people who don't know one another" or the "byproduct of complex social organization."

Part II concludes with Kenneth Himma's chapter, "Ethical Issues Involving Computer Security: Hacking, Hacktivism, and Counterhacking." Himma considers whether and to what extent various types of unauthorized computer intrusions by private persons and groups (as opposed to state agents and agencies) are morally permissible. After articulating a *prima facie* general case against these intrusions, Himma considers intrusions motivated by malicious intentions and by certain benign intentions, such as the intent to expose security vulnerabilities. The final sections of his chapter consider controversies associated with "hacktivism" and "counterhacking" (or hack backs). Himma's chapter can also be read in connection with Dorothy Denning's chapter on the ethics of cyber conflict.

III PROFESSIONAL ISSUES AND THE INFORMATION-RELATED PROFESSIONS

Part III comprises five chapters that examine a diverse set of professional-ethics issues affecting the information and information-related professions—for example, concerns that affect library professionals, software engineering/development professionals, (online) research professionals, medical and healthcare professionals, and business professionals. It opens with Kay Mathiesen and Don Fallis' chapter, "Information

Ethics and the Library Profession." Mathiesen and Fallis note that, in general, the role of the professional librarian is to provide access to information, but they also point out that librarians vary in their activities depending on the goal of such access, and on whether they are corporate librarians, academic librarians, or public librarians. The authors begin their analysis by considering the "mission" of the librarian as an "information provider" and then focus on some of the issues that arise in relation to "the role of the librarian as an information provider." In particular, the authors focus on questions pertaining to the "selection and organization of information," which, in turn, raises concerns having to do with "bias, neutrality, advocacy, and children's rights to access information."

Mathiesen's and Fallis's analysis of ethical challenges facing librarians and the library profession is followed by an examination of controversies affecting open source software development and the computing profession in Frances Grodzinsky's and Marty Wolf's chapter, "Ethical Interest in Free and Open Source Software." Grodzinsky and Wolf begin by comparing free software (FS) and open source software (OSS), and by examining the history, philosophy, and development of each. Next, they explore some important issues that affect the ethical interests of all who use and are subject to the influences of software, regardless of whether that software is FS or OSS. The authors also argue that the distinction between FS and OSS is one that is philosophically and socially important. Additionally, they review some issues affecting the autonomy of OSS software developers and their "unusual professional responsibilities."

Next, Elizabeth Buchanan and Charles Ess examine some professional-ethical issues affecting online research in their chapter, "Internet Research Ethics: The Field and its Critical Issues." Buchanan and Ess begin by noting that Internet research ethics (IRE) is an emerging multi- and interdisciplinary field that systematically studies the ethical implications that arise from the use of the Internet as "a space or locale of, and/or tool for, research." The authors believe that no one discipline can claim IRE as its own. Because Internet research is undertaken from a wide range of disciplines, they argue that IRE builds on the research ethics traditions developed for medical, humanistic, and social science research. For Buchanan and Ess, a "central challenge for IRE is to develop guidelines for ethical research that aim toward objective, universally recognized norms, while simultaneously incorporating important disciplinary differences in research ethics." The authors consider and review a range of the most common ethical issues in IRE, and they offer some suggestions for possible resolutions of specific ethical challenges.

Buchanan and Ess's analysis of IRE-related ethical issues is followed by Kenneth Goodman's chapter, "Health Information Technology: Challenges in Ethics, Science, and Uncertainty." Goodman notes that the use of information technology in the health professions has introduced numerous ethical issues and professional challenges. The three principal issues that Goodman examines in the context of these challenges are (1) privacy and confidentiality; (2) the use of decision support systems; and (3) the development of personal health records.

Part III closes with Bernd Carsten Stahl's examination of some business-related ethical concerns in his chapter, "Ethical Issues of Information and Business." Stahl begins his analysis with a brief definition of the concept of business and then discusses some specific business-ethics issues affecting privacy/employee surveillance, intellectual property, globalization, and digital divides. He considers various approaches to these and related business-ethical issues, drawing on some of the debates in computer and information ethics. Stahl notes that in these debates, different "sets of ethical discourse" have been used. He also notes that in some instances, these "ethical discourses" overlap and have "the potential to inform each other." Stahl's chapter aims at establishing a link between these discourses.

IV RESPONSIBILITY ISSUES AND RISK ASSESSMENT

The five chapters that make up Part IV examine a wide range of topics, each of which touches on one or more aspects of responsibility and risk involving information technology. In the opening chapter, "Responsibilities for Information on the Internet," Anton Vedder begins by noting that issues involving responsibility for Internet service providers (ISPs) are much broader in scope than they are sometimes portrayed in the research literature, where the emphasis has tended to be more narrowly on concerns affecting accountability with regard to illegal content. He then examines some issues affecting the responsibilities involved in the possible negative impact of "the dissemination of information" on the Internet. Here, he focuses mainly on three parties: (1) those who put forward information on the Internet, that is, the *content providers*; (2) the organizations that provide the infrastructure for the dissemination of that information — the *ISPs*; and (3) the receivers or users of the information, that is, the *third parties*.

Vedder's analysis of responsibility for the dissemination of information on the Internet is followed by Philip Brey's chapter, "Virtual Reality and Computer Simulation." Brey argues that virtual reality and computer simulation have not received much attention from ethicists, including ethicists in the computing profession, and that this relative neglect is unjustified because of the important ethical questions that arise. He begins his chapter by describing what virtual reality and computer simulations are and then describes some current applications of these technologies. Brey then discusses the ethics of three distinct aspects of virtual reality: (1) representation in virtual reality and computer simulations, (2) behavior in virtual reality, and (3) computer games. He concludes with a discussion of issues affecting responsibility, such as, responsibility in the development and professional use of virtual reality systems and computer simulations.

Next, Antonio Marturano examines some issues in genetic research that overlap with questions in information ethics. In his chapter, "Genetic Information: Epistemological and Ethical Issues," Marturano first analyzes some basic information-related concepts of molecular biology and then considers the ethical consequences of their misuse. He notes that genetics has utilized many concepts from informatics and that these concepts are used in genetics at different, but related, levels. At the most basic level, for example, genetics has taken the very notion of *information* — central to the field of informatics — to explain the mechanisms of life. Marturano notes that some authors have questioned the application of informational concepts in genetics. He also believes that it is important to understand the way the information-related concepts of molecular biology are interpreted to understand the reason why their "incorrect application—and consequent rhetorical use by geneticists—turns into an ethical failure." In this sense, Marturano's chapter is also concerned with issues affecting responsibility and the use of informational concepts.

In the next chapter, Dorothy Denning examines some ethical aspects of "cyber conflict." Denning believes that there are three areas of cyber conflict where the ethical issues are problematic. The first is "cyber warfare at the state level," when conducted in the interests of national security. One of the questions raised in this context is whether it is ethical for a state to penetrate or disable the computer systems of an adversary state that has threatened its territorial or political integrity. The second area involves "nonstate actors," whose cyber attacks are politically or socially motivated. This domain of conflict is often referred to as "hacktivism," the convergence of hacking with activism. Denning notes that if the attacks are designed to be "sufficiently destructive as to severely harm and terrorize civilians," they become "cyberterrorism" —the integration of cyber attacks with terrorism. The third area involves the "ethics of cyber defense," particularly what is called "hack back," "strike back," or "active response." If a system is under cyber attack, can the system administrators attack back to stop it? What if the attack is coming from computers that may themselves be victims of compromise? Since many attacks are routed through chains of "compromised machines," can a victim "hack back" along the chain to determine the source? Denning's chapter, which raises questions about responsibility and risk issues affecting cyber conflict, can also be read in conjunction with Ken Himma's analysis of security-related issues in Chapter 8.

In the closing chapter of Part V, "A Practical Mechanism for Ethical Risk Assessment—A SoDIS Inspection," Don Gotterbarn, Tony Clear, and Choon-Tuck Kwan examine some specific issues and concerns involving risk analysis. The authors begin by noting that although the need for high quality software may be obvious, information systems are "frequently plagued by problems that continue to occur in spite of a considerable amount of attention to the development and applications of certain forms of risk assessment." They claim that the narrow form of risk analysis that has been used, with its limited understanding of the scope of a software project and information systems, has contributed to significant software failures. Next, the authors introduce an expanded risk analysis process, which goes beyond the concept of "information system risk" to include social, professional, and ethical risks that lead to software failure. They point out that using an expanded risk analysis will enlarge the project scope considered by software developers.

V REGULATORY ISSUES AND CHALLENGES

Part V includes five chapters that examine a diverse set of issues and challenges affecting the regulation of information. It opens with John Weckert and Yeslam Al-Saggaf's chapter, "Regulation and Governance on the Internet," which raises the question: What, if anything, on the Internet should be governed or regulated? Weckert and Al-Saggaf note that we live in a world where people misbehave, and that for groups and societies to function satisfactorily, some restrictions on behavior are required. They also note that even where there is no malicious intent, there can be a need for some centralized body or perhaps "decentralized bodies" to coordinate activities.

Weckert and Al-Saggaf's analysis is followed by David Levy's chapter, "Information Overload." Levy first provides a preliminary definition of information overload and then identifies some of the questions surrounding it. He also discusses the history of the English phrase "information overload" and shows how industrialization and "informatization" prepared the ground for its emergence. In the closing section of his chapter, Levy explores some of the consequences, both practical and ethical, of overload and he considers what can be done in response.

Next, Keith Miller and James Moor examine controversies associated with spam in their chapter "E-mail Spam." The authors begin their analysis with a short history of spam, and they note that not every unwanted e-mail can be defined as spam. Miller and Moor suggest a "just consequentialist" approach to controversies involving e-mail spam — an approach that takes into account several different characteristics that help to differentiate spam from other e-mails. The authors conclude by noting that while the "struggle against unwanted e-mails" will likely continue, ethical analysis can be useful in analyzing spam-related issues provided that ethicists are careful to look at "individual stakeholders as well as systematic stakeholders" (i.e., both micro- and macro-level issues). They also argue that ethical analysis should start with a "clear exposition" of the characteristics of the e-mails that will be considered "spam."

Miller and Moor's analysis is followed by John Snapper's discussion of plagiarism in his chapter, "Plagiarism: What, Why, and If." Snapper defines plagiarism as an "expression that incorporates existing work either without authorization or without documentation, or both," and he points out that plagiarism can occur irrespective of possible copyright violation. Drawing some useful distinctions between plagiarism and copyright violation, Snapper shows how the two can occur simultaneously in some cases but are completely independent in others.

Part V concludes with Richard Spinello's chapter, "Intellectual Property: Legal and Moral Challenges of Online File Sharing." Spinello asks whether we should hold companies such as Napster, Grokster, or BitTorrent morally accountable for the direct infringement of their users, particularly if they intentionally design the code to enable the avoidance of copyright liability. He presents the conflicting arguments on both sides of this provocative debate. Although he focuses primarily on the ethical dimension of this controversy, Spinello claims that we cannot neglect the complex and intertwined legal issues. Taking this point into account, he then discusses the recent *MGM v. Grokster* (2005) case where both kinds of these issues have surfaced.

VI ACCESS AND EQUITY ISSUES

Part VI, the final section of the Handbook, includes four chapters that examine controversies affecting either access or equity, or both, with respect to information technology. It opens with Kay Mathiesen's chapter, "Censorship and Access to Expression," which can also be read in conjunction with some of the chapters in Part V that examine regulatory issues and challenges. But since Mathiesen's chapter includes an important analysis of issues affecting access to information, within her broader discussion of censorship, we decided to include it here. Mathiesen argues that the term "censorship" is commonly used in ways that "go much beyond the strict confines of First Amendment law." She also believes that philosophers, even those who have written much of "freedom of expression," have not tried to provide a conceptual analysis of censorship itself. Mathiesen tries to fill in this gap by providing an acceptable definition of censorship.

Mathiesen's analysis is followed by Alison Adam's examination of access issues affecting gender in her chapter, "The Gender Agenda in Computer Ethics." Adam's chapter is concerned with two interrelated questions: (1) What gender issues are involved in computer ethics? (2) What contribution may feminist ethics offer computer ethics? After briefly introducing the topic of feminist ethics, she reviews existing research on gender and computer ethics. Adam believes that this research falls into two main categories: (i) empirical comparisons of computer ethics decision-making by men and women; and (ii) other aspects of gender and computing that have been considered in ethical terms in the literature (which, she notes, usually involve a consideration of the low numbers of women in computing). She then identifies a number of gaps where extended discussion from a gender perspective would benefit several current problem areas within the purview of contemporary computer ethics; these include topics such as cyberstalking and hacking. In the concluding section of her chapter, Adam speculates that a gender analysis of computer ethics from the perspective of theoretical development of feminist ethics may enable the framing of "the discussion on 'cyberfeminism' as a possible locus for a feminist computer ethics."

Next, Maria Canellopoulou-Bottis and Kenneth Himma examine a different set of access issues affecting information technology in their chapter, "The Digital Divide: A Perspective for the Future." Bottis and Himma argue that the digital divide is not any one particular "gap" between rich and poor or between local and global, but rather includes a "variety of gaps believed to bear on the world's inequitable distribution of resources." They argue that there is a comparative lack of meaningful access to information and communication technologies (ICTs), which can be viewed in terms of several kinds of "gaps": (1) a gap in *access to the ICTs* themselves; (2) a gap in *having the skills needed* to use these technologies; (3) a gap between rich and poor in their *ability to access information* needed to compete in a global economy; and (4) a gap in *education* that translates into a "gap in abilities to process and absorb information." The authors also point out that there are "nondigital gaps" that contribute to the distribution of resources. Himma and Bottis believe that the moral importance of the digital divide as a problem that needs to be addressed is linked to "inequalities between the rich and the poor — especially between wealthy nations and nations in absolute poverty."

In the final chapter of Part VI, and of the *Handbook*, Rafael Capurro examines some intercultural issues in information ethics in his chapter, "Intercultural Information Ethics." He begins with an examination of the foundational debate of morality in general, which is addressed within the background of continental European philosophy (but also "with hints to Eastern traditions"). Next, Capurro presents some ethical questions about the impact of information and communication technologies on different cultures in Asia and the Pacific, Latin America and the Caribbean, Africa, Australia, and Turkey. Then, he addresses special issues such as privacy, intellectual property, online communities, "governmentality," gender issues, mobile phones, health care, and the digital divide.

We believe that the 27 chapters comprising *The Handbook of Information and Computer Ethics* address most of the rich and diverse issues that arise in and, in effect, define the field of information/computer ethics. We hope that the *Handbook* readers will discover for themselves why this field warrants serious attention by both professionals and the general public and why it is becoming one of the most important fields of applied ethics.

PART I

FOUNDATIONAL ISSUES AND METHODOLOGICAL FRAMEWORKS

CHAPTER 1

Foundations of Information Ethics

LUCIANO FLORIDI

1.1 INTRODUCTION

We call our society "the information society" because of the pivotal role played by intellectual, intangible assets (knowledge-based economy), information-intensive services (business and property services, communications, finance, and insurance), and public sectors (education, public administration, health care). As a social organization and way of life, the information society has been made possible by a cluster of information and communication technologies (ICTs) infrastructures. And as a full expression of *techne*, the information society has already posed fundamental ethical problems, whose complexity and global dimensions are rapidly growing and evolving. Nowadays, a pressing task is to formulate an information ethics that can treat the world of data, information, and knowledge,[1] with their relevant life cycles (including creation, elaboration, distribution, communication, storage, protection, usage, and possible destruction), as a new environment, the *infosphere*,[2] in which humanity is and will be flourishing. An information ethics should be able to address and solve the ethical challenges arising in the infosphere.

The last statement is more problematic than it might seem at first sight. As we shall see in some detail in the following sections, in recent years, "Information Ethics" (IE) has come to mean different things to different researchers working in a variety of disciplines, including computer ethics, business ethics, medical ethics, computer

[1]For this distinction, see Floridi (1999b).

[2]*Infosphere* is a neologism I coined years ago (see, e.g., Floridi (1999b) or Wikipedia) based on "biosphere," a term referring to that limited region on our planet that supports life. It denotes the whole informational environment constituted by all informational entities (thus including informational agents as well), their properties, interactions, processes, and mutual relations. It is an environment comparable to, but different from, cyberspace (which is only one of its subregions, as it were), since it also includes offline and analogue spaces of information.

The Handbook of Information and Computer Ethics, Edited by Kenneth Einar Himma and Herman T. Tavani

science, the philosophy of information, social epistemology ICT studies, and library and information science. This is not surprising. Given the novelty of the field, the urgency of the problems it poses, and the multifarious nature of the concept of information itself and of its related phenomena, perhaps a Babel of interpretations was always going to be inevitable.[3] It is, however, unfortunate, for it has generated some confusion about the specific *nature*, *scope*, and *goals* of IE. Fortunately, the problem is not irremediable, for a unified approach can help to explain and relate the main senses in which IE has been discussed in the literature. This approach will be introduced in the rest of this section. Once it is outlined, I shall rely on it in order to reconstruct three different approaches to IE, in Sections 1.2–1.4. These will then be critically assessed in Section 1.5. In Section 1.6, I will show how the approaches can be overcome by a fourth approach, which will be qualified as macroethical. In Section 1.7 two main criticisms, often used against IE as a macroethical theory, are discussed. Section 1.8 concludes this chapter with some brief, general considerations.

The approach mentioned above is best introduced schematically and by focusing our attention on a moral agent A. ICTs affect an agent's moral life in many ways. Recently (Floridi, forthcoming), I suggested that these may be schematically organized along three lines (see Fig. 1.1).

Suppose our moral agent A is interested in pursuing whatever she considers her best course of action, given her predicament. We shall assume that A's evaluations and interactions have some moral value, but no specific value needs to be introduced at this stage. Intuitively, A can avail herself of some information (information as a *resource*) to generate some other information (information as a *product*) and, in so doing, affect her informational environment (information as *target*). This simple model, summarized in Fig. 1.1, may help one to get some initial orientation in the multiplicity of issues belonging to Information Ethics. I shall refer to it as the RPT model.

The RPT model is useful to explain, among other things, why any technology that radically modifies the "life of information" is bound to have profound moral implications for any moral agent. Moral life is a highly information-intensive activity, and ICTs, by radically transforming the informational context in which moral issues arise, not only add interesting new dimensions to old problems, but may lead us to rethink, methodologically, the very grounds on which our ethical positions are based.[4]

At the same time, the model rectifies an excessive emphasis occasionally placed on specific technologies (this happens most notably in *computer* ethics), by calling our attention to the more fundamental phenomenon of information in all its varieties and long tradition. This was also Wiener's position,[5] and it might be argued that the various difficulties encountered in the conceptual foundations of information and computer ethics are arguably connected to the fact that the latter has not yet been recognized as primarily an environmental ethics, whose main concern is (or should be) the ecological

[3]On the various senses in which "information" may be understood see Floridi (2005a).

[4]For a similar position in computer ethics see Maner (1996) on the so-called "uniqueness debate" see Floridi and Sanders (2002a) and Tavani (2002).

[5]The classic reference here is to Wiener (1954). Bynum (2001) has convincingly argued that Wiener may be considered as one of the founding fathers of information ethics.

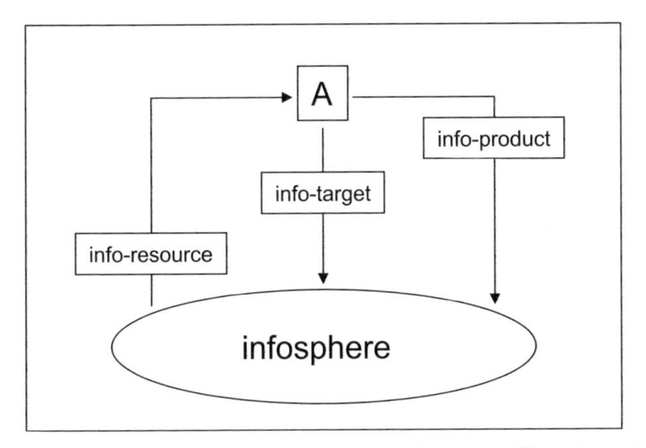

FIGURE 1.1 The "External" R(esource) P(roduct) T(arget) model

management and well-being of the infosphere (see Floridi and Sanders (2002b) for a defense of this position).

Since the appearance of the first works in the eighties,[6] Information Ethics has been claimed to be the study of moral issues arising from one or another of the three distinct "information arrows" in the RPT model. We are now ready to map the different approaches to IE by following each arrow.

1.2 THE FIRST STAGE: IE AS AN ETHICS OF INFORMATIONAL RESOURCES

According to Froehlich (2004),[7] the expression "information ethics" was introduced in the 1980s by Koenig et al. (1981) and Hauptman (1988), who then went on to establish the *Journal of Information Ethics* in 1992. It was used as a general label to discuss issues regarding information (or data) confidentiality, reliability, quality, and usage. Not surprisingly, the disciplines involved were initially library and information science and business and management studies. They were only later joined by information technologies studies.

It is easy to see that this initial interest in information ethics was driven by concern about information as a resource that should be managed efficiently, effectively, and fairly. Using the RPT model, this meant paying attention to the crucial role played by information as something extremely valuable for A's evaluations and actions, especially in moral contexts. Moral evaluations and actions have an epistemic component, as A may be expected to proceed "to the best of her information," that is, A may be expected to avail herself of whatever information she can muster, in order to reach (better) conclusions about what can and ought to be done in some given

[6]An early review is provided by Smith (1996).

[7]For a reconstruction of the origins of IE see also Capurro (2006).

circumstances. Socrates already argued that a moral agent is naturally interested in gaining as much valuable information as the circumstances require, and that a well-informed agent is more likely to do the right thing. The ensuing "ethical intellectualism" analyzes evil and morally wrong behavior as the outcome of deficient information. Conversely, A's moral *responsibility* tends to be directly proportional to A's degree of information: any decrease in the latter usually corresponds to a decrease in the former. This is the sense in which information occurs in the guise of judicial evidence. It is also the sense in which one speaks of A's informed decision, informed consent, or well-informed participation. In Christian ethics, even the worst sins can be forgiven in the light of the sinner's insufficient information, as a counterfactual evaluation is possible: had A been properly informed, A would have acted differently and hence would not have sinned (Luke 23:44). In a secular context, Oedipus and Macbeth remind us how the mismanagement of informational resources may have tragic consequences.[8]

From a "resource" perspective, it seems that the moral machine needs information, and quite a lot of it, to function properly. However, even within the limited scope adopted by an analysis based solely on information as a resource, care should be exercised lest all ethical discourse is reduced to the nuances of higher quantity, quality, and intelligibility of informational resources. The more the better is not the only, nor always the best, rule of thumb, for the (sometimes explicit and conscious) withdrawal of information can often make a significant difference. A may need to lack (or preclude herself from accessing) some information in order to achieve morally desirable goals, such as protecting anonymity, enhancing fair treatment, or implementing unbiased evaluation. Famously, Rawls' "veil of ignorance" exploits precisely this aspect of information-as-a-resource, in order to develop an impartial approach to justice (Rawls, 1999). Being informed is not always a blessing and might even be morally wrong or dangerous.

Whether the (quantitative and qualitative) presence or the (total) absence of information-as-a-resource is in question, it is obvious that there is a perfectly reasonable sense in which Information Ethics may be described as the study of the moral issues arising from "the triple A": *availability*, *accessibility*, and *accuracy* of informational resources, independently of their format, kind, and physical support. Rawls' position has been already mentioned. Since the 1980s, other important issues have been unveiled and addressed by IE understood as an Information-as-Resource Ethics: the so-called *digital divide*, the problem of *infoglut*, and the analysis of the *reliability* and *trustworthiness* of information sources (Froehlich, 1997; Smith, 1997). Courses on IE, taught as part of Information Sciences degree programs, tend to share this approach as researchers in library and information sciences are particularly sensitive to such issues, also from a professional perspective (Alfino and Pierce, 1997; Mintz, 1990; Stichler and Hauptman, 1998).

One may recognize in this original approach to Information Ethics a position broadly defended by Van Den Hoven (1995) and more recently by Mathiesen (2004),

[8]For an analysis of the so-called IT-heodicean problem and of the tragedy of the good will, see Floridi (2006 b).

who criticizes Floridi and Sanders (1999) and is in turn criticized by Mather (2005). Whereas Van den Hoven purports to present this approach to IE as an enriching perspective contributing to the wider debate on a more broadly constructed conception of IE, Mathiesen appears to present her view, restricted to the informational needs and states of the individual moral agent, as the only correct interpretation of IE. Her position seems thus undermined by the problems affecting any univocal interpretation of IE, as Mather correctly argues.

1.3 THE SECOND STAGE: IE AS AN ETHICS OF INFORMATIONAL PRODUCTS

It seems that IE began to merge with computer ethics only in the 1990s, when the ICT revolution became so widespread as to give rise to new issues not only in the management of information-as-a-resource by professional figures (librarians, journalists, scholars, scientists, IT specialists, and so forth) but also in the distributed and pervasive creation, consumption, sharing, and control of information, by a very large and quickly increasing population of people online, commonly used to dealing with digital tools of all sorts (games, mobiles, emails, CD players, DVD players, etc.). In other words, the Internet highlighted how IE could also be understood in a second but closely related sense, in which information plays an important role as a *product* of A's moral evaluations and actions (Cavalier, 2005). To understand this transformation, let us consider the RPT model again.

It is easy to see that our agent A is not only an information consumer but also an information producer, who may be subject to constraints while being able to take advantage of opportunities in the course of her activities. Both constraints and opportunities may call for an ethical analysis. Thus, IE, understood as Information-as-a-Product Ethics, will cover moral issues arising, for example, in the context of *accountability, liability, libel legislation, testimony, plagiarism, advertising, propaganda, misinformation*, and more generally of *pragmatic rules of communication* à la Grice. The recent debate on P2P software provides a good example, but, once again, this way of looking at Information Ethics is far from being a total novelty. Kant's classic analysis of the immorality of *lying* is one of the best known case studies in the philosophical literature concerning this kind of Information Ethics. Cassandra and Laocoön, pointlessly warning the Trojans against the Greeks' wooden horse, remind us how the ineffective management of informational products may have tragic consequences. Whoever works in mass media studies will have encountered this sort of ethical issues.

It is hard to identify researchers who uniquely support this specific interpretation of IE, as works on Information-as-Product Ethics tend to be inclusive, that is, they tend to build on the first understanding of IE as an ethics of informational resources and add to it a new layer of concerns for informational products as well (see, e.g., Moore, 2005). However, the shift from the first to the second sense of IE (from resource to product) can be noted in some successful anthologies and textbooks, which were carefully revised when undergoing new editions. For example, Spinello (2003) explicitly emphasizes much more the ethical issues arising in the networked society, compared to the first edition (Spinello, 1997), and hence a sort of IE that is closer to the sense clarified in this section rather than that in the previous section. And Severson (1997), after the typical introduction to ethical ideas, dedicates a long chapter to respect for intellectual property. Finally, it would be fair to say that the new perspective can be more often found shared, perhaps implicitly, by studies that are socio-legally oriented and in which IT-professional issues appear more prominently.

1.4 THE THIRD STAGE: IE AS AN ETHICS OF THE INFORMATIONAL ENVIRONMENT

The emergence of the information society has further expanded the scope of IE. The more people have become accustomed to living and working immersed within digital environments, the easier it has become to unveil new ethical issues involving informational realities. Returning to our initial model, independently of A's information input (info-resource) and output (info-product), in the 1990s there appeared works highlighting a third sense in which information may be subject to ethical analysis, namely, when A's moral evaluations and actions affect the informational environment. Think, for example, of A's respect for, or breach of, someone's information *privacy* or *confidentiality*.[9] *Hacking*, understood as the unauthorized access to a (usually computerized) information system, is another good example because it shows quite clearly the change in perspective. In the 1980s it was not uncommon to mistake hacking for a problem to be discussed within the conceptual frame of an ethics of informational resources. This misclassification allowed the hacker to defend his position by arguing that no use (let alone misuse) of the accessed information had been made. Yet hacking, properly understood, is a form of breach of privacy. What is in question is not what A does with the information, which has been accessed without authorization, but what it means for an informational environment to be accessed by A without authorization. So the analysis of hacking belongs to what in this section has been defined as an Information-as-Target Ethics. Other issues here include *security* (including issues related to digital warfare and terrorism), *vandalism* (from the burning of libraries and books to the dissemination of viruses), *piracy*, *intellectual property*, *open source*, *freedom of expression*, *censorship*, *filtering*, and *contents control*. Mill's analysis "Of the Liberty of Thought and Discussion" is a classic of IE interpreted as Information-as-Target Ethics. Juliet, simulating her death, and Hamlet, reenacting his father's homicide, show how the risky management of one's informational environment may have tragic consequences.

Works in this third trend in IE are characterized by environmental and global concerns. They also continue the merging process of Information and Computer Ethics begun in the 1990s (Woodbury, 2003), moving toward what Charles Ess has labeled ICE (Weckert and Adeney, 1997). Perhaps one of the first works to look at IE as an ethics of "things" that, as patients, are affected by an agent's behavior is Floridi (1999a) (but see

[9] For further details see Floridi (2005c).

also Floridi, 2003). On the globalization of IE, Bynum and Rogerson (1996) is among the important references (but see also Buchanan, 1999; Ess, 2006), together with the regular publication of the *International Review of Information Ethics*, edited by Rafael Capurro at the International Centre for Information Ethics (http://icie.zkm.de/).

1.5 THE LIMITS OF ANY MICROETHICAL APPROACH TO INFORMATION ETHICS

So far we have seen that the RPT model may help one to get some initial orientation in the multiplicity of issues belonging to different interpretations of Information Ethics. Despite its advantages, however, the model can still be criticized for being inadequate, for at least two reasons.

First, the model is too simplistic. Arguably, several important issues belong mainly but not only to the analysis of just one "informational arrow." The reader may have already thought of several examples that illustrate the problem: someone's testimony is someone's else trustworthy information; A's responsibility may be determined by the information A holds, but it may also concern the information A issues; censorship affects A both as a user and as a producer of information; misinformation (i.e., the deliberate production and distribution of misleading information) is an ethical problem that concerns all three "informational arrows"; freedom of speech also affects the availability of offensive content (e.g., child pornography, violent content, and socially, politically, or religiously disrespectful statements) that might be morally questionable and should not circulate. Historically, all this means that some simplifications, associating decades to specific approaches to IE, are just that, simplifications that should be taken with a lot of caution. The "arrows" are normally much more entwined.

Second, the model is insufficiently inclusive. There are many important issues that cannot easily be placed on the map at all, for they really emerge from, or supervene on, the interactions among the "informational arrows." Two significant examples may suffice: "big brother," that is, the problem of *monitoring and controlling* anything that might concern A; and the debate about information *ownership* (including copyright and patent legislation) and *fair use*, which affects both users and producers while shaping their informational environment.

Both criticisms are justified: the RPT model is indeed inadequate. Yet why it is inadequate is a different matter. The tripartite analysis just provided helps to structure both chronologically and analytically the development of IE and its interpretations. But it is unsatisfactory, despite its initial usefulness, precisely because any interpretation of Information Ethics based on only one of the "informational arrows" is bound to be too reductive. As the examples mentioned above emphasize, supporters of narrowly constructed interpretations of Information Ethics as a *microethics* (i.e., a one-arrow-only ethics, to use our model) are faced with the problem of being unable to cope with a large variety of relevant issues, which remain either uncovered or inexplicable. In other words, the model shows that idiosyncratic versions of IE, which privilege only some limited aspects of the *information cycle*, are unsatisfactory. We should not use the model to attempt to pigeonhole problems neatly, which is impossible. We should rather exploit it as a useful first approximation to be superseded, in view of a more encompassing approach to IE as a *macroethics*, that is, a theoretical, field-independent, applicable ethics. Philosophers will recognize here a Wittgensteinian ladder that can be used to reach a new starting point, but then can be discharged.

In order to climb up on, and then throw away, any narrowly constructed conception of Information Ethics, a more encompassing approach to IE needs to

(i) Bring together the three "informational arrows";
(ii) Consider the whole information cycle; and
(iii) Analyze informationally all entities involved (including the moral agent A) and their changes, actions, and interactions, by treating them not apart from, but as part of, the informational environment, or *infosphere*, to which they belong as informational systems themselves.

As steps (i) and (ii) do not pose particular problems, and may be shared by any of the three approaches already seen, step (iii) is crucial but involves an "update" in the ontological conception of "information" at stake. Instead of limiting the analysis to (veridical) semantic contents—as any narrower interpretation of IE as a microethics inevitably does—an ecological approach to Information Ethics also looks at information from an object-oriented perspective, and treats it as entity as well. In other words, one moves from a (broadly constructed) epistemological conception of Information Ethics—in which information is roughly equivalent to news or semantic content—to one which is typically ontological, and treats information as equivalent to patterns or entities in the world. Thus, in the revised RPT model, represented in Fig. 1.2, the agent is embodied and embedded, as an informational agent, in an equally informational environment.

A simple analogy may help to introduce this new perspective.[10] Imagine looking at the whole universe from a chemical perspective.[11] Every entity and process will satisfy a certain chemical description. To simplify, a human being, for example, will be 90% water and 10% something else. Now consider an informational perspective. The same entities will be described as clusters of data, that is, as informational objects. More precisely, our agent A (like any other entity) will be a discrete, self-contained, encapsulated package containing:

(i) The appropriate data structures, which constitute the nature of the entity in question, that is, the state of the object, its unique identity and its attributes; and

[10] For a detailed analysis and defense of an object-oriented modeling of informational entities see Floridi and Sanders (1999), Floridi (2003, 2004).

[11] "Perspective" here really means level of abstraction; however, for the sake of simplicity the analysis of levels of abstractions has been omitted from this chapter. The interested reader may wish to consult Floridi and Sanders (2004a).

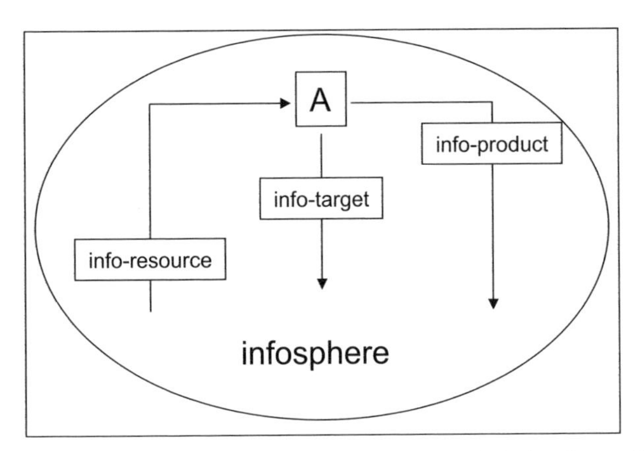

FIGURE 1.2 "Internal" R(esource) P(roduct) T(arget) model: the Agent A is correctly embedded within the infosphere.

(ii) A collection of operations, functions, or procedures, which are activated by various interactions or stimuli (i.e., messages received from other objects or changes within itself) and correspondingly define how the object behaves or reacts to them.

At this level of analysis, informational systems as such, rather than just living systems in general, are raised to the role of agents and patients (senders and receivers) of any action, with environmental processes, changes and interactions equally described informationally.

Understanding the *nature* of IE ontologically, rather than epistemologically, modifies the interpretation of the *scope* and *goals* of IE. Not only can an ecological IE gain a global view of the whole life cycle of information, thus overcoming the limits of other microethical approaches, but it can also claim a role as a macroethics, that is, as an ethics that concerns the whole realm of reality. This is what we shall see in the next section.

1.6 THE FOURTH STAGE: INFORMATION ETHICS AS A MACROETHICS

The fourth interpretation of IE, as a macroethics, may be quickly summarized thus: IE is a *patient-oriented*, *ontocentric*, *ecological* macroethics (Floridi and Sanders, 1999). These are technical expressions that can be intuitively explained by comparing IE to other environmental approaches.[12]

---

[12]For an initial development of Information Ethics and a more technical treatment of some of the themes discussed in this paper see the following papers, available from http://www.philosophyofinformation.net/papers.htm: Floridi (1995, 1999a, 2002, 2003, 2005d, 2005c, 2006a, 2006b, forthcoming), Floridi and Sanders (1999, 2001, 2002b, 2004a, 2004b, 2005).

Biocentric ethics usually grounds its analysis of the moral standing of bioentities and ecosystems on the intrinsic worthiness of *life* and the intrinsically negative value of *suffering*. It seeks to develop a patient-oriented ethics in which the "patient" may be not only a human being, but also any form of life. Indeed, Land Ethics extends the concept of patient to any component of the environment, thus coming close to the approach defended by Information Ethics. Any form of life is deemed to enjoy some essential proprieties or moral interests that deserve and demand to be respected, at least minimally and relatively, that is, in a possibly overridable sense, when contrasted to other interests. So biocentric ethics argues that the nature and well-being of the patient of any action constitute (at least partly) its moral standing and that the latter makes important claims on the interacting agent, claims that in principle ought to contribute to the guidance of the agent's ethical decisions and the constraint of the agent's moral behavior. The "receiver" of the action, the patient, is placed at the core of the ethical discourse, as a center of moral concern, while the "transmitter" of any moral action, the agent, is moved to its periphery.

Now substitute "life" with "existence" and it should become clear what IE amounts to. IE is an ecological ethics that replaces *biocentrism* with *ontocentrism*. It suggests that there is something even more elemental than life, namely *being*—that is, the existence and flourishing of all entities and their global environment—and something more fundamental than suffering, namely *entropy*. The latter is most emphatically *not* the physicists' concept of thermodynamic entropy. Entropy here refers to any kind of *destruction*, *corruption*, *pollution*, and *depletion* of informational objects (mind, not of information), that is, any form of impoverishment of *being*. It is comparable to the metaphysical concept of *nothingness*. IE then provides a common vocabulary to understand the whole reality informationally. IE holds that *being*/information has an intrinsic worthiness. It substantiates this position by recognizing that any informational entity has a *Spinozian* right to persist in its own status, and a *Constructionist* right to flourish, that is, to improve and enrich its existence and essence. As a consequence of such "rights," IE evaluates the duty of any moral agent in terms of contribution to the growth of the *infosphere* and any process, action, or event that negatively affects the whole infosphere—not just an informational entity—as an increase in its level of nothingness (or entropy) and hence an instance of evil (Floridi, 2003; Floridi and Sanders, 1999, 2001).

In IE, the ethical discourse concerns any entity, understood informationally, that is, not only all persons, their cultivation, well-being, and social interactions, not only animals, plants, and their proper natural life, but also anything that exists, from paintings and books to stars and stones; anything that may or will exist, like future generations; and anything that was but is no more, like our ancestors or old civilizations. IE is impartial and universal because it brings to ultimate completion the process of enlargement of the concept of what may count as a center of a (no matter how minimal) moral claim, which now includes every instance of *being* understood informationally, no matter whether physically implemented or not. In this respect, IE holds that every entity, as an expression of *being*, has a dignity, constituted by its mode of existence and essence (the collection of all the elementary proprieties that constitute it for what it is), which deserves to be respected (at least in a minimal and overridable sense) and hence places moral claims on the interacting agent and ought to contribute to the constraint and guidance of his ethical decisions and behavior. This *ontological equality principle* means that any form of reality (any instance of information/*being*), simply for the fact of *being* what it is, enjoys a minimal, initial, overridable, equal right to exist and develop in a way that is appropriate to its nature. The conscious recognition of the ontological equality principle presupposes a disinterested judgment of the moral situation from an objective perspective, that is, a perspective that is as nonanthropocentric as possible. Moral behavior is less likely without this epistemic virtue. The application of the ontological equality principle is achieved whenever actions are impartial, universal, and "caring."

The crucial importance of the radical change in ontological perspective cannot be overestimated. Bioethics and Environmental Ethics fail to achieve a level of complete impartiality, because they are still biased against what is inanimate, lifeless, intangible, or abstract (e.g., even Land Ethics is biased against technology and artifacts). From their perspective, only what is intuitively alive deserves to be considered as a proper center of moral claims, no matter how minimal, so a whole universe escapes their attention. Now, this is precisely the fundamental limit overcome by IE, which further lowers the minimal condition that needs to be satisfied, in order to qualify as a center of moral concern, to the common factor shared by any entity, namely its informational state. And as any form of *being* is, in any case, also a coherent body of information, to say that IE is infocentric is tantamount to interpreting it, correctly, as an ontocentric theory.

The result is that all entities, *qua* informational objects, have an intrinsic moral value, although possibly quite minimal and overridable, and hence they can count as moral patients, subject to some equally minimal degree of moral respect understood as *a disinterested, appreciative, and careful attention* (Hepburn, 1984). As Naess (1973) has maintained, "all things in the biosphere have an equal right to live and blossom." There seems to be no good reason not to adopt a higher and more inclusive, ontocentric perspective. Not only inanimate but also ideal, intangible, or intellectual objects can have a minimal degree of moral value, no matter how humble, and so be entitled to some moral respect.

Deep Ecologists have already argued that inanimate things too can have some intrinsic value. And in a famous article, White (1967) asked "Do people have ethical obligations toward rocks?" and answered that "To almost all Americans, still saturated with ideas historically dominant in Christianity. . .the question makes no sense at all. If the time comes when to any considerable group of us such a question is no longer ridiculous, we may be on the verge of a change of value structures that will make possible measures to cope with the growing ecologic crisis. One hopes that there is enough time left." According to IE, this is the right ecological perspective and it makes perfect sense for any religious tradition (including the Judeo-Christian one) for which the whole universe is God's creation, is inhabited by the divine, and is a gift to humanity, of which the latter needs to take care. IE translates all this into informational terms. If something can be a moral patient, then its nature can be taken into consideration by a moral agent A, and contribute to shaping A's action, no matter how minimally. In more metaphysical terms, IE argues that all aspects and instances of *being* are worth some initial, perhaps minimal, and overridable, form of moral respect.

Enlarging the conception of what can count as a center of moral respect has the advantage of enabling one to make sense of the innovative and epochal nature of ICTs, as providing a new and powerful conceptual frame. It also enables one to deal more satisfactorily with the original character of some of its moral issues, by approaching them from a theoretically strong perspective. Through time, ethics has steadily moved from a narrow to a more inclusive concept of what can count as a center of moral worth, from the citizen to the biosphere (Nash, 1989). The emergence of the infosphere, as a new environment in which human beings spend much of their lives, explains the need to enlarge further the conception of what can qualify as a moral patient. IE represents the most recent development in this ecumenical trend, a Platonist and ecological approach without a biocentric bias, a move from the biosphere to the infosphere. More than 50 years ago, Leopold defined Land Ethics as something that "changes the role of *Homo sapiens* from conqueror of the land community to plain member and citizen of it. It implies respect for his fellow members, and also respect for the community as such. The land ethic simply enlarges the boundaries of the community to include soils, waters, plants, and animals, or collectively: the land" (Leopold, 1949, p. 403). IE translates environmental ethics into terms of infosphere and informational objects, for the land we inhabit is not just the earth.

1.6.1 Moral Agents

We have seen that the whole infosphere counts as a patient of moral action, according to IE. But what sort of moral agents inhabit the infosphere? The short answer is that IE defines as a moral agent any *interactive*, *autonomous*, and *adaptable transition system* that can perform *morally qualifiable actions* (Floridi, 2004). As usual, the rest of this section is devoted to explaining and discussing this definition.

A transition system is *interactive* when the system and its environment (can) act upon each other. Typical examples include input or output of a value, or simultaneous engagement of an action by both agent and patient—for example, gravitational force between bodies.

A transition system is *autonomous* when the system is able to change state without direct response to interaction, that is, it can perform internal transitions to change its state. So an agent must have at least two states. This property imbues an agent with a certain degree of complexity and independence from its environment.

Finally, a transition system is *adaptable* when the system's interactions (can) change the transition rules by which it changes state. This property ensures that an agent might be viewed as learning its own mode of operation in a way that depends critically on its experience.

All we need to understand now is the meaning of "morally qualifiable action." Very simply, an action qualifies as moral if it can cause moral good or evil. Note that this interpretation is neither consequentialist nor intentionalist in nature. We are neither affirming nor denying that the specific evaluation of the morality of the agent might depend on the specific outcome of the agent's actions or on the agent's original intentions or principles.

With all the definitions in place, it becomes possible to understand why, according to IE, *artificial agents* (not just digital agents but also social agents, such as companies, parties, or hybrid systems formed by humans and machines, or technologically augmented humans) count as moral agents that are morally *accountable* for their actions.

The enlargement of the class of moral agents by IE brings several advantages. Normally, an entity is considered a moral agent only if (i) it is an *individual* agent and (ii) it is *human based*, in the sense that it is either human or at least reducible to an identifiable aggregation of human beings, who remain the only morally responsible sources of action, like ghosts in the legal machine. Limiting the ethical discourse to *individual agents* hinders the development of a satisfactory investigation of *distributed morality*, a macroscopic and growing phenomenon of global moral actions and collective responsibilities, resulting from the "invisible hand" of systemic interactions among several agents at a local level. Insisting on the necessarily *human-based nature* of the agent means undermining the possibility of understanding another major transformation in the ethical field, the appearance of artificial agents that are sufficiently informed, "smart," autonomous, and able to perform morally relevant actions independently of the humans who created them, causing "artificial good" and "artificial evil" (Floridi and Sanders, 1999, 2001).

We have seen that morality is usually predicated upon responsibility. By distinguishing between moral responsibility, which requires intentions, consciousness, and other mental attitudes, and moral accountability, we can now avoid anthropocentric and anthropomorphic attitudes toward agenthood. Instead, we can rely on an ethical outlook based not only on punishment and reward (responsibility-oriented ethics) but also on moral agenthood, accountability, and censure. We are less likely to assign responsibility at any cost, forced by the necessity to identify individual, human agent (s). We can stop the regress of looking for the *responsible* individual when something evil happens, since we are now ready to acknowledge that sometimes the moral source of evil or good can be different from an individual or group of humans (note that this was a reasonable view in Greek philosophy). As a result, we are able to escape the dichotomy:

(i) [(responsibility implies moral agency) implies prescriptive action], versus
(ii) [(no responsibility implies no moral agency) implies no prescriptive action].

There can be moral agency in the absence of moral responsibility. Promoting normative action is perfectly reasonable even when there is no responsibility but only moral accountability and the capacity for moral action.

Being able to treat nonhuman agents as moral agents facilitates the discussion of the morality of agents not only in cyberspace but also in the biosphere—where animals can be considered moral agents without their having to display free will, emotions, or mental states—and in contexts of "distributed morality," where social and legal agents can now qualify as moral agents. The great advantage is a better grasp of the moral discourse in nonhuman contexts.

All this does not mean that the concept of "responsibility" is redundant. On the contrary, the previous analysis makes clear the need for further analysis of the concept of responsibility itself, especially when the latter refers to the ontological commitments of creators of new agents and environments. The only "cost" of a "mind-less morality" approach is the extension of the class of agents and moral agents to embrace artificial agents. It is a cost that is increasingly worth paying the more we move toward an advanced information society.

1.6.2 The Responsibilities of Human Agents

Humans are special moral agents. Like demiurges, we have "ecopoietic" responsibilities toward the whole infosphere. So Information Ethics is an ethics addressed not just to "users" of the world but also to producers, who are "divinely" responsible for its creation and well-being. It is an ethics of *creative stewardship* (Floridi, 2002, 2003; Floridi and Sanders, 2005).

The term "ecopoiesis" refers to the morally informed construction of the environment, based on an ecologically oriented perspective. In terms of a philosophical anthropology, the ecopoietic approach, supported by IE, is embodied by what I have termed *homo poieticus* (Floridi, 1999b). *Homo poieticus* is to be distinguished from *homo faber*, user and "exploitator" of natural resources, from *homo oeconomicus*, producer, distributor, and consumer of wealth, and from *homo ludens* (Huizinga, 1970), who embodies a leisurely playfulness, devoid of the ethical care and responsibility characterizing the constructionist attitude. *Homo poieticus* is a demiurge who takes care of reality to protect it and make it flourish. This reality has been defined above as the infosphere.

The ontic powers of *homo poieticus* have been steadily increasing. Today, *homo poieticus* can variously exercise them (in terms of control, creation, or modeling) over himself (e.g., genetically, physiologically, neurologically, and narratively), over his society (e.g., culturally, politically, socially, and economically), and over his natural or artificial environments (e.g., physically and computationally). The more powerful *homo poieticus* becomes as an agent, the greater his duties and responsibilities become, as a *moral agent*, to oversee not only the development of his own character and habits but also the well-being and flourishing of each of his ever-expanding spheres of influence, to include the whole infosphere. To move from individual virtues to global values, an *ecopoietic* approach is needed that recognizes our *responsibilities* toward the environment (including present and future inhabitants) as its enlightened creators, stewards or supervisors, not just as its virtuous users and consumers.

1.6.3 Four Moral Principles

What sort of principles may guide the actions of *homo poieticus*? IE determines what is morally right or wrong, what ought to be done, what the duties, the "oughts," and the "ought nots" of a moral agent are, by means of four basic moral laws. They are formulated here in an informational vocabulary and in a patient-oriented version, but an agent-oriented one is easily achievable in more metaphysical terms of "dos" and "don'ts":

(1) Entropy ought not to be caused in the infosphere (null law);
(2) Entropy ought to be prevented in the infosphere;
(3) Entropy ought to be removed from the infosphere;
(4) The flourishing of informational entities as well as of the whole infosphere ought to be promoted by preserving, cultivating, and enriching their properties.

The basic moral question asked by IE is: what is good for informational entities and for the infosphere in general? We have seen that the answer is provided by a minimalist theory: any informational entity is recognized to be the center of some basic ethical claims, which deserve recognition and should help to regulate the implementation of any informational process involving it. It follows that approval or disapproval of A's decisions and actions should also be based on how the latter affects the well-being of the infosphere, that is, on how successful or unsuccessful they are in respecting the ethical claims attributable to the informational entities involved, and hence in improving or impoverishing the infosphere. The duty of any moral agent should be evaluated in terms of contribution to the sustainable blooming of the infosphere, and any process, action, or event that negatively affects the whole infosphere—not just an informational object—should be seen as an increase in its level of entropy and hence an instance of evil.

The four laws are listed in order of increasing moral value. Their strict resemblance to similar principles in medical ethics is not accidental, since both approaches share an ethics of care. They clarify, in very broad terms, what it means to live as a responsible and caring agent in the infosphere.

On the one hand, a process that satisfies only the null law—the level of entropy in the infosphere remains unchanged after its occurrence—either has no moral value, that is, it is morally irrelevant or insignificant, or it is equally depreciable and commendable, though in different respects. Likewise, a process is increasingly deprecable, and its agent source is increasingly blameworthy, the lower is the number-index of the specific law that it fails to satisfy. Moral mistakes may occur and entropy may increase if one wrongly evaluates the impact of one's actions because projects conflict or compete, even if those projects aim to satisfy IE moral laws. This is especially the case when "local goodness," that is, the improvement of a region of the infosphere, is favored to the overall disadvantage of the whole environment. More simply, entropy may increase because of the wicked nature of the agent (this possibility is granted by IE's negative anthropology).

On the other hand, a process is already commendable, and its agent-source praiseworthy, if it satisfies the *conjunction* of the null law with at least one other law, not the *sum* of the resulting effects. Note that, according to this definition, an action is unconditionally commendable only if it never generates any entropy in the course of its implementation; and the best moral action is the action that succeeds in satisfying all four laws at the same time.

Most of the actions that we judge morally good do not satisfy such strict criteria, for they achieve only a balanced positive moral value, that is, although their performance causes a certain quantity of entropy, we acknowledge that the infosphere is in a better state after their occurrence.

1.7 TWO RECURRENT OBJECTIONS AGAINST IE

Since the nineties,[13] when IE was first introduced as an environmental macroethics and a foundationalist approach to computer ethics, some standard objections have been made that seem to be based on a few basic misunderstandings. The point of this final section is not that of convincing the reader that no reasonable disagreement is possible about the value of IE in general and on IE as a macroethics in particular. On the contrary, several of the theses seen in the previous pages are interesting precisely because they are also open to discussion. Rather, the goal here is to remove some ambiguities and possible confusions that might prevent the correct evaluation of IE in its various interpretations, so that disagreement can become more constructive.

1.7.1 Does it Make Sense to Talk of Informational Entities and Agents?

By defending the intrinsic moral worth of informational entities and the importance of considering artificial agents as moral agents IE does not refer to the moral value of any other piece of well-formed and meaningful data such as an e-mail, the *Britannica*, or Newton's *Principia* (Himma, 2004, Mathiesen, 2004) or some science fiction robot such as *Star Wars'* C3PO and R2D2. What IE suggests is that one adopt an informational approach (technically, a level of abstraction) to the analysis of *being* in terms of a minimal common ontology, whereby human beings as well as animals, plants, artifacts, institutions, and so forth are interpreted as informational entities. IE is not an ethics of the BBC news or some artificial agent à la Asimov. Of course, it remains open to debate whether an informational level of abstraction adopted is correct. For example, the choice and hence its implications have been recently criticized by Johnson (2006) and Capurro (2006) has argued against the ontological options adopted by IE.

1.7.2 Is IE Inapplicable?

Given its ontological nature and wide scope, one of the objections that is sometimes made against IE is that of being too abstract or theoretical (too philosophical in the worst sense of the word) to be of much use when human agents are confronted by very

[13]Fourth International Conference on Ethical Issues of Information Technology (Department of Philosophy, Erasmus University, The Netherlands, March 25–27, 1998); this was published as Floridi and Sanders (1999).

concrete and applied challenges (Mathiesen, 2004; Siponen, 2004). IE would work at a level of metaphysical abstraction too philosophical to make it of any direct utility for immediate needs and applications. Yet, this is the inevitable price to be paid for any foundationalist project. One must polarize theory and practice to strengthen both. IE is not immediately useful to solve specific ethical problems (including computer ethics problems), but it provides the conceptual grounds that then guide problem-solving procedures. Imagine someone who, being presented with the declaration of human rights, were to complain that it is too general and inapplicable to solve the ethical problems she is facing in a specific situation, say in dealing with a particular case of cyberstalking in the company that employs her. This would be rather out of place. The suspicion is that some impatience with conceptual explorations may betray a lack of understanding of how profound the revolution we are undergoing is, and hence how radical the rethinking of our ethical approaches and principles may need to be in order to cope with it. IE is certainly not the declaration of human rights, but it seeks to obtain a level of generality purporting to provide a foundation for more applied and case-oriented analyses. So the question is not whether IE is too abstract—good foundations for the structure one may wish to see being built inevitably lie well below the surface—but whether it will succeed in providing the robust framework within which practical issues of moral concern may be more easily identified, clarified, and solved. It is in its actual applications that IE, as an ethics for our information society, will or will not qualify as a useful approach; yet building on the foundation provided by IE is a serious challenge, it cannot be an objection. It is encouraging that IE has already been fruitfully applied to deal with the "tragedy of the digital commons" (Greco and Floridi, 2004), the digital divide (Floridi, 2002), the problem of telepresence (Floridi, 2005d), game cheating (Sicart, 2005), the problem of privacy (Floridi, 2005b), environmental issues (York, 2005) and software protocols design (Turilli, 2007).

1.8 CONCLUSION

In one of Einstein's letters there is a passage that well summarizes the perspective advocated by IE understood as a macroethics: "A human being is part of the whole, called by us 'universe,' a part limited in time and space. He experiences himself, his thoughts and feelings, as something separated from the rest, a kind of optical delusion of his consciousness. This delusion is a kind of prison for us, restricting us to our personal desires and to affection for a few persons close to us. Our task must be to free ourselves from our prison by widening our circle of compassion to embrace all humanity and the whole of nature in its beauty. Nobody is capable of achieving this completely, but the striving for such achievement is in itself a part of the liberation and a foundation for inner security" (Einstein, 1954). Does looking at reality through the highly philosophical lens of an informational analysis improve our ethical understanding, or is it an ethically pointless (when not misleading) exercise? IE argues that the agent-related *behavior* and the patient-related *status* of informational objects *qua* informational objects can be morally significant, over and above the instrumental function that may be attributed to them by other ethical approaches, and hence that they can contribute to determining, normatively, ethical duties and legally enforceable rights. IE's position, like that of any other macroethics, is not devoid of problems. But it can interact with other macroethical theories and contribute an important new perspective: a process or action may be morally good or bad irrespective of its consequences, motives, universality, or virtuous nature, but depending on how it affects the infosphere. An ontocentric ethics provides an insightful perspective. Without IE's contribution, our understanding of moral facts in general, not just of ICT-related problems in particular, would be less complete and our struggle to escape from our anthropocentric condition, being this Plato's cave or Einstein's cage, less successful.[14]

REFERENCES

Alfino, M. and Pierce, L. (1997). *Information Ethics for Librarians.* McFarland & Co., Jefferson, NC.

Buchanan, E.A. (1999). An overview of information ethics issues in a world-wide context. *Ethics and Information Technology,* 1(3), 193–201.

Bynum, T. (2001). Computer ethics: basic concepts and historical overview. In: Edward, N.Z. (Ed.), *The Stanford Encyclopedia of Philosophy.* http://plato.stanford.edu/entries/ethics-computer.

Bynum, T.W.E. and Rogerson, S.E. (Eds.). (1996). *Global Information Ethics.* Opragen Publications. (A special issue of the journal *Science and Engineering* Ethics, April 1996.)

Capurro, R. (2006). Toward an ontological foundation of information ethics. *Ethics and Information Technology,* 8(4), 175–186.

Cavalier, R.J. (2005). *The Impact of the Internet on Our Moral Lives.* State University of New York Press, Albany, NY.

Einstein, A. (1954). *Ideas and Opinions.* Crown Publishers, New York.

Ess, C. (2006). Ethical pluralism and global information ethics. *Ethics and Information Technology,* 8(4), 215–226.

Floridi, L. (1995). Internet: which future for organized knowledge, Frankenstein or Pygmalion? *International Journal of Human-Computer Studies,* 43, 261–274.

Floridi, L. (1999a). Information ethics: on the philosophical foundations of computer ethics. *Ethics and Information Technology,* 1(1), 37–56. Reprinted, with some modifications, in *The Ethicomp Journal,* 1(1), 2004. http://www.ccsr.cse.dmu.ac.uk/journal/articles/floridi_l_philosophical.pdf. Abridged French translation in *L'Agora,* 5.4 (July–August 1998), pp. 19–20. Polish translation in *Ethos,* the journal of the John Paul II Institute at the Catholic University of Lublin (August–September 2005).

Floridi, L. (1999b). *Philosophy and Computing: An Introduction.* Routledge, London, New York.

[14]This chapter is based on Floridi (1999a, 2003, 2005c), Floridi and Sanders (2001, 2004b, 2005), and Floridi (forthcoming). I am in debt to all the colleagues and friends who shared their comments on those papers. Their full list can be found in those publications. Here I wish to acknowledge that several improvements are because of their feedback, and the editorial suggestions by Kenneth Himma and Herman Tavani.

Floridi, L. (2002). Information ethics: an environmental approach to the digital divide. *Philosophy in the Contemporary World,* 9(1), 39–45.

Floridi, L. (2003). On the intrinsic value of information objects and the infosphere. *Ethics and Information Technology,* 4(4), 287–304.

Floridi, L. (Ed.). (2004). *The Blackwell Guide to the Philosophy of Computing and Information.* Blackwell, Oxford, New York, 40–61.

Floridi, L. (2005a). Information, semantic conceptions of. In: Edward, N.Z. (Ed.), *Stanford Encyclopedia of Philosophy.* http://plato.stanford.edu/entries/information-semantic/.

Floridi, L. (2005b). An interpretation of informational privacy and of its moral value. *Proceedings of CEPE 2005—6th Computer Ethics: Philosophical Enquiries Conference, Ethics of New Information Technologies*, University of Twente, Enschede, The Netherlands.

Floridi, L. (2005c). The ontological interpretation of informational privacy. *Ethics and Information Technology,* 7(4), 185–200.

Floridi, L. (2005d). The philosophy of presence: from epistemic failure to successful observability. *Presence: Teleoperators and Virtual Environments,* 14(6), 656–667.

Floridi, L. (2006a). Four challenges for a theory of informational privacy. *Ethics and Information Technology,* 8(3), 109–119.

Floridi, L. (forthcoming). Information ethics: its nature and scope. In: van den Hoven, J. and Weckert, J. (Eds.), *Moral Philosophy and Information Technology.* Cambridge University Press, Cambridge.

Floridi, L. (2006b). Information technologies and the tragedy of the good will. *Ethics and Information Technology*, 8(4), 253–262.

Floridi, L. and Sanders, J.W. (1999). Entropy as evil in information ethics. *Etica & Politica, Special Issue on Computer Ethics,* 1(2).

Floridi, L. and Sanders, J.W. (2001). Artificial evil and the foundation of computer ethics. *Ethics and Information Technology,* 3(1), 55–66.

Floridi, L. and Sanders, J. (2002a). Mapping the foundationalist debate in computer ethics. *Ethics and Information Technology,* 4(1), 1–9.

Floridi, L. and Sanders, J.W. (2002b). Computer ethics: mapping the foundationalist debate. *Ethics and Information Technology,* 4(1), 1–9. Revised version published in the 2nd edition. In: Spinello, R. and Tavani, H. (Eds.), *Readings in Cyberethics.* Jones and Bartlett, Boston, pp. 81–95.

Floridi, L. and Sanders, J.W. (2004a). The method of abstraction. In: Negrotti, M. (Ed.), *Yearbook of the Artificial-Nature, Culture, and Technology, Models in Contemporary Sciences.* Peter Lang, Bern, pp. 177–220.

Floridi, L. and Sanders, J.W. (2004b). On the morality of artificial agents. *Minds and Machines,* 14(3), 349–379.

Floridi, L. and Sanders, J.W. (2005). Internet ethics: the constructionist values of *Homo Poieticus.* In: Cavalier, R. (Ed.), *The Impact of the Internet on Our Moral Lives.* SUNY, New York.

Froehlich, T.J. (1997). *Survey and Analysis of Legal and Ethical Issues for Library and Information Services*, G.K. Saur, Münich. UNESCO Report (Contract no. 401.723.4), for the International Federation of Library Associations. IFLA Professional Series.

Froehlich, T.J. (2004). A brief history of information ethics. *Textos universitaris de biblioteconomia i documentacio*, http://www.ub.es/bid/13froel2.htm

Greco, G.M. and Floridi, L. (2004). The tragedy of the digital commons. *Ethics and Information Technology,* 6(2), 73–82.

Hauptman, R. (1988). *Ethical Challenges in Librarianship*. Oryx Press, Phoenix. Hepburn, R. (1984). *Wonder and Other Essays*. Edinburgh University Press, Edinburgh.

Himma, K.E. (2004). There's something about Mary: the moral value of things *qua* information objects. *Ethics and Information Technology,* 6(3), 145–159.

Huizinga, J. (1970). *Homo ludens: a study of the play element in culture*. Paladin, first published 1938 London.

Johnson, D.G. (2006). Computer systems: moral entities but not moral agents. *Ethics and Information Technology,* 8(4), 195–204.

Koenig, M.E.D., Kostrewski, B.J., and Oppenheim, C. (1981). Ethics in information science. *Journal of Information Science,* 3, 45–47.

Leopold, A. (1949). *The Sand County Almanac*. Oxford University Press, New York.

Maner, W. (1996). Unique ethical problems in information technology. *Science and Engineering Ethics,* 2(2), 137–154.

Mather, K. (2005). Object oriented goodness: a response to Mathiesen's 'what is information ethics?' *Computers and Society,* 34(4), http://www.computersandsociety.org/sigcas_ofthefuture2/sigcas/subpage/sub_page.cfm?article=919&page_number_nb=911

Mathiesen, K. (2004). What is information ethics? *Computers and Society,* 32(8), http://www.computersandsociety.org/sigcas_ofthefuture2/sigcas/subpage/sub_page.cfm?article=909&page_number_nb=901

Mintz, A.P. (Ed.). (1990). *Information Ethics: Concerns for Librarianship and the Information Industry: Proceedings of the Twenty-Seventh Annual Symposium of the Graduate Alumni and Faculty of the Rutgers School of Communication, Information and Library Studies,* April 14, 1989. Jefferson, NC, McFarland, London.

Moore, A.D. (Ed.). (2005). *Information Ethics: Privacy, Property, and Power.* University of Washington Press, London, Seattle, Washington.

Naess, A. (1973). The shallow and the deep, long-range ecology movement. *Inquiry,* 16, 95–100.

Nash, R.F. (1989). *The Rights of Nature*. The University of Wisconsin Press, Madison, Wisconsin.

Rawls, J. (1999). *A Theory of Justice,* revised edition. Oxford University Press, Oxford.

Severson, R.J. (1997). *The Principles of Information Ethics*. M.E. Sharpe, Armonk, NY.

Sicart, M. (2005). On the foundations of evil in computer game cheating. *Proceedings of the Digital Games Research Association's 2nd International Conference–Changing Views: Worlds in Play,* June 16–20, Vancouver, British Columbia, Canada.

Siponen, M. (2004). A pragmatic evaluation of the theory of information ethics. *Ethics and Information Technology,* 6(4), 279–290.

Smith, M.M. (1996). *Information Ethics: An Hermeneutical Analysis of an Emerging Area in Applied Ethics*, Ph.D. Thesis. The University of North Carolina at Chapel Hill, Chapel Hill, NC.

Smith, M.M. (1997). Information ethics. *Annual Review of Information Science and Technology,* 32, 339–366.

Spinello, R.A. (1997). *Case Studies in Information and Computer Ethics*. Prentice Hall, Upper Saddle River, NJ.

Spinello, R.A. (2003). *Case Studies in Information Technology Ethics and Policy*, 2nd edition. Prentice Hall, Upper Saddle River, NJ.

Stichler, R.N. and Hauptman, R. (Eds.). (1998). *Ethics, Information, and Technology: Readings*. Jefferson, NC, McFarland, London.

Tavani, H.T. (2002). The uniqueness debate in computer ethics: what exactly is at issue, and why does it matter? *Ethics and Information Technology,* 4(1), 37–54.

Turilli, M. (2007). Ethical protocols design. *Ethics and Information Technology*, 9(1), 49–62.

Van Den Hoven, J. (1995). Equal access and social justice: information as a primary good. *ETHICOMP95: An International Conference on the Ethical Issues of Using Information Technology*. De Montfort University, Leicester, UK.

Weckert, J. and Adeney, D. (1997). *Computer and Information Ethics*. Greenwood Press, Westport, Connecticut.

White, L.J. (1967). The historical roots of our ecological crisis. *Science,* 155, 1203–1207.

Wiener, N. (1954). *The Human Use of Human Beings: Cybernetics and Society. Revised Edition.* Houghton Mifflin, Boston.

Woodbury, M.C. (2003). *Computer and Information Ethics*. Stipes, Champaign, Ill.

York, P.F. (2005). Respect for the world: universal ethics and the morality of terraforming. Ph.D. Thesis. The University of Queensland.

CHAPTER 2

Milestones in the History of Information and Computer Ethics*

TERRELL WARD BYNUM

2.1 INTRODUCTION

The academic field of information ethics was born—unintentionally and almost accidentally—in the middle of the Second World War. At that time, philosopher/scientist Norbert Wiener was working with a group of scientists and engineers who were involved with him in the invention of digital computers and radar, and the creation of a new kind of antiaircraft cannon that could (1) perceive the presence of an airplane, (2) gather information about its speed and trajectory, (3) predict its future position a few seconds later, (4) decide where to aim and when to fire the shell, and (5) carry out that decision. All of these steps were to take place almost instantaneously—and without human intervention! With remarkable insight and foresight, Wiener realized that the new science and technology that he and his colleagues were creating would have "enormous potential for good and for evil." He predicted that, after the war, the new information technology would dramatically change the world just as much as the Industrial Revolution had done in the nineteenth and early twentieth centuries. Wiener predicted a "second industrial revolution," an "automatic age," that would generate a staggering number of new ethical challenges and opportunities.

When the war ended, Wiener wrote a book (Wiener, 1948) about the new science of "cybernetics" that he and his colleagues had created. Two years later he followed up

*Published by permission of Terrell Ward Bynum. The Editors would also like to thank Edward N. Zalta, Principal Editor of *The Stanford Encyclopedia of Philosophy,* and its publisher, the Metaphysics Research Lab at Stanford University, for permission to include, in Internet and World Wide Web publication, relevant passages from the following article in *The Stanford Encyclopedia of Philosophy*: Bynum, T., "Computer and Information Ethics: Basic Concepts and Historical Overview", URL http:// plato.stanford.edu/entries/ethics-computer.

The Handbook of Information and Computer Ethics, Edited by Kenneth Einar Himma and Herman T. Tavani

with a second book (Wiener, 1950) about the likely social and ethical impacts of the new information technologies. With these two books—apparently without realizing it—Wiener laid the foundations of information ethics and computer ethics. His thinking was so far ahead of other scholars, however, that many considered him to be an eccentric scientist who engaged in flights of fantasy about the future. No one recognized at the time—not even Wiener himself—the profound importance of his ethical achievements. Nearly two decades would pass before the social and ethical impacts of computing, which Wiener had predicted in the late 1940s, would become obvious to the world. And another decade would go by before Walter Maner coined the name "computer ethics" to refer to the new branch of applied ethics that Wiener had founded.

The present essay begins with a discussion of Wiener's powerful foundation for information and computer ethics, and then it describes a number of additional "milestones" in the history of this new and vital branch of ethics.

2.2 NORBERT WIENER'S FOUNDATION OF INFORMATION ETHICS

Wiener (1950), in his groundbreaking book, *The Human Use of Human Beings*, explored the likely impacts of information technologies upon central human values, such as *life, health, happiness, security, freedom, knowledge, opportunities*, and *abilities*. Even in today's "Internet age" and the search for "global information ethics," the concepts and procedures that Wiener employed can be used to identify, analyze, and resolve social and ethical problems associated with information technologies of all kinds—including, for example, computers and computer networks; radio, television, and telephones; news media and journalism; even books and libraries. Given the breadth of his concerns and the applicability of his ideas and methods to every kind of information technology, the term "information ethics" is an apt name for the field that he founded. *Computer ethics*, as it is typically understood today, is a *subfield* of Wiener's information ethics. The ethical issues that Wiener analyzed, or at least touched upon, decades ago (see Wiener 1948, 1950, 1954, 1964) include computer ethics topics that are still of interest today: computers and unemployment, computers and security, computers and learning, computers for persons with disabilities, computers and religion, information networks and globalization, virtual communities, teleworking, the responsibilities of computer professionals, the merging of human bodies and machines, "agent" ethics, artificial intelligence, and a number of other issues (see Bynum, 2000, 2004, 2005, 2006).

Wiener based his foundation for information ethics upon *a cybernetic view of human nature and of society*, which leads readily to an ethically suggestive account of *the purpose of a human life*. From this, he identified "great principles of justice" that every society should follow, and he employed a practical strategy for analyzing and resolving information ethics issues wherever they might occur.

Although these achievements founded information ethics as a field of academic research, Wiener did not knowingly or intentionally set out to create a new branch of applied ethics. The field simply emerged from the many ethical remarks and examples that he included in his writings on *cybernetics*. Even though he *did* coin the name "cybernetics" (based upon the Greek word for the pilot of a ship) for the new science that he and his colleagues had created, he nevertheless *did not* invent a special term like "information ethics" or "computer ethics" for the new branch of applied ethics that emerged from his work. Even after he published an entire book in 1950 (*The Human Use of Human Beings*) on the social and ethical implications of cybernetics and electronic computers, he did not describe what he was doing as the creation of a new branch of applied ethics, and he did *not* provide metaphilosophical commentary on what he was doing and why. To understand Wiener's foundation for information ethics, therefore, we must *observe* what he does in his works, rather than look for metaphilosophical explanations about his intentions.

Wiener's cybernetic account of human nature emphasized the physical structure of the human body and the tremendous potential for learning and creative action that human physiology makes possible. To explain that potential, he often compared human physiology to the physiology of less intelligent creatures like insects:

> *Cybernetics takes the view that the structure of the machine or of the organism is an index of the performance that may be expected from it.* The fact that the mechanical rigidity of the insect is such as to limit its intelligence while the mechanical fluidity of the human being provides for his almost indefinite intellectual expansion is highly relevant to the point of view of this book . . .. man's advantage over the rest of nature is that he has the physiological and hence the intellectual equipment to adapt himself to radical changes in his environment. The human species is strong only insofar as it takes advantage of the innate, adaptive, learning faculties that its physiological structure makes possible. (Wiener, 1954, pp. 57–58, italics in the original).

On the basis of his analysis of human nature, Wiener concluded that the purpose of a human life is to flourish as the kind of information-processing organisms that humans naturally are:

> I wish to show that the human individual, capable of vast learning and study, which may occupy almost half of his life, is physically equipped, as the ant is not, for this capacity. Variety and possibility are inherent in the human sensorium—and are indeed the key to man's most noble flights—because variety and possibility belong to the very structure of the human organism. (Wiener, 1954, pp. 51–52).

Wiener's understanding of human nature presupposed a metaphysical account of the universe that considered the world and all the entities within it, including humans, as combinations of two fundamental things: matter-energy and information. Everything is a mixture of both; and thinking, according to Wiener, is actually a kind of information processing. Consequently, the brain:

> Does not secrete thought "as the liver does bile," as the earlier materialists claimed, nor does it put out in the form of energy, as the muscle puts out its activity. Information is information, not matter or energy. No materialism which does not admit this can survive at the present day. (Wiener, 1948, p. 155).

Given Wiener's metaphysical view, all things in the universe come into existence, persist, and then disappear by means of the continuous mixing and mingling of information and matter-energy. Living organisms, including human beings, are actually *patterns of information* that persist through an ongoing exchange of matter-energy. Thus, he says of human beings,

> We are but whirlpools in a river of ever-flowing water. We are not stuff that abides, but patterns that perpetuate themselves. (Wiener, 1954, p. 96).
>
> ...
>
> The individuality of the body is that of a flame ... of a form rather than of a bit of substance. (Wiener, 1954, p. 102).

Today we would say that, according to Wiener, humans are "information objects" whose personal identity and intellectual capacities are dependent upon persisting patterns of information and information processing within the body, rather than on specific bits of matter-energy.

2.2.1 Wiener's Account of a Good Life

To live well, according to Wiener, human beings must be free to engage in creative and flexible actions that maximize their full potential as intelligent, decision-making beings in charge of their own lives. This is the purpose of a human life. Different people, of course, have various levels of talent and possibility, so one person's achievements will differ from another's. It is possible, though, to lead a good human life—to flourish—in an indefinitely large number of ways; for example, as a teacher, scientist, nurse, doctor, housewife, midwife, diplomat, soldier, musician, artist, tradesman, artisan, and so on.

Wiener's view of the purpose of a human life led him to adopt what he called "great principles of justice" upon which a society should be built—principles that, he believed, would maximize a person's ability to flourish through variety and flexibility of human action. To highlight Wiener's "great principles," I call them "The Principle of Freedom," "The Principle of Equality," and "The Principle of Benevolence." (Wiener simply stated them, he did not assign names). Using his own words yields the following list of principles (Wiener, 1954, pp. 105–106):

THE PRINCIPLE OF FREEDOM—Justice requires "the liberty of each human being to develop in his freedom the full measure of the human possibilities embodied in him."

THE PRINCIPLE OF EQUALITY—Justice requires "the equality by which what is just for A and B remains just when the positions of A and B are interchanged."

THE PRINCIPLE OF BENEVOLENCE—Justice requires "a good will between man and man that knows no limits short of those of humanity itself."

Wiener's cybernetic account of human nature and society leads to the view that people are fundamentally social beings who can reach their full potential only by actively participating in communities of similar beings. Society, therefore, is essential to a good human life. But a despotic society could be oppressive, and thereby stifle human freedom, so Wiener introduced a fourth principle intended to minimize society's negative impact upon freedom. (Let us call it "The Principle of Minimum Infringement of Freedom.")

> THE PRINCIPLE OF MINIMUM INFRINGEMENT OF FREEDOM—"What compulsion the very existence of the community and the state may demand must be exercised in such a way as to produce no unnecessary infringement of freedom." (1954, p. 106).

Given Wiener's account of human nature and a good society, it follows that many different cultures—with a wide variety of customs, practices, languages, and religions—can nevertheless provide an appropriate context for a good human life. Because of his view that "variety and possibility belong to the very structure of the human organism," Wiener can welcome the existence of a broad diversity of cultures in the world to maximize the possibility of choice and creative action. The primary restriction that Wiener would impose on any society is that it should provide a context where humans can realize their full potential as sophisticated information-processing agents, making decisions and choices, and thereby taking responsibility for their own lives. Wiener believed that this is possible only where significant freedom, equality, and human compassion prevail.

Ethical relativists sometimes cite the wide diversity of cultures in the world—with different values, laws, codes, and practices—as evidence that there is no underlying ethical foundation that can apply everywhere. Wiener could respond that his account of human nature and the purpose of a human life can embrace and welcome a rich variety of cultures and practices while still advocating adherence to "the great principles of justice." These principles offer a foundation for ethics across cultures; and they still leave room for—indeed, welcome—immense cultural diversity.

2.2.2 Wiener's Information Ethics Methodology

When one observes Wiener's way of analyzing and trying to resolve information ethics issues, one finds—for example, in his book *The Human Use of Human Beings*—that he attempts to *assimilate new cases by applying already existing, ethically acceptable laws, rules, and practices*. In any given society, there will be a nexus of existing practices, principles, laws, and rules that govern human behavior within that society. These "policies"—to borrow a helpful word from Moor (1985)—constitute a "received policy cluster" (see Bynum and Schubert, 1997); and in a reasonably just society, they can serve as *a good starting point for developing an answer to any information ethics question*. Wiener combined the "received policy cluster" with his account of human nature, his "great principles of justice," and his critical skills in clarifying vague or ambiguous language; and he thereby achieved a very effective method for analyzing information ethics issues:

(1) Identify an ethical question or case regarding the integration of information technology into society. Typically this will focus upon technology-generated possibilities that could significantly affect (or are already affecting) life, health, security, happiness, freedom, knowledge, opportunities, or other key human values.
(2) Clarify any ambiguous or vague ideas or principles that may apply to the case or issue in question.
(3) If possible, apply already existing, ethically acceptable principles, laws, rules, and practices (the "received policy cluster") that govern human behavior in the given society.
(4) If ethically acceptable precedents, traditions, and policies are insufficient to settle the question or deal with the case, use the purpose of a human life plus the great principles of justice to find a solution that fits as well as possible into the ethical traditions of the given society.

If the traditions, precedents, and policies that one starts with are embedded within a reasonably just society, then this method of analyzing and resolving information ethics issues will likely provide just solutions that can be assimilated into that society.

Note that this way of doing information ethics does not require the expertise of a trained philosopher (although such expertise might prove to be helpful in many situations). Any adult who functions successfully in a reasonably just society is likely to be familiar with existing customs, practices, rules, and laws that govern one's behavior and enable one to tell whether a proposed action or policy would be ethically acceptable. As a result, those who must cope with the introduction of new information technology—whether they are public policy makers, computer professionals, business people, workers, teachers, parents, or others—can and should engage in information ethics by helping to integrate new information technology into society in an ethically acceptable way. Information ethics, understood in this very broad sense, is too important to be left only to philosophers or to information professionals.

In the late 1940s and early 1950s, Wiener made it clear that, in his view, the integration into society of newly invented computing and information technology would lead to the remaking of society—to "the second industrial revolution"—to "the automatic age." It would affect every walk of life, and would be a multifaceted, ongoing process requiring decades of effort. In Wiener's words, the new information technology had placed human beings "in the presence of another social potentiality of unheard-of importance for good and for evil." (Wiener, 1948, p.27) Today, the "information age" that Wiener predicted half a century ago has come into existence; and the metaphysical and scientific foundation for information ethics that he laid down can still provide insight and effective guidance for understanding and resolving many ethical challenges engendered by information technologies of all kinds.

2.3 COMPUTER ETHICS DEVELOPMENTS AFTER WIENER AND BEFORE MANER

The information ethics achievements of Norbert Wiener in the 1950s and early 1960s were much broader than the more specific field of computer ethics, as we think of it today. Thus, Wiener's information ethics ideas and methods apply not only to computer ethics, in the narrow sense of this term, but also to other specific areas such as "agent" ethics, Internet ethics, the ethics of nanotechnology, the ethics of bioengineering, even journalism ethics, and library ethics.

In the specific field of computer ethics, there were some developments that occurred in the 1960s and early 1970s *before* Walter Maner coined the name "computer ethics" and offered a definition of the field. (See the discussion of Maner below.) In the mid-1960s, for example, Donn Parker—a computer scientist at SRI International—began to notice and study unethical and illegal activities of computer professionals. Parker (1968) gathered example cases of computer crimes, and he published the article, "Rules of Ethics in Information Processing," in *Communications of the ACM*. He also headed the development of the first Code of Professional Conduct for the Association for Computing Machinery, which eventually was adopted by the ACM in 1973. Later, he published a number of books and articles on computer crime (see Parker, 1979; Parker et al., 1990).

During the late 1960s, Joseph Weizenbaum—a computer scientist at MIT—created a simple computer program that he called ELIZA. He wanted it to provide a rough imitation of "a Rogerian psychotherapist engaged in an initial interview with a patient." Weizenbaum was surprised and upset by the reactions that people had to his simple program. Some psychiatrists, for example, considered ELIZA to be evidence that computers soon would perform automated psychotherapy; and some computer scientists at MIT, who knew how the ELIZA program worked, nevertheless became emotionally involved with it and shared their intimate thoughts with it. Because of such reactions to his simple computer program, Weizenbaum was concerned that an "information-processing model" of human beings was reinforcing an already growing tendency among scientists, and even the public, to see humans as "mere machines." In response, Weizenbaum (1976) wrote the book *Computer Power and Human Reason* in which he forcefully expressed his worries. This book and some speeches he gave around the country inspired a number of scholars to think about the social and ethical impacts of computing.

2.4 WALTER MANER'S COMPUTER ETHICS INITIATIVE

Until the mid-1970s, notwithstanding the works of Wiener, Parker, and Weizenbaum, the name "computer ethics" was not in use, and no one, it seems, expressed a belief that those scholars had been working in a new branch of applied ethics. This changed in the mid-1970s, when Walter Maner, a faculty member in Philosophy at Old Dominion University, noticed in his medical ethics class that ethical problems in which computers became involved often were made worse or were significantly altered by the addition of computing technology. Indeed, it seemed to Maner that computers might even create new ethical problems that had never been seen before. Additional examination of this phenomenon in areas other than medicine led Maner to conclude that a new branch of applied ethics, modeled upon medical ethics or business ethics, should be recognized by philosophers. He coined the name "computer ethics" to refer to this proposed new field, and he developed an experimental course designed primarily for students of computer science. The course was a success, and Maner started to teach computer ethics on a regular basis.

On the basis of his teaching experiences and his research in the proposed new field, Maner (1978) created a Starter Kit on Teaching Computer Ethics and disseminated copies of it to attendees of workshops that he ran and speeches that he gave at philosophy conferences and computing conferences in America. In 1980, Helvetia Press and the National Information and Resource Center on Teaching Philosophy, headed by the present author, published Maner's computer ethics "starter kit" as a monograph (Maner, 1980). It contained curriculum materials and pedagogical advice for university teachers to develop computer ethics courses. It also included suggested course descriptions for university catalogs, a rationale for offering such a course in a university, a list of course objectives, some teaching tips, and discussions of topics like privacy and confidentiality, computer crime, computer decisions, technological dependence and professional codes of ethics.

During the early 1980s, Maner's Starter Kit on Teaching Computer Ethics was widely disseminated to colleges and universities in America and elsewhere. He also continued to conduct workshops and teach courses in computer ethics. As a result, a number of scholars, especially philosophers and computer scientists, were introduced to computer ethics because of Maner's trailblazing efforts.

2.5 DEBORAH JOHNSON'S INFLUENTIAL TEXTBOOK AND THE START OF THE "UNIQUENESS DEBATE"

As Maner was developing his new computer ethics course, he began to describe the proposed new field as one that studies ethical problems that are "aggravated, transformed or created by computer technology." Some old ethical problems, he said, are made worse by computers, while others are unique problems never seen before computing was invented. A colleague in the Philosophy faculty at Old Dominion University, Deborah Johnson, got interested in Maner's proposed new field of ethical research. She was especially struck by his claim that computer technology generates wholly new ethical problems, because she did not believe that this is true. She did grant, though, that computing technology could alter old ethical problems in interesting and important ways and thereby "give them a new twist." Maner and Johnson had a number of discussions about this proposed new field, and about some allegedly new ethical cases. These early conversations launched a decades-long series of comments and publications on the nature and uniqueness of computer ethics—the "uniqueness debate"—a series of scholarly exchanges that attracted other scholars and led to a number of helpful contributions to computer ethics (see Bynum, 2006, 2007; Floridi and Sanders, 2004; Gorniak-Kocikowska, 1996; Himma, 2003; Johnson, 1999, 2001; Maner, 1980, 1996, 1999; Mather, 2005; Tavani, 2002, 2005).

After leaving Old Dominion University, Johnson joined the staff of Rensselaer Polytechnic Institute and secured a National Science Foundation grant to prepare a set of teaching materials in computer ethics. These turned out to be very successful, and she incorporated them into a textbook, *Computer Ethics*, published in 1985. On page 1 of that book, she noted that computers "pose new versions of standard moral problems and moral dilemmas, exacerbating the old problems, and forcing us to apply ordinary moral norms in uncharted realms." She did not, however, grant Maner's claim that computers create *wholly new* ethical problems. Her book was the first major textbook in computer ethics, and it quickly became the primary text used in university computer ethics courses. It set the research agenda in computer ethics, covering topics such as the ownership of software and intellectual property, computing and privacy, responsibility of computer professionals, and the just distribution of technology and human power. In later editions (1994, 2001), Johnson added new ethical topics, such as "hacking" into people's computers without their permission, computer technology for persons with disabilities, and the Internet's impact upon democracy.

Also in later editions in her textbook, Johnson added to the ongoing "uniqueness debate" with Maner and others regarding the nature of computer ethics issues. For example, she noted that computing technology led to the creation of new types of entities, such as software and electronic databases, and new ways to "instrument" human actions. One can indeed raise new *specific* ethical questions about such innovations—for example, "Should ownership of software be protected by law?" or "Do huge databases of personal information threaten privacy?" but she argued in both later editions that such questions are merely "*new species of old moral issues*" like the protection of human privacy or the ownership of intellectual property. They are not, she said, wholly new ethical problems requiring additions to traditional ethical theories, as Maner had claimed.

2.6 JAMES MOOR'S CLASSIC PAPER AND HIS INFLUENTIAL COMPUTER ETHICS THEORY

The year 1985 was a "watershed year" in the history of computer ethics, not only because of Johnson's agenda-setting textbook, but also because of the publication of Moor's classic paper, "What is computer ethics?", which was the lead article in a special issue of the journal *Metaphilosophy*.[1] Moor's essay contained an account of *the*

[1]That special issue (October 1985) was published as the monograph *Computers and Ethics* (Bynum, 1985), and it contained several other computer ethics articles, including an important one by Pecorino and Maner on teaching computer ethics (Pecorino and Maner, 1985).

nature of computer ethics that was broader and more ambitious than those of Maner or Johnson. It went beyond descriptions and examples of computer ethics problems and offered an explanation of *why* computing technology raised so many ethical questions compared to other technologies. Computing technology is genuinely revolutionary, said Moor, because it is "logically malleable":

> Computers are logically malleable in that they can be shaped and molded to do any activity that can be characterized in terms of inputs, outputs, and connecting logical operations . . .. Because logic applies everywhere, the potential applications of computer technology appear limitless. The computer is the nearest thing we have to a universal tool. Indeed, the limits of computers are largely the limits of our own creativity. (Moor, 1985, p. 269).

Because of logical malleability, computing technology enables human beings to do an enormous number of new things that they never were able to do before. Since no one did them before, the question arises whether one *ought* to do them. In a significant number of cases like this, one may discover that no laws or standards of good practice or ethical rules have been created to govern them. Moor identified such situations as "policy vacuums," some of which might generate "conceptual muddles":

> A typical problem in computer ethics arises because there is a policy vacuum about how computer technology should be used. Computers provide us with new capabilities and these in turn give us new choices for action. Often, either no policies for conduct in these situations exist or existing policies seem inadequate. A central task of computer ethics is to determine what we should do in such cases, that is, formulate policies to guide our actions . . .. One difficulty is that along with a policy vacuum there is often a conceptual vacuum. Although a problem in computer ethics may seem clear initially, a little reflection reveals a conceptual muddle. What is needed in such cases is an analysis that provides a coherent conceptual framework within which to formulate a policy for action. (Moor, 1985, p. 266).

Many people found Moor's account of computer ethics to be insightful and a helpful way to understand and deal with emerging computer ethics issues. His account of the nature of computer ethics quickly became the most influential one among a growing number of scholars across America who were joining the computer ethics research community.

In 1996, in his ETHICOMP96 keynote address (see Moor, 1998), Moor enhanced his computer ethics theory with additional ideas. One of these was his notion of the "informationalization" of a task. This occurs when one uses computers to do an "old" job more efficiently. Eventually, however, computing begins to do the old job in a new way, and information processing becomes an integral part of the task. The resulting informationalization of a task can sometimes alter the meanings of old terms and create conceptual muddles that need to be clarified.

Another significant addition to Moor's theory, also introduced in the ETHICOMP96 address, is the notion of "core values." Some human values—such as *life, health, happiness, security, resources, opportunities,* and *knowledge*, for example—are so important to the continued survival of any community that virtually all communities do value them. If a community did not value these things, it soon would cease to exist. In later papers, Moor used the concept of core values very effectively to address computer ethics issues, such as privacy (Moor, 1997), and to add an account of justice, which he called "just consequentialism" (Moor, 1999), to the rest of his computer ethics theory.[2]

Moor's way of analyzing and resolving computer ethics issues was both creative and very practical. It provided a broad perspective on the nature of the Information Revolution; and, in addition, by using powerful ideas like "logical malleability," "policy vacuums," "conceptual muddles," "core values," and "just consequentialism," Moor provided a very effective problem-solving method:

(1) Identify a policy vacuum generated by computing technology.
(2) Eliminate any conceptual muddles.
(3) Use the core values and the ethical resources of just consequentialism to revise existing, but inadequate policies or to create new policies that will fill the vacuum and thereby resolve the original ethical issue.

2.7 THE PROFESSIONAL-ETHICS APPROACH OF DONALD GOTTERBARN

In the early 1990s, a different understanding of the nature of computer ethics was advocated by Donald Gotterbarn. He believed that computer ethics should be seen as a *professional ethics* devoted to the development and advancement of standards of good practice and codes of conduct for computing professionals. Thus in 1991, in the article "Computer ethics: responsibility regained," Gotterbarn said:

> There is little attention paid to the domain of professional ethics—the values that guide the day-to-day activities of computing professionals in their role as professionals. By computing professional I mean anyone involved in the design and development of computer artefacts .... The ethical decisions made during the development of these artifacts have a direct relationship to many of the issues discussed under the broader concept of computer ethics. (Gotterbarn, 1991).

With this understanding of the nature of computer ethics in mind, Gotterbarn actively created and participated in a variety of projects intended to advance professional responsibility among computing professionals. Even before 1991, for example, Gotterbarn had been working with a committee of the ACM creating the third version of that organization's "Code of Ethics and Professional Conduct" (adopted by the ACM in 1992). Later, he worked with others in the ACM and the Computer Society of the IEEE (Institute of Electrical and Electronic Engineers)

[2]Moor developed his "just consequentialism" theory by combining his notion of "core values" with Bernard Gert's theory of justice in his book *Morality: Its Nature and Justification* (Gert, 1998).

developing licensing standards for software engineers. He became Chair of the ACM Committee on Professional Ethics, and he headed a joint taskforce of the IEEE and ACM to create the "Software Engineering Code of Ethics and Professional Practice" (adopted by those organizations in 1999). In the late 1990s, he created the Software Engineering Research Institute (SEERI) at East Tennessee State University (see http://seeri.etsu.edu/); and in the early 2000s, together with Simon Rogerson from De Montfort University in the UK, Gotterbarn created a computer program called SoDIS (Software Development Impact Statements) to assist individuals, companies, and organizations in the preparation of ethical "stakeholder analyses" to determine the likely ethical impacts of software development projects (Gotterbarn and Rogerson, 2005). These and many other important projects that focused upon *professional responsibility* made Gotterbarn one of the most important thinkers among those who are advancing the professionalization and ethical maturation of computing practitioners (see Anderson et al., 1993; Gotterbarn, 1991, 2001; Gotterbarn et al., 1997).

2.8 COMPUTING AND HUMAN VALUES

A common ethical thread that runs through much of the history of computer ethics, from Norbert Wiener onward, is concern for the protection and advancement of central human values such as life, health, security, happiness, freedom, knowledge, resources, power, and opportunity. Most of the specific examples and cases that Wiener dealt with in his relevant books (Wiener, 1950, 1954, 1964) are examples of defending or advancing such values, for example, preserving security, resources and opportunities for factory workers by preventing massive unemployment from robotic factories, or avoiding threats to national security from decision-making, war-game machines. Moor called such central human values "core values," and he noted that they are crucial to the long-term survival of any community (Moor, 1996).

The fruitfulness of the "human-values approach" to computer ethics is reflected in the fact that it has served as the organizing theme of some major computer-ethics conferences. For example, in 1991 the National Conference on Computing and Values (see the section below on "exponential growth"), was devoted to examining the impacts of computing upon security, property, privacy, knowledge, freedom, and opportunities.[3] In the late 1990s, a new approach to computer ethics, "value-sensitive computer design," emerged based upon the insight that potential computer-ethics problems can be avoided, while new technology is under development, by *anticipating possible harm to human values and designing new technology from the very beginning in ways that prevent such harm.* (see Brey, 2000; Flanagan et al., 2007; Friedman, 1997; Friedman and Nissenbaum, 1996; Introna, 2005a; Introna and Nissenbaum, 2000.)

[3]Materials from that conference can be found on the Web site of the Research Center on Computing & Society: http://www.southernct.edu/organizations/rccs/

2.9 LUCIANO FLORIDI'S INFORMATION ETHICS THEORY

In spite of the helpfulness and success of the "human-values approach" to computer ethics, some scholars have argued that the purview of computer ethics—indeed of ethics in general—should be widened to include much more than simply human beings and their actions, intentions and characters. One such thinker is Luciano Floridi, who has proposed a new general ethics theory that is different from traditional human-centered theories such as utilitarianism, Kantianism, or virtue ethics. This new ethical theory is INFORMATION ETHICS[4] (IE) developed by Floridi (with some of his colleagues in Oxford University's Information Ethics Research Group) in the late 1990s and early 2000s (see Floridi, 1999, 2006; Floridi and Sanders, 2004).

Floridi and his colleagues developed IE as a "foundation" for computer ethics. On the one hand, it is a "macroethics" that is *similar* to utilitarianism, Kantianism, contractualism, or virtue ethics, because it is intended to be applicable to all ethical situations. On the other hand, it is *different* from these more traditional Western ethical theories because it is not intended to replace them, but rather to supplement them with further ethical considerations that go beyond the traditional theories, and that can be overridden, sometimes, by traditional ethical considerations (Floridi, 2006).

The name INFORMATION ETHICS is appropriate to Floridi's theory, because it treats everything that exists as "informational" objects or processes:

> (All) entities will be described as clusters of data, that is, as informational objects. More precisely, (any existing entity) will be a discrete, self contained, encapsulated package containing:
>
> (i) the appropriate data structures, which constitute the nature of the entity in question, that is, the state of the object, its unique identity and its attributes; and
> (ii) a collection of operations, functions, or procedures, which are activated by various interactions or stimuli (that is, messages received from other objects or changes within itself) and correspondingly define how the object behaves or reacts to them.
>
> At this level of abstraction, informational systems as such, rather than just living systems in general, are raised to the role of agents and patients of any action, with environmental

[4]In this essay, I use the convention of writing the name of Floridi's new ethical theory in SMALL CAPS to distinguish it from the much broader, and less rigorously developed, information ethics theory of Wiener. Although there are some similarities, Floridi's theory and Wiener's have very different metaphysical foundations. Thus, Wiener's theory is a kind of materialism grounded in the laws of physics, while Floridi's theory presupposes a Spinozian, perhaps even a Platonic, metaphysics (Floridi, 2006). In Floridi's theory, but not in Wiener's, nonhuman entities such as rivers, databases, and stones have "rights" that ought to be respected. Floridi's "entropy" is not the entropy of physics, as it is in Wiener; Floridi's "information" is not the "Shannon information" of physics, as it is in Wiener, and Floridi's world includes nonmaterial Platonic entities that have no place in Wiener's universe.

> processes, changes and interactions equally described informationally. (Floridi, 2006, 9–10).

Since everything that exists, according to Floridi's theory, is an informational object or process, he calls the totality of all that exists—the universe considered as a whole—"the infosphere." Objects and processes in the infosphere can be significantly damaged or destroyed by altering their characteristic data structures. Such damage or destruction Floridi calls "entropy," and it results in partial "impoverishment of the infosphere." *Entropy in this sense is an evil that should be avoided or minimized*, and Floridi offers four "fundamental principles" of INFORMATION ETHICS:

(i) Entropy ought not to be caused in the infosphere (null law).
(ii) Entropy ought to be prevented in the infosphere.
(iii) Entropy ought to be removed from the infosphere.
(iv) The flourishing of informational entities as well as the whole infosphere ought to be promoted by preserving, cultivating, and enriching their properties.

Floridi's IE theory is based upon the idea that everything in the infosphere has at least a minimum worth that should be ethically respected, even if that worth can be overridden by other considerations:

> IE suggests that there is something even more elemental than life, namely *being*—that is, the existence and flourishing of all entities and their global environment—and something more fundamental than suffering, namely *entropy* .... IE holds that *being*/information has an intrinsic worthiness. It substantiates this position by recognizing that any informational entity has a *Spinozian* right to persist in its own status, and a *Constructionist* right to flourish, that is, to improve and enrich its existence and essence. (Floridi, 2006, p. 11, italics in the original).

By construing every existing entity in the universe as "informational," with at least a minimal moral worth, Floridi is able to supplement traditional ethical theories and go beyond them by shifting the focus of one's ethical attention away from the actions, characters, and values of human agents toward the "evil" (harm, dissolution, and destruction)—"entropy"—suffered by objects and processes in the infosphere. With this approach, every existing entity—humans, other animals, plants, organizations, even nonliving artifacts, electronic objects in cyberspace, pieces of intellectual property—can be interpreted as *potential agents* that affect other entities, and as *potential patients* that are affected by other entities. In this way, Floridi treats INFORMATION ETHICS as a "patient-based" nonanthropocentric ethical theory to be used in addition to the traditional "agent-based" anthropocentric ethical theories like utilitarianism, Kantianism and virtue ethics.

Floridi's IE theory, with its emphasis on "preserving and enhancing the infosphere," enables him to provide, among other things, an insightful and practical ethical theory of robot behavior and the behavior of other "artificial agents" like softbots and cyborgs[5] (see Floridi and Sanders, 2004). His IE theory is an impressive component of a much more ambitious project covering the entire new field of the philosophy of information.

2.10 CONCLUDING REMARKS: THE EXPONENTIAL GROWTH OF COMPUTER ETHICS

The paragraphs above describe some key contributions to "the history of ideas" in computer ethics; but the history of computer ethics includes much more. Just as important to the birth and growth of a discipline is cooperation among a "critical mass" of scholars, as well as the creation of courses to teach, conferences to attend, research centers for planning and conducting research projects, and journals and other places to publish the results of the research. This concluding section of the present essay describes some milestones in that important aspect of the history of computer ethics.

The year 1985 was a pivotal year for computer ethics. Johnson's new textbook, plus the *Computers and Ethics* issue of the journal *Metaphilosophy* (October 1985)—especially Moor's article "What is computer ethics?"—provided excellent curriculum materials and a conceptual foundation for the field. In addition, Maner's earlier trailblazing efforts (and, to some extent, similar efforts by the present author) had created a "ready-made audience" of enthusiastic computer science and philosophy scholars. The stage was set for exponential growth.

Rapid growth of computer ethics occurred in the United States during the decade between 1985 and 1995. In 1987, for example, the Research Center on Computing & Society (RCCS, see www.southernct.edu/organizations/rccs/) at Southern Connecticut State University was created. Shortly thereafter, the Director of RCCS (the present author) joined with Walter Maner to organize a national computer ethics conference that would bring together computer scientists, philosophers, public policy makers, lawyers, journalists, sociologists, psychologists, business people, and others. The goal of the conference was to examine and push forward some of the major subareas of

---

[5]If one assumes, as Floridi does, that everything in the universe deserves moral respect, and things other than humans can damage—through their behavior and interactions with the world—other beings that have moral worth, then ethics needs some way to understand "evil" behavior (i.e., behavior that damages things of value) and "good" behavior (i.e., behavior that causes things of value to flourish, or at least does not damage them). Floridi's IE is an effort to make moral sense of these possibilities. Consider, for example, possible rules of behavior for machines, which already exist today, and which will exist by the millions in a few years, that gather information about the world, make decisions based on that information, and carry out those decisions: Specific examples include machines that automatically inject patients with medicine when the machine detects the need for such medicine; or machines that fly airplanes by constantly checking how the flight is going and making adjustments so that the planes get to their destinations safely. If such machines do this in a way that makes the world a better place and fosters the flourishing of things of value, their behavior is good. If not, their behavior is bad. What are the rules that good machines—indeed good agents of any kind—ought to obey to be considered good rather than evil? This is a vital ethical question, whose importance will grow exponentially in the next few decades. Floridi is developing a carefully argued theory to try to fill this need (and many other needs as well, given the pace and increasing complexity of the "information revolution").

computer ethics, such as computer security, computers and privacy, ownership of software and intellectual property, computing for persons with disabilities, and the teaching of computer ethics. More than a dozen of the most important thinkers in the field joined with Bynum and Maner to plan the conference, which was named "the National Conference on Computing and Values" (NCCV). Funded by grants from the National Science Foundation, NCCV occurred in August 1991 on the campus of Southern Connecticut State University; and it was considered a watershed event in the history of the subject. It included 65 speakers and attracted 400 attendees from 32 American states and 7 other countries. It generated a wealth of new computer ethics materials—monographs, video programs, and an extensive bibliography—that were disseminated to hundreds of colleges and universities during the following two years.

At the same time, professional ethics advocates such as Donald Gotterbarn, Keith Miller, and Dianne Martin—and organizations such as Computer Professionals for Social Responsibility (www.cpsr.org), the Electronic Frontier Foundation (www.eff.org), and the Special Interest Group on Computing and Society (SIGCAS) of the ACM—spearheaded developments relevant to computing and professional responsibility. For example, the ACM adopted a new version of its Code of Ethics and Professional Conduct (1992), computer ethics became a required component in undergraduate computer science programs that were nationally accredited by the Computer Sciences Accreditation Board (1991), and the important annual "Computers, Freedom, and Privacy" conferences (www.cfp.org) began (1991).

In 1995, rapid growth of computer ethics spread to Europe when the present author joined with Simon Rogerson of De Montfort University in Leicester, England to create the Centre for Computing and Social Responsibility (www.ccsr.cse.dmu.ac.uk) and to organize the first computer ethics conference in Europe, ETHICOMP95. That conference attracted attendees from 14 different countries, mostly European, and was a major factor in generating a "critical mass" of computer ethics scholars in Europe. For a decade thereafter, every 18 months, another ETHICOMP conference was held in a different European country, including Spain (1996), the Netherlands (1998), Italy (1999), Poland (2001), Portugal (2002), Greece (2004) and Sweden (2005). In 1999, with assistance from Bynum and Rogerson, the Australian scholars John Weckert and Chris Simpson created the Australian Institute of Computer Ethics (see aice.net.au) and organized AICEC99 (Melbourne, Australia), which was the first international computer ethics conference south of the equator. In 2007 Rogerson and Bynum also headed ETHICOMP2007 in Tokyo, Japan, and an ETHICOMP "Working Conference" in Kunming, China to help spread interest in computer ethics to Asia.

The person most responsible for the rapid growth of computer ethics in Europe was Simon Rogerson. He not only created the Centre for Computing and Social Responsibility at De Montfort University and coheaded the ETHICOMP conferences, but also he (1) added computer ethics to De Montfort University's curriculum, (2) created a graduate program with advanced computer ethics degrees, including the Ph.D., and (3) cofounded and coedited (with Ben Fairweather) two computer ethics journals—*The Journal of Information, Communication and Ethics in Society* in 2003 (www.troubador.co.uk/ices/, Rogerson and Fairweather, 2003); and the electronic *ETHICOMP Journal* in 2004 (www.ccsr.cse.dmu.ac.uk/journal/). Rogerson also served on the Information Technology Committee of the British Parliament, and participated in several computer ethics projects with agencies of the European Union.

Other important computer ethics developments in Europe in the late 1990s and early 2000s included, for example, (1) Luciano Floridi's creation of the Information Ethics Research Group at Oxford University in the mid 1990s; (2) Jeroen van den Hoven's founding of the CEPE (Computer Ethics: Philosophical Enquiry) series of computer ethics conferences, which have occurred alternately in Europe and America since 1997; (3) Van den Hoven's (1999) creation of the journal *Ethics and Information Technology*; (4) Rafael Capurro's creation of the International Center for Information Ethics (icie.zkm.de) in 1999; (5) Capurro's (2004) creation of the journal *International Review of Information Ethics*; and Stahl's (2005) creation of *The International Journal of Technology and Human Interaction*.

In summary, since 1995 computer and information ethics developments have exponentially proliferated with new conferences and conference series, new organizations, new research centers, new journals, textbooks, Web sites, university courses, university degree programs, and distinguished professorships. Additional "subfields" and topics in computer ethics continually emerge as information technology itself grows and proliferates. Recent new topics include online ethics, "agent" ethics (robots, softbots), cyborg ethics (part human, part machine), the "open source movement," electronic government, global information ethics, information technology and genetics, computing for developing countries, computing and terrorism, ethics and nanotechnology, to name only a few examples. (For specific publications and examples, see the list of selected resources below.)

Compared to many other scholarly disciplines, the field of computer and information ethics is very young. It has existed only since the mid-1940s when Norbert Wiener created it. During the first three decades, it grew very little because Wiener's insights were far ahead of everyone else's. In the past 25 years, however, computer and information ethics has grown exponentially in the industrialized world, and today the rest of the world has begun to take notice. As the "information revolution" transforms the world in the coming decades, computer and information ethics will surely grow and flourish as well.

REFERENCES AND SELECTED RESOURCES

Adam, A. (2000). Gender and computer ethics. *Computers and Society*, 30(4), 17–24.

Adam, A., and Ofori-Amanfo, J. (2000). Does gender matter in computer ethics? *Ethics and Information Technology*, 2(1), 37–47.

Anderson, R. Johnson, D., Gotterbarn, D. and Perrolle, J. (1993). Using the ACM code of ethics in decision making. *Communications of the ACM*, 36, 99–107.

Begg, M.M. (2005). *Muslim Parents Guide: Making Responsible Use of Information and Communication Technologies at Home*. Centre for Computing and Social Responsibility, De Montfort University, Leicester, UK.

Brey, P. (2000). Disclosive computer ethics. *Computers and Society*, 30(4), 10–16.

Brey, P. (2006). Evaluating the social and cultural implications of the Internet. *Computers and Society*, 36(3), 41–44.

Brey, P., Introna, L., and Grodzinski, F. (2005). Ethics of new information technologies. *Proceedings of the 6th Computer Ethics—Philosophical Enquiries Conference*. University of Twente, The Netherlands.

Bynum, T.W. (1982). A discipline in its infancy. *The Dallas Morning News, Dallas, TX*, at http://www.southernct.edu/organizations/rccs/resources/research/introduction/bynum_dallas.html.

Bynum, T.W. (Ed.) (1985). *Computers and Ethics*. Blackwell, Oxford, UK. [Published as the October, 1985 issue of the journal *Metaphilosophy*].

Bynum, T.W. (1999). The development of computer ethics as a philosophical field of study. *The Australian Journal of Professional and Applied Ethics*, 1(1), 1–29.

Bynum, T.W. (2000). The foundation of computer ethics. *Computers and Society*, 30(2), 6–13.

Bynum, T.W. (2001). Computer ethics: basic concepts and historical overview. *Stanford Encyclopedia of Philosophy*. Available at http://plato.stanford.edu/entries/ethics-computer/.

Bynum, T.W. (2004). Ethical challenges to citizens of the "automatic age": Norbert Wiener on the information society. *Journal of Information, Communication and Ethics in Society*. 2(2), 65–74.

Bynum, T.W. (2005). Norbert Wiener's vision: the impact of the "automatic age" on our moral lives. In: Cavalier, R.J. (Ed.), *The Impact of the Internet on Our Moral Lives*. SUNY Press, Albany, NY, pp. 11–25.

Bynum, T.W. (2006). Flourishing ethics. *Ethics and Information Technology*, 8(4), 157–173.

Bynum, T.W. (2007). Norbert Wiener and the rise of information ethics. In: van den Hoven, W.J. and Weckert, J. (Eds.), *Moral Philosophy and Information Technology*. Cambridge University Press, Cambridge, UK.

Bynum, T.W. and Schubert, P. (1997). How to do computer ethics—a case study: the electronic mall Bodensee. In: van den Hoven, W.J. (Ed.), *Computer Ethics–Philosophical Enquiry*. Erasmus University Press, Rotterdam, pp. 85–95 (Proceedings of CEPE97).

Bynum, T.W. and Rogerson, S. (Eds.), *Computer Ethics and Professional Responsibility*. Blackwell, Oxford, UK.

Capurro, R. (2007a). Information ethics for and from Africa. *International Review of Information Ethics*, 7, 2–10.

Capurro, R. (2007b). Intercultural information ethics. In: Capurro, R., Frühbauer, J. and Hausmanninger, T. (Eds.), *Localizing the Internet: Ethical Issues in Intercultural Perspective*, ICIE Series, Vol. 4, Fink, Munich.

Capurro, R. (2006). Toward an ontological foundation for information ethics. *Ethics and Information Technology*, 8(4), 157–186.

Capurro, R. (2004). The German debate on the information society. *The Journal of Information, Communication and Ethics in Society*, 2, Supplement, 17–18.

Cavalier, R.J. (Ed.) (2005). *The Impact of the Internet on Our Moral Lives*. SUNY Press, Albany, NY.

Edgar, S.L. (1997). *Morality and Machines: Perspectives on Computer Ethics*. Jones and Bartlett, Sudbury, MA.

Elgesem, D. (1995). Data privacy and legal argumentation. *Communication and Cognition* 28 (1), 91–114.

Elgesem, D. (1996). Privacy, respect for persons, and risk. In: Ess, C., (Ed.), *Philosophical Perspectives on Computer-Mediated Communication*, SUNY Press, Albany, NY.

Elgesem, D. (2002). What is special about the ethical problems in internet research? *Ethics and Information Technology*, 4(3), 195–203.

Elgesem, D. (2007). Information technology ethics. In: van den Hoven, W.J. and Weckert, J. (Eds.), *Moral Philosophy and Information Technology*. Cambridge University Press, Cambridge, UK.

Ess, C. (1996). The political computer: democracy, CMC, and Habermas. In: Ess, C. (Ed.), *Philosophical Perspectives on Computer-Mediated Communication*. SUNY Press, Albany, NY, pp. 197–230.

Ess, C. (2001a). What's culture got to do with it? Cultural collisions in the electronic global village. In: Ess, C. (Ed.), *Culture, Technology, Communication: Towards an Intercultural Global Village*. SUNY Press, Albany, NY, pp. 1–50.

Ess, C.(Ed.) (2001b). *Culture, Technology, Communication: Towards an Intercultural Global Village*. SUNY Press, Albany, NY.

Ess, C. (2005a). Computer-Mediated Communication and Human-Computer Interaction. In: Floridi, L. (Ed.), *The Blackwell Guide to the Philosophy of Computing and Information*. Blackwell, Oxford, UK, pp. 76–91.

Ess, C. (2005b). Moral imperatives for life in an intercultural global village. In: Cavalier, R.J. (Ed.), *The Impact of the Internet on Our Moral Lives*. SUNY Press, Albany, NY, pp. 161–193.

Fairweather, B. (1998). No PAPA: Why incomplete codes of ethics are worse than none at all. In: Collste, G. (Ed.), *Ethics and Information Technology*. New Academic Publishers, Delhi, India.

Flanagan, M., Howe, D., and H. Nissenbaum, H. (2007). Values in design: theory and practice. In: van den Hoven, W.J. and Weckert, J. (Eds.), *Information Technology and Moral Philosophy*. Cambridge University Press, Cambridge UK.

Floridi, L. (1999). Information ethics: on the theoretical foundations of computer ethics. *Ethics and Information Technology,* 1(1), 37–56.

Floridi, L. (Ed.) (2004). *The Blackwell Guide to the Philosophy of Computing and Information*. Blackwell, Oxford, UK.

Floridi, L. (2005a). Information ethics: its nature and scope. *Computers and Society*, 36(3), 21–36

Floridi, L. (2005b). Internet ethics: the constructionist values of *homo poieticus*. In: Cavalier, R.J. (Ed.), *The Impact of the Internet on Our Moral Lives*. SUNY Press, Albany, NY, pp. 195–214.

Floridi, L. (2006). Information technologies and the tragedy of the good will. *Ethics and Information Technology*, 8(4), 253–262.

Floridi, L. and Sanders, J.W. (2004). The foundationalist debate in computer ethics. In: Spinello, R.A. and Tavani, H.T. (eds.), *Readings in CyberEthics*, 2nd edition. Jones, Bartlett, pp. 81–95.

Friedman, B. (Ed.) (1997). *Human Values and the Design of Computer Technology*, Cambridge University Press, Cambridge, UK.

Friedman, B. and Nissenbaum, H. (1996). Bias in Computer Systems. *ACM Transactions on Information Systems*, 14(3), 330–347.

Gert, B. (1998). *Morality: Its Nature and Justification*. Oxford University Press, Oxford, UK.

Gert, B. (1999). Common morality and computing. *Ethics and Information Technology*, 1(1), 57–64.

Gorniak-Kocikowska, K. (1996). The computer revolution and the problem of global ethics. In: Bynum, T. W. and Rogerson, S. (Eds.), *Global Information Ethics*, a special issue of *Science and Engineering Ethics*, 2(2), 177–190.

Gorniak-Kocikowska, K. (2005). From computer ethics to the ethics of global ICT society. In: Bynum, T.W., Collste, G., and Rogerson, S. (Eds.), *Proceedings of ETHICOMP2005,* published on a CD, Linköpings University, Sweden.

Gorniak-Kocikowska, K. (2007). ICT, globalization and the pursuit of happiness: the problem of change. In: *Proceedings of ETHICOMP2007.* Meiji University Press, Tokyo.

Gotterbarn, D. (1991). Computer ethics: responsibility regained. *National Forum: The Phi Beta Kappa Journal*, 71, 26–31.

Gotterbarn, D. (2001). Informatics and professional responsibility. *Science and Engineering Ethics*, 7(2), 221–230.

Gotterbarn, D. (2002). Reducing software failures: addressing the ethical risks of the software development life cycle. *Australian Journal of Information Systems*, 9(2), 155–165.

Gotterbarn, D., Miller, K., and Rogerson, S. (1997). Software engineering code of ethics. *Information Society*, 40(11), 110–118.

Gotterbarn, D. and Miller, K. (2004). Computer ethics in the undergraduate curriculum: case studies and the joint software engineer's code. *Journal of Computing Sciences in Colleges*, 20(2), 156–167.

Gotterbarn, D. and Rogerson, S. (2005). Responsible risk analysis for software development: creating the software development impact statement. *Communications of the Association for Information Systems*, 15, Article 40.

Grodzinsky, F. (1997). Computer access for students with disabilities. *SIGSCE Bulletin*. ACM Press, March 1997.

Grodzinsky, F. (1999). The practitioner from within: revisiting the virtues. *Computers and Society*, 29(2), 9–15.

Grodzinsky, F., Miller, K., and Wolfe, M. (2003). Ethical issues in open source software. *Journal of Information, Communication and Ethics in Society*, (4) 193–205.

Grodzinsky, F. and Tavani, H.T. (2002). Ethical reflections on cyberstalking. *Computers and Society*, 32(1), 22–32.

Grodzinsky, F. and Tavani, H. (2004). Verizon vs. the RIAA: implications for privacy and democracy. In: Herkert, J. (Ed.), *Proceedings of ISTAS 2004: The International Symposium on Technology and Society*. IEEE Computer Society Press, Los Alamitos, CA.

Himma, K.E. (2003). The relationship between the uniqueness of computer ethics and its independence as a discipline in applied ethics. *Ethics and Information Technology*, 5(4), 225–237.

Himma, K.E. (2004a). The moral significance of the interest in information: reflections on a fundamental right to information. *Journal of Information, Communication, and Ethics in Society*, 2(4), 191–202.

Himma, K.E. (2004b). There's something about Mary: the moral value of things qua information objects. *Ethics and Information Technology*, 6(3), 145–159.

Himma, K.E. (2006). Hacking as politically motivated civil disobedience: Is hacktivism morally justified? In: Himma, K.E. (Eds.), *Readings in Internet Security: Hacking, Counterhacking, and Society*. Jones and Bartlett, Sudbury, MA.

Himma, K.E. (2007). Artificial agency, consciousness, and the criteria for moral agency: What properties must an artificial agent have to be a moral agent? *Proceedings of ETHI-COMP2007.* Meiji University Press Tokyo.

Huff, C.W., Fleming, J.F., and Cooper, J. (1991). The social basis of gender differences in human-computer interaction. In: Martin, C.D. (Ed.), *In Search of Gender-Free Paradigm for Computer Science Education*. ISTE Research Monographs, Eugene, OR, pp. 19–32.

Huff, C.W. and Finholt, T. (Eds.) (1994). *Social Issues in Computing: Putting Computers in Their Place*. McGraw-Hill.

Huff, C.W. and Martin, D. (1995). Computing consequences: a framework for teaching ethical computing. *Communications of the ACM*, 38(12), 75–84.

Huff, C.W. (2002). Gender, Software Design, and Occupational Equity. *SIGCSE Bullet: Inroads*, 34, 112–115.

Huff, C.W. (2004). Unintentional power in the design of computing systems. In: Bynum, T.W. and Rogerson, S. (Eds.), *Computer Ethics and Professional Responsibility*. Blackwell, Oxford, UK (Originally presented as a paper at ETHICOMP95).

Huff, C., Johnson, D., and Miller, K. (2007). Virtual harms and real responsibility. In: Brennan, L. and Johnson, V. (Eds.), *Social, Ethical, and Policy Implications of Information Technology*. Idea Group Information Science Publishing, Hershey, PA.

Introna, L.D. (1997). Privacy and the computer: why we need privacy in the information society. *Metaphilosophy*, 28(3), 259–275.

Introna, L.D. (2002). On the (im)possibility of ethics in a mediated world. *Information and Organization*, 12(2), 71–84.

Introna, L.D. (2005a). Disclosive ethics and information technology: disclosing facial recognition systems. *Ethics and Information Technology*, 7(2), 75–86.

Introna, L.D. (2005b). Phenomenological approaches to ethics and information technology. *The Stanford Encyclopedia of Philosophy*. http://plato.stanford.edu/entries/ethics-it-phenomenology/.

Introna, L.D. and Nissenbaum, H. (2000). Shaping the web: Why the politics of search engines matters. *The Information Society*, 16(3), 1–17.

Introna, L.D. and Pouloudi, N. (2001). Privacy in the information age: stakeholders, interests and values. In: Sheth, J.N. (Eds.), *Internet Marketing*. Harcourt College Publishers, Fort Worth, TX, pp. 373–388.

Johnson, D.G. (2001[1985,1994]). *Computer Ethics*, 1st edition 1985, Prentice-Hall, Englewood Cliffs, NJ. 2nd edition 1994, Prentice-Hall, Englewood Cliffs, NJ. 3rd edition 2001, Prentice-Hall, Upper Saddle River, NJ.

Johnson, D.G. (1997a). Ethics online. *Communications of the ACM*, 40(1), 60–65.

Johnson, D.G. (1997b). Is the global information infrastructure a democratic technology? *Computers and Society*, 27(4), 20–26.

Johnson, D.G. (1999) Sorting out the uniqueness of computer-ethical issues. In: Floridi, L. (Ed.), *Etica & Politica* (special issue on computer ethics), Vol. 2. Available at http://www.univ.trieste.it/~dipfilo/etica_e_politica/1999_2/homepage.html.

Johnson, D.G. (2004). Computer ethics. In: Floridi, L. (Ed.), *The Blackwell Guide to the Philosophy of Computing and Information*, Blackwell, Oxford, UK, pp. 65–75.

Johnson, D.G. and Nissenbaum, H.(Eds.) (1995). *Computing, Ethics & Social Values*, Prentice Hall, Englewood Cliffs, NJ.

Maner, W. (1980[1978]). Starter Kit on Teaching Computer Ethics. Self-published in 1978, published in 1980 by Helvetia Press in cooperation with the National Information and Resource Center on Teaching Philosophy, Hyde Park, NY.

Maner, W. (1996). Unique ethical problems in information technology, a keynote address at ETHICOMP95. In: Bynum, T.W. and Rogerson, S. (Eds.), *Global Information Ethics, A Special Issue of Science and Engineering Ethics*, 2(2), 137–154.

Maner, W. (1999). Is computer ethics unique? In: Floridi, L. (Ed.), *Etica & Politica* (special issue on computer ethics), Vol. 2. Available at http://www.univ.trieste.it/~dipfilo/etica_e_-politica/1999_2/homepage.html.

Martin, C.D. and Martin, D.H. (1990). Professional codes of conduct and computer ethics education. *Social Science Computer Review*, 8(1), 96–108.

Martin, C.D., Huff, C., Gotterbarn, D., and Miller, K., (1996a). A framework for implementing and teaching the social and ethical impact of computing. *Education and Information Technologies*, 1(2), 101–122.

Martin, C.D., Huff, C., Gotterbarn, D., and Miller, K. (1996b). Implementing a tenth strand in the computer science curriculum (second report of the Impact CS steering committee). *Communications of the ACM*, 39(12), 75–84.

Mather, K. (2005). The theoretical foundation of computer ethics: stewardship of the information environment. *Contemporary Issues in Governance*. Monash University, Melbourne, Australia. (*Proceedings of GovNet Annual Conference*, Melbourne, Australia, pp. 28–30 November, 2005.)

Miller, K. (2005b). Web standards: Why so many stray from the narrow path. *Science and Engineering Ethics*, 11(3), 477–479.

Miller, K. and Larson, D. (2005a). Agile methods and computer ethics: raising the level of discourse about technological choices *IEEE Technology and Society*, 24(4), 36–43.

Miller, K. and Larson, D. (2005b). Angels and artifacts: moral agents in the age of computers and networks. *Journal of Information, Communication & Ethics in Society*, 3(3), 151–157.

Moor, J.H. (1979). Are there decisions computers should never make? *Nature and System*, 1 217–229.

Moor, J.H. (1985). What is computer ethics? In: Bynum (Ed.), *Computers and Ethics* (a special issue of *Metaphilosophy*), 16(4), 263–275.

Moor, J.H. (1998[1996]). Reason, relativity and responsibility in computer ethics, a keynote address at ETHICOMP96 in Madrid, Spain. Later published in *Computers and Society*, 28 (1), 14–21.

Moor, J.H. (1997). Toward a theory of privacy for the information age. *Computers and Society*, 27(3), 27–32.

Moor, J.H. (1999). Just consequentialism and computing. *Ethics and Information Technology*, 1 (1), 65–69.

Moor, J.H. (2001). The future of computer ethics: You ain't seen nothin' yet. *Ethics and Information Technology*, 3(2), 89–91.

Moor, J.H. (2005). Should we let computers get under our skin? In: Cavalier, R.J. (Ed.), *The Impact of the Internet on our Moral Lives*. SUNY Press, Albany, NY, pp. 121–138.

Moor, J.H. (2006). The nature, importance, and difficulty of machine ethics. *IEEE Intelligent Systems*, 21(4), 18–21.

Nissenbaum, H. (1995). Should I copy my neighbor's software? In: Johnson, D.G. and Nissenbaum, H. (Eds.), *Computers, Ethics and Social Values*, Prentice-Hall, Englewood Cliffs, NJ.

Nissenbaum, H. (1998[1997]). Can we protect privacy in public? Proceedings of Computer Ethics–Philosophical Enquiry 97 (CEPE97), Erasmus University Press,

Rotterdam, The Netherlands, pp. 191–204. Reprinted in *Law and Philosophy*, 1998, 17, 559–596.

Nissenbaum, H. (1998). Values in the design of computer systems. *Computers in Society*, 38–39.

Nissenbaum, H. (2005a). Hackers and the contested ontology of cyberspace. In: Cavalier, R.J. (Ed.), *The Impact of the Internet on our Moral Lives*. SUNY Press, Albany, NY, pp. 139–160.

Nissenbaum, H. (2005b). Where computer security meets national security. *Ethics and Information Technology*, 7(2), 61–73.

Parker, D. (1968). Rules of ethics in information processing. *Communications of the ACM*, 11, 198–201.

Parker, D. (1979). *Ethical Conflicts in Computer Science and Technology*, AFIPS Press, Arlington, VA.

Parker, D., Swope, S., and Baker, B.N. (1990). *Ethical Conflicts in Information & Computer Science, Technology and Business*. QED Information Sciences, Wellesley, MA.

Pecorino, P. and Maner, W. (1985). A proposal for a course on computer ethics. In: Bynum (Ed.), *Computers and Ethics, A Special Issue of Metaphilosophy*, 16 (4), 327–337.

Rogerson, S. (1996). The ethics of computing: the first and second generations. *The UK Business Ethics Network News*.

Rogerson, S. (1998). Computer and information ethics. In: Chadwick R. (Ed.), *Encylopedia of Applied Ethics*. Academic Press, San Diego, CA, pp. 563–570.

Rogerson, S. (2004). The ethics of software development project management. In: Bynum T.W. and Rogerson S. (Eds.), *Computer Ethics and Professional Responsibility*, Blackwell, Oxford, UK, pp. 119–128.

Rogerson, S. and Bynum, T.W. (1995). Cyberspace: the ethical frontier. *The Times Higher Education Supplement*, No 1179, June 9, 1995, p iv.

Rogerson, S., Fairweather, B., and Prior, M. (2002). The ethical attitudes of information systems professionals: outcomes of an initial survey. *Telematics and Informatics*, 19, 21–36.

Rogerson, S. and Fairweather, B. (2003). Entitlement cards. *IMIS Journal*, 13(2).

Rogerson, S. and Gotterbarn, D. (1998). The ethics of software project management. In: Collste G. (Ed.), *Ethics and Information Technology*. New Academic Publishers, Delhi, India, pp. 137–154.

Spinello, R.A. (1997). *Case Studies in Information and Computer Ethics*. Prentice-Hall, Upper Saddle River, NJ.

Spinello, R.A. (2000). *CyberEthics: Morality and Law in Cyberspace*. Jones and Bartlett, Sudbury, MA.

Spinello, R.A. and Tavani, H.T. (2001a). The Internet, ethical values, and conceptual frameworks: an introduction to cyberethics. *Computers and Society*, 31(2), 5–7.

Spinello, R.A. and Tavani, H.T.(Eds.) (2004[2001b]). Readings in CyberEthics, 2nd edition 2004. Jones and Bartlett, Sudbury, MA.

Spinello, R.A. and Tavani, H.T. (Eds.) (2005). *Intellectual Property Rights in a Networked World: Theory and Practice*. Idea Group/Information Science Publishing, Hershey, PA.

Stahl, B.C. (2004a). Information, ethics and computers: the problem of autonomous moral agents. *Minds and Machines*, 14, 67–83.

Stahl, B.C. (2004b). *Responsible Management of Information Systems*. Idea Group, Hershey, PA.

Stahl, B.C. (2005). The ethical problem of framing e-government in terms of e-commerce. *Electronic Journal of E-Government*, 3(2), 77–86.

Stahl, B.C. (2006). Responsible computers? A case for ascribing quasi-responsibility to computers independent of personhood or agency. *Ethics and Information Technology*, 8 (4), 205–213.

Tavani, H.T. (Ed.) (1996). *Computing, Ethics, and Social Responsibility: A Bibliography.* CPSR Press, Palo Alto, CA.

Tavani, H.T. (1999). Privacy and the Internet. *Proceedings of the Fourth Annual Ethics and Technology Conference.* Boston College Press, Chestnut Hill, MA, pp. 114–125.

Tavani, H.T. (2002). The uniqueness debate in computer ethics: What exactly is at issue and why does it matter? *Ethics and Information Technology*, 4(1), 37–54.

Tavani, H.T. (2004). Ethics and Technology: Ethical Issues in an Age of Information and Communication Technology, 2nd edition 2007. John Wiley and Sons, Hoboken, NJ.

Tavani, H.T. (2005). The impact of the Internet on our moral condition: Do we need a new framework of ethics? In: Cavalier, R.J. (Ed.), *The Impact of the Internet on our Moral Lives.* SUNY Press, Albany, NY, pp. 215–237.

Tavani, H.T. (2006). *Ethics, Computing, and Genomics.* Jones and Bartlett, Sudbury, MA.

Turner, E. (2006). Teaching gender-inclusive computer ethics. In: Trauth, I. (Ed.), *Encyclopedia of Gender and Information Technology: Exploring the Contributions, Challenges, Issues and Experiences of Women in Information Technology.* Idea Group, pp. 1142–1147.

Van den Hoven, M.J. (1997a). Computer ethics and moral methodology. *Metaphilosophy*, 28 (3), 234–248.

Van den Hoven, M.J. (1997b). Privacy and the varieties of informational wrongdoing. *Computers and Society*, 27(3), 33–37.

Van den Hoven, M.J. (1998). Ethics, social epistemics, electronic communication and scientific research. *European Review*, 7(3), 341–349.

Van den Hoven, M.J. (1999). Knowledge and democracy in cyberspace. *Etica e Politica* 1(2) at http://www.univ.trieste.it/dipfilo/etica_e_politica/1999_2/index.html.

Volkman, R. (2003). Privacy as life, liberty, property. *Ethics and Information Technology*, 5(4), 199–210.

Vokman, R. (2005). Dynamic traditions: Why globalization does not mean homogenization. *Proceedings of ETHICOMP2005.* Published on CD. Linköpings University, Sweden.

Volkman, R. (2007). The good computer professional does not cheat at cards. *Proceedings of ETHICOMP2007.* Meiji University Press, Tokyo.

Weckert, J. (2002). Lilliputian computer ethics. *Metaphilosophy*, 33(3), 366–375.

Weckert, J. (2005). Trust in cyberspace. In: Cavalier, R.J. (Ed.), *The Impact of the Internet on our Moral Lives.* SUNY Press, Albany, NY, pp. 95–117.

Weckert, J. and Adeney, D. (1997). *Computer and Information Ethics.* Greenwood Press, Westport, CT.

Weizenbaum, J. (1976). *Computer Power and Human Reason: From Judgment to Calculation.* Freeman, San Francisco.

Wiener, N. (1948). *Cybernetics: or Control and Communication in the Animal and the Machine.* Technology Press, John Wiley & Sons, New York.

Wiener, N. (1954[1950]). *The Human Use of Human Beings.* Houghton Mifflin, Boston, 1950, 2nd edition Revised. Doubleday Anchor, New York, 1954.

Wiener, N. (1964). *God & Golem, Inc.: A Comment on Certain Points Where Cybernetics Impinges on Religion.* MIT Press, Cambridge, MA.

CHAPTER 3

Moral Methodology and Information Technology

JEROEN VAN DEN HOVEN

Computer ethics is a form of applied or practical ethics. It studies the moral questions that are associated with the development, application, and use of computers and computer science. Computer ethics exemplifies, like many other areas of applied and professional ethics, the increasing interest among professionals, public policy makers, and academic philosophers in real-life ethical questions. Posing ethical questions about privacy, software patents, responsibility for software errors, equal access, and autonomous agents is one thing; answering them is another. How should we go about answering them, and how can we justify our answers? How should we think about practical ethical issues involving computers and information technology (IT)?

I think the way we ought to proceed in the ethics of IT is not very different from the way we ought to proceed in other departments of ethics of technology and engineering[1], although there are certainly differences between the moral problems occasioned by different types of technology and there are certainly specific properties of computers that need to be accommodated in our moral thinking about them. IT, for example, is (1) *ubiquitous* and *pervasive* (in a way in which even automobiles are not) and IT (2) comes closest to being a "universal technology," because of its "logical malleability."[2] We can use it to simulate, communicate, recreate, calculate, and so much more; it can be applied to all domains of life. IT is also (3) a *metatechnology*, a technology that forms an essential ingredient in the development and use of other technologies. IT may also be called (4) a *constitutive technology*; computing

[1] I will not discuss the ways in which moral problems involving computers and IT can be construed as having a special moral status. The uniqueness question has been extensively discussed; see for example Tavani (2002).

[2] An apt expression introduced by Jim Moor to capture the incredible flexibility of digital computers and Turing machines (Moor, 1985).

The Handbook of Information and Computer Ethics, Edited by Kenneth Einar Himma and Herman T. Tavani

technology coconstitutes the things to which it is applied. If IT is used, for example, in health care, health care will change in important ways as a result of it; if it is used in science and education, science and education will never be the same again; if the Internet and the World Wide Web are introduced into the lives of children, their lives will be very different from the childhood of people who grew up without online computer games, MSN, chat rooms, Hyves, and Second Life. Finally, (5) we tend to forget that IT is about *information.*[3] Information has special properties that make it difficult to accommodate it in conceptual frameworks concerned with tangible, material goods—their production, distribution, and use. Peer-to-peer network environments, for example, make the idea of "fair use" difficult or even impossible to apply. The ubiquitous combination of coupled databases, data mining, and sensor technology may start to cast doubt on the usefulness of our notion of "privacy." Ethical analysis and reflection, therefore, is not simply business as usual. We need to give computers and software their place in our moral world. We need to look at the effects they have on people, how they constrain and enable us, how they change our experiences, and how they shape our thinking. This is how we proceeded in the case of the car, the television, the atom bomb, and this is how we will proceed in the case of ubiquitous brain scanning and use of carbon nanotubes, of artificial agents, and the applications of advanced robotics. The commonalities in the moral questions pertaining to these topics are more important than the differences between them. The properties of IT may require us to revisit traditional *conceptualizations* and *conceptions* of privacy, responsibility, property; but they do not require a new way of moral thinking or a radically new *moral methodology*, which is radically different from other fields of technology and engineering ethics.[4] Neuroscience, nanotechnology, and gene technology will provide us with problems we have not dealt with before, but our moral thinking has revolved in the past, and will revolve in the future, around a familiar and central question: how we ought to make use of technology to benefit mankind, and how to prevent harm to human beings, other living creatures, the environment, and other valuable entities we decide to endow with moral status.[5]

In this chapter I will sketch a conception of method for Ethics and Information Technology, which is different from what we have seen thus far in applied ethics, but which does not call for cataclysmic re-conceptualizations.

First, I will give an overview of some of the main methodological positions in applied ethics relevant for computer ethics. Second, I will sketch the proposed conception of method of ethics of technology, which puts emphasis on design *ex ante* and not on analysis and evaluation *ex post*. It does not focus on *acting with*,

[3]Luciano Floridi has drawn attention to this feature and has provided a detailed and comprehensive treatment of it in his work.

[4]Luciano Floridi's approach to the ethical issues of IT would imply a somewhat different approach. See his contribution to this volume and Floridi (1999).

[5]This is of course a very general and simple way of talking about the ultimate rationale of moral thinking, which does not make it less true. Furthermore, underneath the simplicity lies intricacy; a wealth of distinctions and observations can be made with respect to more specific questions as Kamm (2007), for example, shows in her book.

but on *design* and *production of,* information technology, broadly conceived as a socio-technical system. It aims at making moral values a part of technological design in the early stages of its development, and assumes that human values, norms, and moral considerations can be imparted to the things we make and use (technical artifacts, policy, laws and regulations, institutions, incentive structures, plans). It construes technology as a formidable force, which can be used to make the world a better place, especially when we take the trouble to reflect on its ethical aspects in advance, in a stage of development when we can still make a difference and can shape technology in accordance with our considered moral judgments and moral values. Surely, taking design issues into account *ex ante* will not eliminate the need to evaluate the use of the associated technologies.

I will place this proposed conception of method in the context of the development of applied ethics in the last decades, and finally compare it with other conceptions of method in computer ethics, *Disclosive Computer Ethics* (Brey), *Information Ethics* (Bynum, Floridi), *Hermeneutical method* (Maner), *Professional Codes* (Gotterbarn, Rogerson, Berleur), *Virtue Ethics* (Chuck Huff, Frances Grodzinsky), and *Computational Approach* (Van den Hoven, Lokhorst, Wiegel).

3.1 APPLIED ETHICS

Ethics has seen notable changes in the course of the past 100 years. Ethics was in the beginning of the twentieth century predominantly a metaethical enterprise. It focused on questions concerning the meaning of ethical terms, such as "good" and "ought," and on the cognitive content and truth of moral propositions containing them. Later, ordinary language philosophers continued the metaethical work with different means. In the sixties, however, the philosophical climate changed. Ethics witnessed an "Applied Turn." Moral philosophers started to look at problems and practices in the professions, in public policy issues, and public debate. Especially in the USA, philosophers gradually started to realize that philosophy could contribute to social and political debates about, for example, the Vietnam War, civil rights, abortion, environmental issues, animal rights, and euthanasia, by clarifying terms and structuring arguments. Ever since the sixties, applied ethics has been growing. Every conceivable profession and cluster of issues has established in the meanwhile a special or applied ethics named after itself—from "library ethics" to "sports ethics" to "business ethics." The format of the explicit methodological account provided in many applied and professional ethics textbooks usually refers to the *application* of normative ethical theories, such as utilitarianism, Kantianism, or Rawlsian Justice as Fairness, to particular cases. Textbooks often start with a chapter on deontological theories and teleological theories, which are supposed to be applied to the problems in the field. The application process itself, however, is often left un(der)specified.

There has been a longstanding and central debate in practical ethics about methodology. The debate has been between those who think that general items, for example, rules and principles or universal moral laws, play an important or even central role in our moral thinking (this point of view is often referred to as *generalism*) and those who think that general items play no special or important role in our moral thinking. The latter think that people typically discuss particular and individual cases, articulate contextual considerations, the validity of which expires when they are generalized or routinely applied to other cases. This view is often referred to as *particularism.*

Every form of applied ethics, including computer ethics, needs to position itself in this debate.

3.1.1 Generalism

According to Generalism, "the very possibility of moral thought and judgment depends on the provision of a suitable supply of moral principles."[6] The simplest way to be a generalist is to think that there are fairly accurate general moral rules or principles that may be captured and codified, for example, in codes of conduct, which can be applied to particular cases. According to this simple reading of the generalist view, doing practical ethics is a matter of drafting codes of conduct or formulating moral principles or moral rules and drawing up valid practical syllogisms. However, general rules will necessarily contain general terms, and since general terms have an open texture that gives rise to vagueness, application may create difficulties and ambiguities.

Alternatively, one could say that a case is "subsumed" under a covering moral law or rule to derive an action guiding or an evaluative conclusion.

(i) For all actions x, if Ax, then x is permitted (obligatory)

(ii) Aa,

therefore,

(iii) Action a is permitted (obligatory)

According to this view, let us refer to these two views (code application and subsumption) as *The Engineering View,* in which justification in morality is construed analogous to explanation in physics. In the natural sciences, the so-called "deductive-nomological model" for a long time was the dominant view of explanation. The fact that a piece of metal expands (the *explanandum*) is explained by deducing the sentence or the proposition that expresses this fact from two premises (forming the *explanans*), one being a law of nature (all metal expands when heated) and the other stating the relevant facts (this piece of metal was heated). The presuppositions of this model in ethics would seem to be that ethical theory constitutes a distinct body of knowledge of universally valid moral principles, and secondly that relatively uncontroversial empirical descriptions of cases can be given, and that the application of this body of moral knowledge takes place through logical deduction, which is held to be value neutral and impartial. We can easily see what the shortcomings of this simple "engineering model" in ethics are. One problem is that the deductive application of moral rules and principles to cases is not an adequate account of what is actually done

[6]Dancy (2004), p. 7.

when people, professionals, or philosophers try to clarify practical moral problems or attempt to justify particular prescriptions or evaluations.

Another problem is that we may arrive at contradictory conclusions because two or more competing principles apply to the same case. The ACM Code of Ethics includes, for example, the imperatives "avoid harm to others" and "respect privacy." But, as we know we may avoid harm to Tom by disrespecting Harry's privacy, or respect Harry's privacy by letting Tom come to harm. Both the harm principle and the privacy principle apply, but it is unclear which of the two should have priority. If a lexical ordering of principles could be established, or if priority rules could be drawn up, then the problem of the collision of principles would not occur and purely deductive application could in principle succeed. Unfortunately, no such ordering can be established without raising the justification problem at another level. Therefore, at some point the Engineering Model is bound to fall back on intuitive balancing of conflicting norms, and thus the rigor and appeal of logical deduction disappear.

There are, however, other problems with the Engineering View. They are related to the logic and to the epistemic status of the premises. The logic of the Engineering Model fails to capture the phenomenon of belief revision, exceptions, *ceteris paribus* clauses, and default logic, which characterizes much of ordinary moral discourse. It furthermore fails to address the problem of open-textured concepts and vague notions. For example, if we are discussing *E-democracy*, different conceptions of democracy could be at stake and ever new ones could be thought of (van den Hoven and coworkers, 2005). These different conceptions, between which we may waver or shuttle back and forth, may issue in different constraints on the design of political information systems. Concepts such as "democracy" are sometimes referred to as "essentially contested concepts" to indicate that controversy concerning the meaning of the term has become part of the meaning of the term. The most interesting part of the applied ethics, therefore, lies in the *articulation* of a relevant and interesting conception of democracy and in the *specification* of associated moral constraints on political information systems in the age of information, and not so much in the *application* of a given conception of democracy.

An important criticism of all generalist positions is the objection formulated by Elizabeth Anscombe.[7] She has pointed out that rule-based approaches are all vulnerable to the problem of acting under a description: "An act-token will fall under many possible principles of action (...) how can we tell which act description is relevant for moral assessment?" [8] Should we, Onora O'Neill asks in her discussion of Anscombe's problem, "(...) assess an action under the description that an agent intends it, or under descriptions others think salient, or under descriptions that nobody has noted?" (O'Neill, 2004, p. 306). And how do we evaluate the actions of persons who, according to us, fail to see the morally significant descriptions of what s(he) does? McDowell claims that instead of establishing *rules* of moral salience, as Barbara Herman has suggested, people should have capacities to appreciate salient features of the situation, or "capacities to read predicaments correctly."

---

[7]Anscombe (1958), p. 124.

[8]O'Neill (2004) p. 306.

Bernard Gert gives an example of how the description of the case is of crucial importance.[9] He analyzes Nissenbaum's analysis of moral permissibility of copying software for a friend. Gert remarks that the disagreement about this issue may be caused by the fact that one of the partners to the disagreement has too narrow a description of the kind of violation to launch ethical thinking in the right direction. Some may describe it as "helping a friend," some as "illegally copying a software program," or as "violating a morally acceptable law to gain some benefit." On the basis of the latter description, Gert claims that "no impartial rational person would publicly allow the act" (Gert, 1999, p. 62).

3.1.2 Particularism

Particularists in ethics oppose the search for universally valid moral rules. They consider universally valid principles an intellectual mirage. Jonathan Dancy defines particularism in *Ethics Without Principles* as follows: "The possibility of moral thought and judgment does not depend on the provision of a suitable supply of moral principles."[10] Persons engaged in moral thinking, deliberation, and decision-making typically discuss individual cases; they give examples, tell stories and cautionary tales, and apply their powers of perception and moral judgment to specific individual situations and cases with which they are confronted. They exercise their practical wisdom, the faculty referred to by Aristotle as *phronesis*, which allows one to size up situations and to identify the morally relevant and salient features of particular situations. For particularists, the desideratum of situational adequacy, that is, the regulative ideal of doing justice to situations and persons in a particular historical context, is of paramount importance. The imposition of general principles and abstract concepts is bound to distort the rich, human, historical reality.

There are important objections to particularism, of which two deserve closer examination in this context. The first is that theory and thinking in terms of moral principles and rules seems to be part of our moral practices. Trying to find general principles to match one's judgments and intuitions in a particular case to extend them to other cases, or to explain them to others, seems a natural thing to do and is simply part and parcel of moral life, especially in the public policy and political arena. Our moral thinking in some cases simply depends on our ability to articulate the covering moral rules or principles. Only when an exaggerated distinction between theory (or principle) and practice (or example) is introduced, can one make the latter seem superior at the expense of the former.

Another related problem with particularism is that it "black-boxes" moral justification. It makes it difficult to provide (an account of) public justification of moral judgments. As Robert Nozick has argued, [11] adopting principles can have "symbolic utility" and enhance public scrutiny of the nature and strength of one's commitment to the moral claims one makes. Particularist methodology, in its unwillingness to produce

[9]Gert (1999), p. 57–64.

[10]Dancy (2004), p. 7.

[11]Nozick (1993).

moral justifications in terms of general principles, runs the risk of dissolving in an intransparent and somewhat mysterious, intellectual power of moral intuition, practical wisdom, or the mental faculty of moral perception and judgment. This makes it more difficult to imagine what a *public* particularist justification looks like. One possible particularist reply would be that getting to the right moral solution is better compared to the exercise of skills and abilities to accomplish something, for example, the ability of a marksman to hit the bull's-eye or that of a craftsman's picking the best piece of wood. The obvious objection to this reply is that this type of situation is so different from moral decision-making in public affairs and the professions that the analogy breaks down when stretched beyond moral reflection about intimate and highly personal problems. In the professions and public life, we hold people accountable and may legitimately ask them to explain why they think what they did was right and invite them to provide the general policy or rule they think applies in this case. We like to think of those who are responsible for the well-being of others as committed to certain principles, which limit options open to them to serve their self-interest. According to generalists, we expect them to justify their actions in terms of certain fairly general and self-binding principles. Furthermore, endorsing a principle communicates to others who one wants to be, where one stands, and what others may expect one to do.

Because justification to others requires at least this amount of transparency, it minimally presupposes the truth of the principle of *supervenient application* of moral reasons. This principle states that there can be no moral difference between cases without a relevant empirical difference between them. If, for example, privacy considerations argue against disseminating information about Tom and Dick, then the same privacy considerations forbid the release of information about Harry *ceteris paribus*. Identical cases ought to be treated identically. One's moral judgment in a particular case, therefore, establishes a *pro tanto* principle, namely, that unless one is able and willing to explain how two cases differ in their nonmoral properties, one is committed to judging them in the same way from a moral point of view.

Particularists have argued on the basis of the abundance of exceptions to principles that this is an absurd idea. In real life outside philosophy textbooks, there are no identical cases, situations, persons. An obligation that arises in one case could never carry over to another case, because of the uniqueness of each individual case. This much has to be admitted to the particularist that although the properties that make "blowing the whistle" the right thing to do in Tom's case are also present in Harry's case, this does not imply that blowing the whistle is the right thing to do for Harry, because there may be other properties present in Harry's case that "cancel out" the force of the rightness-conferring properties of Tom's case.

I think the generalist can concede this point without giving up on generalism. The principle of supervenience should not be construed as implying that one's judgment creates an absolute nondefeasible and exceptionless universal principle. Rather it should be construed as implying that each serious moral judgment gives rise to a legitimate expectation on the part of others that the one who makes the moral judgment accepts a commitment to explain why he or she does not apply the same principle, or judge similarly, in a case that seems identical. If one thinks that Tom's information ought not to be made freely available, because that would violate his right to privacy, one thereby does not incur a definite obligation to apply the same reasoning to Harry although their cases seem identical or at least relevantly similar. One *does* incur, however, an obligation to explain why one thinks Harry's case is different so as to make the warranted judgment that privacy constraints are inapplicable to Harry's case. If one agrees, for example, that considerations of national security override privacy of medical data, and that doctors therefore have to provide intelligence officers with access to patient record systems, one is not committed to accepting that the same applies to library loan systems. There is, however, a reasonable presumption that one would apply it in the same way. There is furthermore a legitimate expectation that one is committed to explaining what the differences between patient record systems and library loan systems are, so as to merit their different treatment in terms of providing access for national security purposes. Thus construed, the principle of supervenience of moral reasons constitutes the part that deserves to be salvaged from the generalist position, because it supplies the logic that propels moral dialogue by establishing *prima facie* and *pro tanto* defeasible general rules.

3.1.3 Reflective Equilibrium

There is a methodological alternative to both pure generalism and pure particularism that combines the strengths of both and accommodates in one model the rationale for generalizing modes of moral thinking (supervenience, consistency, transparency, avoidance of self-serving moral strategies, public justification) and the rationale of particularist modes of moral thinking (all moral judgments have exceptions, are only contextually valid, moral situations and persons are unique, people frequently use references to particular and unique features of situations).

The model combines elements of both methodological extremes. It is an approach that is referred to as the "Method of Reflective Equilibrium (RE)." James Griffin[12] observes about this method in his article "How we do ethics now" that "The best procedure for ethics . . . is the going back and forth between intuitions about fairly specific situations on the one side and the fairly general principles that we formulate to make sense of our moral practice on the other, adjusting either, until eventually we bring them all into coherence. This is, I think, the dominant view about method in ethics nowadays." RE incorporates elements of both the universalist and the particularist views. It allows for appeals to considered judgments and intuitions concerning particular cases, and it acknowledges the appropriateness of appeals to general principles that transcend particular cases. It accommodates the particularist objections to the Engineering View of moral justification, without giving up the principle of the supervenient application of moral reasons as explained above. It is dynamic and supports the nonmonotonicity of everyday moral reasoning.

[12]Griffin (1993).

RE was suggested as a method of moral inquiry for the first time by John Rawls. The so-called Wide Reflective Equilibrium (WRE) that Norman Daniels later proposed [13] aims at producing coherence in a broader set of beliefs held by a moral agent or a group of moral agents, namely, (1) a subset of considered moral judgments, (2) a subset of moral principles, and (3) a subset of relevant background theories. The general procedure involved in achieving a mutual fit between them is that of shuttling back and forth between considered moral judgments about a case and our moral principles, adjusting each in the light of the other and in the light of relevant background theories, to arrive at reflective equilibrium. This state is called "reflective" because we know to which principles our judgments conform, and it is referred to as an "equilibrium" because principles and judgments coincide. The shortcomings of the construal of application in the Engineering Model are remedied by the idea of *coherence* or "fit," suitably interpreted. According to a coherence approach of justification, there are no foundations in the sense of absolutely epistemically privileged propositions. The set of our moral beliefs is like a "web of beliefs," to use Quine's expression.[14] All propositions hang together and give mutual support. And as in a web, there is no apparent beginning. There are only relations of noncontradiction, consonance, and connectedness (not necessarily construed in terms of first-order predicate logic). A story told by someone may *fit* perfectly with the experiences of someone else, and they may both *exemplify* the central theme of a novel. Furthermore, no proposition is immune to revision. Some propositions in our web of moral beliefs, however, may be so well entrenched that they will stick forever. We cannot imagine under what circumstances we would, for example, retract the belief that the Nazi ideology was morally perverted; nor can we imagine under what circumstances we would give up the arithmetical proposition $2+2=4$. In this respect there is no difference between particular judgments and general principles. They may all at some point come up for revision.

It is clear that coherence conceptions of moral justification are more congenial to the phenomena of belief revision and defeasible reasoning than are approaches modeled after the engineering model. This is a second important virtue of WRE. It incorporates a doctrine of intellectual responsibility; in matters that are of great importance to us—and moral issues are of great importance to us by definition—we not only want to reduce the chances of failures, misrepresentations, and mistakes, but we also feel that we are under an obligation of a higher order to reduce the number of dangling, loose, and unjustified beliefs, to try to connect them to other relevant belief we have, and to perspicuously represent our attempts to do so. Furthermore, we also want to make clear to others that we consider ourselves fallible and exposed to criticism. The best way to do so is to take a stand and make general claims. In the public domain and the professions, accountability for one's moral judgments is premised on communicative transparency and the attempted articulation of principles.

[13]Daniels (1979).

[14]Quine and Ullian (1970).

3.2 THE DESIGN TURN IN APPLIED ETHICS

A further development in applied ethics takes the methodology of computer ethics beyond these methodological debates concerning generalism and particularism. After a focus on normative ethical theory and its application and justification, emphasis is now placed by some authors on the *design* of institutions, infrastructure, and technology, as shaping factors in our lives and in society.

Until now technology, engineering, and design were treated in moral philosophy as a mere supplier of thought experiments and counter-examples to arguments and theories. Traditional moral philosophy is full of science fiction and adventure, full of lifeboats and runaway trains, brains in vats, android robots, pleasure machines, brain surgery, and pills that will make one irrational on the spot.

Let us look in somewhat more detail at a famous and central thought experiment used in metaethics and normative theory to sharpen moral intuitions. Suppose you are at the forking path of a downhill railway track and a trolley cart is hurtling down and will pass the junction where you stand. There is a lever that you can operate. If you do nothing, the trolley will kill five people who are tied down to the track further downhill. If you pull the lever, the trolley will be diverted to the other track where there is only one person tied to the track. Is it morally permissible to pull the lever?

If an engineer were to remark after a philosophy paper on the trolley problem that one needed a device that would allow one to stop the train before it reached the fork in the track, and sensors to inform one about living creatures on the track, and preferably a smart combination of both, the presenter would probably remark that in that case the whole problem would not arise and the intervention misses the philosophical point of the philosophical thought experiment. The philosopher is right, strictly speaking.

If moral philosophy were to get sidetracked by focusing on these examples in the real world, it would be surely more interesting to try and think about how we could come up with alternative designs of the situation and systems so as to prevent (1) loss of lives and (2) tragic moral dilemmas, instead of looking at actors in tragic and dilemmatic situations where they have to make choices at gunpoint with very little or no relevant information. The trolley problem was indeed designed to raise other and primarily theoretical issues in ethics, but if moral theories are developed on thought experiments that abstract from the design history and the degrees of freedom present at stages before the tragic dilemma came into being, the resulting theories inherit the ahistorical and design orientation that in real-life cases often contains the beginning of the solution.

Dilemmas and thought experiments in medical ethics and traditional computer ethics also suffer from this shortcoming. The professional—medical or IT—is confronted with the following dilemma D; this presents him or her with two options A and B, what should the professional do? The first reaction is to start wracking our brains and start to try utilitarian calculations and Kantian approaches as antidotes to them, but in any case, *the situation is taken as given*. What is suppressed and obfuscated is the fact that the technologically infused situation—often a computer-supported cooperative work setting—and therefore much broader than just the software and hardware together, but more like what Clark refers to as the *Wideware*, that the professional is confronted with is the result of hundreds of design decisions. Every moral analysis of the situation, which abstracts from this historical design and development antecedents, commits the "fallacy of the path-dependent dilemma." Computer ethicists should probe beyond the technical *status quo* and ask how the problem came into being, and what are the designs and architectural decisions that have led up to it.

As far as the institutional dimensions of moral situations are concerned, this type of question is now being addressed more often. The central question now more often than before is which institutional and material conditions need to be fulfilled if we want the outcomes of our applied ethical analyses to be successful in their implementation. How can we increase the chances of changing the world in the direction in which our moral beliefs—held in wide reflective equilibrium—point? How can we design the systems, institutions, infrastructures, and IT applications in the context of which users will be able to do what they ought to do and which will enable them to prevent what ought to be prevented.

This notable shift in perspective might be termed "The Design Turn in Applied Ethics." The work of John Rawls, again I think, gave rise for the first time to talk about design in ethics. Thinking about social justice can in the context of his theory be described as formulating and justifying the principles of justice in accordance with which we should design the basic institutions in society.

Thomas Pogge, Russell Hardin, Cass Sunstein, Robert Goodin, and Dennis Thompson have taken applied ethics a step further down this path. They not only want to offer applied analyses, they also want to think about the economic conditions, institutional and legal frameworks, and incentive structures that need to be realized if our applied analyses are to stand a chance in their implementation and thus contribute to bringing about real and desirable moral changes in the real world. Design in the work of these authors is primarily focused on institutional design, but the Design Turn clearly brings into view the design of Wideware, socio-technical systems, technological artifacts, and socio-technical systems.

At this point an interesting parallel development needs to be noted in IT, and probably also in other engineering disciplines: a shift from attention to technology *simpliciter*, to technology in organizational and human and values context. In the first phase of its development in the fifties, and sixties, the social and organizational context did not matter much in the production of IT applications. Hardly anyone bothered to ask about *users*, *use* and *usability*, and the fit with the organizational context. Computers were a new and fascinating technology—solutions looking for problems. In the second stage of development in the seventies and eighties, after many failed projects, worthless applications, and bad investments, one gradually started to realize that there were human users, with needs and desires, and real organizations with peculiar properties. It occurred to many at that time that it would be wise and profitable to try and accommodate user requirements and conditions on the work floor in the early stages of the development of applications. The social and behavioral sciences came to the aid of IT in this period. But this is still a minimal way of taking the needs and interests of users, organizations, and society into account, namely, as mere constraints on the successful implementation of systems.

In the nineties it gradually started to dawn on the IT profession not only that the reality of real organizations and real users does matter to the development and use of information technology, but also the fact that human beings, whether in their role as consumers, citizens, or patients, have values, moral preferences, and ideals, and that there are moral and public debates in society about liability, about equality, property, and privacy that need to be taken into account. We are now entering a third phase in the development of IT, where the needs of human users, the values of citizens and patients, and some of our social questions are considered in their own right and are driving IT,[15] and are no longer seen as mere constraints on the successful implementation of technology.

3.3 VALUE SENSITIVE DESIGN

These two separate developments in ethics (theory–application–design) and in IT (technology–social and psychological context–moral value) come together in the idea of Value Sensitive Design (VSD).[16] VSD was first proposed in connection with information and communication technology, and that is still its main area of application. There were several important ideas and proponents of those ideas that led up to it.

Work by Terry Winograd, Batya Friedman,[17] John Perry, Ben Shneiderman,[18] and Helen Nissenbaum in Stanford in the early nineties showed that software could easily come to contain biases, arbitrary assumptions, and peculiar worldviews of makers, which could affect users in various ways. Research by Nissenbaum and Introna[19] on biases in search technology is a good example of this approach. Secondly, legal scholars around the same time observed that regulation in society was taking place by means of computer code and software. Code started to function as law and laws would in the future literally be *en-coded*, as Larry Lessig[20] pointed out. Advocates of the so called Privacy Enhancing Technology at the Dutch and Canadian Data Protection Offices observed that this was probably the only way in which we could deal with privacy compliance and law enforcement issues, given the increasing number of laws and regulations and the vast amount of data that are processed in our society. It is impossible to have lawyers check manually whether certain data practices are in breach of or in compliance with the law. The software in the long run would have to take care of that on our behalf, and not only in the privacy area.

---

[15]One of the interesting examples of that approach to date is the Californian Institute CITRIS (Center for IT Research in the Interest of Society) endowed with 320 million U.S. dollars. The CITRIS research agenda is determined by social problems and their solution.

[16]There is a rapidly expanding literature on Values and Design or VSD. See Web sites of Friedman (http://projects.ischool.washington.edu/vsd/), Nissenbaum (http://www.nyu.edu/projects/valuesindesign/), and Camp (designforvalues.org).

[17]Friedman (1998), p. 334.

[18]Shneiderman (2002).

[19]Nissenbaum and Introna, pp. 169–85.

[20]See his comprehensive Web site http://www.lessig.org/

Rob Kling's Social Informatics had been instrumental in making work in social studies in science and technology available in the ICT field and highlighted the social shaping of technology. At Rensselaer Polytechnic, Langdon Winner famously argued that "artefacts have politics," [21] which means that artifacts can be designed in such a way that they serve political purposes, and Deborah Johnson had articulated the ethical issues in computing. All these developments contributed to Value Sensitive Design in IT.

If the discourse on user autonomy, patient centeredness and citizen centeredness, his privacy, her security is to be more than an empty promise, these values will have to be expressed in the design, architecture, and specifications of systems. If we want our information technology—and the use that is made of it—to be just, fair, and safe, we must see to it that it inherits our good intentions. Moreover, it must be seen that to have those properties, we must be able to demonstrate that they possess these morally desirable features, compare different applications from these value perspectives, and motivate political choices and justify investments from this perspective.

Value Sensitive Design assumes that values and normative assumptions can somehow be incorporated, embodied in designs. What does it mean to say that? Let us look at some examples.

(1) **PACS**

ICT applications in hospitals may have unforeseen effects with moral implications. The so-called Picture Archive Systems (PACS) have now been introduced. This may change traditional and robust knowledge practices. Before the introduction of PACS, typically a team of doctors would stand around a neon-lit glass wall from which X-ray photos hang. Colleagues typically provided their interpretation of what they are looking at. One doctor might correct the opinion of a colleague or give his dissenting opinion. The radiologist might tell about the new X-ray equipment. This gathering constituted an epistemic or knowledge practice, and a pretty interesting one for that matter; it allowed for discursive checking, correcting, and supplying information, which can be scrutinized by others. This practice may now become less common because of the introduction of picture archive systems for digitized medical images. The unit cost for high-resolution medical image viewers on the desks of individual doctors is going down. This architectural decision to provide individual doctors with relatively cheap high-resolution viewers may give rise to a different epistemic practice that is highly individualistic, and provides less opportunity for critical valuable intercollegiate discussions. If no one is aware of this change, it will not be compensated for in the design of the relevant Computer Supported Cooperative Work systems that come with the new Picture Archive System.

[21]Winner (1980).

(2) **Sinks**

Jenkins and McCauley (2006) describe a software application where the choice for a particular algorithm has political and moral consequences.[22] In their paper, "GIS, sinks, fill, and disappearing wetlands: unintended consequences in algorithm development and use," they describe how Geographic Information Systems (GIS) software has become an important computational tool in several fields. On the basis of the output from this software, GIS users make important decisions to plan and manage landscapes (e.g., cities, parks, forests) with real consequences for the ecosystems. Jenkins and McCauley discuss a programming decision in a GIS algorithm originally used to discern flow direction in hydrological modeling—the mapping of streams and rivers. Topographic depressions (sinks) are "filled" in the algorithm to map water flow downstream; otherwise, the GIS algorithm cannot solve the problem of accurately calculating and representing the flow direction. Unfortunately, sinks are often "isolated" wetlands that provide essential habitat for many species not commonly found elsewhere. Thus, the algorithmic filling of sinks can make these wetlands "disappear" in GIS output and land-use decisions based on this output. This outcome occasioned by the choice of the algorithm may have potentially devastating real-world consequences for numerous wetlands because land-use plans made in ignorance cannot adequately conserve these unique habitats and the vital ecosystem services that wetlands provide.

(3) **Real-Time Emergency Medicine**

Darcy and Dardalet[23] describe a telemedicine application. An emergency medicine application consisting of a real-time broadband audio-video link realized through a camera mounted in the helmet of firefighters with an emergency medicine center was developed in France. Technically, the system was designed according to the specifications. In the testing it turned out, however, that nothing was done to prepare for the situation where medical professionals were instructing firefighters to perform certain tasks or to prioritize their tasks, in ways that diverged from the firefighters' own conception of their work and responsibilities. Roles and responsibilities need to be accommodated in protocols, and protocols need to be implemented in the telemedicine application. Without a proper initial value analysis and the implementation of the results in the application, the systems are worthless to the users.

If we shift our attention from general moral philosophy to the various fields of applied and practical ethics, such as environmental ethics, engineering ethics, computer ethics, and medical ethics, we find that the bulk of the work that goes on there is

---

[22]Jenkins and McCauley, paper at SAC2006, ACM conference, Dijon.

[23]Darcy and Dardalet (2003)."Rescuing the emergency: Multiple expertise and IT in the emergency field." In: *Methods of Information in Medicine.*

still traditional applied and professional ethics, that is, thinking about codes of conduct, the problem of dirty hands, the many hands, the many dirty hands, utilitarianism, deontological theories, virtue ethics, applied to video games, hacking, spamming, physician-assisted suicide. Now these issues are of course important, but if we focus on them exclusively, we miss exactly the opportunity that Value Sensitive Design perspective brings to the fore, that is, a proactive integration of ethics—the frontloading of ethics—in design, architecture, requirements, specifications, standards, protocols, incentive structures, and institutional arrangements.

3.4 OTHER CONCEPTIONS OF METHOD IN COMPUTER ETHICS

Philip Brey has proposed a conception of method in computer ethics that is related to the Value Sensitive Design conception. Brey is concerned with *disclosing* and *exposing* the values that are embedded in IT systems and software. He proposes "disclosive computer ethics" and contrasts it with what he calls the standard model of applied ethics (what we called the Engineering Model above). Traditional applied ethics focuses on existing moral controversies and practices and on use, although it should also explore uncharted terrain and focus on the technology and its design. Brey cites many examples of studies (and other ones that I have described above) that have exposed hidden or embedded values in information systems or IT applications and proposes a two-tiered approach of disclosive computer ethics. First, a central value that gets a loose and commonsense definition is used to identify a problem. This gives rise to tentative conclusions. In a second stage, more specific theories and conceptions of the relevant value are used to shed light on the problem; these may then be used to inform design of technology and policy. In the first instance the values are used as "fishing nets" or "search lights" and only later are the problems thus collected investigated and analyzed in detail to produce informative conclusions. Disclosive computer ethics takes place, according to Brey, at three levels: In the first disclosure level, the system is analyzed on the basis of a particular value (e.g., privacy); on the second theoretical level it is further developed and refined in the light of the IT case at hand. The third level is the application level where ethical theories are applied. The second stage is the philosophical stage. The disclosive and application stages require detailed domain knowledge. [24]

Walter Maner,[25] Simon Rogerson, Donald Gotterbarn, Jacques Berleur,[26] and Keith Miller[27] have experimented with checklists (referred to by Maner as *Heuristic Method*); checklist-based decisions support systems, steps models, and code of conduct approaches. These are all pragmatic approaches that offer moral guidance to IT professionals and are useful for teaching ethics in a structured way to students and

---

[24]See also Brey (2000).

[25]"Heuristic Method for Computer Ethics." See URL http://csweb.cs.bgsu.edu/maner/heuristics/maner.pdf

[26]Berleur and Brunnstein (1996).

[27]Collins and Miller (1995), pp. 39–57.

practitioners. They may provide tools to sensitize professionals to ethical issues. The problems that were raised against generalism, however, apply to them as they have a tendency to focus on the problems that are relatively obvious and easy to articulate.

Chuck Huff [28] and Frances Grodzinsky[29] have argued for a virtue ethics approach. Huff studied moral exemplars in the IT profession. This should be seen as a modern reinstatement of a neo-Aristotelian approach, which focuses on moral character traits and moral dispositions. Practical wisdom and the ability to intuitively identify salient and morally relevant characteristics of the situation play an important role in Aristotelian ethics. Clearly this accommodates part of the features of our moral lives: we size up the situation and immediately recognize what is important and what not and almost instantaneously form moral judgments. But this approach assumes that there is a relatively robust set of cases—the training set—that allow individuals to become expert judges. Furthermore, in traditional societies, experience in dealing with standard distributive or retributive justice issues can be handed down from generation to generation. When new technology is involved and we are facing new problems every day, the acquisition of and buildup of dispositional properties may become problematic to the point that virtues ethics loses its attractiveness as a methodological approach to practical problems.

Also, nontraditional and nonstandard conceptions of computer ethics have been proposed. Revisionist conceptions of method in computer ethics are implied in the work of Floridi and Bynum.[30] Their work suggests a redescription of moral phenomena as we know them. Floridi[31] has extensively discussed the informational and theoretical dimensions of IT problems. His Information Ethics is presented as an ecological ethics, which is ontocentric in the sense that it construes being/information as the most fundamental and morally relevant category (more important than life). The principle of Entropy (not quite the same as the notion Entropy in Thermodynamics) is more central than suffering. He claims that the moral status of actions with or without IT concerns their informational status. He expands existing theories by arguing that information objects have moral significance and are hence deserving of respect. He says that "all entities, even when interpreted as only clusters of information, still have a minimal moral worth *qua* information objects and so may deserve to be respected." Computer ethics should thus be concerned with finding out what increases entropy and which actions and events counteract it and increase negentropy. Although this conception and redescription of our moral phenomena seems to capture part of what it means to say that we detest and dislike wanton destruction, killing, and pain and what we like about life, organization, order, and structure and system, often we do not consider these categories at all in our moral reflections. We do not think it is wrong to kill our fellow human beings precisely because it fails to counteract the principle of Entropy. This does not enter into our considerations under this description, and it therefore does not, at present, seem to be the reason why we refrain from killing our

[28]Huff and Rogerson.

[29]Grodzinsky (1999).

[30]Bynum (2007).

[31]See for Luciano Floridi's work: URL www.ethicsofinformation.net

neighbors. Even after having taken classes in Thermodynamics, information theory, and cosmology, we would probably resist this redescription of our moral lives. Morality is about the "heartbreakingly human" aspects of the universe, as Cora Diamond has called them.[32]

Van den Hoven and Lokhorst[33] and Wiegel et al.[34] have argued that hybrid deontic logics can be used to model our moral considerations on the one hand and implement them in software on the other. In designing a hospital information system, deontic epistemic action logic (DEAL) can be used to model moral constraints that apply to information flows. Claims such as "it is not permitted to see to it that John knows that p" can be modeled in this way, as can, for example, expressions such as "if p, and John knows that q, then John has an obligation to see to it that all know that q." In this way a privacy-enhancing hospital information system could be designed that would be morally transparent, efficient, auditable, and effective, in the sense in which paper-based policies and human oversight could never be effective in large IT systems with personal data.[35]

These are all very different approaches to computer ethics. They all seem to capture an aspect of our moral thinking about problems in moral lives in the twenty-first century. They all represent valuable contributions to the methodology of computer ethics. The danger lies in thinking that they are uniquely correct and exclude all others. In the context of computer ethics, I think it is important not to repeat the mistakes that have been made in the history of normative ethics, that is, thinking that all moral problems can be solved by means of the application of one theory, one principle; to use artificial, technologically and empirically naive cases for testing normative theories; and also to think that there is only one correct theory or theoretical orientation (Kantianism, utilitarianism, etc.) that provides unique correct answers to all moral questions. We may be able to identify some answers as morally wrong, but that is probably as far as we can get. [Furthermore, in a global context, while dealing with deep cultural and religious divides, ethics probably need to take the form of—as Anthony Appiah observed, following Van Neuman's characterization of learning math as a matter of getting used to it—"getting used to each other's way of life and ways of thinking," without necessarily getting involved in deep and analytically deep arguments with each other.] Also, probably we need to become more of moral entrepreneurs than many computer ethicists have allowed themselves to be in the past.[36] What Bill Joy, Larry Lessig, and Peter Singer have achieved for our thinking about, respectively, converging technologies, freedom in the age of the Internet, and animal rights has no counterpart in the purely academic world. We also need to deal with design and redesign of economic and institutional arrangements as they relate to technology. But first and foremost we need to realize that information technology

---

[32]See for a discussion Himma (2004).

[33]Van den Hoven and Lokhorst (2002). Reprint of *Metaphilosophy*, 33(3), 376–387.

[34]Wiegel et al. (2005).

[35]Others have now also started to experiment with logic-based approaches to Computer Ethics. See, for example, a report on http://www.nyu.edu/projects/nissenbaum/papers/economist.pdf

[36]See Posner (1999).

shapes the spaces of action of people, imposes constraints and affordances, and requires us to address the development and design of technology at a stage when ethics can still make a difference in the light of our ethical beliefs held in a wide reflective equilibrium.

REFERENCES

Anscombe, E. (1958). Modern Moral philosophy. *Philosophy,* 33.

Berleur, J. and Brunnstein, K. (1996). *Ethics of Computing*. Chapman & Hall, London.

Brey, P. (2000). Method in computer ethics: towards a multi-level interdisciplinary approach. *Ethics and Information Technology,* 125–129.

Bynum, T.W. (2007). Flourishing ethics. *Ethics and Information Technology,* 8, 157–173.

Collins, R. and Miller, K. (1995). Paramedic ethics for computer professionals. In: Johnson and Nissenbaum (Eds.), *Computers, Ethics and Social Values*. Prentice Hall, Englewood Cliffs, NJ.

Dancy, J. (2004). *Ethics Without Principles*. Oxford University Press, Oxford.

Daniels, N. (1979). Wide reflective equilibrium and theory acceptance in ethics. *Journal of Philosophy,* 76(5), 256–82. Reprinted in Daniels, N. (Ed.), *Justice and Justification: Reflective Equilibrium in Theory and Practice.* Cambridge University Press, Cambridge, pp. 21–46.

Floridi, L. (1999). Information ethics: on the philosophical foundations of computer ethics. *Ethics and Information Technology,* 1, 37–56.

Gert, B. (1999). Common morality and computing. *Ethics and Information Technology,* 1(1), 57–64.

Friedman, B. (Ed.) (1998). *Human Values and the Design of Computer Technology.* Series: (CSLI-LN), Lecture Notes. Center for the Study of Language and Information, Stanford.

Griffin, J. (1993). How we do ethics now. In: Griffiths, A.P. (Ed.), *Ethics*. Cambridge University Press, Cambridge.

Grodzinsky, F.S. (1999). The practitioner from within: revisiting the virtues. *Computers and Society,* 29.

Himma, K.E. (2004). There's something about Mary: the moral value of things qua information objects. *Ethics and Information Technology,* 6(3).

Huff, C. and Rogerson, S. Craft and reform in moral exemplars in computing. Available at http://bibliotecavirtual.clacso.org.ar/ar/libros/raec/ethicomp5/docs/htm_papers/30Huff,%20Chuck.htm.

Kamm, F. (2007). *Intricate Ethics*. Cambridge University Press, Cambridge.

Lessig, L. Available at http://www.lessig.org/.

Maner, W. Heuristic method for computer ethics. Available at http://csweb.cs.bgsu.edu/maner/heuristics/maner.pdf.

Moor, J.H. (1985). What is computer ethics? *Metaphilosophy,* 16, 266–75.

Nissenbaum, H. and Introna, L. (1999). Shaping the web; why the politics of search engines matters. *Information Society,* 16, 169–85.

Nozick, R. (1993). *The Nature of Rationality.* Princeton University Press, Princeton.

O'Neill, O. (2004). Modern moral philosophy and the problem of relevant descriptions. In: Hear, A.O. (Ed.), *Modern Moral Philosophy.* Cambridge University Press, Cambridge.

Posner, R.A. (1999). *Problematics of Moral and Legal Theory.* Belknal Press, Harvard University Press, Cambridge, MA.

Quine, W.V.O. and Ullian, J. (1970). *The Web of Belief.* Random House, New York.

Shneiderman, B. (2002). *Leonardo's Laptop, Human Needs and the New Computing Technologies.* MIT Press, Cambridge, MA.

Tavani, H.T. (2002). The uniqueness debate in computer ethics: what exactly is at issue, and why does it matter? *Ethics and Information Technology,* 4(1), 37–54.

Van den Hoven, J. and Lokhorst, G.J. (2002). Deontic logic and computer supported ethics. In: Moor, J.H. and Bynum, T.W. (Eds.), *Cyberphilosophy.* Blackwell, Oxford.

Wiegel, V., van den Hoven, J. and Lokhorst, G. (2005). Privacy, deontic epistemic action logic and software agents. *Ethics and Information Technology,* 7(4), 251–264.

Winner, L. (1980). Do artefacts have politics? *Daedalus,* 109, 121–36.

CHAPTER 4

Value Sensitive Design and Information Systems

BATYA FRIEDMAN, PETER H. KAHN JR., and ALAN BORNING

Value Sensitive Design is a theoretically grounded approach to the design of technology that accounts for human values in a principled and comprehensive manner throughout the design process. It employs an integrative and iterative tripartite methodology, consisting of conceptual, empirical, and technical investigations. We explicate Value Sensitive Design by drawing on three case studies. The first study concerns information and control of web browser cookies, implicating the value of informed consent. The second study concerns using high-definition plasma displays in an office environment to provide a "window" to the outside world, implicating the values of physical and psychological well-being and privacy in public spaces. The third study concerns an integrated land use, transportation, and environmental simulation system to support public deliberation and debate on major land use and transportation decisions, implicating the values of fairness, accountability, and support for the democratic process, as well as a highly diverse range of values that might be held by different stakeholders, such as environmental sustainability, opportunities for business expansion, or walkable neighborhoods. We conclude with direct and practical suggestions for how to engage in Value Sensitive Design.

4.1 INTRODUCTION

There is a longstanding interest in designing information and computational systems that support enduring human values. Researchers have focused, for example, on the value of *privacy* (Ackerman and Cranor, 1999; Agre and Rotenberg, 1998; Fuchs, 1999; Jancke et al., 2001; Palen and Grudin, 2003; Tang, 1997), *ownership* and *property* (Lipinski and Britz, 2000), *physical welfare* (Leveson, 1991), *freedom from*

The Handbook of Information and Computer Ethics, Edited by Kenneth Einar Himma and Herman T. Tavani

bias (Friedman and Nissenbaum, 1996), *universal usability* (Shneiderman, 1999, 2000; Thomas, 1997), *autonomy* (Suchman, 1994; Winograd, 1994), *informed consent* (Millett et al., 2001), and *trust* (Fogg and Tseng, 1999; Palen and Grudin, 2003; Riegelsberger and Sasse, 2002; Rocco, 1998; Zheng et al., 2001). Still, there is a need for an overarching theoretical and methodological framework with which to handle the value dimensions of design work.

Value Sensitive Design is one effort to provide such a framework (Friedman, 1997a; Friedman and Kahn, 2003; Friedman and Nissenbaum, 1996; Hagman et al., 2003; Nissenbaum, 1998; Tang, 1997; Thomas, 1997). Our goal in this chapter is to provide an account of Value Sensitive Design, with enough detail for other researchers and designers to critically examine and systematically build on this approach.

We begin by sketching the key features of Value Sensitive Design and then describe its integrative tripartite methodology, which involves conceptual, empirical, and technical investigations, employed iteratively. Then we explicate Value Sensitive Design by drawing on three case studies. One involves cookies and informed consent in web browsers; the second involves HDTV display technology in an office environment; and the third involves user interactions and interface for an integrated land use, transportation, and environmental simulation. We conclude with direct and practical suggestions for how to engage in Value Sensitive Design.

4.2 WHAT IS VALUE SENSITIVE DESIGN?

Value Sensitive Design is a theoretically grounded approach to the design of technology that accounts for human values in a principled and comprehensive manner throughout the design process.

4.2.1 What is a Value?

In a narrow sense, the word "value" refers simply to the economic worth of an object. For example, the value of a computer could be said to be $2000. However, in the work described here, we use a broader meaning of the term wherein a value refers to what a person or group of people consider important in life.[1] In this sense, people find many things of value, both lofty and mundane: their children, friendship, morning tea, education, art, a walk in the woods, nice manners, good science, a wise leader, clean air.

This broader framing of values has a long history. Since the time of Plato, for example, the content of value-oriented discourse has ranged widely, emphasizing "the good, the end, the right, obligation, virtue, moral judgment, aesthetic judgment, the beautiful, truth, and validity" (Frankena, 1972, p. 229). Sometimes ethics has been subsumed within a theory of values, and other times, conversely, with ethical values viewed as just one component of ethics more generally. Either way, it is usually agreed

[1]The Oxford English Dictionary definition of this sense of value is "the principles or standards of a person or society, the personal or societal judgment of what is valuable and important in life" (Simpson and Weiner, 1989).

(Moore, 1903,1978) that values should not be conflated with facts (the "fact/value distinction"), especially insofar as facts do not logically entail value. In other words, "is" does not imply "ought" (the naturalistic fallacy). In this way, values cannot be motivated only by an empirical account of the external world but depend substantively on the interests and desires of human beings within a cultural milieu. In Table 4.1 in Section 4.6.4, we provide a list of human values with ethical import that are often implicated in system design, along with working definitions and references to the literature.

4.2.2 Related Approaches to Values and System Design

In the 1950s, during the early periods of computerization, cyberneticist Wiener (1953, 1985) argued that technology could help make us better human beings and create a more just society. But for it to do so, he argued, we have to take control of the technology. We have to reject the "worshiping [of] the new gadgets which are our own creation as if they were our masters" (p. 678). Similarly, a few decades later, computer scientist Weizenbaum 1972 wrote,

> What is wrong, I think, is that we have permitted technological metaphors... and technique itself to so thoroughly pervade our thought processes that we have finally abdicated to technology the very duty to formulate questions.... Where a simple man might ask: "Do we need these things?", technology asks "what electronic wizardry will make them safe?" Where a simple man will ask "is it good?", technology asks "will it work?" (pp. 611–612)

More recently, supporting human values through system design has emerged within at least four important approaches. *Computer Ethics* advances our understanding of key values that lie at the intersection of computer technology and human lives (Bynum, 1985; Johnson and Miller, 1997; Nissenbaum, 1999). *Social Informatics* has been successful in providing socio-technical analyses of deployed technologies (Johnson, 2000; Kling et al., 1998; Kling and Star, 1998; Orlikowsi and Iacono, 2001; Sawyer and Rosenbaum, 2000). *Computer-Supported Cooperative Work* (CSCW) has been successful in the design of new technologies to help people collaborate effectively in the workplace (Fuchs, 1999; Galegher et al., 1990; Grudin, 1988; Olson and Teasley, 1996). Finally, *Participatory Design* substantively embeds democratic values into its practice (Bjerknes and Bratteteig, 1995; Bødker, 1990; Carroll and Rosson (in this volume); Ehn, 1989; Greenbaum and Kyng, 1991; Kyng and Mathiassen, 1997). (See Friedman and Kahn, 2003 for a review of each of these approaches.)

4.3 THE TRIPARTITE METHODOLOGY: CONCEPTUAL, EMPIRICAL, AND TECHNICAL INVESTIGATIONS

Think of an oil painting by Monet or Cézanne. From a distance it looks whole, but up close you can see many layers of paint upon paint. Some paints have been applied with careful brushstrokes, others perhaps energetically with a palette knife or fingertips, conveying outlines or regions of color. The diverse techniques are employed one on top of the other, repeatedly, and in response to what has been laid down earlier. Together they create an artifact that could not have been generated by a single technique in isolation of the others. This, too, applies with Value Sensitive Design. An artifact or design emerges through iterations upon a process that is more than the sum of its parts. Nonetheless, the parts provide us with a good place to start. Value Sensitive Design builds on an iterative methodology that integrates conceptual, empirical, and technical investigations; thus, as a step toward conveying Value Sensitive Design, we describe each investigation separately.

4.3.1 Conceptual Investigations

Who are the direct and indirect stakeholders affected by the design at hand? How are both classes of stakeholders affected? What values are implicated? How should we engage in trade-offs among competing values in the design, implementation, and use of information systems (e.g., autonomy vs. security, or anonymity vs. trust)? Should moral values (e.g., a right to privacy) have greater weight than, or even trump, nonmoral values (e.g., aesthetic preferences)? Value Sensitive Design takes up these questions under the rubric of conceptual investigations.

In addition, careful working conceptualizations of specific values clarify fundamental issues raised by the project at hand, and provide a basis for comparing results across research teams. For example, in their analysis of trust in online system design, Friedman et al. (2000a), drawing on Baier (1986), first offer a philosophically informed working conceptualization of trust. They propose that people trust when they are vulnerable to harm from others, and yet believe that those others would not harm them even though they could. In turn, trust depends on people's ability to make three types of assessments. One is about the harms they might incur. The second is about the goodwill others possess toward them that would keep those others from doing them harm. The third involves whether or not harms that do occur lie outside the parameters of the trust relationship. From such conceptualizations, Friedman et al. were able to define clearly what they meant by trust online. This definition is in some cases different from what other researchers have meant by the term. For example, the Computer Science and Telecommunications Board, in their thoughtful publication *Trust in Cyberspace* (Schneider, 1999), adopted the terms "trust" and "trustworthy" to describe systems that perform as expected along the dimensions of correctness, security, reliability, safety, and survivability. Such a definition, which equates "trust" with expectations for machine performance, differs markedly from one that says trust is fundamentally a relationship between people (sometimes mediated by machines).

4.3.2 Empirical Investigations

Conceptual investigations can only go so far. Depending on the questions at hand, many analyses will need to be informed by empirical investigations of the human context in which the technical artifact is situated. Empirical investigations are also often needed to evaluate the success of a particular design. Empirical investigations can be applied to any human activity that can be observed, measured, or documented.

Thus, the entire range of quantitative and qualitative methods used in social science research is potentially applicable here, including observations, interviews, surveys, experimental manipulations, collection of relevant documents, and measurements of user behavior and human physiology.

Empirical investigations can focus, for example, on questions such as: How do stakeholders apprehend individual values in the interactive context? How do they prioritize competing values in design trade-offs? How do they prioritize individual values and usability considerations? Are there differences between espoused practice (what people say) compared with actual practice (what people do)? Moreover, because the development of new technologies affects groups as well as individuals, questions emerge of how organizations appropriate value considerations in the design process. For example, regarding value considerations, what are organizations' motivations, methods of training and dissemination, reward structures, and economic incentives?

4.3.3 Technical Investigations

As discussed in Section 4.5, Value Sensitive Design adopts the position that technologies in general, and information and computer technologies in particular, provide value suitabilities that follow from properties of the technology; that is, a given technology is more suitable for certain activities and more readily supports certain values while rendering other activities and values more difficult to realize.

In one form, technical investigations focus on how existing technological properties and underlying mechanisms support or hinder human values. For example, some video-based collaborative work systems provide blurred views of office settings, while other systems provide clear images that reveal detailed information about who is present and what they are doing. Thus, the two designs differentially adjudicate the value trade-off between an individual's *privacy* and the group's *awareness* of individual members' presence and activities.

In the second form, technical investigations involve the proactive design of systems to support values identified in the conceptual investigation. For example, Fuchs (1999) developed a notification service for a collaborative work system in which the underlying technical mechanisms implement a value hierarchy whereby an individual's desire for privacy overrides other group members' desires for awareness.

At times, technical investigations—particularly of the first form—may seem similar to empirical investigations insofar as both involve technological and empirical activity. However, they differ markedly in their unit of analysis. Technical investigations focus on the technology itself. Empirical investigations focus on the individuals, groups, or larger social systems that configure, use, or are otherwise affected by the technology.

4.4 VALUE SENSITIVE DESIGN IN PRACTICE: THREE CASE STUDIES

To illustrate the integrative and iterative tripartite methodology of Value Sensitive Design, we draw on three case studies with real-world applications, one completed and two under way. Each case study represents a unique design space.

4.4.1 Cookies and Informed Consent in Web Browsers

Informed consent provides a critical protection for privacy, and supports other human values such as autonomy and trust. Yet currently there is a mismatch between industry practice and the public's interest. According to a recent report from the Federal Trade Commission (2000), for example, 59% of Web sites that collect personal identifying information neither inform Internet users that they are collecting such information nor seek the users' consent. Yet, according to a Harris poll (2000), 88% of users want sites to garner their consent in such situations.

Against this backdrop, Friedman, Felten, and their colleagues (Friedman et al., 2000b; Millett et al., 2001) sought to design web-based interactions that support informed consent in a web browser through the development of new technical mechanisms for cookie management. This project was an early proof-of-concept project for Value Sensitive Design, which we use here to illustrate several key features of the methodology.

4.4.1.1 Conceptualizing the Value One part of a conceptual investigation entails a philosophically informed analysis of the central value constructs. Accordingly, Friedman et al. began their project with a conceptual investigation of informed consent itself. They drew on diverse literature, such as the Belmont Report, which delineates ethical principles and guidelines for the protection of human subjects (Belmont Report, 1978; Faden and Beauchamp, 1986), to develop criteria for informed consent in online interactions. In brief, the idea of "informed" encompasses disclosure and comprehension. *Disclosure* refers to providing accurate information about the benefits and harms that might reasonably be expected from the action under consideration. *Comprehension* refers to the individual's accurate interpretation of *what* is being disclosed. In turn, the idea of "consent" encompasses voluntariness, comprehension, and agreement. *Voluntariness* refers to ensuring that the action is not controlled or coerced. *Competence* refers to possessing the mental, emotional, and physical capabilities needed to be capable of giving informed consent. *Agreement* refers to a reasonably clear opportunity to accept or decline to participate. Moreover, agreement should be ongoing, that is, the individual should be able to withdraw from the interaction at any time. (See Friedman et al., 2000a for an expanded discussion of these five criteria.)

4.4.1.2 Using a Conceptual Investigation to Analyze Existing Technical Mechanisms With a conceptualization for informed consent online in hand, Friedman et al. conducted a retrospective analysis (one form of a technical investigation) of how the cookie and web-browser technology embedded in Netscape Navigator and Internet Explorer changed with respect to informed consent over a 5-year period, beginning in 1995. Specifically, they used the criteria of disclosure, comprehension, voluntariness, competence, and agreement to evaluate how well each browser in each stage of its development supported the users' experience of informed consent. Through this retrospective analysis, they found that while cookie technology had improved over time regarding informed consent (e.g., increased visibility of cookies, increased options for accepting or declining cookies, and access to information about cookie content), as of 1999 some startling problems remained. For example, (a) while browsers disclosed to users some information about cookies, they still did not disclose the right sort of information, that is, information about the potential harms and benefits from setting a particular cookie. (b) In Internet Explorer, the burden to accept or decline all third-party cookies still fell to the user, placing undue burden on the user to decline each third-party cookie one at a time. (c) Users' out-of-the-box experience of cookies (i.e., the default setting) was no different in 1999 than it was in 1995—to accept all cookies. That is, the novice user installed a browser that accepted all cookies and disclosed nothing about that activity to the user. (d) Neither browser alerted a user when a site wished to use a cookie and for what purpose, as opposed to when a site wished to store a cookie.

4.4.1.3 Iteration and Integration of Conceptual, Technical, and Empirical Investigations On the basis of the results derived from these conceptual and technical investigations, Friedman et al. then iteratively used the results to guide a second technical investigation—a redesign of the Mozilla browser (the open-source code for Netscape Navigator). Specifically, they developed three new types of mechanisms: (a) peripheral awareness of cookies; (b) just-in-time information about individual cookies and cookies in general; and (c) just-in-time management of cookies (see Fig. 4.1). In the process of their technical work, Friedman et al. conducted formative evaluations (empirical investigations) that led to a further design criterion, minimal distraction, which refers to meeting the above criteria for informed consent without unduly diverting the user from the task at hand. Two situations are of concern here. First, if users are overwhelmed with queries to consent to participate in events with minor benefits and risks, they may become numbed to the informed consent process by the time participation in an event with significant benefits and risks is at hand. Thus, the users' participation in that event may not receive the careful attention that is warranted. Second, if the overall distraction to obtain informed consent becomes so great as to be perceived to be an intolerable nuisance, users are likely to disengage from the informed consent process in its entirety and accept or decline participation by rote. Thus, undue distraction can single-handedly undermine informed consent. In this way, the iterative results of the above empirical investigations not only shaped and then validated the technical work, but impacted the initial conceptual investigation by adding to the model of informed consent the criterion of minimal distraction.

Thus, this project illustrates the iterative and integrative nature of Value Sensitive Design, and provides a proof-of-concept for Value Sensitive Design in the context of mainstream Internet software.

4.4.2 Room with a View: Using Plasma Displays in Interior Offices

Janice is in her office, writing a report. She is trying to conceptualize the higher-level structure of the report, but her ideas will not quite take form. Then she looks up from her desk and rests her eyes on the fountain and plaza area outside her building. She

(a)

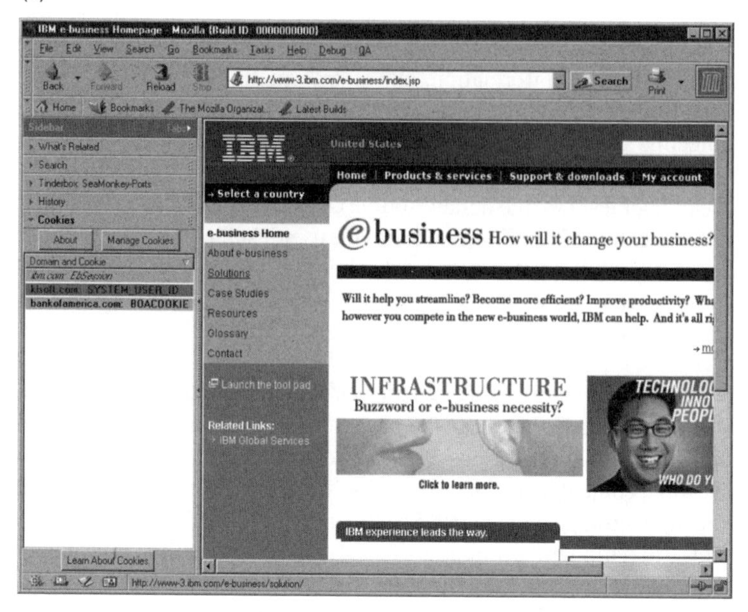

(b)

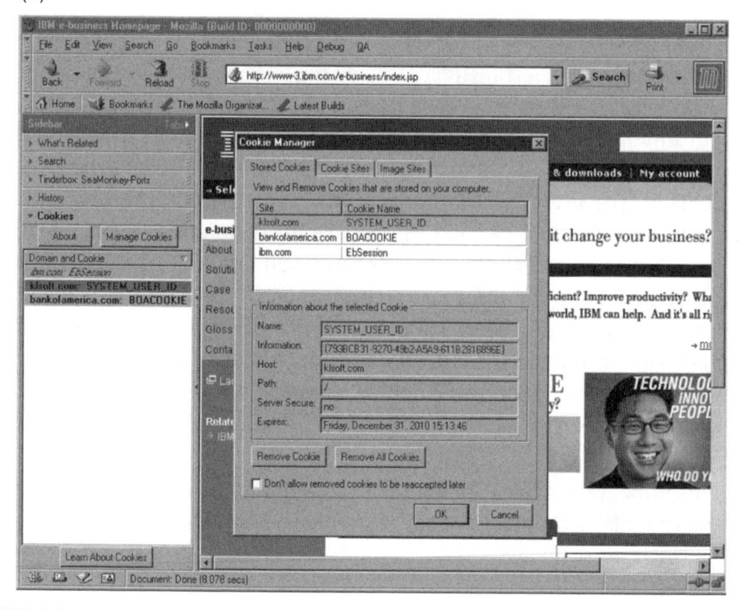

FIGURE 4.1 (**a**) Screen shot of the Mozilla implementation shows the peripheral awareness of cookies interface (at the left) in the context of browsing the web. Each time a cookie is set, a color-coded entry for that cookie appears in the sidebar. Third-party cookies are red; others are green. At the user's discretion, he or she can click on any entry to bring up the Mozilla cookie manager for that cookie. (**b**) Screen shot after the user has clicked on an entry to bring up the just-in-time cookie management tool (in the center) for a particular cookie.

notices the water bursting upward, and that a small group of people are gathering by the water's edge. She rests her eyes on the surrounding pool of calm water. Her eyes then lift toward the clouds and the streaking sunshine. Twenty seconds later she returns to her writing task at hand, slightly refreshed, and with an idea taking shape.

What is particularly novel about this workplace scenario is that Janice works in an interior office. Instead of a real window looking out onto the plaza, Janice has a large-screen video plasma display that continuously displays the local real-time outdoor scene in real-time. Realistic? Beneficial? This design space is currently being researched by Kahn, Friedman, and their colleagues, using the framework of Value Sensitive Design.

In their initial conceptual investigation of this design space, Kahn et al. drew on the psychological literature that suggests that interaction with real nature can garner physiological and psychological benefits. For example, in one study, Ulrich (1984) found that postoperative recovery improved when patients were assigned to a room with a view of a natural setting (a small stand of deciduous trees) versus a view of a brown brick wall. More generally, studies have shown that even minimal connection with nature—such as looking at a natural landscape—can reduce immediate and long-term stress, reduce sickness of prisoners, and calm patients before and during surgery. (See Beck and Katcher, 1996; Kahn, 1999; Ulrich, 1993 for reviews.) Thus Kahn et al. hypothesized that an "augmented window" of nature could render benefits in a work environment in terms of the human values of physical health, emotional well-being, and creativity.

To investigate this question in a laboratory context, Kahn et al. are comparing the short-term benefits of working in an office with a view out the window of a beautiful nature scene versus an identical view (in real time) shown on a large video plasma display that covers the window in the same office (Fig. 4.2a). In this latter condition, they employed a high-definition television (HDTV) camera (Fig. 4.2b) to capture real-time local images. The control condition involved a blank covering over the window. Their measures entailed (a) physiological data (heart rate), (b) performance data (on cognitive and creativity tasks), (c) video data that captured each subject's eye gaze on a second-by-second level, and time synchronized with the physiological equipment, so that analyses can determine if physiological benefits accrued immediately following an eye gaze onto the plasma screen, and (d) social-cognitive data (based on a 50-min interview with each subject at the conclusion of the experimental condition wherein they garnered each subject's reasoned perspective on the experience). Preliminary results show the following trends. First, participants looked at the plasma screen just as frequently as they did the real window, and more frequently than they stared at the blank wall. In this sense, the plasma display window was functioning like a real window. But, when participants gazed for 30 s or more, the real window provided greater physiological recovery from low-level stress as compared to the plasma display window.

From the standpoint of illustrating Value Sensitive Design, we would like to emphasize five ideas.

4.4.2.1 Multiple Empirical Methods Under the rubric of empirical investigations, Value Sensitive Design supports and encourages multiple empirical methods to

FIGURE 4.2 Plasma display technology studies. (**a**) The Watcher, (**b**) The HDTV Camera, (**c**) The Watched .

be used in concert to address the question at hand. As noted above, for example, this study employed physiological data (heart rate), two types of performance data (on cognitive and creativity tasks), behavioral data (eye gaze), and reasoning data (the social-cognitive interview). From a value-oriented perspective, multiple psychological measures increase the veracity of most accounts of technology in use.

4.4.2.2 Direct and Indirect Stakeholders In their initial conceptual investigation of the values implicated in this study, Kahn et al. sought to identify not only direct but also indirect stakeholders affected by such display technology. At that early point, it became clear to the researchers that an important class of indirect stakeholders (and their respective values) needed to be included, namely, the individuals who, by virtue of walking through the fountain scene, unknowingly had their images displayed on the video plasma display in the "inside" office (Fig. 4.2c). In other words, if this application of projection technology were to come into widespread use (as web cams and surveillance cameras have begun to), then it would potentially encroach on the privacy of individuals in public spaces—an issue that has been receiving increasing attention in the field of computer ethics and public discourse (Nissenbaum, 1998). Thus, in addition to the experimental laboratory study, Kahn et al. initiated two additional but complementary empirical investigations with indirect stakeholders: (a) a survey of 750 people walking through the public plaza and (b) in-depth social cognitive interviews with 30 individuals walking through the public plaza (Friedman et al., 2006). Both investigations focused on indirect stakeholders' judgments of privacy in public space, and in particular having their real-time images captured and displayed on plasma screens in nearby and distant offices. The importance of such indirect stakeholder investigations is being borne out by the results. For example, significant gender differences were found in their survey data: more women than men expressed concern about the invasion of privacy through web cameras in public places. This finding held whether their image was to be displayed locally or in another city (Tokyo), or viewed by one person, thousands, or millions. One implication of this finding is that future technical designs and implementations of such display technologies need to be responsive to ways in which men and women might perceive potential harms differently.

4.4.2.3 Coordinated Empirical Investigations Once Kahn et al. identified an important group of indirect stakeholders, and decided to undertake empirical investigations with this group, they then coordinated these empirical investigations with the initial (direct stakeholder) study. Specifically, a subset of identical questions were asked of both the direct stakeholders (the Watchers) and indirect stakeholders (the Watched). The results show some interesting differences. For example, more men in the Watched condition expressed concerns that people's images might be displayed locally, nationally, or internationally than men in the Plasma Display Watcher condition. No differences were found between women in the Watcher Plasma Display Condition and women in the Watched condition. Thus, the Value Sensitive Design methodology helps to bring to the forefront values that matter not only to the direct stakeholders of a technology (such as physical health, emotional well-being, and creativity), but also to the indirect stakeholders (such as privacy, informed consent, trust, and physical safety). Moreover, from the standpoint of Value Sensitive Design, the above study highlights how investigations of indirect stakeholders can be woven into the core structure of the experimental design with direct stakeholders.

4.4.2.4 Multiplicity of and Potential Conflicts Among Human Values Value Sensitive Design can help researchers uncover the multiplicity of and potential conflicts among human values implicated in technological implementations. In the above design space, for example, values of physical health, emotional well-being, and creativity appear to partially conflict with other values of privacy, civil rights, trust, and security.

4.4.2.5 Technical Investigations Conceptual and empirical investigations can help to shape future technological investigations, particularly in terms of how nature (as a source of information) can be embedded in the design of display technologies to further human well-being. One obvious design space involves buildings. For example, if Kahn et al.'s empirical results continue to emerge in line with their initial results, then one possible design guideline is as follows: we need to design buildings with nature in mind, and within view. In other words, we cannot with psychological impunity digitize nature and display the digitized version as a substitute for the real thing (and worse, then destroy the original). At the same time, it is possible that technological representations of nature can garner some psychological benefits, especially when (as in an inside office) direct access to nature is otherwise unavailable. Other less obvious design spaces involve, for example, airplanes. In recent discussions with Boeing Corporation, for example, we were told that for economic reasons engineers might like to construct airplanes without passenger windows. After all, windows cost more to build and decrease fuel efficiency. At stake, however, is the importance of windows in the human experience of flying.

In short, this case study highlights how Value Sensitive Design can help researchers employ multiple psychological methods, across several studies, with direct and indirect stakeholders, to investigate (and ultimately support) a multiplicity of human values impacted by deploying a cutting-edge information technology.

4.4.3 UrbanSim: Integrated Land Use, Transportation, and Environmental Simulation

In many regions in the United States (and globally), there is increasing concern about pollution, traffic jams, resource consumption, loss of open space, loss of coherent community, lack of sustainability, and unchecked sprawl. Elected officials, planners, and citizens in urban areas grapple with these difficult issues as they develop and evaluate alternatives for such decisions as building a new rail line or freeway, establishing an urban growth boundary, or changing incentives or taxes. These decisions interact in complex ways, and, in particular, transportation and land use decisions interact strongly with each other. There are both legal and commonsense reasons to try to understand the long-term consequences of these interactions and decisions. Unfortunately, the need for this understanding far outstrips the capability of the analytic tools used in current practice.

In response to this need, Waddell, Borning, and their colleagues have been developing UrbanSim, a large simulation package for predicting patterns of urban development for periods of 20 years or more, under different possible scenarios (Borning et al., 2008; Noth et al., 2003; Waddell et al., 2002, 2003). Its primary purpose is to provide urban planners and other stakeholders with tools to aid in more informed decision-making, with a secondary goal to support further democratization of the planning process. When provided with different scenarios, packages of possible policies and investments, UrbanSim models the resulting patterns of urban growth and

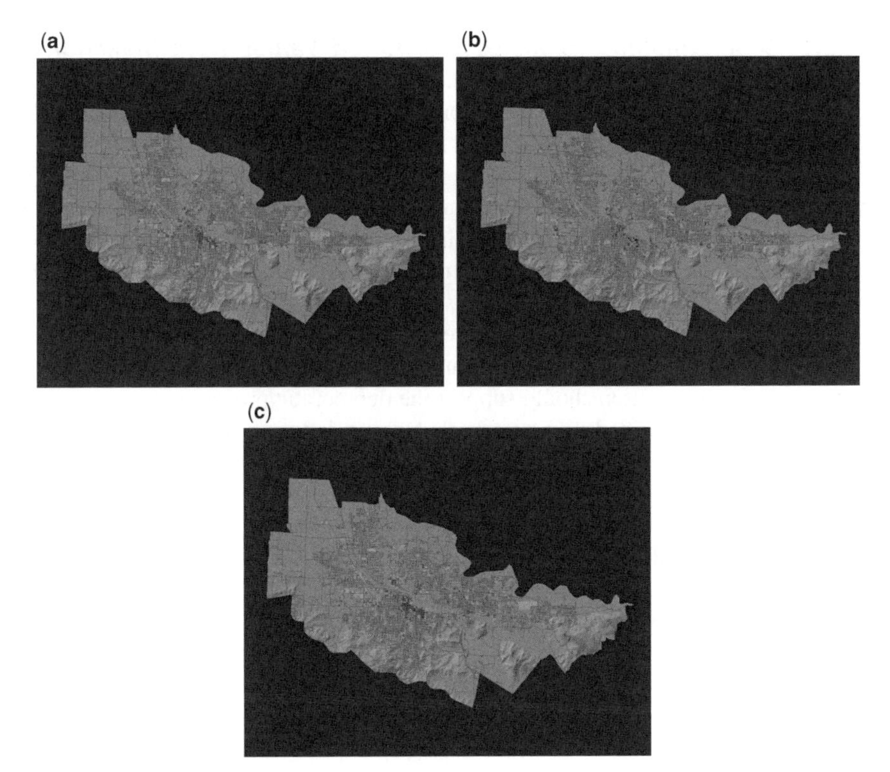

FIGURE 4.3 Results from UrbanSim for Eugene/Springfield, Oregon, forecasting land use patterns over a 14-year period. These results arise from the simulated interactions among demographic change, economic change, real estate development, transportation, and other actors and processes in the urban environment. Map (**a**) shows the employment density in 1980 (number of jobs located in each 150 × 150 m grid cell). Darker red indicates higher density. Map (**b**) shows the predicted change from 1980 to 1994 (where darker red indicates a greater change). Map (**c**) shows the predicted employment density in 1994. In a historical validation of the model, this result was then compared with the actual 1994 employment, with a 0.917 correlation over a one-cell radius.

redevelopment, transportation usage, and resource consumption and other environmental impacts.

As of early 2007, UrbanSim has been applied (either experimentally or in some cases transitioning to operational use) in the metropolitan regions in the United States around Detroit, El Paso, Eugene/Springfield, Oregon (Fig. 4.3), Honolulu, Houston, Salt Lake City, and Seattle; and internationally in Amsterdam, Paris, Tel Aviv, and Zurich. Additional projects have been launched in Burlington, Durham, Phoenix, and San Francisco, and internationally in Melbourne, Australia. Value Sensitive Design has played a central role in the ongoing design and implementation of interactions around UrbanSim indicators. UrbanSim illustrates important aspects of Value Sensitive Design in addition to those described in the previous two case studies.

4.4.3.1 Distinguishing Explicitly Supported Values from Stakeholder Values In their conceptual investigations, Borning et al. (2005) distinguished between explicitly supported values (i.e., ones that they explicitly want to embed in the simulation) and stakeholder values (i.e., ones that are important to some but not necessarily all of the stakeholders). Next, Borning et al. committed to three specific moral values to be supported explicitly. One is fairness, and more specifically freedom from bias. The simulation should not discriminate unfairly against any group of stakeholders, or privilege one mode of transportation or policy over another. The second value is accountability. Insofar as possible, stakeholders should be able to confirm that their values are reflected in the simulation, evaluate and judge its validity, and develop an appropriate level of confidence in its output. The third value is democracy. The simulation should support the democratic process in the context of land use, transportation, and environmental planning. In turn, as part of supporting the democratic process, Borning et al. decided that the model should not *a priori* favor or rule out any given set of stakeholder values, but instead should allow different stakeholders to articulate the values that are most important to them, and evaluate the alternatives in light of these values.

4.4.3.2 Handling Widely Divergent and Potentially Conflicting Stakeholder Values From the standpoint of conceptual investigations, UrbanSim as a design space poses tremendous challenges. The research team cannot focus on a few key values, as occurred in the Web Browser project (e.g., the value of informed consent), or the Room with a View project (e.g., the values of privacy in public spaces and physical and psychological well-being). Rather, disputing stakeholders bring to the table widely divergent values about environmental, political, moral, and personal issues. Examples of stakeholder values are environmental sustainability, walkable neighborhoods, space for business expansion, affordable housing, freight mobility, minimal government intervention, minimal commute time, open space preservation, property rights, and environmental justice. How does one characterize the wide-ranging and deeply held values of diverse stakeholders, both present and future? Moreover, how does one prioritize the values implicated in the decisions? And, how can one move from values to measurable outputs from the simulation to allow stakeholders to compare alternative scenarios?

As part of addressing these questions, the research group implemented a web-based interface that groups indicators into three broad value categories pertaining to the domain of urban development (economic, environmental, and social) and more specific value categories under that. To allow stakeholders to evaluate alternative urban futures, the interface provides a large collection of *indicators*: variables that distill some attribute of interest about the results (Gallopin, 1997). (Examples of indicators are the number of acres of rural land converted into urban use each year, the degree of poverty segregation, or the mode share between autos and transit.) These categories and indicators draw on a variety of sources, including empirical research on people's environmental concepts and values (Kahn, 1999; Kahn and Kellert, 2002), community-based indicator projects (Hart, 1999; Palmer, 1998), and the policy literature. Stakeholders can then use the interface to select indicators that speak to values that are important to them from among these categories.

This interface illustrates the interplay among conceptual, technical, and empirical investigations. The indicators are chosen to speak to different stakeholder values, responding to our distinction between explicitly supported values and stakeholder values in the initial conceptual investigation. The value categories are rooted empirically in both human psychology and policy studies, not just philosophy, and then embodied in a technical artifact (the web-based interface), which is in turn evaluated empirically.

4.4.3.3 Legitimation As we continued our work on VSD and UrbanSim, in our conceptual investigations we identified *legitimation* as a key instrumental value (Borning et al., 2005; Davis, 2006). UrbanSim's legitimacy is crucial for its effective use in the planning process—stakeholders who do not see its use as legitimate may disengage from its use, or if they remain in the process may never accept the analyses that it informs. Our conceptualization of legitimation draws primarily on the work of Habermas (1979, 1984). Since the legitimacy of an urban planning process depends on a huge number of factors—most of which are outside of UrbanSim's scope—we concern ourselves with the *legitimation potential* of the modeling system. Again following Habermas, *communicative action* plays a key role in legitimation potential. The implicit validity claims of an utterance in a communicative act lead to a set of testable design goals for the system regarding comprehensibility, validity, transparency, and freedom from bias, which we then used in structuring our empirical investigations.

4.4.3.4 Technical Choices Driven by Initial and Emergent Value Considerations Most of the technical choices in the design of the UrbanSim software are in response to the need to generate indicators and other evaluation measures that respond to different strongly held stakeholder values. For example, for some stakeholders, walkable, pedestrian-friendly neighborhoods are very important. But being able to model walking as a transportation mode makes difficult demands on the underlying simulation, requiring a finer-grained spatial scale than is needed for modeling automobile transportation alone. In turn, being able to answer questions about walking as a transportation mode is important for two explicitly supported values: fairness (not to privilege one transportation mode over another) and democracy (being able to answer questions about a value that is important to a significant number of stakeholders). As a second example of technical choices being driven by value considerations, UrbanSim's software architecture is designed to support rapid evolution in response to changed or additional requirements. For instance, the software architecture decouples the individual component models as much as possible, allowing them to evolve and new ones to be added in a modular fashion. Further, the architecture separates out the computation of indicator values from the models, making it easy to write new indicators as needed, rather than embedding the indicator code in the component models themselves. For similar reasons, the UrbanSim team uses the YP agile software development methodology (Freeman-Benson and Borning,

2003), which allows the system to evolve and respond quickly to emerging stakeholder values and policy considerations.

4.4.3.5 Designing for Credibility, Openness, and Accountability The credibility of the system is of great importance, particularly when the system is being used in a politically charged situation, and is thus the subject of intense scrutiny. The research group has undertaken a variety of activities to help foster credibility, including using behaviorally transparent simulation techniques (i.e., simulating agents in the urban environment, such as households, businesses, and real estate developers, rather than using some more abstract and opaque simulation technique), and performing sensitivity analyses (Franklin et al., 2002) and a historical validation. In the historical validation, for example, the group started the model with 1980 data from Eugene/Springfield, simulated through 1994, and compared the simulation output with what actually happened. One of these comparisons is shown in Fig. 4.3. In addition, our techniques for fostering openness and accountability are also intended to support credibility. These include using Open Source software (releasing the source code along with the executable), writing the code in as clear and understandable a fashion as possible, using a rigorous and extensive testing methodology, and complementing the Open Source software with an Open Process that makes the state of our development visible to anyone interested. For example, in our laboratory, a battery of tests is run whenever a new version of the software is committed to the source code repository. A traffic light (a real one) is activated by the testing regime—green means that the system has passed all tests, yellow means testing is under way, and red means that a test has failed. There is also a virtual traffic light, mirroring the physical one, visible on the web (www.urbansim.org/fireman). Similarly, the bug reports, feature requests, and plans are all on the UrbanSim project Web site as well. Details of this Open Process approach may be found in Freeman-Benson and Borning (2003).

For interactions around indicators, one project has been carefully documenting the available indicators and their limitations (Borning et al., 2005), including using "live documentation" that directly includes the source code used to compute the indicator values and tests of that source code. Another project has involved partnering with different community organizations to produce "Indicator Perspectives" that provide different views on which indicators are most important, and how they should be evaluated. Finally, Janet Davis's PhD dissertation (Davis, 2006) describes the design and implementation of "personal indicators," which help users answer the question "how will this policy affect me and my family?" in addition to the more region-level results from the existing indicator sets.

Thus, in summary, Borning et al. are using Value Sensitive Design to investigate how a technology—an integrated land use, transportation, and environmental computer simulation—affects human values on both the individual and organizational levels, and how human values can continue to drive the technical investigations, including refining the simulation, data, and interaction model. Finally, employing Value Sensitive Design in a project of this scope serves to validate its use for complex, large-scale systems.

4.5 VALUE SENSITIVE DESIGN'S CONSTELLATION OF FEATURES

Value Sensitive Design shares and adopts many interests and techniques from related approaches to values and system design—computer ethics, social informatics, CSCW, and Participatory Design—as discussed in Section 4.2.2. However, Value Sensitive Design itself brings forward a unique constellation of eight features.

First, Value Sensitive Design seeks to be proactive to influence the design of technology early in and throughout the design process.

Second, Value Sensitive Design enlarges the arena in which values arise to include not only the workplace (as traditionally in the field of CSCW), but also education, the home, commerce, online communities, and public life.

Third, Value Sensitive Design contributes a unique methodology that employs conceptual, empirical, and technical investigations, applied iteratively and integratively (see Section 4.3).

Fourth, Value Sensitive Design enlarges the scope of human values beyond those of cooperation (CSCW) and participation and democracy (Participatory Design) to include all values, especially those with moral import. By moral, we refer to issues that pertain to fairness, justice, human welfare, and virtue, encompassing within moral philosophical theory deontology (Dworkin, 1978; Gewirth, 1978; Kant, 1785/1964; Rawls, 1971), consequentialism (Smart and Williams, 1973; see Scheffler, 1982 for an analysis), and virtue (Campbell and Christopher, 1996; Foot, 1978; MacIntyre, 1984). Value Sensitive Design also accounts for conventions (e.g., standardization of protocols) and personal values (e.g., color preferences within a graphical user interface).

Fifth, Value Sensitive Design distinguishes between usability and human values with ethical import. Usability refers to characteristics of a system that make it work in a functional sense, including that it is easy to use, easy to learn, consistent, and recovers easily from errors (Adler and Winograd, 1992; Nielsen, 1993; Norman, 1988). However, not all highly usable systems support ethical values. Nielsen (1993), for example, asks us to imagine a computer system that checks for fraudulent applications of people who are applying for unemployment benefits by asking applicants numerous personal questions, and then checking for inconsistencies in their responses. Nielsen's point is that even if the system receives high usability scores some people may not find the system socially acceptable, based on the moral value of privacy.

Sixth, Value Sensitive Design identifies and takes seriously two classes of stakeholders: direct and indirect. Direct stakeholders refer to parties—individuals or organizations—who interact directly with the computer system or its output. Indirect stakeholders refer to all other parties who are affected by the use of the system. Often, indirect stakeholders are ignored in the design process. For example, computerized medical record systems have often been designed with many of the direct stakeholders in mind (e.g., insurance companies, hospitals, doctors, and nurses), but with too little regard for the values, such as the value of privacy, of a rather important group of indirect stakeholders—the patients.

Seventh, Value Sensitive Design is an interactional theory—values are viewed neither as inscribed into technology (an endogenous theory) nor as simply transmitted by social forces (an exogenous theory). Rather, the interactional position holds that while the features or properties that people design into technologies more readily support certain values and hinder others, the technology's actual use depends on the goals of the people interacting with it. A screwdriver, after all, is well suited for turning screws, and is also amenable to use as a poker, pry bar, nail set, cutting device, and tool to dig up weeds, but functions poorly as a ladle, pillow, or wheel. Similarly, an online calendar system that displays individuals' scheduled events in detail readily supports accountability within an organization, but makes privacy difficult. Moreover, through human interaction, technology itself changes over time. On occasion, such changes (as emphasized in the exogenous position) can mean the societal rejection of a technology, or that its acceptance is delayed. But more often it entails an iterative process whereby technologies are first invented, and then redesigned based on user interactions, which then are reintroduced to users, further interactions occur, and further redesigns implemented. Typical software updates (e.g., of word processors, browsers, and operating systems) epitomize this iterative process.

Eighth, Value Sensitive Design builds from the psychological proposition that certain values are universally held, although how such values play out in a particular culture at a particular point in time can vary considerably (Kahn, 1999; Turiel, 1998, 2002). For example, even while living in an igloo, Inuits have conventions that ensure some forms of privacy, yet such forms of privacy are not maintained by separated rooms, as they are in most Western cultures. Generally, the more concretely (act-based) one conceptualizes a value, the more one will be led to recognize cultural variation; conversely, the more abstractly one conceptualizes a value, the more one will be led to recognize universals. Value Sensitive Design seeks to work both the concrete and abstract levels, depending on the design problem at hand. Note that this is an empirical proposition, based on a large amount of psychological and anthropological data, not a philosophical one. We also make this claim only for certain values, not all— there are clearly some values that are culture-specific.

The three case studies presented in Section 4.4 illustrate the different features in this constellation. For example, UrbanSim illustrates the goal of being proactive and influencing the design of the technology early in and throughout the design process (Feature 1), and also involves enlarging the arena in which values arise to include urban planning and democratic participation in public decision-making (Feature 2). The cookies work is a good illustration of Value Sensitive Design's tripartite methodology (Feature 3): conceptual, technical, and empirical investigations, applied iteratively and integratively, which were essential to the success of the project. Each of the three projects brings out a different set of human values (Feature 4): among others, informed consent for the cookies work; physical and psychological well-being and privacy in public spaces for Room with a View; and fairness, accountability, and democracy for UrbanSim, as well as the whole range of different, sometimes competing, stakeholder values. The cookies project illustrates the complex interaction between usability and human values (Feature 5): early versions of the system supported informed consent at the expense of usability, requiring additional work to develop a system that was usable and provided reasonable support for informed consent. The Room with a View work considers and takes seriously both direct and indirect stakeholders (Feature 6): the occupants of the inside office (the Watchers) and passersby in the plaza (the Watched). Value Sensitive Design's position that values are neither inscribed into technology nor simply transmitted by social forces (Feature 7) is illustrated by UrbanSim: the system by itself is certainly not neutral with respect to democratic process, but at the same time does not on its own ensure democratic decision-making on land use and transportation issues. Finally, the proposition that certain values are universally held but play out in very different ways in different cultures and different times (Feature 8) is illustrated by the Room with a View project: the work is informed by a substantial body of work on the importance of privacy in all cultures (e.g., the deep connection between privacy and self-identity), but concerns about privacy in public spaces play out in a specific way in the United States, and might do so quite differently in another cultural context.

We could draw out additional examples that illustrate Value Sensitive Design's constellation of features, both from the three case studies presented in Section 4.4 and in other projects but we hope that this short description demonstrates the unique contribution that Value Sensitive Design can make to the design of technology.

4.6 PRACTICAL SUGGESTIONS FOR USING VALUE SENSITIVE DESIGN

One natural question with Value Sensitive Design is, "How exactly do I do it?" In this section we offer some practical suggestions.

4.6.1 Start with a Value, Technology, or Context of Use

Any of these three core aspects—value, technology, or context of use—easily motivates Value Sensitive Design. We suggest starting with the aspect that is most central to your work and interests. In the case of informed consent and cookies, for example, Friedman et al. began with a value of central interest (informed consent) and moved from that value to its implications for web browser design. In the case of UrbanSim, Borning et al. began with a technology (urban simulation) and a context of use (the urban planning process); upon inspection of those two, values issues quickly came to the fore.

4.6.2 Identify Direct and Indirect Stakeholders

As part of the initial conceptual investigation, systematically identify direct and indirect stakeholders. Recall that direct stakeholders are those individuals who interact directly with the technology or with the technology's output. Indirect stakeholders are those individuals who are also impacted by the system, though they never interact directly with it. In addition, it is worthwhile to recognize the following:

- Within each of these two overarching categories of stakeholders, there may be several subgroups.

- A single individual may be a member of more than one stakeholder group or subgroup. For example, in the UrbanSim project, an individual who works as an urban planner and lives in the area is both a direct stakeholder (i.e., through his or her direct use of the simulation to evaluate proposed transportation plans) and an indirect stakeholder (i.e., by virtue of living in the community for which the transportation plans will be implemented).
- An organizational power structure is often orthogonal to the distinction between direct and indirect stakeholders. For example, there might be low-level employees who are either direct or indirect stakeholders and who don't have control over using the system (e.g., workers on an assembly line). Participatory Design has contributed a substantial body of analysis to these issues, as well as techniques for dealing with them, such as ways of equalizing power among groups with unequal power. (See the references cited in Section 4.2.2.)

4.6.3 Identify Benefits and Harms for Each Stakeholder Group

Having identified the key stakeholders, systematically identify the benefits and harms for each group. In doing so, we suggest attention to the following points:

- Indirect stakeholders will be benefited or harmed to varying degrees, and in some designs it is probably possible to claim every human as an indirect stakeholder of some sort. Thus, one rule of thumb in the conceptual investigation is to give priority to indirect stakeholders who are strongly affected, or to large groups that are somewhat affected.
- Attend to issues of technical, cognitive, and physical competency. For example, children or the elderly might have limited cognitive competency. In such a case, care must be taken to ensure that their interests are represented in the design process, either by representatives from the affected groups themselves or, if this is not possible, by advocates.
- Personas (Pruitt and Grudin, 2003) are a popular technique that can be useful for identifying the benefits and harms to each stakeholder group. However, we note two caveats. First, personas have a tendency to lead to stereotypes because they require a list of "socially coherent" attributes to be associated with the "imagined individual." Second, while in the literature each persona represents a different user group, in Value Sensitive Design (as noted above) the same individual may be a member of more than one stakeholder group. Thus, in our practice, we have deviated from the typical use of personas that maps a single persona onto a single user group, to allow for a single persona to map onto multiple stakeholder groups.

4.6.4 Map Benefits and Harms onto Corresponding Values

With a list of benefits and harms in hand, one is in a strong position to recognize corresponding values. Sometimes the mapping is one of identity. For example, a harm that is characterized as invasion of privacy maps onto the value of privacy. Other times the mapping is less direct, if not multifaceted. For example, in the Room with a View study, it is possible that a direct stakeholder's mood is improved when working in an office with an augmented window (as compared with no window). Such a benefit potentially implicates not only the value of psychological welfare but also creativity, productivity, and physical welfare (health), assuming there is a causal link between improved mood and these other factors.

In some cases, the corresponding values will be obvious, but not always. Table 4.1 provides a table of human values with ethical import often implicated in system design. This table may be useful in suggesting values that should be considered in the investigation.

4.6.5 Conduct a Conceptual Investigation of Key Values

After the identification of key values in play, a conceptual investigation of each can follow. Here it is helpful to turn to the relevant literature. In particular, the philosophical ontological literature can help provide criteria for what a value is, and thereby how to assess it empirically. (e.g., Section 4.4.1.1 described how existing literature helped provide criteria for the value of informed consent.)

4.6.6 Identify Potential Value Conflicts

Values often come into conflict. Thus, once key values have been identified and carefully defined, a next step entails examining potential conflicts. For the purposes of design, value conflicts should usually not be conceived of as "either/or" situations, but as constraints on the design space. Admittedly, at times, designs that support one value directly hinder support for another. In those instances, a good deal of discussion among the stakeholders may be warranted to identify the space of workable solutions. Typical value conflicts include accountability versus privacy, trust versus security, environmental sustainability versus economic development, privacy versus security, and hierarchical control versus democratization.

4.6.7 Integrate Value Considerations into One's Organizational Structure

Ideally, Value Sensitive Design will work in concert with organizational objectives. Within a company, for example, designers would bring values to the forefront, and in the process generate increased revenue, employee satisfaction, customer loyalty, and other desirable outcomes for their companies. In turn, within a government agency, designers would both better support national and community values and enhance the organization's ability to achieve its objectives. In the real world, of course, human values (especially those with ethical import) may collide with economic objectives, power, and other factors. However, even in such situations, Value Sensitive Design should be able to make positive contributions, by showing alternate designs that better support enduring human values. For example, if a standards committee were considering adopting a protocol that raised serious

TABLE 4.1 Human Values (with Ethical Import) Often Implicated in System Design

Human Value	Definition	Sample Literature
Human welfare	Refers to people's physical, material, and psychological well-being	Friedman et al. (2003), Leveson (1991), Neumann (1995), Turiel (1983, 1998)
Ownership and property	Refers to a right to possess an object (or information), use it, manage it, derive income from it, and bequeath it	Becker (1977), Friedman (1997b), Herskovits (1952), Lipinski and Britz (2000)
Privacy	Refers to a claim, an entitlement, or a right of an individual to determine what information about himself or herself can be communicated to others	Agre and Rotenberg (1998), Bellotti (1998), Boyle et al. (2000), Friedman (1997b), Fuchs (1999), Jancke et al. (2001), Palen and Dourish (2003), Nissenbaum (1998), Phillips (1998), Schoeman (1984), Svensson et al. (2001)
Freedom from bias	Refers to systematic unfairness perpetrated on individuals or groups, including preexisting social bias, technical bias, and emergent social bias	Friedman and Nissenbaum (1996), cf. Nass and Gong (2000), Reeves and Nass (1996)
Universal usability	Refers to making all people successful users of information technology	Aberg and Shahmehri (2001), Shneiderman (1999, 2000), Cooper and Rejmer (2001), Jacko et al. (1999), Stephanidis (2001)
Trust	Refers to expectations that exist between people who can experience goodwill, extend goodwill toward others, feel vulnerable, and experience betrayal	Baier (1986), Camp (2000), Dieberger et al. (2001), Egger (2000), Fogg and Tseng (1999), Friedman et al. (2000a), Kahn and Turiel (1988), Mayer et al. (1995), Olson and Olson (2000), Nissenbaum (2001), Rocco (1998)
Autonomy	Refers to people's ability to decide, plan, and act in ways that they believe will help them to achieve their goals	Friedman and Nissenbaum (1997), Hill (1991), Isaacs et al. (1996), Suchman (1994), Winograd (1994)

Informed consent	Refers to garnering people's agreement, encompassing criteria of disclosure and comprehension (for "informed") and voluntariness, competence, and agreement (for "consent")	Faden and Beauchamp (1986), Friedman et al., (2000b), The Belmont Report (1978)
Accountability	Refers to the properties that ensure that the actions of a person, people, or institution may be traced uniquely to the person, people, or institution	Friedman and Kahn (1992), Friedman and Millett, (1995), Reeves and Nass (1996)
Courtesy	Refers to treating people with politeness and consideration	Bennett and Delatree (1978), Wynne and Ryan (1993)
Identity	Refers to people's understanding of who they are over time, embracing both continuity and discontinuity over time	Bers et al. (2001), Rosenberg (1997), Schiano and White (1998), Turkle (1996)
Calmness	Refers to a peaceful and composed psychological state	Friedman and Kahn (2003), Weiser and Brown (1997)
Environmental sustainability	Refers to sustaining ecosystems such that they meet the needs of the present without compromising future generations	Hart (1999), Moldan et al. (1997), Northwest Environment Watch (2002), United Nations (1992), World Commission on Environment and Development (1987)

privacy concerns, a Value Sensitive Design analysis and design might result in an alternate protocol that better addressed the issue of privacy while still retaining other needed properties. Citizens, advocacy groups, staff members, politicians, and others could then have a more effective argument against a claim that the proposed protocol was the only reasonable choice.

4.6.8 Human Values (with Ethical Import) Are Often Implicated in System Design

We stated earlier that while all values fall within its purview, Value Sensitive Design emphasizes values with ethical import. In Table 4.1, we present a list of frequently implicated values. This table is intended as a heuristic for suggesting values that should be considered in the investigation; it is definitely not intended as a complete list of human values that might be implicated.

Two caveats: First, not all of these values are fundamentally distinct from one another. Nonetheless, each value has its own language and conceptualizations within its respective field, and thus warrants separate treatment here. Second, as noted above, this list is not comprehensive. Perhaps no list could be, at least within the confines of a paper. Peacefulness, respect, compassion, love, warmth, creativity, humor, originality, vision, friendship, cooperation, collaboration, purposefulness, devotion, loyalty, diplomacy, kindness, musicality, harmony—the list of other possible moral and non-moral values could get very long very quickly. Our particular list comprises many of the values that hinge on the deontological and consequentialist moral orientations noted above: human welfare, ownership and property, privacy, freedom from bias, universal usability, trust, autonomy, informed consent, and accountability. In addition, we have chosen several other values related to system design: courtesy, identity, calmness, and environmental sustainability.

4.6.9 Heuristics for Interviewing Stakeholders

As part of an empirical investigation, it is useful to interview stakeholders to better understand their judgments about a context of use, an existing technology, or a proposed design. A semistructured interview often offers a good balance between addressing the questions of interest and gathering new and unexpected insights. In these interviews, the following heuristics can prove useful:

In probing stakeholders' reasons for their judgments, the simple question "Why?" can go a good distance. For example, seniors evaluating a ubiquitous computing video surveillance system might respond negatively to the system. When asked "Why?" a response might be, "I don't mind my family knowing that other people are visiting me, so they don't worry that I'm alone—I just don't want them to know who is visiting." The researcher can probe again, "Why don't you want them to know?" An answer might be, "I might have a new friend I don't want them to know about. It's not their business." Here the first "why" question elicits information about a value conflict (the family's desire to know about the senior's well-being and the senior's desire to control some information); the second "why" question elicits further information about the value of privacy for the senior.

Ask about values not only directly, but also indirectly, based on formal criteria specified in the conceptual investigation. For example, suppose that you want to conduct an empirical investigation of people's reasoning and values about "X" (say, trust, privacy, or informed consent), and that you decided to employ an interview methodology. One option is to ask people directly about the topic. "What is X?" "How do you reason about X?" "Can you give me an example from your own life of when you encountered a problem that involved X?" There is some merit to this direct approach. Certainly, it gives people the opportunity to define the problem in their own terms. But you may quickly discover that it comes up short. Perhaps the greatest problem is that people have concepts about many aspects of the topic on which they cannot directly reflect. Instead, you will usually be better served by employing an alternative approach. As is common in social cognitive research (see Kahn, 1999, Chapter 5, for a discussion of methods), you could interview people about a hypothetical situation, or a common everyday event in their lives, or a task that you have asked them to solve, or a behavior in which they have just engaged. But, no matter what you choose, the important point is *a priori* to conceptualize what the topic entails, if possible demarcating its boundaries through formal criteria, and at a minimum employing issues or tasks that engage people's reasoning about the topic under investigation.

4.6.10 Heuristics for Technical Investigations

When engaging in value-oriented technical investigations, the following heuristics can prove useful:

Technical mechanisms will often adjudicate multiple if not conflicting values, often in the form of design trade-offs. We have found it helpful to make explicit how a design trade-off maps onto a value conflict and differentially affects different groups of stakeholders. For example, the Room with a View study suggests that real-time displays in interior offices may provide physiological benefits for those in the inside offices (the direct stakeholders), yet may impinge on the privacy and security of those walking through the outdoor scene (the indirect stakeholders), and especially women.

Unanticipated values and value conflicts often emerge after a system is developed and deployed. Thus, when possible, design flexibility into the underlying technical architecture so that it can be responsive to such emergent concerns. In UrbanSim, for example, Borning et al. used agile programming techniques to design an architecture that can more readily accommodate new indicators and models.

The control of information flow through underlying protocols, and the privacy concerns surrounding such control, is a strongly contested area. Ubiquitous computing, with sensors that collect and then disseminate information at large, has only intensified these concerns. We suggest that underlying protocols that release information should be able to be turned off (and in such a way that the stakeholders are confident they have been turned off).

4.7 CONCLUSION

There is a growing interest and challenge to address values in design. Our goal in this chapter has been to provide enough detail about Value Sensitive Design so that other researchers and designers can critically examine, use, and extend this approach. Our hope is that this approach can contribute to a principled and comprehensive consideration of values in the design of information and computational systems.

ACKNOWLEDGMENTS

Value Sensitive Design has emerged over the past decade and benefited from discussions with many people. We would particularly like to acknowledge all the members of our respective research groups, along with Edward Felten, Jonathan Grudin, Sara Kiesler, Clifford Nass, Helen Nissenbaum, John Thomas, and Terry Winograd. This research was supported in part by NSF Awards IIS-9911185, IIS-0325035, EIA-0121326, and EIA-0090832.

REFERENCES

Aberg, J. and Shahmehri, N. (2001). An empirical study of human Web assistants: Implications for user support in Web information systems. In: *Proceedings of the Conference on Human Factors in Computing Systems (CHI* 2000). Association for Computing Machinery Press, New York, NY, pp. 404–411.

Ackerman, M.S. and Cranor, L. (1999). Privacy critics: UI components to safeguard users' privacy. In: *Extended Abstracts of CHI 1999*. ACM Press, pp. 258–259.

Adler, P.S. and Winograd, T. (Eds.) (1992). *Usability: Turning Technologies into Tools*. Oxford University Press, Oxford.

Agre, P.E. and Rotenberg, M. (Eds.) (1998). *Technology and Privacy: The New Landscape*. MIT Press, Cambridge, MA.

Baier, A. (1986). Trust and antitrust. *Ethics*, 231(60), 231–260.

Beck, A. and Katcher, A. (1996). *Between Pets and People*. Purdue University Press, West Lafayette, IN.

Becker, L.C. (1977). *Property Rights: Philosophical Foundations*. Routledge & Kegan Paul, London, England.

Bellotti, V. (1998). Design for privacy in multimedia computing and communications environments. In: Agre, P.E. and Rotenberg, M. (Eds.), *Technology and Privacy: The New Landscape*. The MIT Press, Cambridge, MA, pp. 63–98.

The Belmont Report. Ethical Principles and Guidelines for the Protection of Human Subjects of Research (1978). The National Commission for the Protection of Human Subjects of Biomedical and Behavioral Research.

Bennet, W.J. and Delatree, E.J. (1978). Moral education in the schools. *The Public Interest*, 50, 81–98.

Bers, M.U., Gonzalez-Heydrich, J., and DeMaso, D.R. (2001). Identity construction environments: supporting a virtual therapeutic community of pediatric patients undergoing dialysis. In: *Proceedings of the Conference of Human Factors in Computing Systems (CHI 2001)*. Association for Computing Machinery, New York, NY, pp. 380–387.

Bjerknes, G. and Bratteteig, T. (1995). User participation and democracy: a discussion of Scandinavian research on system development. *Scandinavian Journal of Information Systems*, 7(1), 73–97.

Bødker, S. (1990). *Through the Interface: A Human Activity Approach to User Interface Design*. Lawrence Erlbaum Associates, Hillsdale, NJ.

Borning, A., Friedman, B., Davis, J., and Lin, P. (2005). Informing public deliberation: Value Sensitive Design of indicators for a large-scale urban simulation. In: *Proceedings of the 9th European Conference on Computer-Supported Cooperative Work,* Paris, September 2005, pp. 449–468.

Borning, A., Waddell, P., and Förster, R. (2008). UrbanSim: using simulation to inform public deliberation and decision-making, In: Chen, Hsinchun et al. (Eds.), *Digital Government: Advanced Research and Case Studies*. Springer-Verlag, New York, NY, pp. 439–466.

Boyle, M., Edwards, C., and Greenberg, S. (2000). The effects of filtered video on awareness and privacy. In: *Proceedings of Conference on Computer Supported Cooperative Work (CSCW 2000)*. Association for Computing Machinery, New York, NY pp. 1–10.

Bynum, T.W. (Ed.) (1985). *Metaphilosophy*, 16(4) 263–377.

Camp, L.J. (2000). *Trust and Risk in Internet Commerce*. MIT Press, Cambridge, MA.

Campbell, R.L. and Christopher, J.C. (1996). Moral development theory: a critique of its Kantian presuppositions. *Developmental Review,* 16, 1–47.

Carroll, J.M. and Rosson, M.B. (2000). Dimension of participation in information system design. In: Zhang, P. and Galletta, D. (Eds.), *Human-Computer Interaction and Management Information System*. M.E. Sharpe, Armonk, NY.

Cooper, M. and Rejmer, P. (2001). Case study: localization of an accessibility evaluation. In: *Extended Abstracts of the Conference on Human Factors in Computing Systems (CHI 2001)*. Association for Computing Machinery Press, New York, NY, pp. 141–142.

Davis, J. (2006). Value Sensitive Design of interactions with UrbanSim indicators, Ph.D. dissertation. Department of Computer Science & Engineering, University of Washington.

Dieberger, A., Hook, K., Svensson, M., and Lonnqvist, P. (2001). Social navigation research agenda. In: *Extended Abstracts of the Conference on Human Factors in Computing Systems (CHI 2001)*. Association of Computing Machinery Press, New York, NY, pp. 107–108.

Dworkin, R. (1978). *Taking Rights Seriously*. Harvard University Press, Cambridge, MA.

Egger, F.N. (2000). "Trust me, I'm an online vendor": towards a model of trust for e-commerce system design. In: *Extended Abstracts of the Conference of Human Factors in Computing Systems (CHI 2000)*. Association for Computing Machinery, New York, NY, pp. 101–102.

Ehn, P. (1989). *Work-Oriented Design of Computer Artifacts*. Lawrence Erlbaum Associates, Hillsdale, NJ.

Faden, R. and Beauchamp, T. (1986). *A History and Theory of Informed Consent*. Oxford University Press, New York, NY.

Fogg, B.J. and Tseng, H. (1999). The elements of computer credibility. In: *Proceedings of CHI 1999*. ACM Press, pp. 80–87.

Foot, P. (1978). *Virtues and Vices*. University of California Press, Berkeley and Los Angeles, CA.

Frankena, W. (1972). Value and valuation. In: Edwards, P., (Ed.), *The Encyclopedia of Philosophy*. Vols. 7–8 Macmillan, New York, NY, pp. 409–410.

Franklin, J., Waddell, P., and Britting, J. (2002). Sensitivity analysis approach for an integrated land development & travel demand modeling system. Presented at the Association of Collegiate Schools of Planning 44th Annual Conference, November 21–24, 2002, Baltimore, MD. Preprint available from www.urbansim.org.

Freeman-Benson, B.N. and Borning, A. (2003). YP and urban simulation: applying an agile programming methodology in a politically tempestuous domain. In: *Proceedings of the 2003 Agile Programming Conference*. Salt Lake City. Preprint available from www.urbansim.org.

Friedman, B. (Ed.) (1997a). *Human Values and the Design of Computer Technology*. Cambridge University Press, New York, NY.

Friedman, B. (1997b). Social judgments and technological innovation: adolescents' understanding of property, privacy, and electronic information. *Computers in Human Behavior*, 13(3), 327–351.

Friedman, B., Howe, D.C., and Felten, E. (2002). Informed consent in the Mozilla browser: implementing Value-Sensitive Design. In: *Proceedings of HICSS-35*. IEEE Computer Society, Abstract, p. 247; CD-ROM of full papers, OSPE101.

Friedman, B. and Kahn, P.H. Jr. (1992). Human agency and responsible computing: implications for computer system design. *Journal of Systems Software*, 17, 7–14.

Friedman, B. and Kahn, P.H. Jr. (2003). Human values, ethics, and design. In: Jacko, J. and Sears, A., (Eds.), *The Human-Computer Interaction Handbook*. Lawrence Erlbaum Associates, Mahwah, NJ.

Friedman, B., Kahn, P.H. Jr., and Howe, D.C. (2000a). Trust online. *Communications of the ACM*, 43(12), 34–40.

Friedman, B., Kahn, P.H. Jr., and Hagman, J. (2003). Hardware companions?: What online AIBO discussion forums reveal about the human-robotic relationship. In: *Conference Proceedings of CHI 2003*. ACM Press, New York, NY, pp. 273–280.

Friedman, B., Kahn, P.H. Jr., Hagman, J., Severson, R.L., and Gill, B. (2006). The watcher and the watched: social judgments about privacy in a public place. *Human–Computer Interaction*, 21(2), 235–272.

Friedman, B. and Millett, L. (1995). It's the computer's fault—reasoning about computers as moral agents. In: *Conference Companion of the Conference on Human Factors in Computing Systems (CHI 95)*. Association for Computing Machinery Press, New York, NY, pp. 226–227.

Friedman, B., Millett, L., and Felten, E. (2000b). *Informed Consent Online: A Conceptual Model and Design Principles*. University of Washington Computer Science & Engineering Technical Report 00–12–2.

Friedman, B. and Nissenbaum, H. (1996). Bias in computer systems. *ACM Transactions on Information Systems*, 14(3), 330–347.

Friedman, B. and Nissenbaum, H. (1997). Software agents and user autonomy. *Proceedings of the First International Conference on Autonomous Agents*. Association for Computing Machinery Press, New York, NY, pp. 466–469.

Fuchs, L. (1999). AREA: a cross-application notification service for groupware. In: *Proceedings of ECSCW 1999*. Kluwer Academic Publishers, Dordrecht Germany, pp. 61–80.

Galegher, J., Kraut, R.E., and Egido, C. (Eds.) (1990). *Intellectual Teamwork: Social and Technological Foundations of Cooperative Work*. Lawrence Erlbaum Associates, Hillsdale, NJ.

Gallopin, G.C. (1997). Indicators and their use: information for decision-making. In: Moldan, B., Billharz, S., and Matravers, R. (Eds.), *Sustainability Indicators: A Report on the Project on Indicators of Sustainable Development*. John Wiley & Sons, Chichester, England.

Gewirth, A. (1978). *Reason and Morality*. University of Chicago Press, Chicago, IL.

Greenbaum, J. and Kyng, M. (1991). *Design at Work: Cooperative Design of Computer Systems*. Lawrence Erlbaum Associates, Hillsdale, NJ.

Grudin, J. (1988). Why CSCW applications fail: problems in the design and evaluation of organizational interfaces. In: *Proceedings of the Conference on Computer Supported Cooperative Work (CSCW '88)*. Association for Computing Machinery Press, New York, NY, pp. 85–93.

Habermas, J. (1979). *Communication and the Evolution of Society*. McCarthy, T. (trans). Beacon Press, Boston.

Habermas, J. (1984). *The Theory of Communicative Action*, Vol. 1. McCarthy, T. (trans). Beacon Press, Boston.

Hagman, J., Hendrickson, A., and Whitty, A. (2003). What's in a barcode: informed consent and machine scannable driver licenses. In: *CHI 2003 Extended Abstracts of the Conference on Human Factors in Computing System*. ACM Press, New York, NY pp. 912–913.

Hart, M. (1999). *Guide to Sustainable Community Indicators., 2nd edition*. Hart Environmental Data, North Andover, MA.

Herskovits, M.J. (1952). *Economic Anthropology: A Study of Comparative Economics*. Alfred A. Knopf, New York, NY.

Hill, T.E. Jr. (1991). *Autonomy and Self-Respect*. Cambridge University Press, Cambridge.

Isaacs, E.A., Tang, J.C., and Morris, T. (1996). Piazza: a desktop environment supporting impromptu and planned interactions. In: *Proceedings of the Conference on Computer Supported Cooperative Work (CSCW, 96)*. Association for Computing Machinery Press, New York, NY, pp. 315–324.

Jacko, J.A., Dixon, M.A., Rosa, R.H., Jr., Scott, I.U., and Pappas, C.J. (1999). Visual profiles: a critical component of universal access. In: *Proceedings of the Conference on Human Factors in Computing Systems (CHI'99)*. Association for Computing Machinery Press, New York, NY, pp. 330–337.

Jancke, G., Venolia, G.D., Grudin, J., Cadiz, J.J., and Gupta, A. (2001). Linking public spaces: technical and social issues. In: *Proceedings of CHI 2001,*. pp. 530–537.

Johnson, E.H. (2000). Getting beyond the simple assumptions of organization impact (social informatics). *Bulletin of the American Society for Information, Science,* 26(3), 18–19.

Johnson, D.G. and Miller, K. (1997). Ethical issues for computer scientists and engineers. In: Tucker, A.B. Jr. (Ed.-in-Chief), *The Computer Science and Engineering Handbook*. CRC Press, pp. 16–26.

Kahn, P.H. Jr. (1999). *The Human Relationship with Nature: Development and Culture*. MIT Press Cambridge, MA.

Kahn, P.H. Jr. and Kellert, S.R. (Eds.) (2002). *Children and Nature: Psychological, Sociocultural, and Evolutionary Investigations*. MIT Press, Cambridge, MA.

Kahn, P.H. Jr. and Turiel, E. (1988). Children's conceptions of trust in the context of social expectations. *Merrill-Palmer Quarterly*, 34, 403–419.

Kant, I. (1964). *Groundwork of the Metaphysic of Morals*. (Paton, H.J. trans.). Harper Torchbooks, New York, NY (original work published in 1785).

Kling, R., Rosenbaum, H. and Hert, C. (1998). Social informatics in information science: an introduction. *Journal of the American Society for Information Science*, 49(12), 1047–1052.

Kling, R. and Star, S.L. (1998). Human centered systems in the perspective of organizational and social informatics. *Computers and Society*, 28(1), 22–29.

Kyng, M. and Mathiassen, L. (Eds.) (1997). *Computers and Design in Context*. The MIT Press, Cambridge, MA.

Leveson, N.G. (1991). Software safety in embedded computer systems. *Communications of the ACM*, 34(2), 34–46.

Lipinski, T.A. and Britz, J.J. (2000). Rethinking the ownership of information in the 21st century: ethical implications. *Ethics and Information Technology*, 2(1), 49–71.

MacIntyre, A. (1984). *After Virtue*. University of Nortre Dame Press, Nortre Dame.

Mayer, R.C., Davis, J.H., and Schoorman, F.D. (1995). An integrative model of organizational trust. *The Academy of Management Review*, 20(3), 709–734.

Millett, L., Friedman, B., and Felten, E. (2001). Cookies and web browser design: toward realizing informed consent online. In: *Proceedings of CHI 2001*. ACM Press, pp. 46–52.

Moldan, B., Billharz, S., and Matravers, R. (Eds.) (1997). *Sustainability Indicators: A Report on the Project on Indicators of Sustainable Development*. Wiley & Sons, Chichester, England.

Moore, G.E. (1978). *Principia Ethica*. Cambridge University Press, Cambridge (original work published in 1903).

Nass, C. and Gong, L. (2000). Speech interfaces from an evolutionary perspective. *Communications of the ACM*, 43(9), 36–43.

Neumann, P.G. (1995). *Computer Related Risks*. Association for Computing Machinery Press, New York, NY.

Nielsen, J. (1993). *Usability Engineering*. AP Professional, Boston, MA.

Nissenbaum, H. (1998). Protecting privacy in an information age: the problem with privacy in public. *Law and Philosophy*, 17, 559–596.

Nissenbaum, H. (1999). Can trust be secured online? A theoretical perspective. *Etica e Politca*, 2 (electronic journal).

Nissenbaum, H. (2001). Securing trust online: wisdom or oxymoron. *Boston University Law Review*, 81(3), 635–664.

Norman, D.A. (1988). *The Psychology of Everyday Things*. Basic Books, New York.

Northwest Environment Watch (2002). *This Place on Earth 2002: Measuring What Matters*. Northwest Environment Watch, 1402 Third Avenue, Seattle, WA 98101.

Noth, M., Borning, A., and Waddell, P. (2003). An extensible, modular architecture for simulating urban development, transportation, and environmental impacts. *Computers, Environment and Urban Systems*, 27(2), 181–203.

Olson, J.S. and Olson, G.M. (2000). i2i Trust in e-commerce. *Communications of the ACM*, 43 (12), 41–44.

Olson, J.S. and Teasley, S. (1996). Groupware in the wild: lessons learned from a year of virtual collaboration. In: *Proceedings of the Conference on Computer Supported Cooperative Work (CSCW 96)*. Association for Computing Machinery Press, New York, NY, pp. 419–427.

Orlikowsi, W.J. and Iacono, C.S. (2001). Research commentary: desperately seeking the "IT" in IT research: a call to theorizing the IT artifact. *Information Systems Research*, 12 (2), 121–134.

Palen, L. and Dourish, P. (2003). Privacy and trust: unpacking "privacy" for a networked world. In: *Proceedings of CHI 2003*, pp. 129–136.

Palen, L. and Grudin, J. (2003). Discretionary adoption of group support software: lessons from calendar applications. In: Munkvold, B.E. (Ed.), *Implementing Collaboration Technologies in Industry*. Springer Verlag, Heidelberg.

Palmer, K. (Ed.) (1998). *Indicators of Sustainable Community*. Sustainable Seattle, Seattle, WA.

Phillips, D.J. (1998). Cryptography, secrets, and structuring of trust. In: Agre, P.E. and Rotenberg, M. (Eds.), *Technology and Privacy: The New Landscape*. The MIT Press, Cambridge, MA, pp. 243–276.

Pruitt, J. and Grudin, J. (2003). Personas: practice and theory. In: *Proceedings of DUX 2003*, ACM Press.

Rawls, J. (1971). *A Theory of Justice*. Harvard University Press, Cambridge, MA.

Reeves, B. and Nass, C. (1996). *The Media Equation: How People Treat Computers, Television, and New Media Like Real People and Places*. Cambridge University Press and CSLI Publications, New York, NY and Stanford, CA.

Riegelsberger, J. and Sasse, M.A. (2002). Face it: photos don't make a web site trustworthy. In: *Extended Abstracts of CHI 2002*. ACM Press, pp. 742–743.

Rocco, E. (1998). Trust breaks down in electronic contexts but can be repaired by some initial face-to-face contact. In: *Proceedings of CHI 1998*. ACM Press, pp. 496–502.

Rosenberg, S. (1997). Multiplicity of selves, In: Ashmore, R.D. and Jussim, L. (Eds.), *Self and Identity: Fundamental Issues*. Oxford University Press, New York, NY, pp. 23–45.

Sawyer, S. and Rosenbaum, H. (2000). Social informatics in the information sciences: current activities and emerging direction. *Informing Science*, 3(2), 89–95.

Scheffler, S. (1982). *The Rejection of Consequentialism*. Oxford University Press, Oxford, England.

Schiano, D.J. and White, S. (1998). The first noble truth of cyberspace: people are people (even when they MOO). In: *Proceedings of the Conference of Human Factors in Computing Systems (CHI 98)*. Association for Computing Machinery, New York, NY, pp. 352–359.

Schneider, F.B. (Ed.) (1999). *Trust in Cyberspace*. National Academy Press, Washington, DC.

Schoeman, F.D. (Ed.) (1984). *Philosophical Dimensions of Privacy: An Anthology*. Cambridge University Press, Cambridge, England.

Shneiderman, B. (1999). Universal usability: pushing human–computer interaction research to empower every citizen. ISR Technical Report. 99–72. University of Maryland, Institute for Systems Research, College Park, MD.

Shneiderman, B. (2000). Universal usability. *Communication of the ACM*, 43(5), 84–91.

Simpson, J.A. and Weiner, E.S.C. (Eds.) (1989). "value, *n*." *Oxford English Dictionary*. Clarendon Press, Oxford; *OED Online*, Oxford University Press, 30 May 2003. Available at http://dictionary.oed.com/cgi/entry/00274678.

Smart, J.J.C. and Williams, B. (1973). *Utilitarianism For and Against*. Cambridge University Press, Cambridge.

Stephanidis, C. (Ed.) (2001). *User Interfaces for All: Concepts, Methods, and Tools*. Lawrence Erlbaum Associates, Mahwah, NJ.

Suchman, L. (1994). Do categories have politics? The language/action perspective reconsidered. *CSCW Journal*, 2(3), 177–190.

Svensson, M., Hook, K., Laaksolahti, J. and Waern, A. (2001). Social navigation of food recipes. In: *Proceedings of the Conference of Human Factors in Computing Systems (CHI, 2001)*. Association for Computing Machinery, New York, NY, pp. 341–348.

Tang, J.C. (1997). Eliminating a hardware switch: weighing economics and values in a design decision. In: Friedman, B. (Ed.), *Human Values and the Design of Computer Technology*. Cambridge University Press, New York, NY, pp. 259–269.

Thomas, J.C. (1997). Steps toward universal access within a communications company. In: Friedman, B. (Ed.), *Human Values and the Design of Computer Technology*. Cambridge University Press, New York, NY, pp. 271–287.

Turiel, E. (1983). *The Development of Social Knowledge*. Cambridge University Press, Cambridge, England.

Turiel, E. (1998). Moral development. In: Eisenberg, N. (Ed.), *Social, Emotional, and Personality Development*. In: Damon, W. (Ed.), *Handbook of Child Psychology*, 5th edition, Vol. 3. Wiley & Sons, New York, NY, pp. 863–932.

Turiel, E. (2002). *The Culture of Morality: Social Development, Context, and Conflict*. Cambridge University Press, Cambridge, England.

Turkle, S. (1996). *Life on the Screen: Identify in the Age of the Internet*. Simon and Schuster, New York, NY.

Ulrich, R.S. (1984). View through a window may influence recovery from surgery. *Science*, 224, 420–421.

Ulrich, R.S. (1993). Biophilia, biophobia, and natural landscapes. In: Kellert, S.R. and Wilson, E.O.(Eds.), *The Biophilia Hypothesis*. Island Press, Washington, DC, pp. 73–137.

United Nations (2002). Report of the United Nations Conference on Environment and Development, held in Rio de Janeiro, Brazil, 1992. Available at http://www.un.org/esa/sustdev/documents/agenda21/english/agenda21toc.htm.

Waddell, P. (2002). UrbanSim: modeling urban development for land use, transportation, and environmental planning. *Journal of the American Planning Association*, 68(3), 297–314.

Waddell, P., Borning, A., Noth, M., Freier, N., Becke, M., and Ulfarsson, G. (2003). Microsimulation of urban development and location choices: design and implementation of UrbanSim. *Networks and Spatial Economics*, 3(1), 43–67.

Weiser, M. and Brown, J.S. (1997). The coming age of calm technology. In: Denning, P. and Metcalfe, B. (Eds.), *Beyond Calculation: The Next 50 Years of Computing*. Springer-Verlag, New York, NY, pp. 75–85.

Weizenbaum, J. (1972). On the impact of the computer on society: How does one insult a machine? *Science*, 178, 609–614.

Wiener, N. (1985). The machine as threat and promise. In: Masani, P. (Ed.), *Norbert Wiener: Collected Works and Commentaries, Vol. IV.* The MIT Press, Cambridge, MA, pp. 673–678. reprinted from *St. Louis Post Dispatch*, December 13, 1953.

Winograd, T. (1994). Categories, disciplines, and social coordination. *CSCW Journal*, 2(3), 191–197.

World Commission on Environment and Development (Gro Harlem Brundtland, Chair) (1987). *Our Common Future*. Oxford University Press, Oxford.

Wynne, E.A. and Ryan, K. (1993). *Reclaiming Our Schools: A Handbook on Teaching Character, Academics, and Discipline*. Macmillan & Co., New York.

Zheng, J., Bos, N., Olson, J., and Olson, G.M. (2001). Trust without touch: jump-start trust with social chat. In: *Extended Abstracts of CHI 2001*. ACM Press, pp. 293–294.

PART II

THEORETICAL ISSUES AFFECTING PROPERTY, PRIVACY, ANONYMITY, AND SECURITY

CHAPTER 5

Personality-Based, Rule-Utilitarian, and Lockean Justifications of Intellectual Property

ADAM D. MOORE

5.1 INTRODUCTION: WHAT IS INTELLECTUAL PROPERTY?

Arguments for intellectual property rights have generally taken one of three forms. Personality theorists maintain that intellectual property is an extension of individual personality. Rule-utilitarians ground intellectual property rights in social progress and incentives to innovate. Lockeans argue that rights are justified in relation to labor and merit. While each of these strands of justification has weaknesses, there are also strengths.

In this article, I will present and examine personality-based, rule-utilitarian, and Lockean justifications for intellectual property. Care is needed so that we do not confuse moral claims with legal ones. The brief sketch of Anglo-American and Continental systems of intellectual property below, focuses on legal conceptions and rights, while the arguments that follow—personality based, utilitarian, and Lockean—are essentially moral. I will argue that there are justified moral claims to intellectual works—claims that are strong enough to warrant legal protection.

5.1.1 What Is Intellectual Property?[1]

Intellectual property is generally characterized as nonphysical property that is the product of cognitive processes and whose value is based upon some idea or collection

[1] Intellectual property falls under the umbrella of intangible property. Intangible property is a broader notion including lists of customers, purchasing summaries, medical records, criminal records, and the like. A longer version of this section appears in Moore (2004, 2001, pp. 9–35).

The Handbook of Information and Computer Ethics, Edited by Kenneth Einar Himma and Herman T. Tavani

of ideas.[2] Typically, rights do not surround the abstract nonphysical entity, or *res*, of intellectual property, rather, intellectual property rights surround the control of physical manifestations or expressions. Systems of intellectual property protect rights to ideas by protecting rights to produce and control physical embodiments of those ideas.

Within the Anglo-American tradition intellectual property is protected by the legal regimes of copyright, patent, and trade secret.[3] Copyright protection extends to original works of authorship fixed in any tangible medium of expression.[4] Works that may be copyrighted include literary, musical, artistic, photographic, and cinematographic works, maps, architectural works, and computer software. There are five exclusive rights that copyright owners enjoy and three major restrictions on the bundle.[5] The five rights are the right to reproduce the work, the right to adapt it or derive other works from it, the right to distribute copies of the work, the right to display the work publicly, and the right to perform it publicly. Each of these rights may be parsed out and sold separately. All five rights lapse after the lifetime of the author plus 70 years—or in the case of works for hire, the term is set at 95 years from publication or 120 years from creation, whichever comes first.[6]

The domain or subject matter of patent protection is the invention and discovery of new and useful processes, machines, articles of manufacture, or compositions of matter.[7] Patents yield the strongest form of protection, in that a 20 years exclusive monopoly is granted over any expression or implementation of the protected work.[8] The bundle of rights conferred on patents owners are the right to make, the right to use, the right to sell, and the right to authorize others to sell the patented item.[9] Moreover, the bundle of rights conferred by a patent exclude others from making, using, or selling the invention regardless of independent creation.

[2]For a similar view see Hughes, J. (1997, p. 107).

[3]Trademark and *the law* of *ideas will* not be discussed.

[4]See 17 U.S.C. Section 102 (1988).

[5]The three major restrictions on the bundle of rights that surround copyright are limited duration (17 U.S.C. Section 302), fair use (17 U.S.C. Section 107 and District Judge Leval's opinion in *New Era Publications International v. Henry Holt and Co.*, 695 F.Supp 1493 (S.D.N.Y., 1988)), and the first sale rule (17 U.S.C. Section 109(a)). The first sale rule prevents a copyright holder who has sold copies of the protected work from later interfering with the subsequent sale of those copies.

[6]The U.S. Constitution requires the limited term of copyright and patent. The Constitution empowers Congress to "promote the Progress of Science and useful Arts, by securing *for limited Times* to Authors and Inventors the exclusive Right to their respective Writings and Discoveries" U.S. Const. art. I, Section 8, cl. 8 (emphases added).

[7]See 35 U.S.C. Section 154 (1984 and Supp., 1989). Patents may be granted when the subject matter satisfies the criteria of utility, novelty, and nonobviousness. See 35 U.S.C. Sections 101–107. Unlike copyright, patent law protects the totality of the idea, expression, and implementation. See 35 U.S.C. Sections 101–107.

[8]Patent Act, 35 U.S.C. Section 101 (1988). The 1995 version of the Patent Act has added 3 years to the term of patent protection—from 17 to 20. See 35 U.S.C. Section 154(a)(2).

[9]See 35 U.S.C. Section 154 (1984 and Supp., 1989).

A trade secret may consist of any formula, pattern, device, or compilation of information that is used in one's business.[10] The two major restrictions on the domain of trade secrets are the requirements of secrecy and competitive advantage. Although trade secret rights have no built in sunset, they are extremely limited in one important respect. Owners of trade secrets have exclusive rights to make use of the secret but only as long as the secret is maintained. If the secret is made public by the owner then trade secret protection lapses and anyone can make use of it. Moreover, owner's rights do not exclude independent invention or discovery. Trade secrecy laws rely entirely on private measures, rather than state action, to maintain exclusivity. Furthermore, the subject matter of trade secret is almost unlimited in terms of the content of the information that is potentially subject to protection. Within the secrecy requirement, owners of trade secrets enjoy management rights and are protected from misappropriation.

Continental Systems of Intellectual Property: Article 6 *bis* of the Berne Convention articulates the notion of "moral rights" that are included in continental European intellectual property law.

> Independently of the author's economic rights, and even after the transfer of the said rights, the author shall have the right to claim authorship of the work and to object to any distortion, mutilation or other modification of, or other derogatory action in relation to, the said work, which would be prejudicial to his honor or reputation.[11]

This doctrine protects the personal rights of creators, as distinguished from their economic rights, and is generally known in France as "droits morals" or "moral rights." These moral rights consist of the right to create and to publish in any form desired, the creator's right to claim the authorship of his work, the right to prevent any deformation, mutilation, or other modification thereof, the right to withdraw and destroy the work, the prohibition against excessive criticism, and the prohibition against all other injuries to the creator's personality.[12] Much of this doctrine has been incorporated in the Berne Convention. M.A. Roeder writes,

> When the artist creates, be he an author, a painter, a sculptor, an architect or a musician, he does more than bring into the world a unique object having only exploitive possibilities; he projects into the world part of his personality and subjects it to the ravages of public use. There are possibilities of injury to the creator other than merely economic ones; these the copyright statute does not protect.[13]

---

[10]See The Restatement (Third) of Unfair Competition Sections 39–45 (1995) (containing the most current information about the law of trade secrets).

[11]Berne Convention, Article 6 *bis*.

[12]Generally these moral rights are not recognized within the Anglo-American tradition. See *Crimi v. Rutgers Presbyterian Church*, 194 Misc. 570 (N.Y.S., 1949). Recently, given the inclusion of the United States in the Berne Convention treaty, there has been a move toward indirect recognition. See *Gilliam v. American Broadcasting Companies, Inc.*, 538 F. 2d 14 (2d Cir., 1976), *Wojnarowicz v. American Family Association*, 745 F. Supp. 130 (S.D.N.Y. 1990), and the Berne Convention Implementation Act of 1988.

[13]Roeder, M.A. (1940). The doctrine of moral right: a study in the law of artists, authors, and creators. *Harvard Law Review*, 53, 554.

The suggestion is that individuals can have intellectual property rights involving their personality, name, and public standing.

5.2 PERSONALITY-BASED JUSTIFICATIONS OF INTELLECTUAL PROPERTY

Personality-based defenders maintain that intellectual property is an extension of individual personality. For Hegel the external actualization of the human will requires property. Hegel writes, "The person must give himself an external sphere of freedom in order to have being as Idea."[14] Personality theorists, like Hegel, maintain that individuals have moral claims over their own talents, feelings, character traits, and experiences. Control over physical and intellectual objects is essential for self-actualization—by expanding our self outward beyond our own minds and mixing with tangible and intangible items—we both define ourselves and obtain control over our goals and projects. Property rights are important in two ways according to this view. First, by controlling and manipulating objects, both tangible and intangible, our will takes form in the world and we obtain a measure of freedom. Second, in some cases, our personality becomes infused with an object—moral claims to control feelings, character traits, and experiences may be expanded to intangible works. Josef Kohler echoes this view,

> Personality must be permitted to be active, that is to say, to bring its will to bear and reveal its significance to the world; for culture can thrive only if persons are able to express themselves, and are in a position to place all their inherent capacities at the command of their will.[15]
>
> The writer can demand not only that no strange work be presented as his, but that his own work not be presented in a changed form. The author can make this demand even when he has given up his copyright. This demand is not so much an exercise of dominion over my own work, as it is of dominion over my being, over my personality which thus gives me the right to demand that no one shall share in my personality and have me say things which I have not said.[16]

---

[14]Hegel, G.W.F. (1991). Wood, A. (Ed.). *Elements of the Philosophy of Right,* p. 73. See also Von Humboldt, W. (1969). Coulthard, J. and Burrow J.W. (Eds.) *The Limits of State Action*, Cambridge: University Press, Cambridge; Kant, I. (1983). Von der Unrechtmässigkeit des Büchernachdrucks. In: Macfie, R.A. (Ed.), Copyrights and Patents for Inventions, p. 580 and Kohler, J. (1969). *Philosophy of Law,* In: Albrecht, A. (Ed.). A.M. Kelley, New York.

[15]Kohler, J. (1921). *Philosophy of law,* p. 80.

[16]Kohler, J. (1907). *Urheberrecht an Schriftwerken und Verlagsrecht 15* (quoted in Damich E. (1986). The right of personality, p. 29).

5.2.1 Problems for Personality-Based Justifications of Intellectual Property

There are at least four problems with this view.[17] First, it is not clear that we own our feelings, character traits, and experiences. Although it is true that we have possessed these things or that they are a part of each of us, an argument is needed to establish the relevant moral claims. Second, even if it could be established that individuals own or have moral claims to their personality it does not automatically follow that such claims are expanded when personalities become infused in tangible or intangible works. Rather than establishing property claims perhaps we should view this as an abandonment of personality—similar to the sloughing off of hair and skin cells. Third, assuming that moral claims to personality could be expanded to tangible or intangible items we would still need an argument justifying property rights. Personality-based moral claims may warrant nothing more than use rights or prohibitions against alteration. Finally, there are many intellectual innovations in which there is no evidence of the creator's personality. A list of costumers or a new safety pin may contain no trace of personality. "There may be personality galore in a map of Tolkien's Middle Earth, but not much in a roadmap of Ohio."[18] Thus, personality-based theories may not provide a strong moral foundation for legal systems of intellectual property.

5.2.2 The Personality Theorist's Rejoinder

While acknowledging the force of these worries there does seem to be something intuitively appealing about personality-based theories. Suppose, for example, that Smith buys a painting at a garage sale—a long lost Jones original. Smith takes the painting home and alters the painting with a marker—drawing horns and mustaches on the figures in the painting. The additions are so clever and fit so nicely into the painting that Smith hangs it in a window on a busy street. There are at least two ethical worries to consider in this case. First, the alterations by Smith may cause unjustified economic damage to Jones. Second, and independent of the economic considerations, Smith's actions may damage Jones' reputation. The integrity of the painting has been violated without the consent of the author perhaps causing long-term damage to his reputation and community standing. If these claims are sensible, then it appears that we are acknowledging personality-based moral "strings" attaching to certain intellectual works.

Moreover, personality-based theories of intellectual property often appeal to other moral considerations. Hegel wrote, "The purely negative, but most basic, means of furthering the sciences and arts is to protect those who work in them against *theft* and

[17]Further analysis of the problems for personality-based theories can be found in Hughes (1997, pp. 149–164), Palmer, T. (2005, pp. 143–147) and Schroeder, J.L. (2006). Unnatural rights: Hegel and intellectual property. *University of Miami Law Review*, 60, 453.

[18]Hughes (1997, p. 151).

provide them with security for their property. . ."[19] Perhaps the best way to protect personality-based claims to intangible works is to adopt a more comprehensive system designed to promote progress and social utility. Given this, let us consider incentives based arguments for intellectual property.

5.3 THE RULE-UTILITARIAN INCENTIVES BASED ARGUMENT FOR INTELLECTUAL PROPERTY[20]

In terms of "justification," modern Anglo-American systems of intellectual property are typically modeled as rule-utilitarian.[21] It is argued that adopting the systems of copyright, patent, and trade secret, leads to an optimal amount of intellectual works being produced and a corresponding optimal amount of social utility. These systems or institutions are not comprised by mere rules of thumb. In particular cases, conferring rights to authors and inventors over their intellectual products may lead to bad consequences. Justification, in terms of social progress, occurs at the level of the system or institution. Granting a copyright to Smith and Jones, for example, may not maximize overall social utility, but the system as a whole may yield a better outcome when compared to other systems.

Given that intellectual works can be held by everyone at the same time, cannot be used up or easily destroyed, and are necessary for many lifelong goals and projects it would seem that we have a prima facie case against regimes of intellectual property that would restrict such maximal use. Tangible property, including concrete expressions of intellectual works, is subject to exclusive physical domination in a way that intellectual or intangible property is not. For example, Smith's use of a car excludes my concurrent use, whereas his use of a theory, process of manufacture, or recipe for

[19]Hegel (1991). *Elements of the Philosophy of Right*, pp. 99–100, cited in Balganesh, S. (2004). Copyright and free expression: analyzing the convergence of conflicting normative frameworks. *Chicago-Kent Journal of Intellectual Property*, 4, note 54.

[20]A longer version of this section appears in Moore (2004, 2001, pp. 37–70).

[21]See generally, Oppenheim, C. (1951). Evaluation of the American patent system. *Journal of the Patent and Trademark Office Society* 33; National Patent Planning Commission: *First Report* 783–784 (1943); Report of the President's Commission (1966); Palmer, T. (1997). Intellectual property: a non-Posnerian Law and economics approach. In: Adam, D.M. (Ed.), *Intellectual Property: Moral, Legal, and International Dilemmas*. Rowman and Littlefield, p. 179; Moore, A.D. (Ed.) (2005). Are patents and copyrights morally justified? The philosophy of property rights and ideal objects. *Information Ethics: Privacy, Property, and Power*. University of Washington Press, p. 123; Leonard, G.B. (1989). The University, scientific research, and the ownership of knowledge. In: Weil, V. and Snapper, J. (Eds.), p. 257, Edwin, C.H. (1997). Justifying intellectual property. In: Adam D. Moore (Ed.), *Intellectual Property: Moral, Legal, and International Dilemmas*. Rowman and Littlefield, p. 30; Mackaay, (1990). Economic incentives in markets for information and innovation. *The Harvard Journal of Law and Public Policy*, 12, 867; Miners, R. and Staaf, R. (1990). Patents, copyrights, and trademarks: property or monopoly? *The Harvard Journal of Law and Public Policy* 12, 911; Croskery, P. (1993) Institutional Utilitarianism and intellectual property. *The Chicago-Kent Law Review* 68, 631; and Machlup, F. (1962). *Production and Distribution of Knowledge in The United States*.

success, does not. Thus intellectual works can be seen as nonrivalrous commodities. If this is true, we have an immediate prima facie case against rule-utilitarian justifications of intellectual property rights.[22]

The rejoinder, typically given, is that granting use, possession, and control rights, to both ideas and expressions of ideas is necessary as incentive for the production of intellectual works. Ideas themselves may be independently valuable but when use, possession, and control, are restricted in a free market environment the value of certain ideas increases dramatically. Moreover, with increased value comes increased incentives, or so it is argued.

On this view, a necessary condition for promoting the creation of valuable intellectual works is granting limited rights to authors and inventors. Absent certain guarantees, authors, and inventors would not engage in producing intellectual property. Although success is not ensured by granting rights, failure certainly is, if others who incur no investment costs can seize and produce the intellectual effort of others.

Many utilitarians argue that private ownership of *physical* goods is justified because of the tragedy of the commons or problems with efficiency. Systems of private property are more efficient, or so it is argued, than systems of common ownership. It should be clear that this way of arguing is based on providing incentives. Owners of physical goods are given an incentive to maintain or increase the value of those goods, because the costs of waste, and the like, are internalized.

The incentives based rule-utilitarian argument for systems of intellectual property protection is very similar. In this case, the government grants rights as an incentive for the production of intellectual works, and production of this sort, in turn, maximizes social progress. It is important to note, that on this view, rights are granted to authors and inventors, not because they deserve such rights or have mixed their labor in an appropriate way, but because this is the only way to ensure that an optimal amount of intellectual products will be available for society. A more formal way to characterize this argument is:

Premise 1. Society ought to adopt a system or institution if and only if it leads to or, given our best estimates, is expected to lead to the maximization of overall social utility.

Premise 2. A system or institution that confers limited rights to authors and inventors over what they produce is a necessary incentive for the production of intellectual works.

Premise 3. Promoting the creation and dissemination of intellectual works produces an optimal amount of social progress.

Therefore, Conclusion 4. A system of intellectual property should be adopted.

The first premise, or the theoretical premise, is supported by utilitarian arguments that link theories of the good and theories of the right in a particular way. The rule utilitarian determines a correct moral rule in reference to the consequences of

[22]See Hettinger (1997, p. 30).

everyone adopting it.[23] By adhering to a rule-based component it is argued that the problems that face act utilitarianism, problems of justice,[24] special obligations,[25] integrity,[26] and excessive demands,[27] are circumvented. Moreover, by grounding the theory solely in a consequent component, unlike deontic theories, rule utilitarians argue that the theory is given firm footing. In combining the most promising aspect of act utilitarianism (consequences are all that matter) with the most promising aspect of deontology (its rule following component), rule utilitarians hope to arrive at a defensible moral theory.

The second premise is an empirical claim supported by the aforementioned considerations concerning incentives. The view is that it is empirical fact that authors and inventors will not engage in the appropriate activity unless certain guarantees are in place. What keeps authors and inventors burning the midnight oil, and thereby producing an optimal amount of intellectual works, is the promise of massive profits. The arguments supporting the third premise claim that cultural, technological, and industrial progress is necessary for an optimal amount of social utility. It follows that a system of intellectual property protection should be adopted.

---

[23]This premise could be defended by the act utilitarian in the following way. Consider the adoption of an institution of intellectual property protection as an *act* of congress or government. Members of congress, in voting to adopt some set of rules, are acting so that social utility is maximized—they are adopting a set of rules and attaching sanctions for violating these rules. The sanctions change the consequences of many actions and thus may change what is the correct action for others.
This way of defending the first premise of the argument is not without problems. Although such a view would provide a way to side-step an external critique of rule utilitarianism (see Moore (2004, 2001, pp. 37–70, Chapter 3: "Against Rule-Utilitarian Intellectual Property")), it would not answer any of the internal problems discussed. Moreover, it is not as if, by moving from rule utilitarianism to act utilitarianism, the defender of this view obtains firmer footing—alas there are many damaging criticisms of act-utilitarianism as well. For a lucid account of the many problems with act utilitarianism and rule utilitarianism see Williams, B. (1973). A critique of utilitarianism. *Utilitarianism: For & Against*, pp. 75–150; Rawls, John. (1971). *A Theory Of Justice*, pp. 22–34; McCloskey, H. J. (1984). Respect for human moral rights versus maximizing good. In: Frey, R.G. (Ed.), *Utility and Rights*, pp. 121–136; David, L. (1965). *Forms and Limits of Utilitarianism (1965)*; Nozick, R. (1974). *Anarchy, State, and Utopia*; Smart, J.J.C. (1967). Extreme and restricted utilitarianism. In: Philippa Foot (Ed.), *Theories of Ethics*; and Scheffler, S. (1984). *The Rejection of Consequentialism.*

[24]Generally speaking, the problem of justice for act utilitarianism is found in cases where doing something unjust maximizes overall utility. For example, what if framing an innocent person would lead to the best consequences for everyone affected? Act utilitarianism would seem to require such an unjust act—that is, we would have a moral obligation to frame the innocent person and this seems wrong.

[25]The problem of special obligations is that sometimes we have obligations that stand independent of the consequences. For example, it may be best for all concerned that a teacher give everyone A's, but the teacher has a special obligation to award grades based on merit.

[26]In general terms, the problem of integrity is that act utilitarianism requires individuals to treat their own life-long goals and projects impartially. As a good utility maximizer we each should be willing to abandon our goals and projects for the sake of maximizing overall social utility. The problem is that we cannot be impartial in this way.

[27]The problem of excessive demands is that act utilitarianism demands too much of us. Since everything we do and allow has consequences, every action or inaction is moral or immoral. But this seems wrong. Whether I wake up at 10:00 or 10:05 seems to be outside the realm of morality, assuming of course that I have no prior obligations.

5.3.1 Problems for the Rule-Utilitarian Incentives Based Argument

Putting aside general attacks leveled at rule utilitarianism, which will not be considered in this article, a serious challenge may be raised by questioning the truth of the second premise (hereafter P2). It will be argued that P2 is false or at least highly contentious, and so even granting the truth of the first and third premises, the conclusion does not follow. Given that the truth of P2 rests on providing incentives, what is needed are cases that illustrate better ways, or equally good ways, of stimulating production without granting private property rights to authors and inventors. It would be better to establish equally powerful incentives for the production of intellectual property that did not also require initial restricted use guaranteed by rights.

5.3.2 Alternatives to Patents

One alternative to granting patent rights to inventors as incentive is government support of intellectual labour. This would result in government funded research projects, with the results immediately becoming public property. It is obvious that this sort of funding can and does stimulate the production of intellectual property without allowing initial restricted control to authors and inventors. The question becomes: can government support of intellectual labor provide enough incentive to authors and inventors so that an equal or greater amount of intellectual products are created compared to what is produced by conferring limited property rights? Better results may also be had if fewer intellectual works of higher quality were distributed to more people. If so, then P2 is false and intellectual property rights should not be granted on grounds of utility.

In response to this kind of charge, defenders of the argument based on incentives have claimed that government support of intellectual labor does not and will not create the requisite incentives. It is only by holding out the promise of huge profits that society obtains maximal progress for all. Governments may be able to provide some incentives by paying authors and inventors in advance, but this kind of activity will never approach the incentive created by adopting a system that affords limited monopoly rights to intellectual property.[28]

Another reply typically given, is the standard argument against centralized planning. Governments are notoriously bad in the areas of predicting the demand of future markets, research and development, resource allocation, and the like. Maximizing social utility in terms of optimizing the production of intellectual works is best left in the hands of individuals, businesses, and corporations.[29]

---

[28]For an argument pointing the other direction see Calandrillo, S.P. (1998). An economic analysis of intellectual property rights: justifications and problems of exclusive rights, incentives to generate information, and the alternative of a government-run reward system. *Fordham Intellectual Property, Media, & Entertainment Law Journal,* 9 (Fall, 1998): 301.

[29]For example see Nozick, R. (1974). *Anarchy, State, and Utopia,* p. 45 and Hayek, F. (1940). Socialist calculation: the competitive solution. *Economica,* 7, 125–149.

Rather than a government supported system of intellectual property Steven Shavell and Tanguy Van Ypersele argue for a reward model.[30] Reward models may be able to avoid the problems of allowing monopoly control and restricted access and at the same time provide incentives to innovate. Shavell and Van Ypersele write, "Under a reward system innovators are paid for innovation directly by the government (possibly on the basis of sales), and innovations pass immediately into the public domain."[31] Innovators would still burn the midnight oil chasing that pot of gold and governments would not have to decide which projects to fund or determine the amount of the reward before its "social value" was known. Taxes or collecting percentages of the profits of these innovations may provide the funds necessary to pay the reward.

Two other benefits are also obvious. One criticism of the patent system is that monopoly power allows monopoly prices. Under a reward system, consumers would avoid these prices and likely purchase other goods and services. A second criticism is that patents hinder subsequent innovations and improvements of intellectual works—big firms may be able to control or manipulate an entire industry. "A famous example of this occurred when James Watt, holder of an early steam engine patent, denied licenses to improve it to Jonathan Hornblower and Richard Trevithick, who had to wait for Watt's patent to expire in 1800 before they could develop their high pressure engine."[32] As with monopoly pricing, a reward system avoids this social cost because the intellectual works pass immediately into the public domain.

Certainly the promise of huge profits is part of what drives authors and inventors to burn the midnight oil, but the promise need not be guaranteed by ownership. Fritz Machlup has argued that patent protection is not needed as an incentive for corporations, in a competitive market, to invest in the development of new products and processes. "The short-term advantage a company gets from developing a new product and being the first to put it on the market may be incentive enough."[33] Consider, for example, the initial profits generated by the sales of certain software packages. The market share guaranteed by initial sales, support services, and the like, may provide adequate incentives. Moreover, given the development of advanced copy-protection schemes software companies can protect their investments and potential profits for a number of years.

5.3.3 Alternatives to Copyrights

A reward model may also be more cost effective than copyright protection, especially given the greater access that reward models offer. Alternatively, offering a set of more

[30]Shavell, S. and Van Ypersele, T. (2001). Rewards versus intellectual property rights. *Journal of Law and Economics,* 44, 525–547. See also: Polanvyi, M. (1943). Patent reform. *Review of Economic Studies,* Vol. 11, p. 61.; Wright, B. (1998). The economics of invention incentives: patents, prizes, and research contracts. *American Economic Review,* 73, 1137. Michael Kremer offers an auction model where the government would pay inventors the price that obtains from the public sale of the innovation. See Kremer, M. (1998). Patent buyouts: a mechanism for encouraging innovation. *Journal of Economics Quarterly,* 113, 1137.

[31]Shavell, S. and Van Ypersele, T. (2001). Rewards versus intellectual property rights. p. 525.

[32]Shavell, S. and Van Ypersele, T. (2001). Rewards versus intellectual property rights. p. 543.

[33]Machlup, F. (1962, pp. 168–169).

limited rights may provide the requisite incentives while allowing greater access. Many authors, poets, musicians, and other artists, would continue to create works of intellectual worth without proprietary rights being granted—many musicians, craftsman, poets, and the like, simply enjoy the creative process and need no other incentive to produce intellectual works.

Conversely, though, it may be argued that the production of many movies, plays, and television shows, is intimately tied to the limited rights conferred on those who produce these expressions. But this kind of reply is subject to the same problem that befell patent protection. The short-term advantage a production company gets from creating a new product and being the first to market, coupled with copy-protection schemes, may be incentive enough. And even if the production of movies is more dependent on copyright protection than academic writing or poetry readings, all that can be concluded is that incentives may be needed for the optimal production of the former but not the latter. If correct, a system that afforded different levels of control depending on the subject matter of the intellectual work would likely be better than our current model.

The justification typically given for the "fair use" rule is that the disvalue of limiting the rights of authors is overbalanced by the value of greater access.[34] Perhaps more limitations could be justified in this way—maybe all that is needed is a prohibition against piracy or a prohibition against the direct copying and marketing of intellectual works. Needless to say, even if the incentives argument is correct, the resulting system or institution would be quite different than modern Anglo-American systems of intellectual property.[35]

Another worry that infects copyright, but not reward models, is the conversion of intellectual works into a digital form. A basic rule of rule-utilitarian copyright (and patent law) is that while ideas themselves cannot be owned, the physical or tangible expressions of them can.[36] Ideas, as well as natural laws and the like, are considered to be the collective property of humanity.[37] It is commonly assumed that allowing authors and inventors rights to control mere ideas would diminish overall social utility and so an idea/expression distinction has been adopted.

Nevertheless digital technology and virtual environments are detaching intellectual works from physical expression. This tension between protecting physical expressions and the status of on-line intellectual works leads to a deeper problem. Current Anglo-American institutions of intellectual property are constructed to protect the efforts of authors and inventors and, at the same time, to disseminate information as widely as

[34]17 U.S.C. Section 107. See also, *Folsom v. Marsh*, 9 F. Cas. 342 (C.C.D. Mass. 1841) (No. 4,901).

[35]For arguments calling for the elimination of copyright and patent protection see Palmer, T.G. (2005, p. 123).

[36]17 U.S.C. Section 102(b) (1988) states, "in no case does copyright protection for an original work of authorship extend to any idea, procedure, process, system, method of operation, concept, principle, or discovery, regardless of the form in which it is described, explained, illustrated, or embodied in such work".

[37]See 17 U.S.C. Section 102(b); *International New Service v. Associated Press* 248 U.S. 215, 39 S.Ct. 68, 63 L.Ed. 211 (1918); *Miller v. Universal City Studios, Inc.*, 224 USPQ 427 (1984, CD Cal); and *Midas Productions, Inc., v. Baer* 199 USPQ 454 (1977, DC Cal).

possible. But when intellectual works are placed on-line there is no simple method of securing both protection and widespread access.

The current reaction to these worries has been to strengthen intellectual property protection in digital environments, yet it is unclear that such protection will yield greater social utility. Reward systems or further limiting copyrights would likely avoid the disutility of restricting access in digital environments.

Raymond Shih Ray Ku has argued that copyright is unnecessary in digital environments. "With respect to the creation of music. . . exclusive rights to reproduce and distribute copies provide little if any incentive for creation, and that digital technology make it possible to compensate artists without control."[38] In brief, Shih Ray Ku argues that copyright protects the interests of the publisher—large, up-front distribution costs need to paid for and copyright does the job. Digital environments, however, eliminate the need for publishers with distribution resources. Artists, who receive little royalty compensation anyway, may distribute their work worldwide with little cost. Incentives to innovate are maintained, as they have been, by touring, exhibitions, and the like. Thus, if Shih Ray Ku is correct the incentives based argument would lead us away, not toward, copyright protection for digital intellectual works.

5.3.4 Trade Secret and Social Utility

Trade secret protection appears to be the most troubling from an incentives-based perspective. Given that no disclosure is necessary for trade secret protection, there are no beneficial trade-offs between promoting behavior through incentives and long-term social benefit. From a rule-utilitarian point of view the most promising aspect of granting intellectual property rights is the widespread dissemination of information and the resulting increase in social progress. Trade secret protection allows authors and inventors the right to slow the dissemination of protected information indefinitely—a trade secret requires secrecy.[39] Unlike other regimes of intellectual property, trade secret rights are perpetual. This means that so long as the property holder adheres to certain restrictions, the idea, invention, product, or process of manufacture may never become common property.

5.3.5 What are Long-Term Benefits?

Empirical questions about the costs and benefits of copyright, patent, and trade secret protection are notoriously difficult to determine. Economists who have considered the question indicate that either the jury is out, so-to-speak, or that other arrangements would be better. George Priest claims that "The ratio of empirical demonstration to assumption in this literature must be very close to zero . . . (recently it) has demonstrated quite persuasively that, in the current state of knowledge, economists know almost nothing about the effect on social welfare of the patent system or of other

[38] Ku, R.S.R. (2002). The creative destruction of copyright: Napster and the new economics of digital technology. *University of Chicago Law Review*, 69, 263.

[39] See the Restatement (Third) of Unfair Competition Section 39–45 (1995).

systems of intellectual property."[40] This echoes Machlup's sentiments voiced 24 years earlier and Clarisa Long's view "Whether allowing patents on basic research tools results in a net advance or deterrence of innovation is a complex empirical question that remains unanswered."[41] If we cannot appeal to the progress enhancing features of intellectual property protection, then *the rule utilitarian can hardly appeal to such progress as justification.*

5.3.5.1 The Utilitarian Rejoinder The rule utilitarian may well agree with many of these criticisms and yet still maintain that intellectual property rights, in some form, are justified. Putting aside the last criticism, all of the worries appear to focus on problems of implementation. So we tinker with our system of intellectual property cutting back on some legal protections and strengthening others. Perhaps we include more personality-based restrictions on what can be done with an intangible work after the first sale, limit the term of copyrights, patents, and trade secrets to something more reasonable, and embrace technologies that promote access while protecting incentives to innovate. We must also be careful about the costs of changing our system of intellectual property.

As with personality-based theories of intellectual property there seems to be something intuitive and appealing about rule-utilitarian arguments. If we view rights as rules of thumb or strategic rules—the following of which promotes human flourishing—then we have good moral reasons to adopt legal systems that protect intellectual property. Fine grained empirical evidence may be lacking regarding the benefits and costs of this or that particular rule of copyright or patent law, but there is good evidence that institution of property is better than a "no ownership" or "no protection" view.

Institutions of private property are generally beneficial because the internalization of costs discourages value-decreasing behavior. If Fred forgets to put oil in his car he will pay the costs of his forgetfulness. If Ginger does not market her super efficient electric motor other inventors may produce rival inventions and she will pay the costs of her inactivity—her invention will likely decrease in economic value. Moreover, by internalizing benefits,

> property rights encourage the search for, the discovery of, and the performance of "social" efficient activities. Private property rights greatly increase people's incentives to engage in cost-efficient conservation, exploration, extraction, invention, entrepreneurial alertness, and the development of personal and extra-personal resources suitable for all these activities . . . These rights engender a vast increase in human-made items, the value and usefulness of which tend, on the whole, more and more to exceed the value and usefulness of the natural materials employed in their production.[42]

---

[40]Priest, G. (1986). What economists can tell lawyers about intellectual property. In: Palmer; J. (Ed.), *Research in Law and Economics: The Economics of Patents and Copyrights*, Vol. 8, p. 21.

[41]Machlup, F. (1962); Long, C. (2000). Patent law and policy symposium: re-engineering patent law: the challenge of new technologies: Part II: judicial issues: patents and cumulative innovation. *Washington University Journal of Law and Policy,* 2, 229.

[42]Mack, E. (1995). The self-ownership proviso: a new and improved Lockean proviso. *Social Philosophy & Policy,* 12, 207–208.

If this is true, the upshot is that the rule utilitarian has the resources to argue for specific institutions of property relations. Put another way, it is likely, especially in light of tragedy of the commons problems and the like, that the institution of private property yields individuals better prospects than any competing institution of property relations.[43]

It could be argued that there can be no tragedy of the commons when considering intellectual property. Given that intellectual property cannot be destroyed and can be concurrently used by many individuals, there can be no ruin of the commons.[44] Upon closer examination I think that there can be a tragedy of the commons with respect to intellectual property. To begin, we may ask "What is the tragedy?" Well generally, it is the destruction of some land or other object and the cause of the destruction is scarcity and common access. But the tragedy cannot be the destruction of land or some physical object because, as we all well know, matter is neither created nor destroyed. The tragedy is the loss of value, potential value, or opportunities. Where there was once a green field capable of supporting life for years to come there is now a plot of mud, a barren wasteland, or a polluted stream. If access is not restricted to valuable resources, the tragedy will keep occurring. A prime example is the Tongan coral reefs that were being destroyed by unsavory fishing practices.[45] It seems that the best way to catch the most fish along the reef was to poor bleach into the water bringing the fish to the surface and choking the reef.

The tragedy in such cases is not only the loss of current value but also of future value. Unless access is restricted in such a way that promotes the preservation or augmentation of value, a tragedy will likely result. Now suppose that intellectual works were not protected—that if they "got out" any one could profit from them. In such cases individuals and companies would seek to protect their intellectual efforts by keeping them secret. Contracts, noncompetition clauses, and nondisclosure agreements could be employed to protect intellectual works even within a system of no protection. Secrecy was the predominant form of protection used by Guilds in the middle ages and the result can be described as a tragedy or a loss of potential value. If authors and inventors can be assured that their intellectual efforts will be protected, then the information can be disseminated and licenses granted so that others may build upon the information and create new intellectual works. The tragedy of a "no protection rule" or a system with few protections is secrecy, restricted markets, and lost opportunities.[46] This view is echoed by Roger Meiners and Robert Staaf.

---

[43]Demsetz, H. (1967). Toward a theory of property rights. *American Economic Review,* 47, 347–359, argues that an institution of property rights is the answer to the negative externalities that befall the commons. For general discussions, outside of Demsetz, extolling the virtues of private ownership over various rival institutions see Harden, G. (1968). The tragedy of the commons. *Science,* 162, 1243–48; Anderson and Hill, (1975). The evolution of property rights: a study of the American West. *Journal of Law and Economics,* 18, 163–179.

[44]While intellectual works cannot be destroyed they may be lost or forgotten—consider the number of Greek or Mayan intellectual works that were lost.

[45]The example comes from Schmidtz, D. (1990). When is original acquisition required. *The Monist,* 73, 513.

[46]Not all secrecy is a bad thing. Surely, keeping sensitive personal information to oneself is justified.

> The same story has been told about patents. If inventions lost their exclusivity and became part of the commons, then in the short run there would be over grazing. The inventor could not exclude others, and products that embody previously patentable ideas would now yield a lower rate of return. There would be lower returns to the activity of inventing, so that innovative minds would become less innovative. In the case of open ranges, common rights destroy what nature endows, and in the long run keeps the land barren because no one will invest to make the land fertile. Similarly, common rights would make the intellectual field of innovations less productive relative to a private property right system.[47]

If true, the rule utilitarian has provided the outlines of an argument for protecting the intellectual efforts of authors and inventors. Although this result does not yield a specific set of rules, it does provide a general reply to the epistemological worry that confronts incentives-based justifications of intellectual property.

5.4 THE LOCKEAN JUSTIFICATION OF INTELLECTUAL PROPERTY[48]

A final strategy for justifying intellectual property rights begins with the claim that individuals are entitled to control the fruits of their labor. Laboring, producing, thinking, and persevering, are voluntary and individuals who engage in these activities are entitled to what they produce. Subject to certain restrictions, rights are generated when individuals mix their labor with an unowned object. "The root idea of the labor theory is that people are entitled to hold, as property, whatever they produce by their own initiative, intelligence, and industry."[49] The intuition is that the person who clears land, cultivates crops, builds a house, nurtures livestock, or creates a new invention, obtains property rights by engaging in these activities.

Consider a more formal version of Locke's famous argument.[50] Individuals own their own bodies and labor—that is, they are self-owners. When an individual labors on an unowned object her labor becomes infused in the object and for the most part, the labor and the object cannot be separated. It follows that once a person's labor is joined with an unowned object, and assuming that individuals exclusively own their body and labor, rights to control are generated. The idea is that there is a kind of expansion of rights. We each own our labor and when that labor is mixed with objects in the commons our rights are expanded to include these goods.

---

[47]Miners, R. and Staaf, R. (1990). Patents, copyrights, and trademarks: property or monopoly. *Harvard Journal of Law and Public Policy,* 13, 919.

[48] A longer version of this section appears in Moore, A.D. (2004, 2001, pp. 71–194).

[49]Becker (1977). *Property Rights: Philosophic Foundations* (Routledge & Kegan, Paul, London), p. 32.

[50]There are several distinct strands to the Lockean argument. See Becker, (1977, pp. 32–56).

Locke's argument is not without difficulties.[51] David Hume argued that the idea of mixing one's labor is incoherent—actions cannot be mixed with objects.[52] Nozick asked why doesn't mixing what I own (my labor) with what I don't own a way of losing what I own rather than gaining what I don't?[53] P. J. Proudhon argued that if labor was important why shouldn't the second labor on an object ground a property right in an object as reliable as the first labor.[54] Jeremy Waldron and others have argued that mixing one's labor with an unowned object should yield more limited rights than rights of full ownership?[55] Another worry is what constitutes the boundary of one's labor? If one puts up a fence around 10 acres of land does one come to own all of the land within or merely the fence and the land it sits on?[56] And finally, if the skills, tools, and inventions used in laboring are social products, should not society have some claim on the laborer's property?[57]

Among defenders of Lockean-based arguments for private property, these challenges have not gone unnoticed.[58] Rather then rehearse these points and counterpoints, I would like to present a modified version of the Lockean argument—one that does not so easily fall prey to the objections mentioned above.

Consider the simplest of cases. After weeks of effort and numerous failures, suppose I come up with an excellent recipe for spicy Chinese noodles—a recipe that I keep in my mind and do not write down. Would anyone argue that I do not have at least some minimal moral claim to control the recipe? Suppose that you sample some of my noodles and desire to purchase the recipe. Is there anything morally suspicious with an agreement between us that grants you a limited right to use my recipe provided that you do not disclose the process? Alas, you didn't have to agree to my terms and, no matter how tasty the noodles, you could eat something else.

Here at the microlevel we get the genesis of moral claims to intellectual works independent of social progress arguments. Like other rights and moral claims, effective enforcement or protection may be a matter left to governments. But protection of rights is one thing, while the existence of rights is another.

---

[51] Simmons, A. J. (1992). *The Lockean Theory of Rights*, Princeton University Press, Princeton, 267–269.

[52] Hume, D. (1983). *Treatise of Human Nature*, 3.2.3.: "we cannot be said to join our labor to any thing except in a figurative sense." See also, Waldron, J. (1983). Two worries about mixing one's labor. *Philosophical Quarterly,* 33(37), 40.

[53] Nozick, R. (1974). *Anarchy, State, and Utopia*, p. 175.

[54] Proudhon, P.J. (1966). *What is Property?* Howard Fertig, New York, p. 61, and Plamenatz, J. (1963). *Man and Society.* Longmans and Green, London, p. 247.

[55] Perry, G. (1978). *John Locke*. Allen & Unwin, London, p. 52, and Waldron, (1983) Two worries about mixing one's labor, 42.

[56] Nozick, R. (1974). *Anarchy, State, and Utopia*, p. 174; Mautner, T. (1982). Locke on original appropriation. *American Philosophical Quarterly,* 19, 261.

[57] Rawls J. (1971). *A Theory of Justice*, p. 104; Hettinger, E.C. (1997, pp. 22–26); Grant, R. (1987). *John Locke's Liberalism.* University of Chicago Press, Chicago, p. 112. This worry is addressed in Moore, A.D. (2004, 2001, pp. 169–173).

[58] For example, Simmons A.J. (1992). *The Lockean Theory of Rights,* provides a complex analysis of Lockean property theory and attempts to answer many of these problems.

We may begin by asking how property rights to unowned objects are generated. This is known as the problem of original acquisition and a common response is given by John Locke. "For this labor being the unquestionable property of the laborer, no man but he can have a right to what that is once joined to, at least where there is *enough and as good left for others*."[59] Locke claims that so long as the proviso that enough and as good is left for others is satisfied, an acquisition is of prejudice to no one.[60] Although the proviso is generally interpreted as a necessary condition for legitimate acquisition, I would like to examine it as a sufficient condition.[61] If the appropriation of an unowned object leaves enough and as good for others, then the acquisition is justified. Suppose that mixing one's labor with an unowned object creates a prima facie claim against others not to interfere that can only be overridden by a comparable claim. The role of the proviso is to stipulate one possible set of conditions where the prima facie claim remains undefeated.[62] Another way of stating this position is that the proviso in addition to X, where X is labor or first occupancy or some other weak claim generating activity, provides a sufficient condition for original appropriation.

Justification for the view that labor or possession may generate prima facie claims against others could proceed along several lines. First, labor, intellectual effort, and creation are generally voluntary activities that can be unpleasant, exhilarating, and everything in-between. That we voluntarily do these things as sovereign moral agents may be enough to warrant noninterference claims against others. A second, and possibly related justification, is based on desert. Sometimes individuals who voluntarily do or fail to do certain things deserve some outcome or other. Thus, students may deserve high honor grades and criminals may deserve punishment. When notions of desert are evoked claims and obligations are made against others—these nonabsolute claims and obligations are generated by what individuals do or fail to do. Thus in fairly uncontroversial cases of desert, we are willing to acknowledge that weak claims are generated, and if desert can properly attach to labor or creation, then claims may be generated in these cases as well.

Finally, a justification for the view that labor or possession may generate prima facie claims against others could be grounded in respect for individual autonomy and sovereignty. As sovereign and autonomous agents, especially within the liberal tradition, we are afforded the moral and legal space to order our lives as we see fit. As long as respect for others is maintained we each are free to set the course and direction of our own lives, to choose between various lifelong goals and projects, and to develop our capacities and talents accordingly. Simple respect for individuals would

---

[59]Locke, J. (1690). *The Second Treatise of Government*, Section 27 (italics mine).

[60]Ibid., Section 33, 34, 36, 39.

[61]Both Waldron, J. (1979). Enough and as good left for others. *Philosophical Quarterly*, 319–328, and Wolf, C. (1995). Contemporary property rights, Lockean provisos, and the interests of future generation. *Ethics* 105, 791–818, maintain that Locke thought of the proviso as a sufficient condition and not a necessary condition for legitimate acquisition.

[62]This view is summed up nicely by Wolf, C. (1995). Contemporary property rights, Lockean provisos, and the interests of future generation. 791–818.

prohibit wresting from their hands an unowned object that they acquired or produced. I hasten to add that at this point we are trying to justify weak noninterference claims, not full blown property rights. Other things being equal, when an individual labors to create an intangible work, then weak presumptive claims of noninterference have been generated on grounds of labor, desert, or autonomy.

Suppose Fred appropriates a grain of sand from an endless beach and paints a lovely, albeit small, picture on the surface. Ginger, who has excellent eyesight, likes Fred's grain of sand and snatches it away from him. On this interpretation of Locke's theory, Ginger has violated Fred's weak presumptive claim to the grain of sand. We may ask, what legitimate reason could Ginger have for taking Fred's grain of sand rather than picking up her own grain of sand? If Ginger has no comparable claim, then Fred's prima facie claim remains undefeated. An undefeated prima facie claim can be understood as a right.[63]

5.4.1 A Pareto-Based Proviso

The underlying rationale of Locke's proviso is that if no one's situation is worsened, then no one can complain about another individual appropriating part of the commons. Put another way, an objection to appropriation, which is a unilateral changing of the moral landscape, would focus on the impact of the appropriation on others. But if this unilateral changing of the moral landscape makes no one worse off, there is no room for rational criticism.

The proviso permits individuals to better themselves so long as no one is worsened (weak Pareto superiority). The base level intuition of a Pareto improvement is what lies behind the notion of the proviso.[64] If no one is harmed by an acquisition and one person is bettered, then the acquisition ought to be permitted. In fact, it is precisely because no one is harmed that it seems unreasonable to object to a Pareto-superior move. Thus, the proviso can be understood as a version of a "no harm, no foul" principle.

It is important to note that compensation is typically built into the proviso and the overall account of bettering and worsening. An individual's appropriation may actually benefit others and the benefit may serve to cancel the worsening that occurs from restricted use. Moreover, compensation can occur at both the level of the act and at the level of the institution.

This leads to a related point. Some have argued that there are serious doubts whether a Pareto-based proviso on acquisition can ever be satisfied in a world of scarcity. Given

[63]For a defense of this view of rights see Rainbolt, G. (1993). Rights as normative constraints. *Philosophy and Phenomenological Research*, 93–111, and Feinberg, J. (1986). *Freedom and Fulfillment: Philosophical Essays.* Princeton University Press.

[64]One state of the world, S_1, is Pareto superior to another, S_2, if and only if no one is worse off in S_1 than in S_2, and at least one person is better off in S_1 than in S_2. S_1 is *strongly* Pareto superior to S_2 if everyone is better off in S_1 than in S_2, and *weakly* Pareto superior if at least one person is better off and no one is worse off. State S_1 is Pareto optimal if no state is Pareto superior to S_1: it is *strongly* Pareto optimal if no state is *weakly* Pareto superior to it and *weakly* Pareto optimal if no state is *strongly* Pareto superior to it. Throughout this article I will use Pareto superiority to stand for *weak* Pareto superiority. Adapted from Cohen, G.A. (1995). The pareto argument For inequality. *Social Philosophy & Policy,* 12, 160.

that resources are finite and that acquisitions will almost always exclude, your gain is my loss (or someone's loss). On this model, property relations are a zero-sum game. If this were an accurate description, then no Pareto superior moves could be made and no acquisition justified on Paretian grounds. But this model is mistaken. An acquisition by another may worsen your position in some respects but it may also better your position in other respects. Minimally, if the bettering and worsening cancel each other out, a Pareto superior move may be made and an acquisition justified. Locke recognizes this possibility when he writes,

> Let me add, that he who appropriates land to himself by his labor, does not lessen, but increase the common stock of mankind: for the provisions serving to the support of human life, produced by one acre enclosed and cultivated land, are ten times more than those which are yielded by an acre of land of equal richness lying waste in common.[65]

Furthermore, it is even more of a stretch to model *intellectual* property as zero-sum. Given that intellectual works are nonrivalrous—they can be used by many individuals concurrently and cannot be destroyed—my possession and use of an intellectual work does not preclude your possession and use of it. This is just to say that the original acquisition of intellectual or physical property does not necessitate a loss for others. In fact, if Locke is correct, such acquisitions benefit everyone.

Consider the case where Ginger is better off, all things considered, if Fred appropriates everything compared to how she would have been had she appropriated everything (maybe Fred is a great manager of resources). Although Ginger has been worsened in some respects she has been compensated for her losses in other respects. David Gauthier echoes this point in the following case. "In acquiring a plot of land, even the best land on the island, Eve may initiate the possibility of more diversified activities in the community as a whole, and more specialized activities for particular individuals with ever-increasing benefits to all."[66]

Before continuing, I will briefly consider the plausibility of a Pareto-based proviso as a moral principle.[67] First, to adopt a less-than-weak Pareto principle would permit individuals, in bettering themselves, to worsen others. Such provisos on acquisition are troubling because at worst they may open the door to predatory activity and at best they give antiproperty theorists the ammunition to combat the weak presumptive claims that labor and possession may generate. Part of the intuitive force of a Pareto-based proviso is that it provides little or no grounds for rational complaint. Moreover, if we can justify intellectual property rights with a more stringent principle, a principle that is harder to satisfy, then we have done something more robust, and perhaps more difficult to attack, when we reach the desired result.

---

[65]Locke, J. (1960). *The Second Treatise of Government*, Section 37.

[66]Gauthier (1986). *Morals By Agreement.* Oxford University Press, p. 280.

[67]I have in mind Nozick's Robinson Crusoe case in *Anarchy, State, and Utopia*, p. 185.

To require individuals, in bettering themselves, to better others is to require them to give free rides. In the absence of social interaction, what reason can be given for forcing one person, if she is to benefit herself, to benefit others?[68] If, absent social interaction, no benefit is required then why is such benefit required within society? The crucial distinction that underlies this position is between worsening someone's situation and failing to better it[69] and I take this intuition to be central to a kind of deep moral individualism. Moreover, the intuition that grounds a Pareto-based proviso fits well with the view that labor and possibly the mere possession of unowned objects creates a prima facie claim to those objects. Individuals are worthy of a deep moral respect and this grounds a liberty to use and possess unowned objects.

5.4.2 Bettering, Worsening, and the Baseline Problem

Assuming a just initial position and that Pareto-superior moves are legitimate, there are two questions to consider when examining a Pareto-based proviso. First, what are the terms of being worsened? This is a question of scale, measurement, or value. An individual could be worsened in terms of subjective preferences:satisfaction, wealth, happiness, freedoms, opportunities, and so on. Which of these count in determining moral bettering and worsening? Second, once the terms of being worsened have been resolved, which two situations are we going to compare to determine if someone has been worsened? In any question of harm we are comparing two states—for example, "now" after an acquisition compared to "then" or before an acquisition. This is known as the baseline problem.

In principle, the Lockean theory of intangible property being developed is consistent with a wide range of value theories. So long as the preferred value theory has the resources to determine bettering and worsening with reference to acquisitions, then Pareto-superior moves can be made and acquisitions justified on Lockean grounds. For now, assume an Aristotelian Eudaimonist account of value exhibited by the following theses is correct.[70]

---

[68]The distinction between worsening someone's position and failing to better it is a hotly contested moral issue. See Gauthier (1986). *Morals By Agreement.* Oxford University Press, p. 204; Kagan, S. (1989). *The Limits of Morality.* Oxford University Press. Chapter 3; Harris, J. (1973–1974). The Marxist conception of violence. *Philosophy & Public Affairs,* 3, 192–220; Kleinig, J. (1975–1976). Good samaritanism. *Philosophy & Public Affairs,* 5, 382–407; and Mack, E. (1979–1980). Two articles: Bad Samaritanism and the causation of harm. *Philosophy & Public Affairs,* 9, 230–259; and Causing and Failing to Prevent Harm. *Southwestern Journal of Philosophy,* 7, (1976): 83–90.

[69]This view is summed up nicely by Fressola, A. "Yet, what is distinctive about persons is not merely that they are agents, but more that they are rational planners—that they are capable of engaging in complex projects of long duration, acting in the present to secure consequences in the future, or ordering their diverse actions into programs of activity, and ultimately, into plans of life." Fressola, A. (1981). Liberty and property. *American Philosophical Quarterly,* 18, 320.

[70]For similar views see Rawls J. (1971). *A Theory of Justice.* Harvard University Press, Cambridge. Chapter VII; Aristotle, *Nicomachean Ethics*, bks. I and X; Kant (1785). *The Fundamental Principles of The Metaphysics of Morals,* Academy Edition; Sidgwick (1907). *Methods of Ethics*, 7th edition. Macmillian, London; Perry, R.B. (1962). *General Theory of Value.* Longmans, Green, New York; and Lomasky, L. (1987). *Persons, Rights, and the Moral Community.* Oxford University Press, New York.

(1) Human well-being or flourishing is the sole standard of intrinsic value.
(2) Human persons are rational project pursuers, and well-being or flourishing is attained through the setting, pursuing, and completion of life goals and projects.
(3) The control of physical and intangible object is valuable. At a specific time each individual has a certain set of things she can freely use and other things she owns, but she also has certain opportunities to use and appropriate things. This complex set of opportunities along with what she can now freely use or has rights over constitutes her position materially—this set constitutes her level of material well-being.

Although it is certainly the case that there is more to bettering and worsening than an individual's level of material well-being, including opportunity costs, I will not pursue this matter further at present. Needless to say, a full-blown account of value will explicate all the ways in which individuals can be bettered and worsened with reference to acquisition. Moreover, as noted before, it is not crucial to the Lockean model being presented to defend some preferred theory of value against all comers. Whatever value theory that is ultimately correct, if it has the ability to determine bettering and worsening with reference to acquisitions, then Pareto-superior moves can be made and acquisitions justified on Lockean grounds.

Lockeans as well as others who seek to ground rights to property in the proviso generally set the baseline of comparison as the state of nature. The commons, or the state of nature, is characterized as that state where the moral landscape has yet to be changed by formal property relations. For now, assume a state of nature situation where no injustice has occurred and where there are no property relations in terms of use, possession, or rights. All in this initial state have opportunities to increase their material standing. Suppose Fred creates an intangible work (perhaps a new gathering technique) and does not worsen his fellows—alas, all they had were contingent opportunities and Fred's creation and exclusion adequately benefits them in other ways. After the acquisition, Fred's level of material well-being has changed. Now he has a possession that he holds legitimately, as well as all of his previous opportunities. Along that comes Ginger who creates her own intangible work and considers whether her exclusion of it will worsen Fred. But what two situations should Ginger compare? Should the effects of Ginger's acquisition be compared to Fred's initial state, where he had not yet legitimately acquired anything, or to his situation immediately before Ginger's taking? If bettering and worsening are to be cashed out in terms of an individual's level of well-being with opportunity costs and this measure changes over time, then the baseline of comparison must also change. In the current case we compare Fred's level of material well-being when Ginger possesses and excludes an intangible work to his level of well-being immediately before Ginger's acquisition.

At this point I would like to clear up a common confusion surrounding the baseline of comparison. What if a perverse inventor creates a genetic-enhancement technique that will save lives, but decides to keep the technique secret or charge excessive prices for access? Those individuals who had, before the creation, no chance to survive now have a chance and are worsened because of the perverse inventor's refusal to let others use the technique.[71]

In this case the baseline implies cannot be correct. On this view, to determine bettering and worsening we are to compare how individuals are before the creation of some value (in this case the genetic enhancement technique) to how they would be if they possessed or consumed that value. But we are all worsened in this respect by any value that is created and held exclusively. I am worsened by your exclusive possession of your car because I would be better off if I exclusively controlled the car—even if I already owned hundreds of cars. Any individual, especially those who have faulty hearts, would be better off if they held title to my heart compared to anyone else's holding the title. I am also worsened when you create a new philosophical theory and claim authorship—I would have been better off (suppose it is a valuable theory) if I had authored the theory, so you have worsened me. Clearly this account of the baseline makes the notions of moral bettering and worsening too broad.[72]

A slightly different way to put the Lockean argument for intellectual property rights is:

Step 1: *The Generation of Prima Facie Claims to Control*—suppose Ginger creates a new intangible work—creation, effort, and so on, yield her prima facie claims to control (similar to student desert for a grade).

Step 2: *Locke's Proviso*—if the acquisition of an intangible object makes no one (else) worse off in terms of their level of well-being compared to how they were immediately before the acquisition, then the taking is permitted.

Step 3: *From Prima Facie Claims to Property Rights*—When are prima facie claims to control an intangible work undefeated? Answer: when the proviso is satisfied. Alas, no one else has been worsened—who could complain?

Conclusion: So long as no harm is done—the proviso is satisfied—the prima facie claims that labor and effort may generate turn into property claims.[73]

---

[71]We will also have to suppose that the system of intellectual property protection in this case allows multiple patents assuming independent creation or discovery. If the perverse inventor's intellectual property excluded others from independent creation or discovery then worsening has occurred — the chance or opportunity that someone would find a cure and help will have been eliminated.

[72]This sort of baseline confusion infects Farrelly, C. (2002). Genes and social justice: a reply to Moore. *Bioethics*, 16, 75. For a similar, yet still mistaken, view of the baseline see Waldron, J. (1993, p. 866) and Gordon, W. (1993, p.1564, p.1574).

[73]Ken Himma in correspondence has suggested that this argument could succeed without defending initial prima facie claims to control. "Suppose I have no prima facie claim to X, but my taking X leaves no one worse off in any respect. Since they have no grounds to complain, what could be wrong with my taking it? If, however, there is a prima facie claim on my part, much more would be needed to defeat it than just pointing out that someone is made worse off by it. That's how [moral] claims work it seems to me — and why they're needed: to justify making others worse off." My worry, though, is that without establishing initial prima facie claims to control there would be no moral aspect to strengthen into rights by application of the proviso. In any case, this is an interesting line of inquiry.

If correct, this account justifies moral claims to control intangible property like genetic enhancement techniques, movies, novels, or information. When an individual creates an intangible work and fixes it in some fashion, then labor and possession create a prima facie claim to the work. Moreover, if the proviso is satisfied the prima facie claim remains undefeated and moral claims or rights are generated.

Consider the following case. Suppose Fred, in a fit of culinary brilliance, scribbles down a new recipe for spicy Chinese noodles and then forgets the essential ingredients. Ginger, who loves spicy Chinese food, sees Fred's note and snatches it away from him. On my view of Locke's theory the proviso has been satisfied by Fred's action and Ginger has violated Fred's right to control the collection of ideas that comprise the recipe. We may ask, what legitimate reason could Ginger have for taking Fred's recipe rather than creating her own? If Ginger has no comparable claim, then Fred's prima facie claim remains undefeated.

We can complicate this case by imagining that Fred has perfect memory and so Ginger's theft does not leave Fred deprived of that he created. It could be argued that what is wrong with the first version of this case is that Fred lost something that he created and may not be able to recreate—Ginger herself felt betters, without justification, at the expense of Fred. In the second version of the case Fred has not lost and Ginger has gained and so there is apparently nothing wrong with her actions. But from a moral standpoint, the accuracy of Fred's memory is not relevant to his rights to control the recipe and so this case poses no threat to the proposed theory.[74] Intellectual property rights that are hard to protect have no bearing on the existence of the rights themselves. Similarly, it is almost impossible to prevent that a trespasser from walking on your land has no bearing on your rights to control. In creating the recipe and not worsening Ginger, compared to the baseline, Fred's presumptive claim is undefeated and thus creates a duty of noninterference on others. In both versions of this case Fred has lost the value of control and the control of the value that he created.

Rather than creating a recipe, suppose Fred writes a computer program and Ginger simultaneously creates a program that is, in large part, a duplicate of Fred's. To complicate things further, imagine that each will produce and distribute the software with the hopes of capturing the market and that Fred has signed a distribution contract that will enable him to swamp the market and keep Ginger from selling her product. If opportunities to better oneself are included in the account of bettering and worsening, then it could be argued that Fred violates the proviso because in controlling and marketing the software he effectively eliminates Ginger's potential profits. The problem this case highlights is that what individuals do with their possessions can affect the opportunities of others in a negative way. If so, then worsening has occurred and no duties of noninterference have been created. In cases of competition it seems that the proviso may yield the wrong result.

This is just to say that the proviso, as I have interpreted it, is set too high or that it is overly stringent. In some cases where we think that rights to intellectual property should be justified it turns out, on the theory being presented, that they are not. But surely this is no deep problem for the theory. In the worst light, it has not been shown

[74]If Fred's personality has become infused in the intellectual work, Ginger's taking is even more suspect.

that the proviso is not sufficient but only that it is overly stringent. And given what is at stake—the means to survive, flourish, and pursue lifelong goals and projects—stringency may be a good thing. Nevertheless, the competition problem represents a type of objection that poses a significant threat to the theory being developed. If opportunities are valuable, then any single act of acquisition may extinguish one or a number of opportunities of one's fellows. Obviously this need not be the case every time, but if this worsening occurs on a regular basis, then the proposed theory will leave unjustified a large set of acquisitions that we intuitively think should be justified.

Before concluding, I would like to briefly discuss a strategy for answering the competition problem and related concerns. Continuing with the Fred and Ginger example, it seems plausible to maintain that her complaints are, in a way, illicit. The very opportunities that Ginger has lost because of Fred's business savvy are dependent on the institution of property relations that allows Fred to beat her to market. Moreover, her opportunities include the possibility of others undercutting her potential profits. Contingent opportunities are worth less than their results and so compensation will be less that it would seem. As noted earlier, compensation for worsening could proceed at two levels. In acquiring some object, Fred, himself, could better Ginger's position, or the system that they both operate within could provide compensation. This is just to say that it does not matter whether the individual compensates or the system compensates, the agent in question is not worsened.

5.5 CONCLUSION

In this article, three strategies for justifying intellectual property rights have been presented. Although plausible in some respects, personality-based theories seem the weakest because, I would argue, they are the least well developed. There is something ethically wrong with distorting someone else's intellectual work without consent. But this is just a beginning and a lot of work needs to be done to turn this intuition into a general defense of intellectual property.

Rule-utilitarian incentives-based justifications of intellectual property are stronger, although much depends on empirical claims that are difficult to determine. Nevertheless, the rule utilitarian has the resources to defend moral claims to intellectual works. If these moral claims are to be codified in the law, then we have good reason to adopt a system of intellectual property protection.

For reasons not presented in this article, I would argue that the Lockean justification of intellectual property sketched in the final section is the strongest of the three.[75] If no one is worsened by an acquisition, then there seems to be little room for rational complaint. Locke wrote, "Nobody could think himself injured by the drinking of another man, though he took a good draught, who had a whole river of the same water left him to quench his thirst . . ."[76] Given allowances for independent creation and that

[75]For a critique of rule-utilitarian incentive based justifications of intellectual property see Moore, A.D. (2004, 2001, pp. 36–70); Chapter 3: "Against Rule-Utilitarian Intellectual Property."

[76]Locke, J. (1960). *The Second Treatise of Government;* Chapter 5, Section 33.

the frontier of intellectual property is practically infinite, the case for Locke's water drinker and the author or inventor are quite alike. What is objectionable with the theft and pirating of computer software, musical CD's, and other forms of intellectual property is that in most cases a right to the control something of value and the value of control has been violated without justification. Although the force of this normative claim is easily clouded by replies like, "but they still have their copy" or "I wouldn't have purchased the information anyway" it does not alter the fact that a kind of theft has occurred. Authors and inventors who better our lives by creating intellectual works have rights to control what they produce. How these moral claims take shape in our legal systems is a topic for further discussion.[77]

REFERENCES

Becker, L.C. (1993). Deserving to own intellectual property. *Chicago-Kent Law Review*, 68.

Boyle, J. (1996). *Shamans, Software, and Spleens*. Harvard University Press, Cambridge, MA.

Bugbee, B. (1967). *Genesis of American Patent and Copyright Law*. Public Affairs Press, Washington, DC.

Child, J.W. (1997). The moral foundations of intangible property. In: Moore, A. (Ed.), *Intellectual Property: Moral, Legal, and International Dilemmas*. Rowman and Littlefield, Lanham, MD.

Croskery, P. (1993). Institutional utilitarianism and intellectual property. *Chicago-Kent Law Review*, 68, 631–657.

Gordon, W.J. (1993). Property right in self expression: equality and individualism in the natural law of intellectual property. *Yale Law Journal*, 102, 1533–1609.

Gordon, W.J. (1994). Assertive modesty: an economics of intangibles. *Columbia Law Review*, 94, 2579–2593.

Hettinger, E.C. (1997). Justifying intellectual property. *Philosophy and Public Affairs*, reprinted in: Moore, A. (Ed.), *Intellectual Property: Moral, Legal, and International Dilemmas*. Rowman and Littlefield, Lanham, MD 18, 31–52.

Himma, K. (2006). Justifying intellectual property protection: why the interests of content-Creators usually wins over everyone else's. In: Rooksby, E. (Ed.), *Information Technology and Social Justice*, Idea Group.

Hughes, J. (1997). The philosophy of intellectual property. *Georgetown Law Journal*. Reprinted in: Moore, A. (Ed.), *Intellectual Property: Moral, Legal, and International Dilemmas*. Rowman and Littlefield, Lanham, MD 77, p. 287.

Kuflik, A. (1989). The moral foundations of intellectual property Rights. In: Weil, V. and Snapper, J. (Eds.), *Owning Scientific and Technical Information*. Rutgers University Press, New Brunswick and London.

Machlup, F. (1962). *Production and Distribution of Knowledge in the United States*. Princeton University Press, Princeton.

[77]See Moore, A.D. (2004 [2001], pp. 121–179); Chapter 6: Justifying acts, systems, and institutions and Chapter 7: A new look and copyrights, patents, and trade secrets.

Moore, A.D. (1997). In: Moore, A. (Ed.), *Intellectual Property: Moral, Legal, and International Dilemmas*. Rowman & Littlefield, Lanham, MD.

Moore, A.D. (2004[2001]). *Intellectual Property and Information Control: Philosophic Foundations and Contemporary Issues*. Transaction Publishing, New Brunswick, NJ, 2004, 2001, hardback.

Moore, A.D. (2005). Intangible property: privacy, power, and information control. *American Philosophical Quarterly*, 35, 365–378. Reprinted in: Moore, A. (Ed.), *Information Ethics: Privacy, Property, and Power.* University of Washington Press.

Nelkin, D. (1984). *Science as Intellectual Property.* Macmillan, New York.

Palmer, T.G. (1997). Intellectual property: a non-posnerian law and economics approach. *Hamline Law Review*, 261–304. Reprinted in: Moore, A. (Ed.), *Intellectual Property: Moral, Legal, and International Dilemmas*. Rowman and Littlefield, Lanham MD.

Palmer, T.G. (2005). Are patents and copyrights morally justified? The philosophy of property rights and ideal objects. *Harvard Journal of Law and Public Policy*, 13, 817–866. Reprinted in: Moore, A. (Ed.), *Information Ethics: Privacy, Property, and Power.* University of Washington Press.

Spooner, L. (1971). *The Law of Intellectual Property.* M & S Press (originally published in 1855).

Tavani, H. (2005). Intellectual property rights: from theory to practical implementation (with Richard Spinello). In: Spinello, R.A. and Tavani, H.T. (Eds.), *Intellectual Property Rights in a Networked World: Theory and Practice*. Idea Group/Information Science Publishing, Hershey, PA, pp. 1–65.

Waldron, J. (1993). From authors to copiers: individual rights and social values in intellectual property. *Chicago-Kent Law Review*, 68, 841–887.

Weil, V. and Snapper, J. (Eds.). (1989). *Owning Scientific and Technical Information*. Rutgers University Press, New Brunswick and London.

CHAPTER 6

Informational Privacy: Concepts, Theories, and Controversies

HERMAN T. TAVANI

This chapter examines some key concepts, theories, and controversies affecting informational privacy.[1] It is organized into five main sections. Section 6.1 includes an overview of the concept of privacy in general, whereas Section 6.2 briefly analyzes four distinct kinds of privacy: physical, decisional, psychological, and informational privacy. Section 6.3 critically evaluates some classic and contemporary theories of informational privacy, including the restricted access and control theories. This is followed in Section 6.4 by a consideration of the effect that some specific information technologies have had for four subcategories of informational privacy: consumer privacy, medical privacy, employee privacy, and location privacy. Section 6.5 examines the value of privacy as both an individual and societal good, and it considers some proposals for framing a comprehensive informational privacy policy.

6.1 THE CONCEPT OF PRIVACY

> We demand recognition of our right to privacy, we complain when privacy is invaded, yet we encounter difficulties immediately [when] we seek to explain what we mean by privacy, what is the area, the content of privacy, what is outside that area, what constitutes a loss of privacy, a loss to which we have consented, a justified loss, an unjustified loss.
>
> —H.J. McCloskey (1985, p. 343)

McCloskey identifies some of the difficulties one faces when attempting to give a clear and coherent account of privacy. Yet gaining an adequate understanding of privacy is

[1]This chapter draws from and expands upon material in some of my previously published works on privacy, including Tavani (2004a, 2007a, 2007b, 2007c).

The Handbook of Information and Computer Ethics, Edited by Kenneth Einar Himma and Herman T. Tavani

important because of the role that concept has played and continues to play in social and political thought. For example, DeCew (1997) points out that the concept of privacy has been "central" in most discussions of modern Western life (even though philosophers and legal scholars have only recently tried to elucidate what is meant by privacy). Regan (1995, p. 43) notes that in the West, this concept has its roots in the political thinking of Thomas Hobbes and John Locke and in "the liberal form of democratic government that derived from that thinking."[2] Other authors, such as Flaherty (1989), believe that the roots of privacy as a concept can be traced back to Roman law and Biblical literature. B. Moore (1984) argues that aspects of privacy can also be found in primitive societies and tribes in the non-Western world, whereas Westin (1967) suggests that elements of privacy can even be found in behavior in the animal world.[3]

It would seem that privacy is not simply a static concept, but instead has a dynamic component. Moor (2006, p. 114) argues that privacy is "an evolving concept" and that its "content" is often influenced by the "political and technological features of the society's environment." In a later section of this chapter, we will see how the concept of privacy has evolved significantly in the United States since the eighteenth century. Some theorists, such as Regan, claim that the concept of privacy has existed throughout American history. Yet this concept has also been contested in American jurisprudence, as there is no explicit mention of privacy in the U.S. Constitution. Many Americans now believe that their privacy is severely threatened by the kinds of technologies that have been developed and used in recent years, and some even speak of "the end of privacy."[4] However, Regan (p. 212) argues that the current threat to privacy is not so much the result of recent technologies as it is our failure to "conceptualize privacy" in a way that "sustains public interest and support."[5] But what, exactly, is privacy, and how is it best defined? We briefly examine some answers to these questions that have been proposed by prominent privacy theorists and philosophers.

6.1.1 Unitary, Derivative, and Cluster Definitions of Privacy

Parent (1983a) claims that we cannot frame an adequate definition of privacy unless we are first familiar with the "ordinary usage" of that term. He goes on to argue, however, that this familiarity in itself is not sufficient because our common ways of talking and using language are "riddled with inconsistencies, ambiguities, and paradoxes" (Parent, p. 269). So he believes that while we need a definition of privacy that is "by and large consistent with ordinary language," it must also enable us "to talk

---

[2]Kemp and A.D. Moore (2007) describe some ways in which John Stuart Mill's account of liberty also has influenced contemporary views of privacy in the West.

[3]Westin (p. 8) notes that animal studies have suggested "that virtually all animals seek periods of individual seclusion or social group intimacy . . . in which an organism lays private claims to an area . . . and defends against intrusion by members of its own species."

[4]See, for example, Spinello (1997), who states that our privacy may "gradually be coming to an end."

[5]Solove (2002) also argues that privacy needs to be better conceptualized, or "reconceptualized."

consistently, clearly, and precisely about the family of concepts to which privacy belongs." Yet if one were to consult a dictionary for a definition of privacy, she would likely find that the concept is typically described in terms of notions such as secrecy, solitude, security, confidentiality, and so forth, and often defined in ways that can make it difficult to distinguish privacy from these concepts. So if Parent is correct, privacy needs to be better differentiated from this group of concepts, as well as from those of anonymity, liberty, autonomy, and so forth, before we can frame an adequate definition of privacy. However, one might ask how privacy is distinguishable from this family of related concepts. In other words, how is it possible to define privacy in a way that is sufficiently independent of its cognate concepts? Parent suggests that privacy can be defined as a unitary, or univocal, concept.[6] McCloskey (1985) also believes that a unitary definition is possible. For example, he says that we can "distinguish concepts from privacy, which have been confused with privacy, and thereby make clearer the core notion of privacy" (p. 343).

While Parent and McCloskey believe that privacy can be as understood as a unitary concept that is basic and thus capable of standing on its own, others have argued that privacy is best understood as a "derivative" concept. This view is especially apparent in cases where privacy is conceived of in terms of a right. For example, Thomson (1975) believes that one's "right to privacy" can be derived from other, more fundamental, rights such as one's rights to property and to bodily security. Thomson's defenders, such as Volkman (2003) who argues that matters of privacy can be adequately accounted for by unpacking our natural rights to life, liberty, and property,[7] also describe various ways in which a privacy right can be derived from one or more basic or fundamental rights. But Thomson's critics, including Scanlon (1975) and Parent argue that it is just as plausible to derive other rights from privacy as it is to derive a right to privacy from rights that are alleged to be more basic or fundamental.

Even if Thomson's critics are correct in claiming that privacy is not necessarily derivable from other, more basic, concepts or from other rights, we can still ask whether privacy can be adequately understood as a unitary concept. For example, some theorists reject both the unitary and derivative accounts of privacy. DeCew (p. 61), who argues that it is not possible to give a "unique, unitary definition of privacy that covers all the diverse privacy interests," also believes that the view that privacy is totally derived from other interests is "equally untenable." She suggests that privacy, which is a "broad and multifaceted" notion, is best understood in terms of a "cluster concept."[8]

It is perhaps worth noting at this point that the debate about how privacy is best defined is closely related to the question of whether privacy should be viewed as a full-fledged *right*, or simply in terms of one or more *interests* that individuals have. Perhaps not surprisingly, one's answer to this question can significantly influence one's belief about how privacy should be defined.

---

[6]He defines privacy as "the condition of not having undocumented personal information about us possessed by others" (Parent, p. 269).

[7]Volkman believes that privacy claims are already built into a "neo-Lockean theory of natural rights."

[8]Nissenbaum (2004) and van den Hoven (2004) also endorse a "cluster account" of privacy.

6.1.2 Interest-Based Conceptions Versus Rights-Based Conceptions of Privacy

Whereas Thomson and Scanlon analyze privacy from the perspective of a rights-based concept (as we saw in the preceding section), other philosophers and legal theorists approach privacy in terms of various interests. The U.S. Supreme Court, in its 1977 ruling in *Whalen v. Roe*, recognized privacy as representing two different kinds of interests that individuals have: (i) avoiding disclosure of personal matters (i.e., information) and (ii) independence in making certain kinds of important decisions. DeCew (p. 22), who believes that privacy can be understood as an "umbrella term" for a "wide variety of interests," notes that in recent years there has been "a shift away from reasoning that takes a rights-oriented approach toward more arguments that use a utilitarian cost-benefits analysis which balances the concept of privacy and the benefits to public safety and crime control." Some theorists suggest that it is more useful to view privacy in terms of an interest rather than as a right. However, many who support an interest-based conception of privacy also note that privacy interests must be balanced against many other, often competing, interests. Clarke (1999, p. 60), for example, points out that such competing interests can include those of "the individuals themselves, of other individuals, of groups, and of society as a whole."

What kind of interest is a *privacy interest*, and what does that interest protect? According to Clark (p. 60), privacy protects the interest individuals have in "sustaining a personal space, free from interference by other people and organizations." Alternatively, Alfino (2001, p. 7) argues that privacy protects a "fundamental interest" one has in "being able to lead a rational, autonomous life." In his view, privacy does not simply protect one's "spatial interests," nor does it merely protect one's "reputational interest" from potential harm that might result in a privacy intrusion, it also protects one's interest in being able "to think and plan (one's) conduct. . . and to lead a rational life." Other authors suggest that privacy protects a "tangible property interest" that individuals have with respect to their personal information. For example, Hunter (1995) argues that one way to give individuals control over information about themselves is to vest them with a "property interest" in that information. From this perspective, personal information can be viewed as a kind of property that a person can own and negotiate with in the economic or commercial sphere.[9]

Some who defend an interests-based conception of privacy have suggested that privacy protection schemes can simply be stipulated (as a procedural matter) as opposed to having to be grounded in the kinds of philosophical and legal theories needed to justify rights. Discussions involving privacy in terms of an explicit right—moral, legal, or otherwise—have often become mired in controversy. Following DeCew (p. 27), who leaves open the question of how extensively privacy "ought to be protected" (as a right), we can agree with her claim that privacy is "an interest which also can be invaded." Unfortunately, a detailed analysis of arguments affecting

[9]Branscomb (1994) has defended a "property-interest" conception of personal information, whereas Spinello (2006) describes some of the difficulties that can arise in framing privacy policies based on such an account of personal information. However, we will not pursue this debate here, because it is not central to our objectives in this chapter.

rights-based versus interests-based conceptions of privacy is beyond the scope of this chapter; our purpose in mentioning this distinction here has been to show how these two competing conceptions of privacy often influence definitions of that notion.

6.2 FOUR DISTINCT KINDS OF PRIVACY: PHYSICAL/ACCESSIBILITY, DECISIONAL, PSYCHOLOGICAL/MENTAL, AND INFORMATIONAL PRIVACY

As noted in the preceding section, the meaning of privacy in the United States has evolved considerably since the eighteenth century.[10] Initially, privacy was understood in terms of freedom from (physical) intrusion—as implied, for example, in the Fourth Amendment to the U.S. Constitution. Privacy later came to be interpreted by the U.S. courts as freedom from interference in making important decisions, especially with respect to one's choices involving reproduction and marriage. (A variation of this view interprets privacy as including freedom from interference involving one's ability to think and express one's thoughts without external pressures to conform.) Since the latter half of the twentieth century, privacy has increasingly come to be associated with concerns about protecting one's personal information, which is now easily collected and stored electronically and easily exchanged between electronic databases. Although our main focus on privacy in this chapter centers on issues affecting access and control of personal information – i.e., *informational* privacy – we also briefly examine the three alternative views.

6.2.1 Privacy as Nonintrusion Involving One's Physical Space: Physical/Accessibility Privacy

"Physical privacy" and "accessibility privacy" are relatively recent expressions used to refer to a conception of privacy that emerged in the late nineteenth century in response to an influential article on privacy by Samuel Warren and Louis Brandeis that appeared in the *Harvard Law Review* in 1890. Floridi (1999, p. 52) defines "physical privacy" as the freedom a person enjoys from sensory intrusion, which is "achieved thanks to restrictions on others' ability to have bodily interactions with (that person)." Note the emphasis here on privacy as *physical* nonintrusion. Because the conception of privacy as freedom from unwarranted intrusion focuses on the kind of harm that can be caused through physical access to a person or through access to a person's physical possessions, some have used the expression *accessibility privacy* to describe this view.[11]

As noted in the preceding paragraph, the origin of the physical/accessibility view of privacy is often traced to Warren and Brandeis (1890), who argued that privacy could

---

[10]Etzioni (1999, pp. 188-189) describes three distinct stages of conceptual development regarding privacy in America: (1) pre-1890 (when the concept of privacy was understood primarily in relation to property rights; (2) 1890–1965 (the period in which "the right to privacy" was developed, mainly from tort law; and (3) post-1965 (a period in which the right to privacy has been significantly expanded).

[11]See, for example, DeCew (1997), Floridi (1999, 2006), and Tavani (2004a, 2007a).

be understood as "being let alone" or "being free from intrusion." This conception of privacy was also articulated in a later work by Brandeis (1928), when, as a U.S. Supreme Court Justice, he wrote the dissenting opinion in *Olmstead v. United States*. Justice William Brennan (1972) also appealed to this view of privacy in his majority opinion for *Eisenstadt v. Baird*. Although Warren and Brandeis are often credited with having made the first explicit reference to privacy as a legal right in the United States, DeCew points out that the phrase "the right to be let alone" had been used by Justice Thomas Cooley in his *Treatise on the Law of Torts* in 1880 (approximately one decade before the seminal article by Warren and Brandeis).[12] And Regan notes that as early as 1886, the Supreme Court recognized that the Fourth Amendment protected privacy interests (in its decision in *Boyd v. United States*).[13]

The conception of privacy in terms of physical nonintrusion or being let alone has been criticized because of its tendency to conflate two different concepts that need to be distinguished—namely, having privacy and being let alone. To see this flaw, consider a situation in which one might not be let alone and yet enjoy privacy. If a student approaches her professor, who is on his or her way to teach a class, and asks her professor a question about a previous class assignment, the student has not, strictly speaking, let her professor alone; however, she has also not violated her professor's privacy. Next, consider a situation in which one might be let alone but still not have his or her privacy intact. If a student surreptitiously follows her professor one day and records each of her professor's movements on and off campus, she has in one sense "let her professor alone" (physically) but, arguably, has also intruded upon her professor's privacy. So, even though there may be something intuitively appealing about a view of privacy in terms of protecting against unwarranted physical access to people and to their physical possessions, we can question whether physical/accessibility privacy, based on the notions of physical nonintrusion and being let alone, in and of itself offers an adequate conception of privacy. We next examine a view of privacy in terms of *noninterference* in one's making choices.

6.2.2 Privacy as Noninterference Involving One's Choices: Decisional Privacy

Privacy is sometimes conceived of as freedom from interference in one's personal choices, plans, and decisions; many now refer to this view as *decisional privacy*.[14] Floridi (p. 52) defines decisional privacy as "freedom from procedural interference. . .achieved thanks to the exclusion of others from decisions (concerning, e.g., education, health care, career, work, marriage, faith). . .." In this view, one has decisional privacy to the extent that one enjoys freedom from interferences affecting these kinds of choices. Inness (1992, p. 140) endorses a view similar to this when she

[12]DeCew also notes that the first occurrence of the term "privacy" itself in U.S. jurisprudence can be found in the U.S. Supreme Court case *DeMay v. Roberts* (1881).

[13]Regan (p. 36) also points out that in *Katz v. United States* (which overruled *Olmstead v. United States*), the Court ruled that the Fourth Amendment cannot be translated into a general right to privacy.

[14]See, for example, Floridi (1999, 2006) and Tavani (2004a, 2007a).

defines privacy as protecting a realm of "intimate decisions," as well as "*decisions* concerning matters that draw their meaning and value from the agent's love, liking, and caring" (italics added).

Whereas the nonintrusion account defines privacy in terms of being let alone with respect to invasions involving physical space (including invasions affecting one's home, papers, effects, and so forth), the noninterference view focuses on the kinds of intrusions that can affect one's ability to make important decisions without external interference or coercion. This view of privacy is often traced to the 1965 U.S. Supreme Court case *Griswold v. Connecticut*, which ruled that a married person's right to get counseling about contraceptive techniques could not be denied by state laws. This (privacy) right was later extended to include unmarried couples as well, in *Eisenstadt v. Baird* (1972), where Brennan (1992, p. 453) referred to individuals having the right to be free from unwarranted government interference "into matters so fundamentally affecting a person as the *decision* whether to bear or beget a child" (italics added). The view of privacy as freedom from external interference affecting one's personal choices and decisions has since been appealed to in legal arguments in a series of controversial court cases, such as cases involving abortion and euthanasia. For example, a variation of the noninterference view of privacy was appealed to in the landmark Supreme Court decision on abortion, *Roe v. Wade* (1973), as well as in a state court's decision involving Karen Ann Quinlan's right to be removed from life-support systems and thus her "right to die."

As in the (physical) nonintrusion account of privacy, conceptual difficulties also arise in the noninterference view (i.e., with regard to one's freedom to make choices and decisions). For example, it is possible for someone to be interfered with in making a decision and yet for that person still to enjoy privacy. Imagine that I decide to walk across a busy intersection and someone interferes with me, as I am about to enter the street, to protect me from being hit by an oncoming vehicle that I had not seen. Has my privacy been violated in the process of this interference? It would seem not. Conversely, it is also possible for one not to be interfered with in decision-making, but for that person also not to have privacy. For example, someone who snoops through a trash receptacle of mine to obtain information about my sales receipts has not, strictly speaking, interfered with me, but that person has violated my privacy. So it would seem that the noninterference (or decisional) account of privacy, like that of the (physical) nonintrusion view, is not adequate, and thus neither view in itself captures the breadth of what is covered by the concept of privacy.

6.2.3 Privacy as Nonintrusion/Noninterference Involving One's Thoughts and One's Personal Identity: Psychological/Mental Privacy

What, exactly, is meant by expressions such as "psychological privacy" and "mental privacy"? Regan refers to "psychological privacy" in connection with protecting one's intimate thoughts,[15] whereas Floridi uses the expression "mental privacy" to describe "freedom from psychological interference." According to Floridi (p. 52), an individual, S, has mental privacy when there is a "restriction on others' ability to access and

[15]Regan is particularly concerned with the way that polygraph testing can threaten psychological privacy.

manipulate S' mind." A variation of this view can also be found in Alfino (2001), who believes that the harm caused by a privacy intrusion lies not so much in whether the intruder has gained information about that person but, in the fact that the "very act of the intrusion" prevents the victim from thinking or concentrating on her "life and actions."[16] For example, Alfino asks us to consider the case of someone being stared at or followed, "even as a harmless prank or joke." In this case, he argues that the "violation to our privacy involves the way in which the offending conduct prevents us from concentrating or thinking clearly" (Alfino, pp. 6–7). Rosen (2000, p. 8) seems to endorse a similar conception of privacy when he notes that people need "sanctuaries from the gaze of the crowd" to flourish. And Benn (1971, p. 26) also supports a position similar to this when he argues that "(r)espect for one as a person. . .implies respect for him as one engaged in a kind of self-creative enterprise, which could be disrupted, distorted, or frustrated even by so limited an intrusion as watching."

Some theorists who embrace the psychological/mental view also see privacy as essential for protecting the integrity of one's personality. For example, Freund (1971) defines privacy in terms of an extension of personality or personhood. Warren and Brandeis (1890) seem to suggest a position along these lines when they state that the principle of respect for privacy is that of "inviolate personality."[17] It would also seem that Hofstadter and Horowitz (1964) view privacy in a similar way when they describe it in terms of a right to protect "against unwarranted appropriation or exploitation of one's personality."[18] A related view of privacy, which is concerned with protecting the "core self" and individuality, can be found in Westin (1967) and in McCloskey (1985, pp. 349–350), who states that "the private" relates to the person's "unique self," which is a "thinking, feeling, and self-conscious being, typically aware of its self-identity." For each of these thinkers, one's having privacy is essential either (a) for a person's awareness to think and be conscious of one's self or (b) for the protection of the integrity of one's personality.[19]

Like the physical and the decisional accounts of privacy, the psychological/mental view of privacy has also been criticized. In particular, critics have pointed out flaws in the personality-based aspect of this view. For example, A.D. Moore (2008) notes that although personality-based conceptions of privacy tend to indicate why privacy is valued and why it is considered important, they do not actually define what privacy is.

[16]Alfino believes that our conception of privacy in our everyday thinking is "misplaced" whenever we confuse "practical needs to prevent access to personal information" with "fundamental moral needs to protect our ability to think."

[17]DeCew (pp. 16–17) believes that Warren and Brandeis's discussion of "inviolate personality" indicates their concern for values that today are often described in terms of "personhood" and "self-identity."

[18]Aspects of a personality theory of privacy can also be found in Reiman (1976). Arguably, Floridi's ontological theory of informational privacy (described in Section 6.3) can also be viewed as a personality theory of privacy (Tavani, 2007c).

[19]Some theorists argue that having privacy not only enables an individual to protect his individuality, self-identity, personality, and thoughts, but also allows him to *express* his thoughts. For example, Schoeman (1992) claims that privacy enhances one's control over "self-expression." DeCew (p. 77) refers to this view as "expressive privacy," pointing out that privacy "protects a realm for expressing one's self-identity or personhood."

6.2.4 Privacy as Having Control Over/Limiting Access to One's Personal Information: Informational Privacy

As noted above, our main focus in this chapter is on issues affecting *informational* privacy. Parent (1983b), who rejects accounts of privacy based on nonintrusion and noninterference, has suggested that the only legitimate use of the term "privacy" is to refer to "informational privacy." But what, exactly, is informational privacy? Floridi (p. 52) defines it as "freedom from epistemic interference" that is achieved when there is a restriction on "facts" about someone that are "unknown." Which kinds of personal information are affected in this privacy category? According to DeCew (p. 75), such information can include data about "one's daily activities, personal lifestyle, finances, medical history, and academic achievement." As such, informational privacy concerns can affect personal data that is both stored in and communicated between electronic databases (database privacy) and personal information communicated between parties using e-mail, telephony, and wireless communication devices. Some authors have further differentiated a category of "communications privacy"—see, for example, Johnson and Nissenbaum (1995) and Regan (1995)—to separate the latter kinds of informational privacy concerns from the former. However, in this chapter, we will treat both kinds of privacy concerns as instances of one generic category of informational privacy.[20]

The practice of collecting and using personal information is hardly new. For example, Regan (p. 69) points out that in the eleventh century, William the Conqueror compiled a "Domesday Book" that included data about each of his subjects, which he used to "plan taxation and other state policies." Others note that the practice of gathering personal information for census records by governments goes back at least as far as the Roman era. So, we might ask why the current use of technology by governments and other organizations to collect personal information should be viewed as something controversial. Moor (1997, p. 27) points out that information has become "greased" as it can easily slide to many "ports of call." He also argues that because of the "explosive" generation and use of computer technology, the concept of privacy has now become "informationally enriched" (Moor, 2004, p. 43). However, one still might ask whether current practices involving the collection and use of personal information by governmental and nongovernmental organizations, as well as by individuals and organizations in the commercial sector, have merely exacerbated existing privacy concerns, or whether they have also introduced new kinds of privacy worries that are qualitatively different from traditional concerns. Arguments for both positions have been advanced. We briefly examine some claims that would seem to support each side in this debate.

I have argued elsewhere that the effect that computer/information technology has had for personal privacy can be analyzed in terms of four factors: (1) the amount of personal information that can be collected, (2) the speed at which personal information

[20]Regan (pp. 7–8) points out that concerns about "informational privacy" first appeared in the U.S. Congressional agenda in 1965, in response to a proposal for establishing a National Data Center. Worries about Big Brother and concerns about the emergence of a "dossier society" led to the establishment of the Privacy Protection Study Commission, which eventually resulted in the passing of the Privacy Act of 1974.

can be exchanged, (3) the duration of time that the information can be retained, and (4) the kind of information that can be acquired.[21] Whereas (1)–(3) are examples of privacy concerns that differ quantitatively from traditional worries, (4) introduces a qualitatively different kind of information about persons to merit our concern.

First, consider that the *amount* of personal information that could be collected in the precomputer era was determined by practical considerations, such as the physical space required for storing the data and the time and difficulty involved in collecting the data. Today however, digitized information is stored electronically in computer databases, which takes up very little storage space and can be collected with relative ease. This is further exacerbated by the *speed* at which information can be exchanged between databases. In the precomputer era, records had to be physically transported between filing destinations; the time it took to move them depended upon the transportation systems, for example, motor vehicles, trains, airplanes, and so forth, that carried the records. Now, of course, records can be transferred between electronic databases in milliseconds through high-speed cable lines. With respect to concerns about the *duration* of information, that is, how long information can be kept, consider that before the information era, information was manually recorded and stored in file cabinets and then in large physical repositories. For practical reasons, that information could not be retained indefinitely.

Informational privacy concerns affected by the amount of personal data collected, the speed at which the data is transferred, and the duration of time that data endures can all be viewed as issues that differ *quantitatively* from earlier privacy concerns, because of the degree to which those concerns have been exacerbated. However, the *kind* of personal information that can now be collected, processed, and manipulated via computers and information technology represents a *qualitative* difference regarding concerns about the collection and flow of one's personal information. Consider that every time one engages in an electronic transaction, such as making a purchase with a credit card or withdrawing money from an ATM, "transactional information"[22] about that person is collected and stored in several computer databases. This kind of information, made possible by electronic technology, can also be easily transferred electronically across commercial networks to agencies that request it. Personal information, retrieved from transactional information that is stored in computer databases, has been used to construct electronic dossiers, containing detailed information about an individual's commercial transactions, including purchases made and places traveled—information that can reveal patterns in a person's preferences and habits.[23]

[21]See Tavani (2004a, 2007a). Along somewhat similar lines, Floridi (2005) has argued that informational privacy concerns have tended to be examined in terms of a scheme that he calls the 2P2Q Hypothesis, which focuses on issues affecting *Processing*, *Pace*, *Quantity*, and *Quality*. In this scheme, privacy issues are examined from the perspectives of four characteristics: *processing* capacities, *pace* (or speed), *quantity* of data, and *quality* of data.

[22]See Burnham (1983) for an account of why "transactional information" can be viewed as a qualitatively different kind of information affecting concerns about personal privacy.

[23]For a more detailed analysis of the effect that these four factors have had for informational privacy issues, see Tavani (2007a).

Concerns involving specific technologies that threaten informational privacy, such as, cookies, data mining, and RFID (Radio Frequency Identification), are examined in Section 6.4. First, however, we look at some philosophical theories of informational privacy that aim at helping us to better understand which conditions need to be taken into consideration for adequately protecting an individual's personal information.

6.3 THEORIES OF INFORMATIONAL PRIVACY

Since the latter half of the twentieth century, several informational privacy theories have been advanced. According to Floridi (2005, p.194), these theories tend to fall into two broad categories: the "reductionist" and the "ownership-based" accounts. He points out that in the reductionist scheme, informational privacy is valued because it protects against certain kinds of undesirable consequences that may result from a breach of privacy. Here, privacy is conceived of as a "utility" in that it can help to preserve human dignity. Floridi notes that those who embrace ownership-based theories, on the contrary, tend to believe that privacy needs to be respected because of each person's rights to bodily security and property[24]—that is, because in this view, a person "owns" his or her information. Although Floridi's analysis via these two broad categories may be appropriate for privacy theories in general, it is not clear that they accurately capture the debate about *informational* privacy in the computer ethics literature. Alternatively, I have argued that most analyses of issues affecting informational privacy have invoked variations of the "restricted access" and the "control" theories.[25] We briefly examine both theories, as well as some related informational privacy theories.

6.3.1 The Restricted Access Theory

According to the *restricted access* theory, one has informational privacy when she is able to limit or restrict others from access to information about herself.[26] In this framework, "zones" of privacy (i.e., specific contexts) need to be established to limit or restrict others from access to one's personal information.[27] Variations of this theory have been defended by Bok (1983), Gavison (1980), Allen (1988), and others.[28]

[24]In his analysis of the ownership-based theory, Floridi seems to have Thomson (1975) in mind.

[25]See Tavani (2004a, 2007a).

[26]One might point out that the "accessibility privacy" view examined in Section 6.2.1 also focuses on concerns involving access. However, we should note that whereas accessibility privacy describes issues affecting physical *access to persons* (and their possessions), the restricted access theory of informational privacy is concerned with issues affecting access to *information about persons*.

[27]In writing the majority opinion in *Griswold v. Connecticut*, Douglas (1965) used the expression "zone of privacy" to refer to the areas protected by the right to privacy "emanating" from the Constitution and its Amendments in the "penumbral" right to privacy that he defended.

[28]See, for example, Parent (1983b, p. 269), who seems to endorse a variation of the restricted access theory when he argues that a person's privacy "is diminished exactly to the degree that others possess . . . knowledge about him."

According to Bok (p. 10), "privacy is the condition of being protected from unwanted access by others, "including access to one's personal information." Gavison (p. 428) describes privacy as a "limitation of others' access" to information about individuals. And Allen (p. 3) defines privacy as "a degree of inaccessibility of persons, their mental states, and information about them to the senses and surveillance of others."

Arguably, one of the insights of the restricted access theory is in recognizing the importance of zones and contexts that need to be established to achieve informational privacy. Also, there would seem to be something intuitively appealing in linking one's informational privacy with limited access to information about that individual. However, this theory can be criticized on at least two distinct grounds. First, some critics argue that the restricted access view fails to draw an adequate distinction between "private" and "public" contexts or zones in which access to personal information is restricted. As a result, Elgesem (2004, p. 427) believes that on the restricted access account "we will have to admit that we always have some degree of privacy, since there will always be billions of people who have physically restricted access to us." Second, the restricted access theory can be interpreted as conflating privacy and secrecy, because it suggests that the more one's personal information can be withheld (or kept secret) from others, the more privacy one has.[29] This confusion would seem to be apparent in the account offered by Gavison (p. 428), who argues that an individual, X, enjoys "perfect privacy" when "no one has information about X."

An additional problem for the restricted access theory is that it tends to ignore, or at least seriously underestimate, the role of control or choice that is also required for one to enjoy privacy as it pertains to one's personal information. For example, this theory does not explicitly acknowledge the point that someone who has privacy can elect to grant others access to information about herself, as well as to restrict or limit their access. In other words, one can *control* the flow of her personal information, rather than simply limiting or restricting it.

6.3.2 The Control Theory

According to the *control theory*, one's having privacy is directly linked to one's having control over information about oneself. This view of privacy has been articulated and defended by Fried (1990), Rachels (1975), and many others.[30] For Fried (p. 54), privacy "is not simply an absence of information about us in the minds of others, rather it is the *control* over information we have about ourselves" (italics added). Rachels (p. 297) notes that there is a connection between "our ability to *control* who has access to information about us and our ability to create and maintain different sorts of relationships" (italics added).

---

[29]See my critique of this view in Tavani (2000, 2007b), where I refer to the restricted access account as the "limitation theory of privacy."

[30]See, for example, Westin (1967, p. 7) who seems to endorse the control theory when he describes privacy as the "claim of individuals . . . to determine for themselves when, how, and to what extent information about them is communicated to others," and Miller (1971, p. 25) who describes privacy as "the individual's ability to control the circulation of information relating to him." See also A.D. Moore (2003), who embraces a variation of the control theory.

One of the control theory's principal insights is in recognizing the role that individual choice plays in privacy theory. Consider, for example, the choice that an individual who has privacy can exercise in electing either to grant or to restrict others from access to information about him. So, this account of informational privacy would seem to have an advantage over the restricted access theory. However, the control theory is also flawed; it can be criticized with respect to two points on which it is not altogether clear. First, this theory does not clearly specify the *degree* of control (i.e., how much control) that one can expect to have over his or her personal information. For example, the control theory seems to imply that one must have total or absolute control over one's personal information to have privacy. But this would seem impossible to achieve on practical grounds, because we are often required to disclose certain kinds of information about ourselves in ordinary day-to-day transactions, especially in commercial transactions.

Second, the control theory is not clear with regard to the *kind* of personal information over which one can expect to have control. It would not seem reasonable that one could expect to have control over *every* kind of personal information affecting oneself.[31] For example, suppose I happen to see you at a concert that we are both attending. Here, you have no control over the fact that I have gained information about your attending this public event (even if, for some reason, you do not want me to know this piece of information about you). The fact that I now have this particular information would hardly seem to be a violation of your privacy, as would seem to be implied in the standard interpretation of the control theory of privacy. So, this theory needs to provide a clearer account of both the degree of control and the kind of personal information over which one can expect to have control to enjoy privacy.[32]

Another difficulty for the control theory is that it seems to imply that one could, in principle, disclose every piece of personal information about oneself and yet still claim to have privacy. In this sense, the control theory tends to confuse privacy with autonomy, where someone has autonomously decided to abdicate all informational privacy interests by disclosing all private facts about herself from which she had an interest or right to exclude other people. However, this would seem to run counter to our intuitions about privacy.

We have seen that neither the control nor the restricted access theories provide an adequate account of informational privacy. Yet each theory has something important to say with respect to the question of why privacy protection regarding personal information is important. Can these two theories be reconciled or synthesized in a way that incorporates the respective insights of each into one comprehensive privacy theory? One attempt at doing this has been made by Moor (1997), who has put forth a

---

[31]Perhaps the kind of personal information over which one can expect to have control is limited to "non-public personal information" (or what some now refer to as NPI), which includes information about sensitive and confidential data such as financial and medical records. This kind of information can be contrasted with personal information that is public in nature, or "public personal information" (PPI), such as information about where a person works, lives, shops, dines, and so forth. For a more detailed discussion of some of the key differences between PPI and NPI, see Tavani (2004a, 2007a).

[32]For an expanded discussion of these two challenges for the control theory of privacy, see Tavani (2007b).

framework that we refer to as the Restricted Access/Limited Control (RALC) theory of privacy.[33]

6.3.3 The Restricted Access/Limited Control (RALC) Theory

RALC distinguishes between the *concept of privacy*, which it defines in terms of restricted access, and the *management of privacy*, which is achieved via a system of limited controls for individuals.[34] So, like the restricted access theory, RALC stresses the importance of setting up zones that enable individuals to limit or restrict others from accessing their personal information. Unlike the restricted access theory, and like the control theory, RALC recognizes the important role that individual control plays in privacy theory. Unlike the control theory, however, RALC does not build the concept of control into the definition of privacy; nor does it require that individuals have total or absolute control over their personal information to have privacy. Instead, only limited controls are needed to manage one's privacy.

In the RALC framework, one has privacy "in a *situation* with regard to others" if, in that situation, one "is protected from intrusion, interference, and information access by others" (Moor, 1997, p. 30; italics added). Moor provides several examples of "situations." For instance, he notes that they can include the "storage and access of information" (e.g., information stored or manipulated in a computer), as well as an "activity in a location" (Moor, 1990, p. 76). He also notes, however, that not all situations deserve normative privacy protection. In making this point, Moor draws an important distinction between two kinds of situations, namely, "naturally private" and "normatively private" situations. In naturally private situations, one is "naturally" protected or shielded from observation, intrusion, and access by others. In this type of situation, one's privacy can be lost, but not necessarily violated because no norms—legal or ethical—have been established to protect one's privacy. Here, individuals are shielded or blocked off from others by natural means. For example, if someone is jogging around a track field in a rural area at 3:00 a.m., he might be free from observance by others. If, however, the jogger happens to be seen a few minutes later by someone deciding to take an early morning walk, the jogger could be said to have lost privacy in a descriptive sense. But the jogger's privacy has not been violated in that (naturally private) situation. Next, consider some normatively private situations, which can include "activities" such as voting, "information" such as medical records, and "locations" such as a person's house (where outsiders are expected to knock and get permission to enter) (Tavani and Moor, 2001). In each of these situations, people are normatively protected by privacy policies and laws. In these situations, one's privacy cannot only be lost but also violated.

[33]Initially, Moor (1997) called his theory the "control/restricted access theory." With Moor's permission, I refer to it as the Restricted Access/Limited Control (RALC) framework (based on a later version of that theory articulated in Tavani and Moor [2001]). For a more detailed analysis of RALC, see Tavani (2007b).

[34]Additionally, RALC provides a *justification* for privacy protection. However, we will not examine that feature of RALC in this section.

We noted that although RALC defines privacy in terms of protection from observation, intrusion, and information gathering by others (through situations or zones that are established to restrict access), the notion of "limited control" also plays an important role, namely, in the management of one's privacy. In the RALC framework, an individual enjoys some degree of control with respect to considerations involving *choice*, *consent*, and *correction*. For example, people need some control in *choosing* which situations are and which are not acceptable to them with respect to the level of access granted to others. In managing their privacy, people can also use the *consent process*. For example, they can waive their right to restrict others from access to certain kinds of information about them. The *correction process* also plays an important role in the management of privacy because it enables individuals to access their information with an ability to amend it if necessary.[35] So, in the RALC scheme, individuals can manage their privacy via these kinds of limited controls and thus do not need to have absolute or total control over all of their information.

One way in which limited controls work in the RALC framework can be illustrated in an example involving personal medical information. Moor notes that although individuals do not have complete control over who has access to their information within a medical setting, normative zones of protection have been established to restrict most others from accessing that information. In situations involving one's medical information, doctors, nurses, insurance providers, administrative assistants, and financial personnel may each have legitimate access to various pieces of it (Tavani and Moor, 2001). Others, however, have no legitimate right of access to that information. Thus, normative privacy policies, such as those implemented in medical settings, provide individuals with some control regarding who has access to information about themselves in particular contexts.

In drawing a distinction between naturally private and normatively private situations, we saw that RALC is able to distinguish between a mere loss of privacy (in a natural setting) and a violation of privacy (in a situation that is normatively protected). In this sense, RALC would seem to be an improvement over both the restricted access and control theories. However, critics might challenge certain aspects of the RALC framework. Among the potential objections they might raise, two are worth identifying and briefly describing: (1) the notion of a "situation" is too vague, and (2) the distinction between natural and normative situations can be drawn arbitrarily. With regard to (1), Moor intentionally leaves the concept "situation" broad or "indeterminate" so that it can cover a wide range of contexts. For example, a situation can be as specific as "one's attending a private viewing of a particular work of art in a friend's home," or it can be as broad or general as "individuals attending professional sporting events in large public stadiums." There can even be "situations within situations," such as a situation involving two people having a private conversation within the context of a public restaurant (another situation). In any event, the concept of a situation needs to be sufficiently broad or indeterminate for RALC or any adequate theory of privacy to capture the wide range of human activities and relationships that

[35]For example, consumers need to be able to (a) access information about their credit history and credit scores and (b) challenge and correct any erroneous information contained in those records and documents.

are candidates for being private. Other terms or phrases such as "circumstances" or "states of affairs" or "events" could be used. But these will be equally vague as generic terms and will need to be filled out in detail just as the notion of a situation needs to be described in terms of specific features of a given context.

Regarding (2), one might ask how the RALC theory can tell us which kinds of situations deserve normative protection and which do not – that is, whether there are clearly identifiable standards for determining if a particular *kind* of personal information applicable in a given context deserves normative protection. RALC gives us a framework for understanding what privacy is, but to decide whether a situation should be private requires the application of some ethical analysis. Often this is done by comparing the consequences of requiring or not requiring some situations to be private. Requiring banks to keep their patrons' financial records private from the public has much better consequences than not requiring it. Moor (1999) argues that we justify ethical policies in terms of the goods and evils caused or avoided along with considerations of justice. People of good faith can have disagreements about what should be private and what not. But note that often the disagreements hinge on what they think the expected consequences would be. Also note that the same person might advocate different policies for similar situations that vary by a key ethical difference. For example, Moor (1997) asks us to consider the following question: Should information about college faculty salaries be a normatively private situation? He points out that we can have good reasons for not declaring information about the salaries of professors who teach at large public colleges, especially where the salaries are paid out of taxpayer dollars, to be a normatively private situation; but he also notes that we can have good reasons for declaring that same kind of information as normatively private in other contexts, such as in the case of small private colleges. In this sense, he shows why there is nothing inherent in information pertaining to faculty salaries per se that tells us whether this kind information should be protected as a normatively private situation. In the RALC framework, it will ultimately depend on whether we have agreed that a normatively private situation or zone needs to be established to protect information about individuals, and not on the nature of the particular kind of information itself.

Thus far, we have examined some ways that RALC improves upon the control and restricted access theories of informational privacy by incorporating key elements of both theories. We have also seen how RALC can respond to two kinds of potential objections from critics. However, much more needs to be said about RALC and the role it plays in protecting privacy. We revisit the RALC theory in Section 6.5.3, where we examine some proposals for framing comprehensive policies that protect informational privacy.

6.3.4 Three "Benchmark Theories" of Informational Privacy

Some benchmark theories—that is, outlines or sketches of privacy theories, as opposed to full-fledged theories—have also recently been advanced to address controversies affecting informational privacy. We briefly examine three examples of such theories: (1) "privacy as contextual integrity," articulated by Nissenbaum (2004); (2) an "ontological interpretation of informational privacy" proposed by Floridi (2005); and (3) a framework called "categorial privacy," introduced by Vedder (2004).

6.3.4.1 Privacy as Contextual Integrity Nissenbaum (2004) proposes an "alternative benchmark theory of contextual integrity" that links adequate privacy protection to "norms of specific contexts." Her framework requires that the processes used in gathering and disseminating information are "appropriate to a particular context" and that they comply with the "governing norms of distribution" for that context (Nissenbaum, p. 119). This theory expands upon her earlier work on the problem of "privacy in public" (Nissenbaum, 1997, 1998), where she notes that normative protection does not generally extend to personal information gathered about us in the public sphere. She points out that privacy norms (i.e., explicit privacy laws and informal privacy policies) protect personal information considered to be intimate and sensitive – for example, information such as medical records and financial records – but typically do not extend to personal information about us in public places (such as where we shop, dine, recreate). Nissenbaum (1997) questions an assumption that, she believes, informs many normative privacy theories and policies, namely, "there is a realm of public information in which no privacy norms apply." She notes that in the past, information about what one did in public might not have required the kind of protection it now warrants. However, current technologies make it possible to gather, combine, aggregate, and disseminate that kind of information in ways that were not previously possible. As a result, Nissenbaum believes that public information about persons, which we can no longer assume to be innocuous, needs normative protection.

Many of the issues underlying the problem of privacy in public are further illustrated in her theory of contextual integrity, which offers a procedure for resolving as well as identifying these issues. At the core of Nissenbaum's framework are two principles: (a) the activities people engage in take place in a "plurality of realms" (i.e., spheres or contexts), and (b) each realm has distinct set of norms that govern its aspects. These norms both shape and limit or restrict our roles, behavior, and expectations by governing the flow of personal information in a given context.[36] Nissenbaum (2004) distinguishes between two types of informational norms: norms of appropriateness and norms of distribution. Norms of appropriateness determine whether a given type of personal information is either appropriate or inappropriate to divulge within a particular context – that is, these norms "circumscribe the type or nature of information about various individuals that, within a given context, is allowable, expected, or even demanded to be revealed" (Nissenbaum, p. 138). Norms of distribution, on the contrary, restrict the flow of information within and across contexts. When either norm has been "breached," a violation of privacy occurs; conversely, the contextual integrity of the flow of personal information is maintained when both kinds of norms are "respected" (Nissenbaum, p. 125).

Nissenbaum believes that her theory of contextual integrity contrasts with, and improves upon, alternative privacy theories in two important respects. First, personal

[36]The contextual integrity model proceeds on the assumption that "no areas of life are not governed by norms of information flow" (Nissenbaum, p. 137).

information that is revealed or disclosed in a particular context is always "tagged" with that context and thus never "up for grabs." Nissenbaum argues that this is not the case in alternative accounts of personal information gathered in public places, because those theories lack the appropriate "mechanisms" to prevent the "anything goes" approach to this kind of personal information.[37] Second, the "scope of informational norms" is always "internal to a given context." Nissenbaum (p. 125) notes that in this sense, the norms are "relative" or "non-universal." She also notes that in her benchmark theory of contextual integrity, "context-relative qualifications" can be "built right into the informational norms" of any given context, unlike other normative theories of privacy where these qualifications tend to be treated as "exceptions or tradeoffs" (Nissenbaum, p. 138).[38]

Like the RALC framework (examined in the preceding section), Nissenbaum's theory shows why it is always the context in which information flows, not the nature of the information itself, that determines whether normative protection is needed. As we saw in the example of information pertaining to faculty salaries in our analysis of RALC, there was nothing inherent in the information about the professors' salaries per se that enabled us to determine whether it was appropriate to protect that information; rather, it was the particular context, for example, the norms governing the distribution of information in the context of a large public university versus a small private college, that determined the appropriateness.

Nissenbaum's model can be applied to a wide range of contemporary technologies to determine whether they breach the informational privacy norms that govern specific contexts. For example, I have argued elsewhere (Tavani, 2007b) that Nissenbaum's account of the problem of privacy in public, in conjunction with her framework of contextual integrity, can help us to better understand the kinds of privacy threats posed by data mining technology (described in Section 6.4.2). For an interesting discussion of how Nissenbaum's theory of contextual integrity can be applied to privacy issues involving "vehicle-safety communications technologies," see Zimmer (2005).

6.3.4.2 An "Ontological Interpretation" of Informational Privacy

Floridi (2005) has recently articulated a view that he calls "the ontological interpretation of informational privacy." He grounds this model in his overall framework of Information Ethics (IE), a "macroethical" theory for analyzing particular "microethical" issues in computer ethics (CE) such as privacy (Floridi, 1999).[39] Floridi argues that because information has moral worth, the information entities that comprise the "infosphere" deserve moral consideration (and thus, by

---

[37]Nissenbaum (1998) points out that when it comes to questions about how to protect personal information in public contexts, or in what she calls "spheres other than the intimate," most normative accounts of privacy have a theoretical "blind spot."

[38]In this sense, she believes that her theory allows for the possibility of "context-relative variation" as an "integral part of contextual integrity."

[39]His IE framework includes the concept of the "infosphere," comprising the totality of information entities/ objects. Floridi's infosphere expands upon the notion of the "ecosphere" (or "biosphere") by including inanimate entities (i.e., information entities) as well as biological life-forms.

extension, privacy protection). Unfortunately, it is not possible in the limited space of this section to describe Floridi's IE framework in the detail it deserves.[40] For our purposes, however, we can critically examine Floridi's account of privacy without necessarily having to embrace (or reject) his overall IE framework.[41]

Floridi draws a distinction between "pre-digital" and digital ICTs (Information and Communication Technologies), noting that examples of pre-digital (or older) ICTs include tools such as telescopes and telephones. He believes that pre-digital ICTs have tended to reduce informational privacy in the infosphere because they reduce "ontological friction" (Floridi, 2006, p. 110). Ontological friction refers to "the forces that oppose the information flow within (a region of) the infosphere." For Floridi, "*informational privacy is a function of the ontological friction of the infosphere*" (Floridi, 2005, p. 187, italics in original). Because digital ICTs are "ontologizing devices" that can "re-ontologize the infosphere" by increasing ontologizing friction in it, Floridi believes that digital ICTs can increase informational privacy. He also believes that because digital ICTs can "change the nature of the infosphere," they change the "very nature of privacy" as well as our appreciation of it. Thus, Floridi claims that informational privacy "requires an equally radical reinterpretation," – that is, one that takes into account the "essentially informational nature of human beings and of their operations as informational social agents."

Floridi (1999, p. 53) argues that privacy is "nothing less than the defence of the personal integrity of a packet of information" (i.e., the individual) and that an invasion of an individual's informational privacy is a "disruption of the information environments that it constitutes." He goes on to assert that a violation of privacy "is not a violation of ownership, of personal rights," but instead is "a violation of the nature of information itself." So in Floridi's account, a violation of privacy is not tied to an agent's personal rights (e.g., rights involving control and ownership of information); rather, it is linked to conditions affecting the information environment that the agent constitutes. According to Floridi, an agent "is her or his information." For example, he points out that the sense of "my" in "my information"

> is not the same as 'my' in 'my car' but rather the same as 'my' as in 'my body' or 'my feelings'; it expresses a sense of constitutive belonging, not of external ownership, a sense in which my body, my feelings, and my information are part of me but are not my (legal) possessions (Floridi, 2005, p. 195).

Because each person is "constituted by his or her information," Floridi concludes that a breach of one's informational privacy is "a form of aggression towards one's *personal identity*" (Floridi, 2005, p. 194, italics in original). He also argues that there is no difference between one's "informational sphere" and his or her personal identity, because one is one's information. In equating privacy protection with the protection of one's personal identity, however, Floridi's theory seems to conflate informational privacy with another

---

[40]For an excellent overview of Floridi's IE theory, see the account by Bynum in Chapter 2 of this volume.

[41]I have argued elsewhere (Tavani, 2007c) that Floridi's privacy theory can be analyzed independently of his ontological framework of IE.

kind of privacy, namely, psychological privacy, or what he calls "mental privacy." Furthermore, Floridi tends to undermine a fundamental distinction he makes elsewhere (Floridi, 1999, p. 52) in differentiating informational privacy from mental privacy. So in Floridi's scheme, it would seem that informational privacy and mental privacy reduce to the same thing. If this is the case, Floridi's theory also could be construed to be a "personality theory of privacy" as much as it is an "ontological theory of privacy." In this sense, his privacy theory can be analyzed independently of its connection with his overall IE (macroethical) framework and the ontological concepts that comprise it.

Floridi's ontological theory can be interpreted as providing a fresh and interesting perspective on informational privacy, despite potential objections that might arise in response to concerns about the roles that concepts such as the infosphere and "ontological friction" play in it. For one thing, his theory presents a clear and straightforward case for why a violation of one's privacy need not be tied to one's loss of control/ownership of personal information. However, his theory is less clear with respect to two important points: (1) when a privacy violation (as opposed to a mere loss of privacy) occurs in the infosphere and (2) which kinds of information entities deserve privacy protection. With regard to (1), we saw that Floridi suggests that a "disruption" of the information environment results in a violation of privacy, but he does not tell us what counts as an actual violation of privacy (in a normative sense) versus a loss of privacy (in a descriptive sense) in the infosphere. Perhaps if Floridi were to incorporate into his privacy theory the notion of a "situation" (a la RALC) or a "context" (in Nissenbaum's sense of the term), he could respond to this challenge. Regarding (2), we should note that Floridi describes privacy as a "fundamental right of information entities," but he does not say anything about which information entities enjoy this right and which do not. Perhaps an answer to this challenge can be found in Floridi's account of notions such as "agency" and "informational social agents in the infosphere." However, an analysis of these concepts is beyond the scope of this section.

Perhaps Floridi will respond to the kinds of challenges described in the preceding paragraphs in future work on privacy theory within the context of his overall IE framework.[42] We should note, however, that Floridi may intend for his (ontological) privacy theory to work in conjunction with, and thus supplement, other informational privacy theories rather than replace those theories altogether.[43]

6.3.4.3 Categorial Privacy Vedder (2004) argues that we need a new category of privacy, distinct from both "individual privacy" and "collective privacy", which he calls "categorial privacy." He believes that this new privacy category is needed because of the way individuals can easily become associated with and linked to newly created groups that are made possible by technologies such as KDD (Knowledge Discovery in

---

[42]I have argued elsewhere (Tavani, 2007c) that Floridi's theory would benefit from addressing these challenges.

[43]Floridi (2006) suggests this possibility when he says that a "mature theory" will coordinate with *other* privacy theories. This interpretation is compatible with Floridi's overall IE theory, which he sees as working in conjunction with other macroethical theories rather than replacing them.

Databases) and data mining (described in Section 6.4.2). Vedder argues that individuals who eventually become identified with some of these newly created groups lack adequate privacy protection because current privacy laws protect only "personally identifiable information" that applies to individuals themselves. He points out that these laws do not protect individuals once their personally identifiable information has become aggregated, as it is, for example in the data mining process. Thus, Vedder believes that a new normative category of privacy protection must be established to protect individuals from misuses of their personal data in aggregated form, especially because such data can be used to make important decisions about individuals.

Of course, we can ask whether a new category of privacy protection is necessary, even if Vedder's analysis of current privacy laws as inadequate is correct. We examine Vedder's framework of categorial privacy in detail in Section 6.4.2, in our analysis of privacy threats posed by data mining technology.

In concluding this section, we should note that other kinds of benchmark theories of informational privacy also have been advanced in the information ethics literature. Unfortunately, however, considerations of space will not allow us to examine those theories in the detail that some might warrant. We next examine four categories of informational privacy in terms of some challenges that specific technologies pose for them.

6.4 SOME TECHNOLOGY-BASED CONTROVERSIES AFFECTING FOUR CATEGORIES OF INFORMATIONAL PRIVACY: CONSUMER, MEDICAL, EMPLOYEE, AND LOCATION PRIVACY

Informational privacy concerns can affect many aspects of an individual's life – from commerce to healthcare to work to recreation. For example, we speak of consumer privacy, medical/healthcare privacy, employee/workplace privacy, and (more recently) "location privacy." We briefly examine some examples of each vis-à-vis threats posed by specific information/computer technologies.[44]

6.4.1 Consumer Privacy and the Threat from Cookies Technology

What expectations should consumers have about retaining their privacy when engaging in consumer transactions online, as millions of people currently do in e-commerce activities? For one thing, many online consumers may not realize that when interacting with Web sites they are subject to privacy-related threats posed by cookies technology. Cookies are text files that Web sites send to and retrieve from the computer systems of Web users.[45] Cookies technology enables Web site owners to collect information about users' online browsing preferences when users interact with their sites. This technology is controversial, in part, because of the novel way that

[44]Many examples used in Sections 6.4.1–6.4.4 are taken from my discussion of privacy in Tavani (2007a).

[45]Cookies are not only used by commercial Web sites; users who interact with Web sites solely for recreational (and other noncommercial) purposes are also subject to cookies.

information about users is collected, stored on the hard drives of the user's computer system, and later retrieved from the user's system and resubmitted to a Web site the next time the user accesses that site.

Both critics and defenders of cookies technology have advanced arguments. Critics tend to worry about the kinds of consumer profiles that can ultimately be created via cookies.[46] They also point out that the exchange of data between the user and Web site typically occurs without the user's knowledge and consent and that unlike alternative data-gathering mechanisms, cookies technology actually stores the data it collects about a particular user on that user's computer system. Consequently, many critics believe that practices involving the monitoring and recording of an individual's activities while visiting a Web site, and the subsequent downloading of that information onto a user's computer (without informing the user), violate that individual's privacy. Among those who defend the use of cookies are many owners and operators of online businesses. Some argue that cookies technology actually performs a service for repeat users of a Web site by customizing the user's means of information retrieval. They point out, for example, that cookies provide users with a list of preferences in future visits to that Web site.

It is worth noting that some technology-based solutions to problems generated by cookies have been proposed and implemented. These include privacy-enhancing technologies (or PETs) that assist Internet users who wish to identify, block, or disable cookies on a selective basis. PETs such as PGPcookie.cutter, for example, have been particularly helpful to users with older Web browsers; many current browsers provide users with a built-in feature to disable cookies, so that users can either opt in or opt out of cookies. Of course, this feature is useful only to users who (a) are aware of cookies technology and (b) know how to enable/disable that technology on their Web browsers. Even with this technology, however, users cannot necessarily opt out of cookies for every site they visit. For example, some Web sites will not allow users access unless they accept cookies. So, online consumers may be forced to accept cookies to interact with certain e-commerce sites. We revisit some issues affecting cookies and PETs in Section 6.5.3 in our discussion of informational privacy policies.

6.4.2 Medical/Healthcare Privacy and the Threat from Data Mining Technology

Are concerns about medical and healthcare privacy, especially the protection of a patient's medical records, exacerbated by information technology? If so, are new policies and laws needed to address these concerns? Privacy advocates have raised concerns about the ways in which a person's medical and healthcare data can be manipulated and abused with the aid of information/computing technologies. These concerns have helped to influence the enactment of some medical/healthcare-specific privacy laws such as the Health Insurance Portability and Accountability Act

[46]A controversy arose when DoubleClick.com, an online ad marketing firm that used cookies technology, tried to acquire Abacus, a database company. If DoubleClick had succeeded, it could have merged records in the "off-line" database of Abacus' customers with information about the consumer profiles it had acquired online via cookies. For an analysis of this case, see Tavani (2004a, 2007a).

(HIPAA), which includes standards for protecting the privacy of "individually identifiable health information" from "inappropriate use and disclosure." However, critics, such as Baumer et al. (2006), note that it is not clear whether HIPAA adequately addresses concerns affecting "nonconsensual secondary uses" of personal medical information, which has been exacerbated by technologies such as data mining.

How, exactly, does data mining contribute to privacy concerns affecting the (non-consensual) secondary uses of personal information? In answering this question, we first briefly describe what data mining is and then show why it can be controversial with respect to the manipulation of personal data. Data mining is a technique that searches for patterns in data that can reveal or suggest information about persons that would otherwise not be easily attainable. Unlike the kind of personal data that resides in explicit records in databases, information acquired about persons via data mining is often derived from *implicit* patterns in the data.[47] The patterns, in turn, can suggest the existence of "new groups" that can also include certain "facts," relationships, or associations about the persons included in them. Unlike profiles involving traditional or conventional groups – for example, the group of females, the group of historians, the group of libertarians (and even) the group of people who drive blue cars, and so forth – groups generated from data mining can be based on profiles that include nonobvious information. Because people may not be aware that they have been assigned to such groups, and because decisions about them can be made on the basis of their association with those groups, data mining has been considered a controversial practice.

Most people who are assigned to traditional groups would not be surprised if they were told that they have been identified with such groups or categories (such as the group of females, historians, and so forth). However, individuals assigned to some of the new groups generated by the use of data mining tools might be very surprised to discover their association with these groups and to learn of some "new facts" about them that can be suggested by virtue of their identification with these groups. For example, Custers (2006) notes that a data mining application might reveal an association that suggests a "fact" about a person who drives a red car – namely, that such a person is a member of the group of people who both drive red cars and have colon cancer. In this case, the owner of a red car may be very surprised if he were to learn that he belongs to a group of individuals likely to have or to develop colon cancer merely because of statistical correlation that would seem to have no obvious basis in reality.

Typically, the new groups that are based on arbitrary and nonobvious statistical information, via data mining, are derived from personal data after it has been aggregated. But as Vedder (2004) points out, once an individual's data is aggregated and generalized, the individual has no say in how that data is further processed—for example, how it is used in secondary applications.[48] This is because normative requirements for protecting

[47]For an account of why data mining technology threatens informational privacy in ways that go beyond the kinds of privacy threats posed by earlier computing/information technologies, see Tavani (1999).

[48]One controversy here involves the question of informed consent. If an individual, X, consents to having his personal information used in procedure A, and A implies B (say an aggregation of data involving A), does it follow that X has consented to have his or her information used in B? O'Neill (2002) describes this problem as the "opacity" of informed consent.

personal data apply only to that data *qua* individual persons.[49] Vedder argues that current privacy laws offer individuals virtually no protection with respect to how information about them acquired through data mining activities is subsequently used, even though important decisions can be made about those individuals on the basis of patterns found in the mined personal data. Whereas an individual has a legal right to access and rectify her personal data as it applies strictly to that individual, she does not enjoy the same rights with respect to personal information that is *derived from* that data.[50] For example, that kind of information can be used by others without their having to obtain the explicit consent to do so from the individual whose data has been aggregated.[51] Consider that information about an individual's having a statistical probability of acquiring a certain disease merely because of an association with an arbitrary group profile generated by data mining could result in that individual's being the victim of stigmatization and discrimination. As we noted in Section 6.3.4.3, Vedder has argued that a new category of privacy protection, called "categorial privacy," is needed to protect individuals from the kind of privacy threats made possible by the use of data mining technology (in both medical and nonmedical contexts).

6.4.3 Employee/Workplace Privacy and the Threat from Surveillance Technologies

In the past, it was not uncommon for organizations to hire people to monitor the performance of employees in the workplace. Now, however, there are "invisible supervisors" – i.e., computers – that can continuously monitor the activities of employees around the clock without failing to record a single activity of the employee. Introna (2004) points out that 45% of major businesses in the United States record and review employee communications and activities on the job. These include an employee's phone calls, e-mail, and computer files.[52] Introna believes that surveillance technology, which has become less expensive, has also become "less overt and more diffused." He also points out that current technology has created the potential to build surveillance features into the "very fabric of organizational processes." Consider that this technology is now being used to measure such things as the number of keystrokes a worker enters per minute, the number of minutes he or she spends on the telephone completing a transaction (such as selling a product or booking a reservation), and the number and length of breaks he or she takes.

One controversial practice affecting workplace surveillance has involved the monitoring of employee e-mail. Should employees expect that they can send and

[49]Vedder (p. 463) argues that because personal data is commonly defined and understood as data and information relating to an "identified or identifiable person," our privacy laws are based on "too narrow a definition of personal data."

[50]This factor has prompted Fulda (2004, p. 472) to raise the following question: "Is it possible for data that do not in themselves deserve legal protection to contain implicit patterns that do deserve legal protection . . .?"

[51]For an analysis of some of the problems in achieving "valid informed consent" for research subjects in the context of data mining and genetic/genomic research, see Tavani (2004b).

[52]Introna notes that additional forms of monitoring and surveillance, such as reviews of phone logs or videotaping for security purposes, bring the overall figure on "electronic oversight" to 67.3%.

receive private e-mail messages on an employer's computer system? While many companies have developed explicit policies regarding the use of e-mail, as well as the use of other employer-owned computer system resources, not all organizations have done so. As a result, it is not always clear which kinds of privacy protection employees can expect in e-mail conversations.

Rudinow and Graybosch (2002) describe two controversial cases affecting e-mail monitoring: one involving a major hospital chain in the United States that uses software to scan employee e-mail for keywords that may indicate "inappropriate" content, and another involving a major corporation that uses software to reject incoming e-mail messages that have what that company determines to be "offensive" content. Do practices such as these violate employee privacy rights and expectations? If so, under which conditions (if any) should e-mail monitoring be morally permissible? A thoughtful analysis of these questions is offered by Spinello (2002), who believes that individuals have a "prima facie right" to the confidentiality of their e-mail communications. He also believes, however, that this right must be carefully balanced with a corporation's "information requirements" and its "need to know."

Issues affecting e-mail privacy in particular, and employee/workplace privacy in general, merit much more discussion than can be provided here. Nonetheless, it should be clear from our analysis thus far why the concerns briefly identified here are representative of some of the important challenges for informational privacy in the twenty-first century.

6.4.4 Location Privacy and the Threat from RFID Technology

Recent technologies have made it possible to track a person's locations at any given point in time. For example, Global Positioning System (GPS) technologies can be used to pinpoint a person's location within a few meters. And Highway Vehicle Transportation schemes, such as E-ZPass, record the date and exact time a vehicle passes through a toll plaza. In response to concerns raised by these and related technologies, privacy analysts have recently introduced a new category called *location privacy*. One technology that is closely associated with this kind of privacy concern is RFID, which consists of a tag (microchip) and a reader. The tag has an electronic circuit, which stores data, and an antenna that broadcasts data by radio waves in response to a signal from a reader. The reader also contains an antenna that receives the radio signal, and it has a demodulator that transforms the analog radio signal into suitable data for any computer processing that will be done (Lockton and Rosenberg, 2005).

Although the commercial use of RFIDs was intended mainly for the unique identification of real-world objects (e.g., tracking items sold in supermarkets), which would not seem to be controversial, the tags can also be used to monitor those objects after they are sold. RFID technology can also be used for tracking the owners of the items that have these tags. As Nissenbaum (2004, p.135) worries:

> Prior to the advent of RFID tags, customers could assume that sales assistants, store managers, or company leaders recorded point-of-sale information. RFID tags extend the duration of the relationships, making available to. . .others a range of information about customers that was not previously available.

So, on the one hand, RFID tags can function as "smart labels" that make it much easier to track inventory and protect goods from theft or imitation. On the other hand, these tags pose a significant threat to individual privacy. Thus, some privacy advocates worry about the accumulation of RFID transaction data by RFID owners and how that data will be used in the future. For example, organizations such as the Electronic Privacy Information Center (EPIC) and the American Civil Liberties Union (ACLU) have expressed concerns about the kinds of privacy threats made possible by RFID-based systems.

Many ranchers in the United States now use RFID technology to track their animals by inserting tags into their animals' ears. In the not-too-distant future, major cities and municipalities might require RFID tags for pets. Policies requiring RFID tags for humans, especially for children and the elderly, may also be established in the future. Currently, some nursing homes in the United States provide their patients with RFID bracelets. Adam (2005) notes that chips (containing RFID technology) can be implanted in children so that they can be tracked if abducted; however, she worries that we may come to rely too heavily on these technologies to care for children. So, despite some of the benefits of RFID, a number of social concerns affecting location privacy will need to be resolved as we continue to implement this technology.

Thus far, we have examined some specific kinds of informational privacy threats posed by the development and implementation of a range of diverse technologies – i.e., from cookies to data mining to RFID. In an earlier section of this chapter, we examined some theories of informational privacy that attempt to explain what privacy is, why it is important, and why it should be protected. In the next section, we examine some relatively recent proposals that have been advanced for framing comprehensive policies for protecting informational privacy.

6.5 FRAMING APPROPRIATE POLICIES FOR PROTECTING INFORMATIONAL PRIVACY

Can we establish policies that are sufficiently comprehensive in scope to protect informational privacy, given the threats posed by the kinds of information technologies described in the preceding section? Before answering that question, I believe that it is useful to consider two more basic questions: (1) What kind of value is privacy? (2) Why is privacy valued, and thus worth protecting?

6.5.1 What Kind of Value is Privacy?

Philosophers generally distinguish between intrinsic values—that is, where some things are valued for their own sake—and instrumental values, where things are valued because they provide a means for achieving some end or ends. Van den Hoven (2005), in his analysis of privacy as a value, distinguishes between "intrinsicalist" and "functionalist" accounts. Whereas the former view privacy as valuable in itself, the latter argue that privacy serves other values and that its importance should thus be explained in terms of those values (e.g., security, autonomy, and so forth).

Because it is difficult to interpret and defend privacy as something that has intrinsic value, it would seem that, by default, privacy must be viewed as an instrumental value. However, Fried (1990) and Moor (2004) have both put forth accounts that attempt to straddle the divide between intrinsic and instrumental values of privacy. Fried believes that privacy is unlike most instrumental values, which serve simply as one means among others to achieve some end. Instead, he argues that privacy is essential for certain important human ends, that is, it is necessary to achieve trust and friendship. Fried also notes that we tend to associate instrumental values with contingent, or "non-necessary," conditions. But owing to the fact that privacy is necessary for achieving certain human ends, it cannot be regarded as something that is merely contingent (like most instrumental values). In Fried's scheme, we do not simply value privacy to achieve important human ends; rather, those ends would be inconceivable without privacy.

Moor also suggests that privacy is more than simply an instrumental value, but he uses a strategy that is different from Fried's to make this point. Moor argues that privacy can be understood as the articulation, or expression, of a "core value"—namely, *security*—which he believes is essential across cultures for human flourishing.[53] He also points out that as information technology insinuates itself more and more into our everyday lives, it increasingly threatens our privacy. As a result, individual privacy is much more difficult to achieve in highly technological societies. And because privacy expresses the core societal value of security, the possibility of fully realizing human flourishing in an information society is threatened. Thus, in Moor's scheme, privacy is a value that would seem to have more than mere instrumental worth.

6.5.2 Why is Privacy Valued?

Some privacy theorists, including Westin (1967), believe that privacy is essential for freedom and democracy. Others, such as Henkin (1974) and Johnson (1994), believe that privacy is essential for autonomy.[54] Moor (2006) and DeCew (1997) each use the metaphor of a "shield" to show why privacy is an important human value. For example, Moor (p. 114) describes privacy as a "kind of shield that protects individuals from the harmful demands and idiosyncrasies of society, and in some cases protects other members of society from individuals." DeCew argues that because privacy acts as a shield to protect us in various ways, its value lies in the "freedom and independence" it provides for us. In her view, privacy also serves as a shield that protects us from certain kinds of "intrusions and pressures arising from others' access to our persons and to details about us" as well as from pressure to conform (DeCew, p. 74).

We should also note, however, that some authors have focused their analyses on what they believe to be a "darker side" of privacy. For example, MacKinnon (1989) and Allen (2003), both writing from the feminist perspective, worry that privacy claims can

[53]See also Moor (1999) for an account of the core values needed for human flourishing.

[54]Also, see Benn (1971) for a description of the relationship between privacy and autonomy.

easily be used to shield abusive situations in families.[55] Others who have been critical of privacy as a positive value include Posner (1978) and Etzioni (1999). Posner (p. 22) worries that privacy can be used by individuals to "manipulate the world" around them through the "selective disclosure of facts about themselves," or it can be used as a "cloak" to avoid disclosing information altogether. Etzioni believes that privacy can promote an individualist agenda that can be harmful to the social good.[56]

If these critics are correct, it would seem that privacy is something that mainly benefits individuals, and that it does so at the expense of society as a whole. However, Regan (1995, p. 212) argues that privacy not only is a value to the individual as an individual, but also is important to society in general. She points out that often we mistakenly frame debates about privacy in terms of how to balance privacy interests as *individual goods* against interests involving the larger *social good*. Regan believes that there are three bases for understanding the social importance of privacy. First, privacy is a *common value* in that virtually all societies and "all individuals value some degrees of privacy and have some common perceptions about privacy." Second, privacy is a *public value* in that it has value "not just to the individual as an individual or to all individuals as individuals in common but also to the democratic political system." And, third, privacy is becoming what Regan calls a *collective value* in that "technology and market forces are making it hard for any one person to have privacy without all persons having a similar minimum level of privacy" (Regan, p. 213).

Regan notes that instead of trying to understand the importance of privacy *to society*, we have instead focused on the importance of privacy to the individual. One result of this has been that the philosophical basis that has been used to ground privacy policy has oversimplified the importance of privacy to the individual and has failed to recognize its broader social importance (Regan, p. 221). She believes that if privacy were understood in its broader dimensions, that is, as something not solely concerned with individual goods but as contributing to the broader social good, then privacy would have a greater chance of receiving the appropriate consideration it deserves in policy debates involving the balancing of competing values and interests. Because many of the interests competing with privacy have generally been recognized as "social interests," Regan believes that privacy "has been on the defensive." She concludes that we need policies that take into account privacy's value as an important social good.

6.5.3 Framing a Comprehensive Policy for Protecting Informational Privacy

Generally, policy proposals for protecting informational privacy have called for either (a) strong privacy legislation (and governmental regulation) or (b) better industry self-regulation. Although most privacy advocates have preferred the former, many in the commercial sector have tended to support the latter. Others, however, have argued that

[55]See DeCew (p. 83) for a description of how MacKinnon's views can be interpreted to support the "darker side of privacy."

[56]For a sympathetic analysis of some of Etzioni's views on privacy, especially as they intersect with issues affecting security, see Himma (2008).

a "technological solution" is possible via privacy-enhancing technologies (or PETs), defined in Section 6.4.1. Some proposals see PETs as providing a crucial role because they can "empower" users to take greater control of their privacy. Although not everyone agrees that PETs provide the "silver bullet" that their proponents suggest,[57] these tools have nonetheless played a prominent role in some recent proposals for privacy policies.

A number of comprehensive informational privacy policies have been proposed during the last decade. Most have called for some combination of legislation and industry self-regulation. Additionally, some proposals have called for a privacy oversight commission (see, e.g., Clarke, 1999) similar to those established in countries such as Canada. A proposal by Clarke also calls for the establishment of a "privacy watchdog agency" as well as the use of sanctions, which he believes are needed for his (or any) privacy protection policy to be successful. A similar kind of policy has been proposed by Wang et al. (1998), who believe that governments, businesses, and individuals each have a key role to play. Like Clarke, Wang et al. call for the establishment of independent privacy commissions that would (i) oversee the implementation and enforcement of various privacy laws and (ii) educate the public about privacy issues. Whereas businesses would be responsible for promoting self-regulation for fair information practices and for educating consumers about online privacy policies, individuals themselves would be responsible for using privacy-enhancing technologies and security tools.

Another kind of comprehensive proposal, put forth by DeCew (2006), would require federal guidelines mandating the "priority of privacy." In this scheme, the collection and storage of data require "maximal privacy protection as the default" (DeCew, p. 128). Calling for a "presumption in favor of privacy," the model proposed by DeCew would try to preserve the anonymity of data wherever possible. It would also establish "fair procedures" for obtaining the data. For example, it would require that the personal data collected has both "relevance and purpose," and it would specify the "legitimate conditions of authorized access." This scheme would also require commitments for developing "systematic methods for maintaining data quality," that is, ensuring that "data collected for one purpose not be used for another purpose, or be shared with others without the consent of the subject" (DeCew, pp. 127–128). The retention time of the data would also be limited to what is "necessary for the original purpose of the data collection." DeCew's model would also require that people are educated, consulted, and allowed to give consent or refusal through a process of "dynamic negotiation," where individuals could continuously negotiate with businesses. In this sense, individuals would have some say as to whether or not data about them is collected and, if so, how that data is subsequently used. Unlike the current default practice in many e-commerce transactions, individuals would have to specifically opt in to have their data collected and used, rather than being required to explicitly opt out.

Although none of the models we have examined thus far is easily implemented, each shows why a comprehensive policy is needed to ensure adequate privacy

---

[57]For a critical analysis of PETs, see Tavani and Moor (2001).

protection. Each proposal also supports the view that an adequate privacy policy should be as open and transparent as possible. Perhaps the RALC framework, which we examined in Section 6.3.3, can inform the design of a privacy policy that aspires to be both transparent and comprehensive. In our analysis of RALC, we saw that individuals could only be ensured of having their privacy protected in "situations" or zones that were declared "normatively private." Additionally, RALC requires that the rules for establishing a normatively private situations be "public" and open to debate and that the rules must be known to all those in or affected by a situation (Moor, 1997). These rules are explicitly articulated in RALC's Publicity Principle, which states "*Rules and conditions governing private situations should be clear and known to the persons affected by them*" (Moor, p. 32; italics in original).

Consider how this principle might be incorporated into a privacy policy involving cookies technology (described in Section 6.4.1). An adequate policy would need to spell out clearly the rules affecting the privacy expectations and requirements for users who interact with Web sites that use cookies. First, users would need to be informed that the site they are considering accessing uses cookies. The users would also need to be told whether information about them acquired via cookies could be used in subsequent contexts. At the same time, online businesses might wish to inform consumers about some of the financial benefits, for example, in the form of consumer rebates and discounts, they could expect to receive if they accepted cookies. When all of the information about the potential implications of cookies has been disclosed, users could make an informed decision about whether to accept or reject them. So the Publicity Principle could be an integral component for a privacy policy affecting a technology such as cookies.

Not only can RALC's Publicity Principle help us to frame a clear and explicit privacy policy for cookies technology, it can also be incorporated into policies affecting the use of other kinds of information technologies that threaten personal privacy. For example, the same principle used in determining whether personal information accessible to cookies technology should be declared a normatively private situation (following an open debate in the public forum) can also be applied in the case of technologies such as data mining (described in Section 6.4.2) and peer-to-peer (P2P) computer networks.[58] In this sense, RALC's Publicity Principle can play a key role in informing the kinds of clear, open, and comprehensive policies that are needed to protect informational privacy.

6.6 CONCLUDING REMARKS

We have examined various definitions and conceptions of privacy, with a principal focus on controversies affecting informational privacy. In our analysis of informational privacy, we critically examined some traditional and contemporary theories. We also examined some ways in which some relatively recent information/computer

[58]For a discussion of how RALC can be used to frame a policy affecting data mining, which closely follows the strategy here involving cookies, see Tavani (2007b). And for a description of how RALC can be incorporated into a privacy policy for P2P networks, see Grodzinsky and Tavani (2005).

technologies have introduced informational privacy concerns that affect four broad categories: consumer privacy, medical privacy, employee privacy, and location privacy. Finally, we considered some proposals for comprehensive policies that aim at protecting informational privacy.

ACKNOWLEDGMENTS

I am grateful to Ken Himma and Jim Moor for their helpful comments on an earlier draft of this chapter. I am also grateful to John Wiley & Sons, Inc., for permission to reuse some material from my book *Ethics and Technology* (2004, 2007) in this chapter.

REFERENCES

Adam, A. (2005). Chips in our children: can we inscribe privacy protection in a machine. *Ethics and Information Technology,* 7(4), 233–242.

Alfino, M. (2001). Misplacing privacy. *Journal of Information Ethics,* 10(1), 5–8.

Allen, A. (1988). *Uneasy Access: Privacy for Women in a Free Society.* Rowman and Littlefield, Totowa, NJ.

Allen, A. (2003). *Why Privacy Isn't Everything: Feminist Reflections on Personal Accountability.* Rowman and Littlefield, Lanham, MD.

Baumer, D., Earp, J.B., and Payton, F.C. (2006). Privacy of medical records: IT implications of HIPAA. In: Tavani, H.T. (Ed.), *Ethics, Computing, and Genomics*. Jones and Bartlett, Sudbury, MA, pp. 137–152.

Benn, S. I. (1971). Privacy, freedom and respect for persons. In: Pennock, J.R. and Chapman, J. W. (Eds.), *Nomos XIII: Privacy.* Atherton Press, New York.

Bok, S. (1983). *Secrets: On the Ethics of Concealment and Revelation*. Pantheon Books, New York.

Boyd v. United States (1886). 116 U.S. 186.

Brandeis, L.D. (1928). *Olmstead v. United States* (277 U.S. 438). Dissenting opinion.

Branscomb, A. (1994). *Who Owns Information? From Privacy to Public Access*. Harper Collins, New York.

Brennan, W. (1972). *Eisenstadt v. Baird* (405 U.S. 438). Majority opinion.

Burnham, D. (1983). *The Rise of the Computer State*. Random House, New York.

Clarke, R. (1999). Internet privacy concerns confirm the case for intervention. *Communications of the ACM,* 42(2), 60–67.

Cooley, T. (1880). *A Treatise on the Law of Torts*. Callaghan, Chicago.

Custers, B. (2006). Epidemiological data mining. In: Tavani, H.T. (Ed.), *Ethics, Computing, and Genomics*. Jones and Bartlett, Sudbury, MA, pp. 153–165.

DeCew, J.W. (1997). *In Pursuit of Privacy: Law, Ethics, and the Rise of Technology.* Cornell University Press, Ithaca, NY.

DeCew, J.W. (2006). Privacy and policy for genetic research. In: Tavani, H.T. (Ed.), *Ethics, Computing, and Genomics*. Jones and Bartlett, Sudbury, MA, pp. 121–136.

DeMay v. Roberts (1881). 46 U.S. 160.

Douglas, W.O. (1965). *Griswold v. Connecticut* (381 U.S. 479). Majority opinion.

Eisenstadt v. Baird (1972). 405 U.S. 438.

Elgesem, D. (2004). The structure of rights in directive 95/46/EC on the protection of individuals with regard to the processing of personal data and the free movement of such data. In: Spinello, R.A. and Tavani, H.T. (Eds.), *Readings in CyberEthics*. 2nd edition. Jones and Bartlett, Sudbury, MA, pp. 418–435.

Etzioni, A. (1999). *The Limits of Privacy*. Basic Books, New York.

Flaherty, D.H. (1989). *Protecting Privacy in Surveillance Societies*. University of North Carolina Press, Chapel Hill, NC.

Floridi, L. (1999). Information ethics: on the philosophical foundations of computer ethics. *Ethics and Information Technology,* 1(1), 37–56.

Floridi, L. (2005). The ontological interpretation of informational privacy. *Ethics and Information Technology,* 1(1), 185–200.

Floridi, L. (2006). Four challenges for a theory of informational privacy. *Ethics and Information Technology,* 8(3), 109–119.

Freund, P.A. (1971). Privacy: one concept or many. In: Pennock, J. R. and Chapman, J.W. (Eds.), *Nomos XIII: Privacy.* Atherton Press, New York, pp. 182–198.

Fried, C. (1990). Privacy. In: Ermann, M.D.,Williams, M.B., and Gutierrez, C. (Eds.), *Computers, Ethics, and Society.* Oxford University Press, New York. pp. 51–67.

Fulda, J.S. (2004). Data mining and privacy. In: Spinello, R.A. and Tavani, H.T. (Eds.), *Readings in CyberEthics*, 2nd edition. Jones and Bartlett, Sudbury, MA, pp. 471–475.

Gavison, R. (1980). Privacy and the limits of the law. *Yale Law Journal,* 89, 421–471.

Griswold v. Connecticut (1965). 381 U.S. 479.

Grodzinsky, F.S. and Tavani, H.T. (2005). P2P networks and the *Verizon v. RIAA* case: implications for personal privacy and intellectual property. *Ethics and Information Technology,* 7(4), 243–250.

Henkin, L. (1974). Privacy and autonomy. *Columbia Law Review,* 77, 1410–1425.

Himma, K.E. (2008). Privacy vs. security. *San Diego Law Review*, in press.

Hofstadter, S.H. and Horowitz, G. (1964). *The Right to Privacy.* Central Books, New York.

Hunter, L. (1995). Public image. In: Johnson, D.G. and Nissenbaum, H. (Eds.), *Computers, Ethics and Social Values*. Prentice Hall, Englewood Cliffs, NJ, pp. 293–299.

Inness, J. (1992). *Privacy, Intimacy, and Isolation*. Oxford University Press, New York.

Introna, L. (2004). Workplace surveillance, privacy, and distributive justice. In: Spinello, R.A. and Tavani, H.T. (Eds.), *Readings in CyberEthics*, 2nd edition. Jones and Bartlett, Sudbury, MA, pp. 476–487.

Johnson, D.G. (1994). *Computer Ethics*, 2nd edition. Prentice Hall, Upper Saddle River, NJ.

Johnson, D.G. and Nissenbaum, H. (1995). Privacy and databases. *Computers, Ethics and Social Values*. Prentice Hall, Englewood Cliffs, NJ, pp. 262–268.

Katz v. United States (1967). 389 U.S. 347.

Kemp, R. and Moore, A.D. (2007). Privacy. In: Himma, K.E. (Ed.), *Information Ethics*. Special issue of *Library Hi Tech*, 25(1), 58–78.

Lockton, V. and Rosenberg, R.S. (2005). RFID: the next serious threat to privacy. *Ethics and Information Technology,* 7(4), 221–231.

MacKinnon, C. (1989). *Toward a Feminist Theory of the State*. Harvard University Press, Cambridge, MA.

McCloskey, H.J. (1985). Privacy and the right to privacy. In: Purtill, R.L. (Ed.), *Moral Dilemmas: Readings in Ethics and Social Philosophy*. Wadsworth, Belmont, CA, pp. 342–357.

Miller, A.R. (1971). *The Assault on Privacy*. Harvard University Press, Cambridge, MA.

Moor, J.H. (1990). The ethics of privacy protection. *Library Trends,* 39(1–2), 69–82.

Moor, J.H. (1997). Towards a theory of privacy in the information age. *Computers and Society,* 27(3), 27–32.

Moor, J.H. (1999). Just consequentialism and computing. *Ethics and Information Technology,* 1(1), 65–69.

Moor, J.H. (2004). Reason, relativity, and responsibility in computer ethics. In: Spinello, R.A. and Tavani, H.T. (Eds.), *Readings in CyberEthics*, 2nd edition. Jones and Bartlett, Sudbury, MA, pp. 40–54.

Moor, J.H. (2006). Using genetic information while protecting the privacy of the soul. In: Tavani, H.T. (Ed.), *Ethics, Computing, and Genomics*, Jones and Bartlett, Sudbury, MA, pp. 109–119.

Moore, A.D. (2003). Privacy: its meaning and value. *American Philosophical Quarterly,* 40, 215–227.

Moore, A.D. (2008). Towards informational privacy rights. *San Diego Law Review,* in press.

Moore, B. (1984). *Privacy: Studies in Social and Cultural History,* M.E. Sharpe, Inc., Armonk, NY.

Nissenbaum, H. (1997). Toward an approach to privacy in public: challenges of information technology. *Ethics and Behavior,* 7(3), 207–219.

Nissenbaum, H. (1998). Protecting privacy in an information age. *Law and Philosophy,* 17, 559–596.

Nissenbaum, H. (2004). Privacy as contextual integrity. *Washington Law Review,* 79(1), 119–157.

Olmstead v. United States (1928). 277 U.S. 438.

O'Neill, O. (2002). *Autonomy and Trust in Bioethics*. Cambridge University Press, Cambridge, UK.

Parent, W.A. (1983a). Privacy, morality and the law. *Philosophy and Public Affairs,* 12(4), 269–288.

Parent, W.A. (1983b). A new definition of privacy for the law. *Law and Philosophy,* 2, 305–338.

Posner, R.A. (1978). An economic theory of privacy. *Regulations,* May–June, 19–26.

Rachels, J. (1975). Why privacy is important. *Philosophy and Public Affairs,* 4(4), 323–333.

Regan, P.M. (1995). *Legislating Privacy: Technology, Social Values, and Public Policy.* University of North Carolina Press, Chapel Hill, NC.

Reiman, J.H. (1976). Privacy, intimacy, and personhood. *Philosophy and Public Affairs,* 6(1), 26–44.

Roe v. Wade (1973). 410 U.S. 153.

Rosen, J. (2000). *The Unwanted Gaze: The Destruction of Privacy in America*. Random House, New York.

Rudinow, J. and Graybosch, A. (Eds). (2002). *Ethics and Values in the Information Age.* Wadsworth, Belmont, CA.

Scanlon, T. (1975). Thompson on privacy. *Philosophy and Public Affairs,* 4(4), 315–322.

Schoeman, F. (1992). *Privacy and Social Freedom.* Cambridge University Press, Cambridge, MA.

Solove, D. (2002). Conceptualizing privacy. *California Law Review,* 90, 1087–1155.

Spinello, R.A. (1997). The end of privacy. *America*, 176, 9–13. Reprinted in: Long, R.E. (Ed.), *Rights to Privacy.* H. W. Wilson Co., New York, pp. 25–32.

Spinello, R.A. (2002). Electronic mail and panoptic power in the workplace. In: Rudinow, J. and Graybosch, A. (Eds.), *Ethics and Values in the Information Age.* Wadsworth, Belmont, CA, pp. 303–311.

Spinello, R.A. (2006). Property rights in genetic information. In: Tavani, H.T. (Ed.), *Ethics, Computing, and Genomics.* Jones and Bartlett, Sudbury, MA, pp. 213–233.

Tavani, H.T. (1999). Informational privacy, data mining, and the internet. *Ethics and Information Technology,* 1(2), 137–145.

Tavani, H.T. (2000). Privacy and security. In: Langford, D. (Ed.), *Internet Ethics.* Macmillan, London, pp. 65–95.

Tavani, H.T. (2004a, 2007a). *Ethics and Technology: Ethical Issues in an Age of Information and Communication Technology,* 2nd edition. (2007) John Wiley & Sons, Hoboken, NJ.

Tavani, H.T. (2004b). Genomic research and data mining technology: implications for personal privacy and informed consent. *Ethics and Information Technology,* 6(1), 15–28.

Tavani, H.T. (2007b). Philosophical theories of privacy: implications for an adequate online privacy policy. *Metaphilosophy,* 38(1), 1–22.

Tavani, H.T. (2007c). Floridi's ontological theory of informational privacy: some implications and challenges. In: Hinman L. et al. (Eds.), *Proceedings of the Seventh International Conference on Computer Ethics: Philosophical Enquiry (CEPE 2007).* Centre for Telematics and Information Technology, Enschede, The Netherlands, pp. 385–396. Reprinted in *Ethics and Information Technology,* 10, in press.

Tavani, H.T. and Moor, J.H. (2001). Privacy protection, control of information, and privacy-enhancing technologies. *Computers and Society,* 31(1), 6–11.

Thomson, J.J. (1975). The right to privacy. *Philosophy and Public Affairs,* 4(4), 295–315.

Van den Hoven, J. (2004). Privacy and the varieties of informational wrongdoing. In: Spinello, R.A. and Tavani, H. (Eds.), *Readings in CyberEthics*, 2nd edition. Jones and Bartlett, Sudbury, MA, pp. 488–500.

Van den Hoven, J. (2005). Privacy. In: Mitcham, C. (Ed.), *Encyclopedia of Science, Technology, and Ethics*, MacMillan, New York, pp. 1490–1492.

Vedder, A.H. (2004). KDD, privacy, individuality, and fairness. In: Spinello, R.A. and Tavani, H.T. (Eds.), *Readings in CyberEthics*, 2nd edition. Jones and Bartlett, Sudbury, MA, pp. 462–470.

Volkman, R. (2003). Privacy as life, liberty, property. *Ethics and Information Technology,* 5(4), 199–210.

Wang, H.M., Lee, K.O., and Wang C. (1998). Consumer privacy concerns about internet marketing. *Communications of the ACM,* 41(3), 63–70.

Warren, S. and Brandeis, L. (1890). The right to privacy. *Harvard Law Review,* 14(5), 193–220.

Westin, A.F. (1967). *Privacy and Freedom.* Atheneum Press, New York.

Whalen v. Roe (1977). 429 U.S. 589.

Zimmer, M. (2005). Surveillance, privacy and the ethics of vehicle safety communications. *Ethics and Information Technology,* 7(4), 201–210.

CHAPTER 7

Online Anonymity

KATHLEEN A. WALLACE

The term anonymity has been used to denote a number of related things: namelessness, detachment, unidentifiability, lack of recognition, loss of sense of identity or sense of self, and so on. Anonymity can also be brought about in a variety of ways and there are many purposes, both positive and negative, that anonymity could serve, such as, on the positive side, promoting free expression and exchange of ideas, or protecting someone from undesirable publicity or, on the negative, hate speech with no accountability, fraud or other criminal activity. Anonymity and privacy are also considered to be closely related, with anonymity being one means of ensuring privacy.

7.1 ANONYMITY AS A FEATURE OF COMPLEX SOCIAL STRUCTURES

Before turning to a discussion of the concept of anonymity itself, it is worth noting that recent concerns about anonymity may be an expression of a widespread feature of contemporary social organization and of the extent to which technological features of that organization affect with whom we interact, how we interact with one another, and how we think of ourselves as related, socially and ethically, to others. Anonymity may be deliberately sought or something that occurs spontaneously from the sheer complexity of modern life. The idea of a kind of naturally occurring, "spontaneous anonymity" is embodied in characterizations of someone as a member of an anonymous mass or in expressions such as "the logic of anonymity in modern life." There are two ideas at work here. One is the thought that anonymity could be the byproduct of sheer size as when one is among a throng of people who don't know one another. The other is the thought that anonymity could be the byproduct of complex social organization, where society is organized such that one's social locations are dispersed and not necessarily connected with one another; for example, one's work environment may be disconnected from or not overlapped with one's familial locations. The worry embodied in a phrase like "the

The Handbook of Information and Computer Ethics, Edited by Kenneth Einar Himma and Herman T. Tavani

logic of anonymity in modern life" seems to be that sheer complexity, sheer size, or both render someone, as a particular individual, largely unknown to others and that when individuals are not known as individuals to others this can lead to depersonalization, stereotyping, and prejudicial attitudes.[1] Such spontaneously occurring social anonymity may be involved in the problem of accountability for institutional decisions made by "many hands."[2] At the same time, such anonymity may also enhance the development of individuality in so far as it frees one from the possible stifling effects of close and/or traditional social relations. Anonymity may also occur in more local and intimate contexts too, but one reason that there may be so many contemporary expressions of concern about anonymity and its effects on social relationships, responsible decision-making, and accountability may be due to a worry about the apparent ease with which it is thought to occur spontaneously, so to speak in modern life.

The Internet as a social environment may be of concern in so far as it has the capacity to increase the scope of natural or spontaneous anonymity as a by-product of or endemic to the nature of online communicative relations. For instance, in comparison to face-to-face and telephonic communications, social and person cues are reduced in online communications, and people can easily adopt pseudonyms and personae in chat rooms, blogs, and so on. However, the idea that in "cyberspace" anonymity is given (whereas in "real space" anonymity has to be created), may be mistaken. With the development of modern information technologies, there seems to be an increase in the ease with which anonymity may be assumed and, at the same time, an increase in the ease with which it may be undermined. Examples of the former would be the capacity for anonymizing communication by using e-mail accounts that do not require giving any personal information or which allow giving false information, or by using chained remailers that encrypt and send portions of messages.[3] Regarding the latter, it may be that with techniques of data mining and data correlation, clickstream tracking, or the patterns of users' web-browsing, anonymity in cyberspace is, in some respects, only apparent; in other words, the online user and the user of other electronic devices, such as cell phones with GPS tracking capability, may be easily identifiable to others, although the user may not realize it.[4] In addition, as a new social environment, the Internet may offer new ways of self-identification, expression, and interaction with identifiable others.[5]

To the extent that anonymity is possible, worries about the undermining of identity, self-cohesion and community have been expressed. For example, Dreyfus has

---

[1]This seems to be the kind of anonymity that Natanson (1986) has in mind in his development of Schutz's concept of typification. Gordon (1997), also extending Schutz's notion of typification, articulates a notion of "perverted anonymity" as a way of conceptualizing one kind of prejudice experienced by dark-skinned people of African origin.

[2]Thompson (1987).

[3]See Froomkin (1995), Kling et al. (1999). Lance Cottrell designed one remailer protocol called Mixmaster and has since gone on to found Anonymizer, Inc. See www.anonymizer.com and www.livinginternet.com/i/is_anon.htm (each last visited on September 24, 2006).

[4]As noted by Hayne and Rice (1997) and Spears and Lea (1994).

[5]Sites such as MySpace and Facebook may be instances of such development.

suggested that the anonymity that is possible on the Internet creates the possibility of detached, desituated spectators who are unable to make lasting commitments (Dreyfus, 2001). The thought seems to be that the Internet makes it easy to move from one communicative context or relationship to another with no lasting social ties having been established. In so far as commitments are fundamental to the formation and sustenance of a self, the Internet might be thought to undermine self-identity and involvement in community and social interaction.

Goold, making the following point specifically with respect to closed circuit televisions, has suggested that the anonymity or nonidentifiabilty of an observer or monitor threatens privacy, specifically the dignity and autonomy that depend on some degree of privacy over who knows what about us.[6] Extending that worry to the Internet, monitoring of online communications where the monitor is invisible or anonymous to the user, but the user knows or suspects that such monitoring is occurring, may inhibit free expression and inquiry. The ability to engage in free, political speech may in some contexts, for example, under repressive political regimes, depend on the capacity to make anonymous or pseudonymous utterances. Yet, in some contexts, anonymous monitoring is important for criminal investigation and public security. On the other hand, anonymity for the user is often associated with privacy and the ability to make sensitive inquiries without public exposure, for instance, anonymity may facilitate counseling, online or through other means, such as the telephone, and crisis outreach.

Ferguson (and others) has argued that a right to anonymity is an adjunct to a right to privacy, where privacy is understood as applying to acts or communications that pose no risks to others.[7] von Hirsch has argued that expectations of anonymity, meaning nonidentifiability, protect a sense of dignity and autonomy as we go about our lives in public spheres (von Hirsch, 2000).

7.2 THE CONCEPT OF ANONYMITY

Anonymity has sometimes been taken to mean "un-name-ability" or "namelessness," but that is somewhat too narrow a definition.[8] While a name is often a key and clear identifier of a person, that is not always the case. A name could be ambiguous (because it is not unique), or there may be contexts in which some other tag (e.g., a social security number) is a less ambiguous identifier of a person. Moreover, someone could be clearly and unambiguously identified without naming her, for example, by giving enough other identifying information such that the person can be uniquely picked out even without having been named. For example, in the case of the CIA agent whose

---

[6]Goold's (2002) suggestion is made in the context of a discussion of closed circuit television cameras (CCTV) in public spaces, where CCTV raises the worries articulated by Foucault regarding the disciplinary implications of surveillance and panoptic practices.

[7]Ferguson (2001) defined anonymity as "a condition where actors conceal or withhold their names from an action they have performed" (p. 231), but as I will suggest anonymity should be understood more broadly than only the withholding of a name. See also Feldman (1994, 1997).

[8]Others have made this point as well. See, for example, Nissenbaum (1999).

identity was revealed to the press by someone in the Bush administration: that Valerie Plame's name was not uttered by those who revealed her identity does not mean that they did not clearly identify her.

Anonymity presupposes social or communicative relations.[9] In other words, it is relative to social contexts in which one has the capacity to act, affect or be affected by others, or in which the knowledge or lack of knowledge of who a person is is relevant to their acting, affecting, or being affected by others.[10] A hermit may be "nameless" or unknown, but is not "anonymous." Rather a hermit is an unrelated, socially disconnected agent whose life and actions do not, for the most part, affect or are not affected by the lives of others in a social environment (more on this below). It might be difficult for someone to be a complete hermit in the sense that one's actions are never known by, affect, or are affected by others, but it is conceivable. In any case, while isolation is not necessarily the same as anonymity, it may be a means of achieving anonymity. Isolated existence may be a mechanism for ensuring anonymity[11] with respect to some action, as for instance, in the case of the Unabomber,[12] but is not by itself the same as anonymity.

Marx has offered the following conceptualization of anonymity: "To be fully anonymous means that a person cannot be identified according to any of...seven dimensions of identity knowledge."[13] The seven dimensions singled out by Marx—legal name, locatability, linkable pseudonyms, nonlinkable pseudonyms, pattern knowledge, social categorization, and symbols of eligibility-noneligibility[14]—identify some of the most well recognized ways and methods by which one can be anonymous.

In contrast, I have defined anonymity as "noncoordinatability of traits in a given respect." (Wallace, 1999, p. 25) Anonymity—or noncoordinatability of traits such that the person (or persons) is (are) nonidentifiable—obtains when it is known that someone (or some people) exists, but who it is (or they are) is unknown; the action or trait in virtue of which someone is known or believed to exist is not coordinatable[15]

---

[9]Wallace (1999), Kling et al. (1999).

[10]Wallace (1999). Others make this point as well. For example, "anonymity is fundamentally social. Anonymity requires an audience of at least one person. One cannot be anonymous on top of a mountain if there is no form of interaction with others and if no one is aware of the person" (Marx, 1999, p. 100).

[11]And vice versa. We should perhaps note that either of these could be involuntary or imposed as well as chosen and sought.

[12]The Unabomber, now known to be Theodore Kaczynski, sent bombs to individuals in the United States, most of whom had some association with computer technology, over a 17 year period from 1979 to 1995. At first, it was not even known whether it was a single individual or a group of individuals who was the anonymous actor. He was identified after he succeeded in having some of his writing published in major newspapers under the identity "Unabomber" and his brother recognized the ideas expressed and the writing as those of his brother, Theodore Kaczynski.

[13]Marx (1999).

[14]Marx (1999), p. 100.

[15]While the term "coordinate" might be associated with its use in "coordination problems," I prefer "coordinate" to "associate" because I want to suggest the idea that traits are colocated and co-ordered (although such ordering is not limited to temporal succession or seriality). For example, someone's familial traits (e.g., the fact that she has no siblings) may be coordinated with some personality traits (the traits may each be determinant conditions of the other). Anonymity obtains when others are unable to trace such relations between traits so as to be able to identify, to pick out, the person.

with other traits of the person(s).[16] The "Unabomber" was able to remain anonymous for as long as he did because he was socially isolated. But his anonymity consisted in the lack of coordination between his agency *qua* "sender of bombs to computer scientists" and his other socially recognizable traits and locations in social networks of action. The definition of anonymity that I am offering differs from Marx's in so far as by my definition (1) there has to be some potentially identifying trait that is known, even though it is not otherwise known to whom the trait refers and (2) there is no inherent limit on the kind of trait by virtue of which someone could be anonymous (or, for that matter, known).

By "location" I mean not merely geographical or physical location but "position in a network of social relations." By "trait" I mean any feature, action, or location of a person that can serve to get reference going. One might think of a trait as expressible in a referential use of a definite description;[17] thus, wherever there is anonymity, we have some description that refers to an object, presumably a person, an agent, that cannot be identified in a more specific way that would enable us to identify that person. A trait could be what is sometimes thought of as a property or attribute of an object or person, but it could also be something less strongly characteristic of a person. For example, a number assigned to a person and their blood sample would not typically be thought of as a characteristic of a person, but it could suffice for reference in some context.[18]

Alternatively, a property that is characteristic of a person could be known, but anonymity could still also obtain. Thus, "sender-of-bombs-to-computer-scientists" is an attribute of Theodore Kaczynski (and a trait that could form the content of a definite description referring to the Unabomber), but as long as it was noncoordinatable with other traits, the Unabomber was not identifiable as Theodore Kaczynski. Thus, the Unabomber was initially known only as "sender of bombs to computer scientists," and that trait could not be coordinated with other traits (such as name, address, employment).

---

[16]It is possible to be mistaken in a belief that there is someone who exists. For example, if a pseudonym is used that suggests there is a single person as actor or author when in fact the product or action is produced or performed by many people. In addition, it could also be the case that an action is only indirectly known or experienced because one experiences only the effects of the action and is not aware of the actual cause of those effects and thus is not aware of the agent. This would constitute an additional layer of inaccessibility or noncoordinatability between the agent who seeks to remain anonymous and others.

[17]Russell (1956). In the case of anonymity, the idea is that there is a trait (action, feature, location) of a person (or of persons) that is known and that gets reference going, but without actually identifying the object of reference. (It is possible, of course, to be mistaken about the object of reference, for example, in thinking that it is a single individual, when it might be a group of persons. And, with definite descriptions, it is possible to have a definite description that is false but that still refers to an object, but that is a distinct issue concerning definite descriptions. The main point here is that for a person or group of persons to be anonymous, there is some trait of that person [or group] that gets reference going.) Comments from Ken Himma and Tony Dardis, and discussions on this point with Jeroen van den Hoeven, Eric Steinhart, Tony Dardis, and John Weckert were helpful on this point.

[18]Marx (1999) would categorize this use of numbers, or "opaque identifiers" as nonidentifiability through a pseudonym, either linked or not linked to other identifying information. The degree of nonidentifiability would depend on the degree and kind of linkages.

Anonymity then is nonidentifiability by others by virtue of their being unable to coordinate some known trait(s) with other traits such that the person cannot be identified; it is a form of nonidentifiability by others to whom one is related or with whom one shares a social environment, even if only or primarily by virtue of the effects of one's actions. Anonymity defined as nonidentifiability by virtue of noncoordinatability of traits is a general definition of anonymity that is intended to encompass any specific form of anonymity.

This approach to anonymity presupposes a model of a person (or a group of persons) as a unique combination of interrelated traits; each trait is a position in a network of relations, or, "order."[19] ("Order" is a generic term for designating any such network.) Anonymity is the noncoordination or noncoordinatability of traits such that a person cannot be picked out. To return to the Unabomber example; the Unabomber, *qua* resident of Montana was located in a variety of economic orders, political orders, cultural orders, and so on. The Unabomber as son (brother, uncle, nephew, etc.) is located in a familial order, as bomb maker is located in an order of persons knowledgeable about explosives, as someone fluent in English is located in the order of English language speakers, as writer of his "Manifesto" is located in an order of authors of political tracts, and so on.[20] As Unabomber, he was anonymous because that trait (maker and sender of bombs) could not be coordinated with other traits such that he as a person (as a unique interrelation of traits) was not identifiable to others. However, when his "Manifesto" was published, he was located in another order (of authors of published writings) by which another trait of his became known, and that was coordinatable with his familial traits (through his family's recognition of the ideas); there was then sufficient coordinatability among traits that the Unabomber became identifiable, initially to his family members and then through them to others.

This example illustrates how someone could be anonymous in one respect (e.g., with respect to public authorities) and not in another (e.g., with respect to family members): the Unabomber became identifiable (not anonymous) to family members while remaining anonymous to law enforcement officials and the wider public. It also illustrates how anonymity (noncoordinatability of traits) in a given respect can depend on voluntary cooperation of others to whom one is not anonymous. Thus, if Theodore Kaczynski's family members had chosen to remain silent and not bring their suspicions to the attention of law enforcement officials, the Unabomber may have remained anonymous to those officials and the wider public for some indefinite period of time.

---

[19]For clarification, identical twins are not literally identical even though they have many genetic and biological traits in common, each is a unique combination of traits. Even if there were exactly similar counterparts in other worlds, location in another world is irrelevant to the issue of anonymity that is a kind of identifiability or nonidentifiability relative to a world.

[20]"Being an English speaker" or "being a resident of Montana" is each a trait in so far as each characterizes or is a determinate feature of the person. A trait doesn't have to be unique; it could be trivial, although in some context it could be the salient feature of someone (for example, suppose some context in which the person happens to be the sole English speaker or the sole resident of Montana. In this sense, "being an English speaker" or "being a resident of Montana" is no different as far as being a trait goes from "having brown eyes").

Identifiability depends on there being relations between the traits of a person that are coordinatable by others such that the person can be picked out, identified. In one sense, a person is identifiable by virtue of every trait or location that she or he has. The Unabomber is identifiable as bomber, resident of Montana, son, brother, and so on. Each trait can potentially *identify* a person in so far as it is continuous with (connected to, located in, provides access to) the unique interrelation of traits that constitutes the person. It is these relations that mechanisms of anonymity render noncoordinatable by (and hence, inaccessible to) others. Thus, the Unabomber was anonymous because his social isolation made it practically impossible to coordinate the Unabomber *qua* bomber with enough other traits of the person so as to be able to identify the unique interrelation of traits of that person.[21] Similarly, in anonymous HIV testing, a randomly assigned number linking a person with a blood sample is not coordinated or coordinatable with other traits of the person.[22] The person is identifiable as "number x" in the order of—with respect to—testing of blood samples, but is anonymous in so far as that trait of the person cannot be coordinated with other traits of the person.

Anonymity is achievable because there are ways in which persons can deliberately set up mechanisms by which to block the coordination of their traits with others. But anonymity may also occur "spontaneously," as noted earlier. In some contexts, for instance in complex modern life, where persons may occupy many social orders that do not overlap or are not connected with one another, traits that identify a person in one social order may not be readily coordinatable with traits that are salient in another social order. For example, a trait like familial location is a location or position in a network of kinship relations (schematically represented by family trees). Familial location is typically a clear identifier of a person, that is, provides access to the overall integrity of the person because, typically, it is related to other locations of a person such as first name and surname, place and year of birth, other family members with relations to and knowledge of the person. Hence, knowledge of someone's familial location is likely to provide access to further traits of the person. Typically, this trait can readily pick out a person because it is immediately related to or coordinated with other traits

[21]Note: one doesn't have to know all of someone's traits in order to identify to be able to identify her. What is sufficient for identification may be highly variable, depending on the context, other background knowledge, cognitive abilities of the persons involved, and so on.

[22]This, a randomly assigned number, would be what Nissenbaum refers to as an "'opaque' identifier, a sign linking reliably to a person... that *on the face of it*, carries no information about the person." (Nissenbaum, 1999). One would be further identifiable by one's face were an HIV counselor or technician to recognize one on the street (assuming one had not used a disguise). Facial (and other physical) traits are another type of location. A method of random assignment of numbers for securing anonymity works best either where there is little likelihood of other overlapping locations of the persons involved or where overlapping locations are likely to provide only fleeting encounters that are not easily coordinated with other locations of the persons involved. Thus, it may work well in a large, urban environment but less well in a small town. In electronic communication, such as that which takes place on the Internet, traces of any communication are virtually inevitable. Whether they are coordinatable with other traits of a user/communicator would depend on the scope of linkages established in sets of databases as well as the scope of "data mining." The possibility of systematic coordinatability of traits on the Internet may be stronger than it is in the large urban environment and if it is, then that would pose interesting ethical questions concerning the desirability of linked databases and the practices and methods of data mining.

across a wide range of locations or social relations. However that may not always be the case; for example, it may not be readily coordinated with one's actions in the context of being a New York City subway rider, or with one's actions in the context of employment if that is socially remote from one's family. Conversely, if the only known trait of someone was that she was a NYC subway rider, that trait by itself would most likely be not coordinatable with other locations of the person.[23]

In the United States most persons are located in the order of the social security system, which like familial location and unlike location in the order of English speakers, is immediately related to many other locations of a person. Knowledge of a social security number can provide access to many other locations of a person (e.g., marital, financial, health [records, insurance]). This highly coordinated connection of traits is why identity theft involving social security numbers (or any other tag that has easily coordinated trait connections) is so problematic as far as *identity* is concerned; the other problems with identity theft concern *theft*, fraud, potential damage to one's credit record by virtue of fraudulent purchases, and so on.

It is not the uniqueness of a trait alone that enables the picking out of a person, but rather the coordination of a trait with other traits. For example, "sender-of-bombs-to-computer-scientists" may be a unique trait, but it could not *by itself* pick out a particular person until it was coordinated with other traits of the person. By itself, it did not provide access to other traits of the person until the paths of coordination with other traits became traceable.

This approach to anonymity also allows for the following: that someone could be known in one (some) respect(s) and anonymous in another. For example, an anonymous donor could be known in other ways by the recipient, even though the donor *qua* donor is anonymous to the recipient. Ted Kazinski, we may suppose, was known for many years to his family, before they ever suspected that he might be the Unabomber, as son, brother, eccentric, and so on, even while, unbeknownst to them, he was also the Unabomber. Or, suppose a detective investigating a murder has a best friend, Boris, whom the detective knows well in a variety of respects and contexts. However, suppose, unbeknownst to the detective, Boris is also the murderer in the crime that the detective is investigating.[24] Ted Kazinski was known and identifiable to his family members as son, brother, eccentric, and so on, but was anonymous *qua* Unabomber. Similarly, Boris is known and identifiable to his detective friend as friend, bowling buddy, and so on, but was anonymous *qua* murderer. (The detective knows many things about his friend Boris, that the murderer is anonymous to him, but doesn't know that it is Boris *qua* murderer who is anonymous.)

Whether a trait provides the possibility of access to other traits and hence, to further identification of a person depends on the relations of that trait to other traits of the person *and* the extent to which a prospective identifier has or can have access, or has the

[23]It is also possible that someone could be such an unusual subway rider that atypicality in the order of subway riders would allow for further identification of the person. For example, a regular rider of the Broadway line in NYC might be able to locate some unusual characters for whom that subway line is their "panhandling beat."

[24]I borrow this example from Boër and Lycan (1986), p. 7.

skills to get access, to traits and their connections. Methods of securing anonymity, partial or (nearly) complete, seek to cut off access to other connected traits, seek to isolate a trait or the location of a person from its connections to other traits or locations. The appropriateness of the method in any given context depends on the capabilities of the relevant others, the scope of anonymity sought, and the context(s) in which it is sought. Adults and children may have different cognitive abilities or practices and this may be important to whether anonymity in a particular respect is achievable. For example, an adult at a children's Christmas party who is dressed up as Santa Claus might be anonymous *to* the children, but not to the adults, if the former could not, but the latter could coordinate the traits of the disguise with other traits of the person such as to recognize who the adult is.[25] In such a case, the degree and scope of anonymity could be fairly limited. Cognitive abilities might be relevant differentiating factors in online situations involving children; among adults different degrees of technical ability and familiarity with internet protocols and capacities could be relevant differentiating factors that could contribute to appropriate use and understanding of online anonymity.

In addition, in some contexts, including online contexts, anonymity may also depend on participants' and providers' voluntary observance of normative restraints, for example, legal or moral restraints governing access to confidential information.[26] This would be another way in which anonymity might depend on the voluntary cooperation of others in sustaining the noncoordinatability of traits and hence the nonidentifiability of the person.

Anonymity is never complete unknowability. For anonymity to obtain there is always some knowledge or identifier of the person, even if it is only in virtue of a single trait or location that cannot be coordinated with other locations. As I have noted, an identifier can be relatively opaque, meaning that the known trait, by itself, does not carry much attributively true information about a person, for example, a bank account number or blood sample number. Or, an identifier can be attributively true about a person, but still noncoordinated with other traits, for example, the trait "sender-of-bombs-to-computer-scientists." Anonymity does not mean that there are no other connected traits, but that in the relevant context(s) coordinatability of those traits by others is severed or shielded (even if in some cases, others voluntarily enable the severing by observing normative restraints). The isolation of the trait (or traits) from other traits—the noncoordinatability of traits—is what renders a person anonymous.

I have been arguing that the main characteristic of anonymity is noncoordinatability of traits such that a person is (or persons are) nonidentifiable by others in at least some respect(s).[27] However, anonymity is to be distinguished from simply being unknown. Anonymity applies to an agent or a recipient of an action, when an action, event, or utterance has occurred and either the agent of the action or the recipient of the action cannot be further identified (except as doer or recipient of action, respectively). If there

---

[25]John Weckert proposed this example. In this case, anonymity is achieved through pseudonymity or disguise.

[26]Walter Sinnott-Armstrong raised this point.

[27]Wallace (1999). See also Marx (1999).

is no specifiable respect in which there is a person who is unidentifiable to others, then the person is not anonymous, but simply unknown. For example, if I am sitting in a hotel room in a city where no one knows me, I might be unknown and have all the negative feelings of being unknown (e.g., feeling small, insignificant, invisible), but I am not really anonymous. I become anonymous if, for example, I do something that cannot be traced back to me, where I cannot be identified as the author of the action.[28] Or, as another example, an unknown writer may feel unrecognized, unappreciated, and so on, but being unknown or not knowing that someone exists at all or in some respect (s) is different from being anonymous. An anonymous author has a trait(s) or product (s), for example, a book (that is not coordinatable with other traits of the author) that is located in social orders, whereas the unpublished, and hence, unknown, writer does not. For example, if the trait "author (or writer) of *Waverly*" were not coordinated by the reading public with the person otherwise known as Sir Walter Scott then Scott would have been anonymous to his readers. In so far as Scott's name, familial relations, address, and the like would not be known to readers, readers would be unable to coordinate a known trait ("author of Waverly") with enough other traits of the author to be able to identify the person as such. The extent to which such coordination could not (or could not without a great deal of effort) be made is the extent to which anonymity is sustainable. Thus, anonymity is to be distinguished from total unknowability. With anonymity, typically, there is some trait which is known, for example, that a book has been authored, but the authorship cannot be coordinated with other traits of the author; the identity of the anonymous author is unknown to others except as author of the book.

Another sense in which unknowability is to be distinguished from anonymity is when there are, or were, unknown but accessible facts about someone. So, consider again an unknown writer. Suppose a writer has published something but with an obscure press or publishing house, or the piece has gone out of print, having made no or little continuing impact on any community of readers. In such a case, we would say the writer is unknown, because some facts are unknown or have disappeared in the dross of history, rather than the writer is anonymous. A spy or a "peeping tom" whose action(s) is unknown by anyone except the spy or peeping tom would not be anonymous by this definition of anonymity. If and when the action becomes known by someone, then it is possible to refer to the anonymous (unidentifiable) person who is the agent. Anonymity presupposes some known fact, action, utterance, or event with respect to which and in some context some person(s) is (are) not further identifiable; anonymity is a way of referring to a person(s). The performing of an action such as spying, even when it is unknown and hidden, may create the possibility of the action becoming known, but until it is known reference (to an anonymous agent) can't get going.

It might be thought that anonymity can be sought for its own sake, as in cases of persons who just want to be unknown as a matter of strong psychological preference. This might simply be a way of expressing a strong desire for privacy, or a desire to have a high degree of control over one's social relations, or a desire for minimal (or the ability to minimize) social relations. Social isolation could be a means to anonymity

[28]Thanks to Ken Himma for this example.

(as it was in the case of the Unabomber), but by the approach being suggested here, by itself unknownness by virtue of social isolation is not the same as anonymity.

7.3 ANONYMITY AND ETHICAL ISSUES IN INFORMATION TECHNOLOGY

I will discuss only a few out of the many possible types of cases where anonymity may be importantly relevant to understanding and evaluating the ethics of online activities and the capacities of information technology. See also Bynum (2001). I will then turn to a more general analysis of the purposes and ethics of anonymity.

7.3.1 Data Mining, Tracking, and User's Presumption of Anonymity

Activities of computer users are tracked and compiled in databases, as well as are attributes or characteristics (e.g., names, social security numbers, phone numbers) of individuals. Depending on the extent of coordinatability and routes of access between sets of databases, they can provide sophisticated and detailed ways of identifying individuals for explicit marketing, eavesdropping, and other purposes. One problem in this area is that the user may assume that her activities are anonymous, that is, not coordinatable with her as an individual or with other distinguishing traits of her as an individual. In fact, computer-mediated or online communication may encourage the impression that one is anonymous, even though one's activities may be relatively easily coordinated, leading to identifiability by marketers, researchers, government officials, and so on.[29] The user in such a case presumably desires anonymity as a way of ensuring privacy, that is, nondisclosure of one's person (interests, activities, associates, and the like) to others. One problem here is not with anonymity itself, but that anonymity is not in fact the case when it is presumed to be present. Ethical issues may arise here if deliberate deceptive practices encourage such mistaken presumptions about anonymity and privacy, especially when those presumptions occur in contexts that lead to the unwary user making choices or engaging in actions that make her or him vulnerable to unwanted exposure or intrusive monitoring.

---

[29]Computer-mediated communication can facilitate the impression of anonymity for a variety of reasons: because a user is often isolated, or at least alone in some respect[s] during computer use; because apart from chat rooms and other simultaneous user participation areas there may be an absence of immediate feedback from specific others; because of use of account names or numbers that are not immediately correlatable with one's name; because of ubiquitous cookies; because of many users' lack of sophistication in managing their computer-mediated communications; and so on. The case of AOL Internet User No. 4417749, who was revealed to be Thelma Arnold through detailed compilation of the searches she conducted, is a good example of the user who presumes and acts on the presumption of anonymity. The number was supposed to provide a shield of anonymity, but the record of searches published by AOL provided a discernible route to Ms. Arnold. (The New York Times, August 9, 2006, http://www.nytimes.com/2006/08/09/technology/09aol.html?_r=1&oref=slogin, visited August 12, 2006. AOL's publication of data is a distinct issue from the user's beliefs about her anonymity (and privacy). The AOL publication of search records may itself be a case of a provider failing to observe appropriate normative restraints.)

A related issue concerns encryption technology for the content of messages and interactions. Such technology could protect only the content of messages and interactions, but not patterns of communication. Thus, while one might have privacy with respect to the content of one's messages, one might not be anonymous with respect to the patterns of communications. In tracking terrorists or other criminal activities, the online capabilities for such tracking might be legitimate from the point of view of national security and public safety. At the same time, depending on the scope of such monitoring and tracking, since a good deal could be inferred from the monitoring of such patterns, the privacy of ordinary individuals could be compromised.

This kind of online tracking is similar to other forms of electronic tracking. For example, cellphone users or car owners may be assigned anonymous tracking numbers so that they are not identifiable as the particular individuals that they are. However, through the linking of multiple samples enough path information may be accumulated such that an individual user could be reidentified. This is not to say that accumulation of some path information may not be justified with respect to promoting a public good; for example, it might be helpful in epidemiological research or in improving roadways and vehicles.[30] But, associated with monitoring may be risks to things thought to be of value to an individual good life. For instance, individuals may curtail free and autonomous self-expression, or develop repressive or suspicious relations with others. It may be difficult to formulate a general rule that would sort out when a practice is morally permissible from the point of view of the public good and when a practice is too intrusive from the point of view of an individual's prudential or moral interests in privacy or poses too great a threat to what is taken to be basic to living an individually good life.

There are both technical as well as ethical and political issues in this area that call for further analysis and evaluation. For example, to what extent should "invisible" surveillance be allowed, where the observer is anonymous to the observee and where the observee is unaware of the surveillance or of the possibility of surveillance and personal identification through tracking of path information? If no one were aware of surveillance, then the surveillant would be (as in the previous discussion of the peeping tom) unknown, but not anonymous. If someone, even if not the observee, knows that surveillance is going on but doesn't know who the surveillant is, then the surveillant is anonymous to those who know of the surveillance. So, for instance, members of a law enforcement agency might know that surveillance is going on, but might not know who the surveillant is, and the person(s) being surveiled might not know that they are being surveiled. The surveillant would be anonymous to the members of the law enforcement agency, but would be unknown to the observee. If the observee knows that surveillance is happening, but not who is doing it, then the surveillant is anonymous to the observee as well. On one hand, it might be argued that such surveillance is consistent with a morally legitimate interest that a government has in protecting citizens from potential criminal or terror activity. On the other hand, the individual's morally legitimate interests in free speech, privacy and, anonymity might be sufficiently impacted such that one could argue that they outweigh a government's interests in protecting citizen

[30]See, for example, Gruteser and Hoh (2005).

security. If people know about the practice in general, but not in any specific instance, then in online contexts, this in turn could impact the very freedom of communication that the Internet was thought to encourage by impacting how and with whom one interacts.[31]

Commercial and marketing applications of online tracking might be thought to help businesses, as well as consumers. By establishing a consumer's patterns of preferences, marketers can do selective advertising or make suggestions for purchases that are consistent with a user's expressed paths of interest and buying patterns. Given the robustness of clickstream data, this is readily done. And it is routinely done by businesses and even some political organizations where there are repeat users, some of whom choose to set up well-defined profiles in order to filter advertising and simplify their online experiences. However, this is a contested practice. As much as it may be in a business's self-interest, that interest may be outweighed by individuals' protected moral interests in privacy and freedom of expression, especially where that practice was not chosen or agreed to by the individual. There is also the worry that users' self-selection of online profiles or subscription to highly selective political discussion lists may limit exposure to alternative viewpoints and ideas and thus, undermine the free, critical exchange of ideas that is thought to be essential to robust democratic processes. Moreover, differences in users' cognitive ability and computing skills, as well as their understanding of the implications of setting up such profiles or of subscribing to particular discussion lists, all raise a variety of ethical and social issues regarding how online communication and activity alters the nature of social relations, individual expression, economic activity, and political participation. Finally, to the extent that people tracking capabilities are available to the ordinary user, the scope of "monitoring" is not limited to governmental or commercial interests, and could enhance the capacity for immoral behaviors such as stalking and harrassment. In so far as such information is obtainable by individuals, there is also a trend toward "disintermediation" (Bennett, 2001, p. 207), or "decline of the middleperson," that is, interpretive expert. While in some respects, direct access to information may be a good thing and may promote openness, equality, and democracy, it may also blur real distinctions of power and expertise.

7.3.2 Anonymity and Attribution Bias

The idea that anonymity removes the influence of bias is supported by some practices. For example, the now widespread practice of "blind" auditions for orchestral positions has greatly contributed to more equitable evaluation of applicants such that women and persons of color now have decent chances of being selected for orchestral membership. In computer-mediated communications for group brainstorming, anonymity can contribute to decreased evaluation apprehension, conformance pressure and domination and status competition, and increased private self-awareness, all of

[31] Bennett (2001) argues that due to the surveillance capacities of the Internet, the Internet as a "form of life" is gradually shifting away from the assumptions of anonymity upon which it was originally founded. Bennett also discusses the problem of invisible monitoring and tracking (e.g., at p. 203).

which can lead to increased exploration of alternatives and surfacing of assumptions. Anonymity can also, however, lead to loafing, disinhibition, deindividuation[32] and contribute to decreased effectiveness and satisfaction (Hayne and Rice, 1997, p. 431;[33] Spears and Lea, 1994). For example in a context where high social presence and informational richness is needed or desirable, reduction in social cues associated with computer-mediated communication may be associated with misunderstanding, lack of consensus, feelings of impersonality (Hayne and Rice, 1997, p. 434; citing Kiesler et al. 1984; Rice, 1984) and dissatisfaction with a social or decisional process.

One particular risk to note is that anonymizing practices might inadvertently contribute to bias due to the strong human tendency for attribution, that is, to want to identify who is communicating or acting. The tendency to make attributions of authorship or agency can lead to bias and poor decisions if attributions turn out to be inaccurate. A study by Hayne and Rice (1997) showed that in computer-mediated communications in which participants' comments were rendered technically anonymous (author identifying tags were removed), participants frequently attempted attribution, but were mistaken most (nearly 90%) of the time. The concern that this phenomenon raises is that to the extent that attribution is mistaken but also functions as a basis for participants to evaluate other participants' comments and contributions, the anonymizing practice could lead to biased decision making (even though the bias may itself be based on an error in attribution) or encourage irresponsible (even if unintentional) behavior. A similar sort of problem may occur in other anonymous practices, such as blind review of journal submissions, student papers, and the like, where the community of contributors and evaluators may be small enough such that participants will seek to make attributions, often be mistaken, and the mistaken attributions contribute to bias in the evaluations. Therefore, where anonymity for the sake of eliminating bias is desirable, one cannot assume that technical anonymity by itself guarantees the removal of bias. Rather, if technical anonymity allows for biased

---

[32]In discussions of "deindividuation" the idea is that in a group the individual is "anonymous" and experiences a loss of self-awareness as a distinct individual self and identity. The term "anonymous" in such a context may or may not be coincident with the definition of anonymity as nonidentifiability by virtue of noncoordinatability of traits. If the terms refer to an individual's experienced sense of blurring between individual and group, it would not be coincident. If, however, group membership or participation leads to nonidentifiability in the sense of noncoordinatability of traits by others, then the meaning would be coincident. An example of the latter might be where an agent is not identifiable as the author of an act because group membership makes it impossible or difficult to coordinate authorship of an act with a specific individual; acting as a hooded member of the Ku Klux Klan might be such an instance where group membership (along with artifices of disguise) renders an individual agent anonymous to others outside the group (and perhaps even to other group members), while at the same time contributing to the individual's sense of group identity.

[33]Spears and Lea (1994) have suggested that whether anonymity attenuates or increases group conformity depends on whether it is personal or individual identity that is salient or social identity that is salient: "If group identity is not salient, the isolation of CMC [computer-mediated communications] may further weaken the salience of group identity and thus undermine both conformity to in-group norms, and the tendency to engage in intergroup behavior. In these circumstances CMC may weaken the power of the group over the individual. However, if one identifies with one group or another, the deindividuating conditions of isolation can strengthen the salience of this identification, so that people are more likely to conform to group norms and engage in intergroup behavior" (445).

misattribution, then, at least in some contexts, there may need to be additional steps introduced to mitigate this possibility.

7.3.3 Anonymity and Expression of Self

One function of anonymity might be to allow an individual to act or to express herself in ways that would not be possible or recognized if the identity of the individual were known. For example, a woman writer in taking a male pseudonym (here, a pseudonym might indeed function to ensure anonymity[34]) might enable recognition of her work which would otherwise have not been published at all (e.g., George Eliot, George Sand). Or, as noted above, the ability of women and persons of color to successfully compete for orchestral positions has been considerably advanced by the introduction of blind auditions, whereby applicants play behind a screen and thus, for the purposes of the audition, are anonymous to the conductor or panel of musicians performing the evaluations.

Computer-mediated or online communication may facilitate communication, participation and exploration of the self, or the development of free political speech that might not otherwise be possible or recognized. Regarding the expression of self, the proliferation of personal websites and creation of personal profiles, especially, among teenagers, may be indicative of new avenues of self-expression and risk-free experimentation with adopting various personae in the development of personality. MySpace and Facebook are two such currently prominent examples of how the capabilities of the Internet may ameliorate the "facelessness" of modern life that some find worrisome. The development of such sites and creation of personal profiles are indicative of complicated attitudes that individuals may have about privacy and anonymity in different contexts and perhaps also at different stages of their lives. They are also indicative of some of the ways in which the Internet and online communication are altering social patterns of communication and interaction and restructuring social relations. One consequence might be that relationships become more episodic and one's social contacts more widely geographically dispersed. In addition, while instant messaging allows for speedy and direct response, it may tend to discourage or weed out more thoughtful responses. It may even in some cases lead to misunderstanding. Sometimes the meaning of a communication is not immediately apparent and only becomes so as the recipient of the communication reflects on it and allows interpretation to take place. Instant messaging might tend to displace the occurrence of this more extended reflective, interpretive process. On the other hand, the rapidity of response may allow for speedier clarifications of meaning and hence avoid misunderstandings.

Wide access to the sites containing personal profiles may also expose individuals in ways that may be harmful to them, for example, in the context of job seeking when employers check their personal sites, or if the profiles open them up to cyberstalking. One can imagine how anonymity might figure into both the

[34]A pseudonym need not function as an anonymizing practice; for example, the mystery writier Ruth Rendell uses the pseudonym "Barbara Vine" in order to take on a particular authorial voice and it is widely known that she does so.

facilitation of cyberstalking and protection from cyberstalking. For the agent who is the cyberstalker, the capacity for anonymity with respect to action on the Internet could increase the degree and extent of cyberstalking. For the recipient of such stalking, anonymity on the Internet might be an important device for protection from such stalking. While online communication may not alter the basic ethical issues involved in stalking, it may both enhance methods of protection for the stalked and the scope of the abilities of the stalker.

It might be argued that online communication is detrimental in principle because it represents a decline in or loss of face-to-face interaction. If face-to-face communication is important for human flourishing, then as a moral matter, online communication that diminishes such contact might be morally problematic. Online personal communication and exploration may also be a source of exploitation of intimacy and trust. Another issue might involve equity in access; if it were the case that there were a disparity between different groups of people, for example, between genders, races or ethnicities, in their access to computer-mediated communication (CMC), and that styles of communication in CMC may favor some subcultures and styles over others, then, anonymity could serve to render underrepresented participants even more invisible or be a device for suppressing their individual identities in order to be able to participate or conversely, to avoid unwanted attention (Spears and Lea, 1994, p. 450).

Regarding the expression of free political speech, it may be that in some contexts anonymity enables such political speech without fear of retribution and the capacity to do so via the Internet may be an important development in human rights.[35] (Unfortunately, it also makes it easier for child pornographers and pedophiles to communicate, as well as for racial or ethnic supremacists to form hate groups.) In a global context, the possibility of mass distribution of such speech may also enhance its effectiveness, for example, by engaging citizens in other countries to urge their governments to bring political pressure to bear on oppressive regimes. At the same time, there may be the risk that some behavior or speech that passes as political also serves to inflame passions for irrational and destructive behaviors. There might nonetheless be a presumption in favor of anonymous political speech, at least in the US context, as part of a First Amendment right of freedom of speech and association.[36]

7.3.4 Globalization of Online Activity

Online communication and self-expression may facilitate forms of interaction that could be beneficial to human and cultural experience, albeit in different ways and in different respects. However, there are risks. Given the international character of the Internet, there might also be concerns raised about anonymous speech in so far as it could create problems for enforcement of libel and intellectual property law. For

---

[35]For example, Lance Cottrell established the Kosovo Privacy Project enabling individuals to use anonymizer services to report from within the 1999 Kosovo war zone without fear of retaliation (www.livinginternet.com/i/is_anon.htm, last visited September 24, 2006).

[36]See McIntyre v. Ohio Elections Commission, 115 S.Ct. 1511 (1995); Froomkin (1999).

instance, Internet libel could be spread worldwide and be effectively indelible since it could be stored in innumerable, untraceable computers.[37]

In a global context, the Internet may contribute to the blurring of jurisdictional boundaries—raising such questions as: where does speech, hacking, fraud, or any other type of online activity occur; where do its effects occur; and which government has jurisdiction over what behavior? The Internet may, at least for a time, render legal remedies for illegal online activities difficult to precisely adjudicate and execute. For instance, in light of the international character of the Internet, legal rules imposed by a particular nation may have less and less importance.[38] But the Internet may also be a force that pushes for more international cooperation. One worry might be that too much intergovernmental cooperation is threatening to democratic processes if it entails nonaccountability to an electorate, but discussion of this particular issue is beyond the scope of this article.

7.3.5 Anonymity and Identity Theft

In identity theft, typically some important identifying tags of a person are stolen, for example, social security number, credit card information, banking information, name and address, and so on. The thief may use the stolen identifying tags for her or himself, or may sell or transfer them to others, who then use them. The uses themselves typically involve fraud or theft, for example, using credit card information to make expensive consumer purchases, or using bank information to transfer or withdraw sums of money.

How is this kind of theft related to anonymity? Does it constitute a form of anonymity? The thief might be anonymous (not identifiable) in so far as the theft is performed in such a way that there are no paths of accessibility from the action (the theft) to the person who performed the action (the thief). Thus, *qua* thief the thief is anonymous. This is not different in principle from any other kind of theft or crime, where a criminal aims to leave no traceable trail back to her or himself.[39] The use of such information to perform another act, for example, fraud, credit card purchase, and the like, is a case of deliberately misdirecting identification. That is, the user of such information is able to steal and defraud by, in part, assuming the identity of someone else and hence misdirecting identification away from her or himself to that other party.

Identity theft involves stealing identifying information and, in effect, using it as a pseudonym to misdirect identification and render the actual agent anonymous. In most cases, identity theft is for the sake of theft, access to credit and money, and not for the sake of taking over someone's identity. However, the reason it is called identity theft is precisely because it seeks to accomplish theft by making the transaction appear to be legitimate by virtue of the assumed identity (a kind of pseudonym in order to achieve anonymity). Of course, in physical world transactions, one could often use cash to ensure anonymity with respect to a particular transaction (provided there are also no cameras or other means by which one could

[37]Froomkin (1999), p. 114.

[38]Froomkin (1999), p. 124.

[39]Recall the earlier discussion of the Unabomber.

be identified). But, if the idea is to steal or defraud and to do so online, then identity theft permits thieves to adopt a persona or pseudonym to perform the theft and by doing so allows them to steal anonymously. The extent to which the actual purchaser, the one using someone else's identifying information, is unidentifiable (anonymous) will depend on how well the user makes sure that there are no traceable paths back to the actual user. It could be argued that the basic action of theft and the ethical issues are not different in principle from the use of other ruses or disguises in perpetrating criminal activity, which, by misdirecting identification, provide anonymity (nonidentifiability) to the thief. However, by using someone else's identifying information and thus implicating them in the transaction, and by producing considerable aggravation for the person whose identifying information has been stolen and used, it could also be argued that the action is not only theft, but a kind of false accusation of another. If so, the latter would be a distinct ethical issue from that of ordinary theft. Other cases where identity theft is more transparently a kind of false accusation is where it is performed for the deliberate purpose of setting someone up, either as the "fall guy" for financial or consumer fraud or for some other nefarious activity. One can imagine, for instance, a terrorist using identity theft to misdirect identification so as to enable the terrorist to keep the authorities busy elsewhere, rather than with the actual terrorist, potentially leading to serious harm (arrest, detention, and so on) to an innocent person. The collection of identifying tags, their security in databases and the extent to which innocent people could be inconvenienced, and perhaps quite seriously harmed, by breaches of security are ongoing concerns in computer assisted technology, in so far as any of these collections are vulnerable to theft.

7.4 PURPOSES AND ETHICS OF ANONYMITY

Anonymity may serve or be sought for a variety of purposes: to shield someone from accountability for action, to ensure privacy, to prevent discrimination or stigmatization, to facilitate communication, to avoid reprisals, and so on. It is useful to identify the general categories into which purposes or goals of anonymity can be grouped. I have suggested that there are three (Wallace, 1999):

(1) anonymity for the sake of furthering action by the anonymous person, or agent anonymity;
(2) anonymity for the sake of preventing or protecting the anonymous person from actions by others, or recipient anonymity;
(3) anonymity for the sake of a process, or process anonymity.

Any given case of anonymity may serve any or all of these purposes; in other words, they are not mutually exclusive. In most cases, it will probably be a matter of looking at the *primary* purpose or goal that the anonymity serves.

(1) With anonymity for the sake of enabling action the action could be good, bad, or neutral, and presumably ethical evaluation of any given case will depend on the

particular action or range of action that anonymity enables. For example, the Unabomber presumably sought and maintained anonymity primarily for the sake of being able to send bombs to computer scientists with impunity and thus, would be an example of "agent anonymity." The anonymity aims to prevent punishment for his action, and in that sense it also serves to protect him from actions of others, for example, law enforcement agents. An identity thief and the user of the identifying information of others would be another example of agent anonymity. Other examples of anonymity serving the purpose of enabling or furthering action might include anonymous donors, anonymous authors, anonymous bidders at art auctions, anonymous sources in news reporting, and police work.[40] Internet-based discussion groups and chat rooms may involve both spontaneous anonymity and deliberately assumed anonymity of participants with respect to communicative expression, in which cases the action is speech or expression.[41] In this sort of case, the act (speech or expression) may be ethically valuable (e.g., free political speech that promotes the creation of an informed public) or not (e.g., hate speech).

(2) Anonymity could serve the primary purpose of preventing actions by others or more generally protecting the anonymous person from being the recipient of actions by others, hence, for short, "recipient anonymity." For example, in HIV testing,[42] test results are anonymous for the purpose of protecting the potentially HIV-positive client from stigmatization, loss of employment and health insurance, and so on.[43] (In so far as such anonymity is also for the sake of encouraging testing in high risk populations, its purpose might also be to ensure the reliability of a public health process.)

(3) Anonymity could also be for the sake of preserving the validity of a process. This is the type of case where the anonymity is primarily or also for the purpose of some other goal than enabling or protecting particular (anonymous) person(s). For example,

[40]Again, in any of these cases, anonymity might also serve to protect persons from actions (e.g., public attention) of others, as well as enable the action itself.

[41]An internet chat room pseudonymous identity would be a case where a pseudonym does function as a mechanism for achieving anonymity, even though as we noted earlier, pseudonyms need not function so.

[42]Individuals and their blood samples are given a number and in some cases the individual is asked to leave a first name. It need not be their own name. The individual is told when to return to the testing clinic for her/his results. The individual provides no other personal information, such as name, address, telephone number, social security number, credit card numbers, employer, with which the assigned number could be coordinated. If the individual does not appear to obtain her/his results, she/he cannot be traced by clinic counselors. This arrangement provides for anonymity, at least in a large urban environment, because even when one meets face-to-face with a clinic counselor and hence could theoretically be recognized by a counselor on the street, such chance encounters are highly unlikely when counselor and individual do not already know one another and the HIV testing is their only contact and reason for contact. Since there is no other personal information connected to the blood sample, the test results on that blood sample cannot be traced to the particular individual other than through the number tag.

[43]Such anonymity might, in any instance, also serve to further actions of the person and hence be a case of "agent anonymity" as well. For example, suppose the HIV-positive person was also homosexually active and was more concerned to have that activity be the uncoordinatable trait. In such an instance being HIV-positive would be a strong signifier for such activity and thus anonymity with respect to HIV tests results would also serve the purpose of being anonymous with respect to sexual activity, although this might at the same time allow infected persons to infect others, spreading the disease.

test taking, research studies, peer review of work, anonymous online brainstorming, and the judicial arena are all areas in which anonymity has been thought to serve the purposes of neutrality and impartiality. Thus, in double-blind peer review or blind evaluation of tests, the purpose of anonymity between the participants may be for the primary purpose of maintaining fairness of the process, the objectivity or impartiality of the evaluations, and the like.[44] At the same time, it may serve the purpose of protecting persons from bias (recipient anonymity) in the process of being evaluated (even though, as noted earlier, there may be a risk of attribution bias in some contexts). College classes that set up anonymous discussion "rooms" on a LAN, where students' comments cannot be coordinated with their student authors, may be supposed to promote freedom of expression and freedom to experiment without worry of grading or knowledge of their comments by the instructor. It might be thought that the process of the free exchange of ideas warrants the risks of irresponsible, discriminatory, or even harrassing comments. This would be an instance of process anonymity. Similarly, anonymity may encourage people to post requests for information to public bulletin boards that they might not be willing to make if they could be identified. Anonymity might serve to protect the individual, but it might also serve a larger public health process of providing information about communicable diseases.

It may also be possible to act anonymously and for the action to have effects in the lives of others, without those affected being able to identify the specific action itself or the (anonymous) actor. For example, someone could take steps that are decisive in another person being fired or promoted and the person affected could not be able to correctly identify the causes of her dismissal or promotion. In such a case, the action is not directly known, but is only indirectly experienced.[45] Another example might be the use of secret evidence, where both an accuser and the specific accusation are unknown to a defendant;[46] both the action and the agent would be unknown to the person

[44]Any particular instance of anonymity could serve all three purposes. For example, in test taking, the name of a test taker might be made noncoordinatable by an evaluator with an answer sheet in order to prevent bias that might be harmful to the test taker. In so far as the prevention of bias also serves as a way of ensuring the validity or objectivity of test results, anonymity may also serve to promote impartiality more generally. Evaluators might also be anonymous both to encourage uncontaminated evaluation as well as to protect evaluators from reprisals from disgruntled test takers.

[45]Some action is known, that is, being dismissed or promoted, but some agentially caused event may not be known. If the latter is not known by anyone, then the agent would be unknown, rather than anonymous. If the latter is known in at least some respect by someone (even if not the affected employee) but the agent is not identifiable, then the agent would be anonymous to those who know of the action. Thus, suppose coworkers anonymously submit complaints about another coworker to a manager, and suppose that in light of those complaints as well as other problems with the worker, the manager decides not to renew the worker's contract, but without telling the worker about the complaints. The coworkers would be anonymous to the manager, but unknown to the worker whose contract was not renewed. An example of the former possibility (dismissal or promotion but without anyone knowing of an agential cause) would be if someone secretly planted incriminating information in someone's personnel record or conversely secretly raised someone's evaluations such that the employee was dismissed or promoted and the employee and others never had any idea that the records had been falsified. In such a case, since the action is not known, reference to an anonymous agent can't get going, and therefore, the action and the agent would be simply unknown.

[46]Recent cases of this involve the U.S. Federal Government using secret evidence as grounds for the deportation of immigrants. (See New York Times, 8/15/1998, New England Edition, pp. A1 and A11.)

affected. However, there may be a borderline sense in which these examples might qualify as potentially anonymous. For there is still an action or effects or traces of an action(s), even though the action(s) may not be not clearly identifiable, let alone coordinatable with other traits of the agent. These cases seem to be distinct from cases where there is no agent, or person who is the agent; for example, suppose a computer program were to initiate a series of calculations or actions, unforeseen by programmers, which affected others, *and* it were also not clear exactly what the actions were that the program had taken. This should be distinguished from anonymity that should be understood as applicable to persons and not simply to unidentifiable causes. However, it may be impossible to make precise distinctions, for example, in computer design and programming, where it may be difficult to draw a clear line between the program itself and the action of the person *qua* programmer or designer, especially when programs themselves do some programming.[47]

Groups of persons, as well as individual persons, may seek or have anonymity. Group anonymity can have a variety of purposes. For example, with a group like Alcoholics Anonymous anonymity may serve several purposes simultaneously: it may help provide a "safe" therapeutic setting; it may help to prevent critical or prejudicial judgment (stigmatization) or action by others against those attempting to recover from alcoholism as well as to possibly render noncoordinatable by members of the group their own other locations, again for similar reasons; all these would be instances of recipient anonymity. In a pedophile association or the Ku Klux Klan, anonymity may serve both the purpose of enabling action (agent anonymity) and the purpose of shielding members from accountability and prosecution for their actions (recipient anonymity). Group anonymity can also mean that members are anonymous to one another in so far as their other traits are rendered noncoordinatable with their group membership and activities by the other members of the group. This may also be a characteristic of persons who participate in online chat rooms and the like.

Whatever be the purpose(s) of a given instance of anonymity, it raises the issue of accountability, whatever other issues, such as attribution bias, noted above, may be involved or at stake. We might label the issue of accountability the *Ring of Gyges scenario*. In Plato's parable,[48] Gyges finds a ring that makes him invisible whenever he wears it. (I'm not suggesting that we rely on magical means of anonymizing; rather I am suggesting that Plato's parable makes vivid the risk in anonymity.) Plato's point is that the successfully unjust (immoral or unethical) person seeks and relies on strategies that enable her to avoid accountability for her actions. The Ring of Gyges scenario is when someone's ability to be invisible — anonymous — allows unethical or criminal action with impunity.

Anonymity, then, carries with it the risk of minimizing accountability for action. Just as it might encourage freedom of expression, it might also raise the specter of

[47]Some of these difficulties would involve what Moor calls *invisible programming values* (Moor, 1985). For discussions about whether a program would count as an actor or agent, see Moor (1979), Snapper (1985), Bechtel (1985), Nissenbaum (1996), Szolovits (1996).

[48]Plato, *Republic*, II, 359b–360e.

secret informers and malicious denunciations.[49] The Unabomber or identity thief would be clear cases of anonymity that illustrate the Ring of Gyges scenario (harassment activity by members of a Ku Klux Klan might be a case at a group level). Even when the initial primary purpose is to protect from harmful actions by others or to promote positively valued activity, anonymity also provides space for action with impunity and hence, the Ring of Gyges scenario.[50] This does not necessarily entail that anonymity is something that should be avoided, but it does urge caution and minimally suggests that wherever anonymity is appropriate, there ought to be safeguards to mitigate the risk of the Ring of Gyges scenario.

In addition to accountability, other general ethical issues associated with anonymity may be conceptualized in different ways, depending on the type of ethical theory invoked. Anonymity itself may be problematic for some ethical approaches that stress the importance of recognition of the other in their fullness as a person.[51] A care ethics might be concerned that anonymity depersonalizes the person too much such that she/he wouldn't be an object of care.[52] It might be argued that the Internet, by providing the means for enhanced knowledge and awareness, enhances our ability to care about poverty and suffering persons around the globe. On the other hand, a care ethicist might argue that for a person to be an object of care, it is not sufficient that there be some general sense of "caring about" an issue or population of persons, but that a person be someone for whom one does the work of "caring for," and that can only be done in so far as the person is individuated and their specific care needs recognized. To the extent that the Internet removes us from persons in this way, it may undermine "caring for." From the point of view of such theories that emphasize face-to-face contact, a loss of full presence in the social interaction, or the absence or hiddenness of the person, might be *in principle* undesirable, because the fundamental ethical relation (recognition, care) cannot be reliably established. On the other hand, since anonymity may in some contexts enable recognition of at least some aspects of the person that might otherwise go either unnoticed or be perverted by bias and discrimination, a wholesale condemnation of it on these grounds seems unwarranted. The care ethics worry might be less easily addressed, since the argument there is that having intimate knowledge of the person being cared for is necessary in order to enact the duties of care. Care ethics might also give voice to some concerns about social distancing effects of the Internet. However, care ethics might also both highlight the problem of disparities between those who are able to fully participate in a technology and those who are permanently disabled, and, at the same time, offer ethical justifications for using technology to enhance the care for and capabilities of those who are disabled.

---

[49]Froomkin (1999) makes the same point (p. 114).

[50]For example, presumably an anonymous whistle-blower seeks anonymity in order to protect herself from reprisal, for example, being fired or harassed. At the same time, such anonymity could provide space for slanderous or false charges of incompetence by a disgruntled employee.

[51]Consider, for example, Buber's I-Thou relationship, Hegelian-based recognition ethical theories and Levinas's notion of the Face as the central ethical concept.

[52]Gilligan (1982) coined the term "care ethics" in her seminal work, *In a Different Voice*. For a succinct discussion of the central issues in care ethics see Tronto (1993).

Other ethical theories might interpret anonymity as desirable, permissible, or perhaps even, in some context, obligatory. For example, in a Kantian-based ethics, respecting privacy might be important in so far as doing so embodies a principle of respect for persons and their autonomy. Therefore, to the extent that anonymity is a means to privacy, then an ethics that takes respect for persons as a basic ethical principle would have a place for anonymity in so far as it embodied and cohered with such respect. A moral value of respecting persons might be advanced by anonymity that aims to preserve privacy, where issues of noninvasiveness or autonomy or both might be involved. Thus, it might be argued that others (people, groups, or corporations) do not have a right to have access to some (perhaps many) locations of a person and therefore, that mechanisms that ensure anonymity are devices that promote or ensure minimal conditions of respect of persons.[53]

A consequentialist-based ethics would evaluate practices of anonymity in relation to the desirability of their consequences, that is, the benefits and harms that some particular practice promoted. Such an approach would not have a principled objection to anonymity, but would evaluate more specific instances and practices in terms of their harms and benefits.

There may also be other ethical issues involved that are not directly about anonymity itself. Thus, when anonymity depends on a third party, for example, when someone represents an agent without revealing the identity of the agent, there are contractual and trust issues between the representative and the person(s) represented. Some of these issues may be present in the online setting as well, for instance, when a service provider or a business represents itself to a user as ensuring anonymity or security of personal information.

Because there are many forms of anonymous communication and activity, and a variety of purposes that anonymity may serve, it may be important to distinguish what type of communication or activity is involved, rather than have a single legal policy or ethical stance toward anonymity (Allen, 1999).

REFERENCES

Allen, C. (1999). Internet anonymity in contexts. *The Information Society*, 15, 145–146.

Bechtel, W. (1985). Attributing responsibility to computer systems. *Metaphilosophy*, 16(4), 296–306.

Bennett, C. (2001). Cookies, web bugs, webcams and cue cats: patterns of surveillance on the world wide web. *Ethics and Information Technology*, 3, 197–210.

Boër, S.E. and Lycan, W.G. (1986). *Knowing Who*. MIT Press.

Bynum, T. (2001). Computer ethics: basic concepts and historical overview. In: Edward, N.Z. (Ed.), *The Stanford Encyclopedia of Philosophy (Winter 2001 Edition)*. URL=http://plato.stanford.edu/archives/win2001/entries/ethics-computer/

[53]There is a vast literature discussing the meaning and value of privacy. One suggestion, made by Reiman (1976), is that a right to privacy is essential to the development of personhood and thus derives from a principle of respect for persons.

Dreyfus, H. (2001). *On the Internet*. Routledge, London and New York.

Feldman, D. (1994). Secrecy, dignity or autonomy? Views of privacy as a civil liberty. *Current Legal Problems*, 47(2) 41–71.

Feldman, D. (1997). Privacy-related rights and their social value. In: Birks, P. (Ed.), *Privacy and Loyalty*. Oxford University Press.

Ferguson, K. (2001). Caller ID — whose privacy is it, anyway? *Journal of Business Ethics*, 29, 227–237.

Froomkin, A.M. (1995). Anonymity and its enmities. *Journal of On-line Law*, 4, 1–27.

Froomkin, A.M. (1999). Legal issues in anonymity and pseudonymity. *The Information Society*, 15, 113–127.

Gilligan, C. (1982). *In a Different Voice: Psychological Theory and Women's Development*. Harvard University Press.

Goold, B. (2002). Privacy rights and public spaces: CCTV and the problem of the "unobservable observer". *Criminal Justice Ethics*, 21(1), 21–27.

Gordon, L.R. (1997). Existential dynamics of theorizing black invisibility. In: Gordon, L.R. (Ed.), *Existence in Black: An Anthology of Black Existential Philosophy*. Routledge, New York.

Gruteser, M. and Hoh, B. (2005). On the anonymity of periodic location samples. In: Hutter, D., Ullmann, M. (Eds.), *Security in Pervasive Computing*. Springer-Verlag, Berlin and Heidelberg, pp. 179–192.

Hayne, S. and Rice, R.E. (1997). Accuracy of attribution in small groups using anonymity in group support systems. *International Journal of Human Computer Studies*, 47, 429–452.

Kiesler, S., Siegel, J., and McGuire, T. (1984). Social psychological aspects of computer-mediated communication. *American Psychologist*, 39, 1123–1134.

Kling, R., Lee, Ya-ch., Teich, A., and Frankel, A.S. (1999). Assessing anonymous communication on the internet: policy deliberations. *The Information Society*, 15, 79–90.

Marx, G.T. (1999). Reflections on the sociology of anonymity. *The Information Society*, 15, 99–112.

Moor, J. (1979). Are there decisions computers should never make? *Nature and System*, 1, 217–229.

Moor, J. (1985). What is computer ethics? *Metaphilosophy*, 16(4), 266–275.

Natanson, M. (1986). *Anonymity: A Study in the Philosophy of Alfred Shutz*. Indiana University Press.

Nissenbaum, H. (1996). Accountability in a computerized society. *Science and Engineering Ethics*, 2, 25–42.

Nissenbaum, H. (1999). The meaning of anonymity in an information age. *The Information Society*, 15(2), 141–144.

Reiman, J.H. (1976). Privacy, intimacy and personhood. *Philosophy and Public Affairs*, 6(1), 26–44. Reprinted in: Schoeman, F.D. (Ed.), *Philosophical Dimensions of Privacy*. Cambridge University Press, 1984, pp. 300–316.

Rice, R. (1984). Mediated group communication. In: Rice, R.E. and Associates (Eds.), *The New Media: Communication, Research and Technology*. Sage, Beverly Hills, CA, pp. 129–154.

Russell, B. (1956). On denoting. In: Robert, C.M. (Ed.), *Logic and Knowledge*. Routledge, pp. 39–56.

Snapper, J.W. (1985). Responsibility for computer-based errors. *Metaphilosophy*, 16(4), 289–295.

Spears, R. and Lea, M. (1994). Panacea or panopticon? The hidden power in computer-mediated communication. *Communication Research*, 21(4), 427–459.

Szolovits, P. (1996). Sources of error and accountability in computer systems: comments on "accountability in a computerized society". *Science and Engineering Ethics*, 2, 43–46.

Thompson, D. (1987). The moral responsibility of many hands. *Political Ethics and Public Office*. Harvard University Press, pp. 46–60.

Tronto, J. (1993). *Moral Boundaries: A Political Argument for an Ethic of Care*. Routledge, New York.

von Hirsch, A. (2000). The ethics of public television surveillance. In: von Hirsch, A., Garland, D., and Wakefield, A. (Eds.), *Ethical and Social Perspectives on Situational Crime Prevention*. Hart Publishing, pp. 59–76.

Wallace, K.A. (1999). Anonymity. *Ethics and Information Technology*, 1, 23–35.

CHAPTER 8

Ethical Issues Involving Computer Security: Hacking, Hacktivism, and Counterhacking

KENNETH EINAR HIMMA

This chapter considers whether and to what extent various types of unauthorized computer intrusions by private persons and groups (as opposed to state agents and agencies) are morally permissible;[1] this chapter does not cover other security-related issues, such as issues at the intersection of computer security and privacy, anonymity, and encryption.[2] The first section articulates a *prima facie* general case against these intrusions. The second considers intrusions motivated by malicious intentions and by certain benign intentions, such as the intent to expose security vulnerabilities. The third considers hacktivism, while the fourth considers counterhacking (or hackbacks).

Certain assumptions about "hacker" and related terms should be made explicit. Although these terms were once used to refer to accomplished programmers and their achievements, they are now used to refer to unauthorized computer intrusions and the persons who commit them. Thus construed, "hacking" is used, without moral judgment, to refer to acts in which one person gains unauthorized entry to the

---

Portions of this chapter appeared in my articles "Hacking as politically motivated digital civil disobedience: Is hacktivism morally justified?" and "The ethics of hacking back: active response to computer intrusions," which were originally published in Himma, K.E. (2007) *Internet Security: Hacking, Counterhacking, and Society.* Jones & Bartlett. I'm grateful to the publishers for permission to reprint those portions here.

[1]Public acts (i.e., those performed by state agencies) raise radically different issues. States are frequently permitted to do things that private individuals are not—such as incarcerating persons.

[2]For a discussion of these other issues, see Tavani, H. (2007). "The conceptual and moral landscape of computer security" In: Himma, K.E. (Ed.), *Internet Security: Hacking, Counterhacking and Society.* Jones & Bartlett, Sudbury, MA. In that essay, Tavani also differentiates ethical issues affecting three distinct aspects of computer security: data security, system security, and network security.

The Handbook of Information and Computer Ethics, Edited by Kenneth Einar Himma and Herman T. Tavani

computers of another person, and "hacker" is used to refer to someone who has committed such acts. Although some programmers bemoan the change in meaning, this chapter acquiesces to current usage.

8.1 THE *PRIMA FACIE* CASE AGAINST HACKING

At first glance, it might seem obvious that hacking is wrong. Although the more malicious of these acts involve serious wrongs because of the harm they cause, all are wrong because they constitute a digital trespass onto the property of another person. Unauthorized entry into some other person's computer seems not relevantly different than uninvited entry onto the land of another person. Real trespass is morally wrong, regardless of whether it results in harm, because it violates the owner's property right to control the uses to which her land is put and hence to exclude other people from its use. Similarly, digital trespass is wrong, regardless of whether it results in harm, because it violates the owner's property right to exclude other people from the use of her computer, which, like land, is physical (as opposed to intangible) property.

There are two problems with this argument. First, assuming that hacking is a species of trespass, it doesn't follow that all hacking is wrong because not all trespasses are wrong. It is permissible to trespass onto your land if doing so is the only way to capture a murderer fleeing the crime scene; committing a minor trespass is morally justified as the only way to secure the great good of stopping a killer. If hacking is trespass, then hacking necessary to secure some good that significantly outweighs the evil involved in trespass would also be justified.

Second, and more importantly, it is not clear that the concept of trespass properly applies to digital intrusions. The term "trespass" has largely been reserved—at least in moral usage—to refer to acts in which one person enters upon physical space owned by another, but a hacker is not in any *literal* sense *entering* upon a physical space owned by another person. Perhaps digital intrusion is more like using heat sensors to see what is going on inside a house, which is not usually characterized as "trespass," than like coming into the house without permission.

Even so, it seems clear that digital intrusions impinge upon legitimate interests of computer users. It seems clear, for example, that an unauthorized computer intrusion impinges upon the victim's property rights. Someone who gains access to my computer without my permission is appropriating a physical object in which I have a legitimate property interest; it is, after all, *my* computer—and I have, at the very least, a presumptive moral right to exclude other people from appropriating my computer.

If this is correct, then an unauthorized computer intrusion also impinges upon privacy rights. Someone who hacks into my computer without my permission gets access to something in which I have a legitimate expectation of privacy. If I may legitimately exclude others from my computer, then it is reasonable to regard my computer as a private space in which I can store sensitive information. Indeed, insofar as a computer user has a legitimate expectation of privacy in the contents stored on her computer, an unauthorized intrusion impinges upon the victim's privacy rights—regardless of whether there is, in fact, any sensitive information stored on that machine.

Moral rights are not, however, absolute. If I may trespass to capture a fleeing killer, then a person's right to property can be outweighed by more important rights when they conflict; property rights are weaker, for example, than the right to life. Similarly, a person's privacy rights can be outweighed by other more important interests; a person might, for example, be obligated to disclose sensitive information if needed to ensure another person's safety. Thus, the mere fact that a person has property and privacy rights in her computer does not imply that all unauthorized intrusions are impermissible.

Even so, the burden rests on the hacker to show that a particular intrusion is morally permissible. This will involve showing that any legitimate property or privacy interests are outweighed, as an ethical matter, by interests that can be secured only by committing the intrusion. Insofar as an intrusion involves causing damage to the files of the user, the intrusion can be justified only to the extent that it serves correspondingly greater interests.

Nevertheless, intrusions intended to cause harm out of *malice* are generally wrong. Although it is sometimes permissible to inflict harm on another person when necessary to secure a greater good, a malicious *intention* does not seek a greater good. This is not to say that it is necessarily wrong for one party to hack into another party's computer for the purpose of causing harm. Presumably, it is permissible to hack into the computers of known terrorists to delete files associated with a terrorist plot; given that innocent lives are at stake, this sort of digital harm seems clearly permissible. But these motivations are not *malicious*; the motivation here is to secure the greater good of saving lives, which justifies inflicting the comparatively minor harm of deleting files intended to advance an egregiously wrongful plot.

8.2 OVERCOMING THE *PRIMA FACIE* CASE: HACKING MOTIVATED BY BENIGN PURPOSES

Many hackers believe benign intrusions not calculated to cause damage can be justified on the strength of a variety of considerations. Such considerations include the social benefits resulting from such intrusions; speech rights requiring the free flow of content; and principles condemning waste. These arguments are considered in this section.[3]

8.2.1 The Social Benefits of Benign Intrusions

Hackers point out that benign intrusions have a number of social benefits. First, by gaining insight into the operations of existing networks, hackers develop knowledge that can be used to improve those networks. Second, the break-ins themselves call attention to security flaws that can be exploited by malicious hackers or, worse, terrorists. These are benefits that conduce to the public good and are thereby justified.

[3] I should acknowledge, at the very outset, that the discussion in this section has been deeply influenced by Spafford, E. (1992).

None of these benefits justifies benign intrusions. Even if we assume that privacy and property rights might sometimes yield to such utilitarian considerations, these social benefits can be achieved without infringing upon any moral rights. For example, hackers can develop these techniques and technologies in settings where the consent of all parties has been obtained. In cases where hackers seek entry to the machines of ordinary, noncommercial users, they can solicit the consent of other like-minded individuals to allow them to attempt to circumvent the relevant security measures. In cases where hackers seek entry to the machines of larger commercial users, they can seek employment at those firms or advise persons already employed at those firms. It is wrong to infringe privacy and property interests to achieve social benefits that can be achieved without infringing those rights.

There is, however, a deeper problem with this strategy of argument. If, as is commonly believed, the privacy and property interests of computer owners in their machines rise to the level of moral *rights*, then an appeal to social benefits cannot justify hacking. It is part of the very concept of a right that the infringement (as opposed to violation) of a right cannot be justified solely by an appeal to the desirable consequences of doing so. The mere fact that someone could do a lot of social good by stealing, say, a billion dollars from Bill Gates cannot justify stealing that sum if Gates has a property right to all of that money. As Ronald Dworkin famously puts the point, rights trump consequences.[4]

The social benefits argument, then, fails because it is the wrong kind of argument. The property and privacy rights computer owners have in their machines can justifiably be infringed by an unauthorized intrusion only if required to secure some more important right that outweighs those privacy and property rights. If rights trump consequences, then hackers must identify some stronger reason that justifies an intrusion: the appeal to social benefits, by itself, is insufficient to justify the intrusions.

8.2.2 Benign Intrusions as Preventing Waste

Hackers have also defended benign intrusions on the ground that they make use of computing resources that would otherwise go to waste. On this line of reasoning, it is morally permissible to do what is needed to prevent valuable resources from going to waste; benign hacking activity is justified on the strength of a moral principle that condemns squandering valuable resources in a world of scarcity in which there are far more human wants than resources to satisfy them.

This argument, unlike the social benefits argument, is the right kind of argument because it attempts to identify a moral principle that might limit other rights, like the right to property. Here it is crucial to note that rights are often limited by other moral principles; the right to life, for example, is limited by a moral principle that allows persons to kill if necessary to save their own lives from a culpable threat.

---

[4]See Dworkin, R. (1978). *Taking Rights Seriously.* Harvard University Press, Cambridge, MA.

Nevertheless, the argument fails. If one person has a property right in some object X, it is wrong for other persons to appropriate X without permission to prevent X from being wasted. As Spafford aptly puts this point:

> I am unable to think of any other item that someone may buy and maintain, only to have others claim a right to use it when it is idle. For instance, the thought of someone walking up to my expensive car and driving off in it simply because it is not currently being used is ludicrous. Likewise, because I am away at work, it is not proper to hold a party at my house because it is otherwise not being used. The related positions that unused computing capacity is a shared resource, and that my privately developed software belongs to everyone, are equally silly (and unethical) positions.

If it is wrong to appropriate someone's car without her permission to prevent waste, then there is no general moral principle that justifies infringing property rights to prevent waste and hence none that would justify hacking to prevent waste.

8.2.3 Benign Intrusions as Exercising the Right to a Free Flow of Content

This argument is grounded in the idea that the moral right to free expression entails that there should be no restrictions on the free flow of content; as this latter idea is sometimes put, information (or content generally) wants to be—or, better, ought to be—free. But if restrictions on the free flow of content are wrong in virtue of violating the right to free expression, then security measures designed to keep hackers out of networks violate their rights to free expression because they inhibit the free flow of content.

This argument also attempts to identify a moral principle that might limit other rights—indeed, one grounded in a putatively stronger right than privacy and property rights, namely the right of free expression. So strong is the right of free expression, on this analysis, that it entails a moral principle that would prevent any other right from permitting restrictions on the free flow of content.

It is, however, no more successful than the others in justifying hacking. The claim that there are no morally legitimate restrictions on the free flow of content precludes there being *any* right to informational privacy that entitles persons to exclude others from information in which they have a reasonable expectation of privacy; efforts that exclude others from information, by definition, impede the free flow of content. If we have any right to informational privacy (in, say, our medical records), as seems plausible, then the right to free expression permits restrictions on the free flow of content.

Further, the claim that there are no morally legitimate restrictions on the free flow of content is inconsistent with there being any moral intellectual property (IP) rights. Of course, moral IP rights might be much weaker than the right defined by IP law. But the idea that there are any moral IP rights is inconsistent with the claim that there are no morally legitimate restrictions on the free flow of content. If we have any moral IP right to exclude people from the contents of at least some of our creations, then the right of free expression permits restrictions on the flow of content.

But even if it were true that there are no legitimate restrictions on the free flow of content, it doesn't follow that people have *carte blanche* to get content any way possible. For example, it is clearly wrong to break into someone's house in order to gain information about what websites she visits. Even assuming that the right of free expression entails that other people are entitled to that information, there are limits on what persons can do *to exercise* that right. In ordinary circumstances, one cannot violate another person's property to exercise the right to free expression. If, as seems reasonable, hacking violates the legitimate property interests of a person in her computer (which is physical tangible property), then it is wrong regardless of whether the hacker is otherwise entitled to information on the victim's computer.

8.3 HACKTIVISM: HACKING AS POLITICALLY MOTIVATED ACTIVISM AND CIVIL DISOBEDIENCE

Recently a more plausible justification of hacking as protected free expression has emerged. According to this argument, attacks on government and corporate sites can be justified as a form of civil disobedience (CD) (Manion and Goodrum, 2000). Since CD is morally justifiable as a protest against injustice, it is permissible to commit digital intrusions to protest injustice. Insofar as it is permissible to stage a sit-in in a commercial or governmental building to protest, say, laws that violate human rights, it is permissible to intrude upon commercial or government networks to protest such laws. Thus, digital intrusions that would otherwise be morally objectionable are morally permissible if they are politically motivated acts of electronic CD—or "hacktivism," as such intrusions have come to be called.

8.3.1 CD and Morality

As a conceptual matter, CD involves (1) the open, (2) knowing (3) commission of some nonviolent act (4) that violates a law *L* (5) for the expressive purpose of protesting or calling attention to the injustice of *L*, some other law, or the legal system as a whole. An act need not target a law or system that is unjust as an objective moral matter to be properly characterized as "civil disobedience." It is enough that the actor is motivated by a belief that the law or system is unjust and that the act is contrived to protest it and call attention to its injustice. Since acts of CD are deliberately open so as to call attention to the putative injustice of the law or legal system, they are fairly characterized as "political expression."

It is tempting to think that acts of CD, as political expression, are morally justified as an exercise of the moral right to free expression. On this line of analysis, the right to free expression entails a right to express one's political views about the legitimacy of the law. Since the very point of CD is to call attention to the illegitimacy of the law, it is a morally justified exercise of the right to free expression.

This line of reasoning is problematic. First, the claim that X has a right to express *p* does not imply that expressing *p* is morally permissible. One might have a right to express all sorts of ideas it is morally wrong to express. For example, one might have a right to express racist ideas even though expressing these ideas is wrong. Rights like the right to free expression are negative rights that are constituted by obligations on the part of other persons or entities. My right to free expression, for example, is constituted by an obligation on the part of others not to coercively interfere with my speech. But the claim that X has an obligation not to interfere with my saying *p* does not imply that my saying *p* is morally permissible; it just means that X should not use coercive means to prevent me from saying *p*.

Second, CD might be expressive, but it is primarily conduct. CD, by its nature, involves disobeying something with the status of law. It is one thing to *assert* a law is unjust; it is another thing to deliberately and openly behave in a manner that violates the law; the former is a pure speech act, while the latter is conduct. CD might be expressive conduct, but it is primarily *conduct* and secondarily expression.

Expressive conduct is subject to more stringent moral limits than those to which pure speech is subject. The reason for this has to do with the effects of these different kinds of act. As a general matter, pure speech acts are primarily calculated to affect only mental states. Speech acts intended to advance some view are calculated only to alter or reinforce the belief structure in the audience. In contrast, while conduct might frequently be *intended* to have only such effects, conduct tends to have effects on other important interests. Someone who expresses anger with you by hitting you not only affects your beliefs, but also causes you physical and emotional injury. Injury is just not a reasonably likely outcome from pure speech acts of just about any kind.

This is not to deny that violating the law might sometimes be morally permissible in certain circumstances. Legal and political philosophers are nearly unanimous in believing not only that there is no general moral obligation to obey the law, but also that there is no general moral obligation to obey the law of even reasonably just states; even reasonably just states, like morally wicked states, might sometimes enact legal content so wicked it does not generate a moral obligation to obey.

The circumstances in which one is permitted or obligated to disobey the law, however, will be comparatively rare in a morally legitimate democratic system with a body of law that is largely, though not perfectly, just for a number of reasons. First, citizens have alternative channels through which to express their political views in democratic systems with rights to free speech that can be exercised in a variety of ways—including blogs potentially reaching billions of people. Second, one of the virtues of democracy is that it affords each person an equal voice in determining what becomes law. Someone who violates democratically enacted law arrogates to herself a larger role than what she is entitled to in a democracy. It is *prima facie* problematic to circumvent legitimate democratic procedures in this way. Third, legitimate democratic states have latitude to enforce some unjust laws. Although some laws, like Jim Crow laws, are so unjust that it was wrong for states to coercively enforce them, a state may permissibly enforce some bad laws that do not reach some threshold level of injustice (e.g., tax laws that are not perfectly fair). But the state can be justified in coercively enforcing a law only insofar as a citizen is morally culpable in disobeying the law. This suggests that citizens are sometimes morally obligated to obey even bad laws.

Even so, the idea that there are limits on the scope of a legitimate state's permission to coercively enforce law indicates that CD is sometimes morally justified. In cases where a legitimate state has enacted a sufficiently unjust law that falls outside the scope of its coercive authority, citizens have a qualified moral permission to disobey it. That is, in cases where the state is not justified in coercively enforcing a law, citizens may permissibly disobey that law because it does not give rise to any moral obligation to obey.

But this permission is qualified by a number of factors. The agent should be in cognitive possession of plausible justification for the position motivating the act of CD, and the position itself should be reasonable. Here it is important to note that the mental state of someone who commits an act of CD is not entirely unproblematic from a moral point of view. While such a person's *motivations* might be laudable, she will likely have another mental state that is not unproblematic from a moral point of view. Someone who commits an act of CD is usually acting on the strength of a conviction that is deeply contested in the society—and, indeed, one that is frequently a *minority* position.

This, by itself, can obviously be laudable in many circumstances. The courage to act on one's convictions and the willingness to sacrifice for them are both virtues. We encourage a child, for example, not to follow the crowd when it is wrong or foolish, knowing that such behavior will frequently result in unpleasant social consequences to the child, such as ridicule or ostracism. One who is willing to risk ridicule and ostracism in order to honor her moral convictions is courageous and deserves praise.

There are, however, moral limits on the costs one can impose on innocent third parties on the strength of an even a laudable motivation. After passage of a citizen initiative banning affirmative action by the state in Washington, protesters marched on a Washington highway in order to shut down traffic, something they succeeded in doing for hours. Those protesters *deliberately* caused significant inconvenience to other persons (many of whom voted against the initiative) after having their position rejected at the polls. Although their position might have been the correct one, their willingness to cause such inconvenience to others on the strength of a view that might, or might not, have been particularly well reasoned is morally problematic—even if, all things considered, their conduct was morally permissible.

The mental state of someone who deliberately imposes detriment on innocent third parties on the strength of a moral conviction that lacks adequate epistemic support is morally problematic in at least two possibly related ways. First, it evinces disregard for the interests of innocent third parties, a failure to appreciate the importance of other people. Second, it evinces an arrogant judgment about the importance and reliability of one's own judgments. It seems, at the very least, arrogant for one person to deliberately subject another to a risk of harm on the strength of an idea that lacks adequate support.[5]

---

[5]This primarily applies to individuals; the state is in a somewhat different position because, in many cases, it cannot avoid taking a position on a contested issue by refraining from acting. If it refrains from prohibiting abortion, for example, the absence of a prohibition presupposes (at least to the extent that we presume the state is trying to do what is morally legitimate) that abortion does not result in murder. If it prohibits abortion, the prohibition presupposes either that a woman does not have a privacy right in her body or that abortion results in murder. Citizens are rarely in such a position.

One sign that a moral conviction lacks adequate epistemic support is that it is deeply contested among open-minded, reasonable persons of conscience in the culture. The idea that there are many open-minded reasonable persons of conscience on both sides of an issue suggests that both positions are reasonable in the sense that they are backed by good reasons that lack an adequate rebuttal. Insofar as a disagreement is *reasonable* in this sense, neither side can claim to have fully adequate support.[6]

Agents must also be willing to accept responsibility under the law for their acts of CD. Willingness to accept responsibility goes beyond merely openly defying the law; one might openly defy a law but attempt to evade apprehension by the police by, say, leaving the country. Intuitively, there is a world of difference, for example, between someone who defaces a billboard in front of 15 police officers to protest its content and someone who does so in a clandestine manner hoping to avoid detection. One seems fairly characterized as vandalism while the other, even if ultimately unjustified, does not.

This last factor is especially important in evaluating hacktivism. Many theorists worry that it can be difficult to distinguish hacktivism from cyberterrorism. Huschle (2002) argues that hacktivists should make it a point to accept responsibility for their actions precisely to ensure that their acts are not mistaken for cyberterrorism, which could cause much more disruption than was intended to result from their acts. As Manion and Goodrum (2000) put the point:

> The justification of hacktivism entails demonstrating that its practitioners are neither "crackers"—those who break into systems for profit or vandalism—nor are they cyberterrorists—those who use computer technology with the intention of causing grave harm such as loss of life, severe economic losses, or destruction of critical infrastructure. Hacktivism must be shown to be ethically motivated (pp. 15–16).

The acceptance of responsibility and the legal consequences of disobedience signals that the act is motivated by a principled stand, a feature that operates to legitimize these acts. Moreover, the willingness of the agent to accept responsibility signals that the breach of the public peace is exceptional rather than part of a general pattern of misconduct and hence need not give rise to the feelings of vulnerability and insecurity to which breaches of the public peace typically give rise.

The foregoing discussion suggests a useful framework for evaluating acts of CD that weigh the moral benefits and costs. The following considerations weigh in favor of finding that an act of CD in a legitimate democratic state is morally permissible. First, the act is committed openly by properly motivated persons willing to accept responsibility for the act. Second, the position is a plausible one in play among open-minded, reasonable persons in the relevant community. Third, persons committing an act of CD are in possession of a thoughtful justification for both the position and

[6]Not all disagreement is reasonable, of course. It seems clear, for example, that persons who disagreed in the 1960s and 1970s with the idea that race-based segregation is wrong lacked even minimal support for a position that had been all but conclusively refuted by that juncture. Sometimes there are a lot of unreasonable, narrow-minded persons who simply refuse to see the light.

the act. Fourth, the act does not result in significant damage to the interests of innocent third parties. Fifth, the act is reasonably calculated to stimulate and advance debate on the issue.

In contrast, the following considerations weigh in favor of finding that an act of CD against an otherwise legitimate state is morally wrong. First, the act is not properly motivated or committed openly by persons willing to accept responsibility. Second, the position is implausible and not in play among most thoughtful open-minded persons in the community. Third, the people who have committed an act of CD lack a thoughtful justification for the position or the act. Fourth, the act results in significant harm to innocent third parties. Fifth, the act is not reasonably calculated to stimulate or advance debate on the issue.

The civil rights sit-ins of the 1960s are paradigms for justified acts of CD under the above framework. Someone who refuses to leave segregated lunch counters until police arrive to remove her is clearly committing an open act and is willing to accept the consequences. The view that segregation is wrong was not only in play among open-minded, reasonable persons of conscience, but had pretty much won the day by the time the mid-1960s arrived. The people who committed these acts justified them by reference to a principle of equality that open-minded, reasonable persons of conscience in the culture had nearly universally accepted. Lunch counter sit-ins had significant effects only on the owners who wrongly implemented policies of segregating blacks and whites. These sit-ins helped to call attention to the ongoing racial injustices in the southern United States.

In contrast, acts of vandalism by anarchists during the 1999 World Trade Organization (WTO) protests in Seattle were not justified acts of CD under this framework. Anarchists who broke windows typically fled the scene as soon as the police arrived. Anarchism is not in play among reasonable, open-minded persons in the community. If televised interviews with many of them were any indication, they generally lacked a thoughtful justification for their views; most of them I saw were strikingly inarticulate. The cost of replacing a large plate-glass storefront window is in excess of $10,000—a morally significant cost to innocent store owners. These acts of vandalism tend to alienate people and entrench them further in their opposition to anarchism, rather than provoke reasoned discussion.

8.3.2 What is Hacktivism?

For our purposes, "hacktivism" can be defined as "the commission of an unauthorized digital intrusion for the purpose of expressing a political or moral position." *Qua* digital act, hacktivism is nonviolent in nature. *Qua* activism, hacktivism does not seek to achieve its political purposes, unlike terrorism, by inspiring terror among the population; it attempts to achieve these purposes by stimulating discussion and debate. Hacktivism is thus conceptually distinct from cyberterrorism—though the boundaries, as we will see, sometimes seem to blur. Hacktivism is distinct from other forms of benign hacking (e.g., motivated by a desire for knowledge) in that it is motivated by the laudable desire to protest injustice.

Not all digital activism counts as hacktivism or CD. Posting a Web site in the United States with a petition to end the war in Iraq would be a form of digital activism, on this definition, but would not count as hacktivism because it does not involve an unauthorized digital intrusion. Nor, for that matter, would such an act count as an act of electronic CD because the posting of such content online breaks no laws; CD necessarily involves violating a valid law.

In contrast, the following count as both hacktivism and CD: (1) a denial-of-service (DoS) attack launched against the WTO Web site to protest WTO policies;[7] (2) the altering of the content of a government Web site to express outrage over some policy of that government;[8] and (3) the unauthorized redirection of traffic intended for a KKK Web site to Hatewatch.[9] Each of these acts would involve some unauthorized digital intrusion and hence would, since presumably intended as a piece of political activism, count as hacktivism.

8.3.3 Is Hacktivism Morally Justified as CD?

The issue of whether hacktivism is justified CD must be addressed on a case-by-case basis. Some hacktivists, for example, make no attempt to conceal their identity and accept responsibility, while others conceal their identities to evade detection. Some acts do not involve significant damage to innocent third parties (e.g., defacing a governmental Web site to protest its policies), while others do [e.g., shutting down commercial Web sites with distributed denial-of-service (DDoS) attacks]. Open acts of hacktivism that do not impact innocent third parties have a different moral quality than clandestine acts that harm innocent third parties.

One of the key issues in evaluating whether an act of hacktivism is morally justified is the extent to which the act harms the interests of innocent third parties. In thinking about this issue, it is important to reiterate that the context being assumed here is a morally legitimate democratic system that protects the right of free expression and thus affords persons a variety of avenues for expressing their views that do not impact the interests of innocent third parties.

How much harm is caused depends on whether the target is a public, private, commercial, or noncommercial entity. Attacks on public noncommercial, purely informative Web sites, for example, tend to cause less damage than attacks on private, commercial Web sites. The reason is that attacks on commercial Web sites can result in significant business losses passed on to consumers in the form of higher prices or to employees in the form of layoffs. If the information on a public Web site is nonessential

[7]In 1999, the Electrohippies attacked a WTO Web site for such reasons. For a summary of notable hacker attacks, see "Timeline of hacker history," *Wikipedia*; available at http://en.wikipedia.org/wiki/Infamous_Hacks.

[8]In 1996, hackers changed the content of the Department of Justice Web site, replacing "Justice" with "Injustice."

[9]Anonymous hackers did exactly this in 1999. Intriguingly, a Hatewatch press release characterized the act as "vandalism." See Hatewatch Press Release: Activism versus Hacktivism, September 4, 1999. Available at http://archives.openflows.org/hacktivism/hacktivism01048.html.

(i.e., unrelated to vital interests), an attack on that Web site is likely to result in nothing more serious than inconvenience to citizens who are not able to access that information.[10]

This should not be taken to suggest that hacktivist intrusions upon public entities *cannot* result in significant harm to third parties. One can conceive of a depressingly large variety of acts that might very well cause significant damage to innocent third parties. A digital attack on a public hospital server might very well result in deaths. Of course, these more serious acts are probably not motivated by expressive purposes and, if so, would not count either as CD or as hacktivism as these notions are defined here.

Acts of hacktivism directed at private individuals can also have morally significant effects. A DoS attack, for example, that effectively denies access to a citizen's Web site can impact her moral rights. A DoS attack on a citizen's site impacts her ability to express her views and hence infringes her moral right to free expression. An attempt to gain access to files on a citizen's computer impacts her rights to privacy, as well as her property rights in her computer.

How much harm is done also depends on the nature of the attack.[11] As a general matter, some digital attacks are less likely to cause harm than others. Defacement of a Web site—or "E-graffiti" as sympathetic theorists sometimes call it—seems far less likely to cause significant harm than attacks that simply deny access to a Web site. Changing "Department of Justice" on a government Web site to "Department of Injustice" is not likely to result in significant harm to third-party interests. At most, it will cause embarrassment to the government agency running the site.

This should not, however, be taken to suggest that defacement of a Web site can never result in significant damage to innocent third parties. Publishing sensitive information about individuals, like social security numbers, as "E-graffiti" on a government Web site could obviously result in significant damage to those individuals. As is true of physical graffiti, one must look to the specific circumstances to evaluate the damage caused by digital graffiti to ensure an accurate assessment.

Nevertheless, it is reasonable to think that, as a general matter, DoS and DDoS attacks are likely to cause more damage, other things being equal, than defacement of Web sites. These attacks are calculated to deny access of third parties to the content of a Web site, effectively shutting it down by overwhelming the server with sham requests for information. While there are undoubtedly exceptions to any generalizations about the comparative harm caused by defacement of Web sites and DoS attacks, it seems reasonable to think that shutting down a Web site is a more harmful act than merely defacing it. For this reason, DoS attack will be harder to justify, as a general matter, as permissible electronic CD than defacement.

---

[10]The Electrohippies justify attacks on various public Web sites precisely on such grounds: "Neither the Whitehouse nor 10 Downing Street Web site are [sic] essential services. For the most part they merely distribute the fallacious justifications in Iraq, as well as trying to promote the image of the two prime movers behind war in Iraq: Messrs. Bush and Blair" (Electrohippies, 2003). The idea here is that the harm caused by attacks that ultimately deny access to public Web sites that are not providing essential services results in no significant harm to innocent third parties.

[11]For a helpful discussion of various tactics, see Auty (2004). My discussion in this and the last section owes an obvious debt to Auty's discussion.

Indeed, a coordinated and sustained DDoS attack on the largest commercial Web sites could result in an economic downturn that affects millions of people. Al Qaeda is exploring the possibility of large-scale cyberattacks on public and commercial networks precisely because a large enough attack might suffice to weaken confidence in E-commerce to such an extent as to precipitate a recession—or worse. Here it is worth noting that an increase of the unemployment rate in the United States from 5% to 6% means the loss of approximately one and a half million jobs—a consequence of great moral significance.

Another important factor in evaluating an act of CD is that the persons committing the act are willing to accept responsibility for those acts. Manion and Goodrum (2000), for example, assert that "willingness (of participants) to accept personal responsibility for outcome of actions" is a necessary, though not sufficient, condition for the justification of an act of CD: "In order for hacking to qualify as an act of civil disobedience, hackers must be clearly motivated by ethical concerns, be nonviolent, and be ready to accept the repercussions of their actions" (p. 15).

There is a difference between claiming responsibility for an act and being willing to accept the legal consequences of that act. One can claim responsibility without coming forward to accept the legal consequences of one's act. One can do this by giving some sort of pseudonym instead of one's real name or by attributing the act to a group that protects the names of its members. Although such a claim of responsibility signals an ethical motivation, this is not tantamount to being willing to accept responsibility.

The heroic civil rights activists of the 1960s who staged sit-ins went beyond merely claiming responsibility; they accepted, even invited, prosecution. It was part of their strategy to call attention to the injustice of Jim Crow laws in the South by voluntarily subjecting themselves to prosecution under those very laws. These courageous activists did not anonymously claim responsibility for the sit-ins from a safe distance: they would continue the protests until the police arrived to arrest them.

Some noteworthy hacktivists evince a similar willingness to accept responsibility for their actions. As Manion and Goodrum (2000) observe:

> Examined in the light, the hack by Eugene Kashpureff clearly constitutes an act of civil disobedience. Kashpureff usurped traffic from InterNIC to protest domain name policy. He did this nonanonymously and went to jail as a result (p. 15).

But this is the exception and not the rule. There are a variety of hacktivist groups, including Electrohippies, MilwOrm, and Electronic Disturbance Theatre, but these groups typically claim responsibility for acts *as a group* without disclosing the identities of any members. For example, MilwOrm and another group claimed responsibility for the defacement of approximately 300 Web sites (they replaced the existing content with a statement against nuclear weapons and a photograph of a mushroom cloud), but did not disclose the identities of members who belong to the group. Hacktivists typically attempt to conceal their identities to avoid exposure to prosecution—even when claiming responsibility.

Anonymous hacktivist attacks impose significant costs on social well-being. First, such attacks, regardless of motivation, contribute to an increasing sense of anxiety among the population about the security of the Internet, which has become increasingly vital to economic and other important interests. Second, these attacks require an expenditure of valuable resources, which could be allocated in more productive ways, to protecting computers against intrusions—costs that are passed on to consumers.

In any event, it is worth noting that terrorists typically claim responsibility as a group, but attempt to evade the consequences of their actions by concealing their identities and locations. It is important, of course, not to make too much of this similarity: terrorists deliberately attempt to cause grievous harm to innocent people while hacktivists do not. The point, however, is merely to illustrate that there is a morally significant difference between claiming responsibility and accepting responsibility. Accepting responsibility is, other things being equal, needed to justify an act of hacktivism.

A third factor to consider is that the motivating agenda behind electronic CD, other things being equal, is not as transparent as the motivating agenda behind ordinary CD. Whereas the protesters who shut down the Washington state highway carried signs and alerted the press they were protesting a specific measure, the point of many putative acts of hacktivism is not clear. A DDoS attack, for example, directed against Amazon.com could mean any number of things—some of which have nothing to do with expressing a political view (e.g., a recently discharged employee might be taking revenge for her dismissal). The absence of a clear message is problematic from a moral standpoint.

Acts of hacktivism are frequently motivated to protest the violation of human rights by oppressive nondemocratic regimes and are directed at servers maintained and owned by governmental entities in those regimes. It is worth noting that, strictly speaking, many such acts will not count as CD. The reason is that many of these attacks will be from people who live outside the repressive regime and are not subject to the legal consequences within the regime. But insofar as these legal consequences are draconian and drastically out of proportion to what is morally appropriate, acceptance of responsibility is not necessary for such acts to be justified. Accordingly, these attacks that originate from outside the target nation might be justified hacktivism, but they will not be justified *as CD*.

Other features suggesting that such acts are justified as follows. First, the primary impact of such acts is on the parties culpable for committing violations of human rights. Defacing a governmental Web site that does not provide essential services or information is not likely to have any significant effects on innocent citizens. Second, the targeted regimes do not respect a right of free expression and forcefully repress political dissent. It is reasonable to think that the moral calculus of CD is considerably different in states that systematically deny citizens the opportunity to express dissent without fear of reprisal. Third, such acts of CD are frequently successful in calling attention to the injustice and stimulating debate. Finally, the position is probably a majority position among people in this culture and worldwide. It is fairly clear that, in Western cultures, support for universal human rights is, far and away, a majority position. But it is also reasonable to think that such support is also a majority position in non-Western cultures. In nations where citizens are denied human rights, those citizens frequently demand them. When liberated from oppressive regimes, moreover, citizens tend to behave in ways that were suppressed under those regimes. Women, for example, in Afghanistan adopted a Western style of dress and rejected the oppressive burqa after the Taliban was removed from power. People almost universally want speech rights, equality, and a right to be free from torture or political persecution.

But, unlike the human rights agenda, other positions commonly motivating hacktivism are fairly characterized as fringe positions not generally in play among thoughtful, open-minded members of the community. Consider, for example, the main tenets of the "hacker ethic" as summarized by Levy (1984):

(1) Access to computers should be unlimited and total.
(2) All information should be free.
(3) Mistrust authority—promote decentralization.
(4) Hackers should be judged by their hacking, not by bogus criteria such as degrees, age, race, or position.
(5) You create art and beauty on a computer.
(6) Computers can change your life for the better.

Although tenets 4 through 6 are largely uncontroversial (and so obvious that they do not need to be stated), these are not the tenets that motivate acts of hacktivism. The tenets that are most likely to motivate acts of hacktivism are the first three tenets.

It is hard to know what to say about tenet 3 beyond pointing out that it is overly general (Should all doctors be mistrusted? Always?); however, tenets 1 and 2 are clearly fringe positions not in play among open-minded, thoughtful people. Tenet 1 implies that people have no property rights in their own computers and hence may not permissibly exclude others from their machines—an implausible position that, consistently applied to other forms of property, would vitiate ownership in homes and automobiles. Tenet 2 implies that people have no privacy rights in highly intimate information about themselves. Although many people are rightly rethinking their positions about information ethics in response to the new technologies, tenets 1 and 2 are too strong to be plausible because they are inconsistent with bedrock views about privacy and property rights. For this reason, neither is in play among open-minded, reasonable persons in the community. This operates against thinking hacktivism expressing the hacker ethic is justified.

Nevertheless, it is not enough, according to the framework described above, that an act of hacktivism is motivated by a plausible position in play among thoughtful, reasonably conscientious persons; it is a necessary condition for an act of hacktivism to be justified that the actor be in cognitive possession of a reasonably plausible justification for that position.

As a general matter, hacktivists give little reason to think they are in possession of a reasoned justification supporting the positions they take. Occasionally, they will articulate their position with some sort of slogan, but rarely provide the position with the critical support it needs. Consider, for example, Manion's and Goodrum's (2000) discussion of one such motivation:

> In order to determine the motivations of hacktivists, one place to look is what hacktivists *themselves* say in their motivation . . .. In June, 1998 the hacktivist group "MilwOrm" hacked India's Bhabha Atomic Research Centre to protest against recent nuclear tests. Later, in July of that year, "MilwOrm" and the group "Astray Lumberjacks," orchestrated an unprecedented mass hack of more than 300 sites around the world, replacing web pages with an antinuclear statements(sic) and images of mushroom clouds. Not surprisingly, the published slogan of MilwOrm is "Putting the power back in the hands of people" (Manion and Goodrum, 2000, p. 16).

One should say much more by way of justification for hacking 300 sites than just a vague slogan like "Putting the power back in the hands of people." The victims of such an attack, as well as the public whose peace has been breached, have a right to know exactly what position is motivating the attack and why anyone should think it is a plausible position.

The foregoing argument should not, of course, be construed to condemn all acts of hacktivism. Nothing in the foregoing argument would condemn narrowly targeted acts of electronic CD properly motivated and justified by a well-articulated plausible position that do not result in significant harm to innocent third parties. Acts of hacktivism that have these properties might be justified by the right to free expression—though, again, it bears emphasizing here that such acts will be much harder to justify in societies with morally legitimate legal systems.

But, as a general matter, hacktivists have not done what they should to make sure their acts are unproblematic from a moral standpoint. In their zealousness to advance their moral causes, they have committed acts that seem more problematic from a moral point of view than the positions they seek to attack. If, as Manion and Goodrum (2000) suggest, hacktivists have been misunderstood by mainstream media and theorists, they have only themselves to blame.

8.4 HACKING BACK: ACTIVE RESPONSE TO COMPUTER INTRUSIONS

Victims of digital intrusions are increasingly responding with a variety of "active responses." Some are intended to inflict the same kind of harm on the attacker as the attack is intended to have on the victim. Conxion, for example, overloaded the network from which the Electrohippies staged a DoS attack by redirecting the incoming packets back to the network instead of dropping them at the router.[12] Some, however, are not intended to inflict harm on the attacker's network. The point of a traceback is to identify the parties responsible for the intrusion by tracing its path back to the source.

[12]See Radcliff, D. (2000). Should you strike back? *ComputerWorld.* Available from http://www.computerworld.com/governmenttopics/government/legalissues/story/0,10801,53869,00.html.

This section considers whether and to what extent it is morally permissible for private parties to adopt these active responses.

8.4.1 The Active Response Spectrum

The term "active response" is intended to pick out digital intrusions that come in response to a hacker's intrusion and are intended to counter it; these responses are sometimes called "counterhacking" or "hacking back." As such, active response measures have the following characteristics. First, they are digitally based; assaulting someone who is committing a digital trespass is not active response. Second, they are implemented after detection of an intrusion and are intended to counter it by achieving investigative, defensive, or punitive purposes. Third, they are noncooperative in that they are implemented without the consent of at least one of the parties involved in or affected by the intrusion. Finally, they have causal impacts on remote systems (i.e., those owned or controlled by some other person).

"Benign" responses involve causal interaction with remote systems outside the victim's network, but are neither intended nor reasonably likely to damage those systems. One example of a benign response is a traceback. As noted above, tracebacks attempt to identify the parties responsible for an digital intrusion by following its path in reverse; they causally impact remote systems but without damaging them.

"Aggressive" responses are those calculated to interfere with the availability, integrity, confidentiality, or authenticity of remote systems. Aggressive measures are those intended or highly likely to result in something that the target would regard as harm or damage. An example of an aggressive response is a DoS counterattack of the sort launched by Conxion against Electrohippies.

8.4.2 Relevant Moral Principles

8.4.2.1 A Principle Allowing Force in Defense of Self and Others

It is generally accepted that a person has a moral right to use proportional force when necessary to defend against an attack. If, for example, A starts shooting at B without provocation and B cannot save her own life without shooting back at A, it is permissible for B to shoot at A. The first principle considered here, then, can be stated as follows:

> **The Defense Principle:** It is morally permissible for one person to use force to defend herself or other innocent persons against an attack provided that (1) such force is proportional to the force used in the attack or threat; (2) such force is necessary either to repel the attack or threat or to prevent it from resulting in harm; and (3) such force is directed only at persons who are the immediate source of the attack or threat.

Although the term "force" has traditionally been used to refer to violent physical attacks in which one person attempts to inflict physical harm on another person, it is construed here as applying to both physical *and* digital attacks.

Each element of the Defense Principle states a necessary condition for the justified use of force. First, the Defense Principle justifies the use of no more force than is proportional to the attack. Second, force must be necessary in the sense that the victim cannot either stop the attack or prevent further harm to herself without resorting to its use. Third, the Defense Principle will justify the use of force only against the direct sources of the attack. In limited cases, this might permit the use of force against innocent persons. Many people have the intuition that a person may direct force against an attacker known to be insane and hence not responsible for her actions. Nevertheless, the Defense Principle will never justify directing force against an innocent bystander. Under no circumstances, then, would it allow a person to defend against an attack by interposing an innocent bystander between herself and the attacker.

8.4.2.2 A Principle Allowing Otherwise Wrongful Acts to Secure Greater Moral Good It is also generally accepted that morality allows the infringement (as opposed to violation) of an innocent person's rights when it is necessary to secure a significantly greater good.[13] For example, if A must enter onto the property of B without her permission to stop a murderer from escaping, it is morally permissible for A to do so. Though such an act constitutes a trespass and hence infringes B's property rights, it does not violate B's property rights because it is morally justified. This suggests a second general principle relevant in evaluating an active response:

> **The Necessity Principle:** It is morally permissible for one person A to infringe a right ρ of a person B if and only if (1) A's infringing of ρ would result in great moral value; (2) the good that is protected by ρ is significantly less valuable, morally speaking, than the good A can bring about by infringing ρ; (3) there is no other way for A to bring about this moral value that does not involve infringing ρ; and (4) A's attitude toward B's rights is otherwise properly respectful.

As construed here, the Necessity Principle applies in the context of physical and digital attacks and hence potentially justifies the use of physical or digital force that would ordinarily be impermissible.

Each element of the Necessity Principle states a necessary condition for being justified in doing something that would otherwise be wrong. First, the act is not justified unless it results in a significantly greater good than the interest infringed by the act. Second, the act is justified only if there is no other way to bring about the greater good. Third, the act must be performed with an otherwise respectful attitude.

The Necessity Principle augments the Defense Principle by allowing some action that would infringe the rights of even innocent bystanders: the Necessity Principle seems to allow one person A to infringe the right of an innocent bystander B if

[13]By definition, to say that a right has been "infringed" is to say only that someone has acted in a way that is inconsistent with the holder's interest in that right; strictly speaking, then, the claim that a right has been infringed is a purely descriptive claim that connotes no moral judgment as to whether or not the infringement is wrong. In contrast, to say that a right has been "violated" is to say that the right has been infringed by some act and that the relevant act is morally wrong. Accordingly, it is a conceptual truth that it can be permissible for an individual or entity to infringe a right, but it cannot be permissible to violate a right.

necessary to defend A or some other person from a culpable attack that would result in a significantly greater harm than results from infringing B's right. But insofar as the Necessity Principle requires the achievement of a *significantly* greater good, it will not allow a person to direct at an innocent bystander force that is proportional to the force of the attack.

8.4.2.3 Two Nonstarters: Retaliation and Punishment

It might be thought that victims of an attack have a moral right to retaliate against or punish their attackers by inflicting a proportional harm on their attackers. If, for example, A hits B in the face and then turns and runs away in an obvious attempt to escape, it is morally permissible, on this view, for B to catch A and hit him back in the face; such a measure is permissible either as retaliation or as punishment. Applied to the present context, such a principle would permit the victim of a digital attack to counterattack as a means of "evening the score."

Active response cannot be justified as retaliation. The act of inflicting injury on another person for no other reason than to even the score is "revenge," and revenge is generally regarded as morally wrong because it is no part of the concept of revenge that harm be inflicted to give a person his just deserts. From the standpoint of someone who is retaliating, the point of the retaliatory act is not to restore the balance of justice after it has been disturbed by a wrongful act. Rather, the point is simply to even the score: he did this to me, so I did it back to him. As far as ordinary intuitions go, morality does not allow the infliction of harm on another person without regard for whether that harm is deserved or serves some greater purpose than satisfying a desire for vengeance.

Nor can active response be justified as punishment. In a society with a morally legitimate government, it is morally impermissible for *private citizens* to punish wrongdoing. Mainstream political theorists are nearly unanimous in holding that it is the province of a legitimate government – and not of private persons – to punish wrongdoers after they are found guilty in a fair trial with just procedures. Indeed, vigilantism is universally condemned as morally wrong. Both lines of argument are nonstarters.

8.4.3 An Evidentiary Restriction for Justifiably Acting Under Ethical Principles

As was noted in the discussion of hacktivism, a person must have adequate reason to believe she is justified in acting under a moral principle to be justified in acting under that principle. There is thus another general principle relevant with respect to evaluating an active response—one that is epistemic in character:

> **The Evidentiary Principle:** It is morally permissible for one person A to take action under a moral principle P only if A has adequate reason for thinking that all of P's application-conditions are satisfied.

The Evidentiary Principle implies that one has a duty to ensure that one is epistemically justified in acting under the relevant moral principle. If one person

A takes aggressive action against another person B without sufficient reason to believe the application-conditions of the relevant moral principles have been satisfied, A commits a wrong against B.

Accordingly, the victim of a digital attack can permissibly adopt active response only if she has adequate reason to think the application-conditions of one of the relevant principles are satisfied. Under the Defense Principle, she must have adequate reason to believe that (1) whatever force is employed is proportional to the force used in the attack; (2) such force is necessary either to repel the attack or to prevent it from resulting in harm of some kind; and (3) such force is directed only at persons immediately responsible for the attack. Under the Necessity Principle, she must have adequate reason to believe that (1) the relevant moral value significantly outweighs the relevant moral disvalue; (2) there is no other way to achieve the greater moral good than to do A; and (3) doing A will succeed in achieving the greater moral good.

8.4.4 Evaluating Active Response Under the Relevant Principles

It is important to realize that the risk that active responses will impact innocent persons and their machines is not purely "theoretical." Sophisticated attackers usually conceal their identities by staging attacks from innocent machines that have been compromised through a variety of mechanisms. Most active responses will have to be directed, in part, at the agent machines used to stage the attack. Accordingly, it is not just *possible* that any efficacious response will impact innocent persons, it is nearly inevitable—something that anyone sophisticated enough to adopt an active response is fairly presumed to realize.

Given that innocent persons enjoy a general (though not unlimited) moral immunity against forceful attack, the likelihood of impacting innocent persons with an active response is of special ethical concern. For this reason, the impacts of active responses on innocent parties will occupy a central role in evaluating those responses.

8.4.4.1 Aggressive Measures As a general matter, aggressive active defense cannot be justified by the Defense Principle. Consider, again, Conxion's response to the Electrohippies DoS attack. Instead of simply dropping the incoming packets at the router, Conxion sent those packets back to the Electrohippies' server, overwhelming it. Since dropping the packets at the router would have ended the harmful effects of the attack, Conxion's response was not "necessary" and hence not justified under the Defense Principle.

Additional issues are raised by aggressive response to attacks staged from innocent agent machines. Since the identity of the culpable attacker is unknown in such cases, any aggressive response will invariably be directed at the innocent agents compromised by the attacker, which compounds the harms done to the owners of those machines. Even if it is permissible to use force against innocent attackers, the owners of those machines are not really "attackers" in the sense that an insane person who assaults another person is. If those owners are really "bystanders," the Defense Principle will not allow aggressive response to attacks in which it is evident that the attacker's identity has been concealed.

Moreover, aggressive response will not be justified under the Defense Principle if it is not *necessary* to prevent the harm or to stop the attack. If there is any nonaggressive way for the victim to avoid the attack or the damage caused by it, then an aggressive response cannot be justified under the Defense Principle. This, however, does not mean that the victim is obligated to escape the attack by any means possible. Although it is always possible to escape a digital attack by taking the target offline, such measures can result in significant damage (if, e.g., the target is a web-based business) that the Defense Principle does not require victims to accept. Victims have a duty to escape attacks only insofar as this can be done without incurring injuries that are comparable to those caused by the attack itself.

Aggressive response is problematic under the Necessity Principle for a different reason. The Necessity Principle allows acts that would otherwise be wrong if they are necessary to achieve a significantly greater moral good. Even if we assume that an aggressive response is clearly necessary to achieve the moral good of preventing the damage caused by an attack and that this good significantly outweighs the harms done to the owners of the agent machines, there is an evidentiary problem: for all we can know, an aggressive response might result in unpredictable harms that outweigh the relevant moral goods.

The problem here arises because machines can be linked via a network to one another in a variety of unpredictable ways, making it impossible to identify all the harmful effects of an aggressive response in advance. Suppose, for example, that an attacker compromises machines on a university network linked to a university hospital. If hospital machines performing a life-saving function are linked to the network, an aggressive response against that network might result in a loss of human life. Even worse, suppose an attacker compromises machines used by one nation's government to attack private machines in another nation. If the two nations are hostile toward each other, an aggressive response by the private victim could raise international tensions—a particularly chilling prospect if the two nations are nuclear powers.

The point is not that we have reason to think that these scenarios are likely; rather, it is that we do not have any reliable way to determine how likely they are. A victim contemplating an aggressive response has no reliable way to estimate the probabilities of such scenarios in the short time available to him or her. Since the victim cannot reliably assess these probabilities, she lacks adequate reason to think that the application-conditions of the Necessity Principle are satisfied. Thus, under the Evidentiary Principle, she may not justifiably adopt aggressive measures under this principle.

8.4.4.2 Benign Measures Benign measures are typically concerned with identifying culpable attackers (e.g., tracebacks) and are neither intended nor obviously likely to result in physical damage to affected machines. Even so, benign responses are problematic insofar as they causally impact remote machines. Of course, this does not pose any obvious moral problems when the remote machines are located within the victim's network or when the victim has permission to impact these machines. But unauthorized effects on innocent agent machines are presumptively problematic since they infringe the property rights of an innocent person.

Benign responses do not defend against attacks and hence cannot be justified by the Defense Principle, but they seem to secure an important moral good under the Necessity Principle. Criminal attacks are regarded as offenses against the general public because they violate the legitimate expectations of the public and thereby breach the peace. Like the victim, the public has a compelling reason to ensure that the criminal offender is brought to trial and punished to restore the peace—a good of considerable moral significance. To the extent that tracebacks can reliably be used to identify an attacker, they function to secure the important moral good of restoring the public peace by bringing wrongdoers to justice. Accordingly, responses motivated by such an objective are intended to secure an important moral value.[14]

Moreover, it also seems clear that such goods are important enough to justify comparatively minor infringements of the property rights of innocent persons. If the only way that a private security officer can apprehend a robbery suspect is to commit a trespass against the property of an innocent person, it seems clear that she is justified in doing so under the Necessity Principle. The moral value of bringing the offender to justice and thereby restoring the public peace greatly outweighs the moral disvalue of a simple trespass onto the land of an innocent party.

The problem with benign responses, however, is that it will frequently be unclear whether they are reasonably calculated to succeed in identifying culpable parties. As noted above, any reasonably sophisticated hacker will attempt to conceal her identity by staging the attack from innocent agent machines. Indeed, it is possible for a sophisticated attacker to further insulate herself from discovery by compromising one set of innocent machines to control another set of innocent machines that will be used to stage the attack—a process that can be iterated several times. In such cases, the attacker will interpose several layers of innocent machines between herself and the victim. But the greater the number of layers between attacker and victim, the less likely benign responses will succeed in identifying the culpable party. Although benign response can be highly effective in identifying the culpable parties in attacks staged directly from the hacker's machine, the probability of success drops dramatically with each layer of machines between attacker and victim. Indeed, it is fair to say that the likelihood of identifying the culpable parties in sophisticated attacks by benign responses is morally negligible.

This seems to imply that the victim of a digital attack cannot permissibly adopt benign responses under the Necessity Principle. Unless she has some special reason to think that the attack is being staged directly from the hacker's own machines without the use of benign agent machines or networks, she will not have adequate reason to think that benign measures will succeed in identifying the culpable parties and will not

[14]Not all benign responses are motivated by a desire to prosecute the wrongdoer. Many firms would prefer to avoid prosecution to avoid the unfavorable publicity that might result from the disclosure of security breaches. The above reasoning would not justify benign responses in these cases.

be ethically justified, under the Evidentiary Principle, in acting upon the Necessity Principle.

Nevertheless, it is important to emphasize that the analysis here is limited to current traceback technologies with their limitations. Many researchers are making considerable progress in improving the reliability and efficacy of traceback technologies.[15] Indeed, one might reasonably expect researchers to eventually improve these technologies to the point where they are sufficiently efficacious in identifying culpable parties that they can generally be justified under the Necessity Principle as bringing about the greater moral good of identifying culpable parties to an attack.

8.4.5 The Relevance of Consent

The preceding analysis presupposes that the victim of a digital attack does not have express or implied permission to causally impact the machines of innocent owners. One might think that owners of agent machines somehow consent to being affected by active response measures. If owners of affected machines have consented to such effects, then they have waived any general moral immunity from active response.

Clearly, there is no general reason to think that owners of agent machines have explicitly or expressly consented to either having their machines used for an attack or being targeted by aggressive countermeasures. In the absence of any other reason to think aggressive countermeasures against these machines are permissible, victims would be committing a wrong against the owners under the Evidentiary Principle should they direct aggressive countermeasures at these machines.

In some rare instances, persons can be presumed to have "tacitly" or "impliedly" waived a right on the basis of some nonexpressive behavior not intended to effect a waiver. Indeed, in some instances, a person's failure to object to some act can be treated as tacit consent to that act. For example, there is little disagreement about the justice of the legal rule that treats an attorney's failure to object to something opposing counsel has done as having waived the objection.

Accordingly, one might argue it is reasonable to infer that owners of agent machines used in a digital attack have consented to being targeted by active response. On this line of analysis, the failure of such owners to protect against unauthorized entry with a firewall is reasonably construed as consent to entry in cases where it is needed to investigate or defend against a digital attack staged from their machines. On this line of reasoning, someone who fails to take such precautions is reasonably thought to be sufficiently indifferent about the prospects of intrusions that she may be presumed to consent to them.

There are a couple of problems with this line of reasoning. First, it would not only imply consent to the victim's intrusion, but would also imply consent to the attacker's intrusion—a result that is sufficiently implausible to warrant rejecting any claims that imply it. Second, failure to implement a firewall is no more reasonably construed as consent to entry than failure to lock the door to one's car is reasonably construed as consent to enter one's car. One might forget to take such precautions for any number of

[15]See http://footfall.csc.ncsu.edu, which documents some intriguing advancements in these technologies.

reasons without being indifferent about unwanted entries. Moreover, in the case of computer intrusions, one might simply not know about the available security options.

A somewhat more plausible argument for treating failure to take adequate security precautions is grounded in ethical principles that impute a duty of reasonable care to protect others from foreseeable harm. In cases where one person's negligence potentially puts another person at risk, considerations of fairness require imputing some responsibility or disadvantage to the former person that must ordinarily be voluntarily accepted. If, for example, you negligently disclose my whereabouts to someone who wants culpably to harm me, you might thereby obligate yourself to do something for me that you ordinarily would not be obligated to do—perhaps hide me in your home.

One might, then, argue that persons who fail to take reasonable precautions to prevent unwanted computer intrusions and whose computers are used to stage an attack have tacitly consented to aggressive and benign active defense measures directed at their computers by the victims of those attacks. Since their negligence has wrongfully put innocent persons at risk, they have released the victims of the attacks from any duties they otherwise might have had to refrain from benign or aggressive active defense.

If ordinary intuitions and practices are correct, this line of reasoning will not justify directing aggressive measures at owners of compromised machines. Ethical principles of negligence are not generally thought to justify aggression against negligent parties; they would not, for example, justify me in attacking a person who has negligently injured me or damaging her property. Rather they are thought to require a person to compensate parties for injuries proximately caused by her failure to take reasonable precautions to protect such parties from injury; this, of course, is how such principles are interpreted and applied by the courts under tort law.

Whether ethical principles regarding negligence might justify benign responses is a much more difficult issue that cannot be addressed here. Admittedly, there is little in ordinary practices that would justify an inference that these principles allow benign responses. There are simply no obvious analogs in ordinary practices to digital attacks staged from innocent machines.

Even so, the idea that the owner of a compromised agent machine might have waived any immunity she would have otherwise had to benign responses is not obviously unreasonable. If, for example, it is reasonable to think that such persons have a duty, at the very least, to contribute to compensating the victim of a digital attack for injuries sustained during the attack, it also seems reasonable to think that such persons have a duty to permit victims to commit an intrusion for the purpose of tracing the attack back to its source. If, during the course of an attack, the victim had adequate reason to think that (1) benign responses would successfully identify the attacking party and (2) owners of compromised machines negligently failed to take reasonable precautions to prevent their machines from being used in a digital attack, she might very well be justified in directing benign responses against those machines.

For all practical purposes, however, the argument is moot. Since victims will rarely be able to gather, during the course of a digital attack, adequate evidence for thinking either that benign responses would be successful or that owners of compromised machines have failed to take reasonable precautions, they will not be able to justify adopting these responses under ethical principles of negligence. For this reason, the Evidentiary Principle seems to preclude adopting benign responses on the strength of ethical principles of negligence—assuming, of course, that these principles are even applicable.

8.4.6 The Inadequacy of Law Enforcement Efforts

There is one last argument that can be made in defense of the idea that it is permissible for private individuals to undertake various active defense measures.[16] The argument rests on the idea that the state may legitimately prohibit recourse to self-help measures in dealing with a class of wrongful intrusions or attacks only insofar as the state is providing minimally adequate protection against such attacks. If (1) digital intrusions are resulting in significant harm or injury of a kind that the state ought to protect against and (2) the state's protective efforts are inadequate, then private individuals, on this line of reasoning, are entitled to adopt active defense measures that conduce to their own protection.

Both antecedent clauses appear to be satisfied. Depending on the target and sophistication of the attack, an unauthorized digital intrusion can result in significant financial losses to companies. For example, an extended distributed denial of service attack that effectively takes Amazon.com offline for several hours might result in hundreds of thousands of dollars of business going to one of its online rivals. In the worst-case scenario, these financial losses can result in loss of value to shareholders and ultimately loss of jobs. It seems clear that the harms potentially resulting from digital intrusions fall within a class that the state ought to protect against.

Further, there is good reason to think that the state's protective efforts are inadequate. At this point in time, law enforcement agencies lack adequate resources to pursue investigations in the vast majority of computer intrusions. But even when resources allow investigation, the response might come after the damage is done. Law enforcement simply has not been able to keep pace with the rapidly growing problems posed by digital attackers.

There are a variety of reasons for this. Most obviously, the availability of resources for combating cybercrime is constrained by political realities: if the public is vehemently opposed to tax increases that would increase the resources for investigating cybercrime, then those resources will not keep pace with an increasing rate of intrusions. But, equally importantly, there are special complexities involved in investigating and prosecuting digital intrusions. First, according to Mitchell and Banker, investigation of digital intrusions is resource-intensive: "whereas a typical (non-high-tech) state or local law enforcement officer may carry between forty and fifty cases at a time, a high-tech investigator has a full time handling three or four cases a month." Second, most sophisticated attacks will pose jurisdictional complexities that increase the expense of law enforcement efforts because such attacks will frequently involve crossing jurisdictional lines. For example, an attacker in one country might compromise machines in another country in order to stage an attack on a network in yet a third country.

[16]See Mitchell and Banker (1998).

Although such considerations show that the growing problem associated with digital intrusions demands an effective response of some kind, they fall well short of showing that it is permissible, as a general matter, for private parties to undertake benign or aggressive active defense measures. The underlying assumption is that private individuals can adequately do what the state cannot—namely, protect themselves adequately from the threats posed by digital intrusion.

At this time, however, there is very little reason to think that this underlying assumption is correct. For starters, invasive benign measures intended to collect information are likely to succeed in identifying culpable parties only in direct attacks staged from the attacker's own computer; such measures are not likely to succeed in identifying parties culpable for intrusions that are staged from innocent machines. Since an attacker sophisticated enough to stage an attack likely to result in significant damage is also likely to be sophisticated enough to interpose at least one layer of innocent machines between her and her target, there is little reason to think that invasive investigatory measures are likely to achieve their objectives in precisely those attacks that are likely to result in the sort of damage that the state is obligated to protect against.

Moreover, aggressive measures are not likely to conduce to the protection of the victim in any reasonably sophisticated attack. As noted above, aggressive countermeasures are not usually calculated to result in the cessation of the attack and can frequently result in escalating the attack; for this reason, such countermeasures are not likely to succeed in purely defensive objectives. Unfortunately, they cannot succeed in achieving legitimate punitive objectives in attacks staged from innocent machines. Punitive measures directed at the innocent agents do nothing by way of either punishing the ultimate source of the attack or deterring future attacks. A reasonably sophisticated attacker who knows her target will respond with aggressively punitive measures will simply evade the effects of those measures by interposing an additional layer of innocent machines between her and her target.

REFERENCES

Auty, C. (2004). Political hacktivism: tool of the underdog or scourge of cyberspace? *ASLIB Proceedings: New Information Perspectives*, 56, 212–221.

Huschle, B. (2002), Cyber disobedience: when is hacktivism civil disobedience? *International Journal of Applied Philosophy*, 16(1), 69–84.

Himma, K.E. (2006a). Hacking as politically motivated digital civil disobedience: is hacktivism morally justified? In: Kenneth, E.H. (Ed.), *Readings on Internet Security: Hacking, Counterhacking, and Other Moral Issues*. Jones & Bartlett, Sudbury, MA.

Himma, K.E. (2006b). The ethics of active defense. In: Kenneth, E.H. (Ed.), *Readings on Internet Security: Hacking, Counterhacking, and Other Moral Issues*. Jones & Bartlett, Sudbury, MA.

Himma, K.E. and Dittrich, D. (2006c). Hackers, crackers, and computer criminals. *The Handbook of Information Security.* John Wiley & Sons.

Levy, S. (1984). *Hackers: Computer Heroes of the Computer Revolution*. Delta Trade Paperbacks, New York.

Manion, M. and Goodrum, A. (2000). Terrorism or civil disobedience: toward a hacktivist ethic. *Computers and Society*, June, 14–19.

Mitchell, S.D. and Banker, E.A. (1998). Private intrusion response. *Harvard Journal of Law and Technology*, 11(3), 710.

Spafford, E. (1992). Are computer hacker break-ins ethical? *Journal of Systems Software,* 17(1), 41–48.

Tavani, H. (2007). The conceptual and moral landscape of computer security. In: Himma, K.E (Ed.), *Internet Security: Hacking, Counterhacking, and Society.* Jones & Bartlett, Sudbury, MA.

PART III

PROFESSIONAL ISSUES AND THE INFORMATION-RELATED PROFESSIONS

CHAPTER 9

Information Ethics and the Library Profession

KAY MATHIESEN and DON FALLIS

9.1 INTRODUCTION

Libraries as organized depositories of documents have existed at least from the time of the Sumerians (Fourie and Dowell, 2002, pp. 15–16). Librarianship as a distinct profession, however, is relatively a recent development. With the advent of the printing press, collections of works became larger and more complex, thus creating a greater need for someone devoted to organizing and cataloging such collections. Even then, those who collected, organized, and preserved books and other documents were scholars and their activities as "librarians" were not distinct from their scholarly work. Not until the nineteenth century did librarianship become a separate profession (Gilbert (1994), p. 383).

While in general we can say that the role of the librarian is to provide access to information, librarians vary in their activities depending on the goal of such access. For a corporate librarian, the goal of providing access is to enable and enhance the activities of the corporation. For an academic librarian, the goal of providing access is to enable and enhance the activities of the university community (e.g., research and teaching). For a public librarian, the goal of providing access is to respond to the information needs of all members of the community.

Given the complex communities served by public libraries, many of the most interesting ethical issues arise in the context of the public library. Thus, in this paper, we will largely focus on the public library. Nevertheless, the issues discussed here should be of interest to anyone in the library profession. Indeed, they should be of interest to anyone involved in the activities of collecting, organizing, categorizing, preserving, and providing access to information. This would include, for example,

The Handbook of Information and Computer Ethics, Edited by Kenneth Einar Himma and Herman T. Tavani

many of the people at Google or at Wikipedia, who are traditionally not considered part of the library profession.

We begin this chapter by considering the mission of the librarian as an information provider and the core value that gives this mission its social importance. Of course, librarians face the standard ethical issues that arise in any profession, but our focus here will be on those issues that arise in relation to the role of the librarian as an information provider. In particular, we will be focusing on questions of the selection and organization of information, which bring up issues of bias, neutrality, advocacy, and children's rights to access information.[1] All these issues bring up important challenges to what is commonly seen as the core value of librarianship—intellectual freedom. For example, in providing access to information librarians must make selections among materials. How can they do this in ways that fully enable their patrons' access to information? Some have suggested that the solution is for the librarian to remain neutral. We explore what is meant by neutrality and the benefits and possible costs of taking the "neutral point of view." We then turn to the ethical issues that arise in the organization and categorization of materials. In particular, we discuss the question of what sorts of "labeling" of the content in libraries are appropriate. We end by addressing what is one of the most vexed issues that arises in public libraries. Ought the librarian make any distinctions between children and adults in providing access to information?

9.2 THE CORE VALUE OF THE LIBRARY PROFESSION

In the following sections, we consider the challenges that confront the librarian in carrying out his or her professional duties, in particular with regard to selection of materials and the organization of these materials. Since the ethical obligations of librarians as professionals will at least partly be determined by the role that librarians play in society, we need to understand the mission and values of the librarian in order to understand what those ethical obligations are.[2] Thus, in this section, we characterize the mission of the library profession and discuss the core values related to that mission.

In taking this approach, we are following a number of notable figures in library and information science who have claimed that its values are essential to defining the profession (see Baker, 2000; Finks, 1989; Gorman, 2000; Ranganathan, 1931; Tyckoson, 2000). Many of these authors provide lists of values to which librarianship is committed. Former American Library Association (ALA) President Gorman (2000), for example, suggests the following list of the "Core Values of Librarianship": stewardship, service, intellectual freedom, rationalism, literacy and learning, equity of

[1]We do not, for instance, consider issues of privacy or intellectual property. For a discussion and overview of these issues in relation to the library profession, see Fallis (2007).

[2]There is significant amount of discussion within library and information science about the nature and status of librarianship as a profession (see Arant and Benefiel, 2002; Dilevko and Gottlieb, 2004; Maynard and McKenna, 2005; Walker and Lawson, 1993), particularly in response to the rise of new information technologies (see Diamond and Dragich, 2001). The sociologist Andrew Abbott, who has written on the sociology of professions (Abbott, 1988), provides an insightful analysis of the changing profession of librarianship (Abbott, 1998).

access to recorded knowledge and information, privacy, and democracy. The American Library Association (1999) also provides its own list of core values: "access, confidentiality/privacy, democracy, diversity, education and lifelong learning, intellectual freedom, preservation, the public good, professionalism, service, and social responsibility."

But one drawback of the list approach is that it fails to explain how these different values are related to each other. This is particularly problematic given that it is possible that pursuing some values, such as confidentiality or preservation, may conflict with others, such as access. One would need to understand how preservation is related to access in a structure of values in order to know how such conflicts should be resolved. A second drawback is that some values are not on this list, such as a respect for the intellectual property rights of authors and creators. Finally, in listing values, the theoretical framework that explains and supports these many values is often left out.[3] Thus, such lists may serve as a starting point, but they do not take us very far in serious reflection on the core values of librarianship.

Such serious reflection needs to start with a focus on the point of having libraries and librarians in the first place—that is, the mission of the librarian. Gorman (2000, p. 23) puts it succinctly, "When it comes down to it, libraries exist to make the connection between their users and the recorded knowledge and information they need and want." The imperatives that might follow from the idea of the librarian as a kind of matchmaker between books (or "information") and readers is spelled out most eloquently by Shiyali Ramamrita Ranganathan in his famous *Five Laws of Library Science* (1931):

(1) Books are for use.
(2) Every person his or her book.
(3) Every book its reader.
(4) Save the time of the reader.
(5) The library is a growing organism.

Ranganathan's aphoristic list of these laws is a bit mystifying at first, but their scope and depth reveal themselves as one reads Ranganathan's work. With the first law, "Books are for use," Ranganathan emphasizes that the goal of the librarian should be to make sure that books and other information resources are actually accessed and read. Ranganathan appeals to the first law as a grounding principle for accessibility (e.g., browsability, lack of fees), but also when considering such issues as library location, hours, furniture, and staffing. According to Ranganathan, the primary consideration is whether such features will encourage or discourage "use." This law grounds laws 2, 3, and 4. If our goal is to promote access to books or information more generally, then we will connect particular users to the books they need and want (law 2, "Every reader his/her book"). We will also do what we can to promote people knowing about and using

[3]The ALA (2006a) *Code of Ethics* and *Library Bill of Rights* have similar limitations. For criticisms of the ALA Code of Ethics and Library Bill of Rights on these grounds, see Frické et al. (2000).

the books and information sources that are available (law 3, "Every book, its reader"). And, finally, if there are barriers between a user and the information—if it is too difficult or takes too long to access the information—then persons will be less likely to get the information they need and they most certainly will be likely to get lower quality information[4] (law 4, "Save the time of the reader").

As becomes clear when one unpacks Ranganathan's laws, taking on the job of providing access to information is quite a bit more involved than it might at first appear. It is neither enough to simply fill a building with a bunch of books and other media nor sufficient to just provide an internet connection. If this was all there was to it, there would indeed be little use for librarians at all. Focusing on this point, Ortega y Gasset (1961 [1935]) and Wengert (2001) argue that librarians must not just "provide" access, but must shape the access by carefully selecting information. People do not typically just want or need any old information; they need and want *quality* information, for example, information that is on topic, comprehensible, current, interesting, accurate, and well-written (to name just a few features). And, they want to be able to find this information quickly and easily. This brings up the further goal (beyond just providing access) of evaluating, selecting, and organizing information so as to provide access to quality information.

Ortega y Gasset goes so far as to propose that librarians should serve as "a filter interposed between man and the torrent of books" (1961 [1935], p. 154). His view is rather controversial and is worth describing in more detail. He characterizes the librarian as the individual who handles books for society. Ortega y Gasset points out that books serve the function of preserving the knowledge of other people. Without books, we would all have to start from scratch in learning to deal with the world (cf. Diamond (1997), pp. 215–238).

Books and librarians have, thus, facilitated the progress of society. Ortega y Gasset warns that recently things have reached the point where books have turned against us. There are now too many books (and, according to Ortega y Gasset, too many bad books) for people to sort through for themselves.[5] As a result, librarians need to focus on activities (e.g., carefully organizing and selecting and even publishing books) that save patrons from having to slog through this mass of books for themselves.

Like Ortega y Gasset, Wengert (2001) is concerned with the fact that there is too much (and too much bad) information out there. He also thinks that librarians can play an important role in improving this situation. However, his view is not as extreme as Ortega y Gasset's view. For example, he does not think that librarians should take charge of the production of books. In fact, Wengert seems to think that librarians can ameliorate this problem without restricting access to any materials. Basically, librarians just have to direct people to, and make accessible, those information resources that are likely to be useful to people. Wengert argues that library professionals are "teachers." They are "experts who instruct others on how to

[4]See Mann (1993), on the "principle of least effort."

[5]Ortega was writing prior to the development of the Internet. This problem of "information overload" has clearly become more pressing in the last few decades.

better achieve the projects that they have in mind" (2001, p. 486). They do this by directing people to information resources that are likely to be useful in carrying out these projects. Wengert's view fills in some of what we mean when we say that the mission of the librarian is to provide access to information. In particular, it highlights the fact that merely "exposing someone to data might not provide that person with information" or knowledge.[6] Librarians are particularly well placed to help people gain the knowledge they seek. While librarians may not be "subject specialists" in all of the areas that a patron might have an interest, they are "information specialists." In other words, they have learned how to identify well-reviewed, current, and highly regarded sources of information. Librarians are trained in how to evaluate information resources and they spend their time finding the best information resources so that patrons do not have to sort through everything themselves.[7]

At one time, "moral" quality would have been included in what the librarian should be evaluating information for. Indeed, in his 1908 presidential address to the American Library Association, Arthur Bostwick (1908, pp. 257–259) pictured the librarian as a moral guide whose job is to educate and cultivate the masses. He argued that librarians ought to evaluate works based on their possible "moral teaching or effect" as well as on their "beauty, fitness, and decency." However, librarianship has since moved away from a paternalistic conception of guiding the public taste and morals to a more libertarian conception that people will be better citizens to the extent that they are free to pursue their own intellectual interests.[8] Thus, the goals of the library are more linked to the idea of "intellectual freedom" than of education and "moral uplift." And, indeed, most authors writing on values and librarianship argue that intellectual freedom is the central value of librarianship (ALA, 2006a; Doyle, 2001; Krug, 2003; Oppenheim and Pollecutt, 2000).

A number of authors have provided arguments for why libraries serve an essential function in promoting intellectual freedom. Mark Alfino and Linda Pierce (2001), for example, argue that, "information itself is morally neutral but, in the context of guided inquiry, it supports the development of personal autonomy and personal agency" (p. 481). In response to the view that intellectual freedom is an individualistic value, it is important to emphasize the important *social value* of intellectual freedom. Pierce and Alfino, for example, appeal to what Post (1993) has called the "collectivist" justification of free speech. They argue that "there is an analogy between the ability of a person to become self-governing and the ability of a community to self-govern" (p. 480). Thus, Alfino and Pierce argue that librarians

[6]While we do not wish to take a position on the controversies concerning how to define information, for the purposes of this discussion, we will use information to mean "meaningful data" (see Fetzer (2004)). Thus, we are not limiting "information" to content that is accurate (see Floridi (2005)), for this would exclude too much of what information professionals do.

[7]Indeed, there is much evidence that patrons do not *wish* to sort through all the information for themselves (see Mann (1993)).

[8]Doyle (2001) provides a brief history of intellectual freedom in American libraries. Until fairly recently, libraries were quite paternalistic.

ought to take "an active role" as "public intellectuals," not just "valuing intellectual integrity, personal growth of the patron," but also "the development of their community's reflective skills" (p. 476). Such reflective skills are best promoted in an environment of intellectual freedom where persons are free to express their views and to access the views of others. Libraries, and free public libraries in particular, are fundamental to providing this environment. And, librarianship really is a noble profession insofar as it is devoted to fostering such an environment. Librarians are suited to do this job due to their training in how to find and evaluate information resources. They can promote the development of reflective skills both by providing access to a broad range of quality resources, and by providing information about how to sort through the resources that are available.[9]

The view that the central value of librarianship is intellectual freedom is generally accepted by most writers in library and information science. However, there is quite a bit of debate on how to understand this value. The ALA Office of Intellectual Freedom's (OIF) position, for example, has been criticized by a number of authors as being too extreme.[10] The OIF defines intellectual freedom as "the right of every individual to both seek and receive information from all points of view without restriction. It provides free access to all expressions of ideas through which any and all sides of a question, cause, or movement may be explored" (ALA, 2006b). It further elaborates this definition as requiring absolutely unfettered access to all materials for everyone, including children. As Himma (2004) notes, one may be concerned about having such controversial positions guiding the activities of those employed by "publicly-funded state institutions." He argues that, "The idea that libraries ought to be defending the most expansive conception of free speech is hard to defend on democratic grounds. Most people in this society are in favor of some content-based censorship and believe that obscenity, disclosure of national secrets, corporate and commercial speech, and speech likely to create an imminent threat to public safety are all legitimately restricted by the state" (p. 22). While we share a number of Himma's objections to the OIF's positions on intellectual freedom, we do not think this should lead us to reject the view that intellectual freedom is the core value of librarianship. We can focus on intellectual freedom as the core value of librarianship, without necessarily committing ourselves to the OIF's particular understanding of what intellectual freedom entails.[11]

9.3 SELECTION, BIAS, AND NEUTRALITY

Enabling maximal intellectual freedom might imply providing access to all legally available information. But, since libraries are unlikely to ever be able to provide free

[9]There are a number of ways librarians might do this—via reference interviews or free classes on how to find quality information on the internet, for example.

[10]See Baldwin (1996), Frické et al. (2000), Himma (2004), and Sheerin (1991).

[11]We will note some particular objections to the ALA's position on children's access to information in section IV.

access to all information, some selection decisions must be made.[12] Even if libraries go totally online and thus no longer face issues of space limitations, budget limitations will require that they select some resources and not select others. Atkinson (1996) calls for the use of professional judgment in such decisions. He argues against Asheim (1982) and in support of Ortega y Gasset (1961 [1935]) that "Selection—filtering, to use Ortega's word—far from being an ethical transgression. . .is (and always has been) the core service; indeed, the greatest ethical transgression the library could ever commit would be to avoid selection—that is, not to prescribe" (p. 246).

Given this unavoidable job of selection, ethical issues arise. Are there some rationales for, or procedures for, selection or "deselection" that are professionally objectionable? In particular, how do we avoid selecting in a biased manner that might interfere with patrons' intellectual freedom (by, e.g., excluding works that promote a point of view that the selector disagrees with)?

One might ask, where is the ethical issue? Just give the patrons what they want. Indeed, according to Orr (2003, p. 586), "Most public librarians would agree that their collections should be designed with what might be called a client-centered approach. In other words, the public library collection should fit with the needs of its users." But, this solution is not the panacea that it might appear to be. While no one would argue that patron demand and interest should be ignored, there is also the argument that librarian expertise may find better sources of information on a topic than the public is aware of. Furthermore, even once we have found out what our patrons want in general—mysteries, how-to-books—and in particular—Harry Potter, Anne Coulter—there will still be many selection decisions left to make.

So, librarians must make selections, but in so doing they face the possibility of personal or political bias entering into the process. If collections are biased toward, or away from, certain points of view, the patron may end up being subtly swayed in the direction of the selector's bias, threatening the patron's intellectual freedom. Some have argued that librarians can avoid bias in selection by simply remaining neutral between different points of view. Finks (1989, p. 353), for example, argues that

> At the center of the librarian's commitment to humanity's search for truth and understanding is the goal of remaining always neutral in the battle of competing ideas. No matter how precious to us any faith or philosophy or social movement might be, we have to keep our distance and maintain our impartiality as we help to insure that all the people can hear all the arguments and establish for themselves what is right or true.

This sounds good, but how can we do it? In particular, what does it mean to be "neutral" when making selection decisions? If being a professional means anything, it means making decisions based on one's professional judgment. Thus, any account of neutrality in selection must be distinct from mere nonjudgment. One way to

---

[12]In what follows when we discuss "all information" we will mean all legal information (excluding such things as state secrets, obscene material, etc.). For the purposes of this essay, we will assume that the current law determines the proper boundaries of intellectual freedom. We are indebted to Ken Himma for this suggestion.

understand the call for neutrality is that the selector must put aside or "bracket" his or her own *personal* beliefs and values when evaluating an information source for the purposes of making a selection decision.[13] This fits with the idea that in the context of our professional activities we ought to place our professional obligations over the merely "personal" ones.[14] According to Finks (1991, p. 90), "librarians are obligated to restrain any personal tendencies that are in conflict with the best interests of their agency and occupation." For example, if I am making selections for the religion collection of my public library, I should not use my personal religious beliefs to make a determination about which books to include. One way to understand the idea of "personal" beliefs is that they are beliefs and values that are not part of the organization or community that you serve. If your library does not have a mission to support Christianity, then your selections should not either, even if you are a passionately committed Christian. The concern about the threats to intellectual freedom, which might be posed by the injection of personal bias into the selection decision, can at least be partly corrected by librarians adopting and using explicit collection development policies (cf. Evans (2000), p. 73).

It might be helpful here to compare the neutrality of the librarian to something like a referee's neutrality. The referee is neutral with regard to the teams—he does not make calls based on which team he prefers. But, he is not neutral with regard to the rules of the game—he is a partisan and a defender of the rules. So, we need to determine the rules of the game for selecting information resources. The suggestion of the collection development literature is "information quality."[15] It would not be a "personal" value to refuse to include a book or to remove a book because it is low quality (e.g., because it was factually flawed, dated, unsupported, poorly written, falling apart, vastly inferior to similar texts, etc.). Even the great advocate of neutrality, Finks (1989, p. 353), believed it was appropriate to be an advocate of validity, honesty, accuracy, and quality, "We strive to seize and cherish those items... that speak with validity, that reflect with honesty the heart of our experience, and that reach certain standards of accuracy and quality." In other words, we should select the "good stuff."

But to say we ought to select the good stuff is not enough. The process of selection is not simply about choosing quality individual works, but about providing a quality collection overall. This requires that we be concerned not just about how individual books are selected, but about how a whole collection is designed. Alfino and Pierce (2001, p. 482) argue that, "Part of the librarian's mission might be to model a holistic 'diet' of information, but one that will require substantive judgment, not strict neutrality." This ideal of selection is not one that simply says "pick the good stuff,"

---

[13]Of course, there are serious questions about whether it is possible to be neutral in this way. Our "personal beliefs" may affect our decisions in ways that we are not completely consciously aware of. Thanks to Ken Himma for emphasizing this.

[14]This principle is also embedded in the current ALA Code of Ethics: VII. "We distinguish between our personal convictions and professional duties and do not allow our personal beliefs to interfere with fair representation of the aims of our institutions or the provision of access to their information resources."

[15]In her list of selection criteria, Johnson (2004, p. 107) explicitly includes "quality of scholarship." In addition, she includes several of the traditional dimensions of information quality, such as "currency" and "completeness and scope of treatment" (as well as "veracity").

but one that says "provide a *full range* of information." A model for this might be what the editors of Wikipedia (2006) call the "neutral point of view," which requires that, "where there are or have been conflicting views, these should be presented fairly.[16] None of the views should be given *undue weight* or asserted as being the truth, and all significant published points of view are to be presented, not just the most popular one.[17]

There are clear advantages to taking a "neutral point of view" approach when designing a collection. It promotes both the education and intellectual freedom of those served by the library. By supplying a range of points of view, patrons are more likely to be able to find the works that interest or appeal to them. They are thus free to pursue their intellectual interests through using a library. By providing the range of points of view, the library also allows users to see the range of points of view and beliefs within the culture. This seems to be a natural interpretation of what Finks (1989, p. 353) was after when he argued that librarians should remain "neutral in the battle of competing ideas" and "ensure that all the people can hear all the arguments and establish for themselves what is right or true."

However, it is important to distinguish the "neutral point of view" from the "balance" concept of neutrality. It has been noted by those working in journalism that balance can lead to a false impression. It may simply reinforce the preexisting prejudices of the culture or it may treat a well-established theory or fact as if it was a mere "opinion." In considering Mindich's (2000) history of the origins of objectivity in journalism, Cunningham (2003) writes, "Mindich shows how 'objective' coverage of lynching in the 1890s by *The New York Times* and other papers created a *false balance* on the issue and failed 'to recognize a truth, that African-Americans were being terrorized across the nation.'" More recently, it has been argued that many scientific issues, such as global warming, end up being distorted by a commitment to such "false balance." As Hansen (2006), the Director of the NASA Goddard Institute for Space Studies, points out, "even when the scientific evidence is clear, technical nit-picking by contrarians leaves the public with the false impression that there is still great scientific uncertainty about the reality and causes of climate change."[18] If a librarian thinks that he or she must always balance a book in the collection that says *p* with one that says not *p*, then he or she may be creating a false impression of "equal weight."

Some have argued that, even if one accurately represents the "weight" that various views have within society, one may end promoting the power system of the status quo, such as racism and sexism. If the society as a whole is racist, then accurately representing the balance of views in our library will mean that our library has many

---

[16]The "Neutral Point of View" is one of the content policies that Wikipedia requires their contributors follow (along with "verifiability" and "no original research").

[17]Interestingly, this policy did not originate with Wikipedia. As early as the 17th century, librarians such as Gabriel Naudé in France and John Durie in England recommended that the collection as a whole should be representative of all points of view (Gilbert (1994), p. 384). According to Naudé, "A library arranged for the public must be universal, and it cannot be so if it does not contain all the principal authors . . ." (from his *Advice on Establishing a Library* to Cardinal Richelieu of 1627) [cited in Gilbert ((1994)), p. 384].

[18]See also Mooney (2004).

more racist than nonracist voices. Iverson (1998), for example, writes that, "systemic racism in our society typically limits access to resources to all but the privileged white middle class, by doing so society effectively 'censors' many voices. Consequently, librarians responsible for acquisitions may be recreating racist censorship in their daily practices of selecting from lists of materials produced by mainstream publishing houses and other organizations that perpetuate these patterns." And, as Shera (1970, p. 164) put it, "there are times when silence is not neutrality but assent. . ." The worry, of course, is that in trying to correct for these societal "patterns" the librarian will mold the collection to his or her own views. One might argue that as unfortunate as such societal prejudices are, it is not the job of the librarian to try to correct them via his or her collection development decisions.[19]

9.4 CLASSIFICATION AND LABELING

Libraries by their very nature shape the ways in which we access information. If they did not do this, they would have little use. A big room with all the books and other information stuffed in at random with no way of sorting through it would be relatively useless. As we saw in the previous section, selection is an unavoidable part of the librarian's job; the same is true of organization. However, once one sorts and organizes material or provides particular ways for the information seeker to sort the material herself, one is shaping what information that seeker will get and how the seeker will perceive this information. This shaping may be intentional, or it may simply be an artifact of the way in which the sorting system has been set up. Nevertheless, a library is an intermediary between the person who wishes to access some information and the information. The question is what sorts of shaping are appropriate and which are inappropriate.

In traditional library cataloging and classification, the categories and organization of the information objects are the creations of library professionals. Librarians determine what would be the most useful categories to use in organizing material. This may either be by devising a categorical scheme ahead of time ("faceted classification") or by using the works within the subject area to guide the creation of the categories ("enumerative classification") (Hunter, 1988; Mann, 1993). Not all classifications require such intentional acts of categorization by human beings, however. Classification schemes can also arise from "the ground up," as a result of the ways in which large numbers of people actually label and use information objects. Language itself can be seen as one such classification scheme. The terms used and the categories of objects picked out by those words are the result of the collective activity of human beings. Advances in computer technology have allowed a similar sort of process to create ways to categorize information.[20]

---

[19]This worry about balance and neutrality as opposed to commitment to social justice has also arisen in what has been called the "social responsibility debate" in librarianship (Alcock, 2003; Berninghausen, 1972; Joyce, 1999; Wedgeworth, 1973).

[20]Modern "ground up categorization" via *ad hoc* labeling and tagging systems (e.g., del.icio.us, flickr) has been termed "folksonomy" (Vander Wal, 2007).

Just like selection, organizing and classifying materials shape routes of access. It makes some information easier to find and access than others; it provides descriptors for materials guiding information seekers to particular sources. For example, take two books on space flight. One book may be put in the fiction section and be cataloged under the "science fiction" subject heading, while the other will go into the science section and be cataloged under the "outer space–exploration" subject heading.[21] The ALA characterizes such classification as, "Viewpoint-neutral directional aids," which "facilitate access by making it easier for users to locate materials" (2005). Nevertheless, even though such organization and classification is only intended to aid people in finding information, the organization of information itself shapes how people understand and receive the information that they access. Indeed, studies have shown that labeling or categorizing affects even such seemingly direct perceptual experiences as how things smell or taste (Hardman, 2005).[22]

Given the cognitive effects of categorization, some have argued that we should be particularly careful that our categorizations do not promote or reinforce bias. In *Prejudices and Antipathies: A Tract on the LC Subject Heads Concerning People*, Berman (1993) famously tackled what he saw as bias in the Library of Congress subject headings (LCSH). He argued that racism, sexism, Christocentrism, and other biases were inherent in these subject headings. For example, he notes that in 1966 the LCSH had subject headings for "Negro criminals" and "Jewish criminals," but not for Dutch, Irish, or Italian criminals (p. 35). If one lives in a racist or sexist society, it is not surprising that the categories devised to help people find information will themselves be racist or sexist. As Abbott (1998) notes, "even indexing and retrieval can ultimately be defined as political; like selection, they have a natural slant toward the culturally standard—standard in language, in values, and so on." One may argue that a commitment to social justice gives us a good reason to avoid perpetuating cultural standards that are racist or sexist. Furthermore, there are arguments based solely on providing access to information that would give us a reason to change racist labels; those so labeled within the collection may be turned off and not wish to use the library.

There are worries, however, that the classification system is not the place to try to revise our language. If one uses "politically correct" categories or words that are not commonly or traditionally used, one may make it difficult for persons seeking information to find it either because (a) the person is herself using a "politically *in*correct" term or (b) the person wants to research how people use (or used) this "politically incorrect" term. It may be that two of our goals are in conflict. On one hand, we want people to be able to find the information they want and on the other hand, we want to avoid perpetuating damaging and/or misleading stereotypes. But, when people think in stereotypes, they will be more likely to find the information they want if this information is categorized by such stereotypes. Furthermore, one might

[21]The purpose of such classification is not Aristotle's, that is, to "provide an inventory of everything there is, thus answering the most basic of metaphysical questions: 'What is there?'" (Thomasson, 2004), but to provide "directional aids" for information seekers.

[22]Of course, we do not necessarily recommend eating, or even smelling, the books in the library.

worry that to suggest that librarians correct for such biases is simply an invitation for them to substitute their own (particular) worldview for the "culturally standard" one.[23]

We cannot avoid categorizing materials in order to organize them and better enable access, but it might be argued that we could do more. We could provide patrons with more information about works than merely what is needed to find the works. Is it ever permissible to categorize or label a work in order to give the reader information about the content, which is not merely descriptive or "directional"?[24] The American Library Association says no. The ALA (2005) "opposes labeling as a means of predisposing people's attitudes toward library materials." If, as we noted above, the mere fact of labeling something changes how people perceive it, by labeling materials one is biasing (for good or bad) the reader's perception of that work. Such biasing may interfere with the patron accessing the work (because they would be embarrassed to check out something that said "propaganda" on it, for instance). It might also interfere with the author's ability to "speak for herself" without the intermediary commentary by the librarian.

Several authors have taken a position on labeling contrary to the ALA's position (cf. Hitchcock, 2000; Nesta and Blanke, 1991; Pendergrast, 1988). They suggest that it is permissible to put (nondirectional) labels on library materials when there is a clear risk that these materials will mislead readers. For example, Mark Pendergrast has proposed labeling out-of-date medical texts. Also, Henry Blanke has proposed labeling blatant government propaganda.[25] Such labels would arguably remove the risk of readers being misled without restricting access by removing these books from the collection.

Of course, there are other strategies, which will also make it less likely that readers will be misled. For example, readers might be given instruction on how to evaluate the reliability of information (cf. Fallis, 2004). Such a strategy would arguably be less of a threat to intellectual freedom than informative labeling.[26] However, it is not immediately clear why providing readers with more information about the contents of a book in the form of labels is completely unacceptable. Such information might possibly bias readers for or against that book to some degree. But libraries engage in a number of other activities that have the potential to bias readers. Most notably, as discussed above, biased selection decisions can easily create a biased collection. Nevertheless, recall that in the previous section, we pointed out that librarians can avoid arbitrariness in selection by crafting and abiding by a collection development policy that is

---

[23]Again, we are indebted to Ken Himma for emphasizing this point.

[24]Of course, directional aids, such as subject headings can function as a sort of label, which predisposes people's attitudes toward the content. For example, Holocaust denial literature used to be assigned "Holocaust, Jewish (1939–1945)—*Errors*, inventions, etc." as a subject heading (Wolkoff (1996), p. 92).

[25]Note that it may be clear to experts in evaluating information, such as librarians (e.g., because they note its source, have access to reviews and critiques, can compare it to other more authoritative sources), that a work is propaganda without it being clear to the average person that the work is propaganda.

[26]Of course, even teaching standard evaluation techniques has the potential to bias readers. It is just biases the reader against a whole class of books (e.g., those whose author is anonymous) rather than one specific book.

committed to neutrality. It is not clear why librarians could not adopt similar policies to ensure the neutrality of informative labeling.

9.5 CHILDREN'S ACCESS TO INFORMATION

Some of the most vexed issues that confront public libraries have to do with children's access to information. Much of the American public believes that children ought to be protected from certain sorts of speech. According to a recent poll, "by a 57% to 35% margin, American adults believe that protecting children from indecency is more important than freedom of speech" (Rasmussen Reports, 2007). This concern about the importance of protecting children from certain sorts of speech extends to libraries. According to a not uncommon view, "libraries are supposed to be places where children learn, where they are protected, where responsible adults offer reasonable guidance" (Otis, 2001). And such concerns are borne out in action. Challenges to library materials most commonly concern children's access. The most common initiator of a challenge to materials in public libraries is a parent and the challenges are most commonly in regard to works that the challenger considers sexually explicit, containing offensive language, unsuitable for age group, or too violent (ALA, 2000). It is clearly the desire to "protect" children from such works that motivates adults to suggest that the works be moved or removed.

The American Library Association, however, takes the position that libraries should never impose any limitations to access to information based on age.[27] According to the *Library Bill of Rights*, "a person's right to use a library should not be denied or abridged because of origin, *age*, background, or views. . . The 'right to use a library' includes free access to, and unrestricted use of, *all* the services, materials, and facilities the library has to offer" (2004). The ALA argues that it is the job of the parent, not the librarian, to determine what children should read.

> Librarians and governing bodies should maintain that parents—and only parents—have the right and the responsibility to restrict the access of their children—and only their children—to library resources. Parents who do not want their children to have access to certain library services, materials, or facilities should so advise their children. Librarians and library governing bodies cannot assume the role of parents or the functions of parental authority in the private relationship between parent and child (ALA, 2004).

One might take from this that the ALA's position is a "parents' rights" position. On a parents' right position, the library cannot usurp the role of the parents in shaping their child's access to information. But this parents' rights interpretation would not be born out by the ALA's other positions and statements, in particular, their statement on privacy.

The ALA holds that children have the same rights to privacy in their library records as adults. According to the ALA (2006c), "The rights of minors to privacy regarding

[27]For criticisms of this position, see Etzioni (2001), Frické et al. (2000), and Sheerin (1991).

their choice of library materials should be respected and protected." This right to privacy, according to the ALA, extends to the right to have children's circulation records kept from the parents. While the idea that children have a right to privacy is not controversial, this right is not typically understood as a right children have against their parents. Much information that is typically treated as private, such as medical information and grades, is typically shared with parents. The standard argument for protecting the patron record is that, if a patron knows that others may have access, it will have a chilling effect and make them loath to check out materials of which they think others may disapprove (Garoogian, 1991). If parents have a right to control their children's access to information, then there should be no such worry; so the argument for restricting access in the case of parents and their children does not apply. It thus seems that the ALA must believe that children have a basic right to information regardless of whether the parent wants the child to access such information or not.

Indeed, this view is clearly articulated in the ALA's other positions. The ALA holds that children and adults have the same access rights in relation to third party restrictions (e.g., the government, libraries). In other words, they argue that age is not relevant with regard to what persons should be allowed to access (ALA, 2004). They also hold that in no case should the library restrict access to minors even if the parent requests that such restrictions be put in place. They do hold that parents have a duty to guide their children, but when it comes to restricting access to information this duty is a purely private one. In other words, they argue that it is not the business of anyone else in the society to aid parents in restricting their children's access to information.

So, what can be said in favor of this view? When defending its position, the ALA (2004) frequently appeals to the U.S. Constitution to support their position and state that "children have first amendment rights." This blanket statement, however, fails to capture the more complex state of constitutional law with regard to children's first amendment rights. While the U.S. Supreme Court has determined that children have some first amendment rights, they have also held that these rights are more limited than those of adults. Indeed, the Court has argued that there is a role for the state in "protecting" children from certain sorts of content. So, for example, in *Ginsberg v. New York* (390 U.S. 629 (1968)), the court held that those under 17 "have a more restricted right than that assured to adults to judge and determine for themselves what sex material they may read and see." The court also argued that the state has a role in supporting the parents in carrying out their parental responsibilities. "Constitutional interpretation has consistently recognized that the parents' claim to authority in the rearing of their children is basic in our society, and *the legislature could properly conclude that those primarily responsible for children's well-being are entitled to the support of laws designed to aid discharge of that responsibility.*" Furthermore, it was held that "the state has an independent interest in protecting the welfare of children and safeguarding them from abuses." On this view, there is a legitimate role for the state, and perhaps for such state institutions as the public library, in protecting minors from content that is thought to be harmful to minors.

Merely appealing to the U.S. Supreme Court's decisions, however, does not resolve the philosophical and ethical issues involved. It may be that the Supreme Court is simply wrong about children's information rights. We do wish to point out, however, that contrary to what is implied in the ALA's statement on access to minors, the position that children have full first amendment rights is incorrect. It is our view that while the ALA is welcome to engage in advocacy to forward its view about the importance of intellectual freedom, it owes a duty to its members to provide them with accurate information about the current state of the law.[28]

In any event, we need a more philosophically informed consideration of the question of what limits, if any, there should be to children's access to information. An argument that children have the identical intellectual freedom rights as adults would be difficult to make. First, unlike adults, children have a special need to be protected, even from their own choices. Joel Feinberg, for example, has argued that they have "the right to be protected against harms that befall children because of their childlike vulnerability and whose particular harmfulness is a function of a fact that they befall children" (Archard, 2006). William W. Van Alstyne, a professor at Duke University School of Law and the author of a leading textbook on the First Amendment, puts it a bit more bluntly. "There isn't any doubt that at some age it is preposterous that a child has a right to go to a store and buy matches," he said. "At that point, the child has neither the experience nor gray matter to act in his own best interests" (quoted in Kaplan (2006)). Thus, if some information is harmful to a child, we may have an ethical obligation to protect that child from that information.

Of course, children have a right to protection that can provide a justification for limiting their access to information only if we have good reason to think that some information is likely to be harmful to them. There is not complete agreement that speech such as pornography, violence, and bad language influence the behavior of children, or that such influence generally leads to harm. Many argue that the evidence is sufficient to show the likelihood of harm with regard to violent content, for example (cf. Etzioni (2004)). Others, such as Heins (2004, p. 245) argue that the evidence is not in. She cites a *Lancet* (1999) editorial to the effect that "it is inaccurate to imply that the published work strongly indicates a causal link between virtual and actual violence." Whether limitations on access are justifiable will certainly depend on the empirical evidence.[29]

Second, it is not clear that the arguments for the importance of free access to information fully apply to children. Indeed, no less an advocate of intellectual freedom than Mill (1859/1975 [1859], p. 166) says that his arguments for unfettered access to information only apply to those in the "maturity of their faculties." There is evidence that children's cognitive abilities and capacities for independent decision-making and self-control are significantly limited as compared to adults (Casey et al., 2005). Many of the standard arguments for why persons have a right to access information, such as arguments based on autonomy, free choice, etc., which depend on the sorts of cognitive capacities that children lack, do not easily transfer to the case of children.

---

[28]For a view that the ALA should not take on such advocacy roles as part of their professional duties, see Himma (2004).

[29]For a further discussion of harm as a justification for limiting access to information, see Mathiesen "Censorship and Access to Expression", this volume.

Even so, some argue that we cannot use age as a criterion for what information someone gets access to, because we cannot draw a simple relationship between age and cognitive capacity or level of maturity. According to the ALA (2004), for example, "librarians cannot predict what resources will best fulfill the needs and interests of any individual user based on a single criterion such as chronological age, educational level, literacy skills, or legal emancipation." While there is typically a substantial difference in levels of cognitive development and maturity between a 4 year old, a 12 year old, and a 16 year old, there may also be very large differences between two 16 year olds. Chronological age is only a rough proxy for developmental stage. Interestingly, even those who argue very strongly for children's information rights, such as Heins (2004), agree that the idea of intellectual freedom has "little meaning" before around 7 years of age, or "the age of reason" (252).[30] Of course, many might dispute setting the line at 7 years of age. Clearly there is no one "correct" standard for what is the age at which children are sufficiently adult to merit full intellectual freedom. But, this does not mean that such standards are completely arbitrary. Where we draw the lines may depend on our information about normal cognitive development in children or the age at which children are expected to take on adult roles and responsibilities in a particular society.

But we may argue that, even if children do not have the cognitive development to ground full intellectual freedom, there is still an important justification for trying to provide as full a range of information as possible in support of that children's *future* intellectual freedom. Arguably, in order to develop capacities for understanding, evaluating, and deliberating about information, children need some degree of liberty to make their own information choices as well as access to a broad range of information. Such liberty by itself without adult guidance, however, may inhibit rather than promote the development of an autonomous self.[31] If librarians really care about the current and future intellectual freedom of minors, then merely providing minors with access to all possible content is probably not what is required. This is not to say that individual librarians ought to be making on-the-fly decisions about what individual children should or should not read. What it does say is that promoting the intellectual freedom of children may require quite different sorts of responses than it does in the case of adults.

9.6 CONCLUSION

Since the mid-twentieth century library and information professionals have moved away from a paternalistic conception of their mission as "public uplift" to one of protecting and promoting intellectual freedom (cf. Doyle (2001), pp. 45–50). Librarians continue today to see their central value as the promoting of intellectual freedom.

[30]Nevertheless, she argues that such children would not be able to understand the information that some want to censor, and so such censorship would be pointless.

[31]Indeed, even Heins (2004), an advocate for children's freedom to access information, argues that access to information must be combined with programs that teach children how to evaluate this information.

As we have seen, however, those who wish to promote this value face many challenges. How do librarians provide a "value added service" that selects and guides patrons to quality information while not biasing the collection toward a particular viewpoint? How should librarians respond to the fact of social injustice within the society it serves? What is the appropriate response of the library to the societal desire to protect children from certain types of information? In reflecting on these questions, the ethical theorist must grapple with some very deep underlying questions about the importance to access to information in a complex, democratic society.

REFERENCES

Abbott, A. (1988). *The System of Professions: An Essay on the Division of Expert Labor.* University of Chicago Press, Chicago.

Abbott, A. (1998). Professionalism and the future of librarianship. *Library Trends*, 46(3), 430–444.

Alcock, T. (2003). *Free Speech for Librarians? A Review of Socially Responsible Librarianship*, 1967–1999. [Cited 1/1/07.] Available from http://juteux.net/rory/Alcock.html.

Alfino, M. and Pierce, L. (2001). The social nature of information. *Library Trends*, 49(3), 471–485.

American Library Association. (1999). *Libraries: An American Value*. [Cited 1/6/07.] Available from http://www.ala.org/ala/oif/statementspols/americanvalue/librariesamerican.htm.

American Library Association. (2000). *The 100 Most Frequently Challenged Books of 1990–2000 and Challenges by Initiator, Institution, Type, and Year.* [Cited 7/1/07.] Available from http://www.ala.org/Template.cfm?Section=bbwlinks&Template=/ContentManagement/ContentDisplay.cfm&ContentID=78236.

American Library Association. (2004). *Free Access to Libraries for Minors: An Interpretation of the Library Bill of Rights.* [Cited 1/6/07 2007.] Available from http://www.ala.org/Template.cfm?Section=interpretations&Template=/ContentManagement/ContentDisplay.cfm&ContentID=103214.

American Library Association. (2005). *Labels and Rating Systems: An Interpretation of the Library Bill of Rights.* [Cited 1/20/07.] Available from http://www.ala.org/ala/oif/statementspols/statementsif/interpretations/statementlabeling.htm.

American Library Association. (2006a). *Intellectual Freedom Manual.* American Library Association, Chicago.

American Library Association. (2006b). *Intellectual Freedom and Censorship Q and A.* [Cited 6/8/06.] Available from http://www.ala.org/ala/oif/basics/intellectual.htm.

American Library Association. (2006c). *Questions and Answers on Privacy and Confidentiality.* [Cited 1/6/07.] Available from http://www.ala.org/Template.cfm?Section=interpretations&Template=/ContentManagement/ContentDisplay.cfm&ContentID=34114.

Archard, D.W. (2006). Children's rights. In: Edward, N.Z. (Ed.), *The Stanford Encyclopedia of Philosophy (Winter 2006 Edition).* Available at http://plato.stanford.edu/archives/win2006/entries/rights-children/.

Arant, W. and Benefiel, C. (2002). *The Image and Role of the Librarian.* The Haworth Information Press, Binghamton, NY.

Asheim, L. (1982). Ortega revisited. *Library Quarterly*, 52(3), 215–226.

Atkinson, R. (1996). Library functions, scholarly communication, and the foundations of the digital library: laying claim to the control zone. *The Library Quarterly*, 66(3), 239–265.

Baker, B. (2000). Can library service survive in a sea of change? *American Libraries*, 31(4), 47–49.

Baldwin, G.B. (1996). The library bill of rights—a critique. *Library Trends*, 45(1), 7–27.

Berman, S. (1993). *Prejudices and Antipathies: A Tract on the LC Subject Heads Concerning People*. Scarecrow Press, Metuchen, NJ.

Berninghausen, D. (1972). Antithesis in librarianship: social responsibility vs. the library bill of rights. *Library Journal*, 97, 3675–3881.

Bostwick, A.E. (1908). Librarian as censor. *Library Journal*, 33, 257–259.

Casey, B.J., Tottenham, N., Liston, C., and Durston, S. (2005). Imaging the developing grain: what have we learned about cognitive development? *Trends in Cognitive Science*, 9(3), 105–110.

Cunningham, D. (2003). Rethinking objectivity. *Columbia Journalism Review*, (4). [Cited 1/16/07.] Available from http://cjrarchives.org/issues/2003/4/objective-cunningham.asp.

Diamond, J. (1997). *Guns, Germs, and Steel*. W. W. Norton, New York.

Diamond, R. and Dragich, M. (2001). Professionalism in librarianship: shifting the focus from malpractice to good practice. *Library Trends*, 49(3), 395–414.

Dilevko, J. and Gottlieb, L. (2004). The portrayal of librarians in obituaries at the end of the twentieth century. *The Library Quarterly*, 74, 152–180.

Doyle, T. (2001). A utilitarian case for intellectual freedom in libraries. *Library Quarterly*, 71(1), 44–71.

Etzioni, A. (2001). Suffer the little children. *The Good Society*, 10(1), 67–91.

Etzioni, A. (2004). On protecting children from speech. *Chicago-Kent Law Review*, 79(3), 3–53.

Evans, G.E. (2000). *Developing Library and Information Center Collections*, 4th edition. Libraries Unlimited, Englewood, Colorado.

Fallis, D. (2004). On verifying the accuracy of information: philosophical perspectives. *Library Trends*, 52(3), 463–487.

Fallis, D. (2007). Information ethics for 21st century library professionals. *Library Hi Tech*, 25(1), 23–36.

Fetzer, J.H. (2004). Information: does it have to be true? *Minds and Machines*, 14(2), 223–229.

Finks, L.W. (1989). Values without shame. *American Libraries*, 20, 352–356.

Finks, L.W. (1991). Librarianship needs a new code of professional ethics. *American Libraries*, 22(1), 84–92.

Floridi, L. (2005). Is information meaningful data? *Philosophy and Phenomenological Research*, 70(2), 351–370.

Fourie, D.K. and Dowell, D.R. (2002). Chapter 2: a brief history of libraries. *Libraries in the Information Age: An Introduction and Career Exploration*. Libraries Unlimited, Greenwood Village, CO.

Frické, M., Mathiesen, K., and Fallis, D. (2000). The ethical presuppositions behind the Library Bill of Rights. *Library Quarterly*, 70(4), 468–491.

Garoogian, R. (1991). Librarian/patron confidentiality: an ethical challenge. *Library Trends*, 40(2), 216–233.

Gilbert, P. (1994). Library profession. *The Encyclopedia of Library History*. Garland Publishing, New York.

Gorman, M. (2000). *Our Enduring Values: Librarianship in the 21st Century*. American Library Association, Chicago.

Guns, Lies, and videotape 354(9178). (1999). *Lancet*, [editorial], 525.

Hansen, J. (2006). The threat to the planet. *The New York Review of Books*, 53(12), 12–16.

Hardman, H. (2005). Visual words influence perception of smells. *Medical News Today*. May 19 [cited 6/6/07. Available at http://www.medicalnewstoday.com/medicalnews.php?newsid =24723.

Heins, M. (2004). On protecting children—from censorship: a reply to Amitai Etzioni. *Chicago-Kent Law Review*, 79, 229–255.

Himma, K. (2004). Libraries as political advocates: a critique of the Library Bill of Rights, article III. *ALKI: The Washington Library Association Journal*, 20(2), 21–23.

Hitchcock, L.A. (2000). Enriching the Record. *Journal of Academic Librarianship*, 26(5), 359–363.

Hunter, E.J. (1988). *Classification Made Simple*. Gower, Aldershot.

Iverson, S. (1998/99). Librarianship and resistance. *Progressive Librarian,* 15, 14–20.

Johnson, P. (2004). *Fundamentals of Collection Development and Management*. American Library Association, Chicago, 14–90.

Joyce, S. (1998/1999). A few gates: an examination of the social responsibilties debate in the early 1970s and 1990s. *Progressive Librarian,* 15, 1–19.

Kaplan, C.S. (1998). *Children's First Amendment Rights Lost in the Filtering Debate* 1998 (March 6) [cited 11/05/2006]. Available from http://www.diversityjobmarket.com/library/tech/98/03/cyber/cyberlaw/06law.html.

Krug, J. (2003). Intellectual freedom and the ALA: historical overview. *Encyclopedia of Library and Information Science*. Marcel Dekker, New York.

Mann, T. (1993). *Library Research Models: A Guide to Classification, Cataloging, and Computers*. Oxford University Press, New York.

Maynard, S. and McKenna, F. (2005). Mother goose, spud murphy and the librarian knights: representations of librarians and their libraries in modern children's fiction. *Journal of Librarianship and Information Science*, 37(3), 119–129.

Mill, J.S. (1859/1975). On liberty. *Three Essays*. Oxford University Press, Oxford.

Mindich, D.T.Z. (2000). *Just the Facts: How Objectivity Came to Define American Journalism*. New York University Press, New York.

Mooney, C. (2004). Blinded by science. how 'balanced' coverage lets the scientific fringe hijack reality. *Columbia Journalism Review*, (6). Accessed on-line [05/10/07] at http://cjrarchives.org/issues/2004/6/mooney-science.asp.

Nesta, F. and Blanke, H. (1991). Warning: propaganda! *Library Journal*, 116(9), 41–43.

Oppenheim, C. and Pollecutt, N. (2000). Professional associations and ethical issues in LIS. *Journal of Librarianship and Information Science*, 32(4), 187–203.

Orr, C. (2003). Collection development in public libraries. In: Bates, M.J.M., Niles, M., and Drake, M. (Eds.), *Encyclopedia of Library and Information Science*. Marcel Dekker, New York.

Ortega y Gasset, J. (1961). [1935] The mission of the librarian. *Antioch Review*, 2, 133–154.

Otis, D.S. (2001). *American Library Association Is No Friend of Our Children*. Agape Press, March 23. Available at http://headlines.agapepress.org/archive/3/afa/232001i.asp.

Pendergrast, M. (1988). In praise of labeling; or, when shalt thou break commandments? *Library Journal*, 113, 83–85.

Post, R. (1993). Managing deliberation: the quandry of democratic dialogue. *Ethics*, 103(4), 654–678.

Ranganathan, S.R. (1931). *The Five Laws of Library Science*. Edward Goldston, Ltd., London. Available at http: //dlist.sir.arizona.edu/1220/.

Rasmussen Reports. (2007). *Most Americans Are Very Concerned About Profanity, Sex, Violence on TV* (June 15). [Cited 7/1/2007.] Available from http://www.rasmussenreports.com/content/pdf/4102.

Shera, J. (1970). *Sociological Foundations of Librarianship*. Asia Publishing House, Bombay.

Sheerin, W.E. (1991). Absolutism on access and confidentiality: principled or irresponsible? *American Libraries*, 22(5), 440–444.

Thomasson, A. (2004). Categories. In: Edward, N.Z. (Ed), *The Stanford Encyclopedia of Philosophy (Fall 2004 Edition)*. Available at http://plato.stanford.edu/archives/fall2004/entries/categories/.

Tyckoson, D.A. (2000). Of the people, for the people: public libraries serve democracy. *American Libraries*, 31(4), 40–41.

Vander Wal, T. (2007). *Folksonomy Coinage and Definition*. [Cited 6/6/07.] Available at http://vanderwal.net/folksonomy.html.

Walker, S. and Lawson, V.L. (1993). The librarian stereotype and the movies. *The Journal of Academic Media Librarianship*, 1(1), 16–28.

Wedgeworth, R., et al. (1973). Social responsibility and the library bill of rights: the Berninghausen debate. *Library Journal*, 98(1), 25–41.

Wengert, R. (2001). Some ethical aspects of being an information professional. *Library Trends*, 49(3), 486–509.

Wikipedia. (2006). *Neutral Point of View*. [Cited 9/20/06.] Available from http://en.wikipedia.org/wiki/Wikipedia:Neutral_point_of_view.

Wolkoff, K.N. (1996). The problem of holocaust denial literature in libraries. *Library Trends*, 45, 87–96.

Other Useful Resources

Alfino, M. and Pierce, L. (1997). *Information Ethics for Librarians*. McFarland, Jefferson, NC.

American Library Association. (1981). Evaluating library collections: an interpretation of the library bill of rights. [Cited 1/6/07.] Available from http://www.ala.org/ala/oif/statementspols/statementsif/interpretations/evaluatinglibrarycollections.pdf.

American Library Association. (1990). Diversity in collection development: an interpretation of the library bill of rights. [Cited 1/6/07.] Available from http://www.ala.org/ala/oif/statementspols/statementsif/interpretations/diversitycollectiondevelopment.pdf.

American Library Association. (1991). *Role Definition*. [Cited 7/10/06.] Available from http://www.ala.org/ala/hrdrbucket/3rdcongressonpro/roledefinition.htm.

American Library Association. (2005). *Privacy Took Kit.* [Cited 1/20/07.] Available from http://www.ala.org/ala/oif/iftoolkits/toolkitsprivacy/privacy.htm.

American Library Association. (2006). *Resolution in Support of Online Social Networks.* [Cited 1/6/07.] Available from http://www.ala.org/ala/oif/ifissues/onlinesocialnetworks.pdf.

Asheim, L. (1953). Not censorship, but selection. *Wilson Library Bulletin*, 28, 63–67.

Aviram, A. (1986). The justification of compulsory education: the still neglected moral duty. *Journal of Philosophy of Education*, 20(1), 51–58.

Aviram, A. (1990). The subjection of children. *Journal of Philosophy of Education*, 24(2), 213–234.

Benporath, S.R. (2003). Autonomy and vulnerability: on just relations between adults and children. *Journal of Philosophy of Education*, 37(1), 127–145.

Berman, S. (2001). 'Inside' censorship. *Progressive Librarian*, 18, 48–63.

Berman, S. (2005). Classism in the stacks: libraries and poor people. *Counterpoise*, 9(3), 51–55.

Berninghausen, D. (1975). *The Flight from Reason: Essays on Intellectual Freedom in the Academy, the Press, and the Library.* American Library Association, Chicago.

Bevier, L.R. (2004). Copyright, trespass, and the first amendment: an institutional perspective. *Social Philosophy and Policy*, 21(2), 104–147.

Blanke, H.T. (1989). Librarianship and political values: neutrality or commitment? *Library Journal*, 114(12) 39–43.

Bovens, M. (2002). Information rights: citizenship in the information society. *The Journal of Political Philosophy*, 10(3), 317–341.

Brewerton, A. (2003). The creed of a librarian: a review article. *Journal of Librarianship and Information Science*, 35(1), 47–55.

Bridges, D. (1984). Non-paternalistic arguments in support of parents' rights. *Journal of Philosophy of Education*, 18(1), 89–96.

Buchanan, E.A. (2004). Ethics in library and information science: what are we teaching? *Journal of Information Ethics*, 13(1), 51–60.

Budd, J.M. (2003). The library, praxis, and symbolic power. *Library Quarterly*, 73(1), 19–32.

Burress, L. (1989). *Battle of the Books: Literary Censorship in Public Schools 1950–1985.* Scarecrow Press, Metuchen, NJ.

Calenge, B. (1998). Can we define librarianship? A theoretical essay. *Bulletin des Bibliotheques de France*, 43(2), 8–20.

Callan, E. (1997). The great sphere: education against servility. *Journal of Philosophy of Education*, 31(2), 221–232.

Carbo, T. and Almagno, S. (2001). Information ethics: the duty, privilege, and challenge of educating information professionals. *Library Trends*, 49(3), 510–518.

Carney, S. (2003). Democratic communication and the library as workplace. *Journal of Information Ethics*, 12(2), 43–59.

Coley, K.P. (2002). Moving toward a method to test for self-censorship by school library media specialists. *School Library Media Research*, 5. Accessed on-line [06/01/07] http://www.ala.org/ala/aaslpubsandjournals/slmrb/slmrcontents/volume52002/coley.cfm.

Cotrell, J.R. (1999) Ethics in an age of changing technology: familiar territory or new frontiers? *Library Hi Tech*, 17(1), 107–113.

Couldry, N. (2003). Digital divide or discursive design?: on the emerging ethics of information space. *Ethics and Information Technology*, 5(2), 89–97.

DeCew, J. (2006). *Privacy* (Fall 2006 edition). [Cited 1/16/07.] Available from http://plato.stanford.edu/entries/privacy/.

Dowd, R. (1989). I want to find out how to freebase cocaine; or yet another unobtrusive test of reference performance. *The Reference Librarian*, 25–26, 483–493.

Doyle, T. (2002). Selection versus censorship in libraries. *Collection Management*, 27(1), 15–25.

Doyle, T. (2002). MacKinnon on pornography. *Journal of Information Ethics*, 11(2), 53–78.

Doyle, T. (2004). Should Web sites for bomb-making be legal? *Journal of Information Ethics*, 13(1), 34–37.

Fallis, D. and Mathiesen, K. (2001). Response to "a utilitarian case for intellectual freedom in libraries". *Library Quarterly*, 71(3), 437–438.

Feinberg, J. (1980). A child's right to an open future. In: Aiken, W. and LaFollette, H. (Eds.), *Whose Child? Parental Rights, Parental Authority and State Power*, Littlefield, Adams and Co., Totowa, NJ, pp. 124–153.

Fifarek, A. (2002). Technology and privacy in the academic library. *Online Information Review*, 26(6), 366–374.

Foskett, D.J. (1962). *The Creed of the Librarian: No Politics, No Religion, No Morals*. Library Association, London.

Freeman, L. and Peace, A.G. (2005). *Information Ethics: Privacy and Intellectual Property*. Idea Group, Hershey, PA.

Froehlich, T. (1992). Ethical considerations of information professionals. *Annual Review of Information Science and Technology*, 27, 291–324.

Froehlich, T. (2005). A brief history of information ethics. *Computer Society of India Communications*, 28(12), 11–13.

Harkovitch, M., Hirst, A. and Loomis, J. (2003). Intellectual freedom in belief and in practice. *Public Libraries*, 42 (6), 367–74.

Hauptman, R. (1976). Professionalism or culpability? an experiment in ethics. *Wilson Library Bulletin*, 50, 636–627.

Hauptman, R. (2001). Technological implementations and ethical failures. *Library Trends*, 49(3), 433–440.

Hauptman, R. (2002). *Ethics and Librarianship*. McFarland, Jefferson, NC.

Jaeger, P.T. and Burnett, G. (2005). Information access and exchange among small worlds in a democratic society: the role of policy in shaping information behavior in the post-9/11 United States. *Library Quarterly*, 75(4), 464–495.

Johnson, S.D. (2000). Rethinking privacy in the public library. *International Information and Library Review*, 32, 509–517.

Kleinig, J. (1981). Compulsory schooling. *Journal of Philosophy of Education*, 15(2), 191–203.

Koehler, W. (2006). National library associations as reflected in their codes of ethics: four codes examined. *Library Management*, 27(1/2), 83–100.

Koehler, W.C., Hurych, J.M., Dole, W.V. and Wall, J. (2000). Ethical values of information and library professionals—an expanded analysis. *International Information and Library Review*, 32(3–4), 485–507.

Lancaster, F.W. (1991). *Ethics and the Librarian*. University of Illinois, Graduate School of Library and Information Science, Urbana-Champaign.

Lane, D. (1990). Your pamphlet file supports apartheid. *Library Journal*, 174–177.

Lesk, M. (2003). Copyright enforcement or censorship: new uses for the DMCA? *IEEE Security and Privacy*, 1(2), 67–69.

Lindsey, J.A., and Prentice, A.E. (1985). *Professional Ethics and Librarians*. Oryx, Phoenix, AZ.

Litwin, R. (2002). Neutrality, Objectivity, and the political center. *Progressive Librarian*, 21, 72–74.

Mathiesen, K. (2004). What is information ethics? *Computers and Society*, 32(8). available at http://doi.acm.org/10.1145/1050305.1050312.

McCabe, R. (2001). *Civic Librarianship: Renewing the Social Mission of the Public Library*. Scarecrow Press, Lanham, MD.

McClure, C.R., Joe, R. and John, C.B. (2002). *Public Library Internet Services and the Digital Divide*. Information Use Management and Policy Institute, Tallahassee.

Mintz, A.P. (1990). *Information Ethics: Concerns for Librarianship and the Information Industry*. McFarland & Company, Jefferson, North Carolina.

Moody, K. (2005). Covert censorship in libraries: a discussion paper. *Australian Library Journal*, 54(2), 138–147.

Nagel, T. (1995). Personal rights and public space. *Philosophy and Public Affairs*, 24(2), 83–107.

Oppenheim, C. and Smith, V. (2004). Censorship in libraries. *Information Services and Use*, 24(4), 159–170.

Reid, A. (1998). The value of education. *Journal of Philosophy of Education*, 32(3), 319–331.

Rubin, R.E. (2004). *Foundations of Library and Information Science*, 2nd edition. Neal-Schuman, New York.

Rubin, R.E. and Froehlich, T. (1996). Ethical aspects of library and information science. *Encyclopedia of Library and Information Science*. Marcel Dekker, New York.

Sharman, D. (2001). Intellectual property: a historical perspective on the commodification of information. *Progressive Librarian*, 18, 9-17.

Smith, M. (2001). Global information justice: rights, responsibilities, and caring connections. *Library Trends*, 59(3), 519–537.

Sosa, J.F. (1991). José Ortega y Gasset and the role of the librarian in post industrial America. *Libri*, 41, 3–21.

Sturges, P., Teng, V. and Iliffe, U. (2001). User privacy in the digital library environment: a matter of concern for information professionals. *Library Management*, 22(8/9), 364–370.

Swan, J.C. (1979). Librarianship is censorship. *Library Journal*, 104 (17), 2040–2043.

Tamier, Y. (1990). Whose education is it anyway? *Journal of Philosophy of Education*, 24(2), 161–170.

Trushina, I. (2004). Freedom of access: ethical dilemmas for internet librarians. *Electronic Library*, 22(5), 416–421.

Ward, D.V. (1990). Philosophical issues in censorship and intellectual freedom. *Library Trends*, 39(1 & 2), 83–91.

Weissinger, T. (2003). Competing models of librarianship: do core *values* make a difference? *Journal of Academic Librarianship*, 29(1), 32–39.

West, C. (1983). The secret garden of censorship: ourselves. *Library Journal*, 108(15) 1651–1653.

Westin, A. (1967). *Privacy and Freedom*. Atheneum, New York.

Woodward, D. (1990). A framework for deciding issues in ethics. *Library Trends*, 39(1 & 2), 8–17.

Wong, K.L. (1996). Tobacco advertising and health: the limits of first amendment protection. *Journal of Business Ethics*, 15, 1051–1064.

CHAPTER 10

Ethical Interest in Free and Open Source Software

FRANCES S. GRODZINSKY and MARTY J. WOLF

10.1 INTRODUCTION

Free Software (FS), a concept developed by Richard Stallman in the 1980s, has served as a foundation for important and related movements that have become possible because of the Internet. The most important of these has been the Open Source Software (OSS) movement. OSS, a concept rooted in software methodology and analyzed by Eric Raymond, broke from the FS ethos in 1998. This paper will compare FS and OSS, examining their histories, their philosophies, and development. It will also explore important issues that affect the ethical interests of all who use and are subject to the influences of software, regardless of whether that software is FS or OSS. We will argue that the distinction between FS and OSS is a philosophically and socially important distinction. To make this point we will review the history of FS and OSS with a particular emphasis on four main people: Richard Stallman, Linus Torvalds, Eric Raymond, and Bruce Perens. In addition, we will review the differences between GNU[1] General Public License (GPL) version 2 (v2) and the current draft of the GPL version 3 (v3), and the related controversy in the OSS community. The GPL is the primary mechanism used by the software community to establish and identify software as free software. In section 10.3, we will examine the motivation and economics of OSS developers. We will review issues of quality with respect to OSS, autonomy of OSS software developers, and their unusual professional responsibilities. The final important issue we address is consideration of OSS as a public good.

[1]GNU is a recursive acronym for GNU's not Unix.

The Handbook of Information and Computer Ethics, Edited by Kenneth Einar Himma and Herman T. Tavani

10.2 ON THE DISTINCTION BETWEEN FS AND OSS

10.2.1 The History of Free and Open Source Software

Free software stems from the close ties that early software developers had with academia. As the software industry began to mature, the bond with academia and its ideals of sharing research results weakened. After spending many years as an active participant in the hacker culture, Richard Stallman grew frustrated as more and more software was not free—not free in a financial sense, but free in a way that allowed for its inspection, running, and modification. Stallman took a stand and began the GNU project in 1984. The goal of the project was to establish a software development community dedicated to developing and promoting free software. He established the Free Software Foundation (FSF) to support his plan to create an operating system complete with all of the tools needed to edit, compile, and run software. This effort resulted in a large collection of free software. As part of this work, he codified his notion of free software in the GNU General Public License. Stallman was (and still is) vocal in articulating a moral argument for free software and developing free software as a viable alternative to nonfree software. In the early 1990s, Linus Torvalds was instrumental in further strengthening the viability of free software when he licensed his Linux operating system kernel under the GPL. When this kernel was bundled with GNU's software tools, Stallman's goal of a completely free operating system was achieved.

Although there was an active and productive worldwide community surrounding the GNU/Linux operating system, free software failed to gain much traction in the corporate setting. Eric S. Raymond was instrumental in demystifying many aspects of free software in his essay "The cathedral and the bazaar." This essay motivated Netscape to consider making their browser software free. However, business concerns took hold, and they were unwilling to make the move completely. After consultation with Raymond and others, Netscape released the source code for their browser as *Open Source Software*. It was at this time that Raymond and Bruce Perens founded the Open Source Initiative (OSI). They established a definition of open source software (The Open Source Definition, 2006), distinguishing it from free software. Whereas the two notions are closely related, free software is quite rigid in its definition. There are four basic freedoms, including the freedom to modify and redistribute the software, that cannot be impinged upon. Authors of OSS, however, can place certain restrictions on modifications to and the distribution of the modified software. The history of both free and open software is more fully developed in numerous places, including Grodzinsky et al. (2003). In the next sections we explore deeper distinctions between the free software and open-source software communities. These two communities, although deeply intertwined and closely related, have distinct goals that clearly manifest themselves in the discussion surrounding the release of a draft of the next version (version 3) of the GPL.

10.2.1.1 Free Software

Richard Stallman first articulated the ideals of the Free Software movement in 1985 in *The GNU Manifesto* (Stallman, 1985). In it he articulates his motivations for starting the GNU[2] project and lays the groundwork for the GNU General Public License. In particular, he notes that "[e]veryone will be permitted to modify and redistribute GNU, but no distributor will be allowed to restrict its further redistribution" (Stallman, 1985). It is this notion that served as the foundation for the definition of free software.[3] He argues that all computer users would benefit from the GNU project and the GPL because effort would not be wasted redeveloping software; everyone would be able to make changes to suit his/her own needs; educational institutions would be able to use the software to help students learn about software; and no one would be burdened with the responsibility of deciding who owns which piece of software and exactly what one is allowed to do with it.

Stallman's exact position on the ethics of free software is unclear. He articulates his responsibility as a software developer in the *Manifesto*: "the golden rule requires that if I like a program I must share it with other people who like it" (Stallman, 1985). In a later essay he extends that responsibility by arguing that "programmers have a duty to write free software" (Stallman, 1992). He also seems to articulate a view that selling software is morally wrong. "Software sellers want to divide the users and conquer them, making each user agree not to share with others" (Stallman, 1985). However, in this early paper it is unclear whether he is considering all software or just "infrastructure" software, for example, operating systems, networking software, software development tools, because many of his arguments focus solely on GNU. Later, though, he states his position more pointedly and goes even further, claiming that "proprietary software developers" who obstruct the use of that software by users "deserve a punishment rather than a reward" (Stallman, 1992).

In the *Manifesto*, he also deals with some of the early objections to free software. We mention those with substantial ethical importance here. The first objection centers on programmers being rewarded for their creativity. He makes a distinction between deserving a reward and asking for a reward. He states that "[i]f anything deserves a reward, it is social contribution," and that "[t]here is nothing wrong with wanting pay for work" (Stallman, 1985). However, he insists "the means [of charging for software] customary in the field of software today are based on destruction" (Stallman, 1985). He argues that by asking users to pay for software,[4] certain people will not be allowed to use the software, resulting in reduced benefit to humanity. "Extracting money from users of a program by restricting their use of it is destructive because the restrictions reduce the amount and the ways that the program can be used. This reduces the amount of wealth that humanity derives from the program" (Stallman, 1985). In a later essay he makes the assumption that "a user of software is no less important than an author or even the author's employer" (Stallman, 1992). He acknowledges that not everyone

[2]GNU is a piece of software designed to have the same functionality as Unix and be completely compatible with Unix.

[3]Note that *free* refers to freedom, not price.

[4]Early in this movement, Stallman lacked clarity regarding free software. He often merged the notions of "no cost" and "freedom." Similarly, he seems to confuse the notions of software with restrictive proprietary licenses and charging for software.

may agree with him on this point.[5] However, he argues that those who do are logically required to agree with his conclusions.

A second objection to free software that Stallman deals with in the *Manifesto* is that a programmer has a right to control the results of his/her creative endeavor. Stallman argues that by controlling one's software, one exerts "control over other people's lives; and it is usually used to make their lives more difficult" (Stallman, 1985). When Stallman talks about control here, he is talking about the fact that under proprietary software licenses, users of software are typically restricted from making copies for others and making modifications to the software to meet their own needs. A potential software user, though, needs to weigh the difficulties faced without the software against the difficulties faced when the software is used. Assuming the user purchases the software with full knowledge of the terms and conditions, the user has not been taken advantage of as Stallman suggests.

Stallman brings a social justice bent to this objection as well. "All intellectual property rights are just licenses granted by society because it was thought . . . that society as a whole would benefit by granting them" (Stallman, 1985). Stallman seems to be of the opinion that once you buy a piece of software, you should have rights to control it, much like you would when you purchase a book. He notes that the notion of copyright did not exist in ancient times; it was created in response to technological developments (i.e., the printing press) and was used to prevent businesses from exploiting authors. Society benefited from copyright because authors, knowing they could control the mass production of their works, had sufficient incentive to produce creative works. Stallman sees proprietary software developers taking advantage of the copyright system because the public is unaware of the trade-offs it made in establishing the system. He makes the case that the general population has not examined why it values intellectual property rights. "The idea of natural rights of authors was proposed and decisively rejected when the US Constitution was drawn up. That's why the Constitution only *permits* a system of copyright and does not *require* one; that's why it says that copyright must be temporary. It also states that the purpose of copyright is to promote progress—not to reward authors. Copyright does reward authors somewhat, and publishers more" (Stallman, 1994). His point is that society has not thought thoroughly about copyright for some time, and this issue is not being dealt with honestly by copyright holders. "At exactly the time when the public's interest is to keep part of the freedom to use it, the publishers are passing laws which make us give up more freedom. You see copyright was never intended to be an absolute monopoly on all the uses of a copyright work. It covered some uses and not others, but in recent times the publishers have been pushing to extend it further and further" (Stallman, 2001).

[5]There are those who clearly disagree with Stallman. Himma denies that user interests necessarily win out over creator interests. He argues that content creators invest the most precious resources of their lives, time and effort, in creating content, while the most important interests of users in such content is frequently, but not always, that they merely want the content. Although the fact that someone wants something is of moral significance, Himma argues that, from the standpoint of morality, the content creator's interest in her time and effort (and hence in the content she creates) wins out over mere desires of others (Himma, 2006, 2008).

Ultimately, Stallman sees the social value of an individual modifying and sharing a program as more valuable to society than the author's intellectual property rights.

By the time that version 2 of the GPL was introduced in 1991, Stallman was much clearer in his pursuit of the four freedoms that are essential for free software (although this definition did not appear until 1996):

(1) Freedom to run the program, for any purpose.
(2) Freedom to study how the program works, and adapt it to your needs.
(3) Freedom to redistribute copies so you can help your neighbor.
(4) Freedom to improve the program, and release your improvements to the public, so that the whole community benefits.

In addition to the four freedoms, GPLv2 also introduced a notion called "copyleft." Copyleft is a play on the word "copyright," but, more importantly, it leverages copyright law to propagate the four software freedoms. In particular, it requires that derivative works also be licensed under the GPL. Thus, once a piece of software is made free by the GPL, it and all of its derivative works will always be free. Thus, the GPL is the main mechanism for establishing and propagating software freedom. Later, we will consider the viewpoint that copyleft is coercive because the legal weight of the copyright system is used to force others to propogate free software.

With the clear articulation of the four freedoms (and a mechanism to spread them), Stallman was in a position to argue more clearly for free software. In *Why Software Should Be Free*, he clearly explains the social cost of software having owners (Stallman, 1992). He is careful to separate out the act of creating software from the act of distributing software. He argues that once software is created, society is harmed in three ways when software is not distributed freely: software is used by fewer people, software users are unable to adapt or fix the software, and the software cannot be used to learn from to create new software. He uses a utilitarian argument to suggest that proprietary software is an unethical choice. The purchase of software is zero-sum—wealth is transferred between two entities. "But each time someone chooses to forego use of the program, this harms that person without benefiting anyone" (Stallman, 1992). He goes on to claim that the decision by some not to purchase software harms society because those people do not derive the benefit of that software. Stallman's point is subtle here. Distributing software, unlike distributing material goods, requires no new raw materials or packaging and the incremental distribution costs to allow widespread use are zero or very small.

Stallman also argues that the typical proprietary license damages social cohesion because it restricts one neighbor from helping another. Such a license demands that a person give up the right to copy the software in the event that a neighbor would benefit from its use. The damage comes because people "know that they must break the laws in order to be good neighbors" (Stallman, 1992). He thinks that the software copyright system reinforces the notion that we must not be concerned with advancing the public good. "[T]he greatest scarcity in the United States is not technical innovation, but rather the willingness to work together for the public good" (Stallman, 1992).

A related, weaker claim by Stallman is that proprietary programmers suffer harm in knowing that everyone, quite possibly even themselves, cannot use the software in the case that the owner is the author's employer. Stallman has two additional arguments regarding the social cost of keeping software proprietary. The first is a slippery slope argument that proprietary software begins to destroy the ethic of making contributions to society. This argument is suspect. It is no secret that Bill Gates has made enormous sums of money from proprietary software. The Bill and Melinda Gates Foundation is evidence that the Gates' ethic of making contributions to the greater good is still intact.

Stallman also argues that there is a social cost of frustration and lost capital because of proprietary software. Writing replacement software is frustrating for the programmer and more expensive than modifying and improving existing software (Stallman, 1992). This argument is weak, because there is social value in having two competing pieces of software that are functionally equivalent. It seems that the apparent robustness and security of GNU/Linux has prompted Microsoft to take robustness and security more seriously in the Windows operating system. Also, there are numerous software categories where there are competing FS packages. For example, both KDE and Gnome (both desktop software) have their ardent supporters. Each provides the same functionality (at least on a high level) and allows users and other developers to choose software that most appropriately meets their needs.

In a later essay, *Why Software Should Not Have Owners*, Stallman analyzes arguments for software ownership (Stallman, 1994). He notes that "[a]uthors often claim a special connection with programs they have written" (Stallman, 1994) and because of that special connection, in ethical analyses, software authors' positions should bear more weight. Proponents of this argument claim that this connection comes from extending rights associated with material objects to software. Stallman asserts that material objects are fundamentally different from software and that there is no evidence that software is deserving of the same protection. The fundamental difference stems from the scarcity of material objects relative to the (infinite) abundance of software. As mentioned earlier, it is easy and cheap to make copies of source code without depriving the holder of the source code access to the original copy. Again appealing to an act utilitarian analysis he notes, "[W]hether you run or change a program I wrote affects you directly and me only indirectly. Whether you give a copy to your friend affects you and your friend much more than it affects me" (Stallman, 1994). It might be argued that the last statement is not true. The holder of the original is deprived of the profit that would have been made through selling the software, but this is largely an economic argument.

10.2.1.2 Open Source Software The free software community grew substantially after the introduction of the Internet and Linus Torvalds' contribution of Linux as free software. It quietly made gains, without garnering widespread attention, until 1998, when Eric Raymond and Bruce Perens teamed to create the Open Source Initiative. In 1997, Raymond gave the first thorough analysis of the software development process employed by the free software community in "The cathedral and the bazaar" (Raymond, 2001). He argued that the process is effective at producing superior software and considers numerous reasons that make it effective (which we explore further in Section 10.3). It was at this time that many suggested that most free software developers were either unaware of or motivated by something other than the free software ethos promoted by Stallman. (See Bonaccorsi and Rossi, 2004 and Hertel et al., 2003 for subsequent verification of this observation.) In that same year, Perens published the Debian Social Contract to articulate the developers' commitment to open source software and its users (Perens, 2002). One thing that distinguishes the Debian Social Contract from the GPL is that the needs of the users trump the priority of software freedom as defined by the GPL. Item 4 states, "We will not object to non-free works that are intended to be used on Debian systems." It is clear that Raymond and Perens sought to shape free software into an acceptable choice for businesses by defining open source software so that there are no restrictions on distributing it with proprietary software.

The business case that Raymond and Perens first made was to Netscape, attempting to convince them to make the source code for Netscape Navigator available to the free software community and remove restrictive proprietary licensing terms. In the process it became clear that the business issue was not so much making the source code available to others, but losing control over derivative works. The GPL's copyleft prevented the business from ever "closing" the source code. Raymond, the pragmatist, was motivated by purely practical terms (the widespread distribution of source code is an effective software development technique). Because software freedom was not of particular interest to many free software developers and losing control over derivative works was a risk that business was not willing to take, the requirement to spread software freedom (the notion of copyleft) was weakened and the notion of "Open Source Software" was developed. The Debian Social Contract, which contained the Debian Free Software Guidelines, became the basis of the Open Source Definition. The OSI now publishes licenses that meet the Open Source Definition and declares software distributed under any of these licenses as "OSI Certified."

On philosophical grounds, Stallman is a most ardent critic of Open Source Software. He has two main objections. The first has to do with the weakening of the notion of Free Software. While making the source code available with the executable version will allow a user to achieve most of the four software freedoms, there are ways to license software and the source code that will allow certain users to keep their modifications private (in the sense of source code) while releasing only the executable version. Stallman sees the ability to do this as a violation of the tenets of free software. Putting the needs of any particular user/developer ahead of the concept of software freedom is unacceptable. He is also unabashed in his objection to the use of the term "open source." He argues that obscuring "free software" behind the "open source software" moniker hides the ideals that free software promotes (Stallman, 1998). When people use open source software for pragmatic reasons, there is no reason to believe that they truly understand the ethical importance of free software. Stallman believes that people who use free software and understand the social implications attached to its use and development are much more likely to include the social implications in their deliberations surrounding a switch to proprietary software. As evidence, he recounts a number of incidences where executives in the open source industry publicly indicated a lack of appreciation of the ideal of free software. He attributes this unawareness to the use of the term "open source" rather than the use of the term "free." We do note, however, that Stallman does not object to most of the practices of the Open Source Software community. The fact that the source code on an open source project is available to all is a necessary, but not sufficient, part of software freedom.

Chopra and Dexter also offer analysis of the distinction between FS and OSS (Chopra and Dexter, 2005). They start by noting that software now plays an essential role in the social and political lives of many people and ask the question of whether (open source) software developers are morally obligated to apply copyleft to their work. After taking "as a bedrock principle that freedom is a moral good" and "[t]hat the only justifiable violation of this freedom is the restraint of a person whose actions interfere with the liberty of another," they conclude that FS is the morally superior choice to OSS. Their argument centers around four points. First, they note that the restrictions of copyleft only affect the act of distribution. That is, most freedoms are not affected by copyleft. Next they observe that copyleft does not restrict the ability of the licensee to earn a living, because someone can still be hired to make modifications to copyleft code. Although it is certainly the case that software authors can still make a living developing software, it is not the case that copyleft does not impinge on the methods that they can use to do so. The requirements of copyleft demand that source code be made available, all but ensuring that the author cannot make a living off the distribution of copylefted code. Chopra and Dexter go on to argue that there is no coercion in copyleft.[6] They state that all choices by the original developer and subsequent modifiers of the source code are made with full knowledge of the terms of the license, and, thus, all involved support the notion of free software and perpetuate that notion. Numerous studies of free software developers seem to indicate the contrary (Bonaccorsi and Rossi, 2004; Hertel et al., 2003). Most developers participate for reasons other than promoting free software. It is not unreasonable to conclude that some of them grudgingly contribute to free software. A contributor may be in a position of not wanting to give up distribution rights (as is required by copyleft), yet wanting to make the source code available. Such a contributor must choose one or the other, but not both. Finally, Chopra and Dexter state that OSS developers take the position that developing software is "just engineering" and "free software is not a social or moral imperative." They seem to discount the fact that there may be times when free software may not be worth anything to society. This may be best demonstrated by the fact that FS was largely unknown until the start of the OSS movement. Without the OSS movement, free software might not have moved into the mainstream and business would not have considered it as a viable alternative to proprietary software. By introducing the notion of OSS, the FS community is now in a position to have its ideals considered by a wider audience.

10.2.2 Critiques of Free and Open Source Software

One of the sharpest ethical attacks on free software came from Bertrand Meyer. In the essay "The ethics of free software," Meyer lumps both free software and open source

[6]We include Watson's critique of this point in the next section.

software into the same category (Meyer, 2001). Unfortunately, his analysis begins with some assumptions that are inconsistent with those of Stallman and Raymond. In particular, he assumes that software is the legitimate property of someone and that "free software" is defined in terms of being no-cost, unrestricted in its use, and freely available in terms of the source code. These assumptions make his analysis more indirect. Nonetheless, he has a number of points that require consideration. The first is his assumption that software is the legitimate property of someone. Stallman rejects this notion by arguing that the analogy between real property rights and intellectual creations is weak: "Our ideas and intuitions about property for material objects are about whether it is right to *take an object away* from someone else. They don't directly apply to *making a copy* of something" (Stallman, 1992). Yet Meyer believes we must consider the software developers who have "contribute[d] their time, energy and creativity to free software" (Meyer, 2001). Stallman does not object to the remuneration of the developers. His objection is the restrictions placed on software users because they do not have access to the source code. Meyer sees giving source code away as "an immediate business killer" (Meyer, 2001).

Meyer's second ethical critique stems from the fact that much free software is a "copycat" of some proprietary piece of software. He points out that making software that mimics proprietary software is not unethical, but failing to acknowledge the original proprietary piece of software is an ethical lapse on the part of the developer. Meyer states that "much of the hard work and creativity goes into specifying a system" and that the implementation is really not a place that brilliance is demonstrated. Because, by necessity, the interface is publicly available, it serves as a basis for competitors, both proprietary and free, to begin their work. Although such a lack of attribution may be an ethical lapse, it is not one that speaks to the ethics of free software, but to the ethics of the developer of a particular software package.

Brett Watson offers a critique of both free and open source software that takes an interesting perspective. In *Philosophies of Free Software and Intellectual Property,* he claims that when "one takes the stance that copyright is evil," leveraging copyright to promote the ethical notion of freedom is in itself unethical (Watson, 1999). In particular, Watson objects to the entire notion of copyleft. It becomes a burden, impinging on the freedom of the developer. There should be no requirement of quid pro quo; the fact that a developer is in a position to take advantage of the software written by others does not mean the same developer must return his contributions to the software development community. Watson calls copyleft a coercive system and on those grounds objects to the notion of copyleft. In some sense, he tries to take the argument to a different level. If one is truly concerned about freedom (rather than just software freedom), then one must not try to control the behavior of others. "Advocates of a non-coercive system may themselves dislike being coerced, and by application of 'the golden rule' hence refrain from coercing others" (Watson, 1999).

Watson also considers the use of copyright to promote a noble cause, in this case software freedom that is embodied by the GPL. He argues that there is no reason to limit the promotion to this one noble cause and that we might expect to see other licenses that promote additional noble causes. In fact, he suggests that the only logical conclusions are licenses that include either none or all of an author's noble causes.

Watson does acknowledge the fact that for the GPL to be effective, authors must choose to adopt it. The GPL is designed in such a way that it contains the least restrictive set of clauses needed to promote software freedom. Any more than that and it becomes controversial (as we shall see in the discussion surrounding version 3 of the GPL), and members of the free software community will not adopt it for their software.

Perhaps Watson's most insightful critique of free software is that in some sense copyleft is not really about freedom, it is more about making sure that someone does not earn money off someone else's hard work—even though that person had voluntarily and knowingly given the work away to others. (In some sense copyleft is like preventing the purchaser of a Habitat for Humanity home from selling it at a profit and keeping the money.) Watson points out that if someone takes a piece of software, modifies it, and then tries to sell it without source code, the original piece of software is still available for all to look at and use. The freedom of the original piece of software is unaltered in this process. Stallman would counter that the software modifier in this scenario has done society an injustice by not making the source code for the modified software available. He does not have any qualms with putting legal barriers in place to prevent this sort of antisocial behavior. Watson, on the contrary, sees copyleft as impinging on the autonomy of the software developer, thus reducing freedom. He suggests a scenario in which a developer has no particular attachment to the software and what becomes of it. "[I]t may be flattering to such an author that someone else wishes to create a derived work, regardless of whether that derived work will be free or proprietary" (Watson, 1999).

Watson thinks the world would be a better place if there were no copyright restrictions whatsoever. Without copyright, there is a chance that all software will truly be free—no restrictions at all. However, Watson is a realist in recognizing that this ideal will never be achieved and acknowledging that copyleft is a pragmatic way to maximize most freedoms.

10.2.3 The Controversy Regarding GPL Version 3

In 2006, the Free Software Foundation released drafts of version 3 of the GPL for commentary by the worldwide free software community. In response to technological and legal developments that have occurred since the adoption of GPLv2, GPLv3 articulates the notion of software freedom in a much more nuanced way. One of the more controversial aspects has been the language that deals with the issue of Digital Restrictions Management. The preamble of the first draft of GPLv3 clearly stated that "Digital Restrictions Management is fundamentally incompatible with the GPL" and that it "ensures that the software it covers will neither be subject to, nor subject other works to, digital restrictions from which escape is forbidden" (GPLv3,1st draft, 2006).

It is worth noting that most of the proprietary software industry uses the acronym DRM to refer to "Digital Rights Management." Digital rights management refers to software that copyright holders use to manage creative content and to control the copying of electronic versions of that material. The Free Software Foundation clearly takes a different interpretation of the matter of whose rights are being interfered with by such software. This really comes as no surprise because Stallman has been concerned about this issue for some time. In the essay entitled *Linux, GNU, and Freedom*, he noted some troubling developments with respect to software freedom within the Linux kernel (Stallman, 2002). He stated that the Linux kernel distributions at that time contained nonfree software that probably made it illegal for them to be distributed. This nonfree software came in two forms as part of device driver software. There were special numbers that needed to be placed in the device registers by the driver and a binary form of a substantial piece of software. To deal with this threat of DRM, GPLv3[7] has a section entitled "No denying users' rights through technical measures" with the following terms:

> No covered work constitutes part of an effective technological "protection" measure under section 1201 of Title 17 of the United States Code.[8] When you convey a covered work, you waive any legal power to forbid circumvention of technical measures that include use of the covered work, and you disclaim any intention to limit operation or modification of the work as a means of enforcing the legal rights of third parties against the work's users.

There is another section that deals with the DRM threat by addressing the potential for DRM present in hardware rather than software. The term deals with the requirements for the distribution of source code when object code (the executable version) is the primary distribution mechanism.

> The Corresponding Source conveyed in accord with this section must be in a format that is publicly documented, with an implementation available to the public in source code form, and must require no special password or key for unpacking, reading or copying.

This part of GPLv3 would apply to software that a device (say, a cell phone) manufacturer puts on the device. It would allow the user of the apparatus (or anyone, for that matter) to obtain the source code even though it was not included in the package with the device.

While the language is still fluid as we write this, GPLv3 contains clear language regarding software users' freedoms and suggests pushing those freedoms to other realms. Many within the OSS community are ambivalent about the promotion of software freedom as expressed in GPLv2. However, many OSS developers are concerned about using a software license to push freedom in other realms. Linus Torvalds, the original developer of Linux and holder of the copyright to much of the core of Linux, is quite straightforward about his objection: "The Linux kernel is under the GPL version 2. Not anything else. . . . And quite frankly, I don't see that changing" (Torvalds, 2006a). He is particularly concerned about the first provision noted above. In reference to it, he says, "I believe that a software license should cover the software it licenses, not how it is used or abused—even if you happen to disagree with certain types of abuse" (Torvalds, 2006b). In that same post he goes on to suggest that there are

[7]All quotes from GPLv3 are taken from the second draft, the most current draft at the time of this writing.

[8]This is a reference to the Digital Millennium Copyright Act.

some uses of DRM that might be for the greater good. Curiously, Torvalds has caught Stallman in a contradiction—GPLv3 restricts the use of FS in DRM, yet Software Freedom 0 demands the freedom to run a program for *any* purpose.

The FSF is very clear about the importance of the anti-DRM clauses in GPLv3. Eben Moglen, director of the FSF, in a speech to FSF members said that the anti-DRM clauses in GPLv3 are there because the FSF is "primarily fighting to protect our way of making software" (Moglen, 2006). Because DRM technologies are often used to control access to content other than software (e.g., movies and music), there are those that view the anti-DRM clauses as an attack on the content developer's rights to control the distribution of that content. Essentially, the content developer's right to control his/her creations is "collateral damage" that the FSF is willing to accept in its promotion of software freedom. This view is consistent with Stallman's focus on the rights of all users. In the same speech Moglen says, "What we are playing for is the same thing as always: rights of users." He expresses grave concern about giving up control of many aspects of our personal lives to DRM technology as more and more of the functions within our homes are based on computing technology.

In addition to the controversy surrounding the anti-DRM terms in GPLv3, GPLv3 has raised a practical issue with ethical consequences. In addition to offering GPLv2 for others to use, the FSF offers boilerplate language to use when software is licensed under GPLv2. This language allows a redistributor of GPL'd software to license it with either version 2 or any later version of the GPL. A number of years ago (there is some controversy about exactly when and for which files), Torvalds and other major kernel developers began removing the "or later" clause from the software they wrote. Torvalds has said, "Conversion isn't going to happen" in reference to a possible conversion from GPLv2 to GPLv3 (Torvalds, 2006a). This is a clear demonstration that Torvalds' conviction to software freedom is more pragmatic than that of Stallman's, yet his strong belief in GPLv2 suggests that it is not as pragmatic as Raymond's and Perens'.

10.3 WHY OSS FLOURISHES

The social contract articulated in the Open Source Software Definition is fairly clear about what OSS offers to others. But what do OSS developers expect in return? What motivates developers to contribute to an open source project? Is it altruism, that is, do they consider it a "pro bono" project that contributes to the public good? Is it a reaction against corporate greed? Does it make them feel part of a select community with special talents? Clearly all of these play a part in OSS developer motivation to abide by this contract. Beyond that, however, there is also a sense that developers see their involvement as "enlightened self-interest" (Kollock, 1999).

The analysis of motivations of OSS developers can best be traced through the writings of Eric S. Raymond and Bruce Perens, cofounders of the OSI initiative. The OSI initiative was developed in 1998 and attained general public notice through the publication of Eric Raymond's "The cathedral and the bazaar" (1998, revised 2001), *Homesteading the Noosphere* (2000), and *The Magic Cauldron* (1999, revised 2002).

The impetus for Raymond's initial publication was the emergence of Linux, Linus Torvalds' bazaar-like operating system project. In "The cathedral and the bazaar," Raymond, skeptical that Torvalds' method would work on other large open-source projects, applied it to a project of his own. The results are detailed in the article. The next two articles are further attempts to identify the motivations of OSS developers and the economics associated with the OSS community. Perens in *The Emerging Economic Paradigm of Open Source* (revised 2006) updates the economic analysis started by Raymond in *The Magic Cauldron.* Yochai Benklar in "Coase's Penguin, or, Linux and the Nature of the Firm" is interested in the more general question of "large-scale collaborations in the digital information market that sustain themselves without reliance on traditional managerial hierarchies or markets" (Benklar, 2002).

All of these articles attempt to explain the customs and taboos of the OSS community as well as the sustainability of open source in an exchange market. Although there may be allusions to issues of moral obligation, that is, whether software developers have an obligation to make their source code available, in the explanation of customs and taboos, Raymond declares that he is presenting an economic-utility argument rather than a moral analysis. While Raymond and Perens take the approach of examining software production within the OSS community, Benklar observes the phenomena as a nonpractitioner and extends his observations to other domains.

10.3.1 The Motivations of OSS Developers

In "The cathedral and the bazaar" (1998), Raymond details how the success of GNU/Linux had changed his perception of what the open source community could accomplish.

> The fact that this bazaar style seemed to work, and work well, came as a distinct shock. As I learned my way around, I worked hard not just at individual projects, but also at trying to understand why the Linux world not only didn't fly apart in confusion but seemed to go from strength to strength at a speed barely imaginable to cathedral-builders.

Previously, Raymond envisaged large operating systems as being developed only in the cathedral style of traditional software development. Open source projects, to him, were small and fast, built on rapid prototypes. He decided to test the Linux method of development for himself by developing a POP client for e-mail. He was searching for the motivations that drew hackers into large projects as well as the sustainability of the projects.

Using Torvalds' model of "release early and often," Raymond discovered that there were several features that drew hackers to his project and why this seemingly chaotic model worked. The essay is organized around the lessons learned from the Torvalds model: Primarily, programmers join a community because there is a program that they need for their own personal use and they are willing to put it out to the OSS community at large. "When you start community building, what you need to be able to present is a plausible promise. Your program doesn't have to work particularly well. It can be crude, buggy, incomplete, and poorly documented. What it must not fail to do is convince potential co-developers that it can be evolved into something really neat in the foreseeable future" (Raymond, 2001).

Raymond observes that programmers who participate in this bazaar style of development know the value of reusable code; can start over and throw away the first solution; can keep an open mind and find interesting projects that they want to code; treat users as codevelopers and listen to them; keep the beta test base large so that problems will find a solution; use smart data structures; and can ask a different question, or try a different approach when a wall is hit (Raymond, 2001). These lessons, verified in his own project, confirm Torvalds' methodology. The initial publication of this essay in 1997 drew criticisms that Raymond answered in the 2001 revision. Most traditionalists objected to the dynamic change of project groups in the bazaar style of development. They equated it to a lack of sustainability in the project. Raymond answered these objections by citing the development of Emacs, a GNU editing tool that sustained a unified architectural vision over 15 years (Raymond, 2001).

Ultimately, Raymond concludes that "perhaps in the end the open-source culture will triumph not because cooperation is morally right or software 'hoarding' is morally wrong (assuming you believe the latter, which neither Linus nor I do), but simply because the commercial world cannot win an evolutionary arms race with open-source communities that can put orders of magnitude more skilled time into a problem" (Raymond, 2001). Raymond emphasizes that one of the strengths of the OSS community is that programmers select projects based on interest and skills. He refines this argument in *Homesteading the Noosphere,* where he contrasts the OSS community, a gift culture that is marked by what you give away in terms of time, energy and creativity, with that of an exchange culture that is built on control of the scarcity of materials. In OSS there are always resources of machines and people, and those who try to participate in this culture understand that they are obligated to share their source code. He points out that the culture of OSS only "accepts the most talented 5% or so of the programming population" (Raymond, 2000).

In this article, Raymond explains that although members of the OSS community believe that open-source software is a good and worthy thing, the reasons for this belief vary. He asserts that there are various subcultures within the OSS community: those representing zealotry (OSS as an end in and of itself); those representing hostility to any and all commercial software companies, and any cross product of these two categories (Raymond, 2000). He cites Stallman as an example of a member of the hacker culture who is both "very zealous and very anticommercial" (Raymond, 2000). And, by extension, the FSF supports many of his beliefs. He contrasts the FSF with the pragmatists whose attitudes are only mildly anticommercial and who, in the early 1980s and 1990s, were represented by the Berkeley Unix group. The real shift in power within the hacker culture occurred with the advent of Linux in the early 1990s and the release of the Netscape Source in 1998. When the corporate world took an interest in OSS, the pragmatists became the majority of the hacker culture. By the mid-1990s, this manifested itself in programmers who identified more with Torvalds than with Stallman, and who were less zealous and hostile. The OSS community became more polycentric, developing their own non-GPL licensing schemes.

The OSS development model seems to verify that it is economical and productive to recruit volunteers from the Internet. In general, OSS developers fall into one of two broad categories: hobbyists who enjoy writing software and those who work for a corporation or agency that requires its developers to make contributions to either free or open source software projects. Regardless, there are norms dealing with taboos and ownership customs that act as moral guidelines within the OSS community. The taboos are against forking projects (breaking off and working on another version of the project), distributing changes without the approval of the project owners, and removing someone's name from the credits of a project without prior permission. Owners of OSS projects are likened to the homesteaders of the wild frontier. Homesteaders are those who assume ownership of a project by cultivating the idea and interesting the hacker community, by taking it over from another "owner" who passes it to them, or by picking up a project with no clear chain of ownership and making it their own (Raymond, 2000). In the last case, custom demands that you actively look for the owner and announce that you intend to take over the project. Raymond observes that hackers have been following these norms for years, and that they have evolved. Even in the OSS community there has been movement to "encourage more public accountability, more public notice, and more care about preserving the credits and change histories of projects in ways that (among other things) establish the legitimacy of the present owners" (Raymond, 2000).

In his Lockean analogy, Raymond suggests that the expected return from the programmer's labor comes in the form of reputation among others within the community not only as an excellent programmer, but also as a keeper of the customs associated with homesteading. The recognition of reputation can come only from those already recognized within the culture, and criticism is always directed at the project and not at the person. "The reputation incentives continue to operate whether or not a craftsman is aware of them; thus, ultimately, whether or not a hacker understands his own behavior as part of the reputation game, his behavior will be shaped by that game" (Raymond, 2000). While one could argue that reputation might translate into an economic benefit in a traditional market, what you give away leads to social status within the OSS community. "In the hacker community, one's work is one's statement. There's a very strict meritocracy (the best craftsmanship wins) and there's a strong ethos that quality should (indeed *must*) be left to speak for itself. The best brag is code that 'just works,' and that any competent programmer can see is good stuff" (Raymond, 2001). Therefore, Raymond points out that "the reputation game may provide a social context within which the joy of hacking can in fact become the individual's primary motive" (Raymond, 2000).

The *noosphere* in the essay title refers to "the territory of ideas, the space of all possible thoughts" (Raymond, 2001). Raymond is clear to point out that the noosphere is not cyberspace, where all virtual locations are "owned" by whoever owns the machines or the media. He comments that there is anger against companies, such as Microsoft, because these commercial companies restrict their source code to only their programmers, thereby reducing the noosphere available for development by and for everyone (Raymond, 2001). Within the OSS community itself, there are also certain elements that are not forthcoming with their code, and, in fact, at times will mislead other OSS programmers. These are crackers who do not seem to respect the values and customs of the legitimate OSS community, nor feel morally obligated to participate in the gift culture that OSS embraces. Their philosophy is to hoard rather than share, and they clearly do not embrace the trust that is necessary in a peer review process. For Raymond, sharing good craftsmanship that helps people rewards the developer with personal satisfaction and extends the noosphere.

10.3.1.1 Autonomy One perceived attraction for OSS developers is the autonomy of the programmer. Although developers who embrace OSS do gain a measure of autonomy not available to those working on commercial software, the claim for *complete* autonomy does not appear to be valid. For the most part, OSS developers work as volunteers from the perspective of the project, and can join or quit an effort strictly on their own initiative. These volunteers are not coerced into participation and contribute willingly. Therefore, one might assume that the OSS developer can be depicted as a libertarian ideal, unshackled by corporate controls. However, there are several types of control in OSS, even when no single developer is in charge of an OSS project. An OSS developer cannot be sure that his/her contribution will be accepted into the continuously evolving canonical version. A contribution may be embraced or rejected, and if accepted may later be changed or replaced. The developer is free to contribute or not, but any single developer cannot claim ultimate control over the use of his/her contribution. In *Homesteading the Noosphere*, Eric Raymond states, "the open-source culture has an elaborate but largely unadmitted set of ownership customs. These customs regulate those who can modify software, the circumstances under which it can be modified, and (especially) who has the right to redistribute modified versions back to the community" (Raymond, 2001).

The developers of an open source project must take special care to avoid the symptoms of groupthink. A newcomer to open source development brings very little in terms of reputation when he/she proposes a new piece of code or a new tack on development for a project. Project leaders who are less open to new ideas and ways of doing things may miss the innovation of the newcomer's idea. Not only will the project lose the good idea, but it will also face the potential of losing a good developer. Thus, open source project leaders and developers must show a great willingness to take in new ideas, evaluate them thoughtfully, and respond constructively to nurture both the idea and the developer of the idea.

Project leaders must exercise similar abilities when a subgroup comes with an idea that is controversial. Care must be taken that the larger group does not ride roughshod over the smaller group's idea. Again, in addition to losing out on a good idea and potentially driving people away from the project, doing so will discourage future innovators from taking their ideas forward. Note that the proprietary software development model is not subject to this argument. The innovative developer who meets resistant project leaders or management is typically free to leave the organization, and he/she regularly does. In fact there are social norms that actually encourage this type of behavior; we call these people entrepreneurs.

So it appears that the autonomy experienced by an open source developer is much like the autonomy experienced by a university faculty member—freedom to choose which projects to work on. Thus, an open source developer has increased autonomy compared to a corporate developer. Whereas the corporate developer might find a supportive social structure to take a project in a new direction, the social structure in the Open Source community works to suppress this type of entrepreneurial endeavor.

10.3.2 Economic Foundations for OSS

In *The Magic Cauldron,* Raymond explores the economic foundations of OSS. He continues the discussion touched upon in *Homesteading the Noosphere* of how OSS, largely a gift culture, can economically sustain itself in an exchange economy and presents his analysis from within this context. The motivation for this essay came from the realization that most OSS developers are now working in a mixed economic context. Raymond distinguishes between the "use value" of a program, which is its economic value, and the "sale value," which is its value as a final good. He dispels the myths about the "factory model" of software, in which software is analogous to a typical manufactured good, because most software is not written for sale but rather in-house for specific environments. He states, "First, code written for sale is only the tip of the programming iceberg. In the premicrocomputer era it used to be a commonplace that 90% of all the code in the world was written in-house at banks and insurance companies. This is probably no longer the case—other industries are much more software-intensive now, and the finance industry's share of the total must have accordingly dropped—but we'll see shortly that there is empirical evidence that approximately 95% of code is still written in-house" (Raymond, 2002). Raymond examines the contradiction that software is really a service industry that is masquerading as a manufacturing industry. Consequently, he maintains that consumers lose because price structures replicate a manufacturing scenario even though they do not reflect actual development costs. In addition, vendors do not feel obligated to offer support, as their profit does not come from help center service. Open source offers an economic challenge to this model. "The effect of making software 'free,' it seems, is to force us into that service-fee-dominated world—and to expose what a relatively weak prop the sale value of the secret bits in closed-source software was all along" (Raymond, 2002). If one conceives of OSS as a service model, then consumers would benefit. Raymond cites the example of ERP (Enterprise Resource Planning) systems that base their price structure on service contracts and subscriptions and companies such as Bann and Peoplesoft, that make money from consulting fees (Raymond, 2002).

In seeking to create an economic model for OSS, Raymond tackles the notion of the commons, which at first glance might seem to apply to a cooperative community such as OSS. He rejects the model of the commons, calling OSS an inverse commons where software increases in value as users add their own features, "The grass grows taller when it's grazed upon" (Raymond, 2002). Maintenance costs and risks are distributed among the coders in the project group.

According to Raymond, sale value is the only thing threatened by a move from closed to open source. He cites two models in which developer salaries are funded out of use value: the Apache case of cost sharing and the Cisco case of risk spreading. He demonstrates that by encouraging a group of OSS programmers to work cooperatively to build a better model in a shorter time than one could build it on one's own, companies get more economic value, and mitigate the risk of losing employees who developed the program and might change jobs. With OSS, the company now has a sustainable pool of developers. Because most of this software has no sale value, but rather supports the infrastructure of the company, OSS developers are getting paid to support use value of software. This model has emerged with the Linux for-profit companies such as Red Hat, SuSE, and Caldera (Raymond, 2002). Raymond stresses that a large payoff from open source peer review is high reliability and quality (see Section 10.3.3). Another salient point that Raymond makes is that because the shelf life of hardware is finite, support stops. For those users who continue to use the hardware, having access to the source code makes their lives a lot easier. In a sense, he says, you are "future-proofing" by using OSS.

Bruce Perens takes up the discussion of the economics of OSS in his article *The Emerging Economic Paradigm of Open Source.* Perens examines three paradigms of economic development: the retail paradigm, in which the developer hopes to recover costs from the sale of the finished product; the in-house or contract paradigm, in which programmers are paid for creating custom software; and the open source paradigm. As an advocate for the third paradigm, Perens points out the weaknesses in the other two. For retail, he states that because of its low efficiency (funding software development via retail software purchases is lower than 5%) this paradigm can only be economically feasible if products are developed for a mass market. Whereas the second paradigm is more efficient (50–80%) as it directs "most of each dollar spent toward software development," this software only has a success rate of 50% because the software often fails to meet the customer's goals (Perens, 2006).

These weaknesses disappear with the Open Source paradigm, as contributors are developing a useful product that companies or individuals need. Like Raymond, he emphasizes that the remuneration to the open source software developer may not be as direct as that of the commercial developer; yet, because more than 70% of software is developed as service for customers, there is still a monetary return for OSS programmers (Perens, 2006). Perens distinguishes between technology that makes a company product more desirable to its customers (differentiating) and technology that supports the infrastructure and is general enough so that competitors can know about it (nondifferentiating). Most software is nondifferentiating. Thus, for companies that need software to support their infrastructure, and when that software does not differentiate the business, the OSS community offers reliability, peer review, and sustainability. If a company is not large enough or does not have enough experience to develop software competitively, then open source is a smart alternative. Open source programmers are finding that, more than a hobby, they can get paid by companies such as Red Hat, O'Reilly, and VA Linux Systems to work full time on open source projects. An additional bonus is the absence of advertising costs in OSS software. "The major expense is the time-cost of employee participation" in mature OSS projects (Perens, 2006). In addition, new and creative additions to the software are constantly being developed. Perens describes various ways of using open source within a company: GNU/Linux distribution companies, companies that develop a single open source program as their main product such as MySQL, hardware vendors such as IBM and HP, end-user businesses such as e-Bay, the government, and academics (see Perens). Perens asserts that OSS is self-sustaining because "[i]t is funded directly or indirectly as a cost-center item by the companies that need it" (Perens, 2006). Both Raymond and Perens conclude that companies will be willing to pay for the creation of open source for nondifferentiating software. If a company wishes to produce a retail piece of software, then the OSS paradigm will not work. Because the majority of software is nondifferentiating, there will be opportunity for open source collaboration.

10.3.3 The Quality of OSS

Quality software, in the traditional sense, is software that meets requirement specifications, is well-tested, well-documented, and maintainable (Schach, 2002). Advocates of OSS claim that its developers/users are motivated to do quality work because they are developing software for their own use; their reputations among their peers are at stake. Critics of OSS claim that volunteers will not do professional-quality work if there is no monetary compensation. This has become a rather outdated argument. As we have seen above, there are many who are employed by companies to write open source code and others who are paid to customize it. Critics also claim that documentation and maintenance are nonexistent. Although it is true that documentation and maintenance are concerns, OSS advocates maintain that OSS meets users' requirements, is tested by its developers, and is constantly being upgraded. Documentation evolves as more and more users become interested in the software and use it. For example, books on Linux can be found everywhere.

The question of whether OSS is of higher or lower quality than comparable commercial software is essentially an empirical rather than philosophical question. The answer to this question is not readily available, but we can cite some anecdotal evidence on this issue. The Apache web server is OSS that competes with commercial web servers. The web server market is a potentially lucrative one, and we expect commercial software developers to compete in that market with high-quality software. Yet, despite commercial alternatives, the OSS Apache server is by far the most used web server. Since August 2002, regular surveys have demonstrated that over 60% of web servers on the Internet are Apache (Netcraft, 2006). At least in this market segment, it appears that OSS is sufficiently and consistently high quality for many users. Of course, Apache is free and other servers are not; the cost motivation might explain some of Apache's popularity. But if the Apache server were of significantly lower quality than commercial alternatives, then it would be surprising to see its widespread use. This raises the question of whether market dominance and popularity should be a benchmark for software quality. Does the fact that Microsoft Windows runs on some 90% of home computers assure us of its quality? Popularity and quality might be linked if it can be shown that there is a level of expertise about software quality in the people making the choices. System administrators have more expertise than an average user of a home computer system. Therefore, when a majority of these professionals choose an OSS alternative, it deserves notice.

Another piece of evidence is a study by Coverity, a company whose software is used to detect numerous types of known software defects (Chelf, 2006). In March 2006 the company released a report describing its results of analyzing 32 OSS packages. The defect rate ranged from 0.051 to 1.237 (defects per 1000 lines of code) with an average of 0.434. The defect rate in the better-known packages (Linux, Apache, MySQL, Perl, Python, and PHP) was an even lower at 0.290. Coverity also published the defects that it found, and in a month over 1000 of the initial 7500 defects had been fixed. Since that time, 17 more OSS packages have been added to the analysis, and of the 49 total packages, 11 have none of the defects for which Coverity searches and the remaining 38 packages have an average defect rate of 0.232 (Accelerating, 2006). This is evidence that some OSS developers take code quality seriously and strive to improve it.

The nature of proprietary software makes a fair comparison difficult. Published defect rates for commercial software vary widely, anywhere from 1 to 30, but it seems safe to say that the evidence suggests that at least these popular OSS packages have defect rates that are on par with their commercial counterparts. An earlier study that compared three unnamed proprietary software packages to Linux, Apache, and gcc (the GNU Complier Collection) concluded that the open source projects "generally have fewer defects than closed source projects, as defects are found and fixed more rapidly" (Paulson et al., 2004). A final piece of corroborating evidence is that even though Coverity offers a free analysis for proprietary code that competes with any of the OSS projects, no similar results are available.

10.3.4 The Ethical Responsibilities of Software Developers

Both open source and proprietary developers share the professional ethical responsibility to develop solid, well-tested code. However, the influences on open source software developers to maintain this ethic differ substantially. Proprietary software establishes a strong distinction between developers and consumers. An interesting aspect of OSS is that this distinction can be less pronounced, suggesting that ethical models for analyzing that relationship need to be different.

Most obviously, when developers and users of OSS neither get nor give payment, financial self-interest is no longer a major concern. Developers are not "using" consumers to get their money. Users are not trying to negotiate an unfair deal for software. Instead, both developers and consumers in OSS are cooperating freely in the OSS project.

The social pressure in the open source community to avoid code forking provides incentives for project leaders to ensure that the code is the best it can be. On the contrary, when an open source developer believes there is too much risk associated with a particular piece of code, he/she can rewrite it and release it. Although there is a reputation risk in doing so, there is the opportunity to publicly demonstrate that the forked product is superior.

Because a developer (or group of developers) typically runs an OSS project and is responsible for making decisions about the design of the software and the quality of the code, he/she is ultimately responsible for the "penumbra"[9] (all people who are under the

[9] In the case of proprietary software, software developers and others in the corporate structure share the burden of care for users and the penumbra. In the case of OSS, that responsibility falls entirely upon the software developers.

influence of a piece of software whether they realize it or not (Collins et al., 1994)). Curiously, the interests of the penumbra are closely tied to the life of an OSS project. When a project is in its early stages, the initial users are often the developers and they may be more tolerant of glitches and defects than is acceptable for the penumbra. However, as the project matures, its longevity becomes closely tied to quality. If the quality is not high enough, the project will likely terminate quickly because without any marketing money behind the project, it will not develop the strong user support it takes to make an OSS project successful. As an OSS project increases its market share, those OSS developers are increasingly obligated to consider their responsibilities to the people who use and are affected by the software. A critique of OSS is that sometimes OSS developers have pointed to the low price and claimed, "you get what you pay for" when the software is unreliable. Obviously, the ethical principle of consideration of the public good is clear: OSS developers have professional responsibilities, even though they are different from traditional professionals in how their work is rewarded.

OSS developers have a built-in "informed consent" advantage: by definition, OSS gives users the freedom to examine the source code of the application. Although the source code may only be understandable to some OSS users, this transparency of code (rare in commercial projects) is a fundamentally open stance that encourages a trust relationship between developers and users. OSS literature advocates a level of cooperation and "community" for OSS participants that is not encouraged or observed in, for example, users of shrink-wrapped commercial applications. Thus, the relationship between the developers and the users in OSS is best modeled as a trust relationship between two overlapping groups: the OSS developers on the one hand and the OSS users on the other. Trust is built in two ways. As the user base grows, nonusers can trust that the large group of users who find the software to be of sufficient quality suggests that the software is worthy of consideration. Trust is also built from the lack of financial coercion on either side. Users can explore the software without paying for it; if the users don't find the software reliable, they can choose not to use it without any financial loss. It is in the best interest of the developer to create reliable code in part to sustain his/her reputation within the open source community (see Grodzinsky et al., 2003) and in the user community. Thus, both groups gain when reliability increases and when the groups cooperate in improving the software.

10.3.5 Open Source and Accountability

In her article entitled "Computing and accountability," Helen Nissenbaum cites four barriers to accountability: (1) the problem of many hands, (2) defects, (3) computer as scapegoat, and (4) ownership without liability. She asserts that these barriers can lead to "harm and risks for which no one is answerable and about which nothing is done" (Nissenbaum, 1994). We will examine how OSS may have addressed barriers (1) and (2). Both (3) and (4) are general issues. Number (4) is interesting because almost all software disclaims any warranties. However, OSS does have the advantage of informed consent mentioned above.

"Where a mishap is the work of 'many hands,' it can be difficult to identify who is accountable because the locus of decision-making is frequently different from the mishap's most direct causal antecedent; that is, cause and intent do not converge" (Johnson and Nissenbaum, 1995). When a developer contributes irresponsible code to an open source project, however, it is unlikely to be accepted. In addition, current best open source development practices attribute code to specific authors. So, there is built-in individual accountability for each code segment and the overall software package. Therefore, the many hands problem can be reduced in OSS because parts of code can be ascribed to various developers and their peers hold them accountable for their contributions.

Nissenbaum argues that accepting defects as a software fact of life raises accountability issues. "If bugs are inevitable, then how can we hold programmers accountable for them?" she asks. The open source approach to software development treats the defect problem with a group effort to detect and fix problems. The person who finds a defect in OSS may not be the person to fix it. Because many adept developers examine OSS code, defects are found and corrected more quickly than in a development effort in which only a few developers see the code (Paulson et al., 2004). In this group effort, accountability is not lost in the group, but is instead taken up by the entire group. The question of whether or not this group accountability is as effective as individual responsibility is, again, empirical. The Coverity study offers strong anecdotal evidence that some OSS developers take defects seriously and work diligently to remove them (Chelf, 2006).

Don Gotterbarn is also concerned about issues of professional accountability in OSS (Wolf et al., 2002). In addition to worries about sufficient care in programming and maintaining OSS, Gotterbarn points out that an OSS licensing agreement forces the authors of the software to relinquish control of the software. If someone puts OSS to a morally objectionable use, then the developers have no right to withdraw the software from that use.[10]

Gotterbarn's objection has some theoretical interest, for the OSS licensing agreements clearly state that no one who follows the OSS rules can be blocked from using the software. But if we accept the idea that software developers have a moral duty to police the use of the software they distribute, especially when the software is utility software, we fall into a practical and theoretical thicket. How is a vendor to know the eventual use of software, especially when the software is utility software (such as an operating system or a graphics package)? Are software developers empowered to judge the ethics of each customer or perspective customer? These responsibilities are overreaching ethically, and far too ambitious in a practical sense.

Furthermore, the relinquishment of control argument has practical significance only if existing competing software models include effective control over the use of software. (That is, should OSS be held to a higher standard than commercial software in relation to ethical responsibility for downstream use?) We are unaware of any action by existing commercial software vendors to police the uses to which their software is put. Commercial software vendors are certainly concerned that people who use their software have paid for it. Once paid, vendors slip quietly away.

[10]Curiously, GPLv3 deals with this issue head-on for a single objectionable use: DRM.

10.4 IS OSS A PUBLIC GOOD?

Stallman (2001) notes that the notion of copyright was developed in response to the (corporate) ability to mass-produce creative works, and societies establish copyright laws to promote the production of creative works. He argues that the notion of free software, and by extension open source software, is a return to the pre-printing press days when anyone (with time) could make a copy of a book. A similar notion of freely sharing ideas has also persisted in academia as well. Academia has long had the tradition of sharing ideas without direct payments. Scholarly journals do not pay authors (and in fact may charge them for pages printed). Law has not protected mathematical formulae and formal descriptions of natural laws. Copyright covers the expression of ideas, but not the ideas themselves; patent has (at least traditionally) protected the practical application of ideas, but not the physical laws underlying the ideas. So, if software is viewed as an extended mathematical object, akin to a theorem, then OSS could be a natural extension of the long tradition of free ideas in mathematics. Does that make it a public good?

Peter Kollock, a sociologist at the University of California at Los Angeles, examines the idea of online public goods in his paper entitled "The economies of online cooperation: gifts and public goods in cyberspace" (Kollock, 1999). He defines public goods as those things that are nonexcludable and indivisible. Because the Open Source Definition prohibits discrimination against persons or groups or against fields of endeavor, it supports the definition of a public good being nonexcludable. Public goods in cyberspace can benefit the users of cyberspace irrespective of whether they have contributed to these goods or whether these goods have come from groups or individuals. The fact that one person using OSS does not affect its availability to the whole supports Kollack's idea of indivisibility. He maintains that "[a]ny piece of information posted to an online community becomes a public good because the network makes it available to the group as a whole and because one person's 'consumption' of the information does not diminish another person's use of it" (Kollock, 1999). If a user downloads a copy of GNU/Linux, for example, she does not diminish its availability for other users. So by this definition, we argue that OSS is a public good.

Is there an active interest among developers to create a public good? Are OSS developers actually motivated to do good by contributing software to the public, and by maintaining it in a group effort? Some developers argue that they can customize OSS, and if others find the customizations useful, then they have provided a public good. However, there could be another possible motivation for OSS. It might be a philosophical or instinctive animus toward existing commercial software developers. Bertrand Meyer recites with dismay the many negative statements by OSS advocates about commercial software development and developers (Meyer, 2001). Some see "Microsoft bashing" as a central theme of the OSS movement. Because most Microsoft products compete directly with OSS packages, some friction between OSS advocates and the largest commercial software corporation seems inevitable. But if OSS development is motivated primarily by its opposition to commercial software producers, then its ethical underpinnings are less benign than if OSS is motivated primarily by an altruistic desire to help computer users. Because the OSS movement is, by design, decentralized and evolving, it seems impossible to gauge with any precision the motivations of all its members (Hertel et al., 2003). But the often-repeated disdain for commercial business practices seems more in tune with the hacker culture than with a culture of altruism. So, we would argue that, for the most part, the altruism involved in the creation of a public good in the case of OSS is more of a by-product of developers who are interested in creating tools that are of use for themselves. Customization and expansion of Linux, for example, came from developers who wanted applications for their own use and then shared their code.

Nowhere can OSS be considered more of a public good than in the academic community. Computer Science departments are expected to be on the cutting edge of technology in their curricular offerings. The price of commercial software, even with educational discounts, often straps a department's budget. Academic institutions have strong financial motivations to adopt open source software. GNU compilers, for example, have largely replaced proprietary versions. GNU/Linux is appearing as the operating system of choice, often replacing Solaris. As more and more applications run on GNU/Linux, universities will have less incentive to buy from Unix platform vendors. They will buy cheaper hardware and run GNU/Linux. One caveat to this scenario is the availability of staff who can support the GNU/Linux platform and the availability of documentation for OSS.

Service learning, a concept that is becoming part of the mission of many higher education institutions, also influences the choice between open source software and proprietary software. Consider a scenario in which a software engineering class is to produce a piece of software for a local charity. The choice between open source alternatives and proprietary alternatives is not to be taken lightly. Seemingly, open source software makes good sense for both the students and the charitable organization. The cost is low and, presumably, the quality is sufficient. Yet there are long-term costs that are faced by the charity (as well as any business making such a choice). How expensive will it be to maintain the software? Is there enough open source expertise available to maintain it? And, finally, what documentation and user training can be expected if OSS is the software of choice? Some ongoing support to these charities might be an opportunity for the university to openly support OSS as a public good.

10.5 CONCLUSION

The distinction between Free Software and Open Source Software has had a positive effect on the software development community and on the larger online community as well. Regardless of the motivation of individual developers, it is difficult to find fault with their willingness to give their creative contributions to the world to study and adapt as the world sees fit. Stallman's increasingly clear focus on freedom for all users of software and hardware has forced discussion on issues that many people today have not considered. Elevating discussion of the social purpose of copyright to an international level is valuable. Raymond and Perens' ability to articulate the necessary and sufficient aspects of software freedom that contribute to developing quality software has been an important part of improving the quality of software that society uses. There is some suggestion that, regardless of whether the quality of FS and OSS is high, the mere possibility that it is higher than that of some proprietary software has prompted some proprietary software developers to adopt techniques and processes that lead to better software, again benefiting everyone. Finally, the Free Software movement can be credited with providing an impetus for establishing notions of freedom for other types of digital media, such as the Creative Commons (creativecommons.org).

ACKNOWLEDGMENTS

In composing this chapter, we drew some material from Grodzinsky, F.S., Miller, K., and Wolf, M.J. (2003), "Ethical issues in open source software," *Journal of Information, Communication and Ethics in Society*, I(4), 193–205, Troubadour Publishing, London, and our paper (2006), "Good/fast/cheap: contexts, relationships and professional responsibility during software development," *Proceedings of the Symposium of Applied Computing;* 2006 April.

REFERENCES

Accelerating open source quality. Available at http://scan.coverity.com/. Accessed 2006 July 24.

Benklar, Y. (2002). Coase's Penguin, or, Linux and the Nature of the Firm. *Yale Law Journal,* 112, 369–466.

Bonaccorsi, A. and Rossi, C. (2004). Altruistic individuals, selfish firms? The structure of motivation in Open Source Software. *First Monday,* 9(1). Available at http://firstmonday.org/issues/issue9_1/bonaccorsi/index.html. Accessed 2006 July 25.

Chelf, B. (2006). Measuring software quality: a measure of open source software. Available at http://www.coverity.com/library/index.html, registration required. Accessed 2006 July 24.

Chopra, S. and Dexter, S. (2005). A comparative ethical assessment of free software licensing schemes. *Proceedings of the Sixth International Conference of Computer Ethics: Philosophical Enquiry (CEPE2005)*, Enschede, The Netherlands, July.

Collins, W.R., Miller, K., Spielman, B. and Wherry, P. (1994). How good is good enough? An ethical analysis of software construction and use. *Communications of the ACM*, 37(1), 81–91.

Free Software Foundation (1991). GNU general public license. Available at http://www.gnu.org/licenses/gpl.txt. Accessed 2006 July 19.

GPLv3 1st discussion draft (2006). Available at http://gplv3.fsf.org/gpl-draft-2006-01-16.html. Accessed 2006 August 7.

GPLv3 2nd discussion draft (2006). Available at http://gplv3.fsf.org/gpl-draft-2006-07-27.html. Accessed 2006 August 7.

Grodzinsky, F.S., Miller K., and Wolf, M.J. (2003). Ethical issues in open source software. *Journal of Information, Communication and Ethics in Society,* I(4), 193–205.

Hertel, G., Neider, S., and Herrmann, S. (2003). Motivation of software developers in open source projects: an Internet-based survey of contributors to the Linux kernel. *Research Policy,* 32, 1159–1177.

Himma, K.E. (2006). Justifying intellectual property protection: why the interests of content-creators usually wins over everyone else's. In: Rooksby, E. and Weckert, J. (Eds.), *Information Technology and Social Justice*, Idea Group, pp. 54–64.

Himma, K.E. (2008). The justification of intellectual property rights: contemporary philosophical disputes. Perspectives on Global Information Ethics. *Journal of the American Society for Information Science and Technology,* 59(7).

Johnson, D.J. and Nissenbaum, H. (Eds.) (1995). *Computers, Ethics and Social Values*. Prentice Hall, New Jersey.

Kollock, P. (1999). The economies of online cooperation: gifts and public goods in cyberspace. In: Smith, M. and Kollock, P. (Eds.) (1999), *Communities in Cyberspace*. Routledge, London.

Meyer, B. (2001). The ethics of free software. *Dr. Dobb's Portal*. Available at http://www.ddj.com/dept/architect/184414581. Acessed 2006 July 19.

Moglen, E. (2006). The hardware wars. *The Free Software Foundation Bulletin,* 8 June.

Netcraft (2006). July 2006 Web server survey. Available at http://news.netcraft.com/archives/web_server_survey.html. Accessed 2006 July 24.

Nissenbaum, H. (1994). Computing and accountability. *Communications of the ACM,* 37(1), 72–80.

Paulson, J., Succi, G., and Eberlein, A. (2004). An empirical study of open-source and closed-source software products. *IEEE Transactions on Software Engineering,* 30(4), 246–256.

Perens, B. (2002). *Debian Social Contract*. Available at www.debian.org/social_contract.html. Accessed 2006 July 19.

Perens, B. (2006). *The Emerging Economic Paradigm of Open Source*. Available at http://perens.com/Articles/Economic.html. Accessed 2006 March 1.

Raymond, E.S. (2000). *Homesteading the Noosphere*. Available at http://www.catb.org/esr/writings/homesteading/homesteading/. Accessed 2006 July 19.

Raymond, E.S. (2001). The cathedral and the bazaar. In: Spinello and Tavani (Eds.), *Readings in Cyberethics*. Jones and Bartlett, Sudbury, MA.

Raymond, E.S. (2002). *The Magic Cauldron*, Version 2.0. Available at http://www.catb.org/∼esr/writings/cathedral-bazaar/magic-caulddron/index.html. Accessed 2006 June 06.

Schach, S. (2002). *Object Oriented and Classical Software Engineering*, 5th edition. McGraw Hill, p. 137.

Stallman, R. (1985). *The GNU Manifesto*. Available at http://www.gnu.org/gnu/manifesto.html. Accessed 2006 July 19.

Stallman, R. (1992). *Why Software Should Be Free*. Available at http://www.gnu.org/philosophy/shouldbefree.html. Accessed 2006 July 19.

Stallman, R. (1994). *Why Software Should Not Have Owners*. Available at http://www.gnu.org/philosophy/why-free.html. Accessed 2006 July 19.

Stallman, R. (1998). *Why "Free Software" is Better than "Open Source"*. Available at http://www.gnu.org/philosophy/free-software-for-freedom.html. Accessed 2006 July 19.

Stallman, R. (2001). Copyright versus community in the age of computer networks. Available at http://www.gnu.org/philosophy/copyright-versus-community.html. Accessed 2006 August 8.

Stallman, R. (2002). *Linux, GNU, and Freedom*. Available at http://www.gnu.org/philosophy/linux-gnu-freedom.html. Accessed 2006 July 19.

The Open Source Definition (2006). Version 1.9. Available at http://www.opensource.org/docs/definition.php. Accessed 2006 July 20.

Torvalds, L. (2006a). Re: Linux vs. GPL v3—dead copyright holders. *Linux-kernel mail archives*, January 25, 2006. Available at http://www.ussg.iu.edu/hypermail/linux/kernel/0601.3/0559.html. Accessed 2006 July 20.

Torvalds, L. (2006b). Re: Linux vs. GPL v3—dead copyright holders. *Linux-kernel mail archives*, January 27, 2006. Available at http://www.ussg.iu.edu/hypermail/linux/kernel/0601.3/1489.html. Accessed 2006 July 20.

Watson, B. (1999). *Philosophies of Free Software and Intellectual Property*. Available at http://www.ram.org/ramblings/philosophy/fmp/free-software-philosophy.html. Accessed 2006 July 19.

Wolf, M.J., Bowyer, K., Gotterbarn, D., and Miller, K. (2002). Open source software: intellectual challenges to the status quo. Panel presentation at 2002 SIGCSE Technical Symposium, *SIGCSE Bulletin*, 34(1), 317–318. Available at www.cstc.org/data/resources/254/wholething.pdf.

CHAPTER 11

Internet Research Ethics: The Field and Its Critical Issues

ELIZABETH A. BUCHANAN and CHARLES ESS

11.1 INTRODUCTION TO INTERNET RESEARCH ETHICS: BACKGROUND AND MAJOR ISSUES IN THE LITERATURE

Internet research ethics (IRE) is an emerging multi- and interdisciplinary field that systematically studies the ethical implications that arise from the use of the Internet as a space or locale of, and/or tool for, research. No one discipline can claim IRE as its own, as various disciplines since the 1990s have used the Internet for research and, to some extent, grappled with the ethical implications of such research. Indeed, because Internet research is undertaken from a wide range of disciplines, IRE builds on the research ethics traditions developed for medical, humanistic, and social science research; this means in turn that a central challenge for IRE is to develop guidelines for ethical research that aim toward objective, universally recognized norms, while simultaneously incorporating important disciplinary differences in research ethics—a challenge frequently met in IRE through pluralistic approaches that conjoin shared norms alongside such irreducible differences. Indeed, at the heart of IRE is an intertwined convergence as IRE seeks to draw from the insights of applied ethics, research methods, information and computer ethics, and comparative philosophy, especially *vis-à-vis* the possibility of developing a global IRE that simultaneously preserves irreducible differences defining diverse cultural traditions while developing a more global IRE for Internet research undertaken by researchers around the world.

We begin this chapter by examining the historical emergence of IRE, and then turn to the philosophical dimensions of IRE, including its normative foundations and basic questions. We then review a range of the most common ethical issues in IRE, along with suggestions for possible resolutions of specific ethical challenges. Finally, we consider some of the most current issues in IRE, including the complex interactions

The Handbook of Information and Computer Ethics, Edited by Kenneth Einar Himma and Herman T. Tavani

between methodologies and ethics, and also whether or not a genuinely global IRE may emerge—one that would hold together shared norms alongside irreducible differences between diverse national and cultural ethical traditions.

11.2 IRE: A BRIEF HISTORY

Throughout the 1990s, disparate disciplines began in piecemeal fashion to examine the ethical complexities and implications of conducting research online. In the view of many, it was at best uncertain whether or not such research ethics guidelines as the U.S. *Belmont Report* and such federal human subjects protections as codified in the U.S. Code of Federal Regulations (CFR) (2005) "fit" or were applicable. For example, White (2003) has posed the question, "Representations or People," to highlight and explore how disciplinary perspectives from the arts and humanities, film studies, and cultural studies (as generally oriented toward *representations*) contrast with the human subjects protections models (which stress understanding research participants as *people*) presented through the CFR. Even earlier, broader debates began to take serious academic form when one of the first journal issues devoted entirely to Internet Research Ethics (IRE) appeared in 1996, in a special issue of *The Information Society*, followed by a workshop on IRE in 1999 funded by the National Science Foundation and the American Association for the Advancement of Science; the AAAS/NSF report (Frankel and Siang, 1999) remains a benchmark to which IRE literature refers.

Further evidence of the recognition and development of IRE came through the release of the Association of Internet Researchers (AoIR) Ethics Working Group's report on *Ethical Decision-Making and Internet Research*, chaired by one of the authors of this chapter, Charles Ess (AoIR, 2002). Researchers, policy makers, and such entities as institutional review boards, which were seeing an extraordinary increase in the number of Internet-based research protocols, then continued to build on these foundations, especially as they confronted new ethical challenges in new venues of Internet research (Buchanan, 2003, 2004; Buchanan and Ess, 2003, 2005). Prominent professional societies, too, such as the American Psychological Association, convened a Board of Scientific Affairs Advisory Group on Conducting Research on the Internet, releasing a report in 2004 in *American Psychologist* (Kraut et al., 2004). And finally, three books in the field of IRE were published in 2003 and 2004 (Buchanan, 2004; Johns et al., 2004; Thorseth, 2003). These were all, indeed, important moments in the development of IRE as a discrete research phenomenon, and possibly toward establishment of IRE as a unique field, but more importantly, these publications promoted further serious consideration of the ethical implications of research in online or virtual environments.

11.2.1 Philosophical Foundations: Sources, Frameworks, and Initial Considerations

IRE in Western countries emerged initially from models of human subject research and human subject protections in the life sciences (i.e., medical ethics, bioethics, etc.) and social sciences (e.g., psychology) (Kraut et al., 2004). Moreover, especially the work of the AoIR ethics committee sought insight and guidance from three sources: (1) professional ethics, including codes for computer-related professions, such as the Association for Computing Machinery (1992); (2) ethical codes in the social sciences and the humanities, where our colleagues in the humanities insisted on understanding human beings online as amateur artists or authors who are producing a work that usually needs only copyright protection, in contrast with the social science view of human beings online as "subjects" to be protected in keeping with the standard human subjects protections of anonymity, informed consent, and so on (Bruckman, 2002); and (3) the growing body of information and computing ethics (ICE), (Ess, 2006; Floridi, 2003).

Philosophers who examine extant statements on research ethics from diverse disciplines and diverse countries will recognize that these make use of at least two familiar Western ethical frameworks, namely, deontology and utilitarianism. As a brief reminder, *deontology* is affiliated frequently with Kant and argues for the ethical priority of always respecting human beings as autonomous beings (i.e., free and thereby capable of establishing their own moral norms and rules). In one of the well-known formulations of the Kantian Categorical Imperative, this requires us to never treat others only as a means. Treating human autonomies as ends in themselves then entails a number of rights, duties, obligations, and principles; these include those emphasized in human subjects protections, namely, rights to privacy, confidentiality, anonymity, and informed consent.

Researchers thence have the obligation to respect and protect these rights, regardless of the "costs" of doing so, for example, of having to develop comparatively more complicated and/or costly research design to protect such rights, or even the ultimate cost of giving up an otherwise compelling and potentially highly beneficial research project because it unavoidably violates these basic rights and duties. By contrast, *utilitarianism*, as a species of consequentialism in ethics, seeks to justify ethical choice via a kind of moral accounting that compares potential benefits and costs of a given choice in hopes of thereby promoting maximum human happiness, however defined (e.g., as physical pleasure for sentient beings for Jeremy Bentham, as pleasures both intellectual and physical for J. S. Mill, and so forth). For example, this calculus is expressed in the U.S. CFR (2005) concerning human subjects research, including the requirement that "risks to subjects are reasonable in relation to anticipated benefits, if any, to subjects, and the importance of the knowledge that may reasonably be expected to result" (Section 46.111.a.2): potential benefits can thus be used to justify risking the moral costs of subjects' suffering pain and/or violating subjects' basic rights. Although some deontologists might be able to agree, we will further see that analogous human subjects protection codes from other countries (including the European Union Data Privacy Protection Acts (Directive 95/46/EC, 1995; Directive 2006/24/EC, 2006) and the Norwegian research codes (NESH, 2003, 2006)) rather insist on an absolute protection of basic rights and protections. In this way, they hold to a more strongly deontological view that such rights and protections must be preserved. irrespective of the potential benefits.

Although philosophers will, of course, debate these characterizations of deontology and utilitarianism, these distinctions have proven useful in the development of IRE in two ways. First, for researchers and other nonphilosophers with no formal training in ethics, these distinctions help them "make sense" of their ethical experience and intuitions. And doing so thus helps these colleagues articulate their intuitions in ways that philosophers will recognize as incorporating more objectivist orientations in ethics. These distinctions thus provide a useful (and badly needed) bridge between researchers from disciplines in the humanities and social sciences on the one hand and philosophers and applied ethicists on the other hand. Second, as the examples given above suggest, these distinctions have proven helpful in articulating important differences between national and cultural ethical traditions, and thereby fostering the development of *pluralistic* approaches that most ambitiously will fulfill the goal of a global ICE to foster shared ethical approaches in IRE that at the same time recognize and foster differences essential to specific national cultures and individual identity (Ess, 2006). So, for example, a number of ethicists and political scientists have identified *consequentialist* approaches, including *utilitarianism*, as characteristic of ethical decision-making in the professional ethics and human subjects protections codes in Anglo-American spheres. As we have begun to see, U.S. (and UK) codes characteristically justify research on the basis of its anticipated outcomes, that is, as these promise to benefit society at large in some ways – thereby requiring researchers only to minimize, not eliminate, *risks to research subjects*. But, of course, deontologists point out that "the greatest good for the greatest number" can justify even the most absolute violation of the rights of "the few," as both general examples (e.g., slavery) and specific infamous examples in the history of research (e.g., the Tuskegee Institute syphilis study, Pence, 1990, pp. 184–205) make all too clear. Moreover, the research ethics of countries such as Norway can be accurately characterized as deontological, as they emphasize that the rights of human subjects must never be compromised, irrespective of the potential benefits (NESH, 2003, 2006).[1] Indeed, this contrast between more utilitarian Anglo-American approaches and more deontological European approaches has been noted by earlier researchers (see Burkhardt et al., 2002; Reidenberg, 2000).

Beyond the frameworks of deontology and utilitarianism, contemporary ethical approaches in IRE include feminist and communitarian frameworks that highlight the role and ethical significance of personal relationships and *care* between researchers and their "subjects." Such frameworks in fact offer versions of "the Golden Rule" as they ask researchers to consider how they themselves would *feel* if they found

[1][Editor's note] The distinction being drawn here, primarily for the sake of opening dialogue between philosophers and researchers without extensive philosophical training, is, of course, more complex from an informed philosophical standpoint. For example, an act utilitarian is a consequentialist who indeed would not be able to justify putting human subjects protections and basic rights at risk for the sake of potential research benefits. But a rule utilitarian, by contrast, is a consequentialist who could also insist on a more absolute protection of rights, and so on, in ways attributed here to deontological approaches. The author's experience is that these basic distinctions have worked well in opening up needed dialogue between philosophers and researchers, dialogue that has resulted, in fact, in the development of ethical guidelines (AoIR, RESPECT) endorsed by researchers. But of course, we would hope that further dialogue and debate, including attention to these more sophisticated ethical distinctions, will ensue, ideally with the result of still more philosophically robust codes and guidelines that likewise incorporate the insights and expertise of researchers.

themselves treated in the ways they proposed to treat their subjects. In addition, feminist and communitarian approaches, especially as articulated in Scandinavia, emphasize that the individual "subject" is not simply an autonomous individual, but is rather a human being whose sense of identity and value are intrinsically interwoven with his/her "web of relations." This means that the researcher is not simply obliged to protect, for example, the anonymity and confidentiality of a solitary subject, but rather is required to protect the anonymity and confidentiality of both the subject and of his/her close friends and intimate partner(s) (Johns et al., 2004; NESH, 2006). Manifestly, how we respond to a given ethical issue within research will depend in large measure on which of these diverse ethical frameworks we presume to be primary (cf. AoIR, 2002; Ess, 2003).

For example, although arguments prevailed in the 1990s *against* applying human subjects protection models to online contexts, these models predominate in contemporary discussions of IRE and the three extant ethical guidelines specifically devoted to ethical issues in online research (AoIR, 2002; Kraut et al., 2004; NESH, 2003). These models take both national and international declarations of human rights as their foundation (Michelfelder, 2001; Reidenberg, 2000); they are thus marked by a deontological insistence on protecting the integrity and dignity of human persons first of all by emphasizing *rights* to informed consent, privacy, confidentiality, and anonymity. As we will see more fully in the subsequent paragraphs (**A Global Internet Research Ethics?**), comparable rights to privacy and data privacy protection in Asia, by contrast, rest on much more *utilitarian* arguments that justify privacy and data privacy protection primarily because they contribute to the greater social goods of economic development.

We can see similar contrasts as we now consider some of the most central issues and topics in IRE.

11.2.2 Specific IRE Issues

The IRE literature of the last decade or so has focused on several topics that arise most frequently in online research. We now discuss these "classic" problems in some detail, and then will turn to more recent and emerging issues. For the sake of conceptual clarity, we present these as discrete issues, but the specificity and characteristics of Internet technologies and especially of interdisciplinary research online mean that IRE issues are usually intertwined and consequently more complex.

11.2.2.1 Anonymity/Confidentiality Years ago, an often passed-around cartoon of the Internet read, "On the Internet, no one knows you are a dog." There was a sense of anonymity, of freedom from one's "physical" reality when in the disembodied space of the Internet. Researchers grappled with this newfound reality, while of course acknowledging that a fundamental tenet of research is to protect anonymity and identity. Adequate provisions are designed and embraced to protect the privacy of subjects and maintain confidentiality. Through the process of informed consent, researchers convey their commitment to their subjects or participants to protect their privacy and their identity should revealing something about them cause undue harm, embarrassment, or some other tangible loss. *The Belmont Report*, for instance, demands that privacy of subjects be protected and confidentiality of data be maintained. In online environments, researchers must wonder about the relationships between a screen persona and the onground individual, and to ensure appropriate protections, the researcher should consider a number of issues. Mechanisms must be in place to provide a truly secure online interaction to guarantee anonymity when promised. If this secure environment is not possible, the researcher should explore what type of Internet locations/media are/is *safest*. Questions surround such "anonymous" surveys, as with Survey Monkey or QuestionPro, tools that are growing in use across disciplines.

Ultimately, the researcher must describe how subjects'/participants' identities are protected. One may suggest that encryption is enough. Then, consider how research data are to be stored, given the prevalence of networked computing. Hacking and data corruption are indeed possibilities. These are, of course, data integrity issues, and often researchers do not have the control over an online site needed to secure the interaction from hackers or other forms of data corruption or interference. For instance, a researcher may promise to maintain confidentiality over the data she collects; confidentiality is defined by the U.S. model as pertaining to the treatment of information already revealed. There is an expectation that "the data will not be divulged to others in ways that are inconsistent with the understanding of the original disclosure without permission" (*The Belmont Report*). In online research, an ethical breach may occur not because of researcher negligence or fault, but by or through circumstances beyond his or her control. Online researchers must consider all possibilities of potential breaches, as data may be collected online, and the researcher is not the only one to have access to it; others in an online forum, archiving sites, or other backups may exist that reveal the source of some data. The researcher may not be in control of this.

Moreover, can a research participant be anonymous online? One may have a "different" online identity, but that identity still corresponds to an individual in a physical environment. If an electronic persona is portrayed in research on an electronic support group for a medical condition, will the participant be identifiable? If so, at what risk? Is there the potential for significant harms to the subjects through identification, especially if the topics at hand are sensitive in one or more ways? One could imagine a research report using screen names that can be searched online and identified fairly easily. Thus, though the researcher may have attempted to protect privacy and maintain anonymity and confidentiality, such technological tools as search engines and archives of online discussions may provide enough context and information to make identification not only possible, but harmful. Thus, researchers must consider how they name, describe, or anonymize their online subjects or participants to the extent it could be possible.

11.2.2.2 Copyright A key decision for Internet researchers is often discipline-based, that is, do we treat our "subjects" as subjects (as is characteristic of the social sciences) and thereby invoke familiar human subjects protections, and/or do we treat our "subjects" as posters, as authors (as is characteristic of the humanities)? If the latter is the case, then far from emphasizing the need for anonymity and confidentiality, we are rather dealing with posters who intend to act as public agents online. Moreover,

U.S. law treats anything posted online as protected by copyright. Hence, researchers pursuing more humanistically based approaches will need to consider several questions related to copyright. For example, will any of their citations of these materials count as "fair use" of the materials they encounter online? How will they acknowledge copyright holders, and/or acquire permission for direct citations, especially if the author(s) use pseudonyms or their work is only available on archives whose e-mail addresses are no longer valid ? These considerations force researchers to wrestle with both a set of (deontological) rights (i.e., authorship as protected by copyright, etc.) and costs that must be considered in more utilitarian approaches (e.g., the time and labor required to track down ostensible authors, to certify that they are indeed the authors and thus copyright-holders of a specific text, to acquire consent in ways that overcome the possibility, heightened in the online context, of ensuring that the consent comes from the proper author).

Furthermore, to satisfy review boards or ethics boards, researchers typically state how long they will retain the data. But it is unclear how long e-data lasts, under what circumstances, and in what context. When researchers tell their participants that they will destroy any data after a specified period of time, it may mean nothing in an online context where researchers are not in control of the copyright or ownership.

11.2.2.3 Revealing Identities Online identities are complex, social identities that exist in various forms of online environments. Some choose to use their real names, whereas others choose pseudonyms, screen names, avatars, masks, and so on. Researchers in online settings study these identities through surveys, ethnographies, action research, and participant observation, and consideration has been centered around the representation of such identities in research reports. Should they use screen names that individuals choose for some usually personally significant reason, or should the researcher protect that screen name from potential identification by using a pseudonym of the screen name? This raises questions of ownership and research integrity, trust, copyright, in addition to the more obvious privacy questions. For instance, by changing screen names in a research report, a researcher may detract from the "reality" or "reputation" of the participant. Text searches can reveal more contexts than a researcher may in her reporting, and this raises potential risks. A researcher may allow participants to make this decision about representation, though disciplinary differences in methodology arise in this possibility. As part of the informed consent process, researchers could present options for participants to consider, and participants could be provided the opportunity to review the research report prior to publication. Ultimately, such questions begin to challenge the longstanding process of research, questioning what Forte (2004) has described as scientific takers and native givers. Online research presents an opportunity, then, for greater researcher reflexivity, given the discursive exchange between researchers and researched (Olivero and Lunt, 2004).

11.2.2.4 Public Versus Private Spaces Another more challenging area of IRE for researchers is the differentiation between public and private spaces online. If one contends there is any privacy to online interactions (which may be a big *if*), one can examine whether or not a particular forum, listserve, chat room, bulletin board, and so on, is considered *by its members* to be a public space or a private space. The expectations of privacy held by those members will dictate the role a researcher may occupy there. Many have used the analogy of the public park: What we observe in a public park is available to us as researchers. If the Internet locale is correspondingly public, then the researcher need not seek permission from subjects or participants. However, if we observe something in a house behind closed doors from our space in the public park, that is a private space and therefore unavailable for use by researchers. When considering public and private spaces, a researcher should consider his or her role within that online space: Is he or she an observer, participant, member, or other? Depending on the fora, different expectations arise. In public newsgroups, for instance, one may post something guided by intentions much different than a researcher's. A researcher may use such newsgroup data, removing context and removing the discursive markers of that group, and therefore present an inaccurate presentation of that data. The realities of archived data are challenging from a research ethics perspective. IRBs, for instance, look carefully at the use of preexisting data, as it challenges the process and spirit of informed consent. To protect subjects from harm means researchers must consider the ramifications of using preexisting data from archives, as something from a public space can easily come back to haunt a subject should it be brought to light in research. All of these possibilities quite seriously violate, among many ethical guidelines, the spirit of a consensual research relationship. Sveningsson (2004) has suggested that researchers evaluate this component of IRE along a continuum: public–private and sensitive–nonsensitive. Data falling in the private/sensitive quadrant would be off-limits to researchers, whereas the remaining three quadrants would be used with discretion, according to ethical guidelines and policy.

11.2.2.5 Respect for Persons Human subjects protections models are grounded in respect for persons, as born out of the informed consent process, as well as through a consideration of risks and benefits for the *individual* and for the *larger society*. A careful balance is necessary, though this has certainly not always been the case (e.g., the Tuskegee syphilis study), in protecting individual rights within the greater societal good. Researchers must justify the risk of their studies by the value of potential results. And, through the informed consent process, research participants must clearly understand these risks and benefits, what is taking place in the research, what is expected of them, and what will become of the data. Informed consent must be processual, not a static one-time event.

In online environments, there are many purely practical challenges in obtaining informed consent; these range from fluidity in group membership in short periods of time (for instance, those who log on for a few minutes at a time then quickly log off) to long-time changes in group membership, to individuals who have multiple screen names and identities, to ensuring people receive an informed consent document, to where the informed consent document may be accessible.

Further problems arise in the actual verification of understanding one's role in the research as a participant or subject, which is arguably the cornerstone of informed consent. Disagreement exists over the best process by which to achieve informed consent in online environments. Click boxes or hard copies are options. The inherent nature of some online environments, for instance, chat rooms, defy static informed consent processes. A blanket statement could be used as a corollary to the informed consent document: "I understand that online communications may be at greater risk for hacking, intrusions, and other violations. Despite these possibilities, I consent to participate." Researchers using transaction log data, for instance, have argued that such research is not on human subjects at all, and that it is a truly anonymous online interaction, and therefore, informed consent is unnecessary, as is any review by an ethics board at all.

11.2.2.6 Recruitment In traditional research ethics, the principle of justice demands that subjects have an equal or fair chance of participation—exclusion must be based on some justifiable reason. The federal Office for Human Research Protections (OHRP) in its IRB Handbook, *The Foundations of Human Subjects Protections* in the United States, describes the principle of justice:

> The principle of justice mandates that the **selection of research subjects** must be the result of fair selection procedures and must also result in fair selection outcomes. The "justness" of subject selection relates both to the subject as an individual and to the subject as a member of social, racial, sexual, or ethnic groups.

Thus, justice requires that the benefits and burdens of research be distributed fairly. Individuals must also not be unfairly targeted, as occurred in the Tuskegee experiments, for instance. But many online communities or environments are indeed self-selected based on some quality, and thus questions of justice may not apply in their strict sense; ultimately, in online research, equity/fair representation in the subject pool may not be possible. However, related questions of justice and recruitment emerge: How does the researcher enter the research space to begin recruiting? Many sites, notably proanas (proanorexic/eating disorder sites), for instance, reject researcher presence with notices—researchers are not welcome (see Hudson and Bruckman, 2004). Moreover, Walstrom (2004) has examined participant observation in the face of eating disorder groups online. She explored the possibility where some in a community may consent, whereas others do not. She argues that the researcher must respect the rights of those who do not want to allow their comments, and so on, to be taken as objects and material for research; hence, their privacy, anonymity, and so on, must be protected. At the same time, however, the researcher may proceed to examine the online interactions of those who have given their consent. This proves more difficult online than in on-site research encounters. Clark (2004) has also raised the interesting question of "hybrid" research endeavors: Hybrid research bridges research and researcher presence through groups that have both a physical and a virtual presence simultaneously. Clark (2004, p.247) describes the potential research ethics issue: "Research in a jointly virtual and material context raises unique questions... the primary texts in my research are the community's listserv postings; how will participant perceptions of risk be impacted by physical meetings with me and others after I have written potentially critical things?". One could conceivably consent to participate in one environment but thereby violate confidences from the other locale, intentionally or unintentionally.

11.2.2.7 Research with Minors According to the U.S. model, minors, those under the age of 18, are considered "special populations" (along with pregnant women and fetuses, prisoners, intellectually or emotionally impaired, or handicapped), and as such require special protections. This is codified in the *IRB Guidebook*, which notes that

> The federal regulations require that IRBs give special consideration to protecting the welfare of particularly vulnerable subjects, such as children, prisoners, pregnant women, mentally disabled persons, or economically or educationally disadvantaged persons.

IRBs and review committees take additional precautions when reviewing research involving any of these populations. Boards must have members that are uniquely qualified to represent, and protect, the unique interests of special populations. Moreover, in the case of minors, consent from an adult, in addition to the minor's assent, is required, thus calling for an additional layer in the consent process, whereas in the case of mentally disabled individuals, consent must be given by a guardian. In online environments, research on, or with, minors has raised considerable attention. Both Stern (2004) and Bober (2004) argue that research with minors is fraught with difficulty from a number of perspectives and should be conducted under very special conditions. Online research with minors raises issues of researcher certainty and competence. Researchers must ensure that their online participants are adults, consenting adults, and not a minor in some online forum. As both Stern and Bober have described, securing parental consent is challenging at best, and assent from a child is often easier to secure than parental consent. Furthermore, international agreements differ over the age of consent, contributing to a potential for serious differences across research forums online. Finally, while all researchers assume a level of responsibility to their participants in the research process, researchers dealing with special populations must understand and commit to reporting disturbing or dangerous information, for instance, a child discussing abuse, to an appropriate agency.

11.2.2.8 Emerging Issues Forums such as MySpace and Facebook have become increasingly popular, and of great concern, owing to their inception as popular social networking sites for young adults in particular. From primary to postsecondary schooling, participants of these sites are finding themselves face-to-screen with ethical issues. Notably, these social networking sites provide opportunities for both social and academic conversations, as their usage ranges from finding dates to academic advising. MySpace and Facebook have attracted attention in the popular media, from such instances as online stalking to employers seeking background, and personal, information on applicants.

In particular, these sites occasion questions regarding privacy—first of all, as many of their users seem to assume a level of privacy that is simply not available to them within the network. A number of researchers and ethicists have commented that "newbies" to the Internet, relatively unaware of the intrinsically public nature of most online communication, often assume a level of privacy in their communications that is simply false; for example, many seem to think that an e-mail is like a piece of paper mail, that it travels on the Internet in a sealed container until it reaches its intended destination, and that only the intended receiver may open it. Likewise, many are surprised to discover that their postings to a listserv or chat room can be logged and made publicly available online, most dramatically, for example, when the complete archive of USENET postings were made publicly available on the Web (Sveningsson, 2008). These observations lead to the ethical conundrum of whether or not users' expectations should drive researchers' efforts to protect privacy (cf. the discussion of ethical Good Samaritanism below) even though more legalistic approaches would argue that researchers are bound only by applicable laws and the privacy statements of the sites themselves (despite the fact that most users "click through" these without seriously considering them, thus making their consent of questionable ethical value).

This problem arises in new ways within such environments as Facebook and MySpace; because users must acquire an account and also because users are often affiliated with trusted institutions such as a particular school or organization, many users seem to think that "others"—beginning with their own parents and teachers—are somehow automatically excluded. On the contrary, students are learning, often the hard way, that information they post on their profiles may come back to haunt them in unhappy ways when reviewed by an academic advisor, instructor, prospective employer, and so on. Others, indeed, are finding that they cannot hide from anyone if they have a Facebook profile:

> ...the cops used Facebook to ID a suspected public urinator and bring the pissant to justice. It all started at the University of Illinois at Urbana-Champaign when student Mark Chiles, 22, urinated on a bush in front of a frat house. A cop saw him but the offender ran off before he could be cited. His drinking buddy, Adam Gartner, also 22, was questioned but said he didn't know the miscreant's name. Normally, a cop would give up at this point. But this story has a new digital twist.
>
> The Internet-savvy cop went on Facebook, looked up Gartner's profile and checked out his Facebook friends. Voila! There was the face of the urinator (ZDNET, 2006).

At the time of this writing, Facebook occasioned a number of additional privacy concerns. For example, it introduced a "newsfeed" feature that automatically collects "friends" information and changes to their profiles, displaying this information to a user when first logging in. A form of data mining, this technique was experienced by many Facebook users as a violation of privacy, though, of course, the information collected was perfectly public in the first place. The blogosphere, however, reacted with concern and anger, whereas the U.S. Senate took up a discussion headed by Senators Feingold and Sununu over privacy and data mining.

These discussions are quite new, and nothing can be said with confidence regarding emerging analyses and efforts to resolve these issues. Pending U.S. legislation, specifically, DOPA (Deleting Online Predators Act), would make such networking sites inaccessible in schools and public libraries; great legal and intellectual debate is occurring around this act, but it does clearly indicate that law and ethics are struggling equally with such technological venues. However, it does seem clear that any ethical guidelines and related understandings of privacy, rights to privacy, and so on, in these new venues will have significant effects on legislation and ethics, because these will directly affect a very large number of an important demographic of the Internet, namely, young people—a demographic that, up until now, has largely ignored the ethical dimensions of their participation in such fora. These discussions may be a step toward a much broader and more widely shared sense of ethics online that would be of tremendous help to Internet researchers, whose ethical decisions are often profoundly influenced by users' perceptions and expectations, beginning with privacy online.

11.3 METHODOLOGIES AND ETHICS

As we have already seen in our general discussion regarding the diverse disciplines involved in online research, it is clear that the research ethics implicated by a given ethical problem or difficulty is deeply entwined with and defined by the specific methodology (ies) that shape a specific research project. Indeed, Markham (2006) suggests that all methodological choices are in fact ethical choices.

As a first example, online experiments are a popular research approach—ones that, like their offline counterparts, offer incentives such as a lottery prize, money, or academic credit (to students) to attract and retain participants. At the same time, however, given the multiple ways in which users can mask their identities in online venues, researchers who do not meet with their participants face-to-face may not be able to confirm the offline identity of an online participant. In particular, some experimental designs require that participants remain anonymous. But, of course, delivering incentives promised for participating in an online survey or experiment requires the researcher to know with confidence a given participant's identity and important personal information. Hence, the researcher is faced with the double conundrum of sustaining participant anonymity while also knowing their offline identity to provide them with the promised incentives (cf. Peden and Flashinski, 2004).

Danielle Lawson further highlights a number of ethical problems that emerge in conjunction with specific methodologies. To begin with, research in the social sciences guided by more objectivist methodologies (i.e., ones that seek to emulate as closely as possible the methodologies of the "hard" or natural sciences) presumes such scientific norms as replicatability. To achieve replication, however, would demand publication of relevant data, including gender, age, and so on, of participants. But such publication manifestly puts participant confidentiality and anonymity at risk, especially if accompanied by additional information, such as *verbatim* quotes that can be easily found via a search engine on publicly accessible archives. Other methodologies, for example, those that incorporate Geertz's "thick description" (i.e., a detailed account not only of an event, action, and so on, but also of the *context* needed to understand that event, action, etc.) and/or participant-observation approaches, likewise issue in the need for publishing important details about research participants; but again, doing so only increases the threat to their anonymity and confidentiality (Lawson, 2004). Lawson's own effort to respond to these tensions includes recognizing a range of possible options that may be offered to participants. These begin with maximum privacy protection, as participants may agree to having their nicknames and texts used only for data analysis, but both their names and the texts themselves are not to be included in publication. At the other end of the spectrum, participants are treated much more like authors, as they consent to having their nicknames and texts published, and indeed receive credit for their texts as held by them under copyright as authors (Lawson, 2004, p. 93). Even with the range of options available, however, uncertainty and, more importantly, the role of ethical judgment (what Aristotle would call *phronesis*) cannot be eliminated entirely. Rather, as Lawson observes, researchers will have to make judgments regarding the proper ethical balance between emphasizing human subjects protections (anonymity, informed consent, etc.) and the particular requirements of their chosen methodology regarding publication of relevant information (p. 94).

11.3.1 Participant Observation and Discourse Analysis

A major focus in the IRE literature is on virtual ethnography and its ethical challenges. As a starting point, researchers such as Katherine M. Clegg Smith raise the initial question as to whether a researcher "lurking" (i.e., unannounced and unidentified) in a listserv is more analogous to a researcher taking notes on a public bench, in contrast with doing so while hiding in a bush (Clegg Smith, 2004, 230ff). For Clegg Smith, if a public list provides an introductory message to new members indicating that the list is public and all messages are archived, then the analogy with the researcher on the public bench is salient, so she is not required to ask for informed consent from participants, nor to announce to the list her presence as a researcher "listening" to postings (pp. 231–235). On the contrary, when faced with the question of including potentially sensitive texts as data in her publication, Clegg Smith judged that her posters were more like human subjects than authors, and so chose to keep them anonymous—a decision shared by several of the pioneers in the literature on participant-observation methodologies and IRE (Bromseth, 2002; Markham, 2004; Sveningsson, 2001). Indeed, Olivero and Lunt (2004) argue that the methodologies of participant observation and discourse analysis (i.e., careful analysis of various dimensions of discourse, highlighting not simply the content of discourse but also the styles of discourse, how communicants signal various interactions from turn-taking to seeking dominance, etc.) in online environments *heighten* the importance of privacy, informed consent, and ethical issues surrounding the use of participants' texts. It appears that participant-observation methodology, as it requires researchers to develop a more personal and empathic relationship with participants, *increases* researchers' sensitivity toward participants as coequal human beings; given this heightened sense of engaging in a humane relationship with their participants, researchers may be more likely to accord participants the sorts of privacy protections that researchers would want for themselves. In addition, such heightened attention to protecting privacy, anonymity, and so on, has a strongly pragmatic dimension as well: because participants can easily disconnect from e-mail and other forms of online interviews, researchers have a correlatively greater need to foster and sustain the active engagement of their subjects—in part, by offering the reward of "the gratifying trusted, reciprocal exchange indicated by the feminist perspective" (p. 107). (Such an approach is further argued from feminist and communitarian foundations by Walstrom (2004) and, with the additional appeal to Bakhtin, by Hall et al., (2004).)

Ethicists familiar with the abortion debate will recall that Judith Jarvis Thomson introduces the distinction between "minimally decent" ethics and those actions and choices, such as those of the Good Samaritan, that are admirable precisely because they go beyond our everyday expectations and codes. For Thomson (1971) this means, however, that although such actions and choices are exemplary, they cannot be legally required of everyone in every circumstance. Such ethical Good Samaritanism appears frequently among researchers guided by participant/observation methodology. That is, several researchers have begun with basic ethical and legal requirements of their discipline, but found that these did not go far enough, in their view; for example, these (minimal) requirements did not oblige a researcher to protect the privacy and identity of participants in a listserv as a public space. Rather, these researchers—again, as more directly engaged with their participants as close coequals rather than as distant subjects—have chosen to take the more demanding ethical position of a Good Samaritan. So, for example, several have come to go beyond the minimal demands and insist on protecting participant privacy, even though such protection complicated their research, made greater demands on their time and resources, and so forth (see Clegg Smith, 2004; King, 1996; Reid, 1996).

Given the presence of such "Good Samaritan" ethical choices in the literature of IRE, choices that have shaped foundational work in online research, even once researchers may have achieved clarity regarding what disciplinary guidelines and national law may require, they will further need to consider whether these establish only minimal standards and, if so, if they judge rather to follow a more rigorous "Good Samaritan" approach. (For further discussion of the correlation between distinctive research approaches and their correlative ethical difficulties, see Bakardjieva and Feenberg (2001) and Markham (2003).)

11.3.2 A Global Internet Research Ethics?

The global reach of the Internet means that research participants may be drawn from a wide range of nations and cultures. Coupled with the often international collaborations behind online research, this fact of a global range of participants forces a still more demanding question for ethicists: Given precisely the often significant national differences in the Western world between more deontological and more consequentialist approaches to IRE, how are we to develop a research ethics that is legitimate for researchers and participants from more than one nation, culture, and tradition of ethics and thereby research ethics?

Projects such as the AoIR ethical guidelines and the RESPECT project demonstrate that researchers from a diversity of countries and traditions of ethical decision-making can in fact agree upon a range of basic values and issues, while at the same time preserving local differences in the interpretation and implementation of those values through a strategy of *ethical pluralism.* In the AoIR guidelines, for example, we encountered what at first seemed an irresoluble conflict between U.S. and Norwegian research ethics with regard to the question of whether informed consent would be required for audio and video recordings in public spaces. On the one hand, U.S. approaches argued that such consent is not required, in part because in the U.S. context there is no expectation of having the sort of privacy in a public space that would call for informed consent (Walther, 2002). On the other hand, Norwegian research ethics insisted on precisely such informed consent, in part just because in the Norwegian context there is the expectation that one's privacy will be protected in this way, even in what are otherwise acknowledged to be public spaces (Elgesem, 2002). But in both cases, research ethics begins with a shared focus, namely, just on the expectations of the persons involved. Insofar as both U.S. and Norwegian research ethics acknowledge the normative importance of expectations, while at the same time issuing in contrasting norms regarding informed consent, they thereby articulate a structure of ethical pluralism that holds together shared norms alongside irreducible differences in the interpretation and application of those norms – differences that reflect and thus preserve distinctive cultural identities.[2]

Such pluralisms, in fact, can be discerned in a variety of instances in information and computer ethics (Ess, 2006, 2007). In particular, such pluralisms appear to be at work across East–West boundaries as well, for example, with regard to emerging conceptions of privacy and data privacy protection in China and Hong Kong vis-à-vis Western conceptions. So, for example, in the United States and Germany, rights to privacy and data privacy protection are established on the assumption that privacy is both an intrinsic and instrumental good (e.g., privacy is taken to be a necessary condition for self-development and expression, freedom of opinion and thought, and participation in democratic governance). By contrast, understandings of privacy and data privacy protection codes are justified in China and Hong Kong solely as instrumental goods, that is, these are necessary for a much desired e-economy. Hence, a generally shared notion of "privacy" and the need for data privacy protection is shared across these diverse cultures and nations, but understood and justified within each culture and nation in distinctive ways that directly reflect and preserve their differences (i.e., the differences between a Western emphasis on an atomistic individual as a rights holder in a democratic polity and an Asian emphasis on the individual as a member of a larger community, where community well-being and harmony justify what from a Western perspective would be characterized as more authoritarian regimes; see Ess (2005, 2006, 2007)). Moreover, while discussion of Internet research ethics is very young in Asia, recent examples likewise fit such a pluralistic structure. For example, in a recent study of messages exchanged in a forum, Japanese researcher Tamura Takanori took up what he described as a "more cautious way" in his ethical choices than other Japanese researchers, as he requested consent from the forum coordinator to use forum exchanges; referred to specific authors by way of pseudonyms; and paraphrased rather than using direct quotes (Tamura, 2004). These choices are striking first of all in the light of the clear and significant differences between Western and Japanese understandings of privacy and research practices regarding human subjects protections, differences reflecting still more fundamental differences between Western affirmation of the atomistic individual as a moral autonomy to be

[2]See especially AoIR, 2002, p. 4, including footnotes 6 and 7, as well as "Addendum 2."

affirmed and protected, vis-à-vis especially Buddhist notions of "self" as an illusion that must be overcome if genuine contentment is to be attained (Ess, 2005; Nakada and Tamura, 2005). Such notions support the practices of other Japanese researchers, for example, as they saw no need to protect anonymity, request consent for direct quotes, and so on. Alongside these fundamental differences, however, Tamura's approach deeply resonates with several elements of Western IRE. So, for example, Tamura's insistence on protecting the privacy of the forum participants shows a basic respect for the *expectations* of their authors, a respect that, as we have seen, is a cornerstone especially for deontological approaches to IRE in the West in general and for the issue of informed consent in the United States and Norway in particular (AoIR, 2002, f.7). Moreover, Tamura's "more cautious way" contrasts with the less protective approach of other researchers; in this way, Tamura can be understood to take up a Good Samaritan ethics that goes beyond the requirements of "minimally decent" law and practices, and thereby echoes such ethical Good Samaritanism as we have seen it among Western researchers, especially those following participant-observer methodologies. Finally, Tamura's more cautious way is strikingly consistent with Western, specifically *deontological*, approaches that insist on the rights of the subject, including protection against possible harm, above possible benefits of research.

Tamura's approach thus stands as one side of an East–West ethical pluralism that conjoins shared norms alongside the irreducible differences that define distinctive cultures. To be sure, ethical pluralism will not resolve all cultural differences and conflict in research ethics; nonetheless, these examples suggest that a global IRE may emerge still more fully as information ethics and research ethics traditions in both East and West become ever more developed.

11.4 CONCLUSIONS

We hope this review makes clear that IRE, although a relatively young field at the intersections between applied ethics, information and computer ethics, and professional and research ethics, can now be seen as reasonably well established. It enjoys an extensive literature that helpfully collects historically important analyses along with contemporary considerations of specific issues; this literature, as our discussion of specific topics demonstrates, offers considerable guidance with regard to a range of issues that consistently emerge in the course of online research. Indeed, ethical guidelines such as those established by AoIR and NESH indicate the level of interest in IRE internationally. In addition, as the AoIR guidelines continue to find use in diverse research projects and institutional settings, both in the English-speaking world and beyond, this diffusion suggests that an international consensus regarding online research ethics may well be possible. In fact, at least some examples of apparently contrasting approaches in research ethics, notions of privacy, and IRE proper strongly suggest that a genuinely global IRE may emerge that will conjoin shared norms alongside differences reflecting and fostering irreducible differences between national and cultural ethical traditions, including the most basic differences between Eastern and Western ethical frameworks.

This means that both young and seasoned researchers, as well as oversight institutions responsible for research integrity (e.g., IRBs in U.S. context), now have a considerable range of examples and well-established guidelines to draw from, as well as foundations for a continuing global dialogue aimed toward further developing a global IRE. At the same time, however, the discussion of Facebook and the ongoing efforts to develop a genuinely global IRE make clear that IRE is still very young. Both as new venues and technologies open up new research possibilities, and as researchers and philosophers participate in a growing global dialogue regarding the ethics of online research, philosophers and researchers interested in IRE will confront no shortage of intriguing new examples and issues. We have only just begun ....

REFERENCES

American Association for the Advancement of Science (1999). Ethical and legal aspects of human subjects research in cyberspace. Available at http://www.aaas.org/spp/sfrl/projects/intres/main.htm.

AoIR (Association of Internet Researchers) (2002). Ethical guidelines for internet research. Retrieved, January 10, 2006, from www.aoir.org/reports/ethics.pdf.

Association for Computing Machinery (1992). ACM code of ethics and professional conduct. Retrieved, January 10, 2006, from http://www.acm.org/constitution/code.html.

Association of Internet Researchers Ethics Working Group and Ess, C. (2002). Ethical decision-making and internet research: recommendations from the AoIR Ethics Working Committee. Available at www.aoir.org/reports/ethics. pdf.

Bakardjieva, M. and Feenberg, A. (2001). Involving the virtual subject: conceptual, methodological and ethical dimensions. *Ethics and Information Technology*, 2(4), 233–240.

Bober, M. (2004). Virtual youth research: an exploration of methodologies and ethical dilemmas from a British perspective. In: Buchanan, E. (Ed.), *Readings in Virtual Research Ethics: Issues and Controversies*. Idea Group, Hershey, pp. 288–316.

Bromseth, J.C.H. (2002). Public places–public activities? Methodological approaches and ethical dilemmas in research on computer-mediated communication contexts. In: Morrison, A. (Ed.), *Researching ICTs in Context*. Inter/Media Report 3/2002. University of Oslo, Oslo, pp. 33–61. Available at http://www.intermedia.uio.no/konferanser/skikt- 02/docs/Researching_ICTs_in_context- Ch3- Bromseth. pdf.

Bruckman, Amy (2002). Studying the Amateur Artist: A Perspective on Disguising Data Collected in Human Subjects Research on the Internet. *Ethics and Information Technology*, 4(3), 217–231.

Buchanan, E. (2003). Internet research ethics: a review of issues. Presentation at the Association of Internet Researchers Annual Conference, Toronto, Canada.

Buchanan, E. (Ed.) (2004). *Readings in Virtual Research Ethics: Issues and Controversies*. Idea Group, Hershey.

Buchanan, E. (2005). The IRB review and online research. Presentation at the Association of Internet Researchers Annual Conference, Chicago, IL.

Buchanan, E. and Ess, C. (2003). Understanding internet research ethics: a worskshop. Association of Internet Researchers Annual Conference, Toronto, Canada.

Buchanan, E. and Ess, C. (2005). Internet research ethics workshop. Association of Internet Researchers Annual Conference, Chicago, IL.

Burkhardt, J., Thompson, P., and Peterson, T.R. (2002). The first European congress on agricultural and food ethics and follow-up workshop on ethics and food biotechnology: a US perspective. *Agriculture and Human Values*, 17(4), 327–332.

Clark, D. (2004). What if you meet face to face? A case study in virtual/material research ethics. In: Buchanan, E. (Ed.), *Readings in Virtual Research Ethics: Issues and Controversies*. Idea Group, Hershey, pp. 246–261.

Clegg Smith, K. M. (2004). "Electronic Eavesdropping": the ethical issues involved in conducting a virtual ethnography. In: Johns, M., Chen, S. L., and Hall, J. (Eds.), *Online Social Research: Methods, Issues, and Ethics*, Peter Lang, New York, pp. 223–238.

Code of Federal Regulations (2005). Title 45 (Public Welfare), Part 46, Protection of Human Subjects. Retrieved February 16, 2007, from http://www.hhs.gov/ohrp/humansubjects/guidance/45cfr46.htm.

Directive, 95/46/EC of the European Parliament and of the Council of 24 October 1995 on the protection of individuals with regard to the processing of personal data and on the free movement of such data. Retrieved, February 21, 2007, from http://ec.europa.eu/justice_-home/fsj/privacy/law/index_en.htm.

Directive, 2006/24/EC of the European Parliament and of the Council of 15 March 2006 on the retention of data generated or processed in connection with the provision of publicly available electronic communications services or of public communications networks and amending Directive 2002/58/EC. Retrieved, February 21, 2007, from http://ec.europa.eu/justice_home/fsj/privacy/law/index_en.htm.

Elgesem, D. (2002). What is special about the ethical issues in online research? *Ethics and Information Technology*, 4(3), 195–203. Available at http://www.nyu.edu/projects/nissenbaum/ethics_elgesem.html.

Ess, C. (2003). The cathedral or the bazaar? The AoIR document on internet research ethics as an exercise in open source ethics. In: Consolvo, M. (Ed.), *Internet Research Annual Volume 1: Selected Papers from the Association of Internet Researchers Conferences 2000–2002*. Peter Lang, New York, pp. 95–103.

Ess, C. (2005). "Lost in translation?": Intercultural dialogues on privacy and information ethics (introduction to special issue on privacy and data privacy protection in Asia). *Ethics and Information Technology*, 7, 1–6.

Ess, C. (2006). Ethical pluralism and global information ethics. In: Floridi, L. and Savulescu, J. (Eds.), *Information Ethics: Agents, Artifacts and New Cultural Perspectives*, a special issue of *Ethics and Information Technology*, 8, 215–226.

Ess, C. (2007). Culture and communication on global networks: cultural diversity, moral relativism, and hope for a global ethic? In: Weckert, J. and van den Hoven, J. (Eds.), *Information Technology and Moral Philosophy*. Cambridge University Press, Cambridge, pp. 195–225.

Floridi, L. (Ed.) (2003). *Blackwell Guide to the Philosophy of Information and Computing*. Blackwell, Oxford.

Forte, M. (2004). Co-construction and field creation: website development as both an instrument and relationship in action research. In: Buchanan, E. (Ed.), *Readings in Virtual Research Ethics: Issues and Controversies*. Idea Group, Hershey, pp. 219–245.

Frankel, M. and Siang, S. (1999). Ethical and Legal Aspects of Human Subjects Research of the Internet. A Report presented at the American Association for the Advancement of Science. Retrieved August 1, 2005, from http://www.aaas.org/spp/dspp/sfrl/projects/intres/main.htm

Hall, G. J., Frederick, D., and Johns, M. D. (2004). "NEED HELP ASAP!!!": A feminist communitarian approach to online research ethics. In: Johns, M., Chen, S. L., and Hall, J. (Eds.) (2004). *Online Social Research: Methods, Issues, and Ethics*, Peter Lang, New York, pp. 239–252.

Hudson, J. M. and Bruckman, A. (2004). 'Go Away': Participant objections to being studied and the ethics of chatroom research. *The Information Society* 20(2), 127–139.

King, S. (1996). Researching internet communities: proposed ethical guidelines for the reporting of results. *The Information Society*, 12(2), 119–128.

Kraut, R., Olson, J., Banaji, M., Bruckman, A., Cohen, J., and Cooper, M. (2004). Psychological research online: report of board of scientific affairs' advisory group on the conduct of research on the internet. *American Psychologist*, 59(4), 1–13.

Johns, M., Chen, S. L. and Hall, J. (Eds.) (2004). *Online Social Research: Methods, Issues, and Ethics*. Peter Lang, New York.

Lawson, D. (2004). Blurring the boundaries: ethical considerations for online research using synchronous CMC forums. In: Buchanan, E. (Ed.) (2004). *Reading in Virtual Research Ethics: Issues and Controversies*, Idea Group, Hershey, pp. 80–100.

Markham, A. (2003). Critical junctures and ethical choices in internet ethnography. In: Thorseth (Ed.), Applied Ethics in Internet Research. Programme for Applied Ethics, Norwegian University of Science and Technology, Trondheim, pp. 51–63.

Markham, A. (2004). Representation in online ethnographies: a matter of context sensitivity. In: Johns, M., Chen, S. L., and Hall, J. (Eds.) (2004). *Online Social Research*: Methods, Issues, and Ethics, Peter Lang, New York, pp. 141–155.

Markham, A. (2006). Ethics as method, method as Ethics: a case for reflexivity in qualitative ICT Research. *Journal of Information Ethics*, 15(2).

Michelfelder, D. (2001). The moral value of informational privacy in cyberspace. *Ethics and Information Technology*, 3(2), 129–135.

Nakada, M. and Tamura, T. (2005). Japanese conceptions of privacy: an intercultural perspective. *Ethics and Information Technology*, 7(1: March), 27–36.

NESH (National Committee for Research Ethics in the Social Sciences and the Humanities). (2003). Research ethics guidelines for Internet research. Retrieved, January 10, 2006, from http://www.etikkom.no/Engelsk/Publications/internet03.

NESH (National Committee for Research Ethics in the Social Sciences and the Humanities). (2006). Guidelines for research ethics in the social sciences, law and the humanities. Retrieved, February 21, 2007, from http://www.etikkom.no/English/NESH/ guidelines/

OHRP. *The IRB Guidebook*. Retrieved, January 30, 2007, from http://www.hhs.gov/ohrp/irb/irb_guidebook.htm.

Olivero, N. and Lunt, P. (2004). When the ethic is functional to the method: the case of e-mail qualitative interviews. In: Buchanan (2008). How do various notions of privacy influence decisions in qualitative internet research? In: Markham, A. N. and Baym, N. K. (Eds.) *Internet Inquiry: Conversations About Method*. Sage, Thousand Oaks, CA, pp.101–113.

Peden, B. and Flashinski, D.P. (2004). Virtual research ethics: a content analysis of surveys and experiments online. In: Buchanan (2008). How do various notions of privacy influence decisions in qualitative internet research? In: Markham, A. N. and Baym, N. K. (Eds.) *Internet Inquiry: Conversations About Method.* Sage, Thousand Oaks, CA, pp.1–26.

Pence, G.E. (1990). *Classic Cases in Medical Ethics: Accounts of the Cases that have Shaped Medical Ethics, with Philosophical, Legal, and Historical Backgrounds.*McGraw-Hill, New York.

Reid, E. (1996). Informed consent in the study of on-line communities: a reflection on the effects of computer-mediated social research. *The Information Society*, 12(2), 169–174.

Reidenberg, J.R. (2000). Resolving conflicting international data privacy rules in cyberspace. *Stanford Law Review*, 52, 1315–1376.

Stern, S. (2004). Studying adolescents online: a consideration of ethical issues. In: Buchanan, E. (Ed.), *Readings in Virtual Research Ethics: Issues and Controversies.* Idea Group, Hershey, pp. 274–287.

Sveningsson, M. (2001). Creating a sense of community: Experiences from a Swedish web chat. *Linkoping Studies in Art and Science*, 233.

Sveningsson, M. (2004). Ethics in internet ethnography. In: Buchanan, E. (Ed.), *Readings in Virtual Research Ethics: Issues and Controversies.* Idea Group, Hershey, pp. 45–61.

Sveningsson, M. (2008). How do various notions of privacy influence decisions in qualitative internet research? In: Markham, A. N. and Baym, N. K. (Eds.), *Internet Inquiry: Conversations About Method.* Sage, Thousand Oaks, CA.

Tamura, T. (2004). Internet research ethics in Japan. Unpublished report.

The National Commission for the Protection of Human Subjects of Biomedical and Behavioral Research (1979). *The Belmont Report.* Available at http://ohsr.od.nih.gov/guidelines/belmont.html.

Thomson, J.J. (1971). A defense of abortion. *Philosophy and Public Affairs*, 1(1), 47–66.

Thorseth, M. (2003). *Applied Ethics in Internet Research.* Programme for Applied Ethics, Norwegian University of Science and Technology, Trondheim.

Walther, J. (2002). Research ethics in internet-enabled research: human subjects issues and methodological myopia. *Ethics and Information Technology*, 4(3). Available at http://www.nyu.edu/projects/nissenbaum/ethics_walther.html.

Walstrom, M. (2004). Ethics and engagement in communication scholarship: analyzing public, online support groups as researcher/participant-experiencer. In: Buchanan, E. (Ed.), *Readings in Virtual Research Ethics: Issues and Controversies.* Idea Group, Hershey, pp. 174–202.

White, M. (2003). Representations or people. Available at http://www.nyu.edu/projects/nissenbaum/ethics_whi_full.html.

ZDNET (August 2006). The fuzz wants to add you as a friend. Retrieved, February 18, 2007, from http://education.zdnet.com/?p=411.

CHAPTER 12

Health Information Technology: Challenges in Ethics, Science, and Uncertainty[1]

KENNETH W. GOODMAN

It is sadly and too often the case that many professionals regard ethics as a source of codes for the edification of the not-yet-virtuous, as a place where pointy-headed boffins pass judgment on heathens, as an office to call in search of someone with a horse and a sword to come 'round to smite the evildoers.

Rather, ethics, a branch of philosophy, has the task of studying morality, or (generally) public accounts of the rightness or wrongness of actions. Applied or professional ethics is the analysis of moral issues that arise in, well, the professions. All professions give rise to ethical issues, not necessarily because practitioners do bad things or need to be saved from their many temptations, but because questions of appropriate action arise even in situations in which no one has done anything obviously wrong. That is, professionals encounter ethical issues and challenges in the ordinary course of their work. It is unavoidable. Ought a lawyer advertise? How should an engineer manage complexity? May a scientist build a bomb or clone a sheep? When, if ever, is it appropriate for a physician to use a computer to render a diagnosis? Most importantly, why?

Bertrand Russell made clear that there are no experiments we can do to determine the answers to questions about morality and ethics. Instead, human reason provides the tools for such interrogations and conclusions. It is an exciting enterprise, and has become more so as technology has evolved and given us challenges once unimaginable. This is as true in medicine and biology as in any other domains.

The use of computers or, more generally, information technology in the health professions is indeed a rich source of ethical issues and challenges. Our task here is to

[1]Support for this chapter was provided in part by a grant from the Robert Wood Johnson Foundation® in Princeton, New Jersey.

The Handbook of Information and Computer Ethics, Edited by Kenneth Einar Himma and Herman T. Tavani

identify the largest ones, and point out ways in which applied philosophy provides resources for addressing them. We will look at the following issues:

- privacy and confidentiality,
- use of decision support systems, and
- development of personal health records.

12.1 PRIVACY AND CONFIDENTIALITY

There is arguably no better trigger for reflection on morality and its relationship to the law and society than privacy and its cousin, confidentiality. The demands of privacy are intuitively straightforward and the consequences of its violation obvious. Without a credible promise that privacy and confidentiality will be safeguarded, the task of fostering trust is frustrated. If for instance a patient believes that a physician will disclose interesting or salacious diagnostic data to others, the patient might not disclose information the physician needs to render an accurate diagnosis in the first place. If a patient believes a physician or hospital does not maintain the security of medical records, the patient might similarly be discouraged to tell the truth. And if a patient is dubious about an institution's ability to safeguard data stored in or transmitted by computers and other information systems, then the technology itself will be a source of distrust.

Indeed, there are many reasons to believe that public distrust imperils the growth or expansion of electronic medical records—which, in turn, are seen as needed to replace an aging, fragmented, and inefficient paper-based system. In 2006, the National Committee on Vital and Health Statistics observed that

> as a practical matter, it is often essential for individuals to disclose sensitive, even potentially embarrassing, information to a health care provider to obtain appropriate care. Trust in professional ethics and established health privacy and confidentiality rules encourages individuals to share information they would not want publicly known. In addition, limits on disclosure are designed to protect individuals from tangible and intangible harms due to widespread availability of personal health information. Individual trust in the privacy and confidentiality of their personal health information also promotes public health because individuals with potentially contagious or communicable diseases are not inhibited from seeking treatment... In an age in which electronic transactions are increasingly common and security lapses are widely reported, public support for the [National Health Information Network] depends on public confidence and trust that personal health information is protected. Any system of personal health information collection, storage, retrieval, use, and dissemination requires the utmost trust of the public. The health care industry must commit to incorporating privacy and confidentiality protections so that they permeate the entire health records system (National Committee on Vital and Health Statistics, 2006).

So, without trust—newly imperiled by the belief that computers might constitute new threats to confidentiality—the expansion of health information technology is in jeopardy. Without this expansion, a potentially useful suite of tools for improving health care will be forfeited. Ours is no idle exercise.

Privacy is, most generally, the right entitlement or reasonable expectation people have that they are and will be secure from intrusion. A peeping Tom violates one's privacy, as does a peeping police officer. Because society values both privacy and law enforcement, however, the police officer investigating a crime may and, in fact, must take steps to justify that her official need is worth the intrusion. Put differently, privacy rights are not absolute, but may be balanced against other values. The same is true for *confidentiality*, which applies to information—medical records, for instance. Where privacy is customarily about people, confidentiality applies to information about people. Privacy is also sometimes regarded as including within its scope people's concern about protecting confidentiality. Privacy is a broader concept.

General privacy considerations are addressed in Chapter 7. Issues related to genetics and genomics, including privacy, are assessed in Chapter 22. What is for the most part uncontroversial is the idea that privacy assumes special or unique importance in medical contexts. While we might make privacy claims in banking, education, or Web browsing, for instance, the use of computers to acquire, store, analyze, and transmit medical information applies to information that is generally regarded as more sensitive and more intensely personal than information about other aspects of our lives.

The origins of a physician's duty to ensure confidentiality are customarily traced to the Oath of Hippocrates, whereby one promises that "whatsoever I shall see or hear in the course of my profession, as well as outside my profession in my intercourse with men, if it be what should not be published abroad, I will never divulge, holding such things to be holy secrets."[2] Whatever its origin, there are good reasons for physicians, nurses, psychologists, and others to protect patients' information. One set of reasons is utilitarian. If patients do not trust that their "holy secrets" will in fact be safeguarded they are unlikely to disclose them in the first place or, worse, more likely to deceive clinicians. This erodes the therapeutic relationship, frustrates accurate diagnoses and increases the risk of poor clinical outcomes. Other reasons are rights based. Privacy and confidentiality emerge as entitlements that many people expect by virtue of the (sometimes socially conditioned) desire to control access to their person and representations of or information about their person.

Computers complicate medical privacy and confidentiality in interesting ways. According to a sentinel Institute of Medicine analysis, a number of entities demand

---

[2]Clendening (1960, 1942, p. 15). In fact, few physicians take the oath as written, in part because it invokes several deities no longer usually worshipped. The oath begins: "I swear by Apollo Physician, by Asclepius, by Health, by Panacea, and by all the gods and goddesses, making them my witness, that I will carry out, according to my ability and judgment, this oath and this indenture." So those who are not Apollonians must redact the oath (which also includes prohibitions against abortion and accepting payment for being a medical school faculty member, for instance). Medical students often craft their own oaths, drawing from those of Hippocrates, Maimonides, and others. (It should be noted that some historians debate whether Hippocrates is in fact the author of the oath or any other part of the "Hippocratic corpus.") Clendening contends that the Oath of Hippocrates is not an oath at all, "but an indenture between master and pupil" (p. 13).

patient information to "assess the health of the public and patterns of illness and injury; identify unmet... health needs; document patterns of health care expenditures on inappropriate, wasteful, or potentially harmful services; identify cost-effective care providers; and provide information to improve the quality of care in hospitals, practitioners' offices, clinics, and other health care settings" (Donaldson and Lohr (1994), p. 1. Moreover,

> medical records usually contain a large amount of personal information, much of it quite sensitive. This information is continuous, extending from cradle to grave; it is broad, covering an extraordinary variety of detail; and, with new information technologies, it is accessible as never before. Aside from the patient's name, address, age, and next of kin, there also may be names of parents; date and place of birth; marital status; race; religion; occupation; lifestyle choices; history of military service; Social Security or other national identification number; name of insurer; complaints and diagnoses; medical, social, and family history, including genetic data; previous and current treatments; an inventory of the condition of each body system; medications taken now and in the past; use of alcohol, drugs, and tobacco; diagnostic tests administered; findings; reactions; and incidents (Alpert, 1998; Gellman, 1984).

Computers have been reckoned both to make it easier than paper records to acquire medical information inappropriately (Walters, 1982) and, simultaneously, to provide the means to prevent such inappropriate acquisition (Gostin et al., 1993). This fissure parallels the main challenge underlying the use of computers to manage health information: We want at the same time to make it easy for appropriate users to have access to our medical information and difficult or impossible for inappropriate others to have such access. Several questions follow:

- How should (in)appropriateness of use and user be identified and described?
- How much, if any, privacy/confidentiality should we be willing to trade for improved health care promised by information technology?
- In what circumstances should a patient's privacy/confidentiality rights be overridden?
- What steps should be taken to ensure that computers are not used for inappropriate access?

Some answers to these questions enjoy widespread agreement. For instance, regarding the first question, appropriate access should generally be assigned by patients themselves. As with a paper medical record, there is generally no better judge of this access than the (rational) person to whom the information pertains. To be sure, there are limits on this power. It would be strange and unworkable if a patient did not want anyone, including any physician or nurse, to view the record; on the contrary, a patient might plausibly request that certain hospital staffers, say, should not be able to view the record, and expect to have such a request honored. Requests that seemed to impede treatment could be negotiated.

The second question is most applicable to public or institutional policy. In a teaching hospital, for instance, it will be customary for students and other trainees to view the online record of patients who are of scientific or pedagogic interest. A patient who wanted to restrict all such access should probably seek care at a different institution. More generally and for instance, society values the electronic medical record as a source of research or outcomes data. There are few if any good reasons for a patient to object to such use, especially if the information is anonymized or if the users are trusted researchers. Such a trade-off provides valuable public benefit at modest or no infringement on confidentiality. However, policies must be in place to govern such uses.

The needs of public health provide a suite of reasons for frank infringements of confidentiality in certain restricted cases, including those related to bioterrorism (Goodman, 2003a) and pandemic or syndromic surveillance (Szczepaniak et al., 2006). The field of "emergency public health informatics" raises exquisitely interesting and difficult questions regarding the use of large databases—containing individuals' identifiable personal health information—for collective benefit. With appropriate oversight, this third question invites responses that underscore the importance of public health and invoke standard moral objections in cases in which individuals imperil others for selfish but not unreasonable interests (Goodman, 2003a). So, for instance, the patient with infectious tuberculosis who claims his autonomy is infringed by disclosure of his contagion and forced isolation is correct—but this protest is inadequate in the face of arguments holding that others in his community who do not wish to be exposed unknowingly to this malady have a more powerful claim.

The use of computers in health care provides a solid example of the utility of applied or practical ethics to address challenges raised by the growth of technology. Question 4 here—"What steps should be taken to ensure that computers are not used for inappropriate access?"—is a request for ethically optimized approaches to ensure that access is easy when appropriate and difficult when not. It is perhaps the most important question of all. The U.S. Office of Technology Assessment (OTA, 1993a, 1993b) concluded in a key report that

> all health care information systems, whether paper or computer, present confidentiality and privacy problems ... Computerization can reduce some concerns about privacy in patient data and worsen others, but it also raises new problems. Computerization increases the quantity and availability of data and enhances the ability to link the data, raising concerns about new demands for information beyond those for which it was originally collected. (OTA, 1993a, p. 3; cf. OTA, 1993b; National Research Council, 1997).

What has emerged is for the most part uncontroversial, has elicited widespread support and, for the most part, works. It is this: inappropriate use of computers to obtain confidential information is best reduced by an ensemble of three approaches: institutional policies and federal and state legislation, thoughtful security precautions (including audit trails, for instance), and education about the moral foundations of privacy and confidentiality (Alpert, 1998). Indeed, one might interpret the privacy and security rules under the U.S. Health Insurance Portability and Accountability Act (HIPAA) as embodying just such a three-pronged approach.[3] While HIPAA has met with some resistance and numerous misinterpretations in hospitals and other health care organizations around the country, the first-ever nationwide privacy law has served to raise awareness of the importance of privacy in a networked health environment.[4] (Indeed, one of the motivations for HIPAA was the expectation of increased use of information technology and the belief—to be traced back to the origins of the Hippocratic tradition—that patients would not disclose personal information in the absence of assurances that it would be safeguarded.)

Efforts to establish these measures constitute a collective work in progress, and additional effort is needed (GAO, 2007). But the underlying philosophical and moral challenges have been addressed in a way that is noteworthy for its recognition of an ancient value in the presence of a contemporary technology.

12.2 CLINICAL DECISION SUPPORT SYSTEMS

At least as much as the storing and transmitting of health data and information, which raise issues of privacy and confidentiality, the use of intelligent machines to analyze such data and information raises important questions regarding which uses and users are appropriate and which are not. Indeed, the use of clinical decision support systems (CDSS), including diagnostic expert systems, is arguably one of the most significant—and interesting—ethical issues that arise in the field of computer ethics. What is at stake when intelligent machines are used to augment or even supplant human cognition and decision making is nothing less than the nature of the human-machine relationship and so the extent to which we should be willing to assign complex tasks to such devices. At ground, the issue is this:

> When, by whom, under what circumstances, and with what kinds and levels of oversight, accountability, and responsibility should computers be used to make medical decisions?

The question is delicious, for it precludes nothing while emphatically suggesting that *some* kinds of constraints are appropriate, if not necessary. In the first sustained assessment of the ethical issues raised by computers in medicine, Miller et al., (1985) observed that it might be blameworthy *not* to use an intelligent machine if there were adequate evidentiary warrant to suppose that the machine would improve patient care.

[3]For a comprehensive guide to the federal law in particular, and health data protection in general, see http://privacy.med.miami.edu

[4]Misinterpretations, gleaned from numerous personal communications, include the assertions that HIPAA forbids the use of e-mail to communicate with patients, prevents physicians on a hospital team from communicating with each other without a patient's permission and outlaws research on stored biological samples without explicit consent.

This insight has helped lay out the ethical tension at the heart of the two most common ways computers can be said to assist, augment or even replace human decision making: the use of decision support systems and prognostic scoring systems. The tension has been described in the following way:

> To ask if a computer diagnosis increases (or decreases) the risk of diagnostic or other error is in part to ask whether it will improve patient care. If the answer is that, on balance, the tool increases (the risk of) diagnostic error, then we should say it would be unethical to use it. Significantly, though, what is sought here is an empirical finding or a reasoned judgment—where such a finding is often lacking or even methodologically hard to come by; or where such a judgment is based on inadequate epistemic support, at least according to standards otherwise demanded to justify clinical decisions... This means that we are pressed to answer an ethical question (Is it acceptable to use a decision support system?) in a context of scientific uncertainty (How accurate is the system?). Many challenges in contemporary bioethics share this feature, namely, that moral uncertainty parallels scientific or clinical ignorance. (Goodman, 2007, p. 129).

These issues arise for diagnostic systems and prognostic systems. The latter are used mainly in critical care units to predict the likelihood of a patient surviving (or, conversely, dying); they are said to be most useful, however, in gauging hospital performance, either over time or by enabling comparisons between or among hospitals (Knaus et al., 1991a, b). We should examine each system in a little more detail.

12.2.1 Diagnostic Expert Systems

Computers that can render diagnoses, including artificial-intelligence-based expert systems, have been with us for decades. Indeed, some are already in common use. At the simplest, "reminder systems" serve as a kind of empirical alarm clock to prompt clinicians to perform tests, check results or look for a changes in medical signs and symptoms. More advanced systems issue reminders based not on a clock but on changing clinical input, so a new blood gas value in an electronic medical record, say, would signal clinicians to inspect ventilator settings and adjust as necessary; a change in blood pressure might call for an inquiry into possible drug interactions (where such interactions themselves can be computationally cataloged and vetted). At their most sophisticated, clinical decision support systems (CDSSs) render diagnoses (of, for instance, heart attacks or in general internal medicine) and suggest therapies and treatment plans with more detail and nuance than simple reminder systems (Duda and Shortliffe, 1983; Goodman and Miller, 2006; Miller and Goodman, 1998; Miller and Geissbuhler, 2007).

For health professionals, these devices raise questions related to the nature of professional practice itself. If education, training, and years of experience are reckoned to be necessary conditions for successful practice, what does it mean when a machine can gather data and render a diagnosis or plot a course of therapy? Indeed, CDSSs have been shown to improve outcomes, reduce cost, minimize errors, warn of early adverse drug events and have other salutary effects (Berner and La Lande, 2007). Put differently, they can do some things better than humans. Moreover, if, as above, better outcomes elicit a corresponding increase in duties to use a device, then what is the proper role of a doctor or a nurse?

We should linger here. Few professions are as old and for the most part self-regulating as that of medicine.[5] For several thousands of years, a student would learn medicine, be fledged as a trainee, and eventually be sent into the world, where it was expected that she would remain scientifically up-to-date as an adult learner through continuing education. The practice of medicine has grown steadily more complex as new devices and drugs have become available. Where once a physician could do no more than comfort a dying patient, we now have interventions, treatments, and regimens for a vast array of maladies.

The past half century—a period more or less coextensive with the evolution of the randomized controlled trial—has also seen extraordinary growth of medical information, and in consequence a disconnection between what any individual knows and what can be known. This is in part the impetus for the growth of evidence-based medicine, a movement born partly from the recognition that there must be a better way to move information from successful research trials to clinical practice (Goodman, 2003b). In many respects, the profession evolved from the laying on of hands to a challenge in information management. If only physicians had perfect recall and could wed the world's collective, research-based knowledge with insights gained from every clinical encounter . . . the tasks of medicine would become, well, computational. If that were the case, however, what would be the role of humans? What would be the point of physicians?

But this is too fast, and too much. If the practice of medicine or nursing were merely about data analysis and inference engines, we would be mistaken if we did not allow machines to assume control of diagnosis and treatment. But,

> what is wrong is that the practice of medicine or nursing is not exclusively and clearly scientific, statistical, or procedural, and hence is not, so far, computationally tractable. This is not to make a hoary appeal to the "art and science" of medicine; it is to say that the science is in many contexts inadequate or inapplicable: many clinical decisions are not exclusively medical—they have social, personal, ethical, psychological, financial, familial, legal, and other components; even art might play a role. While we should be thrilled to behold the machine that will make these decisions correctly—at least pass a medical Turing test—a more sober course is to acknowledge that, for the present at least,

---

[5]More recently, of course, nursing has established itself as another health profession with its own codes, standards, and traditions. Nursing informatics likewise has flourished (see Ball et al., 2000). While "medical informatics" and "nursing informatics" name somewhat different subdisciplines, in fact the tools and issues are quite similar for both. For this reason, "health informatics" is often and correctly regarded as the more inclusive—and hence more accurate—term. In this chapter, use of the term "medical" is often one of convenience, and should be construed as applying to all health professions that use computers.

A similar terminological refinement has established the currency of the term "bioethics" as in many contexts preferable to "medical ethics," which once enjoyed broader use. The problem with "medical ethics" is that it excludes nursing, much research, psychology, social work, and other domains, despite the fact that they share many issues.

Also, in "clinical encounter," "clinical practice," and so on, "clinical" means any setting in which physicians, nurses, and others—"clinicians"—practice.

> human physicians and nurses make the best clinical decisions. This entails ethical obligations ... (Miller and Goodman, 1998, p. 111).

What emerges is the need for a sophisticated and nuanced approach to the use of intelligent machines. Such an approach would recognize that a patient-centered approach to technology would not, would never, lose sight of the fact that serving the interests of patients is and ought be the core value. It might very well be the case that a broader use of computers in medicine and nursing will itself foster the interests of patients. (Indeed, a human-centered stance should underlie the use of computers in any setting or profession.) This point will be strengthened if we examine the use of computer programs to predict if patients will die.

12.2.2 Prognostic Scoring Systems

So much information is available and relied on in the modern critical care unit that it would be foolish, hubristic, and dangerous to grow too sentimental about and defensive of the wizened country doc trudging up a hill with his black bag, or the beatific nurse reaching to smooth a brow or apply an unguent. ICUs are wild places, and electronic devices are needed to monitor, alarm, remind, and sift though vast amounts of information. In fact, the very idea of a critical care unit presumes that in a full-court press against death, you want all the information processing wherewithal you can muster.

With enough information, patterns emerge. Once this happens, simple science has an opportunity to identify variables that shape or alter the patterns. The very idea of a prognosis, or medical prediction regarding the course of a malady, is generally an inductive and probabilistic affair, and when the stakes are high it is especially important to have a sense of how things will turn out. A computer system that could sift physiologic and other data and compare a particular patient to the last hundred or thousand or ten thousand would provide potentially very useful information. One might, as above, be able to compare patients in the unit today to those a year ago and see whether, all things being equal, those today were doing better; or one could compare today's patients at Hospital Alpha with those at Hospital Beta to see, all things being equal, which institution was doing a better job (Knaus et al., 1991a, b). Such computations can be quite useful in measuring quality of care, or its improvement.

They would also be of use in determining, all things being equal, whether Mr. Gamma would survive or not. They could, that is, serve as a kind of "computational futility index," an objective measure (based on perfect recall of all other like cases) of whether Gamma is likely to die come what may, that is, no matter how aggressive the treatment (Goodman, 1996). Now, if in fact a clinician believes that a patient is going to die no matter the treatment or effort, does the clinician still have a duty to try to save the patient? Or would that be a waste of time, effort, and resources? How confident should one be before forgoing such treatment? And, most importantly, is the warrant for terminating treatment weaker, about the same or stronger when it comes from a prognostic scoring system?

Those who have asked these or similar questions (e.g., Brody, 1989; Goodman, 1996, 1998a; Knaus, 1993) have tended to arrive at the same answer: treatment should not be withheld or withdrawn based only on a computational score. Perhaps the strongest reasons offered in support of such reluctance to rely on a computer are these:

- The programs and their predictions are imperfect. Accuracy, while initially high, declines over time, meaning that listening to the machine early would foreclose on any chance of success in the future—a chance that is not negligible.
- The programs are statistical and not knowledge based. That is, when an individual patient is compared to thousands in a reference database, a physician who declines to treat based on such computational data is making a decision about a particular patient based *solely* on information about other patients.
- The question whether aggressive treatment is worthwhile is in large part a value judgment about the relative risks and desirability of failing after extensive suffering, succeeding in lengthening a life of unconsciousness, incurring great cost for a brief life extension, and so on. These are neither calculations or computations.

The problem of clinical futility is large, and it has generated an impressive and diverse literature. What is significant is that the notion of withholding or withdrawing care as a statistical or computational decision has enjoyed so little favor.

12.2.3 "The Standard View" and "Progressive Caution"

Ethics under empirical uncertainty is difficult and unavoidable. One could say that the challenge of computers in medicine is not so much how to refine accuracy, but, rather, how to make appropriate use of intelligent machines given their existing limitations. What is needed is an overarching set of principles to guide clinicians.

It turns out that one of the leaders in research and development of CDSSs has also been a leading expositor of the ethical and legal issues raised by decision support. According to Randolph A. Miller, "The Standard View" of appropriate uses of such systems is that they ought not be allowed to trump or over-ride human judgments—at least not yet: "Limitations in man-machine interfaces, and more importantly, in automated systems' ability to represent the broad variety of concepts relevant to clinical medicine, will prevent 'human-assisted computer diagnosis' from being feasible for decades, if it is at all possible" (Miller, 1990). Because the practice of medicine comprises more tasks than rendering diagnoses—asking questions, making observations, recognizing patient variation, formulating treatment plans appropriate to patient preferences, and so on—we should allow Miller's point to apply more broadly. Put this another way: "The Standard View" is a simple acknowledgment of the fact that clinical practice is about much more than induction, even evidence-based induction.

This point should be expanded. Medical students are customarily taught—and fledged physicians told—to take patient lifestyle, preferences, and values into account in making and communicating diagnoses and rendering treatment plans. While such takings-into-account might, in principle, be performed by intelligent machines, it is not yet clear how or when this might be accomplished. Consider a patient who is dying and unlikely to survive more than, say, a few months. Should doctors and nurses attempt to resuscitate this patient if his heart stops? There is a (low) probability the resuscitation will succeed—he is more likely to die or recover with reduced cognitive capacity than to be restored to his (bleak) condition before the heart attack. A rational patient might prefer or disdain the attempted resuscitation, but the decision whether to attempt it should be guided by an understanding of the risks, what they mean, and the alternatives. Does the patient value life so much that he is willing to endure diminished mental capacity for the rest of that life? Has he led a life shaped by an appreciation of experiences enjoyed by intact cognitive function? Is she seeking resuscitation to please a guilty relative? Is he afraid of dying? A good doctor should—according to well-established standards—be able to understand and even empathize with the patient, including his preferences and values, and provide advice about how best to achieve those goals; or the doctor ought to guide the patient who is seeking interventions that are unlikely to work, where "work" is itself a vague concept. It is just not clear that a CDSS will be able to accomplish this any time soon.

As Miller has it in the paper just cited, "people, not machines, understand patients' problems." A set of three ethical principles have been offered as capturing the benefits while mitigating the risks of CDSSs use:

(1) A computer program should be used in clinical practice only after appropriate evaluation of its efficacy and documentation that it performs its intended task at an acceptable cost in time and money.
(2) Users of most clinical systems should be health professionals who are qualified to address the question at hand on the basis of their licensure, clinical training, and experience. Software systems should be used to augment or supplement, rather than to replace or supplant, such individuals' decision making.
(3) All uses of informatics tools, especially in patient care, should be preceded by adequate training and instruction, which should include review of all available forms of previous product evaluations (Goodman and Miller, 2006, p. 383).

This is ultimately a humanistic and well-founded stance. It is not a sentimental appeal to human primacy but an acknowledgment that the best clinicians are not good because of their ability to calculate but because of their ability to incorporate calculations into the vast orbit of patient care. As the technology evolves there might, in principal, be a need to make adjustments to such a policy. The balance to be struck, recall, is between improvements in statistical quality and outcomes and an erosion of an ancient profession's standards while using a tool the benefits of which might be illusory. So a correlate of "The Standard View" might be a principle that seeks benefits from scientific advances while not permitting enthusiasm for the medium to overshadow its limitations.

A candidate for such a principle might be what has been called "progressive caution," or the idea that promotion of scientific and clinical advancement is a value devoutly to be encouraged—at the same time that steps are taken to ensure that evidence and not enthusiasm carries the day. "Progressive caution" has been glossed thus:

> Medical informatics is, happily, here to stay, but users and society have extensive responsibilities to ensure that we use our tools appropriately. This might cause us to move more deliberately or slowly than some would like. Ethically speaking, that is just too bad. (Goodman, 1998b, p. 9).

Implied tacitly by such a principle is the idea that many questions will be resolved, at least a little, by more research. To the extent that ethical challenges are magnified by uncertainty, there is a moral obligation to reduce uncertainty, especially in the professions. Such an approach has as a virtue the fact that it is probably uncontroversial; at the least, it is very difficult to argue against—not that anyone would, which is perhaps another virtue.

12.3 PERSONAL HEALTH RECORDS

For centuries, physicians have kept notes on paper about patient encounters. There is nothing remarkable in this, save that the patient record, medical record or chart, as it is variously known, is an essential part of medical practice. It is needed both for a clinician to refresh her memory about a patient over time and for clinicians to share information about patients within their care. It is a record of signs and symptoms, tests and diagnoses, pharmacologic history and treatment plans. A comprehensive, accurate and accessible record is required for effective medical care. The problem with paper records is that they are often inaccessible; that is, patients are often nowhere near the single copy of their charts.

In the late 1960s, some hospitals created electronic information systems to collect and route medical orders, give clinicians access to laboratory tests, and identify chargeable services (Tang and McDonald, 2006). These early electronic health records (EHR; sometimes electronic medical records [EMR]) have evolved to the point at which a hospital must either have them or be moving forward briskly to adopt such a system lest it be regarded as lagging. EMRs should integrate patient information ranging from lab results to doctors' orders and notes to X-ray images. They should be easily accessible to all who need timely patient information and inaccessible to voyeurs and others. They are essential to the development of decision support systems, for which they provide the data and information for analysis. Obviously, the issues and concerns addressed so far regarding privacy and decision support apply in bold face to EHRs and their uses and users.

But these records and the systems used to maintain, develop, and share them are generally inaccessible to the people whom the information is about—patients. If we accept one of the core values espoused by the HIPAA privacy law, then we accept the idea that patients should generally control access to their information.[6] It is a small step from there to the idea that patients might benefit from being able to access and manage some of their information in ways not dissimilar to the ways in which they access and manage their banking, credit cards, and other personal affairs online. The challenge is that the medical record is not only a valued repository of clinical information needed for treatment, it is also a resource used in legal settings, for research, and to measure hospital outcomes and performance. This means that patients cannot be permitted to alter, redact, or otherwise modify the official record. Moreover, an EHR established in a hospital will not always include information acquired by community physicians, other hospitals, and so on. (Overcoming these discrepancies is a major goal of health information technology research.) Would it not therefore make sense to create resources for patients to monitor and access; for patients to interact with, for instance by entering data or information that might be useful for clinicians on the other end; and for patients to be able to use generally to communicate with doctors and nurses? Such personal health records (PHR) have captured the imagination of leaders in health care. According to one vision,

> PHRs encompass a wide variety of applications that enable people to collect, view, manage, or share copies of their health information or transactions electronically. Although there are many variants, PHRs are based on the fundamental concept of facilitating an individual's access to and creation of personal health information in a usable computer application that the individual (or a designee) controls. We do not envision PHRs as a substitute for the professional and legal obligation for recordkeeping by health care professionals and entities. However, they do portend a beneficial trend toward greater engagement of consumers in their own health and health care. (Markle Foundation, 2006)

On this view, PHRs would contain information from across the life span, from different clinicians and hospitals, from patients' themselves. Patients and clinicians could connect through various online media to explain, share, advise, and otherwise communicate. The growth of PHRs has been rapid, and a number of large employers have announced they will provide PHRs for employees (McWilliams, 2006); this is motivated by the belief that when patients can control such a record it will reduce costs, in part by reducing errors caused or fostered by failures to provide what is called "continuity of care" as well as by impediments to processes designed to ensure that all clinicians have access to an accurate history and a complete, contemporaneous record.

PHRs raise a suite of interesting ethical challenges, and it will be a measure of the utility of the process limned so far if that process can provide ethically optimized guidance to those embracing this new technology. We can itemize the main issues and challenges.

---

[6]To be sure, HIPAA requires that patients have reasonable access to the contents of their medical records, but this requirement so far has been met not by allowing patients to have a look-see at their records on a hospital monitor, or even by making a digital copy of the record and e-mailing it or providing a CD, but by making a paper copy of printouts from the record and then charging patients a per-page fee. This is partly a relic of the fact that many systems remain hybrids containing both paper and electronic records.

Privacy and confidentiality People are more or less computer savvy. But computers remain occult engines for many ordinary people—precisely those expected to make use of personal health records. In balancing ease of use and access with privacy and security, how much responsibility should be assigned to those designing the systems and how much to patients? If a diabetic patient, for instance, is using a personal digital assistant (PDA) or telephone to record and transmit blood glucose levels, say, then what steps should be taken in anticipation of the fact that some of the PDAs or telephones will be lost? Does it matter—and, if so, how much—if the information on the device concerns a behavioral disorder or HIV status or genetic malady? Will family members have access to the devices? Answers to these and other questions have an empirical component, meaning that additional research is needed to determine, for instance, how people use and interact with the devices and which strategies are most successful at maximizing ease of use and minimizing damage from loss or inappropriate disclosure. The role of informed or valid consent plays a role, too, in that if informed, unpressured, competent patients want to use a device and enjoy its benefits (or explore its potential benefits), then it might be patronizing to suggest they are unable to evaluate the risks.

Decision support PHRs can incorporate various kinds and levels of decision support, ranging from reminders ("take your medicine") to proto-diagnoses ("you might be having a bad drug reaction") to advice ("change your diet" or "call 911"). Here, too, is an empirical challenge shaped by moral obligations to reduce or prevent harm. Ethics under uncertainty requires, as ever, that some sort of uncertainty-reduction strategy be undertaken; while such a strategy might fail, and leave us no better off than beforehand, the duty to attempt the reduction is unmitigated. Coupled with "The Standard View" in the context of a "progressive caution" stance, we should be able to see our way clear to exploring and expanding the use of such tools. PHRs do pose a challenge that decision support systems alone generally do not. If we reckon that CDSSs will and ought generally be used by trained clinicians—who then can serve a kind of filter between computer and patient—PHRs might represent an unfiltered conduit between machine and patient. We are largely ignorant of the kinds of advice, say, that might be offered and the attendant risks and benefits. But a number of analogies are available. Many people regularly use books, the World Wide Web, and other means to acquire medical advice. Some of this is helpful, some useless, and some dangerous. Additional research is needed to narrow the epistemological gap between standards and ethics.

Status of the professions Traditionally, physicians practice medicine and nurses nursing by learning about patients' lives and medical histories, conducting examinations and ordering tests and then making diagnoses and developing treatment plans. Might there come a point at which it is not inaccurate to suggest that an intelligent machine is *practicing* medicine or nursing? Moreover, in the event of an affirmative answer to that question, one might reasonably reply, "So what?"—or at least "so what, if patient care is improved?" Such an exchange parallels those in numerous other disciplines concerning the extent of any computer's appropriate use. The challenge for us is the same one that has been a part of the computer ethics firmament for more than a quarter century: Ought there be limits to the kinds of tasks we are prepared to delegate to computers? (Moor, 1979) If a clinician can use a suite of tools, including an interactive PHR, for instance, to improve care, then, as above, it might be blameworthy not to do so.

Throughout this discussion the ethically optimized approach is a process wherein moral analysis is recursively undertaken to incorporate new data and information. To be sure, it should not be otherwise.

12.4 CONCLUSION

We have several opportunities to catalog practical resources for professionals to make ethically optimized use of the tools of information technology. In the health professions, which evolved from ancient struggles to grasp the complexity of human infirmity into an information-rich culture in which some or many of our tools are smarter than we are, at least in some respects, those resources weave the threads of uncertainty reduction, ethically reasonable principles, and professional standards. We value privacy, but share our secrets with healers, lest they fail. We value accuracy and efficiency, but it should be uncontroversial to hypothesize that some people are prepared, in principle, to delegate to machines that which confounds those healers. And we value control over all of this, while hoping that the tools used to manage our health require sacrifices that are not burdensome.

To meet these challenges, we turn to various forms of inquiry: science and ethics. There is of course no alternative. The very idea that use of a tool, in this case a computational tool, might be required or forbidden depending on facts and factors we are unsure of is exhilarating. Applied ethics is too often regarded as consisting in hand-wringing. In fact, it is among the most important things humans do. At our best, we progress: Science and ethics advance in ways that improve the human condition, generally speaking.

To say "a computer is a tool" is inaccurate by understatement if by "tool" we think of a caveman's adze; and a mischaracterization of tools if by "computer" if we think of a science fiction robot, malevolent, and out of control. Humans use tools to do extraordinary things. For the most part, the human brain does a passable job of identifying which uses are good and which are not.

REFERENCES

Alpert, S.A. (1998). Health care information: access, confidentiality, and good practice. In: Goodman, K.W. (Ed.), *Ethics, Computing, and Medicine: Informatics and the Transformation of Health Care*. Cambridge University Press, Cambridge, pp. 75–101.

Ball, M.J., Hannah, K.J., Newbold, S.K., and Douglas, J.V. (Eds.)(2000). *Nursing Informatics: Where Caring and Technology Meet*. 3rd edition. Springer, New York.

Berner, E.S. and La Lande, T.J. (2007). Overview of clinical decision support systems. In: Berner, E.S. (Ed.), *Clinical Decision Support Systems: Theory and Practice*. 2nd edition. Springer, New York, pp. 3–22.

Brody, B.A. (1989). The ethics of using ICU scoring systems in individual patient management. *Problems in Critical Care,* 3, 662–670.

Clendening, L. (1960) [1942]. (Ed.), *Source Book of Medical History.* Dover, New York.

Donaldson, M.S. and Lohr, K.N. (Eds.) (1994). *Health Data in the Information Age: Use, Disclosure, and Privacy.* National Academy Press (Institute of Medicine), Washington, D.C.

Duda, R.O. and Shortliffe, E.H. (1983). Expert systems research. *Science,* 220, 261–268.

General Accounting Office (GAO) (2007). *Health Information Technology: Early Efforts Initiated But Comprehensive Privacy Approach Needed for National Strategy.* GAO-07-238. U.S. Government Accountability Office, Washington D.C.

Gellman, R.M. (1984). Prescribing privacy: the uncertain role of the physician in the protection of patient privacy. *North Carolina Law Review,* 62, 255–294.

Goodman, K.W. (1996). Critical care computing: outcomes, confidentiality and appropriate use. *Critical Care Clinics,* 12, 109–122.

Goodman, K.W. (1998a). Outcomes, futility, and health policy research. In: Goodman, K.W. (Ed.), *Ethics, Computing and Medicine: Informatics and the Transformation of Health Care*. Cambridge University Press, New York, pp. 116–138.

Goodman, K.W. (1998b). Bioethics and health informatics: an introduction. In: Goodman, K. W. (Ed.), *Ethics, Computing and Medicine: Informatics and the Transformation of Health Care*. Cambridge University Press, New York, pp. 1–31.

Goodman, K.W. (2003a). Ethics, information technology and public health: duties and challenges in computational epidemiology. In: O'Carroll, P.W., Yasnoff, W.A., Ward, M.E., Ripp, L.H., and Martin, E.L. (Eds.), *Public Health Informatics and Information Systems*. Springer-Verlag, New York, pp. 251–266.

Goodman, K.W. (2003b). *Ethics and Evidence-Based Medicine: Fallibility and Responsibility in Clinical Science*. Cambridge University Press, New York.

Goodman, K.W. (2007). Ethical and legal issues in use of decision support. In: Berner, E.S. (Ed.), *Clinical Decision Support Systems: Theory and Practice*. 2nd edition. Springer, New York, pp. 126–139.

Goodman, K.W. and Miller, R. (2006). Ethics and health informatics: Users, standards, and outcomes. In: Shortliffe, E.H. and Cimino, J.J. (Eds.), *Biomedical Informatics: Computer Applications in Health Care and Biomedicine*. 3rd edition. Springer, New York, pp. 379–402.

Gostin, L.O., Turek-Brezina, J., Powers, M., Kozloff, R., Faden, R., and Steinauer, D. (1993). Privacy and security of personal information in a new health care system. *Journal of the American Medical Association,* 270, 2487–2493.

Knaus, W. (1993). Ethical implications of risk stratification in the acute care setting. *Cambridge Quarterly of Healthcare Ethics,* 2, 193–196.

Knaus, W.A., Wagner, D.P., and Lynn, J. (1991a). Short-term mortality predictions for critically ill hospitalized adults: science and ethics. *Science,* 254, 389–394.

Knaus, W.A., Wagner, D.P., Draper, E.A., Zimmerman, J.E., Bergner, M., Bastos, P.G., Sirio, C. A., Murphy, D.J., Lotring, T., Damiano, A., and Harrell, F.E. (1991b). The APACHE III prognostic system: risk prediction of hospital mortality for critically ill hospitalized adults. *Chest,* 100, 1619–1936.

Markle Foundation. (2006). *Connecting Americans to Their Healthcare: A Common Framework for Networked Personal Health Information*. New York: The Markle Foundation, available athttp://www.connectingforhealth.org/

McWilliams, G. (2006). Big employers plan electronic health records. *The Wall Street Journal,* 1B. [The headline uses the wrong term, referring not to personal health records but to the already established electronic health (or medical) records.]

Miller, R.A. (1990). Why the standard view is standard: people, not machines, understand patients' problems. *Journal of Medicine and Philosophy,* 15, 581–591.

Miller, R.A. and Geissbuhler, A. (2007). Diagnostic decision support systems. In: Berner, E.S. (Ed.), *Clinical Decision Support Systems: Theory and Practice.* 2nd edition. Springer, New York, pp. 99–125.

Miller, R.A. and Goodman, K.W. (1998). Ethical challenges in the use of decision-support software in clinical practice. In: Goodman, K.W. (Ed.), *Ethics, Computing and Medicine: Informatics and the Transformation of Health Care.* Cambridge University Press, New York, pp. 102–115.

Miller, R.A., Schaffner, K.F., and Meisel, A. (1985). Ethical and legal issues related to the use of computer programs in clinical medicine. *Annals of Internal Medicine,* 102, 529–536.

Moor, J.H. (1979). Are there decisions computers should never make? *Nature and System,* 1, 217–229.

National, Committee on Vital and Health Statistics. (2006). *Privacy and Confidentiality in the Nationwide Health Information Network.* National Committee on Vital and Health Statistics, Washington D.C. Available athttp://www.ncvhs.hhs.gov/060622lt.htm

National Research Council. (1997). *For the Record: Protecting Electronic Health Information.* National Academy Press, Washington D.C.

Office of Technology Assessment (OTA). (1993a). *Report Brief: Protecting Privacy in Computerized Medical Information.* U.S. Government Printing Office, Washington D.C.

Office of Technology Assessment (OTA). (1993b). *Protecting Privacy in Computerized Medical Information.* U.S. Government Printing Office, Washington D.C.

Szczepaniak, M.C., Goodman, K.W., Wagner, M.W., Hutman, J., and Daswani, S. (2006). Advancing organizational integration: negotiation, data use agreements, law, and ethics. In: Wagner, M.W., Moore, A.W., and Aryel, R.M. (Eds.), *Handbook of Biosurveillance.* Academic Press, Boston, pp. 465–480.

Tang, P.C. and McDonald, C.J. (2006). Electronic health record systems. In: Shortliffe, E.H. and Cimino, J.J. (Eds.), *Biomedical Informatics: Computer Applications in Health Care and Biomedicine.* 3rd edition. Springer, New York, pp. 447–475.

Walters, L. (1982). Ethical aspects of medical confidentiality. In: Beauchamp, T.L. and Walters, L. (Eds.), *Contemporary Issues in Bioethics.* 2nd edition, pp. 198–203. Wadsworth, Belmont, Calif. First published in *Journal of Clinical Computing*, 4(1974) 9–20.

CHAPTER 13

Ethical Issues of Information and Business

BERND CARSTEN STAHL

Businesses and the economic system they work in have an important influence on ethical issues arising from information and information and communication technology. This chapter aims at establishing a link between several sets of ethical discourses that concern similar topics. It offers an introduction to some of the current debates in business ethics and considers how information and technology influence the current topics and debates in the area. Drawing on some of the debates in computer and information ethics, the chapter points out areas where these two sets of discourses overlap and where they have the potential to inform each other. The chapter will do so by looking at some prominent examples of issues that arise in business and computer ethics, including privacy and employee surveillance and intellectual property, as well as some macrolevel issues including globalization and digital divides.

13.1 INTRODUCTION

Western industrialized societies are said to be changing into "information societies." The majority of work in these information societies is done in the services sector. Most employees nowadays require large amounts of knowledge and are even called "knowledge workers." Information is becoming increasingly important in most aspects of our lives, and this is particularly true for our economic activities. Recent developments in the way we work and exchange goods and services are highly dependent on information. There are whole industries, ranging from software production and entertainment to education and knowledge brokers, that deal mostly or exclusively in information. But even traditional work such as in agriculture or industrial production gains added value through information. Briefly, modern

The Handbook of Information and Computer Ethics, Edited by Kenneth Einar Himma and Herman T. Tavani

economies require large amounts of information to run, and at the same time they create information in previously unknown quantities.

Business is a central aspect of our lives and as such produces many ethical problems. Information influences and affects many of these problems and creates more in its own right. How are we to address these problems? Many of the problems are hidden within particular business contexts or means of dealing with information. When we talk of information we often refer to information that is made available or accessible via specific technologies, typically summarized as "information and communication technologies" (ICTs). These technologies are central to the way we interact with information and also to the way we organize business. To address the ethics of business and information, we thus need to keep in mind questions of technology. This implies that to discuss these issues we need to consider other disciplines such as computer sciences, software engineering, information systems, and their subspecialties.

Finally, there are scholars who have an interest in some of the combinations of the above issues, such as the relationship of business and ethics or the link between computers and ethics. To comprehensively cover the topic, we would thus have to consider a range of discourses that partly touch on similar issues, but rarely take each other into consideration. The challenge of giving an adequate account of the ethics of business and information is thus considerable.

The chapter will start with a brief definition of the concept of business. This will lead to an introduction of some of the more pertinent approaches to business ethics. The chapter will discuss in some detail the issues of privacy/employee surveillance, intellectual property, globalization, and digital divides. The conclusion will then ask the question of what contribution to the solution of current issues can be expected from ethics, in particular business ethics and computer ethics.

It is important to underline at this early stage that this chapter cannot hope to do the large number of issues and problems justice. Business and the economic constitution of society are at the heart of many ethical problems, and, similarly, information raises new ethical questions. By concentrating on some paradigmatic issues, I attempt to discuss ethical views and possible solutions as well as their shortcomings. The downside of this approach is that I will simply ignore a large number of other issues that are arguably as deserving of attention as the ones that I discuss here. Such neglected issues include the collection, analysis, and sale of customer data, which are integral parts of many businesses and have even spawned new industries and technologies, such as customer relationship management (CRM) activities. I will not touch on the question of quality and reliability of software and hardware, which raise some conceptually new problems in their own right. Furthermore, I will not go into any depth on issues of security and computer misuse and crime, including hacking and counterhacking. I will also assume that economic actors generally move within the confines of the law and ignore legal questions, except where they are pertinent to my topics, including privacy and intellectual property (IP) law. The attempt to unfold a discussion of ethical issues in business and information by taking a somewhat different route than seems to be prevalent in computer and information ethics will hopefully compensate the reader for the fact that not all questions are covered.

13.2 APPROACHES TO ETHICAL ISSUES IN BUSINESS AND INFORMATION

To contextualize ethical issues in business and information, this section will outline some approaches to business ethics. This will require a short introduction of several pertinent aspects of what we mean by business in the first place. The introduction of business ethics will then lead to a comparison of business and computer ethics.

13.2.1 The Concept of Business

Economic activity is a part of every society, and it is arguably one of the most important aspects of current liberal democratic states. Businesses have a large influence on how we live our individual lives and also on how society is regulated. Businesses are social facts, but they are also the objects of theoretical and academic attention. The only introductory remark about business that seems indispensable at this stage of the argument refers to two possible levels of observation of business that will inform the subsequent debate on ethics. The two levels of observation of business and economic activity are the micro- and macrolevels. These are reflected by the distinction between the academic disciplines of economics and business studies. The *foci* of attention of the two levels are different, which is reflected by different methodologies and vocabularies. To address the ethical issues arising from the intersection of business and information, we nevertheless need to consider both levels. Manifest ethical problems, for example, caused by employee surveillance or digital rights management, often occur on the microlevel of the individual business or industry. They cannot be completely divorced, however, from the macrolevel of national and global institutions, which, in turn, are linked to prevalent understanding and theories of economics.

13.2.2 Business Ethics

There is much research and literature in the discipline of business ethics, which deals with the relationship between business and ethics. A possible view might be that business and ethics simply have nothing to do with each other, that the term business ethics is an oxymoron. Immoral behavior of individual market participants, such as high-profile managers or corporations (cf. Enron, WorldCom, etc.), sometimes seems to support this view. However, such a view is not tenable because it overlooks that there are numerous connections between ethics and business and that the two refer to each other in several respects.

Moral norms are important for the functioning of an economic system. If people did not honor contracts, pay their dues, give accurate information about products, and generally follow the moral code of society, economic transactions would become difficult to sustain (De George, 1999; Donaldson and Dunfee, 1999; Hausman and McPherson, 1996; Schwartz and Gibb, 1999; Sen, 1987). At the same time, ethics as the theory of morality plays an important role in justifying the economic system and thus allowing economic agents to feel legitimated in acting within the system. One justification of our current economic system is the utilitarian consideration that free trade creates the goods that allow individuals to satisfy their preferences and live a good life according to their own design (Gauthier, 1986; Goodpaster and Matthews, 1982). Other streams of justification of a market-oriented constitution of society would be the natural rights tradition that can be used to ground a right in personal property (Nozick, 1974). Where personal property is accepted, market mechanisms can easily gain a measure of legitimacy. Markets and free exchange of property can also be justified from a perspective of justice and fairness (Rawls, 2001). Whatever the argument, it is important to note that ethical justification has been a continuous aim of economic theory from Aristotle onward (cf. Keynes, 1994).

The academic discipline of business ethics is now well established in most business schools and recognized as an important part of business studies and research. As in most disciplines, there are a variety of discourses and competing approaches. To a certain degree these reflect the differences in focus outlined above, the difference between the macrolevel of economic activity in society and the microlevel of the corporation. On the macrolevel, business ethicists consider the question of how an economic system can be justified. Since the times of Aristotle, it has been recognized that economic activity is an important part of the "good life." To be able to participate in society and contribute to social interaction, the individual needs material sustenance. Society as a whole requires resources if it is to do the things that are often associated with ethical activity, such as supporting the needy and helping those who cannot help themselves.

Beyond such very general considerations of the ethical foundations of economic activity, there are also more specific issues debated on the macrolevel. Among these we find questions of justice and distribution within and between societies. How we conceptualize justice in a modern society is an important issue often linked to debates surrounding development (cf. Sen, 1987), globalization, and digital divides that will be discussed in more detail below.

While such macrolevel issues are thus of relevance to business ethics and constitute an important part of the theoretical development of the field, many observers view the microlevel analysis of activities of individuals and organizations as the heart of business ethics. The microlevel analysis typically takes the economic framework as given and justified, and considers the question of how an agent is to act morally within this framework and how such moral acts are to be justified. Much of this debate aims at finding useful applications of existing ethical theories to the world of business. Some authors try to compare the most widely discussed ethical theories within business, such as deontology, teleology, virtue, ethics, or care (cf. Velasquez, 1998), whereas others concentrate on a subset and explore in depth individual ethical theories (Bowie, 1999).

Apart from attempts to apply well-established ethical theories to the field of business and economics, some scholars have developed specific theories in business ethics. These include the stakeholder approach, the idea of shared norms or values, and corporate social responsibility. One should also see, however, that there are many further possible approaches and that there is a wealth of ethical traditions and ways of doing business, which are not covered in this debate that is mostly informed by mainstream Western discourses.

13.2.2.1 Shareholders and Stakeholders A widely used approach to business ethics is the stakeholder approach. It is an explicit rejection of the shareholder view of the firm, which holds that decisions of a company, which usually means decisions of management, must concentrate on maximizing the value to the shareholder.[1] This understanding of companies is rather limited and, given the large effects some corporations have on the lives of many people, can be considered too narrow (Koslowski, 2000). Proponents of the stakeholder view of the firm contend that corporations are complex social systems that serve a variety of (sometimes competing) purposes. The shareholder view and its implicit one-dimensional imperative of profit maximization can then be criticized as empirically too narrow, but also ethically insufficient (Gibson, 2000; Hendry, 2001). To overcome the shareholder view, the concept of stakeholder was coined and first used in an internal memorandum of the Stanford Research Institute in 1963 (Kujala, 2001). Stakeholders were originally defined as those groups on whose support a company depends. The dependencies between stakeholders and companies tend to be mutual but not necessarily equal. For example, a company needs employees just as an employee needs employment, which does not imply equality in their relationship. Generally accepted groups of stakeholders include shareowners, employees, customers, suppliers, financial service providers, and society. The definition of stakeholders has broadened over time, and there is no agreement on where or how exactly a line is to be drawn.

The central idea of the stakeholder conception of the company is that the legitimate interests of stakeholders need to be considered when decisions are made. If taken seriously, this has radical consequences for the way companies work. In the market economic systems that we are used to in the contemporary Western world, managers are usually seen as the legitimate representatives of the main interests of the firm (shareholders) and they are free to make decisions they deem appropriate. If those decisions turn out to be wrong, then the shareholders will punish their agents (the managers) by making them redundant. The stakeholder concept of the firm challenges this view and demands from managers that they take a multitude of interests seriously. This poses important epistemological and practical challenges. Even if managers could use an unambiguous definition of stakeholders, they would still have to make sure that they understand the stakeholders' concerns. This would require extreme efforts on the side of the managers, and it might lead to conflicting stakeholder interests that managers would have to balance. Thought through to the end, a stakeholder view of the firm would require a fundamentally different concept of management than the hierarchical, power-oriented one that we currently take for granted (cf. Donaldson and Preston, 1995). The hierarchical model of management cannot work in a stakeholder view for a variety of reasons. The epistemological problem of knowing the stakeholders and their views and legitimate interests precludes a legal solution to the problem where the law would simply require managers to discharge their obligations to

[1]One needs to be aware that this is not the only interpretation of the shareholder concept. An alternative interpretation is that it is meant to strengthen the interest of the shareholders against the power of management. This can be ethically motivated, and the term "shareholder value" then stands for a defense against managerial excesses. In public debate this understanding is not as present as the one outlined above.

stakeholders. Such a law would be too broad, incomplete, or impossible to enforce. More important, the stakeholder view implies a high degree of equality in the relationship between stakeholders and organization. Such a level of equality is difficult to reconcile with a relationship in which one partner retains formal power over decisions and intended outcomes, as is typically the case for management.

It is not surprising that much criticism has been directed at the stakeholder view of the firm. There is conservative resistance from managers who have become accustomed to only taking those into account who have financial value for the firm (Schwarz and Gibb, 1999). However, there are also good economic and philosophical arguments against it. One of these is that the stakeholder view takes away the advantage of the shareholder model, which is simplicity of purpose and reduction of complexity. By concentrating on the profit motive, managers are free to focus their energy and creativity on an achievable goal. A realization of the stakeholder model would politicize the role of managers and thereby lead to slower change and less innovation (Weizsäcker, 1999). It is possible that the practical outcomes of the stakeholder model are less desirable than those of the shareholder model and that it leads to less efficiency (Hank, 2000). At least from a utilitarian point of view that emphasizes the positive contribution of production and by implication of innovation and change, the possible decrease in efficiency resulting from the stakeholder view can count as an ethical counterargument. Finally, it has been argued that the stakeholder model can serve to "sugarcoat" strategic thinking and serve as an excuse when companies are in fact only interested in the bottom line (Gibson, 2000).

13.2.2.2 Shared Norms and Values The stakeholder approach to business ethics says nothing about the norms that stakeholders should follow. These may be implied in the democratic and egalitarian view it offers, but it leaves open the outcomes of stakeholder negotiations. Another set of approaches tries to overcome this perceived weakness by establishing moral norms that the market participants agree on. Much of this thinking can be linked back to Max Weber's (Weber, 1996) observations about the Protestant ethic and the "spirit" of capitalism. On the basis of the observation that wealth was not distributed evenly between different religious groups in several countries in the nineteenth century, Weber argued that moral norms based on religious teachings have an influence on economic success. Specifically, he argued that the Christian Protestant combination of asceticism and a high valuation of work were the historical conditions that allowed Protestants to be successful in market exchange. Hard labor alone does not necessarily lead to the formation of the capital stock, which is important as a starting point of capitalist enterprises. It must be accompanied by the willingness to save the rewards of the labor and renounce immediate gratification by consumption. Such norms can survive the demise of religious faith and take on a meaning of their own.

On the basis of this recognition of the existence of shared and successful norms, some scholars have attempted to identify such norms and use them to ethically justify certain types of economic behaviors. The most salient current example of this are probably the "hypernorms" posited by Donaldson and Dunfee (1999). This idea is embedded in a theory that Donaldson and Dunfee call the "Integrative Social Contracts

Theory" (ISCT). As the name suggests, this theory builds on the contractualist tradition from Hobbes and Rousseau to Rawls. It aims to be descriptively accurate but also normative. Hypernorms are such norms that are shared by all humans and that would become enshrined in the social contract that rational humans would choose in the original position. They are also discoverable in current social practices. They are higher-level norms that can be used to justify lower-level norms. An example of such a hypernorm is efficiency.

Donaldson and Dunfee (1999) are probably the most prominent proponents of such a theory of shared norms, but they are by no means the only ones. De George (1999) identifies shared moral norms that are relevant in markets, as does Stark (1993). The idea for Stark is that business ethics fails to communicate with managers because it uses a language and concepts that are alien to them and have no relevance to their everyday activity. At the same time he posits that there are shared moral norms that business ethics fails to recognize. Such shared moral norms can be extracted from theological teachings, and it is thus not surprising that business ethicists with a theological background promote shared values as a preferred approach to business ethics (cf. Küng, 1997).

13.2.2.3 Corporate Social Responsibility A further approach to ethical issues in business is that of corporate social responsibility (CSR). This approach, which is quite popular in the Anglo-American world, attempts to find an answer to the question of under which conditions the behavior of a corporation as a whole would be considered ethically acceptable. It is based on the view that corporations are social agents and as such they "must assume the responsibility for the effects of their actions [. . .]" (Collier and Wanderley, 2005, p. 169). It overcomes some of the epistemological and practical problems of the stakeholder view (Gonzalez, 2002) and aims to produce practicable solutions rather than philosophical considerations. At the same time, CSR is opposed to the shareholder or stockholder view of the firm, which clearly implies that a company has no social responsibility beyond maximizing its profits within the legal framework that it finds itself in (Hasnas, 1998).

Historically, CSR is closely linked to the attempt to find solutions to the pressing problems of our times, in particular environmental issues. It came to prominence in conjunction with the 1992 Earth Summit in Rio de Janeiro (Wilenius, 2005). Because large corporations have enormous financial resources and are among the biggest users of natural resources, it is plausible that they need to contribute to the solution of the problem of pollution. The CSR approach attempts to facilitate this by extending the democratic principle of participation to corporations and asking what their duties and responsibilities should be. It is based on the recognition that companies benefit from society and thus need to reciprocate. CSR is thus closely linked to the idea of corporate citizenship, which stresses the political side of the corporation's position in society (Husted and Allen, 2000).

CSR is a popular approach, maybe because it does not challenge the existing social and economic order but nevertheless allows the incorporation of ethical issues in corporate decisions. It raises, however, some problems. There is the fundamental philosophical question of whether corporations as collective constructs can be understood as agents and whether they are suitable as possible subjects of responsibility (French, 1979; Werhane, 1985). Much of the criticism of CSR focuses on the problem of instrumental use of the idea of social responsibility. CSR is often described as an integral part of corporate strategy. If this is so, and if the main purpose of the corporation is to create profits, then ethics becomes a tool of profit generation, an idea that many ethicists are uncomfortable with (Husted and Allen, 2000). Moreover, one can view CSR as a strategy that allows companies to appear as ethical entities, which may then have the result of preventing state or international regulation. Or, as Doane (2005) puts it: "CSR has proved itself to be often little more than a public-relations offensive to support business-as-usual."

13.2.3 Business Ethics and Computer Ethics

The above attempt to outline some of the dominant debates of business ethics cannot do such a large field justice. One of its aims, next to introducing some of the concepts and arguments of business ethics, was to provide a platform that will allow us to compare current approaches in business ethics and computer and information ethics.

All of the above theoretical approaches to business ethics find their corresponding view in computer and information ethics. First, there are those scholars who take individual ethical theories, such as utilitarianism, Kantian deontology, ethics of care, or virtue ethics (Grodzinsky, 2001), and apply them to problems raised by information or ICT. Some scholars doubt that abstract ethical theories are useful starting points to discuss ethical problems in ICT and suggest reliance on shared common morality (Gert, 1999, p. 59). This raises some of the problems of hypernorms discussed earlier, such as how we determine which norms are shared and how we decide whether a shared norm is acceptable.

Then there is the stakeholder approach to computer and information ethics. Rogerson (2004), for example, suggests that a stakeholder approach is suitable to cater to the ethical views of the potentially large number of affected parties. This approach is widely accepted by scholars seeking to include ethical considerations in information systems design and use (Heng and de Moor, 2003; Walsham, 1996). Building on such thoughts, Gotterbarn and Rogerson (2005) developed the Software Development Impact Statement (SoDIS), which is a method aimed at providing a structure for an ethically oriented stakeholder analysis.

Just as there are equivalents to the stakeholder approach between business and computer ethics, the same is true for the shared moral norms approach. The widespread use of codes of ethics or codes of conduct can be seen as an indication that there are shared norms that can help ICT professionals to address ethical problems (Anderson et al., 1993; Laudon, 1995; Oz, 1992). The most prominent proponent of a shared values and norms approach is Jim Moor. Moor (1985, 2000, 2001) argues that ICT leads to the development of policy vacua caused by the new properties of such technology. Ethics can be used to address such vacua, and the way to do this in an acceptable fashion is to consider whether and how universally accepted "core values" are affected.

Finally, there are also parallels between corporate social responsibility and computer ethics. Skovira (2003), for example, develops the argument that CSR can be interpreted as a consequence of a contractual understanding of society, and he continues to argue that ICT is changing the nature of this contract and that social responsibility thus needs to explicitly encompass ICT. ICT can effect changes in the nature of interaction in a society and between society and business by increasing transparency of exchanges. Transparency is arguably a moral value where it increases equality of negotiating positions. If one follows this argument, the potential moral impact on the constitution of society warrants explicit duties of ICT specialists such as clarifying the costs and benefits of the technology or implementing it in ways that support accountability.

This section has given an overview of some of the approaches to ethical issues of relevance in business and economic matters. By introducing some of the general topics as well as current streams of business ethics, I have shown that there is general, albeit by no means universal, agreement that ethics and business are related. However, there is much less agreement on how these ethical issues can be addressed. The different views of business ethics have been shown to have corresponding views in computer and information ethics. Based on these theoretical foundations, the chapter can now come to some of the salient ethical issues in business that are related to or caused by information or information technology. This will be done in two sections, one concentrating on individual or microlevel issues, the second targeted at social or macrolevel questions. When discussing some of the most pertinent issues that arise in these areas, I will return to the ethical theories and approaches just outlined and explore how they can contribute to ethically acceptable solutions.

13.3 MICROLEVEL INFLUENCE OF BUSINESS ON ETHICS AND INFORMATION

In this section I will discuss how businesses contribute to the creation and exacerbation of ethical issues arising from information and ICT. The area of interest consists of individual and corporate questions. As examples I will use two of the central ethical issues related to information, namely, privacy and employee surveillance and intellectual property.

13.3.1 The Business Value of Information

Before I can discuss why business has a strong influence on ethical issues of information, I need to lay the groundwork and explain why businesses have an interest in information. The aim of business organizations according to standard economic theory is the maximization of profits. Such organizations will, therefore, aim to minimize cost and maximize revenue. Information has a value for businesses if it can contribute to either of these aims. The business value of information is thus linked to financial gains it can achieve. This is independent of the philosophical debate in information ethics of whether information has an intrinsic value (cf. Himma, 2004).

The answer to the question of the value of information would seem straightforward at first sight. Businesses need to measure the cost of information and subtract this from the added value, and this should give them the required value. Things are not so easy, however. There are some problems surrounding the definition of information in the first place. Putting a money value on it is even more complex. As a result of this, corporations find it very difficult to measure the value of their information technology. Although costs of information technology are sometimes easily measured, the benefits typically are not (Smithson and Hirschheim, 1998; Sriram and Krishnan, 2003; Torkzadeh and Dhillon, 2002).

Despite such distracting voices, most economists and business managers seem to agree that information does have a value for the company, even if it is difficult to quantify. The reaction to this state of affairs is often to collect as much information as possible in the hope that it will be useful. The information collected typically consists of data about business processes, customers, suppliers, or employees. This approach is facilitated by the continuously falling cost of IT and storage facilities. At the same time it has an enormous impact on ethical questions concerning information.

13.3.2 The Impact of Business on Privacy: Employee Surveillance

It is beyond doubt that the question of privacy is a central issue of computer and information ethics. It is less clear what privacy is. The literature has come up with a range of definitions and angles since Warren and Brandeis's (1890) seminal definition as the "right to be let alone." I do not need to discuss the concept and problems of privacy in much depth here because this has been done by other authors in this volume. It will nevertheless be useful to recapture some of the arguments surrounding privacy because this can help us understand the impact of business on ethical issues of privacy. The emphasis here is on informational privacy, which Brey (2001) contrasts with relational privacy. For Brey, relational privacy refers to the freedom from observation and interference, whereas informational privacy concerns control over one's personal information. Floridi (1999) distinguishes between physical, decisional, mental, and informational privacy. As Adam (2005) points out, the different types of privacy are related. If informational privacy is abridged, this can lead to a decrease in other types of privacy as well.

Ethical issues enter the debate when the justifications of a possible right to be left alone (Britz, 1999; Velasquez, 1998) are discussed. In principle, these can be divided into two streams of debate: one that is concerned with data about customers, the other which deals with privacy of employees. Companies usually have more power over their employees than their customers. The arguments in defense of employee privacy are therefore based on stronger ethical concerns. Within the debate about employee privacy, one can distinguish three groups of reasons for its support. They deal primarily with the individual person, with society, and with economic considerations.

Attacks on employee surveillance as the main threat to employee privacy are often strongly grounded in ethics (Weckert, 2005). Violating individual privacy is an ethical problem because it interferes with the development and maintenance of a healthy personality and identity (Brown, 2000; Nye, 2002; Severson, 1997). This is closely linked with personal autonomy (Spinello, 2000), the basis of ethics in the Kantian deontological tradition. A lack of respect for privacy can be interpreted as a lack of respect toward the individual whose privacy is invaded (Elgesiem, 1996; van den Hoeven, 2001). It can hurt the development of trust and security and has the potential to damage the individual's ability to engage in meaningful relationships with others (Introna, 2000). Apart from the effect that a lack of privacy can have on the individual, it can create aggregated problems on the level of society. Democracies require an autonomous and open individual who is willing to engage with others. A lack of privacy can mitigate against the development of these individuals as well as their willingness to engage with others. And finally, there are even economic arguments against limiting employee privacy. These emphasize that missing privacy can hurt labor relations (Bowie, 1999; Weisband and Reinig, 1995) and lead to international legal problems (Culnan, 1993; Langford, 1999; Tavani, 2000). Although most of the above arguments aim at employee surveillance, they can be extended to cover other privacy issues.

One should note that there are a variety of reasons for employers to use surveillance mechanisms on their employees. Private organizations usually give economic reasons for employee surveillance. It is often said that companies lose huge amounts of money because of non-work-related use of company resources (Boncella, 2001; Siau et al., 2002). This seems to be such an important problem that scholars have seen a need to come up with terms such as "cyberslacking" (Block, 2001), "cyberslouching" (Urbaczewski and Jessup, 2002), or "cyberloafing" (Tapia, 2004). The use of surveillance technologies is supposed to limit such personal use of technology and thereby increase worker productivity and company profits.

A related problem is that of legal liability for employee behavior. Companies fear that their staff may abuse their systems and that the company may be held liable for this. Possible problems range from harassment (Spinello, 2000) and negligence in hiring, retention, supervision (Brown, 2000; Panko and Beh, 2002) to cyberstalking and child pornography (Adam, 2005; Catudal, 2001). The solution to all this seems to be to install some sort of technology that will allow managers to know what exactly employees are doing, briefly, surveillance.

There are thus strong economic arguments for employee surveillance. This leaves open the question of the limits of employee surveillance. It seems generally agreed in most Western societies that employers have a legitimate interest in some employee activities that can be subject to surveillance, such as their activity when on the job. At the same time there is agreement that there are limits to employee surveillance, and few would support the installation of surveillance cameras on company toilets. It is not always clear, however, where exactly the limits of legitimate surveillance are. Furthermore, one needs to be aware that such debates always require the background legitimacy of the overall economic system. We generally do not question the right of employers to direct employees, which is usually supported by the argument that employment contracts are entered into on a voluntary basis and that employees consent to employers' positions of power. The underlying assumption of equality of position and bargaining ability between employer and employee can nevertheless be doubted.

What is the relevance of business in this context? Traditionally, people interested in the protection of privacy concentrated on the state as the main threat. This can probably be explained by the historical examples of fascism and communist dictatorships, which used personal information intensively for political purposes. During the 1990s, with its increasing availability and affordability of data collection and processing technology, the threat changed from state to private businesses. Businesses collected a large amount of data on customers as well as employees and competitors, because it was possible and promised financial returns. At some point privacy advocates concentrated on commercial organizations. In conjunction with the 2001 terrorist attacks in the United States and subsequent spread of Islamist terrorism to most other parts of the world, the pendulum swung back toward the state. However, in the wake of the 2001 attacks, states recognized that much information was available from commercial entities and frequently attempted to access this information for security purposes. The main threat to privacy thus seems to be coming from the combination of commercially collected data used by and for purposes of the state (Lessig, 1999, 2001). This means that data collection on employees as well as on customers, which are originally motivated by economic concerns, can no longer be viewed from a purely commercial perspective. The "greased" data (Moor, 2000) that companies produce for their financial aims cannot be confined to these purposes and may lead to privacy problems elsewhere, for example, when consumption patterns get associated with terrorist activities, thus leading to the possible apprehension of terrorists or possibly to unwarranted suspicion of innocent consumers and citizens.

13.3.2.1 The Ethical Response to Employee Surveillance How can the ethical theories outlined earlier help us address the question of privacy and employee surveillance? The first theory is the stakeholder approach. Questions of the ethically justifiable application of ICT for surveillance should in theory be susceptible to the stakeholder approach. Because the issues are likely to arise as employees will object to surveillance, managers can easily identify the stakeholders in question and will find it easy to gather their views. Employees, in turn, can easily identify the problem and know who is in charge or in a position to effect changes. The problem with the stakeholder approach in a company is that there are few incentives for managers to expose themselves to the tedious process of stakeholder participation. A first problem here is the identification of relevant stakeholders and stakeholder groups. If employee surveillance is at issue, then employees themselves are the primary stakeholders. However, there may be others such as trade unions, civil liberty groups, and industry representatives. Indeed, the choice to use ICT for surveillance can be interpreted as an expression of distrust toward employees, which will render it unlikely that managers will be open to incorporating employees' views in their decisions. Financial considerations can thus be one issue to be discussed and may be used as a counterargument against surveillance. There are, however, a multitude of other possible arguments. Some of these will require empirical knowledge, such as whether surveillance hurts employee morale and retention; others will be of a more general and conceptual nature, such as the reach of employers' legitimate interests.

The next approach is that of shared values or norms. Again, privacy seems to be a good contender for a core value or a hypernorm. Privacy is valued in most societies, certainly in all Western industrialized countries. The problem is not a lack of recognition of privacy but a lack of agreement on the limits to which it should be protected.

Whether or not an employee has a right to be unobserved during work time is open to debate. At least in the Anglo-American tradition, such a right to employee privacy tends to be viewed as inferior to the employer's interest in controlling the behavior of employees to ensure compliance with regulations and contractual obligations. The problem of the shared norms approach is thus that it is not clear how they are to be defined or how contentious interpretations can be debated to the point of agreement.

The final example, the corporate social responsibility approach, has little to say on this problem. CSR may be helpful if there are relatively unambiguous issues, such as the contribution to society at large or the preservation of the environment in a general sense, but it becomes difficult to apply if the issue is controversial. Unless a company has chosen to assume responsibility for employee freedom and emancipation, there seem to be few reasons for it to link its CSR stance to a particular view on surveillance. CSR might lead to a variety of corporate views on surveillance, and it is not clear a priori which way a particular company would go.

13.3.3 The Impact of Business on Intellectual Property

Intellectual property (IP), another big issue in information ethics, is also closely linked to business interests. Again, the topic will be dealt with in much more depth elsewhere in this volume. Very briefly, one can distinguish two narratives of justifying the "bundle of rights" (De George, 1999, p. 583) that constitutes intellectual property: the utilitarian and the natural rights approaches. The utilitarian approach emphasizes the overall increase of utility because of the incentives for creators that the protection of intellectual property promises. The natural rights justification of IP distinguishes between Heglian and Lockean approaches (Warwick, 2001), which argue that IP arises directly out of the act of intellectual creation. Intellectual property rights, like most rights, are not absolute but limited by competing rights. In many instances, the statutory protection of IP allows for exceptions, as for example the fair use (U.S.) or fair dealings (UK) exceptions to copyright. Legal protection of IP has changed over time. The opponents of recent change argue that it is driven by the particular interests of IP rights holders to the detriment of IP consumers. The main driving force in this development is represented by the big corporate holders of IP rights, notably the owners of entertainment content (music, films) and software. Although the legal and moral issues are somewhat different between these two, they are united by the fact that corporate interests and lobbying have been successful in extending existing rights (such as the U.S. extension of the length of copyright protection). Moreover, they have been successful in introducing new laws, which criminalize attempts to circumvent IP protection. Finally, much research is invested in the development of new technologies that would allow holders of IP rights to enforce and even extend their rights by circumventing statutory exceptions to their protection. The main case in point here is digital rights management (DRM) technology. Proponents of DRM systems emphasize that their purpose is to help legitimate users enjoy their rights, whereas detractors charge DRM technologies with limiting legal uses (Camp, 2003; George, 2006). A strong argument in this debate refers to the question of fair use or fair dealings, which is the provision that defines acceptable infringements of IP, particularly of copyright law. Given the economic background of copyright protection, most governments have allowed infringements that do not hurt the economic interests of copyright holders while, at the same time, they benefit society as a whole. A prime example of this is the educational use of copyright material, which is generally exempted from copyright enforcement. DRM technologies allocate certain rights that facilitate certain usages (e.g., copying, printing), but they are generally not context aware and sensitive to fair use exceptions. This is not a fundamental problem, and there are ways of designing DRM technologies that allow fair use, but the very use of DRM technologies shifts the power in the direction of IP holders who can use technological means to allocate rights. This can lead to a situation in which a DRM can curtail fair use rights and thus practically limits legal rights that legislatures have given to users.

13.3.3.1 The Ethical Response to Intellectual Property Protection The big debates on IP differ from the employee surveillance issue discussed above in that the main interested parties are not within the same organization but in different segments of society. The IP holders such as the software companies or the content owners such as record or film companies are in conflict with those who use their IP against their wishes. This includes organized crime as well as the individual end-user who downloads a copyrighted MP3 file. An added difficulty is that much IP controversy goes beyond national boundaries, with some of the main violators of IP rights being based in countries where there is little or no IP law. Examples of this are some Southeast Asian countries, such as Vietnam, or China, where Western IP rights are frequently infringed by copying content, software, and also other consumer goods. One justification of this is that these countries simply do not have a strong tradition of IP protection, partly because it is seen as an expression of respect to copy other people's work and partly because the economic tradition in these countries has simply not involved the notion of intellectual property.

Again, we need to ask how ethics can help us understand or even solve the problem. A problem of the stakeholder approach in this context is that there are large numbers of individuals involved and it is not clear whether they can be represented in an acceptable way. Differentiating between different stakeholders is difficult because the views on IP differ vastly. Is the sharing of MP3 files via a peer-to-peer network the same as, or comparable to, the selling of such files by individuals? How does the situation change when such copying is undertaken in countries where IP is not well regulated? What if the main purpose of copying is the generation of profits? On top of such theoretical difficulties, there is the practical issue of achieving consensus even among groups of stakeholders.

The shared norms and values approach runs into similar problems as it did in the case of employee surveillance. Despite some high-profile slogans for the abolition of intellectual property (Stallman, 1995), there is a wide consensus that IP can have positive effects and should be valued. The problem is again not a lack of agreement on the core value of intellectual property, but a lack of agreement on the exact form, limits, and justifications of this value. Current debates are more about how generally accepted exceptions to IP, such as the fair use or fair dealings exceptions to copyright, should be interpreted and applied. A hypernorm such as efficiency is probably useful in justifying IP, but it cannot solve the debate on whether current IP protection is efficient. Whether or not IP is efficient may appear to be an empirical question. I would argue, however, that it is a metaphysical question, which means that one's view of markets and their functions will determine the outcome of any empirical research (Stahl, 2007a).

Finally, a corporate social responsibility view runs into the problem of determining the exact extent of the corporations' responsibility. Would it be an appropriate view of the company to give IP rights to users, or would it act more responsibly by being restrictive and maybe even prosecuting individuals who infringe its rights? There seems no general answer to this, partly due to the fact that because the issue is of different relevance for different industries and even within industries that rely on IP, it will be hard to come to an agreed view. Some companies have IP that is worth huge amounts of money, for example, entertainment giants such as Sony and Disney or software companies such as Microsoft. One should note, however, that this is a self-fulfilling prophecy. These companies have considerable interests and therefore do what they can to stabilize the system, which in turn strengthens their position. A change in IP protection would clearly affect them negatively, but that does not mean that it would be detrimental to society. Counterexamples also exist; for instance, IBM has given up the proprietary off-the-shelf software market and embraced open-source software, positioning itself as a service provider rather than an IT vendor. Such differences in business models and market positions render the question of morally acceptable dealing with IP rights even more difficult.

13.4 THE MACROLEVEL INFLUENCE OF BUSINESS ON ETHICS AND INFORMATION

Privacy and IP are just two examples of the direct influence of business on ethical issues concerning information. I discussed them in some depth because they are central to debates in computer and information ethics. At the same time it is easy to see how business interests directly influence the ethical issues in question. This is not to say, however, that privacy and IP are the only ethical issues on which business interests and information meet. In addition, there are a range of issues that depend on detail, for example, where corporate interests lead to certain design decisions that affect moral views or their ethical justification from the point of view of employees, customers, competitors, or other stakeholders. I will briefly outline some other problems where issues in computer and information ethics are affected by business. All of these have been or are being discussed intensively, but they are rarely brought together as pertaining to the same set of causes.

One central issue is the changing nature of work caused by the growing influence of information and ICT. When computers entered the workplace in ever growing numbers in the 1960s, 1970s, and 1980s, one of the concerns raised was that of replacement of humans and resulting unemployment. This problem is rarely discussed in terms of ICT any more. Where debates about changes of work structures and resulting unemployment surface, they tend to refer to globalization and outsourcing. Outsourcing has had different effects on different sectors. It has increased productivity in some areas and thereby allowed for economic growth. At the same time, employment in other areas, in particular in low-skill and manual work as well as work that can easily be shifted, has suffered. However, it is plain to see that much outsourcing is only possible because of the digitized nature of work and the affordances of ICT. This is obvious in outsourcing of software development and maintenance and mobile service sectors, such as call centers. But ICT is also a driver behind the outsourcing of manual work, for example, the manufacturing of textiles or toys, which are now concentrated in China and other Asian nations. This type of work can only be outsourced because electronic control allows manufacturers to react quickly to markets that are physically located on the other side of the world.

Unemployment is not the only possible effect of ICT on the nature of work. Zuboff (1988) coined the term "to informate" almost 20 years ago to describe the fact that even manual work changes its nature because of the introduction of ICT. Machines can measure and record information about work processes, which fundamentally change the way humans interact with their work. Zuboff gives the example of a paper mill, where the introduction of ICT provided information that allowed not only for partial automation of production processes but also for new control measures and eventually for a different organization of the entire production process. This is certainly not always an ethical problem, but it leads to changes that may produce winners as well as losers. One of the consequences is that even manual work now requires much technological awareness and a high level of education, which changes the nature of the workforce.

One reason why informating is an interesting concept from an ethical point of view is that it is linked to power structures. If data can be automatically collected and simple work structures can be automated, then management can use different means of enforcing corporate views. ICT can thus be used as a simple control mechanism. It differs from traditional control mechanisms, such as punch cards or physical observation of employees, in that it has a longer reach, enabling control and monitoring of every activity, for example, by recording all keystrokes on a computer. This should not be misunderstood to imply a simplistic understanding of ICT as a one-directional means of power exertion. There are many examples in the literature where the attempt to use ICT for the purpose of managerial control was circumvented by employees. Following Foucault (1975), one can argue that all power entails means for resistance. Using a particular control mechanism, say a keystroke logger, will give employees an opportunity to pretend they are doing what is required, although they in fact do something else. And, indeed, the topic of resistance has been explored with regard to ICT in organizations (Doolin, 2004).

Goold (2003) gives an interesting example from the public sector. He investigated the use of CCTV cameras, which are widespread throughout the United Kingdom. The idea behind the introduction of public CCTV cameras was that they would detect and thus deter crime and thereby support police work. Although evidence of the success of CCTV schemes is sketchy and contradictory, Goold made observations that seemed to contradict the rationale of the introduction of the technology. The cameras took pictures not only of criminals and other people they were meant to target, but also of police officers when they were dealing with crime. Police officers became acutely aware of being observed and of pictures being taken of them, which might be used as evidence against them in a court of law. They thus became skeptical of the technology and sometimes tried to avoid being caught on camera doing things they were not supposed to do, such as chat with their colleagues. The reason why this example is interesting in this chapter is that it illustrates that ICT use for control purposes is not simply predictable, but may and often will have effects that are not intended, and that are sometimes contrary to their original aims.

Apart from the changing nature of work and social interaction, there are two interrelated topics of ethical relevance that are of high interest to businesses as well as ICT—globalization and digital divides. I will now discuss these two in some more detail.

13.4.1 Globalization

The term globalization has charged political as well as academic debate for about two decades. The German sociologist Ulrich Beck has called it the "most used, most misused—and most rarely defined, probably most misleading, most nebulous and politically most potent" concept not only of the last few years but also of the coming years (Beck, 1998, p. 42 [translation by the author]). There are many different aspects of globalization, many things that are supposed to become relevant and function on a global rather than a national scale. These include financial markets, corporate strategies, wealth creation, research and development, consumption patterns, and regulatory capabilities (Petrella, 1996, p. 64). All of this is said to contribute to the decreasing importance of national governments. The most pertinent example of globalization is international financial transactions, which have reached staggering dimensions and which seem to move beyond any state or other control (cf. Epstein, 1996). The general idea behind globalization seems to be a worldwide economy in which all countries freely open their markets to competition with other countries and where factors of production as well as products can be exchanged across borders.

There is much debate about the existence and relevance of globalization. Some have argued that the current international exchange is only now coming back to the degree of international cooperation that the world had seen before World War I. Others argue that the current state of affairs may be an aberration, which may come to an end through the next big war or comparable large-scale event. A relatively uncontentious statement is probably that globalization is not a universal phenomenon. There are huge international financial transactions, and the international exchange of goods and services is also growing steadily. This level of globalization is not at all reached by the movement of labor. International migration is an important phenomenon, but labor markets are still fiercely protected by powerful states. It is unclear what the effect of globalization is on communities (Albrow et al., 1997). Also, there are large parts of the world where globalization plays no or only a limited role.

The reason why it is important to include the concept of globalization in this chapter is that it is often seen as a moral problem. On one hand, globalization can be morally positive in facilitating exchange and understanding, thus helping to spread democracy and economic well-being. On the other hand, by taking sovereignty from national governments, the process of globalization leads to consequences that are perceived as problematic. Globalization leads to a movement of employment to lower-paying countries and the lowering of established social standards such as health or unemployment benefits in industrialized countries. Globalization creates winners and losers. The winners are often those who are doing reasonably well in the first place, who are educated, young, healthy, mobile, whereas the losers are those who were disadvantaged before (Bourdieu, 1998). Castells (2000), in his study of the information age, has provided a wealth of data showing that risks and benefits of globalization are unevenly distributed. Globalization can create fears in all parties affected and even lead to the clash of civilizations that Huntington (1993) predicted.

Globalization is driven by economic interests, but it is facilitated by modern information and communication technology. Although attempts at globalizing commercial exchange may be as old as human trade, current technological development allows an immense exchange in scope and scale of such exchanges. The most relevant recent innovation is the Internet. E-commerce and e-business allow consumers to shop regardless of borders. More important, they allow companies to exchange information and develop new ways of collaboration. Furthermore, they facilitate new business models within internationally operating organizations. Globalization as an ethical challenge is thus of high relevance to scholars of computer and information ethics (cf. Johnson, 2000).

13.4.2 Digital Divides

Digital divides are one of the most pertinent ethical issues arising from the globalization of economic activity and ICT. Again, there is little agreement on what constitutes a digital divide, why it is bad, or how it can be addressed (cf. Rookby and Weckert, 2007). One of the reasons why digital divides are perceived as an issue is that they strike most of us as inherently unjust. The reason for this perceived injustice is that they increase and perpetuate the economic inequality within and between nations. A core issue is that some people have an advantage that is linked to their ability to use ICT and thus information whereas others do not. In most cases such digital divides are closely linked to social divides, and those who have few resources off-line are unable to improve their situation because they have even fewer resources online. Business and economic views affect the ethical issues of digital divides in several respects. Businesses have a role to play in overcoming them. Much state effort in both the industrialized world and in less developed countries has been aimed at improving access to ICT. This includes investments in information infrastructure as well as in other conditions such as IT literacy. The costs of this are enormous, and businesses are increasingly called upon to shoulder some of this burden. The justification for this onus on businesses to contribute to overcoming digital divides is that they are likely to profit from a more readily available infrastructure.

Although businesses thus constitute part of the solution, they also exacerbate the problem. The reason is that businesses by definition cater to those who are potential customers, which means those who are financially reasonably well off. Businesses have little interest in poor individuals and few incentives to provide access for them. ICT thus has the potential to worsen social divides by offering the advantages of ICT, for example, e-commerce with its lower transaction costs and wider market range, only to those who are in a favorable position in the first place and by excluding those who are disadvantaged anyway.

13.4.2.1 The Ethical Response to Globalization and Digital Divides

Globalization and digital divides are highly complex phenomena that raise many different ethical and other issues beyond what I could outline here. Owing to the complexity of the problem, the ethical approaches discussed in this chapter are confronted with serious difficulties. The stakeholder approach, which fundamentally aims at the inclusion of outside interests into corporate decision processes, can hardly cope with the complexity. The number of stakeholders and stakeholder groups in the process of globalization is too large to be manageable. An individual company or manager will find it hard to identify relevant stakeholders, much less contact them or consider their views.

Similarly, the shared values or norms approach is problematic in this context because the number of values and norms is simply too great. A pertinent example here may be freedom of speech. Although there is a high level of agreement on the desirability of free speech and one could thus see it as a shared norm (and a human right), there is no agreement on its limits. In Germany, for example, it is an offense to deny that the Holocaust took place, whereas in other countries this would be an unthinkable limitation of free speech. The international uproar concerning the *Jyllands-Posten* Muhammad cartoons controversy is a case in point (Stahl, 2007b). Religious sentiments led to the breakdown of communication because of a number of cartoons that, from my viewpoint as a secular observer, were rather harmless. The point here is that even a relatively straightforward shared value is not really shared when one looks at the details. Stronger disagreements can be expected when one comes to issues such as the ethically relevant ones discussed above, including privacy or intellectual property. Reliance on shared norms, where they can be identified, is likely to bring the problem of shared interpretations to the fore. Then there is the issue of value conflicts where legitimate interests or values of one group conflict with equally legitimate values of others (e.g., free exchange of cultural artifacts vs. cultural self-determination). Finally, the corporate social responsibility view of ethics is not very helpful in this context, either. By definition, CSR aims at the individual company. Companies, in particular large international companies, have without doubt an important role to play in addressing international ethical issues, including globalization and digital divides. In the absence of an international regulatory framework, however, corporations have no clear indication concerning their responsibilities. Whether a company is doing its ethical duty, for example, by building ICT infrastructure or by relying on states to do so, is not easily decided (cf. Stahl, 2007a). The CSR approach relies on the existence of a framework for the ascription of responsibility that arguably does not exist on a worldwide scale.

13.5 CONCLUSION

This chapter aims at providing an overview of the influence of businesses on ethical issues arising in the context of information and ICT. I have concentrated on some salient issues, namely, privacy/employee surveillance and intellectual property on the individual/corporate level, as well as some less clear structural issues including globalization and digital divides on a macrolevel. Using some established approaches to business ethics, which are reflected by computer and information ethics, I have tried to explore what the contribution of ethical thought can be to the issues raised by information and ICT in business and economic contexts.

Perhaps not surprisingly, the chapter does not offer any simple or clear-cut answers. Without doubt, ethical theory can be helpful in raising awareness and shaping views on how ethical problems can be addressed. This is true for the many ethical approaches not discussed here as well as for the three central ones, namely the stakeholder approach, shared values and norms, and corporate social responsibility. One can question whether it is the purpose of ethics to give direct instructions on desirable action in the first place. It is thus probably not too disappointing if the discussed ethical approaches fail to do so in most of the cases discussed here. Ethics is arguably more about raising questions than giving answers.

There nevertheless seems to be a blind spot that the ethical views share and that ethical thinking should aim to overcome. All three of the ethical views discussed in detail, as well as others not introduced here, share as a foundation the acceptance of the status quo. They ask how individual managers or corporations should act in the socioeconomic system they find themselves in. And clearly that is a legitimate question to ask. What they fail to take into consideration, however, is the larger context. They do not question whether and how the economic system is justified, that ascribes intellectual property rights, that gives companies and managers the ability to surveil their employees, or that leads to global disparities and divides. By concentrating on individual or corporate agents as the main focus of attention, such ethical theories thus miss the opportunity to ask how changes in the overall organization of the social, economic, and legal system in which businesses move and use information and ICT can affect ethical questions and possible solutions. One of the aims of this chapter is thus to engage in this debate, to provide a foundation that will not only explain possible views of ethical issues in business and information but also initiate a debate on whether different approaches currently not discussed in business or computer and information ethics might be better equipped to move the debate forward.

REFERENCES

Adam, A. (2005). Delegating and distributing morality: can we inscribe privacy protection in a machine? *Ethics and Information Technology*, 7(4), 233–242.

Albrow, M., Eade, J., Dürrschmidt, J., and Washbourne, N. (1997). The impact of globalization on sociological concepts: community, culture, and milieu. In: Eade, J. (Ed.), *Living the Global City — Globalization as Local Process*. Routledge, London, New York, pp. 20–35.

Anderson, R.E., Johnson, D.G., Gotterbarn, D., and Perrolle J. (1993). Using the new ACM code of ethics in decision making. *Communications of the ACM*, 36(2), 98–106.

Beck, U. (1998). *Was ist Globalisierung? Irrtümer des Globalismus—Antworten auf Globalisierung*, 5th edition, Frankfurt am Main: Edition Zweite Moderne, Suhrkamp Verlag.

Block, W. (2001). Cyberslacking, business ethics and managerial economics. *Journal of Business Ethics*, 33, 225–231.

Boncella, R.J. (2001). Internet privacy: at home and at work. *Communications of the Association for Information Systems*, 7, 1–43.

Bourdieu, P. (1998). *Contre-feux: Propos pour servir à la résistance contre l'invasion néo-libéral*. Paris: Édition, RAISON D'Agir.

Bowie, N.E. (1999). *Business Ethics — A Kantian Perspective*. Blackwell, Oxford.

Brey, P. (2001). Disclosive computer ethics. In: Spinello, R.A. and Tavani, H.T. (Eds.), *Readings in Cyberethics*. Jones and Bartlett, Sudbury, MA, pp. 51–62.

Britz, J.J. (1999). Ethical guidelines for meeting the challenges of the information age. In: Pourciau, L.J. (Ed.), *Ethics and Electronic Information in the 21st Century*. Purdue University Press, West Lafayette, IN, pp. 9–28.

Brown, W.S. (2000). Ontological security, existential anxiety and workplace privacy. *Journal of Business Ethics*, 23, 61–65.

Camp, L.J. (2003). First principles of copyright for DRM design. *IEEE Internet Computing*, 7 (3), 59–65.

Castells, M. (2000). *The Information Age: Economy, Society, and Culture*. Volume I: *The Rise of the Network Society*. 2nd edition. Blackwell, Oxford.

Catudal, J.N. (2001). Censorship, the Internet, and the Child Pornography Law of 1996: a critique. In: Spinello, Richard A. and Tavani, Herman T. (Eds.), *Readings in Cyberethics*. Jones and Bartlett, Sudbury, MA, pp. 170–187.

Collier, J. and Wanderley, L. (2005). Thinking for the future: global corporate responsibility in the twenty-first century. *Futures*, 37(2/3), 169–182.

Culnan, M.J. (1993). "How Did They Get My Name?": an exploratory investigation of consumer attitudes toward secondary information use. *MIS Quarterly*, 17(3), 341–363.

De George, R.T. (1999). *Business Ethics*, 5th edition. Prentice Hall, Upper Saddle River, NJ.

Doane, D. (2005). Beyond corporate social responsibility: minnows, mammoths and markets. *Futures*, 37(2/3), 215–229.

Donaldson, T. and Dunfee, T. (1999). *Ties that Bind: A Social Contracts Approach to Business Ethics*. Harvard Business School Press, Boston.

Donaldson, T. and Preston, L.E. (1995). The stakeholder theory of the corporation: concepts, evidence, and implications. *Academy of Management Review*, 20(1), 65–91.

Doolin, B. (2004). Power and resistance in the implementation of a medical management information system. *Information Systems Journal*, 14, 343–362.

Elgesiem, D. (1996). Privacy, respect for persons, and risk. In: Ess, C. (Ed.), *Philosophical Perspectives on Computer-Mediated Communication*. State University of New York Press, Albany, pp. 45–66.

Epstein, G. (1996). International capital mobility and the scope for national economic management. In: Boyer, R. and Drache, D. (Eds.), *States Against Markets — The Limits of Globalization*. Routledge, London, New York.

Floridi, L. (1999). Information ethics: on the philosophical foundation of computer ethics. *Ethics and Information Technology*, 1(1), 37–56.

Foucault, M. (1975). *Surveiller et punir: naissance de la prison*. Gallimard, Paris.

French, P.A. (1979). The corporation as a moral person. *American Philosophical Quarterly*, 16 (3), 207–215.

Gauthier, D. (1986). *Morals by Agreement*. Clarendon, Oxford.

George, C.E. (2006). Copyright management systems: accessing the power balance. In: Zielinski, C., Duquenoy, P., and Kimppa, K. (Eds.), *The Information Society: Emerging Landscapes (IFIP WG 9. 2 proceedings)*. Springer, New York, pp. 211–222.

Gert, B. (1999). Common morality and computing. *Ethics and Information Technology*, 1(1), 57–64.

Gibson, K. (2000). The moral basis of stakeholder theory. *Journal of Business Ethics*, 26, 245–257.

Gonzalez, E. (2002). Defining a post-conventional corporate moral responsibility. *Journal of Business Ethics*, 39, 101–108.

Goodpaster, K.E. and Matthews, J.B. (1982). Can a corporation have a moral conscience? *Harvard Business Review*, 60(1), 132–141.

Goold, B.J. (2003). Public area surveillance and police work: the impact of CCTV on police behaviour and autonomy. *Surveillance & Society*, 1(2), 191–203.

Gotterbarn, D. and Rogerson, S. (2005). Responsible risk analysis for software development: creating the software development impact statement. *Communications of the Association for Information Systems*, 15, 730–750.

Grodzinsky, F.S. (2001). The practitioner from within: revisiting the virtues. In: Spinello, R.A. and Tavani, H.T. (Eds.), *Readings in Cyberethics*. Jones and Bartlett, Sudbury, MA, pp. 580–592.

Hank, R. (2000). *Das Ende der Gleichheit oder Warum der Kapitalismus mehr Wettbewerb braucht*. S. Fischer Verlag, Frankfurt am Main.

Hasnas, J. (1998). The normative theories of business ethics: a guide for the perplexed. *Business Ethics Quarterly*, 8(1), 19–42.

Hausman, D.M. and McPherson, M.S. (1996). *Economic Analysis and Moral Philosophy*. Cambridge University Press, Cambridge.

Hendry, J. (2001). Economic contracts versus social relationships as a foundation for normative stakeholder theory. *Business Ethics: A European Review*, 10(3), 223–232.

Heng, M.S.H. and de Moor, A. (2003). From Habermas's communicative theory to practice on the internet. *Information Systems Journal*, 13, 331–352.

Himma, K.E. (2004). The question at the foundation of information ethics: does information have intrinsic value? In: Bynum, T., Pouloudi, N., Rogerson, S., and Spyrou, T. (Eds.), *Challenges for the Citizen of the Information Society: Proceedings of the Seventh International Conference on the Social and Ethical Impacts of Information and Communications Technologies (ETHICOMP 2004)*.

Huntington, S.P. (1993). The clash of civilisations? *Foreign Affairs*, 72(3), 22–49.

Husted, B.W. and Allen, D.B. (2000). Is it ethical to use ethics as a strategy? *Journal of Business Ethics*, 27, 21–31.

Introna, L. (2000). Privacy and the computer: why we need privacy in the information society. In: Baird, R.M., Ramsower, R., and Rosenbaum, S.E. (Eds.), *Cyberethics—Social and Moral Issues in the Computer Age*. Prometheus Books, New York, pp. 188–199.

Johnson, D.G. (2000). The future of computer ethics. In: Collste, G. (Ed.), *Ethics in the Age of Information Technology*. Centre for Applied Ethics, Linköping, pp. 17–31.

Keynes, J.M. (1994). Economic model construction and econometrics. In: Hausman, D.M. (Ed.), *The Philosophy of Economics: An Anthology*, 2nd edition. Cambridge University Press, Cambridge, pp. 286–288.

Koslowski, P. (2000). The limits of shareholder value. *Journal of Business Ethics*, 27, 137–148.

Kujala, J. (2001). Analysing moral issues in stakeholder relations. *Business Ethics: A European Review*, 10(3), 233–247.

Küng, H. (1997). *Weltethos für Weltpolitik und Weltwirtschaft*. 3rd edition. Pieper Verlag, München.

Langford, D. (1999). *Business Computer Ethics*. Addison-Wesley, Harlow.

Laudon, K. (1995). Ethical concepts and information technology. *Communications of the ACM*, 38(12), 33–39.

Lessig, L. (1999). *Code and Other Laws of Cyberspace*, Basic Books, New York.

Lessig, L. (2001). The laws of cyberspace. In: Spinello, R.A. and Tavani, H.T. (Eds.), *Readings in Cyberethics*. Jones and Bartlett, Sudbury, MA, pp. 124–134.

Moor, J.H. (1985). What is computer ethics? *Metaphilosophy*, 16(4), 266–275.

Moor, J.H. (2000). Toward a theory of privacy in the information age. In: Baird, R.M., Ramsower, R., and Rosenbaum, S.E. (Eds.), *Cyberethics — Social and Moral Issues in the Computer Age*. Prometheus Books, New York, pp. 200–212.

Moor, J.H. (2001). Reason, relativity, and responsibility in computer ethics. In: Spinello, R.A. and Tavani, H.T. (Eds.), *Readings in Cyberethics*. Jones and Bartlett, Sudbury, MA, pp. 36–50.

Nozick, R. (1974). *Anarchy, State, and Utopia*. Basic Books, New York.

Nye, D. (2002). The 'privacy in employment' critique: a consideration of some of the arguments for 'ethical' HRM professional practice. *Business Ethics: A European Review*, 11(3), 224–232.

Oz, E. (1992). Ethical standards for information systems professionals: a case for a unified code. *MIS Quarterly*, 16, 423–433.

Panko, R.R. and Beh, H.G. (2002). Monitoring for pornography and sexual harassment. *Communications of the ACM*, 45(1), 84–87.

Petrella, R. (1996). Globalization and internationalization: the dynamics of the emerging world order. In: Boyer, R. and Drache, D. (Eds.), *States Against Markets—The Limits of Globalization*. Routledge, London, New York.

Rawls, J. (2001). In: Kelly, K. (Ed.), *Justice as Fairness: A Restatement*. Belknap, Harvard, Cambridge, MA, London.

Rogerson, S. (2004). The ethics of software development project management. In: Bynum, T.W. and Rogerson, S. (Eds.), *Computer Ethics and Professional Responsibility*. Blackwell Publishing, Oxford, pp. 119–128.

Rookby, E. and Weckert, J. (Eds.) (2007). *Information Technology and Social Justice*. Idea Group, Hershey, PA.

Schwartz, P. and Gibb, B. (1999). *When Good Companies Do Bad Things: Responsibility and Risk in an Age of Globalization*. John Wiley & Sons, New York.

Sen, A. (1987). *On Ethics and Economics*. Basil Blackwell, Oxford, New York.

Severson, R.J. (1997). *The Principles of Information Ethics*. M.E. Sharpe, Armonk, NY, London.

Siau, K., Nah, F.F.-H., and Teng, L. (2002). Acceptable internet use policy. *Communications of the ACM*, 45(1), 75–79.

Skovira, R.J. (2003). The social contract revised: obligation and responsibility in the information society. In: Azari, R. (Ed.), *Current Security Management & Ethical Issues of Information Technology*. IRM Press, Hershey, pp. 165–186.

Smithson, S. and Hirschheim, R. (1998). Analysing information systems evaluation: another look at an old problem. *European Journal of Information Systems*, 7, 158–174.

Spinello, R. (2000). *Cyberethics: Morality and Law in Cyberspace*. Jones and Bartlett, London.

Sriram, R.S. and Krishnan, G.V. (2003). The value relevance of IT investments on firm value in the financial services sector. *Information Resources Management Journal*, 16 (1), 46–61.

Stahl, B.C. (2007a). Social justice and market metaphysics: a critical discussion of philosophical approaches to digital divides. In: Rookby, E. and Weckert, C. (Eds.), *Information Technology and Social Justice*. Idea Group, Hershey, PA, pp. 148–170.

Stahl, B.C. (2007b). Truth as the limit of free of speech: a critical perspective. In: Debatin, B. (Ed.), *The Cartoon Debate and the Freedom of the Press: Conflicting Norms and Values in the Global Media Culture*. LIT, Münster, London, pp. 131–136.

Stallman, R. (1995). Why software should be free. In: Johnson, D.G. and Nissenbaum, H. (Eds.), *Computers, Ethics & Social Values*. Prentice Hall, Upper Saddle River, NJ, pp. 190–200.

Stark, A. (1993). What is the matter with business ethics? *Harvard Business Review*, 71(3), 38–48.

Tapia, A.H. (2004). Resistance of deviance? a high-tech workplace during the bursting of the dot-com bubble. In: Kaplan, B., Truex, D.P., Wastell, D., Wood-Harper, A. T., and DeGross, J. (Eds.), *Information Systems Research: Relevant Theory and Informed Practice (IFIP 8. 2 Proceedings)* Kluwer, Dordrecht, pp. 577–596.

Tavani, H. (2000). Privacy and security. In: Langford, D. (Ed.), *Internet Ethics*. McMillan, London, pp. 65–89.

Torkzadeh, G. and Dhillon, G. (2002). Measuring factors that influence the success of internet commerce. *Information Systems Research*, 13(2), 187–204.

Urbaczewski, A. and Jessup, L.M. (2002). Does electronic monitoring of employee internet usage work? *Communications of the ACM*, 45(1), 80–83.

van den Hoeven, J. (2001). Privacy and the varieties of informational wrongdoing. In: Spinello, R.A. and Tavani, H.T. (Eds.), *Readings in Cyberethics*. Jones and Bartlett, Sudbury, MA, pp. 430–442.

Velasquez, M. (1998). *Business Ethics: Concepts and Cases*. 4th edition. Prentice Hall, Upper Saddle River, NJ.

Walsham, G. (1996). Ethical theory, codes of ethics and IS practice. *Information Systems Journal*, 6, 69–81.

Warren, S.D. and Brandeis, L.D. (1890). The right to privacy. *Harvard Law Review*, 5, 193–220.

Warwick, S. (2001). Is copyright ethical? an examination of the theories, laws, and practices regarding the private ownership of intellectual work in the United States. In: Spinello, R.A. and Tavani, H.T. (Eds.), *Readings in Cyberethics*. Jones and Bartlett, Sudbury, MA, pp. 263–279.

Weber, M. (1996). *Die protestantische Ethik und der* "Geist" de Kapitalismus. Edited and introduced by Klaus, L. and Johannes W. 2nd edition, Beltz Athenäum Verlag, Weinheim.

Weckert, J. (Ed.) (2005). *Electronic Monitoring in the Workplace: Controversies and Solutions*. Idea Group Publishing, Hershey, PA.

Weisband, S.P. and Reining, B.A. (1995). Managing user perceptions of email privacy. *Communications of the ACM*, 38(12), 40–47.

Weizsäcker, C.C. (1999). Globalisierung: Garantie für Freiheit und Wohlstand oder Ende der Politik und Abschied vom Staat? Aus ökonmischer Sicht. In: Mangold, H., Weizsäcker, C.C. *Globalisierung—Bedeutung für Staat und Wirtschaft*. Wirtschaftsverlag Bachem, Köln, pp. 9–51.

Werhane, P. (1985). *Persons, Rights, and Corporations*. Prentice-Hall, Englewood Cliffs, NJ.

Wilenius, M. (2005). Towards the age of corporate responsibility? Emerging challenges for the business world. *Futures*, 37(2/3), 133–150.

Zuboff, S. (1988). *In the Age of the Smart Machine: The Future of Work and Power.* Basic Books, New York.

PART IV

RESPONSIBILITY ISSUES AND RISK ASSESSMENT

CHAPTER 14

Responsibilities for Information on the Internet

ANTON VEDDER

14.1 INTRODUCTION

One of the most fascinating aspects of the Internet is that very few accidents happen. This not only holds for the technical infrastructure and maintenance, but also for the communication and information transmitted through the network. Although the many-to-many medium could in principle be abused in so many different ways and on such a large scale, actually only relatively little really goes wrong. This is all the more astonishing as the global phenomenon of the Internet lacks a unique governance core, hierarchy, and central control mechanisms. It is a network that consists of a disparate set of heterogeneous organizations and individuals, ranging from commercial business corporations and private organizations of volunteers, to governmental institutions, universities, and individual citizens. The rise of such a relatively smooth and flawless working and highly influential phenomenon from the voluntary input of so many individuals and organizations is probably one of the happiest developments in the end of the twentieth century.

But what if something nonetheless just goes wrong? What about the accountability and responsibilities involved then? Are there ways of reducing the chances of things going wrong? These questions will become more important in the near future as the speed of data transmission and the accessibility of the network will grow exponentially (Vedder and Lenstra, 2006).

In this essay, I will leave the possibilities of things going wrong, with regard to the infrastructure and maintenance, aside. I will concentrate on the responsibilities involved in the possible negative impact of the dissemination of information on the Internet. I will mainly focus on three parties: those who put forward information on the

The Handbook of Information and Computer Ethics, Edited by Kenneth Einar Himma and Herman T. Tavani

Internet, the so-called content providers (CPs), the organizations that provide the infrastructure for the dissemination of that information, the so-called Internet service and access providers (to which I will refer indifferently as ISPs), the receivers or users of the information, third parties, such as those that deliver quality certificates for Web sites, and others.

Until recently, issues of responsibilities on the Internet have often been discussed in association with specific accountabilities of ISPs with regard to information (including pictures and footage) that are outright illegal or immoral. Think, for instance, of child pornography, illegal weapon sales, the sale of illegal drugs, and the dispersion of hate and discrimination. Typically, in most legal systems, the liabilities of ISPs have been specified with regard to these forms of harmful or offensive information. Member states of the European Union, for instance, have to implement in their laws and regulation the European Directive 2000/31/EC of the European Parliament and of the Council of June 8, 2000 on certain legal aspects of information society services, in particular electronic commerce, in the Internal Market (*Directive on electronic commerce*), articles 12–15. These articles safeguard hosting ISPs from being held liable for the information stored, under the condition that they do not have actual knowledge of the illegal activity or information and, as regards claims for damages, are not aware of facts or circumstances from which the illegal activity or information is apparent. As soon as the ISP obtains such knowledge or awareness, he must act expeditiously to remove or to disable access to the information. The Directive explicitly states, however, that member states are not expected to impose rules of a general character upon ISPs to monitor all of the content that is made available through their services (article 15; compare, however, recitals 47 and 48 that allow of specific monitoring obligations with the help of sophisticated technological tools). The motivating idea behind article 15 may have been that a general obligation of monitoring could affect the effectiveness of the Internet infrastructure on the whole negatively. But of course article 15 may also have been inspired by the awareness of values such as information-related freedoms and privacy of content providers.

In this chapter, I will address a subject that is broader than just ISPs' accountability with regard to illegal content. Much of what I will put forward in this essay will have a bearing on this issue, and in Section 14.3, I will explicitly, but nonetheless concisely, point out what can be said about the responsibilities of the ISPs involved in the dissemination of such content from a moral point of view. I will, however, mainly focus on instances of information that often have not such an immediately clear illegal or immoral character. In doing so, I will also concentrate on other types of actors than ISPs. I will start this essay with mapping out what are normally considered to be the standard conditions of responsibility in moral theory. After dealing with the responsibilities involved in some clear cases of illegal or immoral content, I will continue with an explanation of the different ways in which information on the Internet may have indirect and unintended bad consequences for the users. I will conclude with addressing the different types of responsibilities involved.

14.2 CONDITIONS OF RESPONSIBILITY

In this section, I will elaborate on moral responsibility as it is traditionally conceived of in the everyday moral debate as well as in ethical theory. Normally, the notion of moral responsibility is used in at least two ways that should be carefully distinguished. It can be used in a primarily retrospective sense and in a primarily prospective sense. The former refers to the possibility of rightfully ascribing or attributing actions or consequences of actions to agents. Retrospective responsibility is an equivalent of accountability. The latter refers to duties and obligations that can be imposed upon agents. Having prospective responsibilities is equivalent to having duties and obligations or being bound by these.

For the purposes of this paper, it is important to see that the two cannot be dealt with completely separately. The first cannot be understood adequately without the second (see also: Feinberg, 1970, pp. 187–221; Hart, 1968, pp. 211–230). We only hold people morally responsible (in the retrospective sense) if they had a responsibility (in the prospective sense) to perform or not to perform the action in question at the time when they actually did or did not perform that action. To put it differently, it only makes sense to hold a person responsible, retrospectively, for action or omission X when he or she was under a relevant duty or obligation regarding X. Of course, the presence of a prospective responsibility is just one of the conditions for retrospective responsibility. In order to ascertain the moral responsibility of an agent in the primarily retrospective sense, one has to make sure that three conditions apply.

First, there should be a causal relationship of some kind between the agent and the action or the consequences of the action. This relationship can be direct or indirect, substantial or additional. The relationship need not be the one that can be framed in terms of a sufficient condition or even of a necessary condition as long as it contributes in one way or another to the effect.

Second, the action or its consequences should be performed or produced intentionally. This does not mean that the agent should have or should have had a positive desire to bring about the action or its consequences. The only minimal requirement is that he or she at least did not act or did not refrain from acting in a state of voluntary ignorance regarding the action or the omission and their consequences. Although a thorough discussion about this point would go far beyond the purposes of this chapter, it should be kept in mind that the things that I have said about causality and intentionality are of a rather minimalist vein. What causal relationship and what kind and degree of intentionality should be present depends on the context, the kind of action, and the kind of value that is at stake. Both the character of the causal relationship and the kind and intensity of the intention influence in a complicated way the degree of blame that is imposed on the actor. Whether someone is blamed for doing something wrong and how severely he is blamed depend in part on questions such as: Did he know what was going to happen? Did he consciously want that to happen? Was he negligent with regard to these things? What was his contribution to the effects involved? Was he kept in ignorance about the possible effects? The kind and degree of the causal relationship and the intentionality influence the degree of blame that can be imposed upon an actor in a very complicated way. There is no direct, straight relationship between them. The character and the magnitude of the harm or offense involved are also of importance, as are other aspects of the situation.

The third condition for responsibility leads us back to the relationship between retrospective and prospective responsibility. It should be possible to give a moral qualification of the action or its consequences. There must be some kind of moral principle or value consideration that is applicable to the action or its consequences. At the time of performing the action or producing the consequences for which an agent is held morally responsible, there must be an obligation or duty not to perform or to produce them — at least, not in the way that they have been performed or produced eventually. Would there be no such duty or obligation, then the action and its consequences would be morally indifferent. There would be no need to discover moral responsibility at all.

The connection between prospective responsibility and retrospective responsibility is not only a motivational one; understanding the prospective responsibility involved also focuses our attention on the relevant aspects of a situation when we are deciding whether the first and the second condition of retrospective responsibility have been satisfied. In order to know exactly where to look and find out if the first two conditions are adequately met, it is necessary to know what principle or value consideration is at issue. For an answer to the question what kind of moral responsibility—in the sense of duty or obligation— an agent has in a given situation, one should first of all give careful consideration to all circumstances. Subsequently, one should try to articulate the moral principles or values that call into question these circumstances from a moral point of view. Establishing prospective responsibility in this way enables us to know on what part of the whole machinery of the action and its consequences and from what perspective we have to focus. Doing so in turn enables us to determine the presence of relevant causality relationships and intentionality, and to decide whether these conditions have been met sufficiently or in the degree required. Naturally, this impact of prospective responsibility on the determination of retrospective responsibility is closely tied to the role of the character and dimension of the harm or offense involved in determining the character and degree of the aspects of intentionality and causality.

So, the connection between retrospective and prospective responsibility lies mainly in the need for including some idea of prospective responsibility in the idea of retrospective responsibility. In order to understand fully what retrospective responsibility is, and in order to be able to find out correctly whether someone is responsible in specific situations, we need to have some idea of the types of prospective responsibility that may apply. The converse relationship is not so strong. It makes perfect sense to attribute prospective responsibility to persons without knowing whether these are capable of fulfilling the first and the second condition of retrospective responsibility (causality and intentionality). If, eventually, it turns out that they do not fulfill these conditions sufficiently or to the required degree, they are said to be excused. We would not say that in that case the normative principle invoked does not apply.

Now, all that I have said so far about the conditions of responsibility and the interdependence of retrospective and prospective responsibilities reflects in large part some fundamental tenets of current ascriptive theory, that is, the special sector of moral philosophy that is dedicated to questions concerning the attribution of actions and their consequences to actors; it also reflects broadly shared and deeply held moral convictions, "gut-feelings," of people on a more concrete level. I will return to this in the next section.

First, a few words must be dedicated to our reasons for attributing retrospective responsibility. Why are we interested in doing so? Why should we care to be accurate when we attribute responsibilities? Answers to these questions can be divided into consequentialist ones and Kantian ones. Oddly enough, when asked to give answers to these questions, most people will come up with consequentialist considerations.

Consequentialist reasons for an accurate attribution of retrospective responsibility seem to be more natural than the Kantian ones. Consequentialist reasons have, of course, to do with the clear effects of accurate attribution of retrospective responsibility. Here, one may think of prevention through deterrence or learning. Or one may think of retribution and revenge that may satisfy the preferences or needs of people who have been victims of others' wrongdoings. Kantian reasons are much more complicated. They are about taking persons seriously as individual moral agents. They have to do with respect for the identity and the integrity of the agent, which is rather paradoxically expressed by establishing his responsibility and blaming him for his wrongdoing and lack of integrity. They are also closely connected to concerns about the fairness of judging people and fairness in the distribution of blame and praise. In the next section I will return to these reasons for correctly attributing moral responsibilities. But first, we must have a closer look at the nature of ISPs. Attributing responsibilities to ISPs formed the starting point for the investigation.

14.3 ISPs AND CLEARLY HARMFUL OR OFFENSIVE INFORMATION

Until recently, one of the burning questions in the debate on new information technologies, ethics, and law has been about the responsibilities of ISPs that make the information originally provided by a content provider available to the public. Should ISPs be blamed for the harm or offence caused by, for example, racist expressions and images, slander, offers of drugs, and plagiarism that occur in the contents that are supplied by others? Do they have any kind of obligation to prevent or to compensate for the harm and offense that may be caused by such matters? These are complicated questions. In this section, I will not defend a clear-cut "yes" or "no" to either of them; I will only sketch some preliminaries for the debate on responsibilities of ISPs. In doing so, I will clarify some particularities of the current mainstream in thinking about attributing moral responsibility. As a matter of fact, I will argue that if retrospective responsibilities are to be attributed to ISPs, then the attribution of such responsibilities must be much more motivated by future goals and purposes than is normally the case with regard to the attribution of retrospective responsibilities to individual persons.

Introducing this idea of attributing retrospective responsibilities for future-oriented reasons seems to conflict with some broadly shared and deeply felt intuitions regarding the individuality of responsibility and the relationship between responsibility and guilt. These convictions coincide with some basic ideas in Kantian moral theory and mainstream ascriptive theory. I will explain that the kind of responsibility that perhaps could be attributed to ISPs would better fit in with consequentialist moral theories. Nevertheless, I will also show that, with some adjustments, it may in the end also turn out to be reconcilable with prevailing Kantianism, the moral outlook that hinges on and is dominated by a constitutive ideal of the autonomy and dignity of individual persons.

The position that I will defend differs significantly from the one defended some years ago by Deborah Johnson. Johnson (1994, pp. 124–146) suggests that the relationship between organizations resembling ISPs —she was in fact writing about organizations maintaining electronic billboards—and the negative consequences of the information put forward by content providers can only be evaluated in terms of the legal category of liability. She is of the opinion, however, that this is an exclusively legal matter. As to the moral perspective, she opposes the idea of holding organizations like ISPs liable on the basis of morally normative reasons, that is, reasons regarding the conflict that may arise with information-related freedoms. This seems to be right in many specific cases. Turning this into a general claim seems to go too far. In any case, Johnson is undoubtedly right in assuming that where the question of accountability of ISPs occurs, a notion resembling the one of legal liability is conceptually the most suitable to be applied. Johnson is wrong, however, where she separates the legal and the moral perspectives so strictly, and where she seems to advocate a kind of moral agnosticism regarding the responsibilities of ISPs. In doing so, she was probably led by the predominant Kantianism in current ascriptive theory and the part of the general moral outlook that reflects ascriptive theory. In the subsequent paragraphs, however, I would like to explain, however, that a consequentialist notion of retrospective responsibility can successfully be incorporated in ascriptive theory without compromising basic assumptions. It is of importance to my point that we should not look upon ISP responsibility as just a legal topic that is completely outside the moral domain, but as a moral one fully integrated in that domain.

The first reason for not leaving moral responsibility aside, is that legal liability is very restricted in its possibilities of preventing harm and offence (see also Section 14.4). The second reason has to do with the fact that often ISPs are simply the only ones left to do something in order to prevent harm or offence from happening. With regard to potentially harmful or offensive information, for example, racist phraseology, false incriminations, sale of illegal drugs, and plagiarism, the basic moral responsibilities not to harm and not to offend are, of course, in the first place the responsibilities of the content providers. These are the authors or those who publish the materials on the net. When the content providers do not take their responsibilities seriously, the only ones who can prevent the materials from becoming available or accessible are the ISPs. At least they are the only ones who can try to do so, and who may succeed in doing so to a certain extent. It goes without saying that this has complicated, important technical and financial aspects. Nevertheless, the fact that ISPs sometimes have these possibilities cannot be denied.

Now, in circumstances in which all that can be done to prevent harm or offence from happening can only be done by one (type of) actor, the converse of the well-known adagio "ought implies can" may be true. Sometimes, can implies ought: The sheer ability and opportunity to act in order to avoid or prevent harm, danger, and offense from taking place put an obligation on an agent. This is the case when harm, danger, or offense would be considerable while the appropriate action would not present significant risks, costs, or burdens to the agent, whether it is a natural person or an organization. The absence of other agents with the same kind of abilities and opportunities can make the duty to act even weightier. In the absence of other agents with the same abilities and opportunities, ISPs have weighty duties to prevent harm and offense that may be the effects of publishing materials on the net. The question of how the providers should fulfill their responsibilities exactly cannot be answered here. Instead, I will elaborate somewhat on the urgency of the ISPs taking their responsibilities seriously.

ISPs can be compared with the providers of the traditional mass media. Just like radio, television, or, for that matter, a cable network, the Internet offers opportunities to distribute textual information, images, or sound recordings on an enormous scale. There is little disagreement about the view that the freedom of the more traditional mass media like radio and television to provide information and services should be restricted by certain limiting conditions regarding harmfulness and offensiveness. Many of these traditional media do not produce, themselves, the information and services they transmit or make accessible. In this respect, ISPs do not differ from them. Nevertheless, the traditional mass media are not free to broadcast or distribute whatever textual information, sounds, or pictures are available. They are, for obvious reasons, bound by minimum moral standards concerning the prevention and avoidance of harm and offense. There is no reason at all to think that these standards should not also apply to ISPs.

The similarities between the traditional providers, such as radio and television, and the new providers of the Internet are just one reason to think that they are under a similar moral regime. There is another, and perhaps more important, reason to think that moral restrictions apply to Internet providers. Lack of barriers and easy accessibility of textual information, pictures, and sounds is one of the intriguing characteristics of the Internet. It is relatively easy to disseminate information through the Internet. Publishers, broadcasting companies, printing offices, and production houses can all be left aside. In principle, whatever one likes to publish can be put on the Internet straight from the home, all by oneself. Conversely, it is also very easy to gain access to this information. The recipients need not go to a bookshop and buy their copy of a book or a magazine; they need not wait until the information they want or need is shown on television or broadcast on the radio. They can pick it up at the time they desire, in the way and the circumstances they desire. In short, they are not, or at least much less, hindered by barriers that were formerly present when people tried to get information and materials through media such as newspapers, magazines, books, (propaganda) leaflets, radio, and television.

The fact that such barriers are fading away may, in a certain respect, be considered a good thing. In a sense, easy accessibility advances the equality of opportunities in our societies, where information becomes one of the most important assets and means to obtain welfare and well-being. Nevertheless, it is rather naive to think that all information is useful to the purposes of welfare and well-being. Victims of racist rhetoric, of hatred campaigns or just of the many stupid, undocumented mythological stories on the Internet about diseases such as AIDS or cancer, may testify: not all information is valuable. Before the Internet came into existence, offensive and harmful information was far more difficult to attain. You had to go to a bookshop. You had to await the mailing of the local aberrant political denomination. Or you could switch on your radio or television, fold open your tabloid, and wait for silly information. Now, silliness, bigotry, and sheer hate are just some mouse clicks away from you, to take in when, where, and for as long as you like.

The main argument for attributing responsibilities to ISPs, as put forward in the previous paragraphs, is primarily forward-looking and future-oriented. It is focused on the ISPs' capabilities to prevent harm and offense. Backward-looking ideas about guilt or taking individuals seriously as moral actors and about assigning praise and blame correctly do not play such an important role. This, however, is not completely true. The three elements that must be present for assigning moral responsibility can also be present in the case of ISPs. We can see this once we accept the idea that ability and opportunity can sometimes create obligation and we agree that the complementary contribution of ISPs is of causal relevance to the offensive or harmful effects of publishing certain items on the Internet.

There are, nonetheless, obvious difficulties with assigning such responsibilities. I think, however, that these can be overcome.

First, there is an objection to the attribution of both prospective and retrospective moral responsibilities. This has to do with the fact that ISPs are, for the greater part, private organizations that have to make profits in a context of commercial competition; this could be considered an obstacle to attributing moral responsibilities to them. It is sometimes believed that organizations such as business corporations have no moral responsibilities. Milton Friedman is often cited to explain that business organizations, or rather their managers, have no special competence or expertise concerning social and moral matters. According to Friedman, if they were to have these responsibilities, then these responsibilities might easily conflict with their obligations to make profits for the stockholders (Friedman, 1970).

Friedman, however, did not claim that managers of private organizations, such as business corporations, have no moral responsibilities at all. He held that business is bound by moral norms of minimal decency, meaning that they should avoid and prevent harm. Friedman only wanted to exclude responsibilities or duties of positive beneficence (e.g., funding education and health care for the worst off in the region of the firm). Friedman had a moral reason for not attributing duties of positive beneficence to business. He thought that such activities should be democratically controlled and not decided upon by private persons. According to him, the latter could easily feel tempted to use the enormous power of their corporations for their own, subjective purposes.

But does the argument hold when applied to providers, who often do not produce the information, but are just intermediaries? I do not think so. More often than not, little specific competence or expertise is in fact needed to observe where textual information, images, or sound could be harmful or offensive. Obviously, not every possibility of harm or offense can be understood beforehand. And, of course, harmfulness and offensiveness are matters of degree. This, however, does not mean that clear cases of harm and offense cannot be discovered and need not be tackled. From the difficult and vague cases, we need not at all conclude that every effort to reveal harm and offense and to block further possibilities of harming and offending is useless. Finally, one might consider the fact that ISPs are organizations with the aim to make profits as one more reason to ascribe moral responsibilities. The fact that they can make profits by contributing to the fact that certain people in society are put at the risk of being harmed or offended is just one more reason to hold them responsible.

Secondly, it may be objected that the collectivity of actors prohibits attributing retrospective moral responsibilities of this kind to ISPs. An ISP is an organization, not a person. Many actors, Internet users, or consumers, as well as content providers and the organizations of ISPs, are involved in the process of diffusing information on the Internet. ISPs can only function as providers because they are, as it were, elements in a series connection. The functioning of other providers, in other words, is essential to their own performance. Finally, an ISP accommodates in its systems the information of an enormous number of content providers, among whom are content providers who have subscriptions to other ISPs. Because of all these reasons, attributing responsibility to ISPs cannot be done in the same relatively straightforward sense as attributing responsibility to individual persons.

This, again, does not hold. Although attributing responsibilities to collectivities may be complicated, it is not practically and conceptually impossible. Over the last decades, various studies have been published in which a whole range of arguments have been given for attributing responsibilities to collectivities. Some of these are based on ingenious interpretations of organizations and decision-making procedures in organizations and their resemblance to persons (e.g., French, 1984; May and Hoffman, 1991). Others, such as Goldman (1980), start from consequentialist arguments about the didactic, deterring, or preventive effects of such attributions.

In addition, it should be observed that attributing blame and praise to collectivities such as private organizations, as a matter of fact, is something that happens all the time. People think and talk in terms of attributing responsibilities to organizations and they establish single-issue organizations in order to motivate governments and business companies to take their responsibilities seriously. The law establishes liabilities for organizations. Therefore, one should rather wonder, why in certain regions in the field of moral philosophy, the idea of collective responsibility has still not been accepted.

The underlying reason for this might be a Kantian bias combined with methodic individualism, like the one that seems to be characteristic of the Kantian moral outlook. As seen from the angle of the traditional idea of direct, guilt-related responsibility, it is indeed difficult to understand exactly what it means to hold a collectivity responsible, where this responsibility cannot in any clear way be distributed among the individual members of the collectivity. Nevertheless, attributing such responsibilities just seems to work. Organizations learn from it and change their behavior on that basis. Perhaps, then, it should just simply be admitted that the responsibilities attributed to collectivities, because of their basic future-oriented function, differ only partially from the ones attributed to individuals.

Finally, and most importantly, we come to the objection that to my mind is the most appealing: the apparent irreconcilability of the Kantian and consequentialist moral outlook. Attributing moral responsibility to ISPs is primarily inspired by reasoning of a rather consequentialist kind. Doing so is, in a way, instrumentalist, and may therefore be intuitively felt to be unfair. The categories of blame and guilt are used for purposes that do not relate to the identity and (the lack or restoration of) the integrity of the acting party. This does not seem to do justice to the requirement of respect that we think we ought to pay to the individual persons involved, even by blaming or punishing him or her . . . at least if our morality is of a Kantian vein.

It looks as if this objection is at least in part a question of fundamental outlook, of basic ideological orientation. Nevertheless, it can be argued that attributing responsibilities to ISPs is in large part reconcilable with Kantianism.

Although attributing responsibilities to ISPs is something that is at face value more familiar to consequentialist stances in morality, it is nevertheless closely tied to the ways in which responsibility is traditionally attributed to individuals. Important in this respect is that the idea of a causal relationship — albeit a secondary or additional one — is not completely abandoned. In the case of attributing responsibilities to ISPs, considerations like the practicalities of compensating or preventing losses that result from certain risks for all the parties involved are of importance, but the causality aspect is not completely overlooked. This is so because the requirement of the causal relationship guarantees that it is exactly those who contribute to harm or offense on whom the responsibilities are imposed and who are thereby stimulated to learn from experience and to prevent harm and offense in the future. In this way, even attributing this future-oriented kind of retrospective responsibility pertains to the identity and integrity of the agent. Although, therefore, the consequences of prevention, learning, and deterrence are undoubtedly preponderant among the reasons for attributing responsibilities to ISPs, doing so may have some intuitive appeal to Kantians in so far as it indirectly sees to the identity and integrity of the acting organization.

As I have explained extensively in Section 14.2, even in the traditional views on retrospective responsibility there is a close relationship between retrospective and prospective responsibility. This relationship shows itself in the dependence of the causality and intentionality conditions on the character and the dimension of the harm or offense involved. The significance of the consequences of an action for the requirements regarding the causal relationship and the intentionality, to my mind, at least hints at the functionality of the attribution of retrospective moral responsibility for the ways in which we deal with harm and offense. Put differently, the future-oriented, instrumental approach of moral responsibility, which we might associate with consequentialism, is not at all strange to the traditional idea of retrospective responsibility.

As I already mentioned at the end of Section 14.2, on conditions of responsibility, when asked for reasons for attributing retrospective moral responsibility, most people will come up with consequentialist considerations. The strange thing is that the highly Kantian idea of retrospective moral responsibility, at least in what seems to be a kind of common-sense approach, is embedded in a motivational structure of a highly consequentialist nature. I consider this to be one more reason to assume that the Kantianism of the traditional views and the consequentialism of the views that I have put forward here are ultimately compatible.

Summing up, if ISPs have responsibilities relating to information produced by others but accessible through their services, then these responsibilities are slightly different from the responsibilities that are traditionally attributed to individual persons. They are, however, not completely different. Basic to the traditional idea of responsibility—at least when taken as retrospective responsibility—is the assignment of guilt, which is something that has to do with the identity and the character of an actor. When moral responsibilities regarding negative aspects of information that are accessible through their services, are attributed to ISPs, the primary concern is not so much with guilt but with preventing or compensating for these negative consequences. This, however, is not to say that the question of guilt is completely put aside. I have argued that, whereas this idea may, as such, suit people with a consequentialist moral outlook very well, it may at first and in some respects be difficult to accept for Kantians. The idea does not completely abandon the requirements of a causal relationship and intentionality and, therefore, is not completely alienated from a guilt-centered conception of responsibility.

One may ask whether this whole argument about responsibilities that are better adjusted to consequentialism than to Kantianism and responsibilities that better fit with Kantianism than with consequentialism is not a rather inner-philosophical debate of relatively little importance to everyday life. Is it not a philosophical maneuver aiming at the solution of a problem caused by philosophical idiosyncrasies? Without wanting to be immodest, I do not think so. I think that it is important to update our philosophical conceptual frameworks and vocabularies—and by doing so also our concrete moral concepts and words—frequently in order to adapt them to the new circumstances of our ever-developing societies. Doing so supplies us with conceptual instruments with which we are better fitted to approach contemporary social problems. Reconsidering moral responsibility and introducing a category of responsibility that is oriented toward results and consequences seems all but redundant in an age that witnesses an exponential growth of technologies, the rise of enormous transboundary organizations, and a gradually declining influence of individuals.

14.4 INFORMATION IN GENERAL

As we turn to responsibilities related to information on the Internet in general, the focus shifts even further away from retrospective to prospective responsibilities. As it was mentioned earlier that the debate on information-related responsibilities was until recently restricted to the issue of ISPs' responsibilities with regard to clearly illegal or immoral content. In the remainder of this chapter the discussion will be broadened so as to include responsibilities relating to all kinds of problems that appear in the wake of deficiencies regarding the quality of information and of misperceptions of the quality of the information. The idea of responsibilities for possible negative consequences of (misperceptions of the) quality of information may sound rather vague. For that reason, a large part of the remaining text will be devoted to specifying and articulating what exactly can go wrong with the information on the Internet. These questions gain significance as information on the Internet becomes ever more important in our society.

In comparison with traditional sources of information, such as libraries, books, journals, television, and radio, the Internet makes all kinds of information much more accessible. This phenomenon has often been applauded for its democratizing effects. Unfortunately, there is also a disadvantage. Information that was originally intended for a specific group of people and not in any way processed or adapted to make it fit for a broader audience—"expert information" is the term that I will use to refer in a very loose and broad way to this type of information—can easily be misunderstood and misinterpreted by laymen and, when used as a basis for decisions, lead to unhappy consequences.

Part of the risks of sharing expert knowledge with the general public is caused by the nonexperts' inabilities to recognize and assess the reliability or unreliability of expert information or information that is being presented as expert information. In this section, I will suggest some distinctions and a general conceptual framework, which may offer starting points for nonrestrictive and nonpaternalistic solutions of problems regarding quality of online information.

What exactly is quality of information? It is necessary to ask this question because a clear concept of quality will help to formulate policies for solving the practical problems allegedly caused by flaws of online information. The notion of quality, however, is an ambiguous one. The term is traditionally used to refer to characteristics of an underlying substance, for example, weight, color, and shape, or to properties in general, including formal or supervenient ones. Today, in everyday language, the concept of quality has gained additional or, should we perhaps say, a more specific meaning. Sometimes, quality is simply identified with goodness. More often, however, the term is used in a familiar, though slightly less specific way, that is, to refer to the value of something with respect to its intended use. When applied to data or information, quality is often defined in terms of criteria of truth, accuracy, conformity with facts plus this type of usefulness or functionality. Authors like Frawley et al. (1993) and Berti and Graveleau (1998) already extended their notion of quality to cover the degree of fulfillment of specific interests and preferences of individual users. A common characteristic of both of these accounts is that they do not specify the relationships between the criteria of functionality and the other criteria. The connection between the two types of criteria might, however, shed new light on the problem of quality assessment.

Discussions on issues of quality and quality assessment with regard to information tend to be rather short and clearly aim at particular short-term results. These results can vary from the introduction of new instruments offered by providers to enable users to assess the quality of the information involved, such as certification, to direct efforts to increase different forms of awareness among users of quality issues related to information—media competence or information literacy, as they have been labeled. Deepening the discussion on quality, however, might enable us to find more sophisticated solutions for problems of information quality assessment. It might also give us an opportunity to develop a broader perspective on these issues, which, in turn, might put us in a position to combine and fine-tune a variety of partial solutions. I would like to contend that the discussion can be clarified and deepened with help of what is basically a three-dimensional account of quality. Such an account would be one in terms of reliability, functionality, and significance. In the subsequent paragraphs, I will first expound this account and then turn to the questions of how it may help to solve problems concerning quality assessment and how it may broaden our approach. In doing so, I will emphasize the importance of the user perspective.

Before setting out, I must make a preliminary methodological remark concerning this undertaking. It might be the case that, after ample discussion, I will need to revise certain parts of the proposed account. It might even be the case that disconnecting the concepts of reliability, functionality, and significance might, in the end, make more sense than trying to keep them under the umbrella of quality. What I think is valuable, nonetheless, is the process of analyzing the three dimensions of quality of information and their mutual relationships to clarify quality-related problems and their solutions. What counts is: giving substance to the debate on quality of information and finding starting points for solutions. The exact itinerary is of minor importance.

Reliable information is information that we would be justified in believing. Reliability must be distinguished from truth. Reliable information is not necessarily true, since it is possible that at time t_1 we are justified in believing it, whereas at some later time t_n this information appears to be false: "Discovering that 'a belief is false' does not necessarily mean that, at an earlier time, people were not justified in believing it or that it was wrong to trust it. What is reliable, trustworthy, justified is a matter of what we already know" (Vedder and Wachbroit, 2003, pp. 211).

Assessing the reliability of new information builds on preexisting knowledge. This claim is an epistemologically normative one. It is not to be identified with the factual tendency of many people in everyday life to use the fit or coherence between new information and what they already know as an indication of the reliability of the new information (Vedder, 2002, 2003a). The coherence between new information and previously existing knowledge of one individual can be purely contingent, as long as it is not clear whether his preexisting knowledge is justified. Reliability in the epistemologically normative sense that is under discussion here is a matter of proper justification. In Vedder and Wachbroit (2003), Robert Wachbroit and I distinguish "content criteria" from "pedigree criteria" of reliability.

By "content criteria," we mean the conditions or criteria of reliability that are a function of the content of the information itself. Among these are the criteria of evidence that mostly belong to the domain of experts—people familiar with the subject or with a specific educational background or experience. Other examples of content criteria are logical criteria and, arguably, subject-matter criteria. In general, most people cannot base their assessments of reliability on content criteria. Many determine reliability by pedigree criteria, the conditions or criteria of reliability that relate to the source or intermediary of the information. These have to do with their authoritativeness and having been experienced as credible in the past.

Pedigree criteria are not only used by nonexperts. Experts use them as well. A large part of the training of experts consists in introducing them to the appropriate pedigree criteria applicable in their field of expertise (through courses on how to use libraries, instruments, and sources). Pedigree criteria are established by credibility-conferring institutions. These institutions can be very wide-ranging, from well-organized institutes to broader—sometimes intricate and tangled—networks of cultural and societal arrangements. Perhaps principal ones among the former are the academic institutions such as universities, medical schools, and law schools. Among the broader cultural and societal arrangements are specific conventions and historically grown patterns and traditions of specialization, divisions of labor, and of authority. Here, one may think not only of the traditions that form the cultural basis of the well-organized, credibility-conferring institutions, but also of traditions and conventions that are independently active, for example, certain reputations and small-scale practices and usages, such as the custom of relying on the advice of parents and grandparents in family matters.

Many problems regarding reliability of online information on the Internet are not problems of information lacking reliability, but of receivers misperceiving or not perceiving (un-) reliability. In order to pave the way to discussing this issue, I will, first, give some attention to the dimensions of functionality and significance of information. These dimensions introduce the users' perspective.

Functionality of information should be defined in terms of the connection between the information involved on the one hand, and the purposes of the receivers (including groups and organizations of receivers) on the other. Functionality must not be confused with reliability of information itself. The functionality of information does not influence its reliability. It influences the importance of the information and of its reliability and it affects the degree of urgency of quality enhancing measures. If we say that information is functional, we mean that the information has, in some way or another, a positive bearing on the ways in which the receivers' purposes can be realized. In other words, referring to information as functional information means that the information contributes to the realization of the receivers' purposes. Functionality ultimately depends on the purposes of the receivers. That does not mean that it falls into a totally subjectivist category. In order to ascertain whether information is functional for an individual we need not always know the specific purposes of particular individuals. The purposes of individuals can depend on highly individual tastes and preferences; but they can also be related to the common needs and interests of the human species, communities, and groups.

Some purposes and objectives can be presumed to belong to all or most members of the human species, communities, and groups on the basis of their characteristics and needs. Thus, information can be functional merely for specific individuals, or it can be functional for everyone or for groups of people. Perhaps contrary to ordinary usage, I would like to stipulate the notion of functionality of information as an all-or-nothing notion. Functionality, in my view, should not be considered as a matter of degree. Information is either functional or not. However, it may be useful to make a distinction between functional information as such on the one hand and functional information that is essential—or essential information—on the other. Functional information that is essential is information without which the purpose involved cannot be realized. In the next section, I will specify the notion of essential information further by distinguishing it from significant information. I will also explain that functional information can have different degrees of significance and that essential information can be, but is not necessarily highly significant.

Although reliability of information is not dependent on its functionality, functionality, in a way, is dependent on reliability. In order to be functional, information must, at least in some way or another, be suitable to be grasped and understood by the receivers involved: it must have some structure, some clarity, and must make some sense. This capacity of the information itself, however, must be present whatever the purposes and aims of the receivers might be. In this shallow sense, functionality is dependent on reliability.

Just like functionality, the significance of information has to do with the importance of the information and its reliability. It can also be defined in terms of the connection between the information involved and the purposes of users (including groups and organizations). Significance adds a degree of urgency to functionality. The statement "Information x is functional" just tells us that x is useful with regard to some purpose of a certain individual or a group. "Information x is significant" tells us that knowing x is important because it is functional for a specific purpose that is considered to be important. Significance is a matter of degree. Information can be more or less significant, depending on the importance of the types of purposes for which it is relevant.

It is critical, whether we take what one might call the subjectivist perspective or what one might refer to as the objectivist perspective. In the subjectivist point of view, the significance of information will depend on the individual's appreciation of the purposes for which the information is relevant. The more important the receiver considers his or her purpose to be, the more significant the information will be for this purpose. From the objectivist view, the receiver's exact estimate of the importance of the purposes is irrelevant. The objectivist will measure the importance of the purposes against external standards, such as a certain ranking of basic human needs or a certain view of the good life. For instance: The more a purpose meets an external standard of basic needs of members of the human species, the more significant the information will be considered that enables the user to realize that specific purpose. It would not be very fruitful to try to argue conclusively for or against one of these two conceptions of significance. I consider the restriction to either the subjectivist or the objectivist version of significance as highly artificial. It is far more important to be aware of both interpretations.

Let me finish this part of the argument by explaining the difference between essential information and significant information. Significant information is not necessarily essential, nor is essential information necessarily significant. A piece of information's being essential means that getting to know that piece of information is a necessary condition for the realization of a specific purpose. That purpose, however, can be trivial (according to external standards). In that case, the information, although essential, is also trivial. Of course, the opposite also holds true: When the purposes and the information are significant, the information can, but need not necessarily be, essential.

Now we can take up the thread of the argument again. The distinctions made so far can help us to understand certain problematic phenomena that are related to the assessment of information in general and to the assessment of online information in particular.

As regards problems of reliability there are strictly speaking generally two types:

(a) People lack the necessary expertise to assess information on the basis of content criteria, and they also lack the necessary expertise to assess information on the basis of pedigree criteria. In this case, the problems are due to a lack of competence of the users.
(b) People lack the expertise to assess information on the basis of content criteria, and it is impossible for them to test the information with the help of pedigree criteria. This is the case when the users are, in principle, competent in using pedigree criteria, but the information is presented in such a way that there are no indicators or markers of conformity with pedigree criteria.

Problems with reliability of information can be variations of both themes. The broad accessibility and the many-to-many character of online information, however, put these traditional flaws in a new perspective. Because of the many-to-many character of online information, the very possibility of adequately recognizing pedigree criteria is often lacking where the Internet is concerned (Vedder, 2001). Often, a content provider is anonymous or merely a virtual identity, as the influence of individuals in providing information on the Internet is diminishing, whereas the influence of intelligent systems is increasing. Also, the lack of traditional intermediaries, such as libraries, librarians, and specialized publishers, has a negative influence on the capabilities of information seekers to assess the reliability of information.

These kinds of factors often leave the users without clues or any indication whatsoever about the character, background, and institutional setting of the content provider. An additional complication to the problem is the phenomenon of globalization, which is inherent to online information. Even when the recipient has some information about the content provider, the individual might be unable to estimate the credibility of that provider, simply because the individual will often not be acquainted with the relevant backgrounds and institutional settings from completely different cultures. The recognition procedures and traditions that make up the institutional basis of pedigree criteria may be different in different cultures. A recipient from culture A may not recognize the procedures and traditions of the provider's culture B. It could even be the case that if the recipient from one culture were able to recognize them, he or she would not accept them as credibility-conferring patterns.

The broad accessibility to information also causes different types of reliability-related problems with regard to online information. Information and communication networks like the Internet are media that enormously enhance the accessibility of information. Many people and organizations are able to disperse information through these networks. For many, more information is very easy to find. People do need not to go to libraries anymore; they do not need to order books and journals and lumber a heavy pile home. Complete libraries, books, and journals are available by clicking a mousebutton. The communication channels between experts and specialists (e.g., university libraries, journals) used to be only accessible to these selective groups. Now, these channels are often bypassed as the information is available on publicly accessible Web sites and not on Web pages with specific access requirements such as authentication procedures. This means that many individual users for whom information was not originally intended and for whom that information was traditionally off-limits, are now able to find it. I already referred to the fact that, in practice, many people tend to assess the reliability of information, at least in part, on the basis of the fit or the coherence of the new information with the information that they already have: The degree to which the new information is in accordance with information that is already available, the degree to which the new information reinforces or supports the available information, and vice versa. Depending on whether the person involved is an expert or a nonexpert, the required coherence may concern information on the specific subject of the new information or general background information.

Of course, the degree of fit itself is a reliable indicator of the quality of the information only if the information already present with the user is itself reliable as well.

Nonexperts tend to gain ever more and easier access to information originally intended for an expert audience. That is why problems relating to reliability of information are not exclusively problems that are intentionally or unintentionally caused by content providers or problems inherent to the information. Whereas experts may rightfully use their criterion of fit with regard to this type of information, nonexperts are not able to do the same when they are confronted with information that is originally intended for use by experts.

Similarly, whereas experts may be well equipped to recognize the pedigree criteria that are typical for this specific type of information, nonexperts will be confronted with many more difficulties in recognizing them.

Continuing on this latter point: In order to be able to see whether information satisfies pedigree criteria, we need to have a certain expertise. Depending on the specific type of information, this expertise can be widely shared and consist of experience and an understanding of, for instance, our cultural context. But it can also be the expertise that is typical of certain specialists who have received thorough education or training in a certain field.

There are two causes of the inability to recognize pedigree criteria. It may be the case that the receivers of the information themselves are unable to find and recognize these criteria because they do not know where to look for them. This may be due to the fact that they are not acquainted with the credibility-conferring system behind the criteria or to the fact that they have not been taught where to look. In any case, they lack the required expertise to recognize the markers as markers of reliability.

Another cause of deficient recognition may be more trivial and is situated in the piece of information itself or its presentation due to a deficient visibility of the criterion or, generally, the deficient presentation of the criterion.

14.5 THE RESPONSIBILITIES INVOLVED

The possible causes of problems with regard to the quality of information on the Internet have been identified. What can be done to solve or to prevent these problems?

Unfortunately, the law does not offer many clues as to this question. With regard to information that is not in itself illegal, the possibilities of invoking legal regulation are very limited. Traditionally, the law approaches the problematic consequences of information as a liability problem. As I suggested already in Section 14.3, this approach is insufficient. Liability only arises after the harm and offence have really taken place. Thus, the preventive potential with regard to possible harm and offence and to risks is severely limited. Establishing liability for information is further complicated, because of difficulties of identifying causal relationships, of giving due consideration to the perspectives of content providers and users, and — sometimes — of balancing the good of establishing liability against information freedoms. Furthermore, differences between the liability regimes in different countries may hamper the effective application of liability law to information on the Internet that is, by its very nature, border crossing (Prins and Schellekens, 2004, 2005).

In the previous section, I have distinguished reliability from functionality and significance. With regard to reliability, I have distinguished content criteria from pedigree criteria. I have defined functionality as contributing to the realization of purposes of people. With regard to functionality, I have distinguished between functional information as such and functional information that is also essential, that is, a necessary condition for the realization of the purpose involved. Significant information is functional information that contributes to the realization of important purposes. Significance can be measured against highly individualistic purposes, but also against external standards, for example, those that represent a taxonomy of human needs. With the help of these distinctions, it might be argued that where questions of general policies with regard to quality assessment of online information are concerned, problems regarding significant—as measured against external standards—and essential information should receive priority. It would be useful to elaborate on this point and to draw the rough contours of a typology of different kinds of information that may be considered to represent essential significant information for everyone and for different groups of people. Although this may be a vast project and a cumbersome undertaking—which certainly exceeds the purposes of this paper—it should be kept in mind that even a modest start might already prove to be fruitful, as it could give us a hunch on the directions in which we should seek. As only an indication with regard to the reliability of online medical information, it would probably make sense to say that, in general, medical information should be reliable. More specifically, however, if people look for information on diagnostics or therapeutic treatment because they have a severely ill member of the family or friend, they should be able to feel sure that this information conforms to high standards of reliability.

Interestingly, the typology of essential and significant information is not enough. As we saw, only part of the problems with regard to the assessment of online information are primarily caused by the providers, for example, through the presentation of the information. Often, the initial cause of the problems is the incompetence of users. Therefore, what is necessary is

(1) The creation of new credibility-conferring systems, such as certification systems, allowing us to use pedigree criteria with regard to (online) information, when such systems are lacking.
(2) Raising the visibility of indicators or markers of reliability of information (according to pedigree criteria).
(3) Raising expertise and background knowledge in all users (to enable them to recognize reliability on the basis of pedigree criteria).
(4) Raising the awareness of the varying qualities of information.

With regard to online information, pedigree criteria and the underlying credibility-conferring systems are still largely lacking. In the few cases in which they are already present, they are based on traditional credibility-conferring systems. This is the case, for instance, when well-known brand names are used on the Internet or reference is made to well-known names and titles of newspapers, journals, and broadcasting networks on Web sites. Also some new systems have been developed. There are, for instance, some certification systems that support labels or certificates that appear on Web pages indicating that the information is reliable or that the provider conforms to a self-imposed code guaranteeing reliable information. Generally, an organization or authority that has been especially established, backs up these systems to license information providers to use the label or certificate.

As regards medical information, however, many of these initiatives have been shown to be poor, ineffective, and generally deficient. One of the problems is that the systems supporting these markers are not well established and are too dependent on one form of expressing reliability or, simply, on one licensing authority (Gagliardi and Jadad, 2002). Other problems relate to the intricacies of the systems with which the general public is not familiar, often, the public does not trust the systems to be persistent or viable (Vedder, 2002, 2003a, b). Of course, one must take into consideration that the new media, such as the Internet, lack the long and rich history of credibility-conferring systems that have been developed over the decades and centuries for information dispersed through other media.

With regard to certain types of online information, it may be useful to start thinking anew about credibility-conferring systems and ensuing markers of reliability of information. When developing such a "second generation" of quality systems, it may prove useful to pay more attention to the traditional credibility-conferring systems than seems to have been done in the past. Meticulous study of the complicated patterns and network structure that seem to be characteristic for the traditional systems could be of help when trying to work out systems that will not shut down as soon as one licensing authority disappears. It could also help to find ways of involving experts and the general public and to gain their trust.

The perspectives of the users/receivers of the information should be taken into account in order to decide for what kind of information these markers and basic systems are needed and which kind of information should meet what degree of reliability. The designers of the marker systems should have some sense of the functionalities and the significance that information may have for users. Last, but not least, because in real practice, the degree of fit plays an important role as a criterion for assessing reliability, efforts to introduce new systems for quality assessment run the risk of becoming idle as long as they are not combined with raising the degree of information and education of experts and the general public.

Finally, one may ask: *whose* responsibilities are these? This question, however, is a little premature at this stage. The awareness of the growing dependence on online information is still young. Similarly the search for instruments for maintaining or improving reliability has just started. Perhaps, the safest answer to this question is that, for the time being, responsibilities for safeguarding (the correct perception of) the reliability of online information are responsibilities shared by all parties involved.

ACKNOWLEDGMENTS

The research that lies at the basis of this publication was partially funded by NWO, the Netherlands Foundation for Scientific Research, and the Technology Foundation STW.

REFERENCES

Berti, L. and Graveleau, D. (1998). Designing and filtering online information quality: new perspectives for information service providers. *Proceedings of the Fourth International Conference on Ethical Issues of Information Technology, Ethicomp* 98. EUR, Rotterdam, pp. 79–88.

Feinberg, J. (1970). *Doing and Deserving*. Princeton University Press, Princeton N.J.

Frawley, W.J., Piatetsky-Shapiro, G., and Matheus, C.J. (1993). Knowledge discovery in databases: An overview. In: Piatetsky-Shapiro, G. and Frawley, W.J. (Eds.), *Knowledge Discovery in Databases*. AAAI Press/The MIT Press, Menlo Park, CA.

French, P.A. (1984). *Collective and Corporate Responsibility*. Columbia University Press, New York

Friedman, M. (1970). *The Social Responsibility of Business Is to Increase Its Profits*. The New York Times Magazine, September 13.

Gagliardi, A., and Jadad, A. (2002). Examination of instruments used to rate quality of health information on the internet: chronicle of a voyage with an unclear destination. *British Medical Journal,* 324, 569–573.

Goldman, A.H. (1980). *The Moral Foundations of Professional Ethics*. Rowman and Littlefield, Totowa, N.J.

Hart, H.L.A. (1968). *Punishment and Responsibility*. Oxford University Press, New York/ Oxford.

Johnson, D. (1994). *Computer Ethics*. Prentice Hall, Upper Saddle River, N.J.

May, L. and Hoffman, S. (1991). *Collective Responsibility: Five Decades of Debate in Theoretical and Applied Ethics*. Rowman and Littlefield, Savage, MD.

Prins, C. and Schellekens, M. (2004). The chilling effect of liability law on initiatives to enhance the reliability of online health related information. *European Journal of Health Law,* 11(4), pp. 201–208.

Prins, C. and Schellekens, M. (2005). Fighting untrustworthy internet content: in search for regulatory scenario's. *Information Polity,* 10, 1–11.

Vedder, A. (2001). Misinformation through the internet: epistemology and ethics. In: Vedder, A. (Ed.), *Ethics and the Internet.* Intersentia, Antwerpen, Groningen, Oxford, pp. 125–132.

Vedder, A. (2002). What people think about the reliability of medical information on the Internet. In: Alvarez, I., Ward, B.T., Alvaro de, A.L.J., and Rogerson, S. (Eds.), *The Transformation of Organisations in the Information Age: Social and Ethical Implications.* Universidade Lusiada, Lisbon, pp. 281–292.

Vedder, A. (2003a). Betrouwbaarheid van internetinformatie. In: de Haan, J. and Steyaert, J. (Eds.), *Jaarboek ICT en samenleving. De sociale dimensie van technologie.* Boom/Sociaal Cultureel Planbureau, Amsterdan, pp. 113–132.

Vedder, A. (2003b). Reliability of information. *Computer Ethics, in the Post-September 11 World: Computer Ethics Philosophical Enquiry, Fifth International Conference.* Boston College, June 28, 2003.

Vedder, A. and Lenstra, D. (2006). Reliability and security of information. *Journal for Information, Communication and Ethics in Society,* 4 (1), 3–6.

Vedder, A. and Wachbroit, R. (2003). Reliability of information on the internet: some distinctions. *Ethics and Information Technology,* 5, pp. 211–215.

CHAPTER 15

Virtual Reality and Computer Simulation

PHILIP BREY

15.1 INTRODUCTION

Virtual reality and computer simulation have not received much attention from ethicists. It is argued in this essay that this relative neglect is unjustified, and that there are important ethical questions that can be raised in relation to these technologies. First of all, these technologies raise important ethical questions about the way in which they represent reality and the misrepresentations, biased representations, and offensive representations that they may contain. In addition, actions in virtual environments can be harmful to others and raise moral issues within all major traditions in ethics, including consequentialism, deontology, and virtue ethics. Although immersive virtual reality systems are not yet used on a large scale, nonimmersive virtual reality is regularly experienced by hundreds of millions of users, in the form of computer games and virtual environments for exploration and social networking. These forms of virtual reality also raise ethical questions regarding their benefits and harms to users and society, and the values and biases contained in them.

This paper has the following structure. The first section will describe what virtual reality and computer simulations are and what the current applications of these technologies are. This is followed by a section that analyzes the relation between virtuality and reality, and asks whether virtuality can and should function as a substitute for ordinary reality. Three subsequent sections discuss ethical aspects of representation in virtual reality and computer simulations, the ethics of behavior in virtual reality, and the ethics of computer games. A concluding section discusses issues of professional ethics in the development and professional use of virtual reality systems and computer simulations.

The Handbook of Information and Computer Ethics, Edited by Kenneth Einar Himma and Herman T. Tavani

15.2 BACKGROUND: THE TECHNOLOGY AND ITS APPLICATIONS

15.2.1 Virtual Reality

Virtual reality (VR) technology emerged in the 1980s, with the development and marketing of systems consisting of a head-mounted display (HMD) and datasuit or dataglove attached to a computer. These technologies simulated three-dimensional (3D) environments displayed in surround stereoscopic vision on the head-mounted display. The user could navigate and interact with simulated environments through the datasuit and dataglove, items that tracked the positions and motions of body parts and allowed the computer to modify its output depending on the recorded positions. This original technology has helped define what is often meant by "virtual reality": an immersive, interactive three-dimensional computer-generated environment in which interaction takes place over multiple sensory channels and includes tactile and positioning feedback.

According to Sherman and Craig (2003), there are four essential elements in virtual reality: a virtual world, immersion, sensory feedback, and interactivity. A *virtual world* is a description of a collection of objects in a space and rules and relationships governing these objects. In virtual reality systems, such virtual worlds are generated by a computer. *Immersion* is the sensation of being present in an environment, rather than just observing an environment from the outside. *Sensory feedback* is the selective provision of sensory data about the environment based on user input. The actions and position of the user provide a perspective on reality and determine what sensory feedback is given. *Interactivity*, finally, is the responsiveness of the virtual world to user actions. Interactivity includes the ability to navigate virtual worlds and to interact with objects, characters, and places.

These four elements can be realized to a greater or lesser degree with a computer, and that is why there are both broad and narrow definitions of virtual reality. A narrow definition would only define fully immersive and fully interactive virtual environments as VR. However, there are many virtual environments that do not meet all these criteria to the fullest extent possible, but can still be categorized as VR. Computer games played on a desktop with a keyboard and mouse, like *Doom* and *Half-Life*, are not fully immersive, and sensory feedback and interactivity in them are more limited than in immersive VR systems that include a head-mounted display and datasuit. Yet they do present virtual worlds that are immersive to an extent, and that are interactive and involve visual and auditory feedback. Brey (1999) therefore proposed a broader definition of virtual reality as *a three-dimensional interactive computer-generated environment that incorporates a first-person perspective*. This definition includes both immersive and nonimmersive (screen-based) forms of VR.

The notion of a virtual world, or *virtual environment*, as defined by Sherman and Craig, is broader than that of virtual reality. A virtual world can be defined so as to provide sensory feedback of objects, in which case it yields virtual reality, but it can also be defined without such feedback. Classical text-based adventure games like *Zork*, for example, play in interactive virtual worlds, but users are informed about the state of this world through text. They provide textual inputs, and the game responds with textual information rather than sensory feedback about changes in the world. A virtual world is hence an interactive computer-generated environment, and virtual reality is a special type of virtual world that involves location- and movement-relative sensory feedback.

Next to the term "virtual reality," there is the term "virtuality" and its derivative adjective "virtual." This term has a much broader meaning than the term "virtual reality" or even "virtual environment." As explained more extensively in the following section, the term "virtual" refers to anything that is created or carried by a computer and that mimics a "real," physically localized entity, as in "virtual memory" and "virtual organization." In this essay, the focus will be on virtual reality and virtual environments, but occasionally, especially in the following section, the broader phenomenon of virtuality will be discussed as well.

Returning to the topic of virtual reality, a distinction can be made between *single-user* and *multiuser* or *networked* VR. In single-user VR, there is only one user, whereas in networked VR, there are multiple users who share a virtual environment and appear to each other as avatars, which are graphical representations of the characters played by users in VR. A special type of VR is *augmented reality*, in which aspects of simulated virtual worlds are blended with the real world that is experienced through normal vision or a video link, usually through transparent glasses on which computer graphics or data are overlaid. Related to VR, furthermore, are *telepresence* and *teleoperator systems*, systems that extend a person's sensing and manipulation capability to a remote location by displaying images and transmitting sounds from a real environment that can (optionally) be acted on from a distance through remote handling systems such as robotic arms.

15.2.2 Computer Simulation

A computer simulation is a computer program that contains a model of a particular (actual or theoretical) system. The program can be executed, simulating changes in the system according to certain parameters, after which the output results of the simulation can be analyzed. Computer simulation is also the name of the discipline in which such models are designed, executed, and analyzed. The models in computer simulations are usually abstract and either are or involve mathematical models. Computer simulation has become a useful part of the mathematical modeling of many natural systems in the natural sciences, human systems in the social sciences, and technological systems in the engineering sciences, in order to gain insight into the operations of these systems and to study the effects of alternative conditions and courses of action.

It is not usually an aim in computer simulations, as it is in virtual reality, to do realistic visual modeling of the systems that they simulate. Some of these systems are abstract, and even for those systems that are concrete, the choice is often made not to design graphical representations of the system but to rely solely on abstract models of it. When graphical representations of concrete systems are used, they usually represent only the features that are relevant to the aims of the simulation, and do not aspire to the realism and detail aspired to in virtual reality.

Another difference from virtual reality is that computer simulations need not be interactive. Usually, simulators will determine a number of parameters at the beginning of a simulation and then "run" the simulation without any interventions. In this standard case, the simulator is not himself defined as part of the simulation, as would happen in virtual reality. An exception is an *interactive simulation*, which is a special kind of simulation, also referred to as a *human-in-the-loop* simulation, in which the simulation includes a human operator. An example of such a simulation would be a flight simulator. If a computer simulation is interactive and makes use of three-dimensional graphics and sensory feedback, it also qualifies as a form of virtual reality. Sometimes, also, the term "computer simulation" is used to include any computer program that models a system or environment, even if it is not used to gain insight into the operation of a system. In that broad sense, virtual environments, at least those that aim to do realistic modeling, would also qualify as computer simulations.

15.2.3 Applications

VR is used to simulate both real and imaginary environments. Traditional VR applications are found in medicine, education, arts and entertainment, and the military (Burdea and Coiffet, 2003). In medicine, VR is used for the simulation of anatomical structures and medical procedures in education and training, for example, for performing virtual surgery. Increasingly, VR is also being used for (psycho) therapy, for instance, for overcoming anxiety disorders by confronting patients with virtual anxiety-provoking situations (Wiederhold and Wiederhold, 2004). In education, VR is used in exploration-based learning and learning by building virtual worlds. In the arts, VR is used to create new art forms and to make the experience of existing art more dynamic and immersive. In entertainment, mostly nonimmersive, screen-based forms of VR are used in computer and video games and arcades. This is a form of VR that many people experience on a regular basis. In the military, finally, VR is used in a variety of training contexts for army, navy, and air force. Emerging applications of VR are found in manufacturing, architecture, and training in a variety of (dangerous) civilian professions.

Computer simulations are used in the natural and social sciences to gain insight into the functioning of natural and social systems and in the engineering sciences for performance optimization, safety engineering, training, and education. They are used on a large scale in the natural and engineering sciences, where such fields have sprung up as computational physics, computational neuroscience, computational fluid mechanics, computational meteorology, and artificial life. They are also used on a somewhat more modest scale in the social sciences, for example, in the computational modeling of cognitive processes in psychology, in the computational modeling of artificial societies and social processes, in computational economic modeling, and in strategic management and organizational studies. Computer simulations are increasingly used in education and training, to familiarize students with the workings of systems and to teach them to interact successfully with such systems.

15.3 VIRTUALITY AND REALITY

15.3.1 The Distinction between the Virtual and the Real

In the computer era, the term "virtual" is often contrasted with "real." Virtual things, it is often believed, are things that only have a simulated existence on a computer and are therefore not real, like physical things. Take, for example, rocks and trees in a virtual reality environment. They may look like real rocks and trees, but we know that they have no mass, no weight, and no identifiable location in the physical world and are just illusions generated through electrical processes in microprocessors and the resulting projection of images on a computer screen. "Virtual" hence means "imaginary," "make-believe," "fake," and contrasts with "real," "actual," and "physical." A virtual reality is therefore always only a make-believe reality and can as such be used for entertainment or training, but it would be a big mistake, in this view, to call anything in virtual reality real and to start treating it as such.

This popular conception of the contrast between virtuality and reality can, however, be demonstrated to be incorrect. "Virtual" is not the perfect opposite of "real," and some things can be virtual and real at the same time. To see how this is so, let us start by considering the semantics of "virtual." The word "virtual" has two traditional, precomputer meanings. On the first, most salient meaning, it refers to things almost having certain qualities, or having certain qualities in essence or in effect, but not in name. For instance, if a floor only has a few spots, one can say that the floor is virtually spotless, spotless for all practical purposes, even though it is not formally or actually spotless. Second, virtual can also mean imaginary, and therefore not real, as in optics, where reference is made to virtual foci and images. Note that only on the second, less salient meaning does "virtual" contrast with "real." On the more salient meaning, it does not mean "unreal" but rather "practically but not formally real."

In the computer era, the word "virtual" came to refer to things simulated by a computer, like virtual memory, which is memory that is not actually built into a processor but nevertheless functions as such. Later, the scope of the term "virtual" has expanded to include anything that is created or carried by a computer and that mimics a "real" equivalent, like a virtual library and a virtual group meeting. The computer-based meaning of "virtual" conforms more with the traditional meaning of "virtual" as "practically but not formally real" than with "unreal." Virtual memory, for example, is not unreal memory, but rather a simulation of physical memory that can effectively function as real memory.

Under the above definition of "virtual" as "created or carried by a computer and mimicking a 'real' equivalent," virtual things and processes are simulations of real things, but this need not preclude them from also being real themselves. A virtual game of chess, for example, is also a real game of chess. It is just not played with a physically realized board and pieces. I have argued (Brey, 2003) that a distinction can be made between two types of virtual entities: simulations and ontological reproductions.

Simulations are virtual versions of real-world entities that have a perceptual or functional similarity to them but do not have the pragmatic value or actual consequences of the corresponding real-world equivalent. *Ontological reproductions* are computer imitations of real-world entities that have (nearly) the same value or pragmatic effects as their real-world counterparts. They hence have a real-world significance that extends beyond the domain of the virtual environment and is roughly equal to that of their physical counterpart.

To appreciate this contrast, consider the difference between a virtual chess game and a virtual beer. A virtual beer is necessarily a mere simulation of a real beer: it may look much like a real one and may be lifted and consumed in a virtual sense, but it does not provide the taste and nourishment of a real beer and will never get one drunk. A virtual chess game, in contrast, may lack the physical sensation of moving real chess pieces on a board, but this sensation is considered peripheral to the game, and in relevant other respects playing virtual chess is equivalent to playing chess with physical pieces. This is not to say that the distinction between simulations and ontological reproductions is unproblematic; a virtual entity will be classified as one or the other depending on whether it is judged to share enough of the essential features of its physical counterpart, and pragmatic considerations may come into play in deciding when enough features are present.

Brey (2003) argued that two classes of physical objects and processes can be ontologically reproduced on computers. A first class consists of physical entities that are defined in terms of visual, auditory, or computational properties that can be fully realized on multimedia computers. Such entities include images, movies, musical pieces, stereo systems, and calculators, which are all such that a powerful computer can successfully reproduce their essential physical or formal properties.

A second class consists of what John Searle (1995) has called *institutional entities*, which are entities that are defined by a status or function that has been assigned to them within a social institution or practice. Examples of institutional entities are activities like buying, selling, voting, owning, chatting, playing chess, trespassing, and joining a club and requisite objects like contracts, money, letters, and chess pieces. Most institutional entities are not dependent on a physical medium because they are only dependent on the collective assignment of a status or function. For instance, we call certain pieces of paper money not because of their inherent physical nature but because we collectively assign monetary value to them. But we could also decide, and have decided, to assign the same status to certain sequences of bits that float around on the Internet. In general, if an institutional entity exists physically, it can also exist virtually. Therefore, many of our institutions and institutional practices, whether social, cultural, religious, or economic, can exist in virtual or electronic form.

It can be concluded that many virtual entities can be just as real as their physical counterparts. Virtuality and reality are therefore not each other's opposites. Nevertheless, a large part of ordinary reality, which includes most physical objects and processes, cannot be ontologically reproduced in virtual form. In addition, institutional virtual entities can both possess and lack real-world implications. Sometimes virtual money can also be used as real money, whereas at other times it is only a simulation of real money. People can furthermore disagree on the status of virtual money, with some accepting it as legal tender and others distrusting it. The ontological distinction between reality and virtuality is for these reasons confusing, and the ontological status of encountered virtual objects will often not be immediately clear.

15.3.2 Is the Distinction Disappearing?

Some authors have argued that the emergence of computer-generated realities is working to erase the distinction between simulation and reality, and therefore between truth and fiction. Baudrillard (1995), for example, has claimed that information technology, media, and cybernetics have yielded a transition from an era of industrial production to an era of simulation, in which models, signs, and codes mediate access to reality and define reality to the extent that it is no longer possible to make any sensible distinction between simulations and reality, so that the distinction between reality and simulation has effectively collapsed. Similarly, Borgmann (1999) has argued that virtual reality and cyberspace have led many people to confuse them for alternative realities that have the same actuality of the real world, thus leading to a collapse of the distinction between representation and reality, whereas according to him VR and cyberspace are merely forms of information and should be treated as such.

Zhai (1998), finally, has argued that there is no principled distinction between actual reality and virtual reality and that with further technological improvements in VR, including the addition of functional teleoperation, virtual reality could be made *totally* equivalent to actual reality in its functionality for human life. Effectively, Zhai is arguing that any real-world entity can be ontologically reproduced in VR, given the right technology, and that virtual environments are becoming ontologically more like real environments as technology progresses.

Are these authors right that, in practice if not also conceptually, the distinction between virtuality and reality, and between simulation and reality, is disappearing? First, it is probably true that there is increasingly less difference between the virtual and the real. This is because, as has already been argued, many things are virtual and real at the same time. Moreover, the number of things that are both virtual and real seems to be increasing. This is because as the possibilities of computers and computer networks increase, more and more physical and institutional entities are reproduced in virtual form. There is a flight to the digital realm, in which many believe it is easier and more fun to buy and sell, listen to music or look at art, or do your banking. For many people, therefore, an increasingly large part of their real lives is also virtual, and an increasingly large part of the virtual is also real.

Even if virtuality and reality are not opposite concepts, simulation and reality, and representation and reality, certainly are. Are these two distinctions disappearing as well? Suggesting that they at least become more problematic is the fact that more and more of our knowledge of the real world is mediated by representations and simulations, whether they are models in science, raw footage and enactments in broadcast news, or stories and figures in newspapers or on the Internet. Often, it is not possible, in practice or in principle, to verify the truth or accuracy of these representations through direct inspection of the corresponding state of affairs. Therefore, one might argue that these representations *become* reality for us, for they are all the reality we know.

In addition, the distinction between recordings and simulations is becoming more difficult to make. Computer technology has made it easy to manipulate photos, video footage, and sound recordings, and to generate realistic imagery. Therefore it is nowadays often unclear whether photographic images or video footage on the Internet or in the mass media are authentic or fabricated or enacted. The trend in mass media toward "edutainment" and the enactment and staging of news events has further problematized the distinction.

Yet, all this does not prove that the distinction between simulation/representation and reality has collapsed. People do not get all of their information from media representations. They also move around and observe the world for themselves. People still question and critically investigate whether representations are authentic or correspond to reality. People hence still maintain an ontological distinction, even though it has become more difficult epistemologically to discern whether things and events are real or simulated. Zhai's suggestion that the distinction could be completely erased through further perfection of virtual reality technology is unlikely to hold because it is unlikely that virtual reality could ever fully emulate actual reality in its functionality for human life. Virtual reality environments cannot, after all, sustain real biological processes, and therefore they can never substitute for the complete physical world.

15.3.3 Evaluating the Virtual as a Substitute for the Real

Next to the ontological and epistemological questions regarding distinction between the virtual and the real and how we can know this distinction, there is the normative question of how we should evaluate virtuality as a substitute for reality. First of all, are virtual things better or worse, more or less valuable, than their physical counterparts? Some authors have argued that they are in some ways better: they tend to be more beautiful, shiny, and clean, and more controllable, predictable, and timeless. They attain, as Heim (1994) has argued, a supervivid hyper-reality, like the ideal forms of Platonism, more perfect and permanent than the everyday physical world, answering to our desire to transcend our mortal bodies and reach a state of permanence and perfection. Virtual reality, it may seem, can help us live lives that are more perfect, more stimulating, and more in accordance with our fantasies and dreams.

Critics of virtuality have argued that the shiny, polished objects of VR are mere surrogates: simplified and inferior substitutes for reality that lack authenticity. Borgmann (1999), for example, has argued that virtuality is an inadequate substitute for reality, because of its fundamental ambiguity and fragility, and lacks the engagement and splendor of reality. He also argues that virtuality threatens to alter our perspective on reality, causing us to see it as yet another sign or simulation. Dreyfus (2001) has argued that presence in VR and cyberspace gives a disembodied and therefore false experience of reality and that even immersive VR and telepresence present one with impoverished experiences.

Another criticism of the virtual as a substitute for the real is that investments in virtual environments tend to correlate with disinvestments in people and activities in real life (Brey, 1998). Even if this were to be no loss to the person making the disinvestments, it may well be a loss to others affected by it. If a person takes great effort in caring for virtual characters, he or she may have less time left to give similar care and emotional attention to actual persons and animals, or may be less interested in giving it. In this way, investments in VR could lead to a neglect of real life and therefore a more solitary society. On the contrary, virtual environments can also be used to vent aggression, harming only virtual characters and property and possibly preventing similar actions in real life.

15.4 REPRESENTATION AND SIMULATION: ETHICAL ISSUES

VR and computer simulations are representational media: they represent real or fictional objects and events. They do so by means of different types of representations: pictorial images, sounds, words, and symbols. In this section, ethical aspects of such representations will be investigated. It will be investigated whether representations are morally neutral and whether their manufacture and use in VR and computer simulations involves ethical choices.

15.4.1 Misrepresentations, Biased Representations, and Indecent Representations

I will argue that representations in VR or computer simulations can become morally problematic for any of three reasons. First, they may cause harm by failing to uphold *standards of accuracy*. That is, they may *misrepresent* reality. Such representations will be called *misrepresentations*. Second, they may fail to uphold *standards of fairness*, thereby unfairly disadvantaging certain individuals or groups. Such representations will be called *biased representations*. Third, they may violate standards of decency and public morality. I will call such representations *indecent representations*.

Misrepresentation in VR and computer simulation occurs when it is part of the aim of a simulation to realistically depict aspects of the real world, yet the simulation fails to accurately depict these features (Brey, 1999). Many simulations aim to faithfully depict existing structures, persons, states of affairs, processes, or events. For example, VR applications have been developed that simulate in great detail the visual features of existing buildings such as the Louvre or the Taj Mahal or the behavior of existing automobiles or airplanes. Other simulations do not aim to represent particular existing structures, but nevertheless aim to be realistic in their portrayal of people, things, and events. For example, a VR simulation of military combat will often be intended to contain realistic portrayals of people, weaponry, and landscapes without intending to represent particular individuals or a particular landscape.

When simulations aim to be realistic, they are subject to certain *standards of accuracy*. These are standards that define the degree of freedom that exists in the depiction of a phenomenon and that specify what kinds of features must be included in a representation for it to be accurate, what level of detail is required, and what kinds of idealizations are permitted. Standards of accuracy are fixed in part by the aim of a simulation. For example, a simulation of surgery room procedures should be highly accurate if it is used for medical training, should be somewhat accurate when sold as edutainment, and need not be accurate at all when part of a casual game. Standards of accuracy can also be fixed by promises or claims made by manufacturers. For example, if a game promises that surgery room procedures in it are completely realistic, the standards of accuracy for the simulation of these procedures will be high. People may also disagree about the standards of accuracy that are appropriate for a particular simulation. For example, a VR simulation of military combat that does not represent killings in graphic detail may be discounted as inaccurate and misleading by antiwar activists, but may be judged to be sufficiently realistic for the military for training purposes.

Misrepresentations of reality in VR and computer simulations are morally problematic to the extent that they can result in harm. The greater these harms are, and the greater the chance that they occur, the greater the moral responsibility of designers and manufacturers to ensure accuracy of representations. Obviously, inaccuracies in VR simulations of surgical procedures for medical training or computer simulations to test the bearing power of bridges can lead to grave consequences. A misrepresentation of the workings of an engine in educational software causes a lesser or less straightforward harm: it causes students to have false beliefs, some of which could cause harms at a later point in time.

Biased representations constitute a second category of morally problematic representations in VR modeling and computer simulation (Brey, 1999). A biased representation is a representation that unfairly disadvantages certain individuals or groups or that unjustifiably promotes certain values or interests over others. A representation can be biased in the way it idealizes or selectively represents phenomena. For example, a simulation of global warming may be accurate overall but unjustifiably ignore the contribution to global warming made by certain types of industries or countries. Representations can also be biased by stereotyping people, things, and events. For example, a computer game may contain racial or gender stereotypes in its depiction of people and their behaviors. Representations can moreover be biased by containing implicit assumptions about the user, as in a computer game that plays out male heterosexual fantasies, thereby assuming that players will generally be male and heterosexual. They can also be biased by representing affordances and interactive properties in objects that make them supportive of certain values and uses but not of others. For example, a gun in a game may be programmed so that it can be used to kill enemies but not to knock them unconscious.

Indecent representations constitute a third and final category of morally problematic representations. Indecent representations are representations that are considered shocking or offensive or that are held to break established rules of good behavior or morality and that are somehow shocking to the senses or moral sensibilities.

Decency standards vary widely across different individuals and cultures, however, and what is shocking or immoral to some will not be so to others. Some will find any depiction of nudity, violence, or physical deformities indecent, whereas others will find any such depiction acceptable. The depiction of particular acts, persons, or objects may be considered blasphemous in certain religions but not outside these religions. For this reason, the notion of an indecent representation is a relative notion, barring the existence of universally indecent acts or objects, and there will usually be disagreement about what representations count as indecent. In addition, the context in which a representation takes place may also influence whether it is considered decent. For example, the representation of open heart surgery, with some patients surviving the procedure but others dying on the operation table, may be inoffensive in the context of a medical simulator but offensive in the context of a game that makes light of such a procedure.

15.4.2 Virtual Child Pornography

Pornographic images and movies are considered indecent by many, but there is a fairly large consensus that people have a right to produce pornography and use it in private. Such a consensus does not consist for certain extreme forms of pornography, including child pornography. Child pornography is considered wrong because it harms the children that are used to produce it. But what about virtual child pornography? Virtual child pornography is the digital creation of images or animated pictures that depict children engaging in sexual activities or that depict them in a sexual way. Nowadays, such images and movies can be made to be highly realistic. No real children are abused in this process, and therefore the major reason for outlawing child pornography does not apply to it. Does this mean that virtual child porn is morally permissible and that its production and consumption should be legal?

The permissibility of virtual child porn has been defended on the argument that no actual harm is done to children and that people have a right to free speech by which they should be permitted to produce and own virtual child pornography, even if others find such images offensive. Indeed, the U.S. Supreme Court struck down a congressional ban on virtual child porn in 2002 with the argument that this ban constituted too great a restriction on free speech. The court also claimed that no proof had been given of a connection between computer-generated child pornography and the exploitation of actual children. An additional argument that is sometimes used in favor of virtual child porn is that its availability to pedophiles may actually decrease the chances that they will harm children.

Opponents of virtual child porn have sometimes responded with deontological arguments, claiming that it is degrading to children and undermines human dignity. Such arguments cut little ice, however, in a legal arena that is focused on individual rights and harms. Since virtual child porn does not seem to violate individual rights, opponents have tried out various arguments to the effect that it does cause harm. One existing argument is that virtual child porn causes indirect harm to children because it encourages child abuse. This argument goes opposite the previously stated argument that virtual child porn should be condoned because it makes child abuse less likely. The problem is that it is very difficult to conduct studies that provide solid empirical evidence for either position. Another argument is that failing to criminalize virtual child porn will harm children because it makes it difficult to enforce laws that prohibit actual child pornography. This argument has been used often by law enforcers to criminalize virtual child porn. As Levy (2002) has argued, this argument is, however, not plausible, among other reasons because experts are usually able to make the distinction between virtual and actual pictures.

Levy's own argument against virtual child porn is not that it will indirectly harm children, but that it may ultimately harm women by eroticizing inequality in sexual relationships. He admits, however, that he lacks the empirical evidence to back up this claim. Sandin (2004) has presented an argument with better empirical support, which is that virtual child porn should be outlawed because it causes significant harm to a great many people who are revulsed by it. The problem with this argument, however, is that it gives too much weight to harm caused by offense. If actions should be outlawed whenever they offend a large group of people, then individual rights would be drastically curtailed, and many things, ranging from homosexual behavior to interracial marriage, would still be illegal. It can be concluded that virtual child pornography will remain a morally controversial issue for some time to come, as no decisive arguments for or against it have been provided so far.

15.4.3 Depiction of Real Persons

Virtual environments and computer simulations increasingly include characters that are modeled after the likeness of real persons, whether living or deceased. Also, films and photographs increasingly include manipulated or computer-generated images of real persons who are placed in fictional scenes or are made to perform behaviors that they have not performed in real life. Such appropriations of likenesses are often made without the person's consent. Is such consent morally required, or should the depictions of real persons be seen as an expression of artistic freedom or free speech?

Against arguments for free speech, three legal and moral arguments have traditionally been given for restrictions on the use of someone's likeness (Tabach-Bank, 2004). First, the right to *privacy* has been appealed to. It has been argued that the right to privacy includes a right to live a life free from unwarranted publicity (Prosser, 1960). The public use of someone's likeness, in a particular manner or context, can violate someone's privacy by intruding upon his seclusion or solitude or into his private affairs, by working to publicly disclose embarrassing private facts about him, or to place him in a false light in the public eye. A second argument for restricting the use of someone's likeness is that it can be used for *defamation*. Depicting someone in a certain way, for example, as being involved in immoral behavior or in a ridiculous situation, can defame him by harming his public reputation.

In some countries, like the U.S., there is also a separate recognized *right of publicity*. The right to publicity is an individual's right to control and profit from the commercial use of his name, likeness, and persona. The right to publicity has emerged as a protection of the commercial value of the identity of public personalities, or celebrities, who frequently use their identity to sell or endorse products or services. It is often agreed that celebrities have less of an expectation of privacy because they are public personalities, but have a greater expectation of a right to publicity. In the use of the likenesses of real persons in virtual environments or doctored digital images, rights to free speech, freedom of the press, and freedom of artistic expression will therefore have to be balanced against the right to privacy, the right of publicity, and the right to protection from defamation.

15.5 BEHAVIOR IN VIRTUAL ENVIRONMENTS: ETHICAL ISSUES

The preceding section focused on ethical issues in design and embedded values in VR and computer simulations. This section focuses on ethical issues in the use of VR and interactive computer simulations. Specifically, the focus will be on the question of whether actions within the worlds generated by these technologies can be unethical.

This issue will be analyzed for both single-user and multiuser systems. Before it is taken up, I will first consider how actions in virtual environments take place and what is the relation between users and the characters as that they appear in virtual environments.

15.5.1 Avatars, Agency, and Identity

In virtual environments, users assume control over a graphically realized character called an *avatar*. Avatars can be built after the likeness of the user, but more often they are generic persons or fantasy characters. Avatars can be controlled from a first-person perspective, in which the user sees the world through the avatar's eyes, or from a third-person perspective. In multiuser virtual environments, there will be multiple avatars corresponding to different users. Virtual environments also frequently contain *bots*, which are programmed or scripted characters that behave autonomously and are controlled by no one.

The identity that users assume in a virtual environment is a combination of the features of the avatar they choose, the behaviors that they choose to display with it, and the way others respond to the avatar and its behaviors. Avatars can function as a manifestation of the user, who behaves and acts like himself, and to whom others respond as if it is the user himself, or as a character that has no direct relation to the user and that merely plays out a role. The actions performed by avatars can therefore range from authentic expressions of the personality and identity of the user to experimentation with identities that are the opposite of who the user normally is, whether in appearance, character, status, or other personal characteristics.

Whether or not the actions of an avatar correspond with how a user would respond in real life, there is no question that the user is causally and morally responsible for actions performed by his or her avatar. This is because users normally have full control over the behavior of their avatars through one or more input devices. There are occasional exceptions to this rule, because avatars are sometimes taken over by the computer and then behave as bots. The responsibility for the behavior of bots could be assigned to either their programmer or to whomever introduced them into a particular environment, or even to the programmer of the environment for not disallowing harmful actions by bots (Ford, 2001).

15.5.2 Behavior in Single-User VR

Single-user VR offers much fewer possibilities for unethical behavior than multiuser VR because there are no other human beings that could be directly affected by the behavior of a user. The question is whether there are any behaviors in single-user VR that could qualify as unethical. In Brey (1999), I considered the possibility that certain actions that are unethical when performed in real life could also be unethical when performed in single-user VR. My focus was particularly on violent and degrading behavior toward virtual human characters, such as murder, torture, and rape. I considered two arguments for this position, the argument from moral development and the argument from psychological harm.

According to the argument from moral development, it is wrong to treat virtual humans cruelly because doing so will make it more likely that we will treat real humans cruelly. The reason for this is that the emotions appealed to in the treatment of virtual humans are the same emotions that are appealed to in the treatment of real humans because these actions resemble each other so closely. This argument has recently gained empirical support (Slater et al., 2006). The argument from psychological harm is that third parties may be harmed by the knowledge or observation that people engage in violent, degrading, or offensive behavior in single-user VR and that therefore this behavior is immoral. This argument is similar to the argument attributed to Sandin in my earlier discussion of indecent representations. I claimed in Brey (1999) that although harm may be caused by particular actions in single-user VR because people may be offended by them, it does not necessarily follow that the actions are immoral, but only that they cause indirect harm to some people. One would have to balance such harms against any benefits, such as pleasurable experiences to the user.

McCormick (2001) has offered yet another argument according to which violent and degrading behavior in single-user VR can be construed as unethical. He argues that repeated engagement in such behavior erodes one's character and reinforces "virtueless" habits. He follows Aristotelian virtue ethics in arguing that this is bad because it makes it difficult for us to lead fulfilling lives, because as Aristotle has argued, a fulfilling life can only be lived by those who are of virtuous character. More generally, the argument can be made that the excessive use of single-user VR keeps one from leading a good life, even if one's actions in it are virtuous, because one invests into fictional worlds and fictional experiences that seem to fulfill one's desires but do not actually do so (Brey, 2007).

15.5.3 Behavior in Multiuser VR

Many unethical behaviors between persons in the real world can also occur in multiuser virtual environments. As discussed earlier in the section on reality and virtuality, there are two classes of real-world phenomena that can also exist in virtual form: institutional entities that derive their status from collective agreements, like money, marriage, and conversations, and certain physical and formal entities, like images and musical pieces, which computers are capable of physically realizing. Consequently, unethical behaviors involving such entities can also occur in VR, and it is possible for there to be real thefts, insults, deceptions, invasions of privacy, breaches of contract, or damage to property in virtual environments.

Immoral behaviors that cannot really happen in virtual environments are those that are necessarily defined over physically realized entities. For example, there can be real insult in virtual environments, but not real murders, because real murders are defined over persons in the physical world, and the medium of VR does not equip users with the power to kill persons in the physical world. It may, of course, be possible to kill avatars in VR, but these are, of course, not killings of real persons. It may also be possible to *plan* a real murder in VR, for example by using VR to meet up with a hit man, but this cannot then be followed up by the *execution* of a real murder in VR.

Even though virtual environments can be the site of real events with real consequences, they are often recognized as fictional worlds in which characters merely play out roles. In such cases, even an insult may not be a real insult, in the sense of an insult made by a real person to another real person, because it may only have the status of an insult between two virtual characters. The insult is then only real in the context of the virtual world, but is not real in the real world. Ambiguities arise, however, because it will not always be clear when actions and events in virtual environments should be seen as fictional or real (Turkle, 1995). Users may assign different statuses to objects and events, and some users may identify closely with their avatar, so that anything that happens to their avatar also happens to them, whereas others may see their avatar as an object detached from themselves with which they do not identify closely. For this reason, some users may feel insulted when their avatar is insulted, whereas others will not feel insulted at all.

This ambiguity in the status of many actions and events in virtual worlds can lead to moral confusion as to when an act that takes place in VR is genuinely unethical and when it merely resembles a certain unethical act. The most famous case of this is the case of the "rape in cyberspace" reported by Dibbell (1993). Dibbell reported an instance of a "cyberrape" in LambdaMOO, a text-only virtual environment in which users interact with user-programmable avatars. One user used a subprogram that took control of avatars and made them perform sex acts on each other. Users felt their characters were raped, and some felt that they themselves were indirectly raped or violated as well. But is it ever possible for someone to be sexually assaulted through a sexual assault on her avatar, or does sexual assault require a direct violation of someone's body? Similar ambiguities exist for many other immoral practices in virtual environments, like adultery and theft. If it would constitute adultery when two persons were to have sex with each other, does it also constitute adultery when their avatars have sex? When a user steals virtual money or property from other users, should he be considered a thief in real life?

15.5.4 Virtual Property and Virtual Economies

For any object or structure found in a virtual world, one may ask the question: who owns it? This question is already ambiguous, however, because there may be both virtual and real-life owners of virtual entities. For example, a user may be considered to be the owner of an island in a virtual world by fellow users, but the whole world, including the island, may be owned by the company that has created it and permits users to act out roles in it. Users may also become creators of virtual objects, structures, and scripted events, and some put hundreds of hours of work into their creations. May they therefore also assert intellectual property rights to their creations? Or can the company that owns the world in which the objects are found and the software with which they were created assert ownership? What kind of framework of rights and duties should be applied to virtual property (Burk, 2005)?

The question of property rights in virtual worlds is further complicated by the emergence of so-called virtual economies. Virtual economies are economies that exist within the context of a persistent multiuser virtual world. Such economies have emerged in virtual worlds like *Second Life* and *The Sims Online*, and in massively multiplayer online role-playing games (MMORPGs) like *Entropia Universe*, *World of Warcraft*, *Everquest,* and *EVE Online*. Many of these worlds have millions of users. Economies can emerge in virtual worlds if there are scarce goods and services in them for which users are willing to spend time, effort, or money, if users can also develop specialized skills to produce such goods and services, if users are able to assert property rights on goods and resources, and if they can transfer goods and services between them.

Some economies in these worlds are primitive barter economies, whereas other make use of recognized currencies. Second Life, for example, makes use of the Linden Dollar (L$) and Entropia Universe has the Project Entropia Dollar (PED), both of which have an exchange rate against real U.S. dollars. Users of these worlds can hence choose to acquire such virtual money by doing work in the virtual world (e.g., by selling services or opening a virtual shop) or by making money in the real world and exchanging it for virtual money. Virtual objects are now frequently traded for real money outside the virtual worlds that contain them, on online trading and auction sites like eBay. Some worlds also allow for the trade of land. In December 2006, the average price of a square meter of land in Second Life was L$ 9.68 or U.S. $0.014 (up from L$ 6.67 in November), and over 36,000,000 square meters were sold.[1] Users have been known to pay thousands of dollars for cherished virtual objects, and over $100,000 for real estate.

The emergence of virtual economies in virtual environments raises the stakes for their users, and increases the likelihood that moral controversies ensue. People will naturally be more likely to act immorally if money is to be made or if valuable property is to be had. In one incident that took place in China, a man lent a precious sword to another man in the online game *Legend of Mir 3*, who then sold it to a third party. When the lender found out about this, he visited the borrower at his home and killed him.[2] Cases have also been reported of Chinese sweatshop laborers who work day and night in conditions of practical slavery to collect resources in games like *World of Warcraft* and *Lineage*, which are then sold for real money.

There have also been reported cases of virtual prostitution, for instance on *Second Life*, where users are paid to (use their avatar to) perform sex acts or to serve as escorts. There have also been controversies over property rights. On *Second Life,* for example, controversy ensued when someone introduced a program called CopyBot that could copy any item in the world. This program wreaked havoc on the economy, undermining the livelihood of thousands of business owners in *Second Life*, and was eventually banned after mass protests.[3] Clearly, then, the emergence of virtual economies and serious investments in virtual property generates many new ethical

[1]Source: https://secondlife.com/whatis/economy_stats.php. Accessed 1/3/2007.

[2]Online gamer killed for selling cyber sword. *ABC NewsOnline*, March 30, 2005. http://www.abc.net.au/news/newsitems/200503/s1334618.htm

[3]Linden bans CopyBot following resident protests. *Reuters News*, Wednesday November 15, 2006. http://secondlife.reuters.com/stories/2006/11/15/linden-bans-copybot-following-resident-protests/

issues in virtual worlds. The more time, money, and social capital people invest in virtual worlds, the more such ethical issues will come to the front.

15.6 THE ETHICS OF COMPUTER GAMES

Contemporary computer and video games often play out in virtual environments or include computer simulations, as defined earlier. Computer games are nowadays mass media. A recent study shows that the average American 8- to 18-year old spends almost 6 h per week playing computer games, and that 83% have access to a video game console at home (Rideout et al., 2005). Adults are also players, with four in ten playing computer games on a regular basis.[4] In 2005, the revenue in the U.S. generated by the computer and game industry was over U.S. $7 billion, far surpassing the film industry's annual box office results.[5] Computer games have had a vast impact on youth culture, but also significantly influence the lives of adults. For these reasons alone, an evaluation of their social and ethical aspects is needed.

Some important issues bearing on the ethics of computer games have already been discussed in previous sections, and therefore will be covered less extensively here. These include, among others, ethical issues regarding biased and indecent representations; issues of responsibility and identity in the relation between avatars, users, and bots; the ethics of behavior in virtual environments; and moral issues regarding virtual property and virtual economies. These issues and the conclusions reached regarding them all fully apply to computer games. The focus in this section will be on three important ethical questions that apply to computer games specifically: Do computer games contribute to individual well-being and the social good? What values should govern the design and use of computer games? Do computer games contribute to gender inequality?

15.6.1 The Goods and Ills of Computer Games

Are computer games generally a benefit to society? Many parents do not think so. They worry about the extraordinary amount of time their children spend playing computer games, and about the excessive violence that takes place in many games. They worry about negative effects on family life, schoolwork, and the social and moral development of their kids. In the media, there has been much negative reporting about computer games. There have been stories about computer game addiction and about players dying from exhaustion and starvation after playing video games for days on end. There have also been stories about ultraviolent and otherwise controversial video games, and the ease with which children can gain access to them. The Columbine High School massacre, in 1999, in which two teenage students went on a shooting rampage,

---

[4]Poll: 4 in 10 adults play electronic games. MSNBC.com, May 8, 2006. http://www.msnbc.msn.com/id/12686020/

[5]*2006 Essential Facts about the Computer and Video Game Industry*, Entertainment Software Association, 2006. http://www.theesa.com/archives/files/Essential%20Facts%202006.pdf

was reported in the media to have been inspired by the video game Doom, and since then, other mass shootings have also been claimed to have been inspired by video games. Many have become doubtful, therefore, as to whether computer games are indeed a benefit to society rather than a social ill.

The case against computer games tends to center on three perceived negative consequences: addiction, aggression, and maladjustment. The perceived problem of addiction is that many gamers get so caught up in playing that their health, work or study, family life, and social relations suffer. How large this problem really is has not yet been adequately documented (but see Chiu et al., 2004). There is clearly a widespread problem, as there has been a worldwide emergence of clinics for video addicts in recent years. Not all hard-core gamers will be genuine addicts in the psychiatric sense, but many do engage in overconsumption, resulting in the neglect described above. The partners of adults who engage in such overconsumption are sometimes called gamer widows, analogous to soccer widows, denoting that they have a relationship with a gamer who pays more attention to the game than to them.

Whereas there is no doubt that addiction to video games is a real social phenomenon, there is somewhat less certainty that playing video games can be correlated with increased aggression, as some have claimed. A large percentage of contemporary video games involve violence. The preponderance of the evidence seems to indicate that the playing of such violent video games can be correlated with increases in aggression, including increases in aggressive thoughts, aggressive feeling, aggressive behaviors, a desensitization to real-life violence, and a decrease in helpful behaviors (Bartholow, 2005; Carnagey et al., 2007). However, some studies have found no such correlations, and present findings remain controversial. Whatever the precise relation between violent video games and aggression turns out to be, it is clear now that there is a huge difference between the way that children are taught to behave toward others by their parents and how they learn to behave in violent video games. This at least raises the question of how their understanding of and attitude toward violence and aggression is influenced by violent video games.

A third hypothesized ill of video games is that they cause individuals to be socially and cognitively slighted and maladjusted. This maladjustment is attributed in part to the neglect of studies and social relations due to an overindulgence in video games and to increased aggression levels from playing violent games. But it is also held to be due to the specific skills and understandings that users gain from video games. Children who play video games are exposed to conceptions of human relations and the workings of the world that have been designed into them by game developers. These conceptions have not been designed to be realistic or pedagogical, and often rely on stereotypes and simplistic modes of interaction and solutions to problems. It is therefore conceivable that children develop ideas and behavioral routines while playing computer games that leave much to be desired.

The case in favor of computer games begins with the observation that they are a new and powerful medium that seems to bring users pleasure and excitement, and that seems to allow for new forms of creative expression and new ways of acting out fantasies. Moreover, although playing computer games may contribute to social isolation, it can also stimulate social interaction. Playing multiplayer games is a social activity that involves interactions with other players, and that can even help solitary individuals find new friends. Computer games may moreover induce social learning and train social skills. This is especially true for role-playing games and games that involve verbal interactions with other characters. Such games let players experiment with social behavior in different social settings, and role-playing games can also make users intimately familiar with the point of view and experiences of persons other than themselves. Computer games have moreover been claimed to improve perceptual, cognitive, and motor skills, for example by improving hand-eye coordination and improving visual recognition skills (Green and Bavelier, 2003; Johnson, 2005).

15.6.2 Computer Games and Values

It has long been argued in computer ethics that computer systems and software are not value-neutral but are instead value-laden (Brey, 2000; Nissenbaum, 1998). Computer games are no exception. Computer games may suggest, stimulate, promote, or reward certain values while shunning or discouraging others. Computer games are value-laden, first of all, in the way they represent the world. As discussed earlier, such representations may contain a variety of biases. They may, for example, promote racial and gender stereotypes (Chan, 2005; Ray, 2003), and they may contain implicit, biased assumptions about the abilities, interests, or gender of the player. Simulation games like SimCity may suggest all kinds of unproven causal relations, for example, between poverty and crime, which may help shape attitudes and feed prejudices. Computer games may also be value-laden in the interactions that they make possible. They, for example, may be designed to make violent action the only solution to problems faced by a player. Computer games can also be value-laden in the storylines they suggest for players and in the feedback and rewards that are given. Some first-person shooters award extra points, for example, for not killing innocent bystanders, whereas others instead award extra points for killing as many as possible.

A popular game like The Sims can serve to illustrate how values are embedded in games. The Sims is a game that simulates the everyday lives and social relationships of ordinary persons. The goal of characters in the game is happiness, which is attained through the satisfaction of needs like hunger, comfort, hygiene, and fun. These needs can be satisfied through success in one's career, and through consumption and social interaction. As Sicart (2003) has argued, The Sims thus presents an idealized version of a progressive liberal consumer society in which the goal in life is happiness, gained by being a good worker and consumer. The team-based first-person shooter America's Army presents another example. This game is offered as a free download by the U.S. government, who uses it to stimulate U.S. Army recruitment. The game is designed to give a positive impression of the U.S. Army. Players play as servicemen who obey orders and work together to combat terrorists. The game claims to be highly realistic, yet it has been criticized for not showing certain realistic aspects of military life, such as collateral damage, harassment, and gore. It may hence prioritize certain values and interests over others by presenting an idealized version of military life that serves the interests of recruiters but not necessarily those of the recruit or of other categories of people depicted in the game.

The question is how much influential computer games actually have on the values of players. The amount of psychological research done on this topic is still limited. However, psychological research on the effect of other media, such as television, has shown that it is very influential in affecting the value of media users, especially children. Since many children are avid consumers of computer games, there are reasons to be concerned about the values projected on them by such games. Children are still involved in a process of social, moral, and cognitive development, and computer games seem to have an increasing role in this developmental process. Concern about the values embedded in video games therefore seems warranted. On the contrary, computer games are *games*, and therefore should allow for experimentation, fantasy, and going beyond socially accepted boundaries. The question is how games can support such social and moral freedom without also supporting the development of skewed values in younger players.

Players do not just develop values on the basis of the structure of the game itself, they also develop them by interacting with other players. Players communicate messages to each other about game rules and acceptable in-game behavior. They can respond positively or negatively to certain behaviors, and may praise or berate other players. In this way, social interactions in games may become part of the socialization of individuals and influence their values and social beliefs. Some of these values and norms may remain limited to the game itself, for example, norms governing the permissibility of cheating (Kimppa and Bissett, 2005). In some games, however, like massively multiplayer online role-playing games (MMORPGs), socialization processes are so complex as to resemble real life (Warner and Raiter, 2005), and values learned in such games may be applied to real life as well.

15.6.3 Computer Games and Gender

Game magazines and game advertisements foster the impression that computer games are a medium for boys and men. Most pictured gamers are male, and many recurring elements in images, such as scantily clad, big-breasted women, big guns, and fast cars, seem to be geared toward men. The impression that computer games are mainly a medium for men is further supported by usage statistics. Research has consistently shown that fewer girls and women play computer games than boys and men, and those that do spend less time playing than men. According to research performed by Electronic Arts, a game developer, among teenagers only 40% of girls play computer games, compared to 90% of boys. Moreover, when they reach high school, most girls lose interest, whereas most boys keep playing.[6] A study by the UK games trade body, the Entertainment and Leisure Publishers Association, found that in Europe women gamers make up only a quarter of the gaming population.[7]

[6]Games industry is "failing women." *BBC News*, August 21, 2006. http://news.bbc.co.uk/2/hi/technology/5271852.stm

[7]*Chicks and Joysticks. An Exploration of Women and Gaming.* ELSPA White Paper, September 2004. www.elspa.com/assets/files/c/chicksandjoysticksanexplorationofwomenandgaming_176.pdf

The question of whether there is a gender bias in computer games is morally significant because it is a question about gender equality. If it is the case that computer games tend to be designed and marketed for men, then women are at an unfair disadvantage, as they consequently have less opportunity to enjoy computer games and their possible benefits. Among such benefits may be greater computer literacy, an important quality in today's marketplace. But is the gender gap between usage of computer games really the result of gender bias in the gaming industry, or could it be the case that women are simply less interested in computer games than men, regardless of how games are designed and marketed?

Most analysts hold that the gaming industry is largely to blame. They point to the fact that almost all game developers are male, and that there have been few efforts to develop games suitable for women. To appeal to women, it has been suggested, computer games should be less aggressive, because women have been socialized to be nonaggressive (Norris, 2004). It has also been suggested that women have a greater interest in multiplayer games, games with complex characters, games that contain puzzles, and games that are about human relationships. Games should also avoid assumptions that the player is male and avoid stereotypical representations of women. Few existing games contain good role models for women. Studies have found that most female characters in games have unrealistic body images and display stereotypical female behaviors, and that a disproportionate number of them are prostitutes and strippers.[8]

15.7 VIRTUAL REALITY, SIMULATION, AND PROFESSIONAL ETHICS

In discussing issues of professional responsibility in relation to virtual reality systems and computer simulations, a distinction can be made between the responsibility of developers of such systems and that of professional users. *Professional users* can be claimed to have a responsibility to acquaint themselves with the technology and its potential consequences and to use it in a way that is consistent with the ethics of their profession. The responsibility of *developers* includes giving consideration to ethical aspects in the design process and engaging in adequate communication about the technology and its effects to potential users.

In the development of computer simulations, the accuracy of the simulation and its reliability as a foundation for decision-making in the real world are of paramount importance. The major responsibility of simulation professionals is therefore to avoid misrepresentations where they can and to adequately communicate the limitations of simulations to users (McLeod, 1983). These responsibilities are, indeed, a central ingredient in a recent code of ethics for simulationists adopted by a large number of professional organizations for simulationists (Ören et al., 2002). The responsibility for accuracy entails the responsibility to take proper precautions to ensure that modeling mistakes do not occur, especially when the stakes are high, and to inform users if

[8] *Fair Play: Violence, Gender and Race in Video Games.* Children Now, December 2001. 36 pp. http://publications.childrennow.org/

inaccuracies do or may occur. It also entails the responsibility not to participate in intentional deception of users (e.g., embellishment, dramatization, or censorship).

In Brey (1999), I have argued that designers of simulations and virtual environments also have a responsibility to incorporate proper values into their creations. It has been argued earlier that representations and interfaces are not value-free but may contain values and biases. Designers have a responsibility to reflect on the values and biases contained in their creations and to ensure that they do not violate important ethical principles. The responsibility to do this follows from the ethical codes that are in use in different branches of engineering and computer science, especially the principle that professional expertise should be used for the enhancement of human welfare. If technology is to promote human welfare, it should not contain biases and should regard the values and interests of stakeholders or society at large. Taking into account such values and avoiding biases in design cannot be done without a proper methodology. Fortunately, a detailed proposal for such a methodology has recently been made by Batya Friedman and her associates, and has been termed *value-sensitive design* (Friedman et al., (2006)).

Special responsibilities apply to different areas of applications for VR and computer simulations. The use of virtual reality in therapy and psychotherapy, for example, requires special consideration to principles of informed consent and the ethics of experimentation with human subjects (Wiederhold and Wiederhold, 2004). The computer and video game industry can be argued to have a special responsibility to consider the social and cultural impact of their products, given that they are used by a mass audience that includes children. Arguably, game developers should consider the messages that their products send to users, especially children, and should work to ensure that they develop and market content that is age appropriate and that is more inclusive of both genders.

Virtual reality and computer simulation will continue to present new challenges for ethics because new and more advanced applications are still being developed and their use is more and more widespread. Moreover, as has been argued, virtual environments can mimic many of the properties of real life and, therefore, contain many of the ethical dilemmas found in real life. It is for this reason that they will continue to present new ethical challenges not only for professional developers and users but also for society at large.

REFERENCES

Bartholow, B. (2005). Correlates and consequences of exposure to video game violence: hostile personality, empathy, and aggressive behavior. *Personality and Social Psychology Bulletin,* 31(11), 1573–1586.

Baudrillard, J. (1995). *Simulacra and Simulation*. University of Michigan Press, Ann Arbor, MI. (Translated by Fraser, S.)

Borgmann, A. (1999). *Holding On to Reality: The Nature of Information at the Turn of the Millennium*. University of Chicago Press.

Brey, P. (1998). New media and the quality of life. *Techné: Journal of the Society for Philosophy and Technology,* 3(1), 1–23.

Brey, P. (1999). The ethics of representation and action in virtual reality. *Ethics and Information Technology*1(1), 5–14.

Brey, P. (2000). Disclosive computer ethics. *Computers and Society,* 30(4), 10–16.

Brey, P. (2003). The social ontology of virtual environments. *American Journal of Economics and Sociology,* 62(1), 269–282.

Brey, P. (2007). Theorizing the cultural quality of new media, *Techné. Research in Philosophy and Technology*, 11(1), 2–18.

Burdea, G. and Coiffet, P. (2003). *Virtual Reality Technology*, 2nd edition. John Wiley & Sons.

Burk, D. (2005). Electronic gaming and the ethics of information ownership. *International Review of Information Ethics,* 4, 39–45.

Carnagey, N., Anderson, C., and Bushman, B. (2007). The effect of video game violence on physiological desensitization to real-life violence. *Journal of Experimental Social Psychology*, 43(3), 489–496.

Chan, D. (2005). Playing with race: the ethics of racialized representations in e-games. *International Review of Information Ethics,* 4, 24–30.

Chiu, S., Lee, J., and Huang, D. (2004). Video game addiction in children and teenagers in Taiwan. *Cyberpsychology & Behavior,* 7(5), 571–581.

Dibbell, J. (1993). A rape in cyberspace. *The Village Voice*, December 21, 1993, pp. 36–42. Reprinted in:Trend, D. (Ed.), *Reading Digital Culture*. Blackwell, pp. 199–213.

Dreyfus, H. (2001). *On the Internet*. Routledge.

Ford, P. (2001). A further analysis of the ethics of representation in virtual reality: multi-user environments. *Ethics and Information Technology,* 3, 113–121.

Friedman, B., Kahn, P.H., Jr., and Borning, A. (2006). Value sensitive design and information systems. In: Zhang, P. and Galletta, D. (Eds.), *Human-Computer Interaction in Management Information Systems: Foundations*. M.E. Sharpe, Armonk, New York; London, England, pp. 348–372.

Green, C.S. and Bavelier, D. (2003). Action video game modifies visual selective attention. *Nature,* 423, 534–537.

Heim, M. (1994). *The Metaphysics of Virtual Reality.* Oxford University Press, New York.

Johnson, S. (2005). *Everything Bad Is Good for You: How Today's Popular Culture Is Actually Making Us Smarter.* Riverhead Books, New York.

Kimppa, K. and Bissett, A. (2005). The ethical significance of cheating in online computer games. *International Review of Information Ethics,* 4, 31–37.

Levy, N. (2002). Virtual child pornography: The eroticization of inequality. *Ethics and Information Technology,* 4, 319–323.

McCormick, M. (2001). Is it wrong to play violent video games? *Ethics and Information Technology,* 3(4), 277–287.

McLeod, J. (1983). Professional Ethics and Simulation. In: Roberts, S., Banks, J., and Schmeiser, B. (Eds.), *Proceedings of the 1983 Winter Simulation Conference*, pp. 371–373.

Nissenbaum, H. (1998). Values in the design of computer systems. *Computers and Society,* 1998, 38–39.

Norris, K. (2004). Gender stereotypes, aggression, and computer games: an online survey of women. *Cyberpsychology & Behavior,* 7(6), 714–727.

Ören, T.I., Elzas, M.S., Smit, I., and Birta, L.G. (2002). A code of professional ethics for simulationists. *Proceedings of the 2002 Summer Computer Simulation Conference*. San Diego, CA, pp. 434–435.

Prosser, W. (1960). Privacy. *California Law Review,* 48(4), 383–423.

Ray, S. (2003). *Gender Inclusive Game Design: Expanding The Market*. Charles River Media.

Rideout, V., Roberts, D., and Foehr, U. (2005). *Generation M: Media in the Lives of 8–18 Year-Olds*. A Kaiser Family Foundation Study, March 2005. http://www.kff.org/entmedia/entmedia030905pkg.cfm

Sandin, P. (2004). Virtual child pornography and utilitarianism. *Journal of Information, Communication & Ethics in Society,* 2(4), 217–223.

Searle, J. (1995). *The Construction of Social Reality.* MIT Press, Cambridge, MA.

Sherman, W. and Craig, W. (2003). *Understanding Virtual Reality. Interface, Application, and Design*. Morgan Kaufmann Publishers.

Sicart, M. (2003). Family values: ideology, computer games & sims. *Level Up Conference Proceedings*. University of Utrecht, Utrecht (CD-ROM).

Slater, M., Antley, M., Davison, A., Swapp, D., Guger, C., Barker, C., Pistrang, N., and Sanchez-Vives, M. (2006). A virtual reprise of the Stanley Milgram obedience experiments. *PLoS ONE,* 1(1), e39 (open access).

Tabach-Bank, J. (2004). Missing the right of publicity boat: How Tyne v. Time Warner Entertainment Co. threatens to "Sink" the First Amendment. *Loyola of Los Angeles Entertainment Law Review,* 24(2), 247–288.

Turkle, S. (1995). *Life on the Screen. Identity in the Age of the Internet*. Simon & Schuster.

Warner, D. and Raiter, M. (2005). Social context in massively-multiplayer online games (MMOGs): ethical questions in shared space. *International Review of Information Ethics,* 4, 47–52.

Wiederhold, B. and Wiederhold, M. (2004). *Virtual Reality Therapy for Anxiety Disorders: Advances in Evaluation and Treatment*. American Psychological Association, Washington, D.C.

Zhai, P. (1998). *Get Real: A Philosophical Adventure in Virtual Reality.* Rowman & Littlefield Publishers.

CHAPTER 16

Genetic Information: Epistemological and Ethical Issues[1]

ANTONIO MARTURANO

16.1 INTRODUCTION

Genetics has utilized many concepts from informatics. These concepts are used in genetics at two different, albeit related levels. At the most basic level, genetics has taken the very notion of information, central to the field of informatics, to explain the mechanisms of life. An example is the famous "Central Dogma of Genetics," which Crick (1958) describes as follows:

> The transfer of information from nucleic acid to nucleic acid or from nucleic acid to proteins may be possible . . . but transfer from protein to protein or from protein to nucleic acid is impossible. Information means here the *precise* determination of sequence, either of bases in the nucleic acid and or of amino acid residues in the protein.

At a higher level, molecular biologists claim that cells and molecules are machinery similar to computers; this cell-machinery actually contains devices useful to build up unique biological beings starting from the information stored in a DNA.

Different authors (i.e., Griffiths, 2001; Lewontin, 1992; Mahner and Bunge, 1997; Marturano, 2003, and others) have questioned the application of informational concepts in genetics. Some authors have claimed that such concepts were very useful

[1]The present essay is a development of my research presented to the XV Internordic Philosophical Symposium, Helsinki, May 13–15, 2004, and the 8th Annual Ethics & Technology Conference, Saint Louis University, 24–25 June 2005. I am indebted to the remarks that my colleagues made during these two meetings. I thank Ken for his insightful comments.

The Handbook of Information and Computer Ethics, Edited by Kenneth Einar Himma and Herman T. Tavani

to establish the new science of molecular biology (and bioinformatics) around common basic concepts; but later on the same concepts were manipulated with an ideological sense for business and political purposes (Lewontin, 1992). Nonetheless, "the idea that 'biology is an information technology' is . . . widely accepted . . . partly because of the central role of information in the contemporary scientific world-view" (Griffiths, 2001; see also Fox Keller, 1995).

In this essay, I will first analyze some basic information-related concepts of molecular biology and then elucidate the possible ethical consequences of their misuse. It is indeed very important to understanding the way the information-related concepts of molecular biology are interpreted (we will see they use "information" sometimes in a *metaphorical* sense and sometimes in a *literal* sense, and often the way in which biologists use this concept is very ambiguous; see Griffiths, 2001) to figure out the reason why their possible incorrect application, and consequent rhetorical use by some geneticists (notably, Nobel Laureate Walter Gilbert), might lead to ethical failures.

16.2 INFORMATION THEORY AND THE NOTION OF GENETIC INFORMATION

16.2.1 The Concept of Information

Historically, since its first formulation by Shannon (1947), information theory has been subject to further refinements. The central paradigm of Shannon's classical information theory was the technical problem of the transmission of information (that is a sequence of signs) over a noisy channel without any reference to what is expressed by that sequence. Therefore, Shannon's theory of information was exclusively a physical-mathematical theory of signs (see also Bar-Hillel, 1955). In other words, mathematical information theory studies only the quantity of information in a physical system. The quantity of information in a system can be understood roughly as the amount of order in that system, or the inverse of the entropy (disorder) that all closed physical systems accumulate over time. This measure says nothing about the content of information. Shannon's theory is now labeled as the *syntactical theory of information* (Brehmer and Cohnitz, 2004). In genetics, Shannon's theory stemmed out of the *causal theory of information* (Griffiths, 2001).

Successively, Shannon's theory of information was developed by Bar-Hillel and Carnap (Bar-Hillel, 1955; Brehmer and Cohnitz, 2004), which focused on the meanings attached to signs. According to Rossi (1978, pp. 152 and following), the information theory becomes synonymous with a theory of knowledge, and it includes two different aspects of the word "information"; that is, "data organization" and "command" (or behavioral modification). However, it is very important to underline that semantical information theory presupposes the syntactical one. Following this stream, recently, information was seen as "true propositional content," as only a true propositional content can "resolve uncertainty in the objective sense of 'uncertainty'" (Himma, 2005). In genetics, semantical theories of information are related to the so-called teleosemantic theories, according to which a sign represents whatever evolution designed it to represent (Griffiths, 2001).

In the subsequent section we will discuss in more depth the idea of genetic information following these conceptual distinctions.

16.2.2 The Notion of Genetic Information

We have learned in the previous paragraph one way in which genetics has utilized the notion of information. In Crick's quote above, indeed, genetic information was defined as "the *precise* determination of sequence, either of bases in the nucleic acid and or of amino acid residues in the protein" (Crick, 1958). It seems to me, Crick is relying on the classical mathematical notion of information as proposed by Shannon and Weaver (1948). Shannon and Weaver information theory holds that "an event carries information about another event to the extent that it is causally related to it in a systematic fashion. Information is thus said to be conveyed over a "channel" connecting the "sender" [or signal] with the "receiver" when a change in the receiver is causally related to a change in the "sender" (Gray, 2001, p. 190).[2]

The idea of "genetic information" (Fig. 16.1) is that genes containing an amount of information (the so-called TACG amino acids sequence) and able to build a human being up is today a seldom challenged triviality. This idea is fundamental to the so-called "Central Dogma" of genetics. The "Central Dogma", as originally formulated by Crick, is a negative hypothesis, which states that information cannot flow downward from protein to DNA. Its complement, the "Sequence Hypothesis," is often conflated with the "Central Dogma". Under it, DNA is transcribed to RNA, and RNA is translated into protein. More abstractly, information flows upward from DNA to RNA, to proteins, and, by extension, to the cell, and, finally, to multicellular systems. In the ensuing years, many scientists have merged the two hypotheses and refer to them collectively as the "Central Dogma." We will use the term in this latter collective, conjunctive sense. So, this enlarged notion of "Central Dogma", or—according to

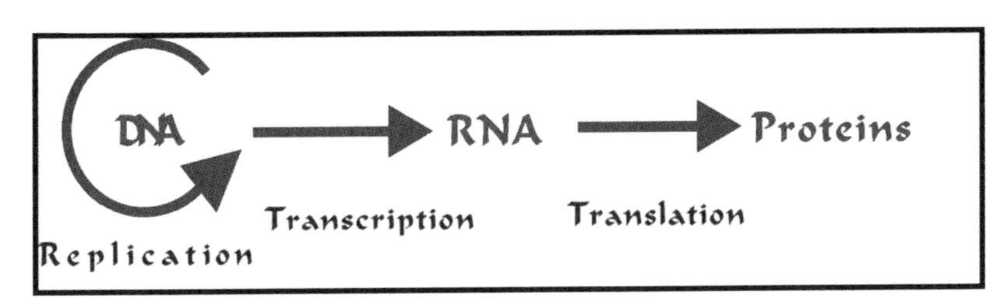

FIGURE 16.1 The "Central Dogma" of genetics.
(*Source*: http://library.thinkquest.org/C0122429/intro/genetics.htm)

[2]Sterelny and Griffiths (1999, p. 102) illustrate how the causal information concept could work in the context of molecular biology: "The idea of information as systematic causal dependence can be used to explain how genes convey developmental information. The genome is the signal and the rest of the developmental matrix provides channel conditions under which the life cycle of the organism contains (receives) information about the genome."

Berlinski (1972) who uses the term in the sense of Kuhn—paradigm "encompasses an account of the cell's ability to store, express, replicate, and change information. These are the fundamental features of life, no less; and a schema that says something interesting about them all has at least a scope to commend it."

Readers who are more expert will understand the characterization of the "Central Dogma" as based on the so-called broadcasting theory of communication in which we have just only one information sender and multiple information recipients, and information flows one way from a receiver to recipients.

However, the idea of genetic information, or better, of a code script into the cell is ascribed to Schroedinger (1944); the code script would be "a sort of cellular amanuensis, set to record the gross and microscopic features of the parental cell and pass the information thus obtained to the cell's descendant."

Other authors reject the idea that the concept of information does apply to DNA because it presupposes a genuine information system, which is composed of a coder, a transmitter, a receiver, a decoder, and an information channel in between. No such components are apparent in a chemical system (Apter and Wolpert, 1965). Even if there were such a thing as information transmission between molecules, this transmission would be nearly noiseless[3] (i.e., substantially nonrandom), so that the concept of probability central to the physical-mathematical theory of information[4] does not apply to this kind of alleged information transfer (Mahner and Bunge, 1997).

16.3 A SEMANTIC OR A SYNTACTIC THEORY OF GENETIC INFORMATION

Several authors have argued that molecular biology developed at the same time as computer technology and information theory; these two parallel processes have remained parallel. The biological notion of "information" has developed independently from the one advance by Shannon (in computer science). The expression "genetic information," used for the first time in Watson and Crick (1953b), has a metaphorical connotation—as we have seen before—without any particular reference to the nature of "code." As Crick explained later, for information they intended the specification of the amino acids sequence in the proteins; in Crick's mind such a notion, so to speak, was an "instructive" one (see Fox Keller, 1995) rather than a "selective" one as it is in Shannon's theory.

[3]In Shannon and Weaver, "Noise" means anything that corrupts information as it moves along a communication channel. More recently, it has been used to refer to the degradation of information, not only during transmission, but also during storage, whether in magnetic memory or in molecules of DNA (Johnson, 1987).

[4]As Warren Weaver summarizes in his famous article in *Scientific American* (1949, "The word information is in relation not so much with what you say as with what you could say. The mathematical theory of communication deals with the carriers of information, symbols and signals, not with information itself. That is, information is the measure of your freedom of choice when you select a message" (p. 12). In short, there is not a semantic dimension included in this notion of information theory.

According to the historian Kay (2000, p. 328), "Up until around 1950 molecular biologists. . .described genetic mechanisms without ever using the term *information*." "Information" replaced earlier talk of biological "specificity."[5] Watson and Crick's second paper of 1953, which discussed the genetical implications of their recently discovered (Watson and Crick, 1953a) double-helical structure of DNA, used both "code" and "information": ". . .it therefore seems likely that the precise sequence of the bases is the *code* which carries the genetical *information*. . ." (Watson and Crick, 1953b, p. 244, emphasis added). In other words, specificity and information had become synonymous terms in biological literature; they were based on the concept of uniqueness of the sequence as a condition for an organism's self-replication at the molecular level.

Corbellini (1998) believes that this model overcomes some applicative limits of Shannon's original model, with respect to the importance of the meaning attached to transmitted information. The importance of a semantical understanding of the biological notion of information, Corbellini claims, was underlined by Hutten (1973) who saw in such a notion of specificity an adequate formal definition to explain the transmission of genetic information from nucleic acid to proteins, which has encapsulated the meaning of an order. In other words, although Shannon's idea of information should be understood as physical-mathematical, in biology the semantical aspect of communicative interactions plays a more fundamental role. During the 1960s, despite these characterizations of the nature of information in terms of specificity, a metaphorical notion of information become fully absorbed into the vocabulary of molecular biology, whether in the context of the nucleic acids or in that of proteins.

But it was not until 1977 that robust and generally applicable sequencing methods were developed, and even then the modern bioinformatics techniques of gene discovery was still years away. Although the development of information/processing by computers proceeded contemporaneously with progress in research into biological and biochemical information processing, the trajectories of these two initiatives were never unified even if they sometimes overlapped at various points.

According to Castells (2001, p. 164), modern science relies largely on computer simulations, computational models, and computational analyses of large data sets. Although genetics is considered to be a process that is entirely independent from microelectronics, it is not really so independent. First, Castells argues that genetics technologies are obviously information technologies, because they are focused on the decoding and eventually the reprogramming of DNA (the information code of living matter). More important, "Without massive computing power and the simulation capacity provided by advanced software, the HGP [Human Genome Project], would not have been completed—nor would scientists be able to identify specific functions and the locations of specific genes" (Castells, 2001, p. 164). Sulston (2002) seems to agree with Castells, "The future of biology is strongly tied to that of bioinformatics, a

[5]Biological specificity is the principle that defines the orderly patterns of metabolic and developmental reactions giving rise to the unique characteristics of the individual and of its species.

field of research that collects all sort of biological data, tried to make sense of living organisms in their entirety and then make predictions."

Others have argued that there is an intrinsically special value in genetic information, which is said to differ either from (a) other kinds of health information, or (b) other kinds of information in general. According to Holm (1999), there is little support for the claim that genetic information has one or more special features that distinguish it from other health-related information in any morally relevant way.

The idea that genetic information has a special relevance is linked to the claims of "genetic essentialism." Genetic essentialism was roughly expressed in the statement that "we are our genes," or as Walter Gilbert put it, "once we will be able to pull a CD out of one's pocket and say, 'Here's a human being: it's me!'" Haraway (1997, p. 247), on the other hand, describes the human represented by the Human Genome Project (HGP)[6] as follows:

> Most fundamentally, ... the human genome projects produce entities of a different ontological kind than flesh-and-blood organisms, 'natural races,' or any other sort of 'normal' organic being. ... the human genome projects produce ontologically specific things called databases as objects of knowledge and practice. The human to be represented, then, has a particular kind of totality, or species being, as well as a specific kind of individuality. At whatever level of individuality or collectivity, from a single gene region extracted from one sample through the whole species genome, this human is itself an information structure...

In other words, according to Nagl (2007), this data structure is a construct of abstract humanness, without a body, without a gender, without a history, and without personal and collective narratives. It does not have a culture, and it does not have a voice. This electronically configured human is an acultural program. And in this very construction, it is deeply culturally determined—we find ourselves confronted with a "universal human," constructed by science as practiced in North America at the close of the twentieth century. This version of "human unity in diversity" is not liberatory but

[6]Completed in 2003, the Human Genome Project was a 13-year project coordinated by the U.S. Department of Energy and the National Institutes of Health. During the early years of the HGP, the Wellcome Trust (UK) became a major partner; additional contributions came from Japan, France, Germany, China, and others. Project goals were to

- *identify* all the approximately 20,000–25,000 genes in human DNA,
- *determine* the sequences of the 3 billion chemical base pairs that make up human DNA,
- *store* this information in databases,
- *improve* tools for data analysis,
- *transfer* related technologies to the private sector, and
- *address* the ethical, legal, and social issues (ELSI) that may arise from the project (Available at http://www.ornl.gov/sci/techresources/Human_Genome/home.shtml).

Genetic essentialism was the epistemological framework of the HGP endeavor.

deeply oppressive. To achieve such a vision in a positive sense, culture cannot be separated from biology.

Genetic information would be, therefore, information about the very essence of a person, whereas other nongenetic information would be only about accidental attributes. In other words, for genetic essentialism and the Human Genome Project (which absorbs such methodological background), I am who I am because of those precise genetics characteristics, whereas medical information about whether or not I got flu is accidental. We would still need to be able to distinguish between genetic and nongenetic information, but if we could do that, genetic information would surely be special.

Genetic essentialism[7] is widely attacked (i.e., Lewontin, 1999) and this view has strongly influenced the public perception of genetics (Nelkin and Lindee, 1995). According to Lewontin (1999, p. 63), however, it takes more than DNA to make a living organism and its history. A living organism at every moment of its life is the unique consequence of a developmental history that results from the interaction of and determination by internal (genetics) and external (environmental) forces. Such external forces are themselves partly a consequence of the activities of the organism itself, produced by the conditions of its own existence. Reciprocally, the internal forces are not autonomous, but act in response to the external. Part of the internal chemical machinery of a cell is manufactured only when external conditions demand it. Therefore, genetic essentialism, which assumes the uniqueness and independence of genetic information, does not give us a plausible argument for treating genetic information in a special category. Another claim that genetic information is special compared with other kinds of health-related information is sometimes based on a further claim that there is some other kind of genetic information that makes it different. Some have pointed out that genetic information is predictive; but it is also worth pointing out that, on the contrary, a lot of genetic information is nonpredictive and much of nongenetic health-related information is predictive. Knowing, for example, the LDL-cholesterol level in the blood of an individual can also predict that person's risk of coronary heart disease. According to Holdsworth (1999), the convergence of the computational notions of information with the biological notion enables us to see that there really are not two different kinds of information; information has the same meaning in information technology and in molecular biology. In each of these two contexts, we find signals that are expressed by ordering of the states of physical substrates. Viewed in this light, the nature of substrates, whether silicon or carbon, is irrelevant; and it is irrelevant even if, in real-world situations, there seem to be contingent reasons for drawing differences.

On the contrary, Maynard Smith (2000a,b) argues that the word "information" is used in two different contexts: "It may be used without semantic implication; for example, we may say that the form of a cloud provides information about whether it will rain. In such cases, no one would think that the cloud had the shape it did in

[7]Griffiths (2001) claims that "The present atmosphere, in which information talk is only applied to genes, makes that way of talking highly misleading. . . I also believe that the asymmetrical use of information talk partly explains the persistence of genetic determinism."

provided information. In contrast, a weather forecast contains information about whether it will rain, and it has the form it does because it conveys that information. The difference can be expressed by saying that the forecast has intentionality whereas the cloud does not." In this unfortunate example, Maynard Smith does not support the claim that there are two different uses of "information." A proposition is information and a weather report expressly asserts propositions; a cloud, on the other hand, does not express propositions, but there are other ways of gleaning information from the world than interpreting sentences. The claim that a weather forecast contains information is compatible with the claim that a cloud provides it—under the same definition of "information." A sad face is not information, but one can make informative inferences about the person from her expression.[8]

The notion of information as it is used in biology, he argues, "is of the former kind: it implies intentionality. It is for this reason, that we speak of genes carrying information during development and of environmental fluctuation not doing so" (Maynard Smith, 2000a). How, then, can a genome be said to have intentionality? Maynard Smith (2000a, 2000b) suggests that "the genome is as it is because of millions of years of selection, favouring those genomes that cause the development of organisms able to survive in a given environment. As a result, the genome has the base sequence it does because it generates an adapted organism. It is in this sense that genomes have intentionality." It is very difficult to understand what intentionality means here. Maynard Smith's statement is a sloppy and confused one. Strictly speaking, only conscious entities have intentionality; or, in other words, intentionality and having consciousness are synonymous; it is, therefore, very odd to understand DNA or molecules as "having consciousness."[9] It is not quite surprising that advocates of so-called "Intelligent Design" in evolution took this idea seriously (Dembski, 1999); conceptions of biological intentionality such as Maynard Smith's presuppose, indeed, a kind of consciousness that can be only externally given.

Moreover, Maynard Smith's idea is a consequence of what we have called before the semantic conception of information, and in particular of that biological interpretation of semantics that is labeled as *teleosemantics* (see also Griffiths, 2001). Teleosemantics is a philosophical program aiming at reducing meaning to biological function (teleology) and then reducing teleology to natural selection. According to Griffiths (2001), although there is considerable controversy about whether such reductions can be successfully carried out, teleosemantics still remains one of the most popular programs for naturalizing intentionality.[10]

[8]I am indebted to Ken Himma for this insightful counterexample.

[9]Claiming that DNA has consciousness has a similar value to the homeopathic idea that water has memory. Homeopathy holds that water is capable of retaining a "memory" of particles once dissolved in it. This memory allows water to retain the properties of the original solute even when there is literally no solute left in the solution. Though some scientific experiments and studies have shown such an effect, double-blind repetitions of many of these experiments have shown that no such effect exists (Maddox et al., 1988).

[10]Discussion about teleosemantics would go beyond the scope of this essay; for further information about how teleosemantics is used as a semantic theory of information in biology see Griffiths, 2001.

16.4 THE CELL AS COMPUTER MACHINERY

16.4.1 Berlinski: Bacterial Cell as Automata

Berlinski (1972) explores another interesting analogy between languages and the genetic code (Wendell-Waechtler and Levy, 1973). More interestingly, he wishes to identify a mechanical instrument, which would "pass" for coding into DNA all and only those well-formed formulae, and rules out those which are not—changes that resulted in strings that could not be generated, would simply not arise, or, failing that, would arise without effective genetic expression.

Some scholars, such as Berlinski (1972), tried to map units and structures in the genetic space directly to units and structures in the linguistic space. It was based on the observation that genetic information was organized as a sequence of discrete nucleotides[11] forming DNA molecules. The particularity of this organization is that there is a molecular system for reading the sequence, in which nucleotides are read three by three (these groups are called *codons*) and command the formation of a corresponding sequence of amino acids, which will then fold up and form a three-dimensional protein. The association between triplets of nucleotides and amino acids seems to be largely arbitrary. The discovery of these structures led researchers to think of nucleotides as an alphabet, codons as words, and genes as sentences whose meaning would be that proteins are associated to the genes (Berlinski, 1972). This led to the proposal that the genetic code itself was a language (Searls, 2002). Nevertheless, this mapping is controversial and a number of strong arguments against the relevance of these detailed correspondences at the levels of units were developed (Oudeyer and Kaplan, 2007).[12]

Berlinski, indeed, suggests that concepts such as code, information, language, control and regulation, and translation come easy to the biologist as he describes the cell according to lights provided by the central dogma (see above) in which the same terms are widely used. These, according to Berlinski (1972), are distinctly theoretical notions and such notions would provide a formal model for the bacterial cell. But, code theory says nothing about the organization that may be evident in the nucleic acid; this deals with problems that arise when information is passed along a channel of communication. According to Berlinski, automata theory would be a better "source for the most natural models of the bacterial cell" (1972).

---

[11]A nucleotide is one of the structural components, or building blocks, of DNA and RNA. A nucleotide consists of a base (one of four chemicals: adenine, thymine, guanine, and cytosine) plus a molecule of sugar and one of phosphoric acid (Available at http://www.genome.gov/glossary.cfm?key=nucleotide).

[12]Wendell-Waechtler and Levy have criticized Berlinski analogy between DNA and a language; they argue that "The trouble with the basic analogy is that it is seriously incomplete." Wendell-Waechtler and Levy suggest that such analogy is wrong because it presupposes the English language as input, but formal code theory does not require such restriction as it includes cases in which the sequences to be encoded are neither complete sentences nor portions of any natural language; they can be sequences of letters of an alphabet or any manner of abstract object. They conclude that Berlinski is not justified in assuming that DNA is a language in the usual sense just because DNA is encoded into proteins (Wendell-Waechtler and Levy, 1973).

In its more general form, automata theory treats the conversion of inputs to outputs via a device that admits states. The conversion is effected between items that are discrete and combinatorial: numerals, words, sentences, and letters. The bacterial cell seems similar to this sort of treatment: sequences of nucleotides resemble sentences, codons, and words. These very intricate processes of control and organization, Berlinski (1972) claims, evoke *computing machinery* of various kinds, in fact, in both cases, the relevant mechanism is transmitting and processing entities that are fairly characterized as expressing or being information.

This idealization of a cell thus appears as automation; the associated programs are designed to push the machine through movements that look vaguely biological. Often the machines turn out to have very strong computational capabilities. The idea that a cell is a machine was very attractive for biologists; J. Monod, for example, has confessed that it is just in the discovery of the machinelike nature of cell that modern biology has had its most impressive triumphs[13] (Monod, 1972, Chapter 4). In particular, Berlinski (1972) claims that cells are somewhat potentially infinite machines; that is, *pushing down storage automata* (PDSA). A PDSA is a machine, which accepts or rejects input strings of nucleotides; a given string of nucleotides is accepted if the machine reads it when the stack is empty. This treatment, as readers can see, is very close to Turing machines. The bacterial cell, Berlinski continues, suffers some idealization when set as a PDSA; bacterial systems quite obviously must complete their computations within close limits of time and space, they not only don't go on processing forever, they cannot go on doing so. Even though constructing a biological automaton would be quite difficult, biological PDSA embody algorithms for the conversion of codons into strings of nucleotides. They thus control the nature of proteins that are sequenced and the order in which they are synthesized. Therefore, states of the bacterial cell can be identified with the states of a PDSA; the full set of codons is formed from a nucleotide alphabet with its set of input symbols.

An automata-theoretic approach satisfied many algorithmic intuitions about the cell; the feeling is that the growth and regulation of the cellular machinery is basically a recursive process in which a finite set of elements are teased into a complex construction—a class of machines with fixed properties and limitations but no specific computational powers. But, according to Berlinski's (1972) analogy between DNA

[13]A machine's "internal structure is given" (Birch and Cobb, 1981, p. 79). It is this fixed or static character of machines that is at the heart of all classical mechanism. Birch and Cobb assert that "the ultimate mechanical model involves 'dissecting the organism down to its constituent controlling mechanisms and building it up from these building blocks" (p. 69). For a philosophy of mechanism, these building blocks are assumed to be static either in being or structure (or at least to have static essential parts), and it is from such fixed structure that the mechanist seeks to predict the nature of the whole, normally by the laws of mechanics. Jacques Monod, for example, is listed as a contemporary mechanist who believes that the DNA molecule is the controlling building block of living systems (Birch and Cobb, 1981, p. 70). Birch and Cobb object to Monod's view because of evidence concerning the nature of DNA: it can self-replicate, it assists in its own synthesis outside the cell, it is counterentropic, it determines necessary conditions for the development of organisms and not sufficient conditions (pp. 81 and following). In short, it is conditioned (in a very complex way) by its environment, as well as conditioning its environment (Henry, 1983).

and codes, the central dogma requires something quite different—a system in which elements of the base vocabulary change at random and thus form output sequences, unlike any that the machine is prepared to manage, as the exchange of information within a cell is virtually noiseless (see above).

Those kinds of biological automata, Berlinski (1972) argues, would be marvelous machines, sorting out among the proteins and instantiating a definition of life itself. Berlinski concludes that the original invocation of automata as models for the bacterial cell carried with it is a conceptual baggage of thought that biological automata not only arranged the affairs of the cell, but also fixed in recursive form computational powers that were sufficient to segregate the viable from the unviable proteins.

16.4.2 Maynard Smith: Eggs as Computer Machineries

In developmental genetics, the model of information has moved even farther John Maynard Smith, one of the champions of this branch of molecular biology, starting from the Mahner and Bunge criticism, expressed the idea that cells are actually a kind of information system. Maynard Smith (2000a) claims that indeed there are such things like coder, transmitter, receiver, coder, or information channel; "if there is 'information' in DNA, copied to RNA, how did it get there?". Maynard Smith concludes that natural selection plays the role of the coder in biological systems: "In human speech, the first 'coder' is the person who converts a meaning into a string of phonemes... In biology, the coder is natural selection". It would be worth remembering that Maynard Smith believes that natural selection is responsible for the selection of such code as he was a supporter of the controversial teleosemantics program (see above). What of the claim that a chemical process is not a signal that carries a message? Maynard Smith argues that this hypothesis is false because of the idea that the same information can be transmitted by different carriers, or better, that information is carrier-free, so information can be carried by chemical systems. Finally, Maynard Smith rejects the idea that probability does not play a fundamental role in biology just on the basis that genetic transmission of information is virtually noiseless. Rather, he argues, difficulties in applying information theory to genetics arise principally not in the transmission of information, but in its meaning.

Maynard Smith (2000a) claims that the analogy between the genetic code and human-designed codes is too apparent to require a justification, and he points our attention to the fact that the genetic code is symbolic, but we have machinery in the cell to process information; there is a decoding machinery (i.e., tRNA), a translating machinery (ribosomes, tRNAs etc.), and, finally, genes that contain coded information.

Finally, he claims "Yolk is just a store of nutrients: it no more carries information than the petrol in your petrol tank... An egg must also contain the machinery—ribosomes etc.—needed to translate the genetic message. The machinery is provided by the mother, and coded for by her genes. It is perhaps the classic example of the chicken and the egg paradox: no coding machinery without genes, and no genes without coding machinery" (Maynard Smith, 2000b). "What is inherited is not the dark pigment itself, but the genetic machinery causing it to appear in response to sunlight" (Maynard Smith, 2000a). Thus arises a new analogy: an egg is similar to a computer in the sense that it contains all the machinery able to process that information useful to build up a new individual with his/her characteristics. The informational metaphor is thus expanded: not only do we have "genetic information," but we also can talk of cells as "computational machines" in which the role of each computational element is defined.

According to Griffiths (2001), the idea that biology is an information technology (as Maynard Smith and Berlinsky seem to suggest) is a weak argument: it can be represented as follows:

(1) There is a genetic code.
(2) In a molecular developmental biology, there is talk of signals, switches, master control genes, and so forth.
(3) Therefore, the information flowing in (2) is information in the code of (1).

In this blunt form, the argument sounds merely frivolous (or even a fallacy of equivocation over the concept "information"). But many discussions of molecular biology, especially those for a nontechnical audience, insinuate something very close to it (see criticisms by Griffiths, 2001; Lewontin, 1992, 1999; Sarkar, 1996).

16.5 USE AND MISUSE OF MODELS

We have just seen two ways in which the informational and computer models were imported by philosophers of biology into the biological and genetic realm. However, these models were widely used and even stressed to other biologists such as W. Gilbert, making that a common sense. Ordinary use of the model, today, has unfortunately collapsed the distinction between a model and the theory for which it is a model. There is, indeed, a one-to-one correlation between the propositions of the theory and those of the model; propositions that are logical consequences of propositions of the theory have correlates in the model, which are logical consequences of the correlates in the model of these latter propositions in the theory and vice versa. But the theory and the model have different epistemological structures: in the model, the logically prior premises determine the meaning of the terms occurring in the representation of the calculus of the conclusions; in the theory the logically posterior consequences determine the meaning of the term occurring in the representation of the calculus of the premises. The widely used realist vocabulary has collapsed this fundamental distinction, turning the informational model into the theory, or, even worse—as in the Gilbert example—into the actual ontological base of genetics. In other words, as Lewontin (2000b, p. 4) suggests, "We cease to see the world *as if* it were like a machine and take it *to be* a machine." The result, he rightly argues, is that "the properties we ascribe to our object of interest and the questions we ask about it reinforce the original metaphorical imagine and we miss the aspects of the system that do not fit the metaphorical approximation" (Lewontin, 2000a, p. 4; author's emphasis).

The idea that models were an epistemologically dangerous tool is a well-known topic in philosophy of science. According to Braithwaite (1953, p. 93 and following) a danger in the use of model is that

> the theory will be identified with a model for it, so that the objects with which the model is concerned—the model-interpretation of the theoretical terms... of the theory's calculus—will be supposed actually to be the same as the theoretical concepts of the theory. To these theoretical concepts will then be attributed properties which belong to the objects of a model but which are irrelevant to the similarity in the formal structure, which is all that is required of the relationship of model to the theory... Thinking of scientific theories by means of models is always *as-if* thinking; hydrogen atoms behave (in certain respects) as if they were solar systems each with an electronic planet revolving round a protonic sun. But hydrogen atoms are not solar systems; it is only useful to think of them as if they were such systems if one remembers all the time that they are not. *The price of employment of models is eternal vigilance* (my emphasis).[14]

In much the same sense, the notion of information in biology was a fundamental operational instrument (that is, an *as-if* thinking, according to Braithwaite) that helped and even boosted genetic research. Unfortunately, at a point, it "collapsed" or was "naturalized" (or, in Braithwaite's own terminology, it led to an identification between the model-interpretation of theoretical terms and the theoretical concepts of the theory), reducing a quite powerful heuristic model into the very research object.

According to Buiatti (1998), those who defended that model have even turned it into an untouchable dogma (Lewontin (1992), for example, says that "Molecular Biology is now a religion, and molecular biologists are its prophets"), so that biological organisms were reduced to "living computers," in particular to the way computers were understood in the early 1960s, and thereby transferred the logical necessity of some of the features of the chosen model into the theory, as Braithwaite warned.[15]

In one of the following sections we will see how this mere epistemological problem might turn into an ethical one as from the result of the combined forces of academic status quo and business interests.

16.6 ETHICAL PROBLEMS OF GENETIC INFORMATION

There are several ethical concerns regarding genetic information. A popular one has led to calls for a genetic privacy law all over the world because of the frequent genetic information disputes arising between individuals. This has commonly been characterized as the "right to know debate" (see for a general discussion Chadwick et al., 1997). Given the nature of genetic information, situations can arise where a genetic test result for one individual will also have a significant effect on his or her siblings and

[14]The same emphasized statement can be found in Rosenblueth and Wiener (1951) who address their concerns in the way the concept of information might be used.

[15]In this very sense Floridi is hypostatizing the concept of information: "We have seen that a person, a free and responsible agent, *is* after all a packet of information." (Floridi, 1999; my emphasis).

wider family circle (Ngwena and Chadwick, 1993). Often there will be no difficulty and the information in question will be passed on. However, there may be many reasons, rational and irrational, why individuals will not want to share the results of their genetic diagnosis. It is this situation that has, at least in part, given rise to the question of whether individuals should have the right to know the results of a third party genetic test to make significant future life decisions for themselves. Equally there will be those, who aware of the potential financial, social, and emotional difficulties that such a diagnosis can carry, will have no desire to learn the results of a genetic diagnosis. These individuals argue that they have a right not to know information that has a direct or indirect reference to their genetic health (Suter, 1993).

In the subsequent sections, the problems of scientific honesty in genetic research, the problem of data access and patenting, and intellectual property rights are discussed.

16.6.1 Ideological Use of a Model and Ethical Issues in Fund-raising

There are many reasons why the information model in genetics was hypostatized (or naturalized); on the one hand, it has provided a powerful research strategy—a kind of reference guide—that has grounded and organized a new discipline: molecular biology. On the other hand, it provided a useful ideological tool for scientists to fund major research programs such as the Human Genome Project.

This raises a moral question about whether molecular biologists used the "collapsed model" in a correct (or honest) way for fund raising or, as Lewontin seems to suggest, it was rather an ideological weapon to monopolize media interests and capitalize on future patent rights. More importantly media drum banging around the HGP might steal room and funds for research with a narrow focus and thus interesting for just a smaller audience. According for Vicedo (1992), "the first task from a moral point of view is for the scientists to inform society about the development of the initiative and its implications. Lack of (or—I would add—biases in) information always raises suspicion, and leads to misunderstanding... (Geneticists should) to assess realistically the value of the project, and to avoid making empty promises."

We have discussed before the fundamental contribution of informational concepts to the central dogma of genetics. That dogma was the principle that gave rise to the biotech industry. Although debate and discussions continue about the truth of the central dogma[16] (see Caruso, 2007), such a large biotechnology industry funded by a

[16]A consequence of the central dogma is that genes operate independently from each other, that is, each gene in living organisms, from humans to bacteria, carries the information needed to construct one protein ("one gene, one protein" principle). In popular science literature, it is often depicted as the existence of a "smoker's gene," a "criminality gene," a "drunkenness gene," and so on. Postgenomic research has shown that genes do not work in isolation; rather, for instance, a disease is caused by the interplay among multiple genes (Caulfield, 2002). This idea was widely accepted amongst scholars, but it was kept hidden from the public by supporters of the HGP. This state of affairs slowed down genomic research as it was the main cause, for example, of the missed unfolding of the so-called junk DNA. Junk DNA is a collective label for portions of the DNA sequence of a chromosome or a genome for which no function has yet been identified. About 80–90% of the human genome has been designated as "junk."

massive infusion of venture capital (based on gene patenting) and an equally significant amount of capital from large, often multinational, pharmaceutical, companies has become an established force and interested only in patentable, working genes (Rabinow, 2000, p. 3).

16.6.2 Cooperation and Public Access of Data

According to Vicedo (1992), one of the main problems arising at the beginning of the HGP was ensuring the coordination of the different tasks and the cooperation among all research groups. She points out: "Some regulatory guidelines could be established to secure the smooth functioning of the project, but the scientists concerned hold different views on this issue. J. Watson, for example, thinks that the groups will develop rules to co-ordinate their efforts as the investigations proceed. Other researchers, such as Walter Gilbert (Harvard), think that clear rules should provide all participating members access to the results. Others suggest that the need for groups to communicate to obtain mutual benefits will force them to co-operate." Elke Jordan believes that the HGP's goals will be unattainable unless it is "built on teamwork, networking and collaboration." In his opinion, "This makes sharing and co-operation an ethical imperative." As Vicedo's remarks suggest, cooperation was a fundamental concern since the beginning of the HGP. One cause for concern arose because of the so-called emerging patenting-and-publish system between researchers and backed by the pharmaceutical and biotechnologies industries. This factor influenced the merging of scientific research with business interests.

This ethical problem is not directly related to the way biologists use the notion of information, nonetheless this problem is related to data banks in which genetic results are stored.[17] The controversy between Celera and the public HGP consortium would provide an example. Indeed, according to HGP researcher John Sulston: "The Human Genome Project and Celera were not working toward a common goal, since only the former generated a public sequence. Like everyone else, Celera had free access to all our assembled sequence. But Celera also asked us for a personal transfer of individual nematode sequence reads. To comply would have been a major distraction from our [HGP] work" (Sulston quoted in Koerner, 2003).

The paper Celera Genomics published in *Science* detailed the results of data sequencing and how this data would be used by the academic community. The material transfer agreement stated that academic users would be able to download up to one megabase per week from the Celera Genomics Web site, subject to a nonredistribution clause; if academics wanted to download more data, they would have to get a signature

[17]Recently an editorial in the very authoritative journal *Nature* (2004, p. 1025) has shown how the problem of genetic data accessibility is related to information; "Increasingly, it is easiest to make [genetic] materials available in the form of information, but even this imposes significant challenges, as high-dimensional biology generates very large files. We currently insist that sequences be deposited in databases such as GenBank and EMBL and, at least for expression data, in the microarray databases GEO and ArrayExpress according to MIAME criteria. But it is time to develop community standards for new kinds of large datasets, and we would welcome suggestions about how to proceed with array CGH, methylation, ChIP on chip and other epigenomic datasets."

from a senior member of their institution guaranteeing that the data would not be redistributed (Sulston and Ferry, 2002, p. 234). Members of the HGP community vigorously protested this agreement. Michael Ashburner, a former reviewing editor for *Science*, led the protest. He explained that such a strategy would be problematic for the future of genetics, because if the strategy employed by Celera Genomics was similarly adopted by other researchers in the field, "the data will fragment across many sites and today's ease of searching will have gone, and gone forever. Science will be the MUCH poorer, and progress in this field will inevitably be delayed" (Ashburner quoted in Moody, 2004, p. 112). Others felt outraged that one of the fundamental principles of scientific progress, the publication and free access of data, should be undermined by the way Celera Genomics wished to keep its data proprietary so that the complete database (including volumes of data on genetic variability in humans and the genomes of animals critical to biomedical research) could be available for mining to any pharmaceutical company in exchange for money. Therefore, in Venter's mind, "Celera would be the definitive source of genomic information in the world, in much the same way that Microsoft had early on made its DOS operating system the standard for personal computers" (Shreeve, 2004, p. 220). Sean Eddy of Washington University and Ewan Birney of the European Bioinformatics Institute claimed, "The genome community has established a clear principle that published genome data must be deposited in the international databases, that bioinformatics is fueled by this principle, and that *Science* therefore threatens to set a precedent that undermines bioinformatics research" (quoted in Moody, 2004, p. 112). Many genome researchers agreed with Eddy and Birney that *Science* had acted unethically by publishing the Celera Genomics paper when Celera Genomics had not entered its data in an international database.

For genome researchers who objected to the proprietary practices of Celera Genomics, the open-source regime offered a welcome alternative, one that not only provided ready access to scientific to the research methodology behind the data, but also one that would highlight "the importance of sharing materials, data and research rights, and requiring [a] fair global access" (Taylor, 2007).

16.6.3 Sequence Patenting and the Open Source Challenge

Another ethical problem linking information with genetics is that of gene patenting. According to Caulfield (2002), "the legal theory supporting gene patents is rooted in the notion that they aren't patents on the naturally occurring gene per se, but on an isolated and purified form of the gene. That language dates from a 1911 case, *Parke-Davis v. H.K. Mulford*, decided by the 2nd U.S. Circuit Court of Appeals." In the modern world of genomics, that logic is a distinction without a difference. If you accept this argument, Caulfield (2002) rightly argues, "then would it not follow that you could patent a human heart once you removed it and preserved it? A human gene is created first in nature, the same way other parts of human bodies are, and the fact that it's isolated, cloned and purified doesn't change that root of origin." He concludes, "Moreover, with the sequencing of the human genome, finding a new gene is now a process that involves little invention. Powerful supercomputers have replaced the lab bench as the source of discovery of genes, so the patents issued are really on information."

But, one value of patenting a gene sequence, according to Lewontin (2000a), "lies in its importance in the production of targeted drugs, either to make up for the deficient production from a defective gene or to counteract the erroneous or excessive production of an unwanted protein." Alternatively, Lewontin continues, . . .the cell's production of a protein code by a particular gene, or the physiological effect of the genetically encoded protein, could be affected by some molecule synthesized in an industrial process and sold as a drug. The original design of this drug and its ultimate patent protection will depend upon having rights to the DNA sequence that specified the protein on which the drug acts (Lewontin, 2000a, p. 181). Lewontin concludes, were the patent rights to the sequence in the hand of a public agency like the NIH, a drug designer and manufacturer would have to be licensed by that agency to use the sequence in its drug research, and even if no payment were required the commercial user would not have a monopoly, but would face possible competition from other producers (Lewontin, 2000a, p. 182). In our view, however, an international competition would be difficult, because, as Lewontin claims, it wants "the NIH to patent the human genome to prevent private entrepreneurs, and especially foreign capital, from controlling what has been created with American funding" (Lewontin, 1999, p. 75). After the successes of Celera Genomics, led by Craig Venter, it could be argued that the actual patenting strategy seems focused now on protecting the interest of the few corporations working in this field. According to Sulston: "The Human Genome Project and Celera were not working toward a common goal, since only the former generated a public sequence. Like everyone else, Celera had free access to all our assembled sequence. But Celera also asked us for a personal transfer of individual nematode sequence reads. To comply would have been a major distraction from our work" (Koerner, 2003). On the contrary, the adoption of open-source philosophy promises to shift, according to Raymond, to a "gift economy," where status among peers is achieved by giving away things that are useful to the community.

In particular, social aspects of science work in a similar way; activities such as publishing papers, giving talks, and sharing results help scientists to obtain status among scientific peers. Science, in this sense, is a sort of gift economy of ideas; the open-source model thus gets at the basic nature of the old and originary way (or imaginary) of scientific research. Indeed, according to Cukier (2003), before the draft of the genome was completed (helped along, controversially, by the private sector company Celera Genomics), the Human Genome Analysis Group at the Sanger Institute in Britain even contacted the father of the free software movement, Stallman (1994) to get advice. Soon, draft license agreements and implementation plans were circulated, followed by a round of legal reviews. A "click-wrap contract" was drawn up so that if a party refined a sequence by mixing the HGP's public draft version with extra sequence data, they would be obliged to release it. "Protecting the sequence from someone taking it, refining it and then licensing it in a way that locked everyone in, was the primary objective," says Hubbard (Cukier, 2003). Allowing patents in DNA is inconsistent with the old model of research in which one scientist is free to build on the work of another, because no one has any intellectual property (IP) rights in earlier work that would preclude further development of the ideas in the work. But assigning IP rights in DNA or sequences effectively precludes scientists who do not belong to the organization holding the patent from advancing the work. There are a couple of ethical problems here worth noting—the gift-economy model respects the expressive and speech rights of scientists. IP thus inhibits speech rights, and, also, that would seem to slow the development of therapies that would conduce to the common good.

But, Cukier concludes, as the industry advances, there is a growing call among researchers to redraw the lines of intellectual property. Instead of simply learning to live with the current system, they want to upend it. In addition to graduate degrees, they are armed with moral arguments, evidence of economic efficiency, and a nascent spirit of solidarity, which is renewing the traditional ethos of cooperation, found in the sciences and the academy. And the approach that is gaining momentum comes from the neighboring industry of open-source information technology. Its underlying principles are the communal development of technology, complete transparency in how it works, and the ability to use and make improvements that are shared openly with others. Where proprietary software's underlying source code is forbidden to be modified (and normally even inspected) by customers, open-source products encourage users to develop it further. The parallel in life sciences are things like the HGP that represent a "common good," says Sulston (2002), corecipient of the 2002 Nobel Prize. "Progress is best in open source," he concludes (Cukier, 2003).

REFERENCES

Apter, M.J. and Wolpert, L. (1965). Cybernetics and development I: information theory. *Journal of Theoretical Biology*, 8, 244–257.

Bar-Hillel, Y. (1955). An examination of information theory. *Philosophy and Science*, 22, 86–105.

Berlinski, D. (1972). Philosophical aspects of molecular biology. *Journal of Philosophy*, 12, 319–335.

Birch, C. and Cobb, J.B. Jr (1981). *The Liberation of Life: From the Cell to the Community.* Cambridge University Press, Cambridge.

Braithwaite, R.B. (1953). *Scientific Explanation. The Tanner Lectures.* Cambridge University Press, Cambridge.

Brehmer, M. and Cohnitz, D. (2004). *Information and Information Flow. An Introduction.* Ontos Press, Frankfurt am Mein.

Buiatti, M. (1998). L'analogia informatica del 'dogma centrale' e le conoscenze attuali in biologia. In: Continenza, B. and Gagliasso, E. (Eds.), *L'informazione nelle Science delta Vita,* FrancoAngeli, Milano. pp. 100–117.

Caruso, D. (2007). A challenge to gene theory, a tougher look at biotech. *New York Times*, July 1. Accessed from http://www.nytimes.com/2007/07/01/business/yourmoney/01frame.html?ex=1188360000&en=4bf18830b9d16a6c&ei=5070#, 17/09/2007.

Castells, M. (2001). Informationalism and the network society. In: Himanen, P. (Ed.), *The Hacker Ethic and the Spirit of the Information Age*. Vintage, London, pp. 155–178.

Caulfield, B.A. (2002). Why we hate gene patents. Accessed from http://www.law.com/jsp/article.jsp?id=1039054490790#, 17/09/2007.

Chadwick, R., Shickle, D., and Leavitt, M. (Eds.), (1997). *The Right to Know and the Right Not to Know*. Avebury Press, Aldershot.

Continenza, B. and Gagliasso, E.(Eds.), (1998). *L'informazione nelle Scienze della Vita*. FrancoAngeli, Milano.

Corbellini, G. (1998). La definizione informazionale della specificita' biologica. In: Continenza, B. and Gagliasso, E. (Eds.), L'informazione nelle Scienze delta Vita, FrancoAngeli, Milano. pp. 66–99.

Crick, F. (1958). Central dogma of molecular biology. *Nature*, 227, 561–563.

Cukier, K. (2003). Open source biotech: can a non-proprietary approach to intellectual property work in the life sciences. *The Acumen Journal of Life Sciences*, 1(3).

Dembski, W.A. (1999). Intelligent Design: The Bridge Between Science and Theology, InterVarsity Press, Downers Grove (II).

Editorial (2004). 'Good citizenship' or good business? *Nature*, 36(10), 1025.

Floridi, L. (1999). Information ethics: on the philosophical foundations of computer ethics. *Ethics and Information Technology*, 1(1), 33–52.

Fox Keller, E. (1995). *Refiguring Life: Metaphors of Twentieth Century Biology, The Welleck Lectures*. Columbia University Press, New York.

Gray, R.D. (2001). Death of the gene: developmental systems strike back. In: Griffiths, P.E. (Ed.), *Trees of Life: Essays in Philosophy of Biology*. Kluwer, Dordrecht. pp. 165–210.

Griffiths, P.E. (2001). Genetic information: a metaphor in search of a theory. *Philosophy of Science*, 68(3), 394–412.

Haraway, D.J. (1997). *Modest_Witness@Second_Millennium.FemaleMan ©_Meets_OncoMouse™*. Routledge, New York.

Henry, G.C. Jr. (1983). The image of a machine in the liberation of life. *Process Studies*, 13(2), 143–153.

Himma, K.E. (2005). Information and intellectual property protection: evaluating the claim that Information should be free, (August 12, 2005). *Berkeley Center for Law and Technology. Law and Technology Scholarship (Selected by the Berkeley Center for Law & Technology*, Paper 12. http://repositories.cdlib.org/bclt/lts/12

Holdsworth, D. (1999). The ethics of the 21st century bioinformatics: ethical implications of the vanishing distinction between biological information and other information. In: Thompson, A.K. and Chadwick, R. (Eds.), *Genetic Information: Acquisition, Access and Control*. Kluwer/Plenum, New York. pp. 85–98.

Holm, S. (1999). There is nothing special about genetic information. In: Thompson, A.K. and Chadwick, R. (Eds.), *Genetic Information: Acquisition Access and Control*. Kluwer/Plenum, New York. pp. 97–104.

Hutten, E.H. (1973). Information theory in physics and in biology. *Ann Ist Super Sanita*, 9(4), 335–361.

Johnson, H.A. (1987). Thermal noise and biological information. *The Quarterly Review of Biology*, 62(2), 141–152.

Kay, L. (2000). *Who Wrote the Book of Life? A History of the Genetic Code*. Stanford University Press, Stanford.

Koerner, B. (2003). Attacking venter capitalism. *Wired Magazine*, 6(6), Accessed from http://www.wired.com/wired/archive/11.06/view.html?pg=3, 17/09/2007.

Lewontin, R. (1992). The dream of the human genome. *The New York Review of Books*, May 28.

Lewontin, R. (1999). *The Doctrine of DNA, Biology As Ideology*. Penguin, London.

Lewontin, R. (2000a). *It Ain't Necessarily So*. Granta, London.

Lewontin, R. (2000b). *The Triple Helix*. Harvard University Press, Cambridge.

Maddox, J., Randi, J., and Stewart, W.W. (1988). 'High-Dilution' experiments a delusion. *Nature*, 334, 287–290.

Mahner, M. and Bunge, M. (1997). *Foundations of Biophilosophy*, Heidelberg, Berlin.

Marturano, A. (2003). Molecular biologists as hackers of human data: rethinking IPR for bioinformatics research. *Journal of Information, Communication & Ethics in Society*, 1(4), 207–216.

Maynard Smith, J. (2000a). The concept of information in biology. *Philosophy of Science*, 67, 177–194.

Maynard Smith, J. (2000b). Reply to commentaries. *Philosophy of Science*, 67, 214–218.

Monod, J. (1972). *Chance and Necessity*. Vintage, London.

Moody, G. (2004). *Digital Code of Life: How Bioinformatics is Revolutionizing Science, Medicine, and Business*. Wiley, Hoboken, NJ.

Nagl, S. (2007). Genetic essentialism and the discursive subject. *Paideia*. Accessed from http://www.bu.edu/wcp/Papers/Bioe/BioeNagl.htm, 17/08/2007.

Nelkin, D. and Lindee, M.S. (1995). *The DNA Mystique: The Gene As A Cultural Icon*. W.H. Freeman and Co., New York.

Ngwena, C. and Chadwick, R. (1993). Genetic diagnostic information and the duty of confidentiality: ethics and law. *Medical Law International*, 1(1), 73–95.

Oudeyer, P-Y. and Kaplan, F. (2007). Language evolution as a darwinian process: computational studies. *Cognitive Processing*, 8(1), 21–35.

Rabinow, P. (2000). *French DNA: Trouble in Purgatory*. Chicago University Press, Chicago.

Rosenblueth, A. and Wiener, N. (1951). Purposeful and non-purposeful behaviour. *Philosophy of Science*, 17(4), 318–326.

Rossi, P.A. (Ed.) (1978). *Cibernetica e teoria dell'informazione*. Brescia, La Scuola.

Sarkar, S. (1996). Biological information: a sceptical look at some central dogma of molecular biology. In: Sarkar, S. (Ed.), *The Philosophy and History of Molecular Biology: New Perspectives*. Kluwer, Dordrecht. pp. 187–232.

Searls, D.B. (2002). The language of genes. *Nature*, 420, 211–217.

Schroedinger, E. (1944). *What is Life. The Physical Aspect of the Living Cell*. Cambridge University Press, Cambridge.

Shannon, C. (1947). A mathematical theory of communication. *The Bell System Technical Journal*. 27, 379–423, 623–656.

Shannon, C. and Weaver, W. (1948). *The Mathematical Theory of Communication*. University of Illinois Press, Urbana, Illinois.

Shreeve, J. (2004). *The Genome War: How Craig Venter Tried to Capture the Code of Life and Save the World*. Alfred Knopf, New York.

Stallman, R. (1994). Why software should not have owners. Accessed from http://www.gnu.org./philosophy/whyfree.html 03/04/2003.

Sterelny, K. and Griffiths, P.E. (1999). *Sex and Death*. University of Chicago Press, Chicago.

Sulston, J. (2002). Heritage of humanity. *Le Monde Diplomatique*. December, 2002. Accessed from http://mondediplo.com/2002/12/15genome 15/02/2004.

Sulston, J. and Ferry, G. (2002). *The Common Thread: A Story of Science, Politics, Ethics, and the Human Genome*. Joseph Henry Press, Washington, DC.

Suter, S.M. (1993). Whose genes are these anyway? Familial conflict over access to genetic information. *Michigan Law Review*, 91(7), 1854–1908.

Taylor, P.L. (2007). Research sharing, ethics and public benefit. *Nature Biotechnology*, 25, 398–401.

Thompson, A.K. and Chadwick, R.(Eds.) (1999). *Genetic Information: Acquisition, Access and Control*. Kluwer/Plenum, New York.

Vicedo, M. (1992). The human genome project: towards an analysis of the empirical, ethical and conceptual issues involved. *Biology and Philosophy*, 7, 255–277.

Watson, J.D. and Crick, F. (1953a). Molecular structure of nucleic acids. *Nature*, 171, 737–738.

Watson, J.D. and Crick, F. (1953b). Genetical implications of the structure of deoxyribonucleic acid. *Nature*, 171, 964–967.

Weaver, W. (1949). The mathematics of communication. *Scientific American*, 181(1), 11–15.

Wendell-Waechtler, S. and Levy, E. (1973). More philosophical aspects of molecular biology. *Philosophy of Science*, 2, 180–186.

CHAPTER 17

The Ethics of Cyber Conflict

DOROTHY E. DENNING

17.1 INTRODUCTION

At least on the surface, most cyber attacks appear to be clearly unethical as well as illegal. These include attacks performed for amusement or bragging rights, such as web defacements conducted "just for fun" and computer viruses launched out of curiosity but disregard for their consequences. They also include attacks done for personal gain, such as system intrusions to steal credit card numbers and trade secrets; denial-of-service attacks aimed at taking out competitor Web sites or extorting money from victims; and attacks that compromise and deploy large "botnets" of victim computers to send out spam or amplify denial-of-service attacks.

There are, however, three areas of cyber conflict where the ethical issues are more problematic. The first is cyber warfare at the state level when conducted in the interests of national security. Some of the questions raised in this context include: Is it ethical for a state to penetrate or disable the computer systems of an adversary state that has threatened its territorial or political integrity? If so, what are the ground rules for such attacks? Can cyber soldiers attack critical infrastructures such as telecommunications and electric power that serve both civilian and military functions? If a nation is under cyber assault from another country, under what conditions can it respond in kind or use armed force against the assailant? Can it attack computers in a third country whose computer networks have been compromised or exploited to facilitate the assault?

The second area with ethical dilemmas involves nonstate actors whose cyber attacks are politically or socially motivated. This domain of conflict is often referred to as "hacktivism," as it represents a confluence of hacking with activism. If the attacks are designed to be sufficiently destructive as to severely harm and terrorize civilians,

The Handbook of Information and Computer Ethics, Edited by Kenneth Einar Himma and Herman T. Tavani

they become "cyberterrorism"—the integration of cyber attacks with terrorism. Although cyberterrorism is abhorrent and clearly unethical, hacktivism raises ethical questions. For example: Is it ethical for a group of hackers to take down a Web site that is being used primarily to trade child pornography, traffic in stolen credit card numbers, or support terrorist operations? Can the hacktivists protest the policies or practices of governments or corporations by defacing Web sites or conducting web "sit-ins?" Can they attack vulnerable machines in order to expose security holes with the goal of making the Internet more secure?

Finally, the third area involves the ethics of cyber defense, particularly what is called "hack back," "strike back," or "active response." If a system is under cyber attack, can the system administrators attack back in order to stop it? What if the attack is coming from computers that may themselves be victims of compromise? Since many attacks are routed through chains of compromised machines, can a victim "hack back" along the chain in order to determine the source?

This paper explores ethical issues in each of these areas of cyber conflict. The objective is not to answer the questions listed above, but rather to offer an ethical framework in which they can be addressed. Examples are used to illustrate the principles, but no attempt is made to reach a final ethical decision. To do so would require a much more thorough analysis of the nature of a particular cyber attack and the context in which it is used.

The framework presented here is based on the international law of armed conflict. Although this law was developed to address armed attacks and the use of primarily armed force, some work has been done to interpret the law in the domain of cyber conflict. The law has two parts: *jus ad bellum*, or the law of conflict management, and *jus in bello*, or the law of war. Despite being referred to as "law," both of these parts are as much about ethical behavior as they are rules of law.

The international law of armed conflict applies to nation states, and thus concerns cyber warfare at the state level. The paper will extend this framework to politically and socially motivated cyber attacks by nonstate actors, and compare this approach with some previous work on the ethics of cyber activism and civil disobedience. It will also apply the international law of armed conflict to the domain of cyber defense, and show how it ties in with the legal doctrine of self-defense and relates to other work on hack back.

Thus, for all three areas, the paper builds on the ethical principles encoded in the international law of armed conflict, and interpretation of those principles in the cyber domain. In this way, the paper approaches the three areas of cyber attack more as domains of conflict, especially international conflict, than as domains of crime—even though the acts themselves may also violate criminal statutes.

There are several areas of cyber conflict that the paper does not address. Besides cyber attacks conducted for pleasure or personal gain, the paper does not consider revenge attacks by insiders—all of which are generally regarded as unethical. In addition, the paper does not address methods of cyber conflict other than cyber attacks, for example, messages transmitted for the purpose of psychological operations or deception. Although other types of activity raise important ethical issues, their treatment is beyond the scope of this paper.

17.2 CYBER WARFARE AT THE STATE LEVEL

The law of international conflict consists of two parts: *jus ad bellum*, or the law of conflict management, and *jus in bello*, or the law of war. Both are concerned with the use of force, particularly armed forces, but the former specifies *when* that force may be applied, while the latter specifies ground rules for *how* it should be applied. Both are about ethical principles as much as they are about "law," and indeed, international law does not carry the same weight as domestic law. Under international law, states, as sovereign entities, assume international legal obligations only by affirmatively agreeing to them, for example, signing a treaty or agreeing to abide by the Charter of the United Nations. They are free to decline participation, and they are free to back out later. By contrast, under domestic laws, the citizens of a country are vulnerable to prosecution for violating any laws, regardless of whether they agree with them, and regardless of whether the laws are even just.

The law of international conflict is designed to promote peace and minimize the adverse effects of war on the world. As a general rule, states are not permitted to attack other states, except as a means of self-defense. Where conflict does arise, the law is intended to ensure that wars are fought as humanely as possible, minimizing collateral damage (harm to civilians and civilian property). Thus, the international law of armed conflict tends to prescribe widely accepted ethical principles.

17.2.1 *Jus ad Bellum*—The Law of Conflict Management

The law of conflict management is primarily concerned with the application of force, particularly armed force. It is codified in the United Nations Charter and specifies the conditions under which member states may apply force against other states. The most relevant parts of the Charter are Articles 2(4), 39, and 51.

Article 2(4) prohibits states from using force against other states:

> All Members shall refrain in their international relations from the threat or use of force against the territorial integrity or political independence of any state, or in any other manner inconsistent with the Purposes of the United Nations.

Although the nature of this force is left somewhat open, it may include more than just the use of armed force, as other parts of the Charter explicitly refer to armed force. However, it is not so broad as to cover generally lawful activity such as boycotts, economic sanctions, severance of diplomatic relations, and interruption of communications (Wingfield, 2000, p. 90).

Article 39 assigns the UN Security Council responsibility for responding to threats and acts of aggression:

> The Security Council shall determine the existence of any threat to the peace, breach of the peace, or act of aggression and shall make recommendations, or decide what measures shall be taken in accordance with Articles 41 and 42, to maintain or restore international peace and security.

Although a wide variety of acts can plausibly be interpreted as a "threat to the peace," the term "aggression" is defined in a UN General Assembly resolution as "the use of armed force" by a member or nonmember state. It includes invasions, attacks, bombardments, and blockades by armed military forces and other groups including mercenaries. Article 41 refers to responses other than armed force, for example, "complete or partial interruption of economic relations and means of communication, and the severance of diplomatic relations." Article 42 refers to the use of air, sea, and land forces, including demonstrations, blockades, and other operations.

Although Article 2(4) prohibits states from launching offensive attacks, Article 51 acknowledges a right to self-defense against armed attacks:

> Nothing in the present Charter shall impair the inherent right of individual or collective self-defense if an armed attack occurs against a Member of the United Nations, until the Security Council has taken measures necessary to maintain international peace and security. Measures taken by Members in the exercise of this right of self-defense shall be immediately reported to the Security Council and shall not in any way affect the authority and responsibility of the Security Council under the present Charter to take at any time such action as it deems necessary in order to maintain or restore international peace and security.

Although Article 51 states that defensive measures, including the use of force, are allowed after a state has been attacked, it is generally understood that states also have a right of "anticipatory self-defense," that is, they can take preemptive action to avert a strike. They are also permitted to exercise "self-defense in neutral territory." This means they can use force against a threat operating in a neutral state when that state is unwilling or unable to prevent the use of its territory as a base or sanctuary for attacks (DoD OGC, 1999, p. 14).

In summary, the UN Charter prohibits states from using force against other states (Article 2(4)), except when conducted in self defense (Article 51) or under the auspices of the Security Council (Article 39). The Charter effectively encodes an ethical principle of *just cause* for attacking another state that most people would accept. States have a moral right to defend themselves against acts and threats of aggression, but they do not have the right to engage in unprovoked aggression. The use of force is permissible only as a means of defending against aggression.

In order to apply these legal/ethical principles to cyber warfare, we must first determine whether cyber attacks constitute the use of force. If they do, then they would fall under the UN Charter along with armed force, implying that cyber attacks at the state level would be justified only as a means of defense. But if they are not considered to be a form of force, the ethical issues regarding their application are more ambiguous, falling closer to the issues raised by "softer" forms of coercion such as trade restrictions and severance of diplomatic relations.

17.2.2 When Does a Cyber Attack Constitute the Use of Force?

Not all cyber attacks are equal. The impact of a cyber attack that denies access to a news Web site for 1 hour would be relatively minor compared to one that interferes with air traffic control and causes planes to crash. Indeed, the effects of the latter would be comparable to the application of force to shoot down planes. Thus, what is needed is not a single answer to the question of whether cyber attacks involve the use of force, but a framework for evaluating a particular attack or class of attacks.

For this, we turn to the work of Michael Schmitt, Professor of International Law and Director of the Program in Advanced Security Studies at the George G. Marshall European Center for Security Studies in Germany. In a 1999 paper, Schmitt, a former law professor at both the US Naval War College and the US Air Force Academy, offered seven criteria for distinguishing operations that use force from economic, diplomatic, and other soft measures (Schmitt, 1999). For each criterion, there is a spectrum of consequences, the high end resembling the use of force and the low end resembling soft measures. The following description is based on both Schmitt's paper and the work of Thomas Wingfield, author of *The Law of Information Conflict* (Wingfield, 2000, pp. 120–127).

(1) **Severity.** This refers to people killed or wounded and property damage. The premise is that armed attacks that use force often produce extensive casualties or property damage, whereas soft measures do not.

(2) **Immediacy.** This is the time it takes for the consequences of an operation to take effect. As a general rule, armed attacks that use force have immediate effects, on the order of seconds to minutes, while softer measures, such as trade restrictions, may not be felt for weeks or months.

(3) **Directness.** This is the relationship between an operation and its effects. For an armed attack, effects are generally caused by and attributable to the application of force, whereas for softer measures there could be multiple explanations.

(4) **Invasiveness.** This refers to whether an operation involved crossing borders into the target country. In general, an armed attack crosses borders physically, whereas softer measures are implemented from within the borders of a sponsoring country.

(5) **Measurability.** This is the ability to measure the effects of an operation. The premise is that the effects of armed attacks are more readily quantified (number of casualties, dollar value of property damage) than softer measures, for example, severing diplomatic relations.

(6) **Presumptive Legitimacy.** This refers to whether an operation is considered legitimate within the international community. Whereas the use of armed force is generally unlawful absent some justifiable reason such as self-defense, the use of soft measures are generally lawful absent some prohibition.

(7) **Responsibility.** This refers to the degree to which the consequence of an action can be attributed to a state as opposed to other actors. The premise is that armed coercion is within the exclusive province of states and is more susceptible to being charged to states, whereas nonstate actors are capable of engaging in such soft activity as propaganda and boycotts.

To see how these criteria could apply to a cyber attack, consider an intrusion into an air traffic control system that causes two large planes to enter the same airspace and collide, leading to the deaths of 500 persons on board the two aircraft. In terms of severity, the cyber attack clearly ranks high. Immediacy is also high, although the delay between the intrusion and the crash may be somewhat longer than between something like a missile strike and the planes crashing. With respect to directness, let us assume the reason for the crash is clear from information in the air traffic control computers and the black boxes on board the planes, so directness ranks high. Invasiveness, however, is moderate, requiring only an electronic invasion rather than a physical one. Measurability, on the other hand, is high: 500 people dead and two planes destroyed. Presumptive legitimacy is also high in that the act would be regarded as illegitimate, akin to a missile attack (the high end of the spectrum corresponds to high illegitimacy). Responsibility comes out moderate to high. In principle, the perpetrator could be anyone, but the level of skill and knowledge required to carry out this attack would rule out most hackers, suggesting state sponsorship. In summary, five criteria (severity, immediacy, directness, presumptive legitimacy, and measurability) rank high, while two rank at least moderate (invasiveness and responsibility). Thus, the attack looks more like the application of force than a softer, more legitimate form of coercion.

Now, consider a massive distributed denial of service (DDoS) attack against a key government Web site that exploits a botnet of hundreds or thousands of compromised computers (zombies) and makes the site inaccessible for 1 day. This would likely rank low to moderate on severity, but high on immediacy. Directness would be moderate to high. Although the effects could as easily be attributed to hardware or software malfunction, network monitoring and inspection of Internet logs would show the problem to be caused by a massive onslaught of traffic. Invasiveness would be about the same as in the previous scenario, namely moderate owing to the electronic penetration. Measurability would be high, as it is easy to determine the downtime of the target Web server. Presumptive legitimacy would also score high, as DoS attacks, like force, are generally regarded as illegitimate and in violation of laws. Responsibility would be low to moderate. Some skill is required, but attribution would be difficult and many hackers would be capable of pulling it off. In summary, the attack looks less like force than the one causing the plane crash in terms of severity and responsibility, but neither does it resemble legitimate measures.

Wingfield suggests assigning a score for each criterion, say from 0 to 10. The idea is that high scores resemble force, whereas low scores resemble the softer measures such as economic and political ones. Under a "primary Schmitt analysis," the seven scores are summed and the average taken. For a "secondary Schmitt analysis," the criteria are assigned weights and the weighted average computed. This would allow severity, for example, to count more than the other criteria. An example with graphs showing the results of both primary and secondary Schmitt analyses is given in (Michael et al., 2003).

To the extent that a particular cyber attack looks like the application of force, its application would violate Article 2(4), possibly triggering an Article 39 response from the UN Security Council or an Article 51 application of force in self-defense by the target. However, under Articles 39 and 51, cyber attacks that resemble force would be allowed as a means of defense against aggressors who use either physical or cyber force.

On the contrary, if the attack looks more like legitimate, soft measures, than the use of force, then its application should not constitute a violation of Article 2(4). Moreover, if not deemed serious, it would likely not trigger an Article 39 response by the UN Security Council, as it would not be interpreted as a threat to the peace or act of aggression. Nor would it provide grounds for the target country to use force in its self-defense under Article 51. Of course, all this is theory. In practice, a nation that is the victim of a cyber attack may perceive it as an act of force worthy of a physical (or cyber) response, regardless of how the perpetrators score it under Schmitt's criteria or any others.

The ethical implications are that cyber attacks that resemble force are, like the use of physical force, morally justified only when they adhere to Articles 2(4), 39, and 51 of the UN Charter; that is, they are inherently defensive in nature. Unprovoked acts of aggression in cyberspace that resemble the use of force are not legally permissible.

Cyber attacks that fall below the Article 2(4) threshold for force are more likely to be ethical than attacks that cross the threshold, but they are not necessarily morally right. Their ethical implications must be examined like any other government action, for example, economic sanctions. However, in general it should be easier to justify cyber operations on ethical grounds as those operations move away from force on the spectrum of violence.

17.2.3 *Jus in Bello*—The Law of War

Whereas the *jus ad bellum* provides a legal framework for determining the lawfulness of a use of force, the *jus in bello* specifies principles governing how that force may be applied during armed conflict. It applies to all parties of the conflict, including the aggressors as well as states operating out of self-defense under Article 51 or in support of a UN operation under Article 39.

Under the *jus in bello*, the legal—and ethical—question regarding a cyber attack is not whether it looks like force, because armed force is permissible, but whether the attack adheres to commonly accepted principles. These principles are embodied in treaties, including Hague Regulations and Geneva Conventions, plus what is called "customary international law." The latter consists of those practices that are so widely adhered to that they are considered to be legally binding.

The U.S. Department of Defense summarizes the law of war with the following seven principles: (DoD OGC):

(1) **Distinction of Combatants from Noncombatants.** Only members of a nation's regular armed forces may use force, and they must distinguish themselves and not hide behind civilians or civilian property.
(2) **Military Necessity.** Targets of attack should make a direct contribution to the war effort or produce a military advantage.
(3) **Proportionality.** When attacking a lawful military target, collateral damage to noncombatants and civilian property should be proportionate to military advantage likely to be achieved.

(4) **Indiscriminate Weapons.** Weapons that cannot be directed with any precision, such as bacteriological weapons, should be avoided.
(5) **Superfluous Injury.** Weapons that cause catastrophic and untreatable injuries should not be used.
(6) **Perfidy**. Protected symbols should not be used to immunize military targets from attack, nor should one feign surrender or issue false reports of cease fires.
(7) **Neutrality.** Nations are entitled to immunity from attack if they do not assist either side; otherwise, they become legitimate targets.

The first three principles essentially state that wars are to be conducted by military forces, and that attacks, whether kinetic or cyber, should be aimed at military targets rather than civilian ones. Cyber attacks against critical infrastructures such as civilian energy distribution, telecommunications, transportation, and financial systems would be permitted only if they did not cause unnecessary or disproportionate collateral damage to noncombatants and civilian property.

The first principle also says that military forces should identify themselves when they engage in attacks, thereby taking responsibility for their actions. Part of the motivation for this is so that targets will not blame innocent civilians or other states for attacks and then take actions against them. Applying this to cyberspace, this means that military cyber soldiers should not attack anonymously in a way that leaves open the possibility that they are operating as civilians or on behalf of another state. Because most attacks are conducted so as to avoid attribution, achieving this objective would require novel means and methods, for example, cyber weapons and attacks that carry a government logo or "flag," or are clearly traceable to a military source. More fundamentally, it would also require a change in perspective, away from the notion that cyber attacks are necessarily covert operations toward one that favors open operations. Governments might oppose this, as it would leave them more open to counterattack.

Although computer intrusions and denial-of-service attacks can be delivered with precision, some cyber weapons could be prohibited on the grounds of being indiscriminate. Most viruses and worms would fall under this category, as they are designed to spread to any vulnerable machine they can find. Viruses and worms might still be used, but they would have to be coded in a way that restricted their spread, say, to a target subnet.

As for cyber weapons causing superfluous injury, there may not be any at this time. However, one could envision a cyber attack that caused such injury, for example, by altering the behavior of a surgical robot during an operation.

There would be ample opportunity for committing perfidy in cyberspace. For example, one could hide Trojan horses on a bogus Web site that bore the Red Cross logo or place a fake notice of surrender from a wanted terrorist leader on Web sites used by him to distribute messages. Under the law of war, such acts are not allowed.

The principle of neutrality protects neutral states from attack. To illustrate, suppose an adversary's cyber attack packets travel through the telecommunications network of a neutral country. It would not be permissible to attack that network to stop the attack as long as the services are offered impartially to both sides and the neutral country is doing nothing more than relaying packets without regard to their content. On the contrary, if the adversary penetrated computers in the neutral country and used them to launch its strike, it would be permissible to launch a counter attack against those machines if the neutral country refused or was unable to help.

In general, then, cyber attacks against an adversary during war could be considered ethical if they follow the above principles. Indeed, they may be less destructive than many kinetic attacks, and thereby preferred on humanitarian grounds. Rather than dropping bombs on a computing center in order to shut down a particular service, thereby causing extensive property damage and possibly loss of life, one might instead penetrate or disrupt the computer systems in a way that accomplishes the same military objectives but with fewer damages and long-term side effects.

17.3 CYBER ATTACKS BY NONSTATE ACTORS

Although the law of information conflict concerns state actors and the application of armed force, its general principles can be applied to nonstate actors who conduct cyber attacks for political and social reasons. This domain of conflict includes hacktivism, which is the convergence of hacking with activism and civil disobedience, and cyberterrorism, which uses hacking as a means of terrorism. In both cases, the objective is change of a political or social nature, but whereas the activist generally avoids causing physical injury or property damage, the terrorist seeks to kill and destroy.

To apply the international law of armed conflict to this domain, recall that the *jus ad bellum* specifies what types of operations are generally considered illegitimate, namely, operations that use force, and the conditions under which these otherwise illegitimate operations can be conducted—conditions that provide a lawful basis for engaging in otherwise prohibited behavior. The *jus in bello*, on the contrary, offers legal principles for the conduct of otherwise illegitimate operations in the face of conflict. The following discusses how each of these applies to hacktivism.

17.3.1 Just Cause for Hacktivism

Just as *jus ad bellum* specifies operations that states are not allowed to initiate against each other during the normal course of events, namely operations that use force, domestic laws specify operations that nonstate actors are not allowed to conduct. In the United States, the laws governing cyber attacks are embodied primarily in Title 18, Section 1030 of the U.S. Code (at the federal level) and in state computer crime laws. These laws generally prohibit most cyber attacks, including denial-of-service attacks, web defacements, network intrusions, and the use of malicious code such as viruses, worms, and Trojan horses.

Jus ad bellum allows states to engage in otherwise illegitimate operations that use force in order to defend themselves or, under the auspices of the UN, other states that are threatened. Domestic legal doctrine also incorporates a notion of self-defense that allows victims to use force that otherwise would be unlawful. Since the use of cyber attacks as a means of self-defense is covered later, this section focuses on other conditions that might provide ethical grounds for politically and socially-minded hackers to engage in cyber attacks.

One area where hacktivism may be morally justified is civil disobedience, which is the active refusal to obey certain laws and demands of a government through nonviolent means. Civil disobedience is conducted to protest and draw attention to laws, policies, and practices that are considered unjust or unethical. It employs such means as peaceful demonstrations, blockades, sit-ins, and trespass. Civil disobedience involves breaking laws, but it is an area where violating a law does not necessarily imply immoral behavior. When Rosa Parks refused to give up her seat on the bus, she committed an act of civil disobedience that was morally permissible as well as courageous. However, acts of civil disobedience are not necessarily ethical. For example, it would be unethical to block the entrance to a hospital emergency room in order to protest the government's health care policy.

The concept of civil disobedience was extended to cyberspace in the mid-90s. Stefan Wray (1998), founder of the New York-based Electronic Disturbance Theater (EDT), credits the Critical Arts Ensemble, which produced two documents, "Electronic Disturbance" in 1994 and "Electronic Civil Disobedience" in 1996. According to Wray, the Critical Arts Ensemble argued that activists needed to think about how they could apply blockade and trespass in digital and electronic forms.

EDT promoted the application of electronic civil disobedience, mainly through "web sit-ins," which were viewed as virtual forms of physical sit-ins and blockades. Each sit-in targeted one or more Web sites at a specified date and time, and was announced in advance in a public forum. To participate, activists would go to a Web site and select a target. This would cause a Java applet called Flood Net to be downloaded onto their computers and generate traffic against the selected Web site. Although the traffic generated by a single participant would have little effect on the performance of the target Web site, when thousands participated, as they did, the combined traffic could disrupt service at the target. EDT initially used their web sit-ins to demonstrate solidarity with the Mexican Zapatistas and protest Mexican and U.S. government policies affecting the Chiapas, but later went on to support numerous other causes. The concept was also picked up by other activists, including the U.K.-based Electrohippies. As web sit-ins became popular, the groups also developed more sophisticated flooding software, including software that could be downloaded in advance and run directly from participant machines, and software that required active involvement on the part of the participant (e.g., moving the mouse around).

To assess the lawfulness of web sit-ins and other forms of hacktivism, Schmitt's criteria for determining whether a cyber attack resembles the use of force versus softer, more legitimate measures are useful. In the domain of activism, legitimate measures include such things as letter writing campaigns, petitions, lobbying, publications, and speaking out. These forms of protest generally come out low on Schmitt's criteria. They do not cause damage and hence are not severe. Their effects are not immediate or direct, and they are hard to measure. They are not particularly invasive and are often carried out at a distance (e.g., public writing and speaking). They are presumed legitimate. Finally, they are low on responsibility since they can be performed by anyone.

One justification for following this approach is that there is a class of crimes called "violent crimes" that are singled out for their gravity. These crimes use or threaten to use violent force against their victims, and include murder, rape, robbery, and assault. In addition, the concept of civil disobedience expressly calls for the use of "nonviolent" means. Thus, it seems reasonable to evaluate forms of hacktivism in terms of the degree to which they resemble the application of violent force, which is effectively the same as armed force in the domain of *jus ad bellum*. An alternative approach would be to compare acts of electronic civil disobedience with physical acts of civil disobedience such as trespass and blockades. However, this begs the question of whether the physical acts themselves are ethical. It would be instructive to use Schmitt's criteria to assess such physical acts of civil disobedience, but that is beyond the scope of this paper.

Using Schmitt's criteria, let us consider a web sit-in that is publicly announced in advance, is scheduled to last 1 hour, and produces a noticeable degradation in service. In terms of severity, it would likely rank low, assuming the target is not providing some critical service. Immediacy, however, would be fairly high, as the effects, if noticed at all, would arise once a critical mass joined the sit-in. Directness would also be high. Although impaired performance at the Web site could be attributed to network problems or increased interest in material posted on the Web site, the prior announcement of the sit-in all but rules out other explanations. Invasiveness is moderate, but measurability is high, as it is straightforward to measure the performance degradation at the target. Presumptive legitimacy is low to moderate. Even though it is generally against the law to intentionally disrupt service, the effects produced by any individual participant are neither particularly disruptive nor clearly illegal, and the effects as a whole may be minor (indeed, many sit-ins have produced no noticeable effects). Finally, responsibility is moderate. Although it may be easy to determine the group responsible for organizing the sit-in from the public announcement, it would be difficult to determine individual participants. In sum, one measure is low (severity), two are high (immediacy and measurability), and four are in the middle (directness, invasiveness, presumptive legitimacy, and responsibility). Thus, web sit-ins do not look all that legitimate, falling somewhere between lawful measures and the illegal use of force.

Indeed, their legitimacy has been questioned by other activists. Following the EDT's sit-ins against the Mexican president's Web site in 1998, for example, the Mexican civil rights group Ame la Paz objected, saying that the use of hacking tools was counterproductive and dangerous. Another group, the Cult of the Dead Cow, criticized the Electrohippies for their sit-ins, arguing that they violated their opponents' rights of free speech and assembly. For their part, the E-Hippies justified their actions on the grounds that they substituted their opponent's forced deficit of speech with broad debate on the issues. They also attempted to justify a planned web sit-in as part of their April 2000 "E-Resistance is Fertile" campaign against genetically modified foods by asking visitors to their Web site to vote on whether to carry out the planned web sit-in. When only 42% voted in support, they cancelled the action.

However, they did not offer this option with other web sit-ins, including a massive 3-day sit-in against the World Trade Organization in late 1999 in conjunction with the Seattle protests.

Another form of hacktivism is the web defacement. Although most web defacements are not conducted for political or social reasons, they have become a popular tool of protest, accounting for tens of thousands of digital attacks. Outrage over the Danish cartoons of the prophet Mohammed alone generated almost 3000 defacements of Danish websites (Waterman, 2006).

One of the earliest defacements took place in 1996 against the U. S. Department of Justice Web site. The hackers used the attack to protest the Communications Decency Act (CDA), which made it illegal to make indecent material available to minors on the Internet. The defaced Web site was retitled "U.S. Department of Injustice" and displayed the message "this page is in violation of the Communications Decency Act!" It also included pornographic images and information about the First Amendment and the CDA (Attrition, 1996). By displaying pornographic material on a Web site accessible to children, the cyber attack violated the very act that was considered unjust. Considering that the CDA was subsequently struck down as unconstitutional by the Supreme Court, one might argue that the defacement was a reasonable response. However, the defacement also violated computer crime laws, making it much more difficult to justify.

Most web defacements violate computer crime statutes. Examining them in terms of Schmitt's criteria, they score high on four: immediacy, directness, measurability, and presumptive legitimacy. Severity may be low, as Web sites generally can be readily restored from backups, but it could be high if the defacement, for example, causes visitors to the site to use erroneous medical information or give up bank account information, or if it severely undermines confidence in the organization owning the Web site. Invasiveness is moderate, and responsibility is low, in that few countries claim responsibility for such actions. Not everyone is capable of defacing a Web site, but there are tens of thousands of hackers who are. In sum, defacements look even less legitimate than web sit-ins, and indeed are scorned by many hacktivists.

Other forms of hacktivism can be examined through Schmitt's criteria. In general, those actions that violate computer crime statutes come out moderate to high, implying their general illegitimacy, laws aside. These include cyber attacks to take down Web sites that traffic in child pornography, and attacks aimed at exposing—and correcting—security vulnerabilities. Even though the ends may be worthy, the means are questionable at best. Acts that would qualify as cyberterrorism would come out high in severity at the very least.

By comparison, cyber actions that do not violate computer crime laws come out low by Schmitt's criteria. Examples include E-Hippies' development and use of software to facilitate letter writing campaigns and Hacktivismo's development of software for getting information censored in China past China's firewalls. These activities are lawful (at least in the United States) and do not resemble the use of force.

Cyber attacks that fall in the middle to upper ranges of Schmitt's criteria are not necessarily unethical, but they are harder to justify. One factor that might be useful is whether the activist's objectives could be achieved by lawful means. For example, consider again the defacement protesting the CDA. The hacker could have displayed his message on his own Web site or, with permission, another party's Web site, and doing so would have given it a longer "shelf life." Defaced sites are rarely up for more than a short time, although they may be mirrored in an archive, as was the case here. The defacement got press coverage that otherwise would have been unlikely, but the criminal act is hard to justify given that civil liberties groups were working hard to overturn the CDA through the courts (as they succeeded in doing). Indeed, the defacement could have undermined the legal efforts by linking the civil liberties objectives to illegal hacking.

17.3.2 Conduct of Hacktivism

The seven principles of *jus in bello* provide guidance for using force and, by extension, for engaging in cyber attacks that resemble force.

The first principle, distinction of combatants from noncombatants, states that only members of a nation's regular armed forces may use force, and that they must distinguish themselves from civilians and not hide behind civilian shields. This principle would prohibit activists from engaging in any form of cyber attack that resembles force. If we interpret web sit-ins and defacements as something less than force, then they might be allowed, but only if the activists identify themselves or their sponsoring organization so that any response is not directed at innocent parties, including governments. Indeed, the organizers of EDT used their real names and talked about their philosophy and actions in public forum. The E-Hippies were also fairly open, and both groups openly acknowledged responsibility for the web sit-ins they organized. Although the tens of thousands of people who participated in their sit-ins did not individually identify themselves by name, participation in the sit-in itself implied an affiliation of sorts with the sponsoring organization. Web defacers also identify themselves, although typically by hacker group names and individual aliases that are not explicitly linked to their real names. But the level of identification is sufficient for an observer to see that the action was performed by a particular group of hackers and not a government.

The second principle, military necessity, requires that the amount of force employed not exceed the requirements of a lawful strike against a legitimate target. Given that most web sit-ins are conducted against the government agency or company whose policies are the target of protest, they could be interpreted as being consistent with the objective of avoiding collateral damage. However, there have been exceptions. For example, within their broad mission to help the Mexican Chiapas, EDT conducted a web sit-in against the Frankfurt Stock Exchange on the grounds that it represented capitalism's role in globalization, which they claimed was "at the root of the Chiapas' problems" (Denning, 2001). Although the connection seems far-fetched, the sit-in did raise this as an issue, which the EDT might have thought necessary to their mission. Indeed, EDT subsequently sponsored several web sit-ins over globalization issues.

A web sit-in can be viewed as a relatively mild form of denial-of-service (DoS) attack that affects its target directly. However, there are other types of DoS attacks that leverage third party computers to amplify their affects. For example, in a distributed denial-of-service (DDoS) attack, thousands of third party computers may be compromised and instructed to attack the target. As the compromised machines serve as a shield to protect the source of attack, this would violate the principle of distinguishing combatants from noncombatants.

Many web defacements have been directed only against the government or organization that was the subject of complaint. For example, the Department of Justice, which was the target of the CDA protest mentioned earlier, had supported and defended the CDA. However, numerous other defacements have been against targets that had little if any direct connection to the grievance. Of the almost 3,000 Danish websites defaced in conjunction with the protest against the Danish newspaper that published the cartoons and the government's response, most belonged to civilian organizations and companies that had nothing to do with the newspaper or government action. However, the attacks did generate press coverage, in part because of their magnitude, likely drawing greater attention to the complaint than simply defacing one or two government sites would have done. Roberto Preatoni, founder and administrator of Zone-h, which recorded the defacements, said that "This is the biggest, most intense assault" he had ever seen (Waterman, 2006). In general, hackers might justify their defacements of civilian Web sites on two grounds: first, because the civilian sites were the only ones they could successfully hack, and second, by hacking more sites, they could generate more publicity.

The principle of proportionality requires that any unintentional but unavoidable injury to noncombatants or damage to their property be proportionate to mission benefits. Returning to the EDT example above, the Frankfurt Stock Exchange reported that it was aware of the protest but believed it had not affected its servers (Denning, 2001). Hence, the sit-in could be considered proportionate to benefits achieved, which arguably were small. By comparison, DDoS attacks affect potentially thousands of noncombatant computers without necessarily meeting mission objectives any better than a sit-in, which does not harm third party computers. Similarly, web defacements against noncombatant servers produce noticeable effects and take time to repair. Besides removing the vulnerability that was exploited and, restoring the home page, system administrators must check for other damage and remove any backdoors and malicious code left behind by the hackers. It is harder to argue that such defacements are proportionate to the protestors' gains.

Hacktivists have employed indiscriminate computer viruses and worms to disseminate protest messages. These would violate the general principle of avoiding indiscriminate weapons. However, one of the earliest worms, Worms Against Nuclear Killers (WANK), stayed within the network of NASA, the target of the protest. The protestors objected to the nuclear power unit for the Galileo probe.

There do not appear to be cases of hacktivists causing superfluous injury or violating the principle of perfidy. Cyber criminals, however, have exploited protected symbols, including the Red Cross logo, for financial gain (e.g., through bogus fund raisers).

The principle of neutrality implies that activists should not launch cyber attacks against neutral states or third parties. While sponsoring web sit-ins to protest the

Mexican government's treatment of the Chiapas, the EDT conducted sit-ins against U.S. government sites as well as Mexican ones. However, they justified the U.S. sit-ins on the grounds that U.S. policies supported the Mexican government at the expense of the Chiapas.

17.3.3 Other Ethical Frameworks for Hacktivism

Mark Manion and Abby Goodrum offer five necessary conditions for acts of civil disobedience, and by extension electronic civil disobedience, to be ethically justified (Manion and Goodrum, 2000). They are as follows:

(1) No damage done to persons or property
(2) Nonviolent
(3) Not for personal profit
(4) Ethical motivation—that is, the strong conviction that a law is unjust, unfair, or to the extreme detriment of the common good
(5) Willingness to accept personal responsibility for outcome of actions

Manion and Goodrum's analysis of several acts of hacktivism suggests they regard web sit-ins, defacements, and some other forms of ethically motivated cyber attacks to be justifiable. However, their analysis ignores their first condition of no damage. Defacements in particular cause information property damage that is analogous to physical property damage (both require resources to repair).

The overall approach taken by Manion and Goodrum differs substantially from the law of war approach taken in this paper. The first principle of *jus in bello*, which states that combatants distinguish themselves, is similar to their fifth condition of accepting responsibility, but the other six principles of *jus in bello*—necessity, proportionality, indiscriminate or superfluous weapons, perfidy, and neutrality—are left out. Instead, Manion and Goodrum rely mainly on the ethical motivations of the hacktivists, taking an "ends justifies the means" approach, at least as long as the attack does not fall in the domain of cyberterrorism.

Kenneth Himma also offers five conditions that weigh in favor of acts of civil disobedience being ethically justified (Himma, 2006a):

(1) The act is committed openly by properly motivated persons willing to accept responsibility for the act.
(2) The position is a plausible one that is, at the very least, in play among open-minded, reasonable persons in the relevant community.
(3) Persons committing the act are in possession of a thoughtful justification for both the position and the act.
(4) The act does not result in significant damage to the interests of innocent third parties.
(5) The act is reasonably calculated to stimulate and advance debate on the issue.

Himma's conditions are stronger than Manion and Goodrum's, examining means (Condition 4) as well as end objectives. However, although Himma's fourth condition relates to several *jus in bello* principles, it offers fewer distinctions.

Neither framework appeals to *jus ad bellum* for assessing just cause and comparing cyber attacks with acts of force, which are generally forbidden by state as well as nonstate actors. On the contrary, both frameworks offer an additional consideration for determining just cause, namely ethical motivation. Further, Himma goes further and asks that activists provide justification for their position and actions; that the position itself be considered plausible by open-minded, reasonable persons in the relevant community; and that the actions be designed to foster debate. Himma's framework is complementary to the law of war framework offered by this paper.

17.4 ACTIVE RESPONSE AND HACK BACK

"Hack back" is a form of active response that uses hacking to counter a cyber attack. There are two principal forms. The first involves using invasive tracebacks in order to locate the source of an attack. The second involves striking back at an attacking machine in order to shut it down or at least cause it to stop attacking.

17.4.1 The Doctrine of Self-Defense

At the state level, the doctrine of self-defense is based on *jus ad bellum* and *jus in bello*, which together allow states to use force in self-defense, but constrain how that force is applied.

An analogous legal doctrine of self-defense allows nonstate actors to use force in order to protect themselves from imminent bodily harm or, under some circumstances, to protect their property from damage. According to Curtis Karnow, formerly Assistant U.S. Attorney in the Criminal Division, the test is whether:

(1) There is an apparent necessity to use force.
(2) The force used is reasonable.
(3) The threatened act is unlawful.

The necessity condition requires that there be a good faith subjective, and objectively reasonable, belief that there were no alternatives to the counterstrike. The reasonableness condition requires that the harm produced by the counterattack be proportional to the harm avoided (Karnow, 2003). Reasonableness would also encompass other principles from *jus in bello*, including neutrality, indiscriminate weapons, superfluous injury, and perfidy, as counterstrikes that violated these principles would seem unreasonable. Karnow observes that while self-defense is a privilege of state rather than federal law, it might protect the defender from prosecution under the federal computer crime statute, which prohibits unauthorized access, on the grounds that self-help provides the requisite authorization.

Karnow also suggests that the legal doctrine of nuisance could justify a counterstrike against cyber nuisances such as viruses and worms. Under nuisance law, a person affected by a nuisance can, as a last resort, use force or other means of self-help to abate or stop it (Karnow, 2003).

The doctrine of self-defense does not justify retaliatory strikes that are motivated by revenge or a desire to get even. The response must be necessary to counter the threat. To illustrate, in the midst of the Electrohippies' 3-day web sit-in against the World Trade Organization's Web site in 1999, the ISP hosting the WTO site, Conexion, conducted a counterstrike against the Electrohippies' site. Conexion's server was configured to retransmit all of the attack packets back to the Electrohippies' Web site, from where they had originated, thereby shutting it down. Himma argues that the strike back was retaliatory and unnecessary, as Conexion could have simply dropped the incoming attack packets (Himma, 2006b). Further, the response had a side effect of motivating the E-hippies to develop sit-in software that could be launched directly from their participants' computers, without the need to go through a central portal. Arguably, this made it more difficult for victims of future sit-ins to defend themselves, as there is no central source for the attack; indeed, such sit-ins more closely resemble DDoS attacks. As another example, the U.S. Department of Defense engaged in active response against a web sit-in conducted by the EDT in 1998. In their case, they redirected the browsers of participants using the EDT portal to a Web page with a hostile applet, which caused the participant's computers to go into an endless loop trying to reload a document (Denning, 2001). The counterstrike raised legal and ethical issues (some participants claimed they lost data), and the Department of Defense did not deploy similar measures in response to future sit-ins.

Besides self-defense and retaliation/punishment, Himma considers an ethical principle for active response based on the need to secure a significantly greater common good, which might justify aggressive measures. However, he cautions that such justification can be problematic because of potential unanticipated side effects. He also argues that persons engaging in active response are morally bound to have sufficient reason to believe they are acting on ethical principles (Himma, 2006b).

17.4.2 Hack Back and Force

For both state and nonstate actors, the doctrine of self-defense allows the application of force against force and threats of force. In general, offensive operations that use less than force call for responses that use less than force. However, even when the offensive act uses force, defensive responses that use less than force are generally preferred over those that use force. Thus, it is useful to know the extent to which active response resembles force versus more legitimate means, the latter being easier to justify on ethical grounds.

To determine the degree to which a particular means of active response resembles force, we again turn to Schmitt's criteria. Consider first an invasive traceback such as the one conducted by Shawn Carpenter in Titan Rain. Carpenter traced an intrusion into Sandia Labs and Department of Defense computers back to a province in China.

Although the details have not been made public, for the purpose of analysis, assume he had to hack back through computers that were not directly responsible for the intrusion in order to locate the source, as this is typical in cyber attacks.

In terms of Schmitt's criteria, severity is low. Indeed, the owners of intermediate machines may not observe any effects or even know of the traceback, especially if they had not noticed the intrusion from China in the first place. Given that the effects could go unnoticed unless and until system logs are examined, it seems reasonable to rate immediacy low as well. Measurability is also low in that there is not much to measure. Directness is low to moderate, as it could be hard to attribute the effects of the intrusive traceback to an active response (vs. some other computer intrusion). Invasiveness is moderate, as in all cyber attacks. Responsibility is also moderate, as some skill is required to conduct an effective traceback, but attribution is difficult.

To assess presumptive legitimacy, we need to know who is conducting the invasive traceback and who owns the machines being hacked. If the traceback is conducted by a state actor against foreign systems, presumptive legitimacy should be low in that the entire operation falls in the domain of foreign intelligence collection, which is generally considered legitimate. If the traceback involves accessing a domestic computer, the state may need additional authorities to access the system. However, if the traceback is conducted by a nonstate actor, the operation likely violates computer crime statutes, although the offense may be minor if no sensitive information was downloaded or files damaged. But even if we rate presumptive legitimacy moderate or high, the invasive traceback as a whole looks less like force than the cyber attacks examined earlier in this paper. This is consistent with Himma's argument that tracebacks are not properly characterized as force (Himma, 2004).

Next, consider an operation that aims to stop a machine from participating in a DoS attack. Suppose that the attacking machine is not even the source of the attack, but rather a victim itself of an earlier compromise. Finally, suppose that the method of stopping the machine from engaging in the attack involves removing malicious code that had been planted on the machine. Severity is low—indeed, removing the malicious code should improve the state of the machine. Immediacy, however, is high: once the malicious code is deleted, the attack packets stop. Directness is moderate, as the attack packets could stop for other reasons. (e.g., the malicious code could be programmed to only attack for 1 hour on a particular day). Invasiveness is also moderate. Measurability is high, as the before and after attack packets can be counted. Presumptive legitimacy is high, as it is normally illegal to tamper with other peoples' machines. Finally, responsibility is moderate. In sum, this operation looks more like force than an intrusive traceback, with at least three criteria (immediacy, measurability, and presumptive legitimacy) ranking high. As a result, it would seem harder to justify on ethical grounds. Even though the attack may appear noble—after all, malicious code is removed—it is also more dangerous. Deleting code can introduce problems, as anyone who has had difficulty uninstalling software has learned the hard way. By comparison, one is less likely to cause damage during traceback.

17.4.3 Conduct of Hack Back

Consider again the traceback operation from the perspective of *jus in bello* and the legal doctrine of self-defense. In both cases, a critical question is whether the traceback is necessary for self-defense. Clearly, the operation itself will not stop the attack. Indeed, the most effective way of stopping most attacks is through improved security. However, traceback may be necessary to find and stop a perpetrator who is exploiting an undetermined vulnerability, as the solution would be unknown. Although the machine could simply be disconnected from the Internet, the effect could be worse than the attack itself, resulting in lost productivity and income. In addition, traceback is necessary to find and then stop the perpetrator from going after other targets and causing greater damage. Furthermore, at the state level, traceback may be necessary to identify the source of foreign intelligence collection against one's own country. Finding that source may be important for national security.

An alternative to traceback is to hand the problem over to law enforcement, but it may be months before law enforcement can even get to the case, let alone solve it. Furthermore, the perpetrator of the attack may have exploited computers in several countries before eventually attacking a particular target, and getting law enforcement agencies in these countries to all participate in the investigation is challenging at best. Moreover, by the time law enforcement responds, the perpetrator may have conducted additional, more serious attacks that could have been averted with a more timely response. Thus, a reasonable argument can be made that at least in certain circumstances, invasive traceback is necessary for a prompt response.

For similar reasons, a traceback that involves invading computers belonging to a neutral country or organization could be warranted if the neutral party is unable or unwilling to stop its own systems from being exploited in the cyber attack in a timely manner.

With respect to proportionality, the seriousness of the cyber attack must be considered along with whether any collateral damage from the traceback is proportional to the harm averted. Whereas traceback may not be justified to defend against a web defacement, it may be appropriate for locating an intruder who has been penetrating a network and downloading sensitive information for months or surreptitiously tampering with or deleting critical data.

With respect to the principles of indiscriminate weapons, superfluous injury, and perfidy, a traceback operation would seem to be in compliance. However, it would not satisfy the principle of distinguishing combatants from noncombatants if the traceback is conducted surreptitiously with the goal of avoiding detection and attribution. To satisfy the principle, the traceback would have to be conducted openly, ideally with permission.

Although the above suggests that invasive tracebacks could be ethically justified in accordance with the principles of self-defense, Himma argued that they are not (Himma, 2004). He based his conclusion on the grounds that they did nothing to either repel or prevent an attack. He further reasoned that tracebacks can locate the source only in direct attacks staged from the hacker's machine, and, therefore, are unlikely to achieve the greater good of identifying the culpable parties. His argument assumes that a traceback identifies the source of a particular IP packet, but not necessarily the source of the attack. In a later paper, Himma observed that improvements in traceback technologies that allow source identification could lead to a different conclusion (Himma, 2006b).

Now, consider the hack back to remove malicious code from the victim machine engaged in the DoS attack. A case for necessity is harder to make, as an alternative course of action would be to notify the owner of the machine of the attack and ask that the machine be taken off the network until the code is repaired. Since most owners would not want to risk being held liable for damages caused by their machines, this approach should be effective, although some effort might be required to determine the machine's owner or get an ISP to notify the owner of a machine on its network. Another course of action would be to get the machines' ISP to block the attack packets, which at least would stop the immediate attack.

It is also harder to make a case for satisfying the principle of proportionality, given that the hack back to remove the malicious code could potentially damage the victim machine beyond that already caused by the presence of the code, and the operation has no effect on eliminating the original source of the attack. The perpetrator could find another victim and resume the DoS attack from the new base of operation.

With respect to the principle of neutrality, the hack back is also difficult to justify if the victim machine is in a neutral country or owned by a neutral third party. The alternative of notifying the machine's owner or ISP would be a better choice.

The hack back does not involve the use of indiscriminate or superfluous weapons. Nor does it involve perfidy. However, unless done openly, it would fail to distinguish combatants from noncombatants. The owner of the victim machine would not know who had hacked the machine. In sum, the hack back to remove code appears less consistent with the doctrine of self-defense than the invasive traceback, and thus harder to justify on moral grounds.

17.5 CONCLUSIONS

This paper has explored the ethics of cyber attacks in three domains of conflict: cyber warfare at the state level, hacktivism conducted by nonstate actors, and active response. It has reviewed how the international law of armed conflict has been interpreted to cover cyber actions in the context of state-level conflict, and then showed how the resulting framework can be applied to nonstate actors and active response.

The framework requires making two determinations: first, whether a particular cyber attack resembles force, and second, whether the attack follows the principles of the law of war. In general, the less an attack looks like force and the more it adheres to the law of war principles, the easier it is to justify ethically. However, attacks that look like force are generally permissible for defensive purposes, so they cannot be ruled out.

To determine the degree to which a particular cyber attack resembles force, the framework uses criteria identified by Michael Schmitt and promoted by Thomas Wingfield. These criteria were developed to distinguish operations that use armed force from softer, more legitimate forms of influence at the state level.

The framework is not intended as a sole instrument for making ethical judgments, but rather as a starting point based on well-established principles. Others have proposed additional considerations that can inform ethical decision making.

ACKNOWLEDGMENTS

I am grateful to Kenneth Himma, Tom Wingfield, and Matt Bishop for helpful suggestions on earlier versions of this paper.

REFERENCES

Attrition (1996). Available at: http://attrition.org/mirror/attrition/1996/08/18/www.doj.gov/. Accessed May 18, 2006.

Denning, D.E. (2001). Activism, hacktivism, and counterterrorsm. In: Arquilla, J. and Ronfeldt, D. (Eds.), *Networks and Netwars*. RAND Santa Monica, CA, Chapter 8, pp. 229–288.

DoD OGC (1999). *An Assessment of International Legal Issues in Information Operations*. 2nd edition, November, Department of Defense, Office of General Counsel, Arlington. Available at: http://www.cs.georgetown.edu/~denning/infosec/DOD-IO-legal.doc. Accessed May 11, 2006.

Himma, K.E. (2004). The ethics of tracing hacker attacks through the machines of innocent persons. *International Journal of Information Ethics*, 2(11),1–13.

Himma, K.E. (2006a). Hacking as politically motivated digital civil disobedience: is hacktivism morally justified? In: Himma, K.E. (Ed.), *Readings on Internet Security: Hacking, Counterhacking, and Other Moral Issues*. Jones & Bartlett, Boston.

Himma, K.E. (2006b). The ethics of "hacking back": active response to computer intrusions. In: Himma, K.E. (Ed.), *Readings on Internet Security: Hacking, Counterhacking, and Other Moral Issues*. Jones & Bartlett, Boston.

Karnow, C.E.A. (2003). Strike and counterstrike: the law on automated intrusions and striking back. *BlackHat Windows Security 2003*, February 27. Seattle, WA.

Manion, M. and Goodrum, A. (2000). Terrorism or civil disobedience: toward a hacktivist ethic. *Computers and Society*, June. ACM Special Interest Group on Computers and Society, New York.

Michael, J.B., Wingfield, T.C., and Wijesekera, D. (2003). Measured responses to cyber attacks using Schmitt analysis: a case study of attack scenarios for a software-intensive system. *Proceedings in Twenty-Seventh Annual International Computer Software and Applications Conference*, November. IEEE, Dallas, TX.

Schmitt, M.N. (1999). Computer network attack and the use of force in international law: thoughts on a normative framework. *Columbia Journal of Transnational Law*, 37, 885–937.

Waterman, S. (2006). Muslim hackers deface Danish web sites. *The Washington Times*, February 24.

Wingfield, T. (2000). *The Law of Information Conflict*. Aegis Research Corporation, Falls Church, VA.

Wray, S. (1998). On electronic civil disobedience. *Presented to the 1998 Socialist Scholars Conference*, March 21–23. New York. Available at: http://www.thing.net/~rdom/ecd/oecd.html. Accessed May 23, 2006.

CHAPTER 18

A Practical Mechanism for Ethical Risk Assessment—A SoDIS Inspection

DON GOTTERBARN, TONY CLEAR, and CHOON-TUCK KWAN

18.1 INTRODUCTION

> The availability of high-quality software is critical for the effective use of information technology in organizations.

Although the need for high-quality software is obvious to all and despite efforts to achieve such quality, information systems are frequently plagued by problems (Ravichandran, 2000). These continued problems occur in spite of a considerable amount of attention to the development and applications of certain forms of risk assessment (which will be discussed in Section 18.2). The narrow form of risk analysis and its limited understanding of the scope of a software project and information systems has contributed to significant software failures. Section 18.3 will introduce an expanded risk analysis process that expands the concept of information system risk to include social, professional, and ethical risks that lead to software failure. Using an expanded risk analysis will enlarge the project scope considered by software developers. This process is further refined by incorporating it into an inspection model and illustrated by its application to a national information system. A tool to develop Software Development Impact Statements (SoDIS) is also discussed.

Informaticians have been evolving and refining techniques to mediate risks of developing software products that meet the needs of their clients. The risks focused on include missed schedule, over budget, and failing to meet the system's specified requirements (Boehm, 2006; Hall, 1998; Jones, 1994). This focus was later expanded to address software security as the highest risk (Stoneburner et al., 2002). In spite of this attention to risks, a high percentage of software systems

The Handbook of Information and Computer Ethics, Edited by Kenneth Einar Himma and Herman T. Tavani

are being delivered late, over budget, and not meeting all requirements, leading to software development being characterized as a "software crisis" and a mistrust of software systems.

18.2 EVOLVING PRACTICES FOR RISK ASSESSMENT

18.2.1 Generic Standards for Risk Analysis Models

Generic standards for software project risk management are available from many professional societies. For example, an American/European standard (IEEE, 2001) and an Australasian standard (AS/NZS, 1999) provide similar systematic approaches to risk management

> ... for establishing the context, identifying, analyzing, evaluating ... risks associated with any activity ... that will enable organizations to minimize losses and maximize opportunities (AS/NZS, 1999, p. 2).

Risk management generally consists of an iterative series of steps like the ones shown in Fig. 18.1. We will use this model to look at each stage of the generic risk analysis model.

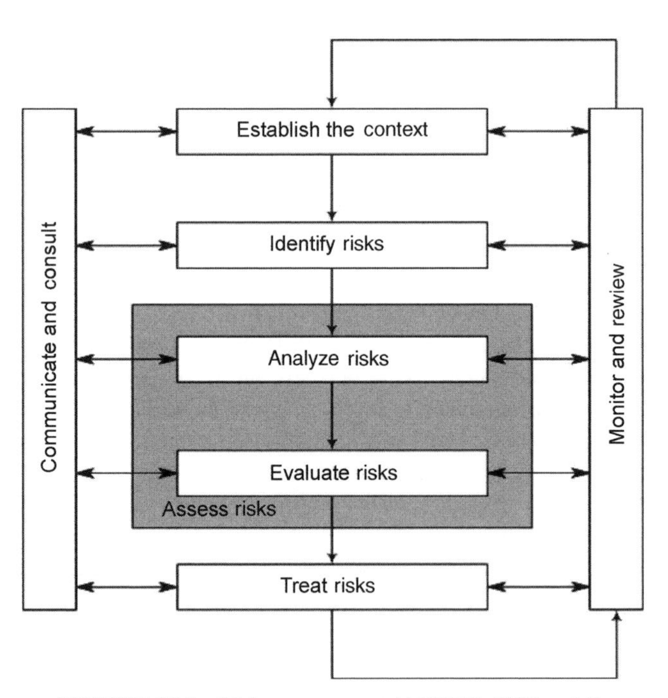

FIGURE 18.1 Risk management (AS/NZS, 1999, p.16).

18.2.1.1 The Context The context referred to in the top box—the context in which the project is being developed—includes the organizational structure and its competitive and political position as well as its risk management structure. This advocates that the scope of the risk analysis should include those things that may derail or interfere with the completion of the project. This model is consistent across most models of software risk analysis in the past 50 years. This defines the area the risk analyst will focus on.

18.2.1.2 Risk Identification The risk identification process identifies potential negative impact on the project and its stakeholders. AS/NZS lists potential negative areas of impact such as

> Asset and resource base of the organization, Revenue and entitlements, Costs, Performance, Timing and schedule of activities, and Organisational behaviour (AS/NZS, 1999, p. 39).

Jones (1994) categorizes software risks by project sector (Table 18.1) and organizes them by frequency of occurrence to help guide risk identification.

The types of risk identified generally include those that have the potential to negatively affect project development (DeMarco and Lister, 2003). Boehm (2006) says the top ten software risks are "personnel shortfalls, unrealistic schedules and budgets, developing the wrong functions, developing the wrong user interfaces, gold-plating, continuing stream of requirements changes, shortfalls in externally-performed tasks, shortfalls in externally-furnished components, real-time performance shortfalls, and straining computer science capabilities."

There are many models of software development, ranging from highly preplanned methods such as the waterfall model to the newer highly adaptive models called agile models of software development. Nevertheless, in all of these models the generic risks addressed are those that may derail the project. Many software development textbooks describe risk simply as the "problem that X will occur and have a negative effect on

TABLE 18.1 Most Common Risk Factors for Various Project Types

Project Sector	Risk Factor	Percentage of Projects at Risk
MIS	Creeping user requirements	80%
	Excessive schedule pressure	65%
	Low quality	60%
	Cost overruns	55%
	Inadequate configuration control	50%
Commercial	Inadequate user documentation	70%
	Low user satisfaction	55%
	Excessive time to market	50%
	Harmful competitive actions	45%
	Litigation expense	30%

some aspect of the development process" (Pfleeger, p. 70). Rook (1993) severely limits the scope of risk identification. For something to be a risk event, it must "create a situation where something negative happens to the project: a loss of time, quality, money, control, understanding, and so on. . .The loss associated with a risk is called the risk impact."

Jones (1994) and Hall (1998) address a waterfall approach. Boehm (1988) has developed a software development model that spreads the analysis of risk throughout the development process. Following Boehm's spiral model, software is developed in a series of incremental releases. Each iteration through the spiral includes tasks related to customer communication, planning, risk analysis, engineering the development of the next level, construction and release, and customer evaluation and assent. Each incremental element of the product that passes through these phases has undergone analysis for these risk types. Although this model introduces a focus on risks, those risks are limited to the risks identified above. The method is also limited in that it assumes all stakeholders are equal and that they will be equally aware of and able to describe their own win conditions (for more on the relation between this model and the SoDIS, see Gotterbarn, 2004). When discussing risks in the Agile model of software development, Highsmith (2002, pp. 57–58) divides risk into three major categories: technical, organizational, and business. The focus of these agile methods is on an iterative approach to improve delivery speed and return on investment.

18.2.1.3 Risk Analysis Once these potential risk effects have been identified, they are prioritized in the risk analysis phase to help order when and if they will be addressed. The risk analysis process divides the identified risks by their severity and the likelihood that they will occur, producing a given level of risk. The analysis of the risk severity is put in either qualitative or quantitative terms. Kerzner (2002) says when doing project risk analysis, those items to be considered are cost evaluation, schedule evaluation, and technical evaluation. Once these are analyzed they are converted into a prioritized schedule (Kerzner, 2002, pp. 669–670), either by quantitative analysis or a limited qualitative analysis that still uses cost and project derailment as the major form of categorization. This is evident in the standard set of risk ratings given by Kerzner (p. 670) (Table 18.2).

Two forms of exposure are commonly calculated. The first method using quantitative risk analysis provides quantitatively expressed assessment of the negative

TABLE 18.2 Risk Rating (Kerzner, 2002, p. 670)

Risk Level	Description
High	Substantial impact on cost, schedule, or technical. Substantial action required to alleviate.
Moderate	Some impact on cost, schedule, or technical. Special action may be required to alleviate issue.
Low	Minimal impact on cost, schedule, or technical. Normal management oversight is sufficient.

consequences of an event as the outcome of an event, for example, "A delay of one day will cost $3000 in sales." The second method uses qualitative risk analysis to address risks that are not readily quantifiable other than by describing the degree of risk, for example, "The delay will upset our distributors, causing significant loss of goodwill."

Quantitative Risk Analysis Generally, qualitative analysis is often used "first to obtain a general indication of the level of risk . . . or where the level of risk does not justify the time and effort for a quantitative analysis . . ." (AS/NZS, 1999, p. 14). The role of quantitative analysis primarily is to characterize and identify the impact of a risk generally assessed in terms of dollars. The risk level or severity is generally determined by using some quantifiable value such as cost or time and statistical or mathematical method. This level of risk is generally determined with statistical analysis or calculations with fault trees and event trees. A typical calculation is "risk exposure," a metric derived by multiplying the anticipated costs by the probability of the event occurring. There are obvious kinds of problems with this type of analysis because the prioritization of the risk types may be industry- or country specific, but there are formal methodologies to harmonize these risk rankings (Kerzner, 2002, p. 693).

Sometimes we find the cost/benefit analysis of this form of risk analysis troubling. A safety calculation in large engineering projects includes a calculation of cost/benefit ratio for an acceptable level of risk for construction workers. These calculations when used in terms of death benefits reduce a person's worth to the number of potential earning years lost multiplied by their expected income during those years. These and similar questions raise concerns about a purely quantitative approach to risk analysis.

Qualitative Risk Analysis As a support for a quantitative risk analysis, a qualitative analysis is sometimes used.[1] Surprisingly, in standard risk methodologies the qualitative risk approach typically looks at quantifiable data that can be easily prioritized and facilitates analysis. Qualitative analysis uses descriptive scales, such as those in Table 18.3, indicating the degree of the risk.

These descriptions are used to prioritize risks and determine the amount of corporate resources devoted to their mitigation. Notice how each of these descriptive levels has an easily quantifiable description.

Even the generic form of qualitative risk analysis is limited in scope to the success of the project, which may include the satisfaction of the customer. Hilson (2004) says, "... if [qualitative risk identification is] done properly it should ensure that all foreseeable risks are listed, representing any uncertain event, or set of circumstances that, if it occurs, would have a positive or negative effect on the project ...."

[1]Qualitative risk analysis should not be confused with the Qualitative management movement of Deming and others, which used an iterative method of process modification (Deming, 2000).

TABLE 18.3 Qualitative Measures of Consequence or Impact (AS/NZS, 1999, p. 42)

Level	Description	Detailed Description
1	Insignificant	No injuries, low financial loss
2	Minor	First aid treatment, on-site release immediately contained, medium financial loss
3	Moderate	Medical treatment required, on-site release contained with outside assistance, high financial loss
4	Major	Extensive injuries, loss of production capability, off-site release with no detrimental effects, major financial loss
5	Catastrophic	Death, toxic release off-site with detrimental effect, huge financial loss

18.2.1.4 Limitations of the Generic Standards The Association for Information Systems defines "system quality" in terms of currency, response time, turnaround time, data accuracy, reliability, completeness, system flexibility, and ease of use (AIS, 2005). Even after using these generic models of risk analysis, information systems have been produced that have significant negative social and ethical impacts. The risks of these impacts are not traditionally included in the tripartite concept of software failure: over-budget, late, or not meeting stated functions.

On the basis of business considerations, some have extended the generic risk analysis to include negative impacts of software to include the impacts on those who have a financial interest in the project (Agle et al., 1999).

Some have extended this generic risk analysis further to include safety critical issues related to the distribution of the software (Leveson, 1995). Their analysis extends those stakeholders considered in the risk analysis process from developer and client/customer to also include those who use the software. Others (Hilson, 2004; Stoneburner et al., 2002) make the data to a stakeholder and make maintenance of the data and security the primary goal of risk analysis. For Hilson (2004), the primary stages of risk analysis over a software life cycle consist of the specification and testing of system security requirements (Fig. 18.2).

Unfortunately, even with all of these modifications to risk analysis, there are still considerable problems with software. Consider a simple case that occurred in

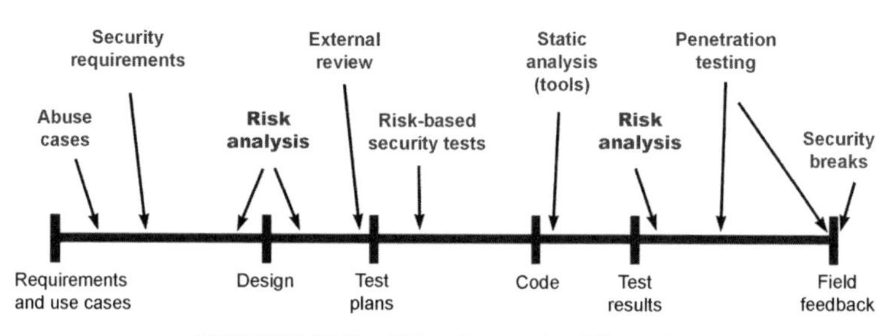

FIGURE 18.2 Hilson's security life cycle.

New Zealand in 2003. Software was developed to allow people to remotely start their cars with the same device that allows them to remotely unlock their cars. There was a complete risk analysis done on the development of this program, and it even included an analysis of risks to those who had a financial stake in the project and to the users of the software. What was not considered in the design of the system was its use in a manual transmission car that was left in gear when it was parked. Nor was the pedestrian considered who was hurt while walking between two cars when one of the cars was started remotely. None of these expanded analyses considered the pedestrian, a stakeholder who was significantly affected by the failure in the design to limit its application to cars with automatic transmission. The problem is that these quantitative forms of analysis miss a whole range of ethical and social issues, which are not issues of project development.

Generic qualitative analysis has a similar weakness. Frequently ethical concerns addressed in a risk analysis are turned into quantitative judgments, for example, the Ford decision not to redesign the Pinto based on the lower cost of law suits from the injured compared to the cost of redesigning the car. Even in qualitative analysis the risk is reduced to some utilitarian calculus, to some cost benefit analysis. The inadequacy of such calculations is seen when the "cost of the US occupation of Iraq is simply reduced to the number of American dead," or the risk assessment of the September 11 attacks is reduced to the number of people who died on that day. These generic types of risk analysis miss significant ethical and social issues.

18.2.1.5 Ethical Risks The ethical stakeholders in developed software are all those who are affected by it even though they are not directly related to the use or financing of a system. The political candidate who is not elected because of a difficult voting machine interface is a stakeholder in the development of that voting machine. The person who suffers identity theft because of a flaw in the security for an information system is a stakeholder in that information system. The developer's obligations to these stakeholders are not included in the generic concept of software failure.

These systems may have been a success in terms of being developed within budget and delivered on schedule, but they were a failure because they failed to take into account the conditions in which they were used. The user interface that met specifications had a significant impact on the lives of others. The system used to record dosages of pediatric medicine correctly handled negative interactions of dosages, but it was awkward to use in emergency situations, resulting in 3 medication errors out of every 100 (Pediatrics, 2006).

Contributing Factors Two interrelated factors related to system stakeholders contribute to missing these professional and ethical failures. The first of these is that limiting the consideration of system stakeholders to just the customer/client, software developer, and those who have a financial stake in the system ignores the needs of other relevant stakeholders.

Some have realized that the focus on technical risks is too narrow, but, unfortunately, the risk focus only expands to other internal issues related to the development of the system. For example, Ravichandran (2000) says, "Research in software quality has focused largely on the technical aspects of quality improvement, while limited attention has been paid to the organizational and socio-behavioral aspects of quality management." This is similar to the position maintained in the Australasian risk analysis model mentioned above.

A second factor is limiting the scope of software risk analysis just to the technical and cost issues. A complete software development process requires (1) the identification of all relevant stakeholders and (2) enlarging risk analysis to include social, political, and ethical issues. A complete risk analysis requires a process to help identify the relevant stakeholders and broaden the scope of risks anticipated.

To meet the goal of quality software, developers focus on particular risks, including project and schedule slips, cost increases, technical and quality risks, the timeliness of the product, and risks that the final product will not fit the business for which it was designed. Nevertheless, developers use the quantitatively assessed risk exposure to help them focus on the most critical risks. The use of easy-to-read fonts or an easy-to-use backup system may be ignored in an effort to get a product out in time or at lower cost.

This quantitative approach is utilitarian. The risks that are addressed are those with the highest risk exposure. All consequences are given dollar values. Even qualitative risks are turned into a numerical hierarchy (McFarland, 1990). The resulting risk of the September 11 disaster was calculated in terms of the number of deaths that occurred on that day or lifetime dollar earnings potential of those who died.

The negative effects that need to be addressed in risk analysis include both overt harm and the denial or reduction of goods. An automated surgical system that randomly moves inches instead of centimeters, hurting patients, would have a negative effect, as a pay phone system that disables all usage, including 911, without an approved credit card would also have a negative effect preventing the report of an accident. These stakeholders, patients and someone hurt in a fire, are not normally considered. The scope of a project needs to be identified in terms of its real stakeholders.

This enlargement of the domain of stakeholders has been implicitly endorsed by professional societies in the paramountcy clause—"Protect public health, safety, and welfare"—in their codes of ethics. This extension has been explicitly adopted in several legal decisions in the United States. This extended domain of stakeholders includes users of the system, families of the users, social institutions that may be radically altered by the introduction of the software, the natural environment, social communities, informatics professionals, employees of the development organization, and the development organization itself.

Including a broader range of stakeholders will also broaden the types of risks considered. The systems we develop perform tasks that affect other people in significant ways. The production of quality software that meets the needs of our clients and others requires both the carefully planned application of technical skills and a detailed understanding of the social, professional, and ethical aspects of the product and its impact on others.

18.3 SODIS AUDIT PROCESS

A process developed by Gotterbarn and Rogerson (2005) using Software Development Impact Statements can mitigate some of these problems and improve software quality by ensuring that the needs of all project stakeholders have been properly considered, thereby broadening the types of risks considered at the outset of a project. The resulting first-cut requirements documents reflect a more comprehensive vision of potential threats to a project's success. The SoDIS process steps analysts through a systematic preaudit of the factors that govern a "typical" project's management and deployment. The results of a SoDIS preaudit can then be used to develop a refined set of requirements, which in turn can be written into the Request for Proposal (RFP) documents common in outsourced software development. Development contracts may also be modified to stipulate subsequent SoDIS project audits or inspections at later stages in a project's life cycle.

The original SoDIS concept, as developed by Gotterbarn and Rogerson (Rogerson and Gotterbarn, 1998), was based on two sets of findings from multiple software development projects. One of the findings showed that software project failures were largely because of defective risk analyses, that is, analyses that failed to consider a system's impact on all who might be affected by that system's deployment. This narrow consideration of stakeholders contributed to a limited view of project scope. The other findings characterized ways in which software development projects could have significant negative impacts on society and its citizens. From these findings they developed a hypothesis about a way to mitigate social and ethical software disasters—a strategy that uses a preliminary analysis of software development plans to alert the developer to a broader range of stakeholders and expand the range of risks considered for these stakeholders. In turn, this would have a positive impact on the development of the software and thereby reduce the negative impact of the software developed. They developed methods to test the efficacy of this strategy and developed the SoDIS process to do the preliminary project auditing. They tested this process on software projects in industry and academe. For example, an application of the SoDIS process in a blind parallel test with Keane Incorporated (Boston) in 1998 led to significant modifications to the SoDIS analysis process and to the development of a prototype tool (called the SoDIS Project Auditor, SPA) to apply the SoDIS process to software project plans.

18.3.1 Software Development Impact Statement[2]

The Software Development Impact Statement, like an environmental impact statement, is used to identify potential negative impacts of a proposed system and specify actions that will mediate those impacts. A SoDIS is intended to assess impacts arising from both the software development process and the more general obligations to various stakeholders.

[2]The description of the SoDIS Audit process is primarily based on Gotterbarn, 2004.

At any point in the development of a system, there are stated system goals and a list of tasks needed to complete that stage of development. The SoDIS process uses that list of tasks and the system goals as its primary input. The goal of the SoDIS process is to identify significant ways in which the completion of individual tasks may negatively affect stakeholders and to identify additional project tasks needed to prevent any anticipated problems.

The process of developing a SoDIS encourages the developer to think of people, groups, or organizations related to the project (stakeholders in the project) and how they are related to each of the individual tasks that collectively constitute the project. Although all software projects have some unique elements, there are significant similarities between projects so that a generic practical approach can be taken to refocus the goal of a project to include a consideration of all ethically as well as all technically relevant stakeholders.

To aid with the major clerical task of completing this process for every task and for every stakeholder, a tool—the SoDIS Project Auditor (SPA)—was developed. The SoDIS Project Auditor is a software tool that keeps track of all decisions made about the impact of project tasks on the relevant project stakeholders, and it enables a proactive way to address the problems identified. A review of the tool will help explain and demonstrate the SoDIS process.

18.3.2 Stakeholder Identification

A preliminary identification of software project stakeholders is accomplished by examining the system plan and goals to see who is affected and how they may be affected. When determining stakeholders, an analyst should ask whose behavior, daily routine, work process will be affected by the development and delivery of this project; whose circumstances, job, livelihood, community will be affected by the development and delivery of this project; and whose experiences will be affected by the development and delivery of this product. All those pointed to by these questions are stakeholders in the project.

Stakeholders are also those to whom the developer owes an obligation. The imperatives of the several software codes of ethics define the rights of the developer and other stakeholders. These imperatives can be used to guide the stakeholder search. The process of identifying stakeholders also identifies their rights and the developers' obligations to the stakeholders. These imperatives have been reduced and categorized under five general principles in the SoDIS process and incorporated into the SoDIS Project Auditor.

On a high-level, the SoDIS process can be reduced to four basic steps: (1) the identification of the immediate and extended stakeholders in a project, (2) the identification of the tasks or work breakdown packages in a project, (3) for every task, the identification and recording of potential ethical issues violated by the completion of that task for each stakeholder, and (4) the recording of the details and solutions of significant ethical issues that may be related to individual tasks and an examination of whether the current task needs to be modified or a new task created to address the identified concern.

A complete SoDIS process (1) broadens the types of risks considered in software development by (2) more accurately identifying relevant project stakeholders.

18.3.3 SoDIS Stakeholders Identification

The identification of stakeholders must strike a balance between a list of stakeholders that includes people or communities that are ethically remote from the project and a list of stakeholders that only includes a small portion of the ethically relevant stakeholders.

The SoDIS process provides a standard list of stakeholders that are related to most projects. This standard list of stakeholder roles changes with each change of project type. For example, a business project will include corporate stockholders, while a military project will not have stockholders in a standard stakeholder role. The system also enables the SoDIS analyst to add new stakeholder roles.

The stakeholder identification form (Fig. 18.3) contains a Statement of Work that helps remind the analyst of the project goals and facilitates the identification of relevant stakeholders. The stakeholder form and the SoDIS analysis form are dynamic

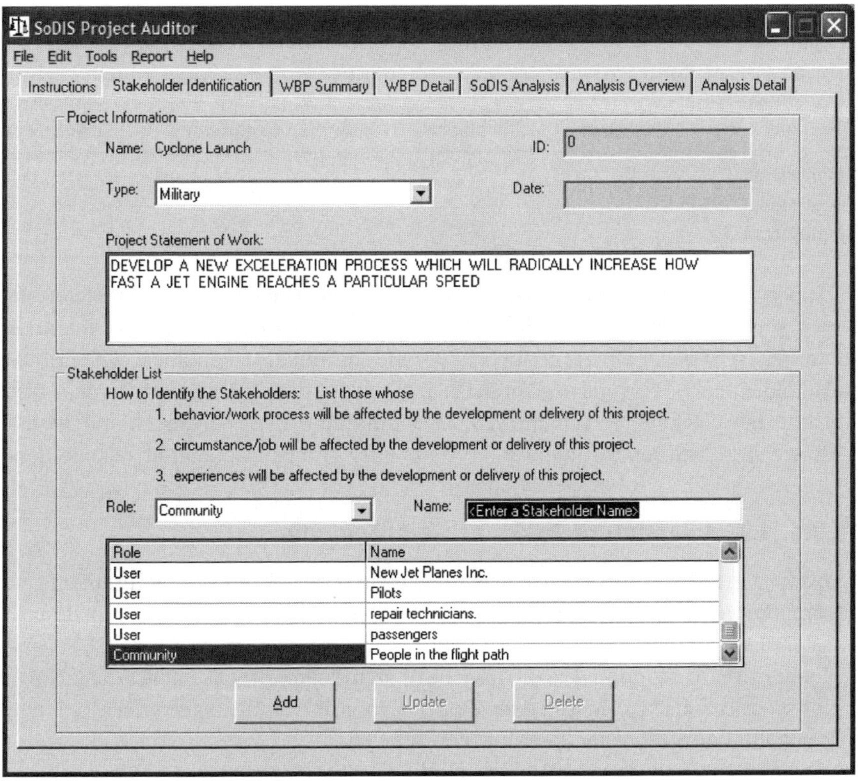

SoDIS Project Auditor
File Edit Tools Report Help
Instructions | Stakeholder Identification | WBP Summary | WBP Detail | SoDIS Analysis | Analysis Overview | Analysis Detail
Project Information
Name: Cyclone Launch
ID: 0
Type: Military
Date:
Project Statement of Work:
DEVELOP A NEW EXCELERATION PROCESS WHICH WILL RADICALLY INCREASE HOW FAST A JET ENGINE REACHES A PARTICULAR SPEED
Stakeholder List
How to Identify the Stakeholders: List those whose
1. behavior/work process will be affected by the development or delivery of this project.
2. circumstance/job will be affected by the development or delivery of this project.
3. experiences will be affected by the development or delivery of this project.
Role: Community
Name: <Enter a Stakeholder Name>

Role	Name
User	New Jet Planes Inc.
User	Pilots
User	repair technicians.
User	passengers
Community	People in the flight path

Add Update Delete

FIGURE 18.3 SoDIS stakeholder identification.

and enable the iterative process. If while doing an ethical analysis one thinks of an additional stakeholder, he/she can shift to the stakeholder identification form, add the stakeholder, and then return to the SoDIS analysis that will now include the new stakeholder.

Rogerson and Gotterbarn (1998) proposed a method to help identify stakeholders based on Gert's moral rules (Gert, 1988). Gert gives 10 basic moral rules. These rules include as follows: Don't kill, Don't cause pain, Don't disable, Don't deprive of freedom, Don't deprive of pleasure, Don't deceive, Don't cheat, Keep your promises, Obey the law, and Do your duty.

A matrix can be set up for each ethical rule such as "Don't cause harm." The column headers of the "Don't cause harm" matrix are the stakeholders, such as the "developer" and the "customer," and there is a row for each major requirement or task. The SoDIS analyst then visits each cell in the matrix, asking for each requirement whether meeting this requirement violates that obligation to the stakeholder. Because the analysis as described is organized by particular software requirements or tasks, it will be easy to identify those requirements that generate a high-level of ethical concern. Thus, the list will also be used to determine if particular requirements have to be modified to avoid significant ethical problems. This method can be used to give a composite picture of the ethical impact of the entire project from the point of view of these stakeholders.

Might the completion of this requirement cause harm to the stakeholder? ("Y" indicates that the task may cause harm to the stakeholder group.)

Requirement/ Stakeholder	Customer	Developer	User	Community	Additional Stakeholders
Requirement 1	N	N		N	
Requirement 2	N	N		Y	
Requirement 3	Y	N		Y	

This process can be used both to identify additional stakeholders and to determine their rights. The first phase of the stakeholder identification should have identified some areas of broader ethical concern and some additional stakeholders. The primary stakeholder analysis is repeated for these newly identified stakeholders. Even if there were no new stakeholders identified, at a minimum the analysis should include software users and related cultural or community groups as potential stakeholders.

18.3.4 Identification of Tasks or Requirements

At every stage of system development there are a series of tasks or requirements that decompose the development into its component parts. These individual task descriptions are used in the reviewing and monitoring of the project.

Each of these individual tasks may have significant ethical impact. The SoDIS Audit process is used to help the developer responsibly address the ethically loaded risk potential of each of the tasks or requirements.

The SoDIS analysis process also facilitates the identification of new tasks or modifications to existing tasks that can be used as a means to mediate or avoid identified concerns. The early identification of these software modifications saves the developer time and money and leads to a more coherent and ethically sensitive software product.

18.3.5 Identify Potential Ethical Issues

This stakeholder identification process has been modified in the SoDIS Project Auditor. Gert's ethical principles have been combined with ethical imperatives from several computing codes of ethics to reflect the professional positive responsibility of software developers. These principles have been framed as a set of 31 questions related to stakeholders in a software system and to generalized responsibility as an informatics professional. These questions are placed in the bottom frame of the SoDIS Analysis screen (Fig. 18.4).

There may be some special circumstances that are not covered by these 31 questions, so, the system enables the SoDIS analyst to add questions to the analysis list. When the analysis is complete, there are several usage statistics reports that give various snapshots of the major ethical issues for the project.

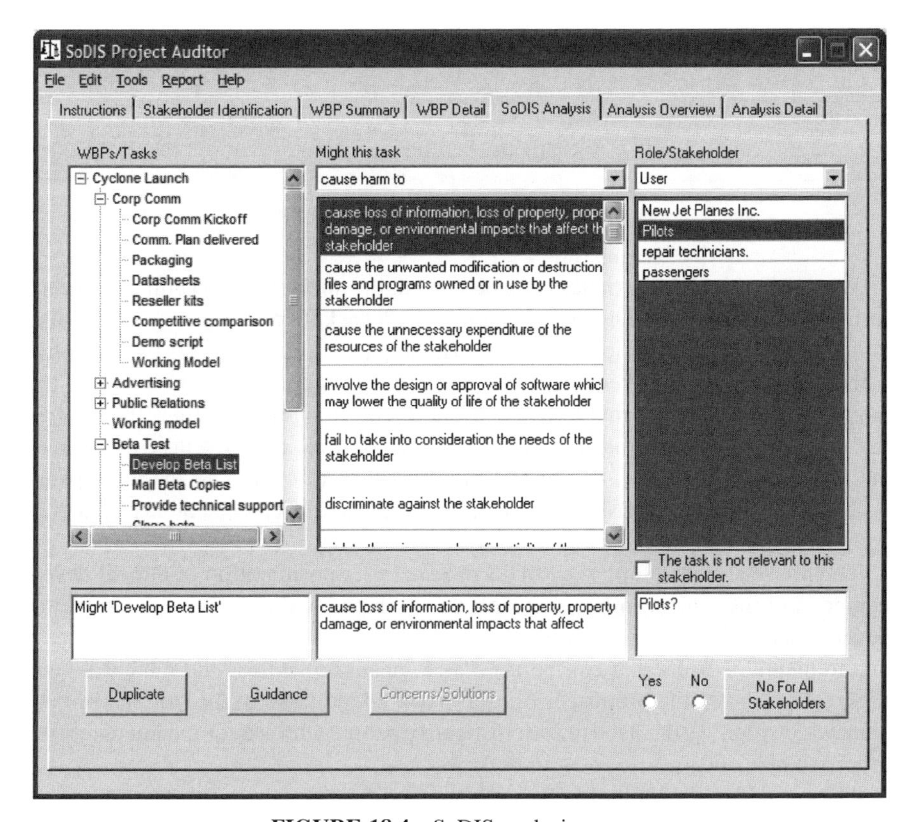

FIGURE 18.4 SoDIS analysis screen.

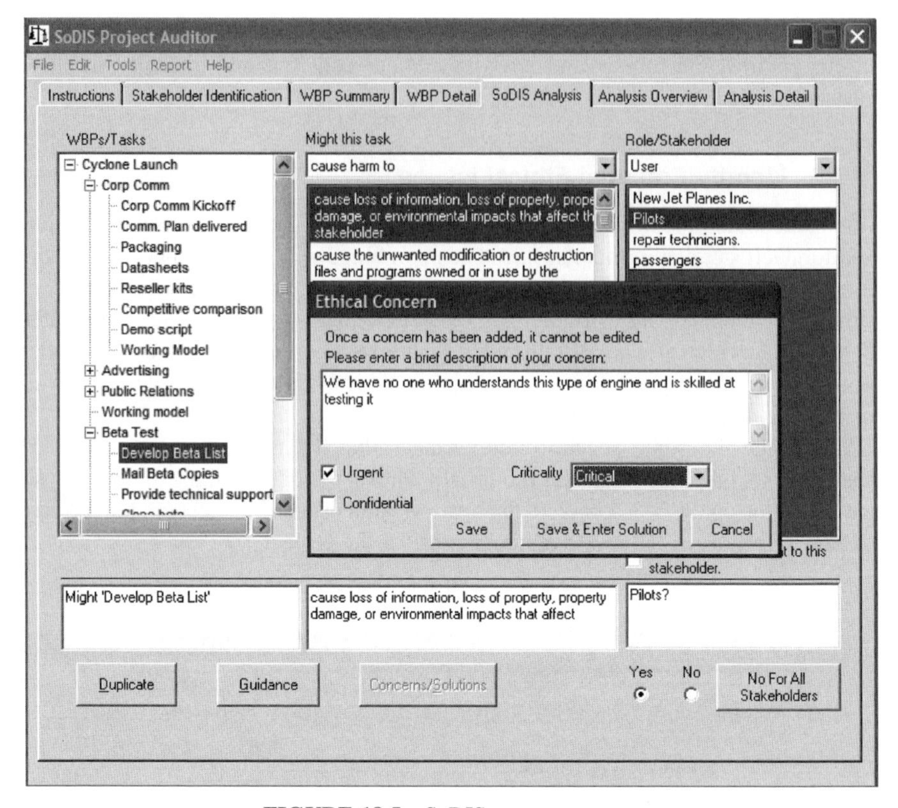

FIGURE 18.5 SoDIS concern screen.

When an ethical concern has been identified, the analyst gets an ethical concern form (Fig. 18.5) to record his/her concern with the task for that stakeholder and also to record a potential solution. The most critical part of this process is on this form, where the analyst is asked to assess the significance of the concern with the requirement/task being analyzed. If the problem is significant then the analyst must determine whether the problem requires a modification of the task, deletion of the task from the project, or the addition of a task to overcome the anticipated problem. It is these adjustments to the software requirements or task list that complete risk analysis.

The process of developing a SoDIS requires the consideration of ethical development and the ethical impacts of a product—the ethical dimensions of software development. The SoDIS analysis process also facilitates the identification of new requirements or tasks that can be used as a means to address the ethical issues. Figure 18.6 shows the proposed solution—two added tasks to the original project plan, which identify the need to start to identify people who are competent to test the proposed new software.

The early identification of these software modifications saves the developer time and money, and leads to a more coherent and ethically sensitive software product.

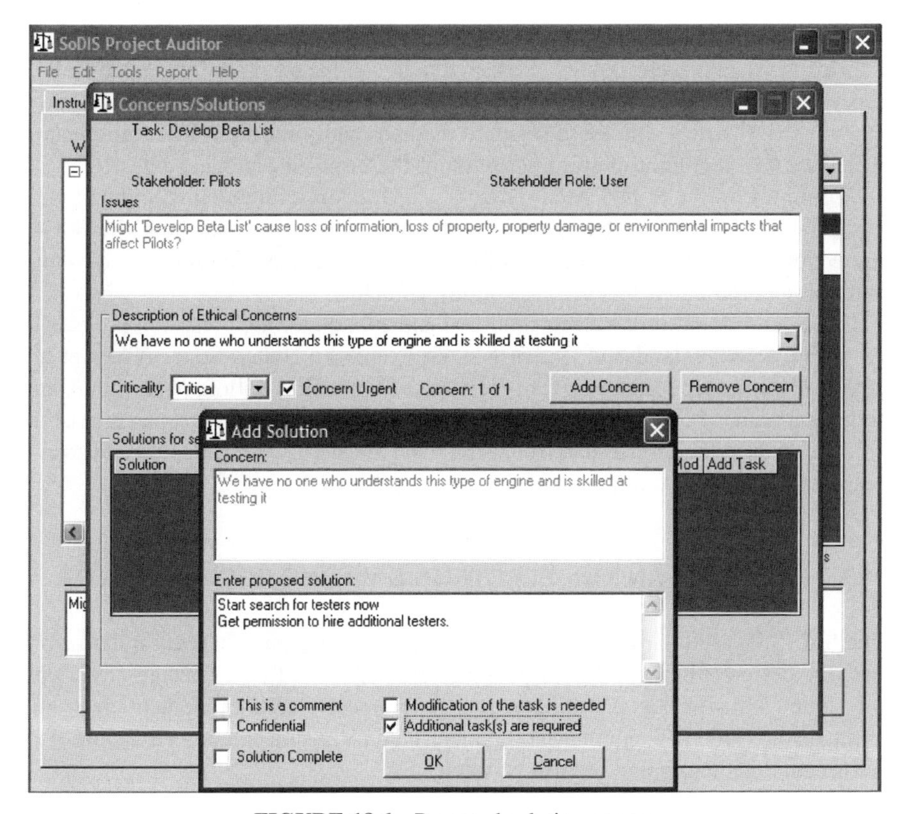

FIGURE 18.6 Proposed solution screen.

When the developers arrive at the point of testing the jet engine in the above scenario, they will have experienced testers.

A SoDIS Audit, in short, represents a promising mechanism for addressing the inherent weaknesses of generic risk analysis. As shown here and elsewhere (Clear et al., 2004; Gotterbarn, 2004; Gotterbarn and Clear, 2004; Gotterbarn and Rogerson, 2005; Koh, 2003; McHaney, 2004), application of the SoDIS Audit in projects has improved the quality of the project scoping, requirements analysis, project management, and risk assessment processes.

18.4 SODIS INSPECTION MODEL

18.4.1 Improve SoDIS Audit with an Inspection Model

The results of previous research on failed projects and our initial use of the SoDIS analysis (Gotterbarn and Rogerson, 2005) led us to make significant modifications to the generic form of risk analysis. Our results clearly indicate that these generic approaches are significantly limited. The application of the SoDIS Audit to specific details of systems development had some limitations, where issues about the application in its context were missed. Research was conducted on an expanded SoDIS process, a SoDIS inspection. The SoDIS Inspection process as a risk analysis method has some similarities with these standard methods and some significant differences, which we will highlight in our description of the inspection model we tested.

We show how the SoDIS Audit process was developed into an inspection model[3] based on work with the UK government and then demonstrate how the SoDIS inspection process has been successfully applied to an outsourced software development project through a specific case described below.

The SoDIS Audit process was modified and applied through research projects in Australia, New Zealand, the United Kingdom, and the United States (Clear et al., 2004). The research followed a four-step cycle of planning, action, observation, and reflection used in both the UK and NZ research projects reported here, and consistent with the dual cycle action research of McKay and Marshall (2001) aimed at addressing both practice and research concerns.

18.5 THE SODIS AND UK ELECTRONIC VOTING REQUIREMENTS

The SoDIS inspection model (cf. Section 18.3) was developed in part through work with the UK government. This work began in 2002, when Rogerson contracted with the United Kingdom to apply SoDIS to a government plan to implement electronic voting in the United Kingdom by 2005. Electronic voting is a paradigm example of an information gathering and reporting system.

This contract outlined a set of deliverables that included assessments of technical requirements for the UK e-voting system. In tandem, both the SoDIS Inspection process itself and the supporting SoDIS Project Auditor case tool were modified in a research context built upon a coherent information ethics framework. Early subsets of the research goals from Table 18.4 (1–5, 9–10) were drivers of the research work. Thus, a set of practice interests and research interests were to be jointly addressed in this project. The research findings from this project formed a basis for the subsequent reflection and formalization of the SoDIS Inspection reported below, whereas the practice findings directly informed the UK government's e-voting implementation plan.

The study began by identifying the technical and social issues related to the electronic voting project. Meetings were held between the SoDIS Team, general election policy specialists, and technical specialists responsible for outsourcing the project. These meetings helped the SoDIS team to gain a high-level understanding of the nature of the application and the key social and technical issues, in an effort to inform the RFP process.

[3]Like Gilb and Graham (1993), we extend the concept of "inspection" (Fagan, 1976) to include informal reviews of all software development artifacts, rather than limiting the concept to "examining a program in detail" (Parnas and Lawford, 2003).

TABLE 18.4 Elements of a Research Intervention – Use of the SoDIS Process in Software Development Projects

Research Element	Description of SoDIS Element
Framework	SoDIS process (Rogerson and Gotterbarn, 1998) Gert's moral rules (Gert, 1988) Codes of professional ethics (Gotterbarn et al., 1998) Practical action research (Carr and Kemmis, 1983)
Research method	Practical action research (Carr and Kemmis, 1983)
Problem-solving method	Process consultancy, SoDIS project preaudits, SoDIS inspections
Problem situation of interest to the researcher	1. How to develop better-quality software 2. How to systematically apply qualitative risk analysis to software development 3. How to develop ethically sensitive software that considers the legitimate interests of all stakeholders 4. How to apply the SoDIS process in project contexts 5. How to apply the SoDIS Project Auditor CASE tool in project contexts 6. Can use of the SoDIS process reduce the risks in software development projects? 7. In what situations is the SoDIS process most applicable? 8. How can the SoDIS process be implemented in particular domain contexts? 9. How should SoDIS inspections be conducted to best effect? 10. How should the SoDIS process be adapted for different types of projects?
A problem situation in which we are intervening (of interest to the practitioner)	1. Developing quality software 2. Developing a structure for software qualitative risk analysis 3. Improving software development risk assessments 4. Reducing the risk of projects failing 5. Improving software acceptability to stakeholders 6. Reducing the risk of failure in outsourced software development projects

Following these initial meetings, the SoDIS team developed a set of high-level strategies for implementing the application. These strategies were developed, in part, by evaluating technical options that included the location of polling, the means of authentication, the user interface, the network communication interface, and the collection and processing infrastructure (Fairweather and Rogerson, 2002). The team also identified 13 of the system's potential stakeholders grouped under five different stakeholder role categories (cf. Table 18.5), and produced a set of 10 generic requirements.

TABLE 18.5 Stakeholders in UK Electronic Voting Project

Stakeholder Role	Name
Customer	Central government
	Local government
	Those seeking election
Community	Minority groups
	Those overseas
	Those with disabilities
	Those with linguistic constraints
	Those from minority ethnic groups
	Those belonging to fringe political parties
	Those living in rural areas
User	Citizens as voters
Vendor	Suppliers of technological elements
Developer	Systems developer

As the focus was on the needs and obligations of the public, the SoDIS Team singled out for detailed analysis each of the 10 requirements for each of the 8 types of "Community" and "User" stakeholders shown in Table 18.5. This evaluation was driven by a set of 32 socioethical-related questions designed to uncover any negative impact that satisfying a given requirement might have on a given stakeholder. Potential problems were addressed by examining the possible modifications to the project, including the use of a particular technical option that would minimize the adverse impact on the stakeholders.

The SoDIS Project Auditor CASE tool (Gotterbarn and Rogerson, 2005) was used to conduct this evaluation. Analysts who attempted to use this tool in isolation, unfortunately, found the work tedious, and it produced incomplete analyses. The subsequent use of pairs and teams of analysts proved more effective.

The final step in the initial analysis involved the use of the SPA CASE tool to generate a Software Development Impact Statement—a register of prioritized concerns and the actions for their potential resolution. In all, a total of 103 concerns with the high-level requirements were identified. To make this list of concerns more manageable, common threads running through the list were identified and labeled. This grouping helped in communications for all involved. In the case of the UK project, this taxonomy was used for communications with governmental officials and politicians in the Office of the Deputy Prime Minister's report (ODPM, 2003).

The team then used these outputs as a guide to repeatedly apply the SPA to the requirements list. This iterative process was conducted until no new concerns were identified.

The concerns identified through the SoDIS process as well as their potential solutions later became the basis for a series of stipulations by the Office of the Deputy Prime Minister (ODPM) for prospective vendors to meet when formulating a Request for Proposal (RFP) for outsourcing the design and development of the system. Some of the key concerns that were identified during the SoDIS process and included in the RFP had to do with designing the system to ensure voter secrecy and safety, equity of access, system performance, and data integrity and security. The analysis also highlighted usability concerns for minority groups and those with disabilities, and also a concern that the means of authentication should not be cost-prohibitive or result in an unacceptable violation of privacy (ODPM, 2003).

The mapping of the concerns raised by the analyst team using the SoDIS Audit process with the system specifications as subsequently issued by the UK Government is detailed in Appendix A.

The application of the SoDIS preaudit process to a high-level set of requirements supported the hypothesis that the SoDIS process is useful in identifying potential problems with requirements. The application of a project preaudit reduced the number and degree of the problems related to this software project and provided at the earliest opportunity warnings of poor development and strategy.

18.5.1 Lessons Learned from the UK Election

The process used in the UK analysis is as shown in Fig. 18.7.

The process showed that the initial methdology of simply using the SoDIS Project Auditor on a list of tasks or requirements was inadequate. This led to several adjustments during the analysis process. Lessons learned from the UK process relevant to this chapter are shown in Table 18.6 below.

These results led to the development of an improved SoDIS process grounded in a formalized context called a "SoDIS Inspection," which could be conducted at chosen points in a project life cycle. It is believed that the SoDIS Inspection could resolve or mitigate four tensions in software development between the forces for change based upon an evolving vision; commercial certainty and cost; project management delivery; and professional quality software (Clear, 2003).

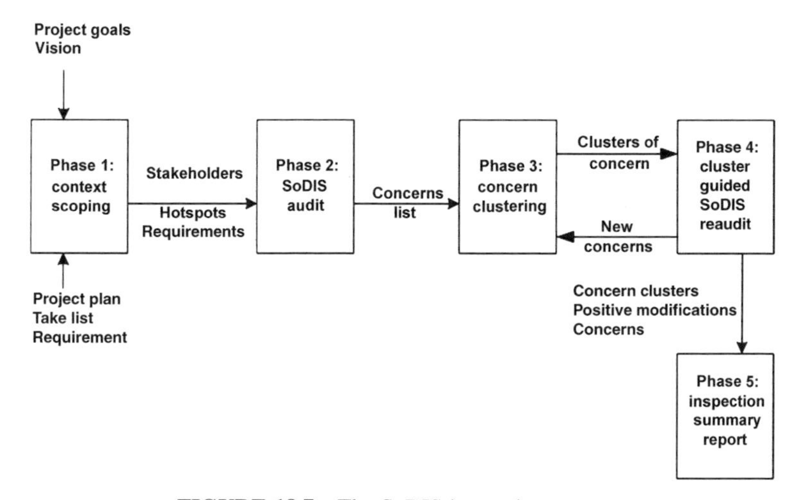

FIGURE 18.7 The SoDIS inspection process.

TABLE 18.6 Results from Applying the SoDIS Process to the UK Electronic Voting Project

No.	Key Results
1.	Tasks can be described at too a high level to yield useful results, particularly if their context was not well understood.
2.	The analysis needs to be conducted on a developer (technical) level and a stakeholder level.
3.	The number of negative questions generated causes analysts to occasionally change focus and think of positive improvements for the project.
4.	A SoDIS analysis can identify a large number of specific low-level issues. These issues can be made easier to understand and help developers maintain focus on the issues by grouping them into categories of problems related to system development.
5.	The process was iterative at every phase.

18.5.2 Research Insights from UK e-Voting Analysis

The lessons learned from this original analysis addressed the following research interests from Table 18.4: (1) developing better quality software; (2) applying systematic risk analysis; (3) ethically considering interests of all stakeholders; (4) applying SoDIS in project contexts; (5) applying SPA in project contexts; and (8) applying SoDIS in particular domain contexts. Other concerns that were addressed in part include (10) adapting SoDIS for different project types; and, from the practitioner's viewpoint (6), the role of SoDIS in reducing risk in outsourced software development projects—upon subsequent reflection. These are all reflected in the key results of Table 18.6, and the subsequent actions of the client in modifying the planned e-voting implementation.

18.5.3 Insights Related to Practice from UK e-Voting Analysis

The questions from Table 18.4 addressed in our research by identifying potential ethical and social risks were about developing quality software, structuring qualitative software risk analysis, improving software development risk assessments, reducing the risk of failed projects, improving software acceptability to stakeholders, and reducing the risk of failure in software development projects. The significant findings from the study and their incorporation into the planning and requirements for e-voting in the UK (cf. Appendix A) demonstrated the efficacy of the process.

The insights drawn from the UK study and reflection upon the lessons learned led to further modification of the SoDIS Inspection process.

18.6 RESEARCH PROJECT

The lessons learned from the UK project and our examination of risk analysis models were brought together in another research project related to a commercial partner's high-level requirements. We present first the SoDIS Inspection model developed as a result of the previous research and then the results of our applying it with a commercial partner.

18.6.1 The Inspection Process

The SoDIS Inspection model developed is directly grounded in the five results described in Table 18.6 (cf. Section 18.5) based on the analysis of UK e-voting outsourced requirements. The inspection process we tested has five phases: (1) context scoping; (2) SoDIS Audit; (3) concerns clustering; (4) cluster guided SoDIS review; and (5) analysis summary. Each phase may be iterated as necessary. The following sections demonstrate how the SoDIS Inspection process links both to the results from the UK election study identified in Table 18.6 and to the steps in a generic risk analysis.

18.6.1.1 SoDIS Inspection Phase 1—Context Scoping UK Result 1 indicated a need for an initial understanding of the project. Phase 1 involves understanding critical elements in the context, including the identification of tasks and overlooked stakeholders. The type of task considered varies with the stage of the development process inspected. For example, "tasks" can be detailed steps in test plans, work breakdown structures in project plans, analysis and design documents, or high-level requirements.

Phase 1 consists of at least two meetings. One meeting identifies the "context of concern" from a technical and project development point of view. The participants are technicians (developers/program managers). The project manager's presentation of scenarios of the product's use in various environments aids in the identification of the people and organizations affected by the software, as well as the identification of the risks for these stakeholders. The other type of meeting also uses scenarios. It identifies the initial context of concern from the perspective of a business analyst, a user, and an affected stakeholder. The focus of this second type of meeting is not the technical development of the software but the impact of the completed project. In this sense it differs from the generic risk assessment step of "establish the context" from Fig. 18.1. The concern for impact here is not the traditional "organization-centric" strategic context, with limited stakeholder models of customer and developer considered in the course of a project. This context of concern meeting adopts a "society-centric" perspective in which the deployment of the project, its impact on society, and all those affected by the software are taken into account.

The structure of the context of concern meeting follows a typical software review. The meetings have a moderator who reports on project scope, available resources, and project goals. There are two parts to these meetings: the first focuses on conveying an understanding of the project proper and the second on highlighting areas of concern. The meeting is held with the business manager/analyst and customer (if it is an internal project) or with the customer, a user representative (if it is an external project), and someone representing the position of the external stakeholders. In many projects these people are not available, in which case a developer is assigned the role of stakeholder representative.

The success of this stage is dependent upon the analyst's and customer's understanding of where difficulties may occur. A scenario technique is employed by the SoDIS analyst, who asks for stories about how the system will be used and in what contexts. The customer is asked about how various stakeholders may be related to the software. The concerns and stakeholder roles identified are recorded and entered into a context of concern form. These descriptions heighten a project's and an analyst's environmental sensitivity and help the analysts and the developers to focus on a broader range of stakeholders. This helps the developer begin to look beyond the purely technical side of development. As in any inspection, the meeting should not be used to resolve any of these issues, or address the details of the concerns identified. Later in verifying the analysis, these context of concern forms will be verified with the stakeholders.

If similar projects have gone through a SoDIS Inspection, then the analysis summary (from Phase 5) for that project should be used as a cross-reference during these meetings. During the analysis, particular hotspots that are critical should be indicated. Hotspots represent places where there is a real danger of negative impact from the completed project. The hotspot's focus may be on particular tasks or stakeholder groups.

This context scoping process helps resolve the problem of missing critical issues because of high-level requirements descriptions (Result 1 of the UK project). Context scoping provides an organization's context and some preliminary directions about where to focus the initial SoDIS Audit. Bringing the user and stakeholder perspective into this phase also starts to address the second UK research result.

18.6.1.2 SoDIS Inspection Phase 2—The SoDIS Audit In Phase 2, the results of the prior "context scoping" phase provide a starting point for the SoDIS analyst's selection of tasks and determine the number and types of questions produced by the SoDIS Project Auditor. The goal of Phase 2 is to search in a structured way for potential concerns related to the project's development, delivery, or use. Based on the context of concern scoping, the analysts select a set of tasks to start their analysis.

The SoDIS Inspection process incorporates elements from pair programming (Cockburn and Williams, n.d.) to improve the efficacy of the analysis. One goal of pair programming is a synergy that produces better software design, continual review of each other's work leading to more effective defect removal, an enhanced problem-solving ability, and an ability to stay focused for longer periods of time. We experimented in several workshops with having the analysts work in pairs. When analysts work in pairs using the SoDIS Audit process (Phase 2 of the inspection model), the results of the analysis are more complete and potential solutions are fuller. One analyst operates the SPA and reads the questions aloud. Both analysts respond to analysis questions. The reader should maintain this role for a maximum 30 min before turning the keyboard and reader role over to the other analyst. The analysis session should last no more than 2 h. This helps reduce the tedium of answering the questions, keeps the analysts focused, and has been shown to generate more "insightful" results.

During the SoDIS Audit process, the SPA forces the analysts to first identify potential stakeholders for the project. The SPA aids the process by providing a partial list of stakeholder types that have been associated with that type of project. Once the stakeholders have been identified, the analysts examine questions that can be either task (hotspot) focused or stakeholder focused. In answering the questions the analysts seek to identify and note potential negative consequences for the identified stakeholders or for the project and, where possible, suggest solutions for the identified items.

This audit is repeated because (1) during the audit, new stakeholders are identified, generating new questions; (2) answering the questions generates a new and more complete picture of the project, which helps clarify issues analysts addressed earlier; and (3) the suggested solutions to earlier concerns may in fact introduce new concerns.

The result of this audit phase (Phase 2) is a Software Development Impact Statement that enumerates potential concerns for the project and project impacts on citizens and organizations. Although the pair analyst approach reduces the tedium of the analysis, it does not reduce the felt need to escape from the negative nature of the analysis. One of the ways by which analysts have escaped from this negative atmosphere is to devise strategies for making a project better while analyzing that project's problems. We modified the SoDIS process to capture these ideas by adding a Positive Modification Form (PMF) to the inspection and asking analysts to record ideas for improvements as they think of them, without respect to implementation, cost, or resource issues. Not only does this capture positive creative thought, but it also reduces the sense of negativism identified in the UK election study (Table 18.5, Result 3).

The audit results in two types of lists: a list of concerns and potential solutions and a list of potential improvements to the project and their anticipated impacts. These lists are the input into the next phase.

18.6.1.3 SoDIS Inspection Phase 3—Concerns Clustering In Phase 3, SPA-generated reports are used to identify trends in the analysis data. The goal of Phase 3 is to provide high-level abstractions of the identified concerns. These abstractions help with further SoDIS analysis and provide high-level risk categories that developers can use in reviewing their projects. This phase meets a need identified in the fourth lesson learned in the electronic vote analysis (cf. Table 18.5—grouping detailed concerns into clusters).

It is difficult to address a large number of ethical issues without careful consideration, clear planning, and resolute action. Phase 3 uses a broader strategy for analyzing risk than the strategy employed in classic risk management (cf. Fig. 18.1). In keeping with the broader notion of risk introduced in Section 18.2, this broadened analysis seeks to identify categories of risks that extend simply beyond the two standard kinds of qualitative risk identified in Table 18.1—namely, physical injury and financial impact. For instance, such critical software development issues as stakeholder disenfranchisement (e.g., because of the development of an unsuitable user interface) are not addressed by these two generic categories. The cluster analyses that are done in Phase 3 support the identification of new perspectives on risk, together with the rebuilding of the project in ways that are consistent with these new perspectives.

In Phase 3, analysts are directed to look for common classes of identified risk, and then to cluster or group the individual concerns identified during the audit. This clustering of risk is based on similarities and differences between concerns identified in the SoDIS Audit phase. For example, a type of problem that emerges with undue frequency may reflect a weakness in the system development plan. Each issue is analyzed to identify common key word descriptors such as "privacy," "access," or "trust." In the UK project, the analysis for several different tasks contained the word "accessibility." Further analysis showed that this was a "concern cluster" that cut across the entire scope of the project.

These risk clusters can convey clear meaning to all types and levels of development project staff. They provide a basis for undertaking practical action to address the project's identified ethical risks. The cluster list is also used to guide subsequent analyses.

Again, incorporating an approach learned from Agile methods research (Cockburn, 2004) and requirements bounding and defining viewpoint analysis (Kotonya and Summerville, 1996), individual SoDIS analysts identify clusters individually and then meet to compare their results. Together they create a single clustering taxonomy based on their individual results.

The elements of each cluster should be evaluated to identify the commonalities and points of difference among the elements that make up the cluster. If a cluster is very large, it is reanalyzed to identify any subclusters. If a cluster addresses more than three types of issues, that cluster should be divided into smaller, more cohesive clusters.

Clusters should be prioritized on the basis of two factors: the priority of individual issues within the cluster and an overall view of the cluster's relative importance or criticality. In this way the project is redefined as a cluster breakdown structure. Cluster analysis enables the project to be rebuilt along a set of new perspectives, thereby addressing Result 4 from the UK election research: the need to group low-level issues in a way that helps system developers to identify new risks.

The cluster analysis that is done during Phase 3 supports the communication of concerns to key stakeholders who stand outside of the project development. The clusters also serve as a filter to determine the completeness of the analysis in Phase 4. The Cluster Analysis Document is used as input to the next phase.

18.6.1.4 SoDIS Inspection Phase 4—Cluster-Guided SoDIS Review In Phase 4, the cluster breakdown structure developed in Phase 3 is used to validate the analysis performed in Phase 2 (Fig. 18.8). The analysts use the new perspective on the project to reengage the task list and the stakeholder list. These lists are compared against the cluster list to identify any issues or stakeholders that may have been overlooked during the earlier analysis. The analysts also reengage the stakeholder list looking for new stakeholders and any issues they may have missed. Access to the Positive Modification Form is also maintained in this stage.

This process is repeated until no new clusters are identified and the analysts are satisfied that the project tasks as redefined do not generate unidentified issues in the

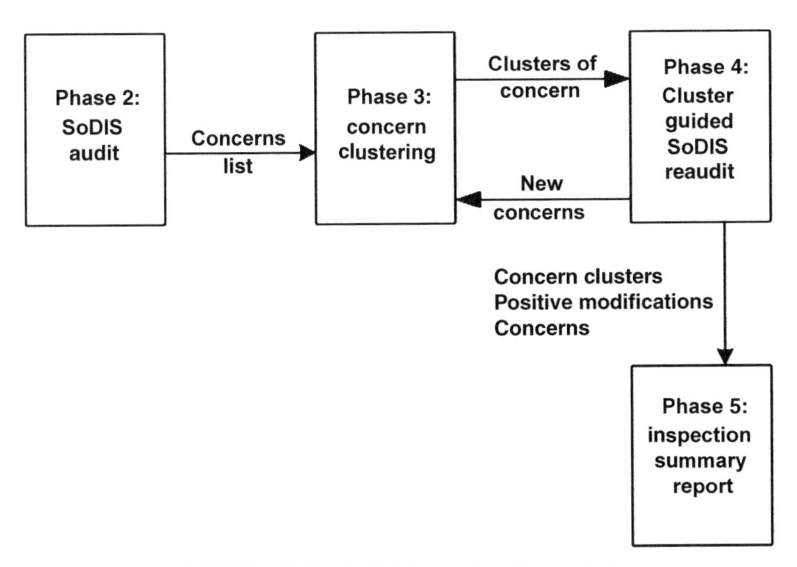

FIGURE 18.8 SoDIS inspection Phases 2–5.

resultant cluster list. This is the phase gate to the development of the SoDIS Inspection Analysis Summary document in Phase 5.

18.6.1.5 SoDIS Inspection Phase 5—Analysis Summary

When the review of tasks and stakeholders using the cluster analysis is completed, a *SoDIS Inspection Analysis Summary* is produced. This document is an overview of the results of the inspection showing the cluster structure and indicating the priority of the cluster issues. This summary functions as an early warning of project directions and tasks that need to be modified to mitigate potential negative impacts of the delivered product. This document is also used to determine whether a project is to continue. This document becomes part of the project library. The information in this document is a historic record that can be used on similar projects.

During this phase the Positive Modification Form is revisited to tidy up the positive suggestions before turning the document over to management for further review. The results of each inspection are used to modify the inspection process details for subsequent inspections. The addition of identified concerns to the inspection model for a particular sector, project, or context is consistent with and supports a continuous process improvement strategy such as the CMM or CMMI (Paulk, 1995).

18.6.2 Iteration

Although this process has been described in a linear fashion, work at any one phase is likely to uncover new information relevant to a previous phase. That phase should be revisited to analyze the impact of the new information on the project and the impact on the project's stakeholders (Table 18.3, Result 5).

18.7 APPLYING THE PROCESS TO OUTSOURCED REQUIREMENTS

The SoDIS Inspection process can be used at most phases of software development. The development of requirements in one form or another is common to all forms of information systems development. The normal risks of requirements gathering are exacerbated by outsourcing. The following case reports the results of research on the SoDIS Inspection model applied to a commercial project in which requirements were developed by a customer and outsourced to diverse developers.

18.7.1 Applying the New SoDIS Inspection Model

18.7.1.1 Background A research study was undertaken with a company in New Zealand, using students and university staff as SoDIS analysts. The team had limited knowledge of the project domain. This use of the SoDIS Inspection process by people unfamiliar with the particular business sector, if successful, would indicate that the process itself and not the analysts' background knowledge was responsible for the results.

The project involved the highest potential for requirements problems because the requirements were developed in-house and their implementation was to be outsourced to multiple parties. If the SoDIS could help mitigate potential problems in this worst case, then we might have a higher degree of confidence in the efficacy of the process for all classes of software development projects.

The Company "NZ*" (the company information has been anonymized but cluster and system issues identified are accurate) gathers information from various sources and locations in New Zealand and makes statistical summaries of that information available to its members and to requesting government agencies. Anyone who works in the NZ* business sector must also be a member of NZ*. The members use the information made available by NZ* in their business. The member organizations are a variety of types and sizes.

NZ*'s paper-intensive system has inputs in diverse formats. Because of the data entry formats, the wide range of IT systems employed by members, and the types of member organizations submitting information, the statistical data is sometimes significantly out of date. There is no way to verify the data in the current paper system before it is used to generate the statistical information for its members. NZ* is responsible for recording and tracking the complaints against its members.

NZ* is planning an automated replacement for the current system. The database design and security will be outsourced to one company and web development and web security to another company. The Content Management System to process additions, alterations of the data, and member input and retrieval formats will be a purchased off-the-shelf system. The hardware technology is leased from a U.S. company.

18.7.1.2 The Inspection and NZ* Prior to the inspection phases we now discuss, the process and its goals were communicated to the research subjects and their consent and buy-in for the engagement was gained. This preliminary step was not a part of the inspection process, but it is a requirement for the method's success.

Phase 1—Context Scoping In their initial meetings, the SoDIS analysis team learned about NZ*'s domain, whereas NZ* was learning about the inspection process. The analysis team took notes and asked few questions because of their lack of prior familiarity with the domain and our inspection process. The enormity of NZ* replacing its entire manual system was being treated with great care by NZ*. NZ* described its business and characterized its primary stakeholders as its members and did not mention those affected by the information. Because the NZ* representative presented a technical view of their systems, the inspection team's questions represented the interests of the user community. The NZ* team's high-level presentation of its plan mistakenly presupposed that the SoDIS analysis team was familiar with its system. Nevertheless, the analysts identified most of the project's stakeholders through the explanations (scenarios in Phase 1) of how the system would be used. A team debriefing was held, and the team then developed context scoping documents listing initial concerns and potential stakeholders affected by the system.

A second meeting with NZ* was held to verify and complete the inspection team's understanding of NZ*'s plan and proposed system. The team presented concerns and raised questions that were not addressed during the first meeting. Specific questions were asked about stakeholders that were not identified in the initial project plan. The team focused its questions during this meeting on how these additional stakeholders would use the system. The scenarios about how these stakeholders would use the system facilitated the revision of the context scoping documents with an expanded list of concerns and those affected by the system.

Phase 2—Guided SoDIS Audit The inspection team used context scoping documents to isolate selected tasks for analysis. The choice of tasks was also partially influenced by the order in which NZ* was going to structure the project. This influence diminished during the analysis as the team came to believe that some tasks scheduled for the later stages of the project had to be addressed early.

OBSERVATIONS ON THE PHASE 2 PROCESS As the team worked through the SPA they did not record some of their observations because they wrongly assumed that the NZ* project manager would surely have thought of these things. The team had not used SPA before and did not answer some questions because they did not understand the NZ* domain well enough. This problem should be addressed by incorporating mechanisms in the SPA for highlighting questions that require further information to answer.

The initial use of the SPA resulted in primarily technical concerns from our inspection team (of technical people). As they felt more relaxed with the process, the domain of concerns expanded to include additional stakeholders. There was still a large number of low-level technical concerns that could not be answered at this stage of the development. This indicates a need to preserve these low-level issues and bring them (as a concern scoping) to an inspection of the more detailed design.

The SoDIS process generates a significant number of questions. Part of its value is that its systematic structure leads an analyst to address questions she may have ignored. However, much software development is often guided by a hysterical rush to finish quickly. When the SoDIS Inspection is used in such situations (Table 18.4, items 9 and 10—lightweight analysis and conducting SoDIS inspections to best effect), care must be taken to avoid haphazard skipping of sections of the process. During the audit phase, analysts will also have an Analysis Decision Document (ADD) available. During the analysis, strategic decisions are made about where to start and where to focus attention. The analysts will generally have special knowledge that allows them to give cursory treatment to some sections of the analysis. To remove the potential for skipping questions out of fatigue, this ADD document is maintained, which records the justification for not analyzing some elements of the project. It also reduces the tendency to stop the analysis just because one or two big issues were identified. Both in the hysterical (light) mode and in the complete analysis mode analysts will be more satisfied and likely to continue if they start with those questions where they have the greatest initial concern, those likely to have the greatest return for the time spent.

Phase 3—Concern Clustering The concerns identified in Phase 2 were reviewed and clustered together. A sample of some clusters identified is listed below:

(1) **Modeling the Existing System**
   (a) In NZ* there was an acceptance of inevitable inaccuracy or out-of-date data in the manual system. The initial design of the new system reflected that acceptance in the manual system. In the manual system, data was examined for accuracy only if it represented an extreme data value that was noticed by the data entry staff. No additional verification was planned. The team felt this unsatisfactory, because the new system would provide for greater distribution of this data to many more stakeholders. The positive modification form also suggested a potential way to mitigate this verification difficulty by returning the collected data electronically to the person who originally entered it for verification at the source.
   (b) In the NZ* manual system, complaints were handled by a data entry person who entered the complaint that had been faxed or phoned in. There was no verification of the accuracy of the complaint description or the tone of the complaint. The computer system facilitates complainant review of the recording and facilitates further complaint tracking. The review process identified the need for a new set of tasks for facilitating complainant review and yet maintaining consistency with existing privacy laws. The complainant had not been viewed as a stakeholder in the system.

(2) **Data Integrity** In the original system, data were submitted by computer, fax, mail, or telephone and then entered into the system. This same mix of input methods was in the new system to provide for input from members with different types of computing facilities. Significant concerns identified here were related to the lack of verification and a risk of data loss, because of an accidental uploading of an incorrect FTP file, for example, an older file that had accidentally overwritten a newer file. This risk could be eliminated by

only allowing a single FTP for each statistical cycle or by asking for member verification of data at the close of the FTP window.

(3) **User Interface** The original, manual system supported a single interface: one for data entry clerks in the home office. NZ* had outsourced the design of the interface. The analysis team identified a cluster of concerns arising from the need to allow a wider range of stakeholders/users to use the system to enter or retrieve data from several different types of system. These concerns identified a need to expand the interface requirements to specify the needs of each of the distinct user groups to the outsourced developers.

(4) **Authorizing Users** The database security was being managed by the group designing the database. They needed to be informed of the different levels of access for each group of users.

(5) **Project Management: Postponing Decisions** A generic issue emerged related to the overall development strategy. NZ* followed a generic outsourcing strategy; each element went to a different specialist outsource agent for development and their results were to be delivered in a linear fashion. First the hardware platform to hold the data would be developed, followed by the database design to hold the data. These deliverables would mimic the existing manual system. After the database was established, the user interface and reports would be developed. Once the software and database were functioning, the issues of security and access were then to be addressed. The way in which the system was contracted was also being used as a model for the design of the system. This is a very risky process. Interface design and data foundations are interdependent, but in the NZ* development model, only after the data foundations are built do the design issues of user interface and security get raised. As was revealed in the previous clusters, the interfaces and the security issues are very different for the proposed system, which would require significant rework of the NZ* hardware and database configurations. The SoDIS Inspection revealed sufficient necessary and desired differences from the manual system to warrant abandoning NZ*'s linear approach for system development.

(6) **Testing** In the process of developing clusters sometimes new insights emerged.

  (a) During Phase 1 of the inspection, the testing method had been described to the team as being done at a client's site a short distance away from the main office. The analysis for clusters 1–4 mentioned above sensitized the team to the variety of types and skills of users, interfaces, levels of authorization, types of computing equipment, and wide distribution of those using the system. The testing cluster, which overlapped most of the other clusters, showed that testing restricted to a local user at a single level would be inadequate. Any test plan proposed by an outsource agent had to be expanded to satisfy these broader concerns.

  (b) The lack of a single integration and testing authority emerged as a second concern. Because of the linear disparate development model being used, each outsource agent was to be responsible for his own product. Individual outsource agents were required to define individual test plans, but no one was assigned to develop and be responsible for the final product meeting the acceptance tests.

OBSERVATIONS ON THE PHASE 3 PROCESS As expected, some large clusters, for example, modeling the existing system, were identified. These were too abstract to be useful as a filter in Phase 4, so they were broken into subclusters: privacy-complaints, data verification, multiple user interfaces, and database integrity.

One of the common problems in the cluster identification and analysis phase is the tendency to collapse similar clusters into a single cluster. Sometimes this causes a loss of significant information. For example, there was an attempt to collapse privacy and integrity. Such a collapse would have lost the distinction between accessing information and preventing its unauthorized modification. Information can be protected from corruption and yet be available to the public. Great care has to be exercised in subsuming one cluster into another. When there is a significant disparity between the concerns identified, this type of collapse will lessen the effectiveness of the cluster and the inspection process.

In a review of the analysis team's initial clustering with NZ*, NZ* accepted the team's revised list of seven stakeholders, and added two further stakeholders. Most of the cluster issues were agreed to, but NZ* trivialized the seriousness of some clusters by claiming that they would be addressed at a later stage of the project or that "the outsource agency would surely take care of that."

Phase 4— Cluster-Guided Revisit of the SoDIS Audit Phase In this phase the analysts used the cluster analysis document to examine the NZ* task list for unidentified concerns that fit into the defined clusters, or that indicated the need for additional clusters or subclusters. This review succeeded in identifying some additional subclusters. For example, the data accuracy cluster had missed the possibility that the system would not recognize that information was missing. This became a subcluster of the data accuracy cluster. Several places where reviews were planned did not allow any time for corrections that might be identified. A subcluster "review time" was added to the project management cluster. After the project task review, the analysts used a stakeholder perspective to guide a further review of the clusters. It was discovered that two significant stakeholder groups had been left out of the development plan, and if they were not considered early then the primary database would have to be redesigned or an arbitrary table added to the database, which would significantly diminish its design quality.

Phase 5— Analysis Summary The results of Phases 2–4 are the foundation of the SoDIS Inspection Analysis Summary report, which provides the project manager with an early list ordered by criticality of potential issues identified in the inspection. The NZ* analysis identified a number of critical issues as observed in the six clusters discussed above. Key among these were the development process related to the database and user interface design, aligning the responsibilities and roles of outsourcing parties, mechanisms for ensuring data integrity, and addressing some resulting negative impacts for the user community.

18.7.1.3 Lessons Learned from the NZ* Study The lessons learned from this analysis more comprehensively addressed the scope of the overall research program than did the UK research program. The second study focused on an expanded and more consciously articulated set of research questions in tandem with the consultancy role. Because the work was undertaken as an unpaid consultancy in the context of a research project, the nature of the engagement may also have supported this, owing to the fact that the research interests were inherently to the fore. Nonetheless, as discussed below, the set of practitioner concerns were also addressed in the course of the study.

18.7.1.4 Research Insights from NZ* Analysis The lessons learned from this analysis addressed the full set of research interests from Table 18.4: 1–10, namely, developing better-quality software; applying systematic risk analysis; ethically considering interests of all stakeholders; applying SoDIS in project contexts; applying SPA in project contexts; determining whether SoDIS reduces risk in outsourced software development projects; determining those situations to which the SoDIS process best applies; applying SoDIS in particular domain contexts; designing lightweight versions of the SoDIS process; determining how SoDIS Inspections should best be conducted; and adapting SoDIS for different project types.

Specifically, in addition to the obvious results of the effectiveness of the process in addressing a broader range of risks and stakeholders, the results of the research showed that

> Stakeholders are not necessarily those people who are **directly involved**, either financially or technically, **in** the development of a project. Stakeholders are also those people and organizations who will be **impacted by** the development and implementation of the project.

In applying the SoDIS Inspection method, we found that it helped in applying systematic risk analysis and ethically considering interests of all stakeholders. Contrary to the generic risk assessment standards discussed in Section 18.3, this research showed that our qualitative risk approach is a necessary condition for a complete risk analysis and not merely an intellectual exercise where the level of risk does not justify the time and effort for a quantitative analysis.[4] The findings identified significant areas of stakeholder impact, which if left unaddressed would have resulted in a lower-quality software system.

We learned more about the flow and logic of a SoDIS Inspection process; its applicability at different stages of a project—initial phases or key milestones; the steps within each phase and the extent to which the method was inherently iterative rather than linear; how to use the SPA to support the process; some needed enhancements to the SPA (discussed under practice insights below); and how to apply the SPA in an outsourced software development project.

The inspection team noted two issues that will need further work. First, different stages in a project should focus on different sets of questions. For instance, early in the

---

[4]The results of a separate research project demonstrated time efficiency of the SoDIS Inspection process.

project life cycle, questions more related to project management, functionality, and domain concerns are significant. Later in the development, questions related to design and technical implementation become important. The SPA and the SoDIS inspection process may need modification to emphasize different foci at different stages of development. Second, different people may be involved in each of these areas of concern. Earlier in the project business analysts may take the lead, whereas later in the more technical stages of a project designers and developers are more prominent. Should these people work together in conducting any SoDIS Inspections?

18.7.1.5 Insights Related to Practice from NZ* Analysis The practice questions from Table 18.4 addressed in the research were the complete set of 1–6, namely, developing quality software; structuring qualitative software risk analysis; improving software development risk assessments; reducing the risk of failed projects; improving software acceptability to stakeholders; and reducing the risk of failure in outsourced software development projects.

We believe that the findings for our client demonstrate the efficacy of the process in practice. In subsequent discussions with the project manager, she expressed her surprise that the team produced such a specific set of findings, given that we had limited information and limited background to work from. This demonstrates the value of a SoDIS Inspection as a relatively "lightweight" risk analysis technique. She also observed that the principles of taking stakeholders needs into consideration are applicable to almost all projects.

18.8 CONCLUSION

The work reported here reflects an ongoing program of research into developing and refining the SoDIS process. It has resulted in the new concept of a SoDIS Inspection. Guidelines for conducting SoDIS Inspections are being developed, with further work required to better determine how far to extend the analysis and when to terminate the process.

Results from trials with commercial partners have shown significant improvement in project planning and requirements identification activities. In the UK electronic voting case, the original goal of e-voting by 2005 (ODPM, 2002) has been tempered by the large number of issues identified through SoDIS analysis, and a more measured program of pilots to be conducted at local body election level has now been adopted (ODPM, 2003). The process has also identified specific issues for responding vendors to address in their e-voting proposals. The issues identified became filters for the UK voting requirements, but these issues once captured are now available for incorporation into any future inspection of e-voting systems. Thus, the use of the process has the potential to improve future inspections. Each use of the SoDIS process provides opportunity for continuous improvement of the process itself by adding domain- or sector-specific stakeholders, questions, clusters, or risks. These are captured via the SPA and other outputs from the process for use in subsequent projects.

In the NZ* case, several critical issues were identified that broadened the scope of the specified requirements to include the needs of additional stakeholders, and suggested significant changes in the outsourcing strategy. Given the higher requirements risk inherent in the case of outsourced software development projects, the value demonstrated by the SoDIS process in these cases indicates its efficacy for more general application to software development projects of all kinds.

ACKNOWLEDGMENTS

The authors gratefully acknowledge all the participants in this program. This research was partially funded by NSF Grant No. 9874684, and the work in New Zealand has been supported by grants from NACCQ, AUT Faculty of Business, AUT School of Computing and Mathematical Sciences, and Kansas State University. East Tennessee State University supported Professor Gotterbarn's work during his sabbatical term as Visiting Professor of Software Engineering Ethics at AUT. Phillip E. Pfeiffer made significant contributions to the clarity of this paper.

APPENDIX A: MAPPING OF CONCERNS IDENTIFIED IN THE UK ELECTRONIC VOTING PROJECT TO THE REQUIREMENTS STIPULATED IN THE RFP DATED NOVEMBER 2002

Cluster Identified in SoDIS	Description of SoDIS Concerns	RFP Requirements that Address the Concerns
Safety	Lack of secrecy could cause danger to voters.	Services must guarantee the confidentiality of the vote until it is counted.
	Individuals might be at risk through the physical stealing of authentication instruments.	A formal risk analysis and documented threat model must be developed and must cover physical, procedural, personnel, and technical security measures to counter the full range of threats identified. Threat of theft or forgery of election details electronically or from the postal system must be taken into account in this analysis.
	A system that involves authentication instruments that can be physically stolen may result in voters being at risk. If biometrics form part of the preventative measure, there may be privacy and health and safety issues relating to individual voters.	- as above -

Privacy	Certain types of system failure might reduce the security infrastructure such that voter data might be accessible to unauthorized parties.	*System Reliability Principle* ensuring that the system is free from malicious codes and accidental bugs. The system must have failover and fully redundant high-availability systems. *System Access Control Principle* ensuring that users and administrators can only access those parts of the system and assets necessary to perform the authorized task.
	Failure to achieve a sufficient degree of anonymity may result in an unacceptable violation of privacy.	*Voter Anonymity Principle* ensuring that the voter must not be associated with voter identity, unless warranted by law.
	The likelihood of identification disclosure increases as costs of searching through ballots decreases. This is a particular concern for political minorities.	*Voter Anonymity Principle* ensuring that the voter must not be associated with voter identity, unless warranted by law. *Data Confidentiality Principle* ensuring that the vote is secret.
	Employers have a legitimate interest in computer and network activity, which may conflict with secrecy of voting at work. Regulation of Investigatory Powers Act obligations may conflict with voter secrecy.	- as above -
	Privacy violations may result from inadequate or inappropriate protection of secrecy.	- as above -
	Most minority groups have an increased risk of privacy violation where specialist interfaces are in use.	- as above -
	Those with linguistic constraints may need support from others if interfaces or authorization tokens do not support any language they are fluent in, which may cause privacy violations.	The system should have the capacity to display/process in different languages.
	The audit process may capture details of voter profiles.	*Open Auditing and Accounting and Public Verifiability Principles* – The system shall provide audit information that can be verifiable by a third party without contravening the requirement for voter anonymity.
	To have effective audit of technologies used by disabled voters, voter identity may be revealed.	- as above -

	Measures to prevent personation may result in a loss of privacy. For example, the procedure may require additional identification to be presented or input at the point of voting.	*Voter Authenticity Principle* ensuring that users are properly identified before being granted permission to vote and that multiple and false identities cannot be registered. *Voter Anonymity Principle* ensuring that the voter must not be associated with voter identity, unless warranted by law.
	Loading of the voting software might result in a voter incurring costs either financial or time or both.	The system must not require specialized applications at the client to secure the system. No constraints are to be placed on the state of voters' equipment as a condition for e-voting service access.
	Some authentication methods may result in an extra time cost because preregistration might be required.	- as above -
Anony-mity	Anonymity may not be attainable and automation removes or changes some of the practical solutions to anonymity attainment.	
	Identifying votes that have been cast with interfaces designed specifically for a special need could result in small groups of voters being identified implicitly or through an amalgamation of data.	*Voter Anonymity Principle* ensuring that the voter must not be associated with voter identity, unless warranted by law.
	Those at the intersection of several minority groups might be easily identifiable.	- as above -
	Voting through specialized interfaces may result in small subsets, making identification of individuals possible implicitly or through an amalgamation of data.	- as above -
	Secrecy may not be attainable particularly when solely dependant on technology.	*Personnel Integrity Principle* ensuring that those developing and operating the voting system have unquestionable records of behavior. *Operator Authentication and Control Principle* ensuring that those operating and administering the system are authenticated and have functional access on the system strictly controlled.

	Voting not under the direct supervision of a polling official cannot guarantee secrecy of ballot.	*Voter Anonymity Principle* ensuring that the voter must not be associated with voter identity, unless warranted by law. *Data Confidentiality Principle* ensuring that the vote is secret.
	For some minority groups, the family culture may make it difficult to vote in secret within the home environment.	
Usability	Technologically assisted voting is inevitably less simple than traditional methods.	E-voting platforms should be as easy to use as possible.
	Existing system specifications may be altered to aid simplicity of the voting procedure, for example, turning off typematic.	The specific needs of disabled people must be taken account of. User trials with people with a diverse range of impairments should be conducted.
	Oversimplification may reduce choice, for example, having the ability to spoil a ballot paper.	
	Overcomplication of the system may prevent those, for example, with learning difficulties, voting independently.	The specific needs of disabled people must be taken account of. User trials with people with a diverse range of impairments should be conducted.
	Providing multilingual interfaces is costly and could increase the complexity of the interface.	E-voting platforms should be as easy to use as possible.
	Design could lead to added complexity in the voter interface to realize the desired level of reliability	- as above -
	Effective prevention of multiple voting may increase complexity unacceptably.	- as above -
	Biometric identification as a means of prevention may be inappropriate for some voters. For example, retinal scanning cannot be used by voters with some visual impairment.	The specific needs of disabled people must be taken account of. User trials with people with a diverse range of impairments should be conducted.
	Effective prevention of personation may increase complexity unacceptably.	E-voting platforms should be as easy to use as possible.
	Provision of alternative interfaces might result in extra stages in accessing the voting process.	- as above -
Access	For those with access to voting solely through the Internet (e.g., overseas) disruption during the polling period will eliminate their ability to vote.	The Internet as a public domain network is assumed to be accessible to potential threat agents and to provide a transmission capability with no service quality assurance.

	A formal risk analysis and documented threat model must be developed and must cover physical, procedural, personnel, and technical security measures to counter the full range of threats identified.
Overcomplication of the system may prevent those, for example, with learning difficulties, voting independently.	E-voting platforms should be as easy to use as possible.
	The specific needs of disabled people must be taken account of. User trials with people with a diverse range of impairments should be conducted.
Some minority groups require specialist interfaces of some description; therefore, failure of such interfaces could lead to discrimination.	*System Reliability Principle* ensuring that the system is free from malicious codes and accidental bugs. The system must have failover and fully redundant high-availability systems.
By requiring a high-level of secrecy in the voting process, some more popular evoting options may be excluded. This may mean that many, if not all, will fail to benefit from the e-voting process.	
Personation by family members.	*Voter Authenticity Principle* ensuring that users are properly identified before being granted permission to vote and that multiple and false identities cannot be registered.
Voters could lose their opportunity to vote through personation.	- as above -
The need to have an unofficial proxy for those with linguistic constraints may be curtailed through antipersonation measures.	The system should have the capacity to display/process in different languages.
Lack of access to appropriate interfaces could lead to some forms of discrimination.	A wide range of optional e-voting devices is catered to.
Particular interface technologies may exclude groups of disabled voters. For example, the telephone interface excludes those with hearing impairments, and ATMs could exclude those with mobility difficulties.	The specific needs of disabled people must be taken account of.
	User trials with people with a diverse range of impairments should be conducted.

		Compliance to international standards is specified to cater to those with mobility difficulties.
	Lack of equity of access may disenfranchise some voters.	A wide range of optional e-voting devices is catered to.
	Some minority groups, for example, rural and socioeconomic, have less access to appropriate technologies.	- as above -
	Members of minority groups with low rates of uptake of relevant technologies and who are also disabled will not be able to benefit from accessibility features built into the voting system.	- as above -
Performance	Complete reliability is probably unattainable or an overemphasis on reliability reduces effort in other equally important aspects.	
	Inclusion of safeguards in system design may result in degradation of system performance.	System must achieve a high level of service availability, capacity, and performance.
	Promotion of equality of access may result in computer system problems owing to the "extra" resource requirement.	- as above -
Misuse	These people may seek to alter votes to change the outcome of an election.	System must have appropriate security to prevent data corruption, loss, sabotage, etc.
	Failure to prevent or detect multiple voting may result in incorrect election results leading to danger to public.	*Voter Authenticity Principle* ensuring that users are properly identified before being granted permission to vote and that multiple and false identities cannot be registered.
	Result of election might be effected by successful personation.	- as above -
Audit	Audit must only consider the efficacy of the process and not capture any details of voter profiles - a precise definition of audit needs to be developed - It is an issue about the nature of the audit and associated trail.	The system shall provide audit information that can be verifiable by a third party without contravening the requirement for voter anonymity.
	The conflict of interest between audit and citizens as voters is aggravated for certain minority groups.	
Data Integrity	Voters could lose their ability to vote or their votes once cast.	System must have a mechanism (electronic or otherwise) by which a person who appears to have voted already may continue to cast their vote, but that vote will be recorded separately.

		System must have appropriate security and DBMS features to prevent data corruption, loss, sabotage, etc., and to preserve data integrity during update operations.
	Software defects could cause the loss of data files.	- as above -
	Tallying defects could result in errors in who is elected, the impact of which could be significant.	The system must deliver 100% accuracy in ballot count.
	If proprietary software is used (directly or indirectly) as any part of the voting system, it is extremely difficult to guarantee it free from vote tallying defects (black box concept).	System must have appropriate security and DBMS features to prevent data corruption, loss, sabotage, etc., and to preserve data integrity during update operations. Accuracy and reliability testing are to be conducted as part of overall quality assurance strategy.
	Some methods used to prevent multiple voting may result in the inappropriate modification of data files.	System must have appropriate security and DBMS features to prevent data corruption, loss, sabotage, etc., and to preserve data integrity during update operations.
Data Security	Tension between open source and the need to safeguard software from disruption whatever the threat.	A formal risk analysis and documented threat model must be developed and must cover physical, procedural, personnel, and technical security measures to counter the full range of threats identified.
	Issue of technical limitation and being able to anticipate the potential threat.	- as above -
Environ-ment	Safeguards against will require redundancy to be built into the system, which may result in additional environmental damage.	
	Distribution of authentication instruments may have an adverse environmental impact.	
Attitude	An ease of development focus may result in simplicity of voting process at the expense of the demotion of equally important considerations.	A comprehensive set of system requirements has been specified.
	Inadequate concern regarding simplicity of voting.	E-voting platforms should be as easy to use as possible.
	There is a temptation to suggest that the system is more reliable and secure than it really is.	Compliance to a comprehensive set of security requirements and standards has been specified.
	Inadequate concern regarding reliability.	A comprehensive set of reliability requirements has been specified.

	Public ignorance of anonymity limitations could be perpetuated in electronic voting.	
	Inadequate concern regarding anonymity.	*Voter Anonymity Principle* ensuring that the voter must not be associated with voter identity, unless warranted by law.
	Inadequate concern regarding secrecy of ballot.	*Data Confidentiality Principle* ensuring that the vote is secret.
	Inadequate concern regarding vote tallying.	The system must deliver 100% accuracy in ballot count.
	Inadequate concern or inappropriate implementation regarding audit.	*Open Auditing and Accounting and Public Verifiability Principles*— The system shall provide audit information that can be verifiable by a third party without contravening the requirement for voter anonymity.
	Inappropriate levels of concern regarding multiple voting.	*Voter Authenticity Principle* ensuring that users are properly identified before being granted permission to vote and that multiple and false identities cannot be registered.
	Some systems may be unable to achieve a level of security against personation which satisfies public expectation.	- as above -
		Open Auditing and Accounting and Public Verifiability Principles— The system shall provide audit information which can be verifiable by a third party without contravening the requirement for voter anonymity.
	Inappropriate levels of concern regarding personation.	- as above -
	Inadequate concern regarding equity of access.	A wide range of optional e-voting devices is catered to.
	There may be a tendency to focus on the needs of the unexceptional citizen at the expense of those in the minorities.	The specific needs of disabled people must be taken account of.
		User trials with people with a diverse range of impairments should be conducted.
Cost & Access	The minimum accessing system requirement may be greater than the specification of the system available to a voter.	The system must not require specialized applications at the client to secure the system.
		No constraints are to be placed on the state of voters' equipment as a condition for e-voting service access.

REFERENCES

Agle, B., Mitchell, R. and Sonnenfeld, J.A. (1999). Who matters to CEOs? An investigation of stakeholder attributes and salience, corporate performance, and CEO values. *The Academy of Management Journal*, 42(5), *Special Research Forum on Stakeholders, Social Responsibility, and Performance* (Oct., 1999), 507–525.

AIS (2005). Definition of system quality. Available at http://business.clemson.edu/ISE/html/system_quality.html 2005. Accessed 2007 Jan 1.

AS/NZS (1999). *Risk Management, Standard 4360–1999, Standards Australia, Standards New Zealand.*

Boehm, B. (1988). Using the WINWIN spiral model: a case study. *Computer*, July.

Boehm, B. (2006). Software risk management: principles and practices. in: Reifer, D. J. (Ed.), *Software Management*, 7th edition. Wiley-IEEE Computer Society Press.

Carr, W. and Kemmis, S. (1983). *Becoming Critical: Knowing Through Action Research.* Deakin University Press, Melbourne.

Clear, T. (2003). The waterfall is dead: long live the waterfall. *SIGCSE Bulletin*, 35, 13–14.

Clear, T., McHaney, R., and Gotterbarn, D. (2004). SoDIS SEPIA: Collaborative partnerships in software engineering research. *NZ Journal of Applied Computing and IT*, 8(1), 2–7.

Cockburn, A. (2004). *Crystal Clear.* Addison-Wesley.

Cockburn, A. and Williams, L. (n.d.). The costs and benefits of pair programming. Available at http://collaboration.csc.ncsu.edu/laurie/Papers/XPSardinia.PDF. Accessed 2007 Jan 1.

DeMarco, T., and Lister, T. (2003). *Waltzing with Bears: Managing Risk on Software Projects.* Dorset House, March.

Deming, W.E. (2000). *Out of the Crisis*, MIT Press.

Fagan, M., (1976). Design and code inspections to reduce errors in program development. *IBM Systems Journal*, 15(3), 182–211.

Fairweather, B. and Rogerson, S. (2002). *Technical Options Report.* Available at http://www.communities.gov.uk/index.asp?id=1133612. Accessed 2007 Jan 1. Office of the Deputy Prime Minister, London.

Gert, B. (1988). *Morality.* Oxford University Press, Oxford.

Gilb, T. and Graham D. (1993). *Software Inspection.* Addison-Wesley Longman Ltd., Essex, England.

Gotterbarn, D. (2004). Reducing software failure with software development impact statements. In: Spinello, R. and Tavani, H. (Eds.), *Readings in CyberEthics*, 2nd edition. Jones and Bartlett Publishers, Sudbury, MA.

Gotterbarn, D. and Clear, T. (2004). Using SoDIS as a risk analysis process: a teaching perspective. *Conferences in Research and Practice in Information Technology*, 30, 83–90.

Gotterbarn, D., Miller, K., and Rogerson, S. (1998). Software engineering code of ethics. *Communications of the ACM.* Available at http://www.acm.org/serving/se/code.htm. Accessed 2007 Jan 27.

Gotterbarn, D. and Rogerson, S. (2005). Responsible risk analysis for software development: creating the software development impact statement. *Communications of the Association for Information Systems*, Vol. 15, Article 40.

Hall, E. M. (1998). *Managing Risk: Methods for Software Systems Development*. Addison-Wesley, Reading, MA.

Highsmith, J. (2002). *Agile Software Development Ecosystems*. Addison-Wesley, Boston.

Hilson, D. (2004). *Effective Opportunity Management for Projects: Exploiting Positive Risk*. CRC Press.

IEEE (2001). Standards for software life cycle process: Risk Management. *IEEE Std. 1540–2000*.

Jones, C. (1994). *Assessment and Control of Software Risks*. PTR Prentice-Hall, Englewood Cliffs, NJ.

Kerzner, H. (2002). *Project Management: A Systems Approach to Planning, Scheduling, and Controlling*, 8th edition. John Wiley & Sons, New York.

Koh, D. (2003). *Using SoDIS for Target Audience Analysis: a Fresh Field Application*. Paper presented at the NACCQ Conference, 6–9 JULY, Palmerston North.

Kotonya, G. and Summerville, I. (1996). Requirements engineering with viewpoints. *BCS/IEE Software Engineering Journal*, 11(1), 5–18.

Leveson, N. (1995). *Safeware: System Safety and Computers*. Addison-Wesley.

McFarland, M. (1990). Urgency of ethical standards intensifies in computer community. *Computer*, March.

McHaney, R. (2004). Special issue on the SoDIS SEPIA symposium. *Bulletin of Applied Computing and IT*. Available at http://www.naccq.ac.nz/bacit/0202/index.html#sodis. Accessed 2007 Jan 26.

McKay, J. and Masshall, P. (2001). The dual imperatives of action research. *Information Technology and People,* 14(1), 46–59.

ODPM (2002). *Implementing electronic voting in the UK*. Available at http://www.odpm.gov.uk/index.asp?id=1133595. Accessed 2005 Nov 13. Office of the Deputy Prime Minister, London.

ODPM (2003). *The government's response to the electoral commission's report: the shape of elections to come—a strategic evaluation of the 2003 electoral pilot schemes* (No. Cm 5975). Office of the Deputy Prime Minister, London.

Parnas, D. and Lawford, M. (2003). Inspection's role in software quality assurance. *IEEE Software*, 20(4), 16–20.

Paulk, M.C. (1995). *The Capability Maturity Model: Guidelines for Improving the Software Process*. Addison-Wesley, Reading, MA.

Pediatrics (2006). Available at http://pediatrics.aappublications.org/cgi/content/abstract/118/5/1872.

Pfleeger, S. (2004). *Software Engineering: Theory and Practice*. 2nd edition. Prentice Hall, p. 70.

Ravichandran, T. (2000). Total quality management in information systems development: key constructs and relationships. *Journal of Management Information Systems*, 1999/2000, 16(3).p119, 37p.

Rogerson, S. and Gotterbarn, D. (1998). The ethics of software project management. In: Collste, G. (Ed.), *Ethics and Information Technology*. New Academic Publishers, Delhi.

Rook, P. (1993). *Risk Management for Software Development*. ESCOM Tutorial.

Stoneburner, G., Goguan, A., and Feringa, A. (2002). NIST special publication 800–30: *Risk Management Guide for Information Technology Systems*. National Institute of Standard, Technology Administration, U.S. Department of Commerce.

PART V

REGULATORY ISSUES AND CHALLENGES

CHAPTER 19

Regulation and Governance of the Internet

JOHN WECKERT and YESLAM AL-SAGGAF

19.1 INTRODUCTION

> Internet governance is the development and application by Governments, the private sector and civil society, in their respective roles, of shared principles, norms, rules, decision-making procedures, and programs that shape the evolution and use of the Internet (WGIG, 2005).

This is a working definition used by the Working Group on Internet Governance (WGIG), set up by the Secretary-General of the United Nations (UN). There are narrow definitions of Internet governance that incorporate the administration and management of the technical infrastructure and broader definitions that also incorporate political and policy issues (King, 2004, p. 243). The above is of the latter type and the kind that is of interest here. There are at least two different questions involved in the governance of the Internet: who, if anybody should be in charge, and what, if anything, should be governed or regulated. The definition above does not say who, if anyone, should be the ultimate authority, nor how the authority should be decentralized, if it should be. Some argue that there should be a global ultimate authority, at least in some areas (see Al-Darrab, 2005; King, 2004), and the UN has set up the WGIG to investigate Internet governance (Drake, 2005). This view has some plausibility given the international character of the Internet and the fact that national borders mean little (or so it is often thought) in this context. Others opt for an approach in which there is cooperation between nations where necessary, but that much can be left to self-regulation by the private sector (see Hassan, 2005; ICC, 2004).

The Handbook of Information and Computer Ethics, Edited by Kenneth Einar Himma and Herman T. Tavani

Although the question of who should be in charge of Internet regulation is an important and interesting one, it is not the one of the main concerns in this chapter. Control of standards and protocols is probably best handled by one centralized body, but content and activity regulation we will assume can be done at a national level, something to which we will return throughout the chapter. The primary concern here is what, if anything, should be governed or regulated? We live in a world where people misbehave and in order for groups and societies to function satisfactorily some restrictions on behavior are required, and even where there is no malicious intent there can be a need for some centralized body or perhaps decentralized bodies, to coordinate activities.

There are a number of different issues raised here, not all of which can be discussed in detail in this chapter. Some of these are issues of political morality, while others are issues of prudential rationality: (1) whether it is legitimate, as a matter of political morality, for use of the Internet to be subject to restrictions; (2) if so, what kinds of restrictions would be morally justified (e.g., would censorship of pornography? Or would limiting use of the web by businesses for commercial purposes, as has been demanded by some hacktivists?); (3) what kinds of enforcement mechanisms for otherwise justified restrictions would be morally permissible (e.g., the usual coercive enforcement mechanisms like threats of incarceration or code-based mechanisms that make the violation of these restrictions technologically impossible); (4) what sorts of enforcement mechanisms are likely to be most effective and hence conduce maximally to these prudential interests; and (5) what sorts of restrictions are prudentially justified as being in everyone's best interest. These questions and issues are all relevant and important and shed light on what follows, but here we will paint with a broader brush and focus on the question, What, if anything, can be justifiably regulated on the Internet?

At the least controversial level, for the Internet to be functional and useful there need to be standards relating to Internet Protocols (IP), including IP addresses, the unique numbers given to all computers connected to the network. Standards are also required for domain names, and some governing body needs to ensure that all names are unique. Various organizations, such as the Internet Corporation for Assigned Names and Numbers (ICANN), the World Wide Web Consortium (W3C), and the Internet Architecture Board (IAB), coordinate different aspects of the Internet. (For discussions of these and other relevant organizations see Vincent and Camp, 2005.) Assuming that there is good governance at this level (and assuming that the physical infrastructure is sound), is there now a well-functioning Internet? Not necessarily. If anyone can do anything on the Internet it is not so well-functioning, at least not if it is considered as part of society. Consider the case of sites that make music freely available for downloading. Although this does not inhibit the functioning of the Internet (apart from potentially degrading its performance for periods), it does raise problems for intellectual property and there is legislation to prevent this activity because of its harm to electronic commerce. The prevalence of child pornography on the Internet is another concern, and there are efforts to control this. Efforts to control these and other criminal activities, including terrorist activities, raise issues about personal privacy and also about freedom of speech and expression on the Internet.

19.2 CONTENT REGULATION

Proposals by governments to regulate content on the Internet are often hotly contested, at least in liberal democratic countries (depending on what the content is, but more on that later). This occurred in the United States of America around 1995 with the introduction of the Communications Decency Act, and in Australia in 1999 (BSA, 1999). On the face of things, this is a little puzzling. There are regulations governing the content of television, radio, newspapers, magazines, movies, and books, so why not the Internet as well? A number of types of arguments against Internet content regulation are advanced. Some arguments are general ones that apply to all media, relating to the principle of rights to freedom of speech, expression, and information (see Hurley, 2004). Others are more specific to the Internet (see Catudal, 2004; Spinello, 2001; White, 2004). One argument is that the Internet is different from all other media and so must be treated differently (Bick, 2006; Goodwin, 2003; Jorgensen, 2001) and another that it is more like books, say, so the regulations applied to it should not be like those applied to television (Graham, 2003). Still others will argue that Internet content regulation should be resisted because it is an extension of government control (Anderson and Rainie, 2006; irrepressible.info, 2007; The OpenNet Initiative, 2004). Not only do governments want to control the other media, but now they also want to control the Internet as well. Then there is the pragmatic argument that has two strands. One is that it is pointless for one country alone to attempt regulation (Zizic, 2000). The Internet is global, so regulation, to be effective, must also be global (Nielsen, 2003). Not only is it pointless, but it can also be harmful economically to that country (Litman, 2002) because many valuable electronic commerce sites may move elsewhere. The other strand is that it can create intolerable situations for individuals who create sites in any country. The material on their sites may be legal in their own country where the site is located, but illegal in another (Bick, 2006). Finally, and importantly, there is the argument that because of the technology itself, the Internet cannot be effectively regulated (Anonymous, 2005; NetAlert, 2006; Zizic, 2000).

Now, however, the Internet is not spoken about only as a type of medium but often as a living space in which people work, play, shop, socialize, and so on. So to some extent the argument about controls on Internet content has shifted a little. Although there is still discussion of pornography, hate language, and the like, there is also discussion of, for example, controlling Internet gambling and downloading music and movies. So the discussion now is partly of Internet activity and partly of Internet content. On the Internet, it must be noted, this distinction is not sharp because most activities apart from simple communication, for example, e-mail, involve web pages that have content. So to that extent, controlling the content controls the activity. Still, it is a useful distinction.

19.3 EFFECTIVE REGULATION

There are two broad issues regarding the regulation of the content on the Internet: *can* content on the Internet be regulated effectively, and *should* it be regulated? At this level we can say that if it cannot be, then the second question does not arise as a practical issue. If it should not be, it does not matter if it cannot be. But the situation is more fine-grained. In the following section it will be seen that Internet content and activity can be regulated to some extent but there are questions about how effective this regulation can be without introducing draconian measures. Even allowing this, there are questions about effective regulation that relate to the second question. To what extent, if any, should there be regulation if that regulation is not, or not very, effective? Avoiding paying income tax on amounts earned over a certain amount is generally illegal, but avoiding such tax is possible by various means, including locating offshore to "tax havens." In this and many other cases regulation is justified even though it is not as effective as most would like. Perhaps a better example is that of the legislation against the distribution and use of many drugs. Although many users and distributors are caught, it is not obvious that worldwide the war on illicit drugs is being won. But these might not be good analogies. It appears to be impossible, or at least very difficult, to effectively ban the use of performance-enhancing drugs in sport, but that is not taken as a reason for not regulating their use. The health dangers as well as the unfairness warrant the regulations. They demonstrate society's disapproval of such drug use and hopefully limit their use to some extent. Other cases of ineffective regulation seem not to be justified, say certain sorts of sexual behavior between consenting adults in private. It all depends on how serious the issue is, and the same of course applies to the Internet. It might be justified to regulate certain kinds of content or activity even if this is not very effective, but in other cases it might not be.

19.4 REGULATION: TECHNICAL ISSUES

To what extent does the technology allow for effective regulation? Many opponents of content regulation in Australia argued that because of the nature of the Internet technology, content regulation is not possible in any practical sense (NetAlert, 2006). Various strategies are available for blocking Internet content (see McCrea et al. (1998) for an early discussion and Goldsmith and Wu (2006) for a recent one). Web pages and ftp files can be blocked by Internet Service Providers (ISPs) with the use of proxy servers. Requests by the ISP's clients go through the proxy server, where each is checked to see if the requested URL is on its black list, the list of forbidden URLs. However, this method is not foolproof. Forbidden Web pages and ftp files can easily be given new names, URLs can be set up to return the contents of a different URL, domain names can be bypassed if the IP address is known, and push technologies bypass proxy filters. In addition there are various costs with employing proxy servers that could make it difficult for small ISPs to remain viable.

An alternative method would be to use routers to block content at the packet level, where the source address of each packet is checked against a black list. To be anywhere near efficient, this would need to be done at the Internet gateways to Australia operated by Backbone Service Providers (BSPs). One problem with this method is that it provides only very coarse filtering. If a site contains some material deemed offensive then the whole site will be blocked, including all the harmless and useful material. Additionally, sites can be renumbered to bybass blocking, and *tunneling*, that is, enclosing an IP packet within another IP packet, can be employed. It is argued, too, that not only is packet blocking not efficient, but it can also create considerable problems, particularly with respect to information going through one country to other countries.

What follows from this for the moral argument that Internet content regulation is justified? Ought implies can, so there cannot be much force in an argument that says that the Internet ought to be regulated if it cannot be in any effective way. Some care needs to be taken here. As we saw in the previous section, it does not follow necessarily that because something cannot be regulated effectively and efficiently it should not be regulated at all. There are important benefits to the community in general in having most people paying tax and in having few drug users, but are there similar benefits to be gained from Internet regulation? That there are benefits in reducing the amount of material that can harm the innocent and vulnerable is obvious. Whether this can be achieved to any significant extent given current technology is not. Given that neutralizing the use of proxy servers requires some effort in terms of renaming Web page and ftp sites, using IP addresses instead of URLs, and so forth, it is likely that the amount of material considered offensive will be reduced to some extent, but perhaps not significantly. But again, perhaps this slight reduction is enough in itself to justify regulation. These benefits, however, must be weighed against the costs, for example, of greater unreliability of Internet access using proxy servers, the possibility of adverse effects on some applications, and the existence of black lists of URL, which could become valuable commodities. The cost of packet blocking could be even higher.

There is no doubt that to some extent the Internet can be and is being regulated by blocking certain material. Before considering arguments about the extent to which Internet content and activity can be justifiably regulated, we will briefly take a quick look at the current situation in various parts of the world because that will be useful for the light that it sheds on what can be done.

19.5 THE CURRENT SITUATION

Different countries differ radically in terms of their position with respect to Internet content regulation. Although countries like China and Saudi Arabia, for example, exercise tight control over Internet content, many European Union countries, on the contrary, take a more liberal attitude toward regulating online content. China is at the top of the list of the countries that implement strict online censorship, and their attitude to content regulation, according to a report by Reporters Without Borders (2006), is one of the harshest toward freedom of expression.[1] Any material that contains pornography, criticism of the regime, or information about Taiwanese independence, the Tiananmen Square protests, or the Dalai Lama is blocked in China (Villeneuve, 2006). Access to online content is controlled at the Internet backbone level, near international gateway points. That is, requests for "blocked" content are blocked before the requests leave China's

[1]See Table 19.1 below for more information on China and other countries.

TABLE 19.1 Country Profiles[c]

Country	Population (Millions)	Number of Internet Users (Millions)	Human Development Index	Press Freedom Index
China	1313	111	85	159[a]
Saudi Arabia	27.019	2.54	77	154
The Netherlands[b]	16.491	10.8	12	5
Australia	20.67	14	3	31
United States	298.444	205.326	10	56

[a]167 being the last in the list and representing the least press freedom.
[b]Selected randomly as a representative of the EU member-states.
[c]The source of the information in Table 19.1 is the United Nations (2005); Human Development Report (2006); CIA *World Fact Book* (2006); Australian Bureau of Statistics (2006); Reporters Without Borders (2006).

backbone lines and enters the international lines (i.e., the outside internet). Content that flows within the country (i.e., inside internet) is also subject to repressive filtering by local ISPs, enforced, of course, by the government through strict regulations that they have to abide by (Human Rights Watch, 2006). The regime also strongly encourages content providers and users to censor their own material (Reporters Without Borders, 2006). Those who violate the rules end up in prison. Forty-nine of the 59 people in prison at present (for something they posted online) are in China alone, which makes it the world's biggest prison for cyber-dissidents.

Another country that exercises very strict control over the internet is Saudi Arabia, which blocks any material that contains pornographic or anti-Islamic content or contains criticism of Saudi Arabia, the Royal Family, or other Gulf states (Internet Services Unit, 2006). The Internet access for the whole country is controlled by a single node, which makes the government the ultimate arbiter on what is permissible to view online. The Communication and Information Technology Commission[2] filters all the web traffic that flows to the country by implementing country-level proxy servers. These proxy servers contain massive databases of banned sites (Internet Services Unit, 2006). Unfortunately, filtering the Internet in this way not only stopped pornographic, anti-Islamic or antigovernment sites from arriving to users' computer screens, but also stopped material, for example, of a medical nature. Any Web site that contains, for instance, the word "breast" is banned. It is believed that many users, including in this case students studying medicine or anatomy, are deprived from accessing this kind of material because of this filtering and as a result deprived from accessing the knowledge contained within these sites. This earned Saudi Arabia a reputation for being one of the countries that allow the least freedom of expression (Reporters Without Borders, 2006).

[2]The government body that controls access to the Internet in the country.

Even the United States, arguably the most free society and where the Internet and the telephone were invented, has tried a number of times in the past to censor the Internet. The first attempt to censor the Internet was the federal Communications Decency Act (CDA), which criminalized the sending of "indecent" material over the Internet (Jorgensen, 2001). Fortunately the CDA, which the Congress passed into law in 1996, was overturned by the Supreme Court because it found it unconstitutional since it went against the free speech principle expressed in the First Amendment, which guarantees the rights of citizens to speak freely. The second attempt to censor the Internet was the Child Online Protection Act (COPA) of 1998, which required information providers to ensure that minors are not granted access to material deemed harmful to them (Depken, 2006). "Material harmful to minors" was to be determined in accordance with "community standards." Again, the law was struck down by the Supreme Court as unconstitutional. The third attempt to regulate Internet content in the United States was the Children's Internet Protection Act (CIPA), a U.S. federal law passed in 2000 (Tavani, 2007). The law required public libraries, as a condition for receiving federal funding (for Internet access), to install filtering software on their computers to stop users from viewing images of obscenity and child pornography, and also to prevent minors from viewing material harmful to them. Although in the beginning the law was rejected by a lower court, on the ground that Internet filters would not only stop material deemed harmful to minors but also innocent material, later in 2003 the law was upheld by the Supreme Court. Unlike the above legislations, which aimed at regulating pornographic content, the USA Patriot Act that was enacted in 2001 (just 1 month after the September 11 attacks on the United States) was aimed at monitoring content relevant to terrorist activities on the Internet (Wikibooks, 2006). The law allows the FBI (with the permission of a special court) to install software on the ISP's machines to monitor the flow of e-mail messages exchanged between potential terrorists and store records of web activity by people suspected of having contact with terrorists.

The picture in many of the EU member-states is completely different when compared with non-Western democracies but not so different if compared with the one in the United States. Although the EU Commission responsible for Information Society and Media is concerned about the well-being of minors, they also feel strongly about freedom of expression and freedom of the media. Freedom of expression for them is a right that is enshrined not only in the Charter of Fundamental Rights of the EU, but also in the European Convention on Human Rights (Reding, 2006). The EU model for regulating the Internet, which appears to be influenced by their desire to protect minors, consists of self-regulatory and coregulatory regimes. The self-regulation is based on three elements: "first, the involvement of all the interested parties (Government, industry, service and access providers, user associations) in the production of codes of conduct; secondly, the implementation of codes of conduct by the industry; thirdly, the evaluation of measures taken" (EU Council, 2006). But self-regulation, according to the Commissioner for Information Society and Media, Viviane Reding, works best when it has a clear legal framework to support it, which is where the coregulatory framework comes in. In this model, while the public authorities leave the task of protecting the minors, for example, in the hands of the self-regulatory mechanisms and codes of conduct, they reserve the right to step in, in case self-regulation is not sufficient.

In Australia the commonwealth law, which is at the federal government level and applies to Internet Content Hosts (ICHs) and ISPs, requires them to remove any content from their servers that is deemed "objectionable" or "unsuitable for minors" on receipt of a take-down notice from the government regulator, the Australian Communications & Media Authority (ACMA) (EFA, 2006a). Content becomes "objectionable" or "unsuitable for minors" only after it has been complained about to the ACMA and determined to be "objectionable" or "unsuitable for minors." The federal government also requires all ISPs to provide filters at cost or below cost to consumers to allow them to exercise filtering at their own ends to protect their children. At the state and territory levels, which apply to content providers/creators and ordinary Internet users, the laws vary from one state to another. Although some states, for example, prosecute ordinary Internet users for making available material that is deemed "objectionable" or "unsuitable for minors" and/or for downloading content that it is illegal to possess, others do not. It should be noted that ISPs in Australian are not required by law to install filtering or blocking software, nor do they have to block access to any site. Also, users are not required by law to use filtering software, nor do they have to buy any such product even if it is offered to them by ISPs at a low cost (EFA, 1999, 2006a). A recent investigation conducted by NetAlert[3] found that accessing the Internet through a content filter at the ISP level is ineffective and leads to a significant reduction in network performance. As a result, they recommended a scheme that focuses strongly on education and voluntary filtering by ordinary Internet users (NetAlert, 2006).

Here again the question of *effective* regulation is raised. That regulation is possible is clear, but the extent to which regulation degrades Internet performance is an important factor in assessing effectiveness. This is obviously of more concern in some countries than in others. It is a moot point whether strict regulation is effective regulation. It is certainly effective in the sense of denying access, but it could be argued that it is not effective if it denies legitimate access, for example, to medical students or researchers.

19.6 ACROSS BORDERS

Individual countries, or political entities within countries, apart from blocking, can also control content through legislation, and this can be relatively effective in certain circumstances despite the fact that the Internet crosses national borders. Consider the recent example of Internet gambling in the U.S. Online gambling has been prohibited in the U.S. (Humphrey, 2006), but at least one prominent gambling business, Party Gaming, is based on Gibraltar, where it is immune from U.S. legislation (Cullen, 2006).

[3] An independent government body that provides advice to the community about managing children's access to the Internet.

The U.S. government, however, has made it illegal for American banks and credit card firms to process payments to gambling sites.

> "No person engaged in the business of betting or wagering may knowingly accept, in connection with the participation of another person in unlawful Internet gambling (1) credit, or the proceeds of credit, extended to or on behalf of such other person (including credit extended through the use of a credit card); (2) an electronic fund transfer, or funds transmitted by or through a money transmitting business, or the proceeds of an electronic fund transfer or money transmitting service, from or on behalf of such other person" (The Unlawful Internet Gambling Enforcement Act of 2006).

Perhaps a more interesting case is that of Yahoo and its Nazi web pages selling Nazi memorabilia. This material was illegal in France but not in the U.S., and Yahoo initially refused to close down the pages despite a French court ruling. Eventually, however, they did, the reason being, it seems, that they had assets in France. (Goldsmith and Wu (2006) discuss this case at length. See also their discussion of the Gutnick case that raises similar cross-border issues.)

The point of these examples is that even without international laws, the national laws can have an effect on Internet content internationally, a point made strongly by Goldsmith and Wu (2006).

Is there anything wrong as such with individual countries regulating the content of the Internet in their country? Content includes things such as libelous material, material protected by intellectual property laws or privacy laws, cultural material (should there be free trade in cultural objects?), political material, and pornographic material. We will consider some problems with this later.

19.7 INTERNET REGULATION: NORMATIVE ISSUES

So to some extent the Internet can be regulated, both through technical means and by legislation, but should it be? We will now consider this in more detail. There are a number of questions: should it be regulated in general; what, if anything, should be regulated; and should it be regulated in any one country in the absence of cooperation of others? We will examine these in turn. It should be noted first that there is already some regulation of Internet content (as has been discussed in the current situation section above). Various things that are illegal in other media are illegal on the Internet as well. One cannot, for example, release state or trade secrets on the Internet (Turner, 2006), and that which constitutes defamation on other media also does so on the Internet (EFA, 2006b). This regulation is covered by laws that do not target the Internet specifically, and it is not these that opponents of Internet regulation oppose most strongly, nor are they the major concern here. In Australia, for example, defamation laws cover the Internet as well as other media, but there have been no objections to this. The regulation of primary concern in this chapter is that which attempts to restrict content or activity specifically on the Internet.

There is nothing new about regulating different media differently. In Australia, for example, there are many more restrictions on free-to-air television than there are on books and magazines, or even on movies. Should the Internet be treated like books and magazines (as some argue, see page 477), or like television? It has aspects of both. In addition, it has aspects that closely resemble the postal and telephone services, and these are subject to very little content regulation. Some of these differences and similarities will be discussed later. But what if, as suggested earlier, the Internet is not really a medium at all in the sense that television is but more like a living space? Then perhaps the question of its regulation is more akin to questions of regulation in society in general.

19.8 CENSORSHIP

Article 19 of *The Universal Declaration of Human Rights* states: Everyone has the right to freedom of opinion and expression; this right includes freedom to hold opinions without interference and to seek, receive, and impart information and ideas through any media and regardless of frontiers (UDHR, 1948).

This implies that censorship is a violation of a *right*. However, commonly consequentialist arguments are brought to bear against censorship, and some of the most compelling of these come from Mill (1975, Chapter 2). The first is that an opinion that is not allowed to be heard might just be true, and the second that it might contain some truth. Therefore restrictions on the freedom of opinion can, and most probably will, deprive the world of some truths. His third reason is that unless beliefs and opinions are vigorously challenged, they will be held as mere prejudices, and finally, those opinions are themselves in danger of dying if never contested, simply because there is never any need to think about them.

Mill has a further argument. His conception of a good human life is one in which we think, reflect, and rationally choose for ourselves from different beliefs and lifestyles according to what seems most true or meaningful to us. This is shown in his arguments for the freedom of expression. His central tenet here is that people ought to be allowed to express their individuality as they please "so long as it is at their own risk and peril" (Mill, 1975, p. 53). The basic argument is that the diversity created has many benefits. One is that "the human faculties of perception, judgment, discriminative feeling, mental activity, and even moral preference, are exercised only in making a choice" (Mill, 1975, p. 55). And exercising this choice makes it less likely that we will be under the sway of the "despotism of custom" (Mill, 1975, p. 66). We will be able to lead happier and more fulfilled lives. And again, if there is this diversity, each human will be more aware of the various options available, and so more competent to make informed choices in lifestyle and self-expression.

These and other such arguments for freedom of speech and expression do support the claims for lack of restrictions and control of material in the media in general. However, the support is qualified for several reasons. Some of the arguments, particularly the first mentioned, only apply where there is propositional content, that can enhance the search for truth, and second, one person's right to freedom of speech or expression can infringe on another's rights, and can clash with other goods. For example, my freedom to talk openly of your financial or medical situation would infringe your rights to privacy, and I clearly cannot be allowed the freedom to express myself through torturing you.

There is little sense in the idea of complete freedom of expression for all. So the issue now becomes one of where to draw the lines for this freedom. A common criterion is harm to others, a criterion endorsed by Mill. Admittedly this is not without problems, because many actions, which appear to be self-regarding in Mill's sense, can harm others indirectly, although he qualifies this by saying that he is talking of action that harms others "directly, and in the first instance," but even this does not make the distinction clear and sharp. However, it is a useful criterion for all that. Many distinctions that are more or less vague are useful, for example, that between day and night.

The freedom of speech or expression of one person can cause harm to another, so some restrictions need to be placed on how and to what extent a person can be allowed free expression. That there should normally be some restrictions placed on harming others, other things being equal, is pretty uncontroversial. There are all sorts of restrictions on what can be said, and in general there is little opposition to this. There are libel and defamation laws and laws against perjury, blasphemy, abusive language, disclosing personal information, and so on. There is debate about what should and what should not be allowed, but little argument that anything and everything ought to be. The value in having some restrictions on what may be said seems just too obvious. Mill also recognized this, and he claimed that if some kinds of utterances are likely to cause riots, for example, there ought to be restrictions placed on them (Mill, 1975, p. 53).

One way to explicate the claim that language can harm is to draw on the speech act theory of Austin (1962). He distinguished between *locutionary* acts, that is, expressing propositions, *illocutionary* acts, that is, expressing beliefs, and *perlocutionary* acts, the creation of some effect on listeners. Consider, for example, the following case of "hate" language: "People of race X are mentally and morally inferior." The locutionary act here is the proposition that people of race X are mentally and morally inferior, the illocutionary act is the expression of the belief that this is the case, and the perlocutionary act might be to incite racial hatred or even violence. Considered from this perspective, the claim for freedom of speech entails the claim for freedom to perform any sort of perlocutionary act, but now it is a claim that looks decidedly weaker. Although it might be argued that it is the violence that causes the harm and not the language, if the language is highly likely to incite the hatred and violence that leads to harm, a strong case can be made on Millian grounds that there should be restrictions on the language (as Mill does in the example above).

19.9 REGULATION OF THE INTERNET: MORAL ARGUMENTS

The above suggests that there are grounds for content regulation of the media in general, and this is fairly widely accepted. If some action harms others there might be legitimate reasons for regulating actions of that kind. The next question is whether the same reasons for regulation apply to the Internet. Concerns about material on the

Internet can roughly be grouped into three areas; pornography, hate language, and information to aid harmful activities.

Questions of free speech and censorship probably arise most frequently in connection with pornography. Although anything available on the Internet would also be available elsewhere, or at least material of the same type would be, the situation is slightly different, simply because it is so much more difficult to control the material put on the Internet, and then to control its distribution. Although there can be effective controls, as was seen early, these call for strong measures that limit the value of the Internet.

Anybody can put anything on, and with varying degrees of difficulty almost anybody can have access to it. In addition, gaining access to pornography on the Internet may be a very private affair. Locked in one's room, one can browse and search to one's heart's content. There is no need to face the possible embarrassment of detection in buying or hiring material from a newsagent or video shop, or even by the interception of mail, if acquiring material by mail order. As a consequence, it is much more difficult to restrict its consumption to adults.

The second main area of concern is hate language, usually racist language. Particular groups, especially white supremacy groups, spread their massages of hate, free from any control, in a way not normally possible using other media.

The third area is the imparting of information designed to cause harm to other people. A common example often mentioned is information on how to construct bombs. Another is advice on how to abduct children for the purpose of molestation. It might be argued that this is nothing new. This information is available anyway, and possibly in the local public or university library. This may be so, but again it is much easier to get it in the privacy of one's room rather than in a public place.

Mill's arguments for freedom of speech, as compelling as they might be in other contexts, give little support to freedom of speech on the Internet in at least two of the areas just mentioned. Pornography has nothing to do with the freedom to express opinions (although it might express a view about the worth of women), and neither does giving information on activities designed to harm people.

Hate language, or racial vilification, may be the expression of an opinion, and so might be supported by Mill's principles above, but it falls foul of the harm principle. These three areas of pornography, hate language, and harmful information might all, perhaps, be protected by Mill's argument for freedom of expression, although it is difficult to see how any of them could in any way assist people in living a good life (in Mill's sense). But in any case, again the harm principle would come into play if they in fact incite or are highly likely leads to harmful behavior.

So the complete nonregulation of the Internet does not get great support from Mill. We can forsake consequentialist arguments and talk instead of rights, but this does not help much. I might have the rights to freedom of speech and expression, but my rights can clash with the rights of others to privacy, to be respected as persons, and so on. It would also be no easy task to show that we indeed have the rights to express ourselves through pornography, hate language, and the circulation of potentially harmful information.

19.10 OTHER ARGUMENTS

A different argument is given by White (2004), who argues against content control with regard to obscene material. The argument is essentially that it is extremely difficult to define the obscene and therefore it is difficult to know what content should be censored, and that on the Internet it is not possible to resort to community standards because there is no community in the relevant sense. Although it is true that what is considered obscene varies enormously between societies (and between individuals), as we have seen controls within various countries can be effective to some extent and even in some cases have effects between countries. Such controls might not be thought desirable and might raise other problems, but it seems that national borders are important enough for some community-based definitions of the obscene to be plausible.

A different kind of argument again is that Internet content regulation is an unwarranted extension of government power (Anderson and Rainie, 2006; irrepressible.info, 2007; The OpenNet Initiative, 2004). However, it can plausibly be claimed that it represents no extension of power or regulation. As activities shift away from other media and to the Internet, if there is no regulation of the Internet, then there is a diminishing of regulation. (This may be good, but that is a different argument.) Consider home entertainment, for example. What can be shown on free-to-air television is quite severely restricted. When entertainment in the home shifts to the unregulated Internet, there is much less regulation on what can be experienced as entertainment in that environment. Regulation of Internet content is thus maintenance of the *status quo*, rather than an extension of regulation.

Yet another argument is that the Internet is different from other forms of media and therefore ought to be treated differently. Although this claim is true, it implies nothing about whether or not there should be Internet regulation. The most that it shows is that there should be a different type or degree of regulation. The Internet has some characteristics of television, but it is not intrusive in the same way (although with the increase of unsolicited advertising, this difference is decreasing). As the television set is on, material is entering my home and I have little control over it. Certainly the set can be turned off or the channel changed, but that action may come too late to stop children seeing inappropriate material. In contrast, content must be *found* on the Internet. In this way it has a greater resemblance to printed media. So it could be argued that if there is to be regulation of Internet content that regulation should be more akin to that applied to printed media than to free-to-air television. This has some plausibility, although there are differences too. As noted earlier, one can "surf" the Internet in the privacy of one's room and thereby avoid the possible embarrassment of being seen purchasing material of which one is not entirely proud!

19.11 REGULATION AND EFFICIENCY

A variation of the above argument is that the Internet is much more important than other media, and so should be left free of regulation. It provides, so the argument runs, enormous benefits: repressed peoples can make their plight known, the isolated can communicate more widely than previously, there is much more access to information than ever before, vast markets for goods are opened up through electronic commerce, and so on. Granted that all these benefits exist, it would still need to be shown that they could not exist with regulation. In fact, perhaps at least some of them could exist to a greater degree with some regulation.

That some degree of regulation improves efficiency is hardly news. Transportation would hardly be more efficient without regulation, and the same is true of at least some areas of Internet activity, especially electronic commerce. The example of the online auction site eBay, illustrates this well:

> The eBay auction system, as libertarian as its origins may have been, depends on an oft-hidden virtue of government power to deter those who would destroy the system (Goldsmith and Wu, 2006, p. 136).

Goldsmith and Wu discuss in some detail how the success of the eBay business depends on strong government regulation in a variety of areas.

An objection to this line of argument is that the fact that regulation is justified in some areas does not show that it is in others. Regulating the environment for electronic commerce is one thing, regulating expression of opinions, pornography, and so on quite another. This is true, but what it means is that talk of regulation of Internet content is not fine-grained enough on its own. One must specify what content is meant.

In principle, then, at least some aspects of the Internet can be justifiably regulated. There are moral justifications for regulation of the media and for regulation of business in general, and there seem to be no good reasons why these justifications should not also be applied to the Internet. This much might be conceded, but the claim could still be that regulation in one country should be resisted. This argument was a common one against the current legislation in Australia. The Internet is global, up to a point (the vast majority of users are in only a few countries), so not only is it futile but also probably damaging to any one country that attempts to introduce regulation. Australia would become a "laughing stock," business would move offshore, and the growth of the Australian Internet economy would be restrained, were common claims.

These claims may all have been true, but they are not to the point if the moral justification is strong enough and if the legislation will be effective. If all countries had economies based on slavery, the repeal of this practice in one country alone would be the moral thing, even if the economy of the country did suffer (it is interesting to note that this economic argument was in fact used in defense of slavery, see Drew, 1963). The argument is not that opposing Internet content regulation is on the same level as supporting slavery; the point is merely that economic argument do not necessarily carry much weight against moral ones. In order for the economic argument to have force it would need to be shown that the economic losses would be such that the innocent suffered. Then of course these arguments would also be moral ones. In the case of the Internet, the economic argument would require showing that the suffering caused would be of a magnitude that outweighed the benefits of the protection of innocents afforded by regulation (this is a central point to which we will return later).

19.12 REGULATION ACROSS LEGAL JURISDICTIONS

There is, however, one worrying aspect to individual countries regulating Internet content, and this relates to the supposed irrelevance of international borders to the Internet, and to the question asked earlier as to whether there is anything wrong with individual countries regulating the Internet in their own countries. Suppose that I develop a Web site in my country with material that is uncontroversially legal. Unbeknownst to me that material is illegal in another country, and of course can be downloaded in that country. What is my legal position relative to that second country? Should I be extradited, or arrested if I travel there? This may seem an unlikely scenario, but consider the following recent case. An Australian citizen had material on his Web site located in Australia where it was legal, but this same material was illegal in Germany. He was arrested while traveling in Germany and charged, among other things, for distributing prohibited material on the Internet (Nixon, 1999). Although he was eventually not convicted on that charge, this case does show that there is a problem and that it is probably only a matter of time before someone is convicted in these circumstances. But this seems rather unfair. Ignorance of the law is no defense, and clearly I have an obligation to know the law (although not in detail) in my own country, but this obligation can hardly extend to a knowledge of the law in all countries that have Internet access. It is one thing to perhaps have some familiarity with the laws of those countries that one visits or does business with, but quite another to have the same familiarity with all of those from which my Web page might be accessed (It should be noted that in the example mentioned there was no ignorance of German law. The material concerned the holocaust, and the relevant laws were well understood.)

This situation *is* a worry if individual countries regulate Internet content, but it does not show that such regulation is necessarily wrong. What it does show is that there must be international agreements that clarify the legal situation. Ideally, in most cases, nobody should be charged if the "offense" took place in a country in which it was legal, but failing that, the countries with content regulation should make their positions very clear and widely known. This latter position is certainly far from ideal, but at least some clarity would be introduced. (There may be exceptions in extreme cases, for example, that of killing Jews in Nazi Germany where it was legal, but just to prosecute Nazis as war criminals.)

The final moral argument to be considered relates to whom the regulations should apply, the content provider, the user, or the provider. Given the ease of moving sites offshore, content providers are difficult to regulate, and attempting to regulate users would involve massive intrusions on privacy. The best way then has appeared to be the regulation of Internet Service Providers (ISPs). But, as has been pointed out, this could be argued to be unfair and a case of "shooting the messenger." On the contrary, one might argue that if one is going into the business of ISP, then one has a special obligation to learn the laws of other countries and to structure one's practices accordingly. And even if something is not illegal in some country, say child pornography, but is clearly immoral, then holding ISPs responsible does not appear unfair. Although it might appear at first sight unfair to hold carriers rather than content providers responsible for content, there might be situations in which it is justifiable.

Suppose that I am given a parcel by a stranger to deliver to another country. I do not know the contents of the parcel, and do not ask, but given that I am making the trip anyway and the parcel is small and there is little inconvenience to me, I agree to take it. In this situation I would almost certainly be held responsible for carrying drugs if that is what the parcel contained, and justifiably so, because I should have known better than to agree to take the parcel. Graham (1999, p. 110), to support the opposing position, says that airline companies are not held responsible for what their passengers carry. This is true, but only up to a point. The airlines are expected, for example, to ensure that passengers do not carry weapons, and it is not difficult to imagine situations in which they could be held responsible for ensuring that passengers did not carry other items as well. Such expectations are not necessarily unjust if there are no other practical ways of avoiding or minimizing harm to innocent people.

There is precedent too for holding people legally responsible for actions that they did not perform. Vicarious liability is "the imposition of liability on one person for the actionable conduct of another, based solely on a relationship between the two persons (or the) indirect or imputed legal responsibility of acts of another..." (Black, 1990, p. 1566). Although ISPs may not be vicariously liable for the actions of content providers, that is not the point. The point is simply that it is not enough to say that ISPs should not be held responsible for content on the grounds that they did not create that content. People can, in the right circumstances, be responsible for the actions of others. It may be of course that vicarious liability itself is not morally justifiable, but the point here is only that there is precedent for having to take responsibility for the actions of others in certain circumstances and that this is generally not thought to be unreasonable.

Vicarious liability is a legal and not a moral term, but the idea of taking responsibility for actions that I did not commit can easily be extended into the moral realm. I can be held morally responsible for the actions of another if I could reasonably have been expected to have prevented those actions. Even if I did not know about them, if my position is such that I should have known, I can still be morally responsible. This is a well-established (if not much adhered to) principle in the Westminster system of government, where cabinet ministers are (or were) expected to take full responsibility for the actions of their subordinates. (The argument is not that this is a case of vicarious liability.)

The argument here falls short of demonstrating conclusively that it is fair and just to hold carriers responsible rather than content providers. It does, however, indicate that it is not obviously unjust. A similar conclusion is reached by Vedder (2001), who argues that Internet access and service providers can justifiably be held responsible for the consequences of content not produced by them by extending the notion of strict liability into the moral sphere (a person can be strictly liable and therefore responsible for harm even if that person is not culpable). His argument is not that this is, or should be, a case of strict liability, merely that the idea of being responsible but not culpable can be applied in the moral sphere.

A frequently cited alternative to regulating Internet content in general is the use of filtering of Web sites, particularly in the home but also in schools, libraries, and other public places. There are, however, a number of objections to filtering. For those who object to certain kinds of material, for example, pornography or advertisements for Nazi memorabilia, filtering is not enough. The material is still available for those who do not filter, so it is not a solution. That the material is available at all is the issue. Several other worries are expressed by Rosenberg (2001). He argues that there are free speech concerns if the filtering is carried out at schools, libraries, or other public places. Part of the problem lies in the fact that filters are not particularly good at blocking just what is meant to be blocked and will normally also stop users viewing completely inoffensive and possibly useful information. So filters are not a satisfactory solution. (For discussion of problems with filtering, see US v. ALA, 2002, which discusses the U.S. Children's Internet Protection Act that required public libraries to install filters as a condition for receiving certain federal funding.)

The argument of this section is primarily that many of the common arguments against Internet content regulation (and thereby activity regulation) do not stand scrutiny. In a previous section of this chapter it was argued that some regulation of the content of media in general is justified in order to protect the innocent and vulnerable. If some regulation in general is justifiable, and if the special arguments regarding the Internet are not sound, then the tentative conclusion must be that Internet content regulation is justifiable.

19.13 CONCLUSION

A strong moral case can be made for regulating the content of the Internet, but there is also a strong case that such regulation cannot be very effective and comes at a price in Internet performance. These last two factors together constitute an argument of considerable weight against attempting to control Internet content through legislation. So what should be done? On balance, a case can be made for content regulation, although that case is probably not as strong as proponents would wish. That the case can be made can be seen by looking a little more closely at the two opposing factors just mentioned. First, while in general laws that are not enforceable to any great extent are to be avoided, in certain instances they can be useful. Consider illicit drugs, for example. The laws banning their use and distribution are not particularly effective, but they are still considered worthwhile by many because they give the message that using those substances in not a good thing. A similar argument could be mounted for content regulation of the Internet. Second, degrading Internet performance will not obviously harm many people very much, depending of course on the degradation. Most of us could wait a little longer when searching or downloading without much of a diminution of our living standards. There may be some problems with electronic commerce if Internet performance is slower, but that will not affect too many people, at least not in the short term, given that this form of commerce is not being accepted by consumers particularly quickly. And in any case, it is not uncontroversially accepted that the benefits of electronic commerce will outweigh its disadvantages (Rogerson and Foley, 1999).

The argument in this chapter has been that Internet content regulation is justifiable, but the problems are recognized. To overcome them, there will need to be more research into technological methods for blocking content, and there must be international cooperation in the formulation and enforcement of laws, practices, and standards. A long-term solution suggested in a recent report is this: It is proposed that Australia participate in international fora to create the necessary infrastructure, so that organizations that host content would be able to determine the jurisdiction of the client software making the request. Having determined the jurisdiction, the server can find out whether the requested content is legal in the client's jurisdiction (McCrea et al., 1998). This proposal might not be the ideal solution, but it is one possibility that ought to be seriously investigated. Internet content can harm, and some regulation is morally justified. Given the benefits of the Internet, however, we do not want to throw the baby out with the bathwater.

ACKNOWLEDGMENTS

Parts of this paper are based on John Weckert, "What is so bad about Internet content regulation?" *Ethics and Information Technology*, 2, 2005, 105–111. We are grateful to Kenneth Himma for his many helpful comments on an earlier version of this paper.

REFERENCES

Al-Darrab, A.A. (2005). The need for international Internet governance oversight. In: Drake, W.J. (Ed.), *Reforming Internet Governance: Perspectives for the Working Group on Internet Governance (WGIG)*, United Nations Information and Communication Technologies Task Force, New York. pp. 177–184. Retrieved (March 18, 2007) from http: //www.wgig.org/docs/book/WGIG_book.pdf

Anderson, J.Q. and Rainie, L. (2006). *The Future of the Internet II*. Pew Internet and American Life Project, Washington, DC. Retrieved (January 15, 2007) from http: //www.elon.edu/e-web/predictions/2006survey.pdf

Anonymous (2005). *Are Australia's current Internet censorship laws effective?* Retrieved (January 14, 2007) from http://www.cs.mu.oz.au/343/2005/essays/58.ps

Austin, J. (1962). *How to do Things With Words*. Oxford University Press, Oxford.

Australian Bureau of Statistics (2006). *Australia's Current Population*. Retrieved (November 14, 2006) from http://www.abs.gov.au/ausstats/abs@.nsf/94713ad445ff1425ca 25682000192 af2/1647509ef7e25faaca2568a900154b63!OpenDocument

Bick, J. (2006). Overseas Courts Limit American Internet Speech. *New Jersey Law Journal*, 185 (1).

Black, H.C. (1990). *Black's Law Dictionary: Definitions of the Terms and Phrases of American and English Jurisprudence, Ancient and Modern*, 6th edition. West Publishing, St. Paul, MN.

BSA (1999). Broadcasting Services Amendment (Online Services) Bill. Retrieved (July 6, 1999) from http://www.aph.gov.au/parlinfo/billsnet/bills.htm

Catudal, J.N. (2004). Censorship, the Internet, and the child pornography law of 1996: a critique. In: Spinello, R.A. and Tavani, H.T. (Eds.), *Readings in CyberEthics*, 2nd edition, Jones and Bartlett Publishers, Sudbury, MA, pp. 196–213.

CIA *World Fact Book* (2006). Retrieved (September 29, 2006) from www.cia.gov

Cullen, D. (2006). Party Gaming axes interim dividend. *The Register*. Retrieved (March 18, 2007) from http://www.theregister.co.uk/2006/10/03/party_gaming_suspends_dividend/

Depken, C. (2006). Who supports Internet censorship? *First Monday*, 11(9), Retrieved (January 15, 2007) from http://firstmonday.org/issues/issue11_9/depken/index.html

Drake, W.J. (Ed.). (2005). *Reforming Internet Governance: Perspectives for the Working Group on Internet Governance (WGIG)*. United Nations Information and Communication Technologies Task Force, New York. Retrieved (March 18, 2007) from http: //www.wgig.org/docs/book/WGIG_book.pdf

Drew, T.W. (1963). Review of the Debate in the Virginia Legislature. In: McKitrick, E.L. (Ed.), *Slavery Defended: the Views of the Old South*, Prentice-Hall, Englewood Cliffs, NJ, pp. 20–33.

EFA (1999). *Clairview Internet Sheriff: An Independent Review*. Retrieved (July 6, 1999) from http://efa.org.au/Publish/report_isheriff.html

EFA (2006a). *Internet Content Filtering and Blocking*. Retrieved (November 14, 2006) from http://www.efa.org.au/Issues/Censor/cens2.html#austlaw

EFA (2006b). *Defamation Laws & the Internet*. Retrieved (January 15, 2007) from http://www.efa.org.au/Issues/Censor/defamation.html

EU Council. (2006). *Protection of Minors and Human Dignity Recommendation*. Retrieved (November 14, 2006) from http://ec.europa.eu/comm/avpolicy/reg/minors/index_en.htm

Goldsmith, J. and Wu, T. (2006). *Who Controls the Internet? Illusions of a Borderless World*. Oxford University Press, New York.

Goodwin, M. (2003). *CyberRights: Defending Free Speech in the Digital Age*. MIT Press, Cambridge, MA.

Graham, G. (1999). *The Internet: A Philosophical Inquiry*. Routledge, London.

Graham, I. (2003). *Australian Internet Censorship Laws: Frequently Asked Questions*. Retrieved (January 14, 2007) from http://libertus.net/liberty/faq_auscens1.html

Hassan, A. (2005). Internet governance: strengths and weaknesses from a business perspective. In: Drake, W.J. (Ed.), *Reforming Internet Governance: Perspectives for the Working Group on Internet Governance (WGIG)*. United Nations Information and Communication Technologies Task Force, New York, pp. 117–128. Retrieved (March 18, 2007) from http: //www.wgig.org/docs/book/WGIG_book.pdf

Human Development Report (2006). *Beyond Scarcity: Power, Poverty and the Global Water Crisis* . http://hdr.undp.org/en/media/hdr06-complete.pdf

Human Rights Watch (2006). *How Censorship Works in China: A Brief Overview*. Retrieved (September 29, 2006) from http://www.hrw.org/reports/2006/china0806/3.htm

Humphrey, C. (2006). *Internet Gambling Funding Ban*. October 13, 2006. Retrieved (January 15, 2007) from http://www.gambling-law-us.com/Federal-Laws/internet-gambling-ban.htm

Hurley, S. (2004). Imitation, media violence, and freedom of speech. *Philosophical Studies*, Springer, The Netherlands, Vol 117, pp. 165–218.

ICC (2004). *International Chambers of Commerce issues Paper on Internet Governance*, Prepared by ICC's Commission on E-Business, IT, and Telecoms.

Internet Services Unit (2006). *User's survey*. Retrieved (November 14, 2006) from http://www.isu.net.sa/surveys-&-statistics/new-user-survey-results.htm

irrepressible.info (2007). *An Amnesty International campaign for freedom of expression online*. Retrieved (January 15, 2007) from http://www.amnesty.org/

Jorgensen, R. (2001). *Internet and Freedom of Expression*. Master Thesis. Raoul Wallenberg Institute. Retrieved (January 14, 2007) from http://www.digitalrights.dk/DRfile66.htm

King, I. (2004). Internationalizing Internet governance: does ICANN have a role to play? *Information and Communications Technology Law*, 13, 243–258.

Kitney, G. (1999). *Trial Sparks Internet Racism Fears*, Friday 12 November. Sydney Morning Herald, Sydney, Australia.

Litman, J. (2002). Electronic commerce and free speech. In: Elkin-Koren, N. and Netanel, N.W. (Eds.), *The Commodification of Information*. Kluwer Academic Publishers, The Hague. pp. 23–42.

Menestrel, M.L., Hunter, M., and de Bettignies, H.C. (2002). Internet e-ethics in Confrontation with an Activists' Agenda: Yahoo! on Trial. *Journal of Business Ethics*, 39(1–2), 135–144. Springer, The Netherlands.

Mill, J.S. (Ed.) (1859). *On Liberty: Annotated Text, Sources and Background Criticism*. W.W. Norton & Company, New York. (*On Liberty*, page citations to edition of David Spitz).

NetAlert. (2006). *Education the Best Filter for Young Australians on the Internet.* Retrieved (November 14, 2006) from http://www.netalert.net.au/03004-Education-the-Best-Filter-for-Young-Australians-on-the-Internet.asp

Nielsen, R. (2003). *Bioterrorism and Science: The Censorship of Scientific Journals Will Do More Harm than Good*. Retrieved (January 15, 2007) from http://www.colorado.edu/PWR/occasions/articles/bioterrorism.html

Nixon, S. (1999). Holocaust critic held in Germany. *The Age*, Saturday 10 April, 1999, Melbourne, Australia.

McCrea, P., Smart, B., and Andrews, M. (1998). *Blocking Content on the Internet: A Technical Perspective*. Prepared for the National Office for the Information Economy, CSIRO, Mathematical and Information Sciences, June. Retrieved (July 6, 1999) from http://www.cmis.csiro.au/projects+sectors/blocking.pdf

Reding, V. (2006). Co-regulation and media literacy: cornerstones for an efficient protection of minors in the European Union. *ICRA Roundtable Brussels "Mission Impossible."* Retrieved (November 14, 2006) from http://europa.eu.int/rapid/pressReleasesAction.do?reference=SPEECH/06/374&format=HTML&aged=0&language=EN&guiLanguage=en

Reporters Without Borders (2006). *Internet Annual report*. Retrieved (September 25, 2006) from http://www.rsf.org/print.php3?id_article=17177

Rogerson, S. and Foley, P. (1999). Internet Electronic Commerce: The Broader Issues. In Institute of Chartered Accountants in Australia. *Ethics and Electronic Commerce: A Collection of Papers*, 9–12.

Rosenberg, R.S. (2001). Controlling access to the Internet: the role of filtering. *Ethics and Information Technology*, 3, 35–54.

Spinello, R.A. (2001). Code and moral values in cyberspace. *Ethics and Information Technology*, 3(2), 137–150.

Stoll, M.L. (2006). Infotainment and the Moral Obligations of the Multimedia Conglomerate. *Journal of Business Ethics*, 66(2–3), 253–260.

Tavani, H.T. (2007). *Ethics and Technology: Ethical Issues in an Age of Information and Communication Technology*, 2nd Edition, United States of America, Wiley.

The OpenNet Initiative (2004). *Documenting Internet Content Filtering Worldwide*. Retrieved (January 15, 2007) from http://www.opennetinitiative.net/modules.php?op=modload&name=Sections&file=index&req=viewarticle&artid=1

The Unlawful Internet Gambling Enforcement Act of 2006. Retrieved (January 15, 2007) from http://www.gambling-law-us.com/Articles-Notes/internet-gambling-prohibition-2006.pdf

Turner, D.D. (2006). Court Hears Final Arguments in Apple Trade Secret Suit. *Eweek*, April 20, 2006. Retrieved (January 15, 2007) from http://www.eweek.com/article2/0,1895,1952111,00.aspeWEEK.com

UDHR. (1948). *The Universal Declaration of Human Rights*. United Nations.

United Nations (2005). *Human Development Report*. Retrieved (November 14, 2006) from http://hdr.undp.org/reports/global/2005/pdf/HDR05_complete.pdf

US v ALA (2002). *United States et al v American Libraries Association Inc et al*. Retrieved (February 6, 2007) from http://www.supremecourtus.gov/opinions/02pdf/02-361.pdf

Vedder, A. (2001). Accountability of Internet access and service providers – strict liability entering ethics? *Ethics and Information Technology*, 3, 67–74.

Villeneuve, N. (2006). The filtering matrix: integrated mechanisms of information control and the demarcation of borders in cyberspace. *First Monday*, 11(1), Retrieved (September 29, 2006) from http://firstmonday.org/issues/issue11_1/villeneuve/index.html

Vincent, C. and Camp, J. (2005). Looking to the Internet for models of governance. *Ethics and Information Technology*, 6(3), 161–173.

WGIG (2005). *Report of the Working Group on Internet Governance* Château de Bossey, June 2005 http://www.wgig.org/docs/WGIGREPORT.pdf

White, A. (2004). The obscenity of Internet regulation in the United States. *Ethics and Information Technology*, 6 (2), 111–119.

Wikibooks (2006). *Legal and Regulatory Issues in the Information Economy/Censorship or Content Regulation*, 20 December 2006. Retrieved (January 15, 2007) from http://en.wikibooks.org/wiki/Legal_and_Regulatory_Issues_in_the_Information_Economy/Censorship_or_Content_Regulation

Zizic, B. (2000). *Copyright Infringement Occurring over the Internet: Choice of Law Considerations*. Master of laws thesis. Queen's University. Kingston, Ontario, Canada. Retrieved (January 15, 2007) from http://www.collectionscanada.ca/obj/s4/f2/dsk2/ftp01/MQ54498.pdf

CHAPTER 20

Information Overload

DAVID M. LEVY

20.1 INTRODUCTION

Is it any wonder that information overload is such a common complaint these days? Given the prevalence of cell phones, voice mail, e-mail, and instant messaging, as well as endless sources of academic, commercial, governmental, and personal information on the Web, it should hardly be surprising if complaints about a flood, a fire hose, or a blizzard of information are not only common but increasing. Googling "information overload" in early 2007 yielded more than six million hits for the phrase. The first of these was the entry in Wikipedia, which states that information overload "refers to the state of having *too much* information to make a decision or remain informed about a topic."[1] Indeed Google, the wildly successful Internet search service, and Wikipedia, the free, online, collaboratively produced encyclopedia, are themselves attempts at taming the fire hose of information by helping to organize it and make it more manageable.

On the face of it, information overload would seem to be a straightforward phenomenon (an excess of information) with a straightforward cause (the recent explosion of information technologies). Yet closer inspection reveals a number of subtleties, questions, and concerns: What does it mean to have too much information, and, for that matter, what exactly is this substance (information) we can apparently have too much of? Is the phenomenon as recent as our anxious complaints suggest, or does it have a longer history? How does it relate to other seemingly related notions, such as data overload, information anxiety, information pollution, technostress, data smog, and information fatigue syndrome? What exactly are the negative consequences, both practical and ethical, and how can we possibly say without pinning down the phenomenon more carefully?

[1]Retrieved (June 11, 2007).

The Handbook of Information and Computer Ethics, Edited by Kenneth Einar Himma and Herman T. Tavani

I cannot hope to address, or certainly answer, all these questions in this relatively brief article. But I do intend to identify some of the major issues, to wrestle with some of them, and along the way to chart some of the dimensions of the phenomenon. I will proceed as follows: In the next section, I will provide a preliminary definition of information overload and will identify some of the questions surrounding it. In Section 20.3, I will discuss the history of the English phrase "information overload," and in Section 20.4, I will show how industrialization and informatization prepared the ground for its emergence. Finally, in Section 20.5, I will explore some of the consequences, both practical and ethical, of overload, and in Section 20.6, I will briefly consider what can be done in response.

20.2 WHAT IS INFORMATION OVERLOAD?

Information overload: Exposure to or provision of too much information; a problematic situation or state of mental stress arising from this. [OED Online, retrieved (June 11, 2007)].

Information overload . . . refers to the state of having *too much* information to make a decision or remain informed about a topic. Large amounts of historical information to dig through, a high rate of new information being added, contradictions in available information, a low signal-to-noise ratio make it difficult to identify what information is relevant to the decision. The lack of a method for comparing and processing different kinds of information can also contribute to this effect. [Wikipedia, retrieved (June 11, 2007).]

Information overload, according to these two definitions, is a condition in which an agent has—or is exposed to, or is provided with—too much information, and suffers negative consequences as a result (experiences distress, finds itself in a "problematic situation," is unable to make a decision or to stay informed on a topic, etc.). In a famous episode of "I Love Lucy," the 1950s American television show, Lucy has a job at a candy factory. Her job is to wrap chocolates as they come by on a conveyor belt. She has no trouble doing this at first, so long as the chocolates arrive at a moderate rate. But as the conveyor belt begins to accelerate, Lucy finds it progressively more difficult to keep up, until finally she is doing anything to get rid of the chocolates, including stuffing them in her mouth, her hat, and her blouse. If we think of the chocolates as morsels of information, then as the belt speeds up, Lucy begins to suffer from information overload.

At the heart of this understanding of the phenomenon is a fairly simple conception of human information processing—a three-stage model consisting of *reception*, *processing*, and *action*. In the first stage, information is received in some manner; it is a system input. It may arrive with little or no effort on the part of the person, as when someone opens an e-mail folder to discover that a number of new messages have arrived; or it may be actively sought after, as when someone performs a literature search and discovers a number of (potentially) relevant sources. In the second stage, the person processes these inputs cognitively to absorb, interpret, and understand. Exactly what this consists of will depend on the nature of the inputs and the uses to which they will be put. In the case of e-mail, for example, processing may include scanning, skimming, reading, and organizing (categorizing, deleting). In the third stage, the person takes some action in response. In Lucy's case, each chocolate arrives in front of her on the conveyor belt (reception), she recognizes it for what it is (processing), and wraps it (action).

While this might seem like a straightforward and unproblematic notion, it raises a variety of questions, concerns, and complexities. I will briefly mention four.

20.2.1 What is Information?

How can one decide if one is suffering from information overload without knowing the range of phenomena encompassed by the word "information"? In both of the definitions above—as in many of the discussions of information overload in the popular press and academic literatures—the meaning of the word is assumed to be understood. This is presumably because the notion of information is unproblematic in the public mind: we require no explanation because we know what it means. When people say they are suffering from information overload, they most likely mean that they are feeling overwhelmed by the number of information goods or products (such as books, e-mail messages, telephone calls, or some combination of these) they are faced with.[2]

Among scholars who concern themselves with information from a theoretical perspective, however, there is little agreement about the notion. In search of stable footing, some have borrowed the information theoretic approach of Shannon and Weaver, which identifies the number of bits of information content in a signal, despite the fact that this is an abstract measure of channel capacity and not of meaning.[3] (Knowing how many bits of information are contained in the transmission of an issue of the *New York Times* over the Internet tells you nothing about the amount of meaningful, humanly interpretable content.) Others take information to be propositional content, facts, or some other postulated unit of meaningful content. Within library and information science, Michael Buckland's article, "Information as thing" (Buckland, 1991), identifies artifacts (physical documents, for example) as one of the three primary senses of the concept. Among the more radical approaches is Agre's suggestion (Agre, 1995) that information is a political and ideological construct rather than a natural kind; or Schement and Curtis' focus on the role that *the idea of*

[2]They might also mean that they are feeling overwhelmed by the information content of these products, which in their mind would probably amount to the same thing.

[3]Nunberg (1996) quotes Warren Weaver: "The word information, in this theory, is used in a special sense that must not be confused with its ordinary usage. In particular, information must not be confused with meaning. In fact, two messages, one of which is heavily loaded with meaning and the other of which is pure nonsense, can be exactly equivalent, from the present viewpoint, as regards information."

information plays within society, an approach that sidesteps the question of what information is (Schement and Curtis, 1995).[4]

20.2.2 More than Information

The two definitions above seem to suggest that information overload is simply concerned with an excess of information. But excess is a relational notion. As Kenneth Himma puts it,"One has too much of something relative to some (normative) standard that defines what is an appropriate amount" (Himma, 2007). If we have too much information, it must be relative to the tasks or purposes it is meant to serve, or the standards we are expected, or expecting, to meet.

Central to judgments of excess is some notion of *capacity*: a person or organization can only handle so much information in a given period of time. And here, it has been usefully pointed out that human beings have a limited attentional capacity; it is this capacity that is stressed in the face of excess. As Warren Thorngate, a Canadian psychologist, has noted: "Information is supposed to be that which informs, but nothing can inform without some attentional investment. Alas, there is no evidence that the rate at which a member of our species can spend attentional resources has increased to any significant degree in the past 10,000 years. As a result, competition for our limited attention has grown in direct proportion to the amount of information available. Because information has been proliferating at such an enormous rate, we have reached the point where attention is an extremely scarce resource, so scarce that extreme measures—from telemarketing to terrorism—have proliferated as fast as information just to capture a bit of it" (Thorngate, 1988, p. 248). Thorngate proposes a set of principles of "attentional economics." These include the principle of Fixed Attentional Assets, which states that attention is a fixed and nonrenewal resource, and the principle of Singular Attentional Investments, which states that attention can, in general, be invested in only one activity at a time. (Time is in a sense a surrogate for attention. If we can we only attend to so much in a given unit of time, then increasing the amount of time available potentially increases the amount of attention that can be applied.)

20.2.3 Perception or Reality?

Is information overload simply a question of one's subjective state—that one *feels* overloaded—or must there be some objective reality to it? Watching Lucy unable to

[4]Schement and Curtis (1995, p. 2) say: "We propose that the idea of information forms the conceptual foundation for the information society. By that we mean a perspective in which information is conceived of as thing-like. As a result, messages are thought to contain more or less 'information.' Marketplaces exist for the buying and selling of 'information.' Devices are developed for the storage and retrieval of information. Laws are passed to prevent theft of information. Devices exist for the purpose of moving information geographically. Moreover, and equally important, the thingness of information allows individuals to see diverse experiences, such as a name, a poem, a table of numbers, a novel, and a picture, as possessing a common essential feature termed 'information.' As people endow 'information' with the characteristics of a thing (or think of it as embodying material characteristics) they facilitate its manipulation in the world of things, for example, in the marketplace."

wrap all the chocolates coming down the conveyor belt, there is no question that objectively she is failing to keep up. But in the case of a human being or an organization contending with (possibly too much) information, we may not be able to monitor directly and transparently the coping process, and thus may have no way to decide what constitutes too much. On the other hand, with human beings and their social units we can receive direct reports that, at the very least, describe their subjective state: *This is more than I can handle*. The relation between perception and reality is particularly important when trying to decide, for example, if we are *more* overloaded by information today than we were in the past. Is it enough to determine that more people today complain about the problem (if this is true), or must we have some independent way to measure the "load," and therefore the overload, on our human systems?

20.2.4 A Novel, Recurrent, or Ever-Present Phenomenon?

It has been suggested that we conceive of information overload today as a unique phenomenon, "an immediate phenomenon—without a history—that the current generation uniquely faces for the first time" (Bowles, 1999, p. 13). Yet complaints about an overabundance of books stretch back at least as far as the sixteenth century. Indeed, an entire issue of the *Journal of the History of Ideas* was recently devoted to exploring "early modern information overload."[5] The editor of this issue, Daniel Rosenberg, notes "the [strange] persistence of the rhetoric of novelty that accompanies so old a phenomenon" (Rosenberg, 2003, p. 2). Bowles has called this perception of novelty the "myth of immediacy."

While it seems wise not to assume that information overload is an entirely new phenomenon, we should also be careful not to assume that the sixteenth century experience of excess is (exactly) the same phenomenon we are dealing with today. What's more, since the phrase "information overload" is less than 50 years old, its application to earlier times is quite literally anachronistic, a point not lost on Rosenberg: "It is worth noting the terminological anachronism deployed by this group of historians in the application of the rubric of 'information overload' to the early modern period. The word 'information' itself appears little if at all in the sources to which these historians refer. Indeed, the use of 'information' to mean something

[5]In the introduction to the issue, Daniel Rosenberg asserts that "During the early modern period, and especially during the years 1550–1750, Europe experienced a kind of 'information explosion.'. . .There is ample evidence to demonstrate that during this period, the production, circulation, and dissemination of scientific and scholarly texts accelerated tremendously" (Rosenberg, 2003, p. 2). In the same issue, Ann Blair quotes Conrad Gesner complaining in 1545 about the "confusing and harmful abundance of books" and Adrien Baillet worrying in 1685 that "we have reason to fear that the multitude of books that grows every day in a prodigious fashion will make the following centuries fall into a state as barbarous as that of the centuries that followed the fall of the Roman Empire" (Blair, 2003, p. 11). Nearly a hundred years after this second complaint, Diderot was still expressing similar concerns in his "Encyclopédie": "As long as the centuries continue to unfold, the number of books will grow continually, and one can predict that a time will come when it will be almost as difficult to learn anything from books as from the direct study of the whole universe" (Rosenberg, 2003, p. 1).

abstract and quantifiable (rather than particular knowledge) does not appear until the early twentieth century, and the usage 'information overload' is even later" (Rosenberg, 2003, p. 7).

20.3 A BRIEF HISTORY OF THE PHRASE

Having briefly explored the concept of information overload, including some of the nuances and questions that surround it, I now want to examine the history of the phrase.

The Wikipedia entry on "information overload" claims that it was coined in 1970 by Alvin Toffler in his book *Future Shock*.[6] Although Toffler, as we will see, did write about the subject, a search of the academic and popular literature of the time reveals that the phrase was already in use by the early 1960s. Consider the following examples:

- In 1962, Richard L. Meier, published a scholarly monograph, *A Communications Theory of Urban Growth*, which applied then-current ideas in information and communication theory to the organization of cities. In a section entitled "The Threat of Information Overload" (Meier, 1962, pp. 132–136), he pointed out that the "expanded flow rates" of communications in dense urban settings would likely lead to "widespread saturation in communications flow... within the next half century" (Meier, 1962, p. 132).
- In an article published the same year, "Operation Basic: the retrieval of wasted knowledge," Bertram M. Gross, a professor of political science at Syracuse University, argued that the accelerating rate of publication was threatening science and society with an "information crisis." There was a "flood" of new publications as a consequence of which, "Nobody... can keep up with *all* the new and interesting information" (Gross, 1962, p. 70). Indeed, thanks to international efforts "to 'educate' the people of lesser-developed economies," these unfortunate citizens would find themselves moving from "the frying pan of information scarcity to the fire of an information overload" (Gross, 1962, p. 70).
- In 1966 the economist Kenneth Boulding published a journal article in which he declared that "management science... is an alternative defense against information overload" (Boulding, 1966, p. B169). The following year, in the same journal, *Management Science*, Russell Ackoff claimed that "most managers receive much more data (if not information) than they can possibly absorb even if they spend all of their time trying to do so. Hence, they already suffer from an information overload" (Ackoff, 1967, p. B148).
- Also in 1967, a paid advertisement appeared in the *New York Times* announcing the publication and sale of a book titled *EDUNET: Report of the Summer Study on Information Networks*, a report by an academic coalition outlining a plan "for a vast interuniversity communications network." The ad bore the title

[6]Retrieved (June 11, 2007).

"EDUNET: Is it the answer to the information overload in our schools and colleges?" The body of the ad copy read:

> "Our contemporary problem is not lack of knowledge or information. Far from it. / Every day there are more and more schools, colleges, and universities. /. . . More books, more journals, more learned papers, more data banks, more symposia, more international meetings, more conference calls—more knowledge and information, in fact, than our present system of scholarly communications can reasonably process. / This is the 'information overload' problem—a problem which EDUNET is designed to relieve" ("EDUNET: Is it the answer to the information overload in our schools and colleges?" 1967).

- In 1969, an article on rock music—"All You Need is Love. Love is All You Need"—appearing in the *New York Times* noted: "The wild images in the song are not, in fact, connected in any logical way. There is no normative meaning, no sirloin for the watchdog. But there is a larger, more important meaning: The quick flashing of disjointed phenomena reproduces the chaos of sensations in our world of information overload" (Murphy and Gross, 1969).

Although Alvin Toffler neither coined the phrase nor introduced it into the culture, his vastly popular book may well have been the first to discuss the topic at some length in a form that was accessible to the general public. Information overload for Toffler was just one manifestation of a period of unprecedented change Western society was then entering, a movement beyond industrialization to what he dubbed "super–industrialism." In this new era, he claimed, "we have not merely extended the scope and scale of change, we have radically altered its pace. We have, in our time, released a totally new social force—a stream of change so accelerated that it influences our sense of time, revolutionizes the tempo of daily life, and affects the very way we 'feel' the world around us" (Toffler, 1970, p. 17). Human beings, he claimed, were ill-prepared for this new, accelerated rate of change, and were beginning to feel the effects of "future shock," an inability to respond comfortably and successfully to the "overload of the human organism's physical adaptive systems and its decision-making processes" (Toffler, 1970, p. 326).

In the last third of the book, Toffler addressed "the limits of adaptability," arguing that the human organism remains "a biosystem with a limited capacity for change" (Toffler, 1970, p. 342). In Chapter 15, "Future Shock: The Physical Dimension," he described how the stress of change could lead to physical illness as "each adaptive reaction extracts a price, wearing down the body's machinery bit by minute bit, until perceptible tissue damage results" (Toffler, 1970, p. 342). In the next chapter, "Future Shock: The Psychological Dimension," he explained that future shock could manifest itself in psychological as well as physical ways: "Just as the body cracks under the strain of environmental overstimulation, the 'mind' and its decision processes behave erratically when overloaded" (Toffler, 1970, p. 343). Toffler identified three forms of overstimulation: sensory, cognitive, and decisional. Cognitive overstimulation occurs when a person's environment is changing so quickly that he doesn't have sufficient time to think about what is happening, to "absorb, manipulate, evaluate, and retain information" (Toffler, 1970, p. 350). Toffler pointed to work in psychology and information science concerning human beings' "channel capacity," which suggested "first, that man has limited capacity; and second, that overloading the system leads to serious breakdown of performance" (Toffler, 1970, p. 352). (The title of this subsection is "Information Overload," but nowhere in the five-page section did he use the phrase in a sentence.)

It is a striking feature of these early citations that authors apparently felt no need to define what they meant by "information overload." For Meier writing in 1962 phrases such as "the overloading of communications channels" (Meier, 1962, p. 2), "communications load" (Meier, 1962, p. 79), and the "threat of information overload" (Meier, 1962, p. 132) emerge fluidly and unproblematically to advance his narrative. Although the 1967 EDUNET ad appears to explain what it means by information overload in the body of the advertisement (more books, journals, papers, data banks, symposia. . .), still, the copywriters were comfortable including the phrase in the headline. And even though Toffler devoted five pages to explaining what he meant by cognitive overstimulation (a high-sounding phrase), he felt comfortable labeling the subsection "Information Overload." In the 1960s, we might conclude, the meaning of the words "information" and "overload" were unproblematic, and the meaning of the conjoined phrase could be straightforwardly extrapolated from them.

Much of this linguistic transparency seems to have come from the meaning of the word "overload," and its root word "load." The Oxford English Dictionary defines "load" as "That which is laid upon a person, beast, or vehicle to be carried; a burden. Also, the amount which usually is or can be carried; e.g., *cart-load*, *horse-load*, *wagon-load*," and it defines "overload" as "An excessive load or burden; too great a load; the condition of being overloaded." These core meanings speak to the recognition that a person or an animal (or by extension, a device such as a cart or wagon) is capable of carrying physical materials and second, that this capacity is finite and can be exceeded.

But the OED also identifies various extensions or specializations of these core meanings. In a use stretching back as far as Shakespeare, "load" has been used to mean something specifically negative: "A burden (of affliction, sin, responsibility, etc.); something which weighs down, oppresses, or impedes." And in more recent times, it has been used to designate "an amount of work, teaching, etc., to be done by one person." Particularly noteworthy is the application of "overload" to electrical circuits to mean "an electric current or other physical quantity in excess of that which is normal or allowed for," a use the OED dates to the early 1900s. Moving from the overloading of an electrical system to the overloading of a communication system does not seem like much of a stretch.

The word "information" has a much more complex history. (For an exploration of the word's history and current meanings, see Nunberg (1996); for an exploration of "The concept of information," see Capurro and Hjorland (2003).) Yet the ease with which the word was bandied about in these citations from 1962 to 1970 suggests that its meaning was clear—or at least clear enough—to audiences of the time.

The period from the 1970s to the present has witnessed the explosive growth of information and communication technologies, including personal computers, e-mail,

TABLE 20.1 Number of Articles Referring to "Information Overload" in Several Literatures

	ProQuest Business Press	ProQuest Science & Technology Press	*New York Times*
2000-present	549	172	86
1990–1999	521	150	99
1980–1989	164	21	41
1970–1979	1	3	7
1960–1969	1	0	2
Total	1266	346	235

instant messaging, the World Wide Web, and cell phones. And the use of the phrase "information overload" has equally expanded, in apparent synchrony with these developments. As an example of this growth, Table 20.1 indicates the occurrence of the phrase in three separate bodies of literature, grouped by decade.[7] Data for the first column, "ProQuest Business Press," was collected by searching the following ProQuest databases: ABI/INFORM Dateline, ABI/INFORM Global, ABI/INFORM Trade & Industry, Accounting & Tax Periodicals & Newspapers, Banking Information Source, ProQuest Asian Business and Reference, ProQuest European Business. Data for the second column, "ProQuest Science & Technology Press," was collected by searching the ProQuest Computing and ProQuest Science Journals databases. Data for the third column was collected by searching the *New York Times* online.

Finally, it should be noted, other phrases are in current use that relate to, or overlap with, "information overload" in various ways, including "data overload," "information anxiety," "information pollution," "information fatigue syndrome," "data smog," and "technostress." While it is beyond the scope of this article to examine the meanings of these phrases, it is worth observing that none of them appears to enjoy anything like the frequency of "information overload."[8]

20.4 CAUSES OF INFORMATION OVERLOAD

For the phrase "information overload" to have been introduced so casually and unproblematically in the 1960s, the groundwork must have been laid earlier. In the early twentieth century, as I noted above, scientists began to talk about the "overloading" of certain kinds of technical systems, namely electrical circuits; from

[7]These literature searches were performed in early 2007. This means that the counts for the period from the decade beginning in the year 2000 are small relative to the counts for earlier decades since the current decade is as yet incomplete.

[8]In Spring 2007, a Google search for each of these phrases found that "information overload" was mentioned in ten times as many web sites as its closest competitor: information overload (1,140,000), data overload (116,000), technostress (87,400), information anxiety (74,100), data smog (42,500), information pollution (34,800), information fatigue syndrome (812).

here it would seem to be a small step to viewing a human being or a human organization as a system that also could be overloaded.

Concerns specifically about an excess or overabundance of information were being expressed in this period, too. Following World War I, Burke (1994) has observed, scientists and librarians began to worry about a "library crisis." Such concerns stimulated various scientists and technologists, including Vannevar Bush, who in 1945 published an influential article in The Atlantic Monthly, "As We May Think" (Bush, 1945), which proposed developing a personal information device called the "memex" for storing and searching one's personal library of microfilmed books and articles. His description of the problem he hoped the memex would solve is information overload by another name: "There is a growing mountain of research. But there is increased evidence that we are being bogged down today as specialization extends. The investigator is staggered by the findings and conclusions of thousands of other workers—conclusions which he cannot find time to grasp, much less to remember, as they appear. Yet specialization becomes increasingly necessary for progress, and the effort to bridge between disciplines is correspondingly superficial."

By the 1960s, then, the language was ripe for use, and the perception existed that there was more information than could be properly handled. While this analysis provides some background for the emergence of the phrase in the 1960s, I would suggest that we look back even further in history—to the Industrial Revolution and its shaping of the Information Society—for insight into the phenomenon. The idea of the information society—or, as it is sometimes called, the knowledge society—can be dated to the 1960s and 1970s. In *The Production and Distribution of Knowledge in the United States* (Machlup, 1962) published in 1962, Fritz Machlup, an economist, noted an increasing reliance on knowledge production, as compared with physical production, in the U.S. economy. (Toffler uses the term "superindustrialism" in much the same spirit.) In the late 1970s, Marc Porat labeled the American economy as an "information economy," and declared American society to be an "information society."[9] Also, in the late 1970s Daniel Bell published *The Coming of Post-Industrial Society* (Bell, 1976). Together these works advanced, for both scholars and the general public, the idea that the balance of the American economy was shifting from the production, distribution, and consumption of physical goods to the provision of information products and services. Many details of the information society idea continue to be debated (see Frank Webster's 1995 book, *Theories of the Information Society* (Webster, 1999)), but its central premise—that there has been a major quantitative and qualitative shift in the use of information products, services, and technologies—is not in dispute.

But when did this shift begin and how did it come about? In his book, *The Control Revolution* (Beniger, 1986), James Beniger argues that the Information Society emerged in the attempt to solve certain problems created by the Industrial Revolution. The Industrial Revolution was essentially a radical transformation of the Western economic system. Steam power made it possible to produce, distribute, and consume

[9]See Porat's *The Information Economy: Definitions and Measurement* (Porat, 1977) and "Communication Policy in an Information Society" (Porat, 1978).

material goods faster and in greater quantity than had ever been possible before. Beniger says:

> "Never before had the processing of material flows threatened to exceed, in both volume and speed, the capacity of technology to contain them. For centuries most goods had moved with the speed of draft animals down roadway and canal, weather permitting. This infrastructure, controlled by small organizations of only a few hierarchical levels, supported even national economies. Suddenly—owing to the harnessing of steam power—goods could be moved at the full speed of industrial production, night and day and under virtually any conditions, not only from town to town but across entire continents and around the world." (Beniger, 1986)

But by the mid-nineteenth century, mass production and accelerated rates of distribution and consumption had precipitated what Beniger calls a "control crisis": both human and technical systems for managing the increased flow were found to be inadequate. JoAnne Yates illustrates one dimension of this crisis in her book *Control Through Communication* (Yates, 1989): the problem of creating organizational structures and management methods that were up to the task of controlling manufacturing and transportation at increased speed. Through the mid-nineteenth century and beyond, most commercial establishments had relied on flat organization and oral, face-to-face communication. But in the new economic conditions, organizations were larger and were more geographically distributed—the railroads were one major instance—and the old management methods no longer worked.

The solution to this control crisis, Beniger explains, was a "control revolution": the development of a whole series of innovations in information technologies and practices. In the case that Yates explores, gaining control of the new industrial organizations meant developing new, more sophisticated bureaucratic methods of management and accountability, as well as new technologies (the typewriter, vertical files, carbon paper) and document genres (the memo, the executive summary, graphs, and tables) to support these methods. Since the late nineteenth century, a stunning series of technical innovations has not only helped to control the material economy but to further accelerate it. This has led to a radical informatization of Western culture—a much greater presence of information products, services, and practices serving a control function; it is this process of radical informatization that led to, and in fact constitutes, the Information Society.

As a result of this transformation, American society, as well as other industrial economies, has witnessed a fairly steady increase in production: more products circulating more quickly. There has been a steady increase in traditional material goods, such as cars, washing machines, and refrigerators. And there has been a steady increase in the production of information goods, which have served two purposes: as end-user commodities, such as books, newspapers, movies, and television shows, and as agents of control in the sense Beniger identifies, such as telephone calls, e-mail messages, and advertisements. (Some of these information goods have served both functions at once.)

Hanging over the economy, however, has been the inevitable double fear—of overproduction and underproduction. Beniger's account provides a useful way of seeing how these threats have been dealt with. In the 1920s, for example, concern grew among American business leaders that the market was becoming saturated—that more goods were being produced than consumers wanted or needed; the very success of the late nineteenth century in learning how to manage accelerated flows of goods seemed to be leading to a crisis of overproduction. There was serious talk of reducing working hours and slowing production. But fortunately (from the perspective of the business leaders) another solution was found, which was nicely summarized in the findings of a national Committee of Recent Economic Changes, chaired by Herbert Hoover while he was secretary of commerce: "The survey has proved conclusively what has long been held theoretically to be true, that wants are almost insatiable; that one want satisfied makes way for another. The conclusion is that economically we have a boundless field before us; that there are new wants which will make way endlessly for newer wants, as fast as they are satisfied" (quoted by Hunnicutt, 1988, p. 44). From this understanding, the modern advertising industry was born, an industry devoted to producing new forms of information goods—advertisements—that would stimulate desire, and thereby consumption.

There is an irony in the increased production and use of information goods, however, an irony that has become increasingly clear throughout the more than century-long process of informatization. Information goods—whether newspapers, television advertisements, books, or e-mail messages—are themselves circulating products; they too need to be managed and controlled, just as much as do traditional material goods. But the more information that is produced to manage and control other forms of production, distribution, and consumption, the greater the need to manage this new information as well. These new needs have led, seemingly inevitably, to further innovations in information practices and technology, which have produced more, and new forms of information, which have led to further needs for control, and so on.

The World Wide Web provides an interesting example of this phenomenon. Writing in 1945, Vannevar Bush hoped that the memex would solve the post-war problem of information overload. The most innovative feature of the memex was to be the establishing of "associative indices" between portions of microfilmed text—what we now call hypertext links—so that researchers could follow "trails" of useful information through masses of literature. Bush never foresaw that a worldwide system of digital links would itself become a source of further overload, or that whole new methods and technologies would be required to manage the links that he thought would help solve the original problem.

From this perspective, then, information overload is simply an inevitable consequence of certain economic conditions and the philosophy of life that underlies them—a philosophy I have elsewhere called "more-faster-better" (Levy, 2006), which gives priority to producing and consuming more and more material and information goods ever faster. Now that digital video is within the reach of ordinary consumers, for example, we are seeing an explosive growth of its production and distribution (witness the YouTube phenomenon), and new technical schemes to manage this explosion (how do you index or search video?); it won't be long before these capabilities are increasingly used within corporations and other complex organizations to effect what Yates calls "control through communication." But the larger point is that systems of control for increased quantities and new forms of information will inevitably lag behind, leading in many cases to a sense of excess and an inability to handle it all.

What then is truly novel about today's sense of information overload and what is simply the recurrence, or the ongoing presence, of an older phenomenon? There is no question that people at other times and in other places have experienced a problematic excess of information goods. Roger Chartier (Chartier, 1994), the French historian of the book, for example, has suggested that after the invention of the printing press, it took an "immense effort motivated by anxiety" "to put the world of the written world in order." This vast, centuries-long effort, he points out, included the development of title pages, cataloging schemes, and the invention of the author. Here, it would seem, is a case that parallels our own, in which innovations in the control systems and practices needed to manage an increased production of information goods (in this case of books) were one step behind innovations in production. In this respect, today's experience of overload might be seen as a recurring phenomenon.

But at the same time, we might note certain features of the current experience that may be unique to this period of informatization. Never before have people had such widely available and powerful tools for communicating and managing affairs at a distance; we now live lives *mediated by* information and communication technologies to a degree that is clearly unprecedented in human history. Over the course of the last 150 years, the use of such technologies has spread to most, if not all aspects of human life: commerce, health care, education, family life, entertainment, and so on. It isn't uncommon for many of us to spend the better part of our days conducting many—in some cases, most—aspects of our life through e-mail, instant messaging, cell phones, and the Web.

What's more, many aspects of these interactions are bureaucratic in nature: filling in forms, pushing buttons to interact with voice mail phone trees, sending task-oriented e-mail messages around the globe. Bureaucratic methods born in the late nineteenth century to effect communication and control within large, distributed organizations have become the norm for communicating with friends and colleagues, for shopping, for interacting with corporations and other large and impersonal organizations. Memos, "fill-in" forms, and elaborate charts and tables were invented to systematize communication *within* organizations. But over the decades these same methods have been adopted for communication *with* corporate organizations, and among citizens. Today, one is just as likely to communicate with one's phone provider or gas company via a Web form or an automated phone tree as by talking with a customer service agent. Many people now find that a detailed calendar is an essential tool for managing not only their professional but their personal activities.

So it may be that some of the concerns being registered as information overload don't simply have to do with the amount of information we are dealing with, or the inability of our current control systems to handle them. Rather, or in addition, they may be the expression of discomfort with our *mode* of operating in the world through information technology. We may feel frustrated not just with the amount of e-mail we receive but with the fact that we are living so much of our lives through such a medium. To this might be added one further irony of the current phenomenon: Thanks to the very media and technologies that are contributing to our sense of overload—newspapers and magazines, radio and television, e-mail, blogs, and so on—we now have more opportunities than ever before to give public voice to such complaints.

20.5 CONSEQUENCES OF INFORMATION OVERLOAD

Information overload, as we have seen, involves more than just the exposure of an agent to excessive amounts of information: that agent must also suffer certain negative effects as a result. One of the most obvious, and straightforward, consequences is a failure to complete the task at hand, or to complete it well. As the conveyor belt speeds up, Lucy is unable to wrap *all* the chocolates passing her station (which is a clear criterion of success); by the end, she isn't wrapping *any* of them.

Faced with more information than we can handle in an allotted time, we find various ways just to get by. Steven J. Bell, library director at Philadelphia University, has noted how students, having pulled up hundreds of articles in response to a Google search, "print out the first several articles—making no effort to evaluate their quality—and then run off to write their papers" (Carlson, 2003). Barry Schwartz, a professor of psychology at Haverford, presents various heuristics people use to choose among a large range of options, and argues that as the number of choices increase, decisions require more effort and errors become more likely (Schwartz, 2004). Paradoxical though it may seem, having access to more information may lead us at times to be *less well* informed, and to make *less* effective choices.

Information overload may have consequences not only for the task but for the well-being of the person performing it, who may experience a diminished sense of accomplishment and a heightened degree of stress. In 1970, Toffler noted the link between increased stimulation and stress, making reference to Hans Selye's ground-breaking work on stress as a contributing factor in illness (Selye, 1956). In the intervening years, scientific evidence has incontrovertibly demonstrated that stress is a contributor to both physical and psychological ailments, including high blood pressure, depression, and anxiety. A certain amount of stress, of course, is inevitable in life, and not all of it is bad. The tensing of major muscle groups, along with increased heart rate and respiration, was, and still is, an appropriate response to physical threat. But living essentially full-time in fight-or-flight mode is bound to wreak havoc on health. As Peter C. Whybrow, a UCLA neuropsychiatrist and the author of *American Mania* (Whybrow, 1989), has observed: "today, it is no longer the life-threatening chance encounter that triggers physiological stress. Now stress is tied largely to social relationships and to the way in which our technology aids or hinders those relationships. The mechanisms of bodily defense that once gave short-term physical advantage are not well suited to the time-starved chronic competition of the Fast New World" (Whybrow, 1989, p. 79). The consequences of this unchecked and unbalanced striving may include "a competitive, unstable workplace, diminished time for family and community life, fragmented sleep, obesity, anxiety, and chronic stress"[10] (Whybrow, 1989, p. 106).

It should hardly be surprising if information overload also has negative consequences for ethical behavior. Nearly 30 years ago, Herbert Simon, a Nobel laureate in economics, observed: "In a world where information is relatively scarce and where problems for decision are few and simple, information is always a positive good. In a world where attention is a major scarce resource, information may be an expensive luxury, for it may turn our attention from what is important to what is unimportant. We cannot afford to attend to information simply because it is there" (Simon, 1978, p. 13). Simon's concern might be understood as simply applying to the cases just discussed above, where inadequate time and attention diminish the likelihood of choosing the most appropriate car or writing the best literature review. But Simon's choice of words —"turn[ing] our attention from what is important to what is unimportant"—might also alert us to the risks that information overload poses to ethical action as well; for an excess of information and the frenzy to which it contributes may easily distract us from adequately taking ethical concerns into account. Josef Pieper, a Roman Catholic philosopher and theologian, makes just this point when he argues that it is the workaholic who may be lazy insofar as his engagement in "the restlessness of a self-destructive work-fanaticism" (Pieper, 1998, p. 27) distracts him from his deeper responsibilities as a human being.

The overwhelming amount of information now presented through media outlets, for example, can make it difficult to stay in touch with those stories that require compassionate action on our part—as when the news about Anna Nicole Smith drives the dire circumstances in Darfur off the front page and out of public consciousness. Within corporate, and even scientific, contexts, it isn't hard to imagine that information overload and deadline pressure may lead individuals, or entire groups, to gloss over the ethical implications of their actions. Indeed, a recent study of scientific research observed that "our respondents were clearly worried about the quality of their own and their colleagues' data but they were *not* overly concerned with data that are purposively manipulated. Rather, they were troubled by problems with data that lie in what they see as a 'gray area,' problems that arise from being too busy or from the difficulty of finding the line between 'cleaning' data and 'cooking' data" (DeVries et al., 2006, p. 46).

Clearly then, there are times when information overload may lead people to fail to respond to, or to notice, what is important ethically. Pressed for time, overwhelmed and overbusy, any one of us may fail to reflect carefully and respond empathically. But there is another ethical effect of information overload, which may be even more consequential: when individuals, groups, or even entire societies become so enmeshed

[10]Also see the *New York Times* article, "Always on the Job, Employees Pay with Health" (Schwartz, 2004), which begins: "American workers are stressed out, and in an unforgiving economy, they are becoming more so every day. Sixty-two percent say their workload has increased over the last six months; 53 percent say work leaves them 'overtired and overwhelmed.'. . .Decades of research have linked stress to everything from heart attacks and stroke to diabetes and a weakened immune system. Now, however, researchers are connecting the dots, finding that the growing stress and uncertainty of the office have a measurable impact on workers' health and, by extension, on companies' bottom lines."

and overwhelmed by the amount of information and the accelerating pace of life that they fail to adequately develop the habits and character traits from which ethical action springs. The novelist Richard Ford, for example, has called the pace of life "morally dangerous," suggesting that in today's fast-paced, overloaded world "vital qualities of our character [may] become obsolete: our capacity to deliberate, to be patient, to forgive, to remain, to observe, to empathize..." (Ford, 1998).

Whybrow covers some of the same territory. Human empathy, he points out, "functions as the immune system of civil society" (Whybrow, 1989, p. 218). "It is empathy that transcends the interests of the 'selfish' self, promoting shared values among individuals and shaping the collective behaviors we call culture. In short, empathic understanding provides the lifeblood—the psychic immune system—that is the humanity of the civil society" (Whybrow, 1989, p. 219). But human empathy, he goes on to say, is a "delicate" commodity; it requires a kind of ongoing cultivation that is in short supply in a world, where, to quote the subtitle of Whybrow's book, "more is never enough."

A recent scientific study reported in the *New York Times* (Carey, 2006) offers evidence that when people who hold strong politically partisan views are given contravening information, they reject those views quickly and unconsciously using a part of their brain more associated with emotional activity than with reasoning. They never fully hear what doesn't fit their beliefs. "It is possible to override these biases," one of the scientists involved in the study is quoted as saying, "but you have to engage in ruthless self-reflection, to say, 'All right, I know what I want to believe, but I have to be honest." Can there be any question that human beings need time and attention to cultivate the depths of their humanity—to develop their capacity for mature thinking and listening, for insight and empathy? While it would be too simple to suggest that information overload is the sole cause, it must surely be counted as a factor insofar as it is both a contributor to and a symptom of the complex of attitudes and behaviors I have called "more-faster-better."

20.6 CONCLUSION: WHAT CAN BE DONE?

In the analysis I have presented here, information overload is one of the side effects of an information society operating under a "more-faster-better" philosophy of life. For a variety of reasons—some economic, some social, and some spiritual[11]—our society's sense of progress and achievement is tied to the accelerated production of material and information goods. Some of these information goods are end-products (films and video games and newspapers), while others are agents of control (advertisements and e-mail messages) that help to manage the accelerating processes of production and consumption. The result is that more and more information products are being produced faster and faster, and attempts to manage these flows lead to the production of yet more information.

[11]See David Loy's assertion that the Western economic system is a "religion of the market" (Loy, 1997).

Can nothing be done then to stem the tide? In fact, many things have been done, and will continue to be done, to reduce people's sense of information overload. Some of these interventions are technological in nature: the development of e-mail filters, for example, to automatically categorize incoming e-mail and to identify and isolate spam. Other interventions are social, as when people decide to take themselves off certain e-mail listservs to reduce their inbox clutter, or to take "Blackberry-free" vacations. Still others involve the law: recent U.S. legislation instituting "don't call" lists has apparently been very effective in eliminating unwanted telemarketing phone calls to people's homes.

How successful these different forms of technical, social, and legal intervention have been, or will be, depends largely on the specific circumstances within which they are embedded. In some cases—"don't call" lists, for example—the intervention may solve the immediate problem, while in other cases—such as e-mail filters or calendaring programs—the solutions may make the problem worse in the long run, because the ability to handle more information more efficiently may contribute to further acceleration that then leads to the production of more information, and so on. All of which suggests that so long as a "more-faster-better" attitude governs the production of material and information goods, we can expect that information overload—along with more and less successful attempts to mitigate it—will be a regular feature of postmodern life.

REFERENCES

Ackoff, R.L. (1967). Management misinformation systems. *Management Science*, 14(4), B147–B156.

Agre, P.E. (1995). Institutional circuitry: thinking about the forms and uses of information. *Information Technology and Libraries*, 14(4), 225–230.

Bell, D. (1976). *The Coming of Post-Industrial Society: A Venture in Social Forecasting*. Basic Books, New York.

Beniger, J.R. (1986). *The Control Revolution: Technological and Economic Origins of the Information Society*. Harvard University Press, Cambridge, MA.

Blair, A. (2003). Reading strategies for coping with information overload ca. 1550–1700. *Journal of the History of Ideas*, 64(1), 11–28.

Boulding, K.E. (1966). The ethics of rational decision. *Management Science*. 12(6), B161–B169.

Bowles, M.D. (1999). *Crisis in the Information Age? How the Information Explosion Threatened Science, Democracy, the Library, and the Human Body, 1945–1999*. Case Western Reserve University, Cleveland, OH.

Buckland, M. (1991). Information as thing. *JASIS*, 42(5), 351–360.

Burke, C. (1994). *Information and Secrecy: Vannevar Bush, Ultra, and the Other Memex*. Scarecrow Press, Metuchen, NJ.

Bush, V. (1945). As we may think. *The Atlantic Monthly*, 176, 641–649.

Capurro, R., and Hjorland, B. (2003). The concept of information. *Annual Review of Information Science & Technology*. 37, 343–411.

Carey, B. (2006). A shocker: partisan thought is unconscious. *New York Times*. p. D1.

Carlson, S. (2003). Has Google won? A librarian says students have more data than they know what to do with. *Chronicle of Higher Education*.

Chartier, R. (1994). *The Order of Books*. Cochrane, L.G. (trans). Stanford University Press, Stanford, CA.

DeVries, R., Anderson, M.S., and Martinson, B.C. (2006). Normal misbehavior: scientists talk about the ethics of research. *Journal of Empirical Research on Human Research Ethics*. 1(1), 43–50.

EDUNET: Is it the answer to the information overload in our schools and colleges? *New York Times*, June 4, 1967, p. BR14.

Ford, R. (1998). Our moments have all been seized. *New York Times*, p. 9.

Gross, B.M. (1962). Operation Basic: the retrieval of wasted knowledge. *Journal of Communication*, 12(2), 67–83.

Himma, K.E. (2007). A preliminary step in understanding the nature of a harmful information-related condition: an analysis of the concept of information overload. *Ethics and Information Technology*, 9(4).

Hunnicutt, B.K. (1988). *Work Without End: Abandoning Shorter Hours for the Right to Work*. Temple University Press, Philadelphia.

Levy, D.M. (2006). More, faster, better: governance in an age of overload, busyness, and speed. *First Monday*, Special Issue #7: Command Lines: The Emergence of Governance in Global Cyberspace http://www.firstmonday.org/issues/special11_9/.

Loy, D.R. (1997). The religion of the market. *Journal of the American Academy of Religion*. 65, 275–290.

Machlup, F. (1962). *The Production and Distribution of Knowledge in the United States*. Princeton University Press, Princeton.

Meier, R.L. (1962). *A Communication Theory of Urban Growth*. MIT Press, Cambridge, MA.

Murphy, K. and Gross, R. (1969). All you need is love. Love is all you need. *New York Times*, p. SM36.

Nunberg, G. (1996). Farewell to the information age. In: Nunberg, G. (Ed.), *The Future of the Book*. University of California Press, Berkeley.

Pieper, J. (1998). *Leisure: The Basis of Culture* (G. Malsbary, Trans.). St. Augustine's Press, South Bend, IN.

Porat, M. (1977). *The Information Economy: Definitions and Measurement*. Department of Commerce, Office of Telecommunications, Washington DC.

Porat, M. (1978). Communication policy in an information society. In: Robinson, G.O. (Ed.), *Communications for Tomorrow*. Praeger, New York.

Rosenberg, D. (2003). Early modern information overload. *Journal of the History of Ideas*, 64(1),1–9.

Schement, J.R. and Curtis, T. (1995). *Tendencies and Tensions of the Information Age: The Production and Distribution of Information in the United States*. Transaction Publishers, New Brunswick, NJ.

Schwartz, B. (2004). *The Paradox of Choice: Why More Is Less*. Ecco, New York.

Schwartz, J., (2004). Always on the job, employees pay with health. *New York Times*.

Selye, H. (1956). *The Stress of Life*. McGraw-Hill, New York.

Simon, H. (1978). Rationality as process and as product of thought. *American Economic Review*, 68, 1–16.

Thorngate, W. (1988). On paying attention. In: Baker, W.J., Mos, L.P., Rappard, H.V., and Stam, H.J. (Eds.), *Recent Trends in Theoretical Psychology*. Springer-Verlag, New York. pp. 247–263

Toffler, A. (1970). *Future Shock*. Bantam Books, New York.

Webster, F. (1999). *Theories of the Information Society*. Routledge, London.

Whybrow, P.C. (2005). *American Mania: When More Is Not Enough*. W.W. Norton, New York.

Yates, J. (1989). *Control Through Communication: The Rise of System in American Management*. Johns Hopkins University Press, Baltimore.

CHAPTER 21

Email Spam

KEITH W. MILLER and JAMES H. MOOR

21.1 INTRODUCTION

A fundamental problem with any philosophical discussion of email spam is definitional. Exactly what constitutes spam? Published definitions by some major players differ dramatically on which emails should be identified as spam. Some emphasize the importance of "consent"; others require the emails to be commercial in nature before they are called spam; still others focus on the number of identical messages that are sent as spam. At least one Web site (Spam Defined, 2007) is soliciting signatories to settle on the definition of spam.

The conceptual muddles about defining spam have immediate philosophical and legislative consequences. These muddles have, for example, made it difficult to write effective legislation regarding spam, and the laws that exist have not been successful at significantly reducing what many people consider a significant problem in cyberspace. At this writing, some estimates show that over 80% of email traffic is spam. Although the definitional and technical challenges of these estimates make it difficult to verify their accuracy, few Internet users doubt that spam emails are a significant and persistent occurrence. In this short article, we'll look at the short history of spam and then describe a just consequentialist analysis of email spam, an analysis that takes into account several different characteristics that help to differentiate spam from other emails.

21.2 A SHORT HISTORY OF THE TERM "SPAM"

If you take a broad view of spam as "unsolicited electronic messaging," the first spam mentioned in Wikipedia (2007) is a telegram sent in 1904. However, the term "spam" wasn't used until the 1980s, when some participants in interactive MUDs (Multi-User

The Handbook of Information and Computer Ethics, Edited by Kenneth Einar Himma and Herman T. Tavani

Dungeons) would use one tactic or the other to flood the interface, often with repeated messages (Northcutt, 2007). Some of the repetitions used the word "spam," in homage to a Monty Python skit (Detritus.og, 2001), and the name caught on.

The first large commercial spam occurred in newsgroups. In 1994, Laurence Canter and Martha Siegel, two lawyers trying to solicit business, posted identical messages to all existing Usenet newsgroups. These messages were ads for their business of "assisting" foreign nationals to apply to a U.S. immigration lottery. (In truth, all that was necessary to enter the lottery was a postcard, but the ad didn't mention that.) (Everett-Church, 1999.)

Spam has rapidly spread to many forms of electronic communication. For example, "spim" is instant messaging spam (Bitpipe.com, 2007). Some Web sites are now called "spam." But most people today associate spam with certain kinds of email. The rest of this chapter will use "spam" to refer to email spam; although there are some ethical nuances that might be different in different kinds of electronic spam, we will focus on email spam.

21.3 SEARCHING FOR A CHARACTERIZATION OF "SPAM"

With a few notable exceptions (see Hayes, 2006), there are few public defenders of spam. People attacking spam are legion and vocal. But it would be a mistake to think that "spam" can be defined simply as "unwanted email." An email that is from an unsolicited, commercial, bulk emailing, often considered spam, may provide a receiver with just the information that he/she does want. And some email, such as email informing someone that he/she is fired, is unwanted but not spam. Nevertheless, in general spam is viewed negatively as something one does not want to receive. But, the reasons for this vary. Different spam critics are offended by different features. In this section, we'll survey some of the characteristics of emails that influence receivers to call the senders "spammers." We do not claim that the list is exhaustive, nor is there an implied priority in the ordering. Many of these aspects interact and overlap; however, each of them has received some attention in the literature on spamming.

21.3.1 Content of the Email

If an email is a short message, perhaps an invitation to a local event, it is unlikely to be labeled "spam," especially if it comes from someone the recipient knows. But, if the content of an email includes yet another routine advertisement for improving one's sexual prowess, it will likely be classified as "spam." Even worse, if it includes a disguised virus that erases the recipient's hard drive, that email will almost certainly be labeled spam by the enraged recipient. This designation will be made regardless of other characteristics below. Notice, however, that an email message that contains a virus can be sent to a specific individual (not part of a mass email) innocuously, by a friend, without any commercial aspect, and with a valid return address. Sometimes mass emailing itself will be enough to count as spam regardless of content, as happens in attempts to flood servers with messages to overwhelm them (denial of service).

21.3.2 Intent of the Sender

The sender's intent is relevant both to receivers and to our analysis. Imagine two emails with identical virus-laden content, one sent to you by a malicious hacker whose purpose is to invade your system and the other sent by a friend unaware of the email's hidden virus. Both emails would affect your system in the same way, but your attitude toward the person who sent the email is likely to be different in the two cases. (Your attitude toward the originator of the virus is likely the same.)

21.3.3 Consequences to the Receiver

Regardless of the intent of the sender, the actual consequences to the receiver often influence the receiver's classification of an email. Every email has an effect. Even emails that are blocked by a spam filter still have a consequence on the performance of the network and the receiver's system. Emails that lure the receiver into revealing personal and financial information, "phishing attacks," and damaging virus attacks can have devastating effects. However, some emails that some think of as spam, some might welcome as useful advertising or helpful announcements. The more detrimental the consequences of an email, the more likely it is that the receiver will label the email as spam.

21.3.4 Consent of the Receiver

If a receiver has given the sender explicit consent for the sender's emails, the receiver is less likely to consider the emails spam. The word "unsolicited" is often used in describing spam, and that term suggests a lack of consent. It should be noted, however, that most emails are "unsolicited," unless they are a direct reply from Y to X after a previous message from X to Y. In case of commercial bulk email advertisements, we can infer that at least some receivers find the advertisements acceptable absent consent, since if absolutely no sales resulted from such emails, it seems unlikely that they would continue to be sent. (A counterargument is that the developers and vendors of antispam software have an incentive to keep sending spam, so that billion-dollar industry continues; however, hardcore evidence is by far lacking that antispam companies are the sources of spam.)

21.3.5 Relationship Between the Sender and the Receiver

The relationship (or lack of relationship) between the sender and the receiver of an email affects the receiver's classification of the email. The relationship may be personal, professional, or commercial; the "relationship" might be based on a shared interest or on a desire to make a transaction; and the relationship can be positive or negative. A close, positive, personal relationship between the sender and the receiver tends to improve the receiver's attitude toward the email. If the receiver perceives no relationship, or a negative relationship, then the receiver is likely to label the email as spam. For example, imagine a "get out the vote" email from a political party. Such an email from an organization for which the receiver has sympathy will be less likely to be labeled spam than a similar email from an organization that the receiver detests, even if the content in both the messages is nearly identical.

21.3.6 The Accountability of the Sender and the Degree of Deception

If the sender of an email message includes an authentic return email address that in truth is the sender's address, and if the sender's address is consistent over a reasonable amount of time, then the receiver is less likely to label the email as spam. There are technical approaches to controlling spam called "blacklisting" and "whitelisting" that depend for their effectiveness on knowing where email comes from. Because of this, emailers wishing to send multiple messages to the same address often use return addresses that are not their own permanent email addresses. The substitutes can be a spoof (someone else's legitimate email address), temporary (used to send a group of bulk emails and then abandoned), or fake (not a real email address). Handshaking protocols have been proposed to distinguish between authentic permanent email return addresses and those substitutes, but none has so far been implemented on a wide scale. The issue of the return email address is only the most visible portion of the larger issue of accountability. Another issue is to what degree the sender is trying to deceive the receiver. Some emails have a subject line that has nothing to do with contents. Other emails masquerade as coming from an acquaintance or from system administrators. Phishing emails fraudulently present themselves as banking security notices. The more receivers detect these attempts at deception, the more they label the emails as "spam," and the angrier they get at the spammers.

21.3.7 Number of Identical Emails Sent

The more identical emails are sent from a sender, the more likely the recipients are to perceive it as spam. The number of identical emails that are sent is not typically known by a recipient, but recipients often get a suggestion about this number from the content of the email. If the email is targeted at a narrow group of which the recipient is not a member (e.g., a customer of a particular bank or someone interested in altering a body part), the email is assumed to be sent in bulk, in an indiscriminating way. System administrators may be able to make a better estimate of this number, since they can detect the influx of many identical messages to their system.

21.3.8 Illegality

One way to distinguish spam from other emails is to determine if the email violates the law. In the United States, the CAN-SPAM Act of 2003 (Controlling the Assault of Nonsolicited Pornography and Marketing Act) defines illegal spam as commercial emails that have false or misleading header information (to, from, and routing information) or deceptive subject headings. The act requires that a commercial email must identify itself as an advertisement, provide an opt-out mechanism, and include a valid physical address (FTC, 2007). Many critics think that CAN-SPAM is flawed and

TABLE 21.1 Examples of Characteristics That Can Range from "Not a Problem" to "Serious Problem" for Emails

Dimension That Matters	Benign or Better	Bad or Worse
Content of message	Birthday party invitation	Debilitating virus
Sender's intent	Inform friends	Steal identity
Consequences to receiver	Gain valuable information	Lose life savings
Consent of receiver	Voluntarily joined group	Desperately wants out
Sender/receiver relationship	Close personal friend	Predator/victim
Accountability and deception	Genuine return address	Spoofed address
Number of emails sent	Less than ten	Several millions
Legality	Legal meeting notice	Illegal bank fraud
Size of a message	Less than a kilobyte	More than a gigabyte

ineffective, and several states have passed their own laws, with their own definitions, against spam (Scanlan, 2005).

Any laws about emails suffer the problems of Internet regulation implemented by a land-based government. The Internet is trans-national, and such laws stop at the border. However, studying these laws is instructive in understanding what emails are considered (at least by some) as spam. Furthermore, if a sender is in a territory that has a law about spam, there is some motivation to follow the law. A counterargument is that such laws are unjust, and should therefore be resisted. Another counterargument is that such laws are impractical and sporadically enforced, and can therefore be ignored (Kulawiec, 2004).

21.3.9 Size of the Message

An unwanted email that is relatively small is less of a problem for a recipient and for the Internet than an unwanted email that is relatively large. Although users probably would not consider the size of a message as determining whether the email is spam or not, users and particularly administrators might very well decide that an email is either "minor, annoying spam" or "major, dangerous spam" based on the number of bytes in such an email (Table 21.1).

21.4 ENVISIONING THE SPAM SPACE: SPECIFIC EMAILS EXHIBIT COMBINATIONS OF CHARACTERISTICS

Given the observations in the last section, one might be pessimistic about ever becoming clear on what "spam" means. The term "spam" is no doubt vague and ambiguous. It is vague because there are borderline cases about which we are uncertain whether to classify the item as spam or not. Do 100 similar emailed messages advertising the office holiday party count as spam or not? And the term "spam" is ambiguous depending upon which of the characteristics one regards as the defining conditions. We might understand "spam" not so much as a technical term but a common language term used in a technical arena. Expecting precise, unambiguous meanings for common language terms is unrealistic and not necessary. Many common language terms are vague but useful. "Night" is a vague term in that it is not possible to identify the exact moment when day stops and night begins, but "night" is obviously a useful term. "Plane" is an ambiguous term, but the context of use often disambiguates whether we are talking about an aircraft or a carpenter's tool.

Our discussion of spam demonstrates that it is difficult to identify a set of necessary and sufficient conditions to define "spam." But again, "spam" is a common language term used in a technical arena. The philosopher Ludwig Wittgenstein argued rather persuasively in his *Philosophical Investigations* that many common language terms do not lend themselves to analysis in terms of necessary and sufficient conditions. The meaning of terms such as "game" or "chair" are resistant to analysis in terms of necessary and sufficient conditions and yet are perfectly useful terms. Wittgenstein suggested that we should look for the use of a term in order to locate its meaning. The items picked out by a term may be united not by a set of necessary or sufficient conditions but through a family resemblance. Thus, we recommend understanding what "spam" means by regarding it as a family resemblance term and looking closely at its various uses. To establish the foundation for our analysis, we introduce the notion of spam space.

If we imagine that each of the characteristics mentioned in the last section (as well as additional characteristics that others might suggest) defines a continuum from good to bad, then we could, in theory, measure each characteristic and locate an email in a multidimensional space. The extremes of this space are relatively easy to imagine. Assume that email E1 includes a short invitation to a birthday party, sent to only six friends by the sender S1; assume further that it includes no viruses, includes S1's correct physical and email addresses, and is sent to personal (nonbusiness) email accounts. Although E1 is unsolicited and is sent out to multiple recipients, few of the recipients would call it spam, and probably none of the friends would object to E1. Indeed, email that tends toward the "good" end of all the characteristics is unlikely to cause recipients annoyance or rage, and is unlikely to be labeled with the pejorative "spam."

At the other extreme of our multidimensional space, we can imagine a second email message, E2. The content of this message includes a virus, spyware, and a phishing attack. The message is nearly a gigabyte in size. The sender intends to defraud the unwary recipients and to attack their systems as well. When the multiple attacks are successful, the recipients will have their identities stolen, their bank accounts drained, and their systems debilitated. If recipients knew the true origin of this email, none of them would consent to it. There is no relationship between the sender and the receiver except that the sender considers the receivers prey. The return address in E2 is spoofed, and any identifying information in the email is a cleverly designed fraud. Millions of emails identical to E2 have been sent out nearly simultaneously by an organized crime syndicate. Although it is not exactly commercial, since no legitimate business is involved, E2 is still labeled as illegal by the CAN-SPAM act, as well as all relevant state laws and international treaties. Anyone receiving this message is likely to immediately consider it spam (if they aren't fooled by it), or will consider it spam after they discover they have been deceived.

There is a great deal of email traffic that will fall somewhere between the benign invitation E1 and the spam from hell, E2. We can envision unsolicited commercial bulk emails that include wonderful bargains that many recipients would like to consider. We can envision persistent, frequent newsletters from a small organization that manage to annoy subscribers so much that they consider it spam even though they voluntarily signed up for the newsletter emails.

A broad variety of emails is possible within the "spam space" defined by these dimensions. We suggest that it is essential in any ethical analysis of spam to be explicit about what subset of this space is being analyzed. In each of the sections that follow, we'll explore one such subset. Our collection of subsets cannot be exhaustive or authoritative. We present our collection as merely illustrative of the kinds of analysis that can be accomplished by limiting the scope of emails considered. After we consider these subsets, we will, in the following section, consider some of the proposed countermeasures against spam.

Before examining some specific cases from spam space, a few general comments about ethics are useful. In considering the appropriate use of email, we are looking for policies that guide us about what to do and what not to do. To evaluate this, we consider the good and the evil consequences that will ensue if everyone knows that a particular policy for action is permitted and in principle can act on it. A policy for one's ethical actions must be a public policy. We are required ethically to be impartial about the ethical policies that we adopt. If it is acceptable for A to deceive B in an email, it must be acceptable for B to deceive A in similar circumstances. The policy must be acceptable to us if we were ignorant about the role we might play under the policy. To ethically evaluate our ethical policies requires *both* a consequentialist analysis of the outcomes of following the policies and a justice analysis (Moor, 1999).

Notice that the test for ethical policies is not to ask what if everyone did it? Instead the test should be, what if everyone knew they were allowed to do it? Sometimes, it is argued that spam should not be sent because, if everyone sent it, the email system would collapse. Such a test is too strong because it would rule out too many actions. For example, what if everyone flew on a plane on Mondays? Hence, the better question to ask is what if everyone were allowed to send spam? The answer to that will depend upon the kind and likelihood of the spam that would be sent. Finally, in ethical analysis it is important to separate the intention of an actor on which we praise or blame him from the consequences of the actor's action. If the actor intends to cause unjustified evil for his own benefit then he is blameworthy regardless of the actual outcomes of the action.

21.4.1 Deceptive Emails Meant to Defraud Are Condemned, Spam or Not

If a sender uses an email to attempt to defraud any recipient or recipients for the sender's selfish gain, the sender is blameworthy. Many of the characteristics described above seem insignificant compared to the intent of these emails. Whether such an email is sent to one recipient or to millions, it is difficult to imagine an ethical justification for the sender. This is not to suggest that deception in email is always ethically condemned. We can imagine deceptive emails engineered by law enforcement, with a warrant, to gather evidence against predators who use the Internet illegally for the purpose of child molestation. Deception is an evil, but in the latter case it is counterbalanced by the greater evil it avoids. In considering the policies for these actions from an impartial point of view, we would reject a policy of outright fraud for selfish gain, but accept a policy that protected children from molestation even though some deception is involved.

21.4.2 Emails Between Well-Meaning Friends Are Probably Not Spam

When people who know each other on a personal basis send each other emails in good faith (i.e., with no intent to harm each other), the term "spam" seems inappropriate, even if identical messages are broadcast to a group of friends. There is typically an implicit consent shared among people who know each other that "unsolicited" emails are allowed, as long as that consent is not abused. This is not to suggest that senders do not have responsibilities toward receivers when the receivers are the sender's friends. Surely, there should be a good faith effort to avoid forwarding emails that contain viruses, phishing attacks, offensive content, or excessively large attachments. Friends shouldn't inundate their friends' email boxes with messages, no matter how well intentioned. But among friends, such problems can be negotiated on a personal basis, and we think the term "email spam" should be reserved for less intimate email encounters.

21.4.3 Unsolicited Commercial Bulk Emails (UCBE)

As pointed out frequently, UCBE have become a significant portion of Internet traffic for simple economic reasons. The incremental cost of one more email in a large UCBE project is negligible. Thus, as long as there is a prospect for increased sales with more emails sent, UCBE projects are likely to grow in scope and frequency. The cost of sending emails is borne by the sender, but the cost of dealing with the emails sent is born by the recipients and the network infrastructure. As increasing amounts of spam (however it is defined) become the majority of the message traffic on the Internet, this imbalance of costs and incentives is a technical and economic situation within which any discussion of UCBE must be framed (Hoanca, 2006).

Another required "framing" is the question of whether or not *any* commercial messages should be allowed on the Internet. This is an interesting theoretical question, worthy of ethical analysis, but in this paper, we will assume that this question has been rendered moot. At its inception, the Internet did not include commercial messages, but it seems unlikely that the Internet (at least in its current state) will ever return to this "pristine" state. This might be a live issue for new initiatives such as Internet2 (2007), but at least for the purposes of this paper, we will assume that commercial email is a given on the Internet.

The characteristic "accountability and deception" should be critical in any ethical analysis of UCBE. The more deceptive the content and the less accountable the sender, the more blameworthy the sender becomes. If the sender of UCBE takes pains to disguise the message and the sender's identity, there is a strong indication that the sender is "up to no good," and blameworthy. However, there are important exceptions to this rule of thumb; for example, a whistle-blower might send a message to a large commercial mailing list, but for the purposes of alerting the recipients to a danger or to an injustice. The whistle-blower may have justifiable reasons for seeking anonymity and for emailing broadly. In the case of whistle-blowing, the sender's intent dominates.

Much of the popular discussion about UCBE is confused and degraded because it doesn't distinguish between two sets of emails: UCBE that is deceptive, intended to harm, with unaccountable senders, and UCBE that is not deceptive, intended for encouraging commerce, with accountable senders. (Some may argue that all advertisements are fundamentally biased and likely to be inherently deceptive; that may be true, but we think it is possible to distinguish between advertisements that misrepresent and are fraudulent and advertisements that present information in a favorable light. In this section we will assume such a distinction.) We will label emails that are fraudulent as F-UCBE, and emails that are not fraudulent as NF-UCBE. Analysts may differ about which category a particular email belongs to, but most analysts will agree that there exist some emails that belong to each category.

NF-UCBE can include emails from a company X that is contacting customers who have previously done business with X. NF-UCBE can include emails from a company Y that has purchased a list of potential customers that Y hopes will be interested in doing business with Y. An email in the category NF-UCBE is analogous to physical "junk mail" and ads in a newspaper, on TV, or on radio; the seller is looking for buyers, and the buyers can accept or reject the offer. A major difference between NF-UCBE and physical junk mail is the relatively insignificant incremental cost for emailing a new customer as opposed to the incremental cost of sending another physical piece of mail or buying another ad. Two differences with ethical significance between NF-UCBE and other advertisements are who pays the real costs of NF-UCBE and the technical consequences of the (somewhat perverse) economic incentives of NF-UCBE. (These incentives are also important for F-UCBE, but we will discuss that aspect separately below.)

Since NF-UCBE is offered in good faith, it seems similar to other advertisements in any fundamental ethical analysis. The technical detail of different cost/benefit ratios for different kinds of advertisements does not, in our opinion, alter the ethical stance of a seller communicating with potential buyers. It could be argued that Internet users are so upset about spam (of all sorts) that sellers using NF-UCBE are generating so much ill-will for themselves that NF-UCBE is increasingly counterproductive. But the determination of whether or not NF-UCBE is, on balance, a good selling strategy seems best left to the sellers. If the small percentage of buyers that NF-UCBE generates is worth the costs, both for sending the emails and generating ill-will, then NF-UCBE will continue to be sent until the costs rise. Perhaps the costs *should* rise, and that possibility will be discussed below.

Users and Internet providers employ filters, whitelists, and blacklists to reduce the amount of spam users receive, including both NF-UCBE and F-UCBE. However, these measures have drawbacks that will be discussed in the following section.

F-UCBE, in our opinion, is not as complex an ethical analysis as NF-UCBE. It is difficult to imagine a situation in which one could impartially advocate a policy that supported F-UCBE. Moral and legal strictures against fraud are not relevant for an analysis of NF-UCBE, but *are* relevant for an analysis of F-UCBE.

21.5 THE ETHICS OF ANTISPAM MEASURES

In ethical analysis, there is a three-part division of policies: some that virtually everyone would impartially accept as an ethical public policy (e.g., sending a friendly birthday email to one's parents), some that virtually nobody would accept as an ethical public policy (e.g., sending a virus in an email that destroys the receiver's hard disk), and some about which there is disagreement as to whether they should be accepted as ethical public policy (e.g., NF-UCBE). It is this last group to which we turn now. We need to consider what the possible policies are. In order to analyze several different suggestions about how spam should be dealt with, we find it convenient to separate the suggestions into two broad categories: suggestions to reduce the number of spam emails sent and suggestions to reduce the number of spam emails received after they are sent.

21.5.1 The Ethics of Doing Nothing

Although it is not much discussed, there is an option of a widespread policy of ignoring this problem, at least for a time. It might be expected that, unfettered by filters and legislation, spam would increase and email would become unusable for many people, and would then become abandoned by an increasing number of users. Perhaps private intranets, isolated from the Internet, would increase in popularity as more and more people refused to use Internet email because of the high percentage of spam messages. At least for a while, other Internet services (like the Web) would slow down because of the heavy spam email traffic. Eventually (at least theoretically), spam would become unproductive because not enough potential customers and victims would be checking their email. It might happen that spammers would slowly leave the business because it would no longer be profitable. If spam lost its luster, the Internet might become more usable again.

This scenario appears to be extremely unlikely. Despite experiments like Internet2 and despite investments in local intranets, the current investments in the Internet seem too substantial to abandon. Furthermore, many users are enraged by spammers, and this kind of "surrender" to spam will be difficult for many Internet users to accept.

Finally, even if such a strategy temporarily succeeded, and users started to return to the Internet because spammers had largely given up, then as soon as sufficient numbers of nonspammers started using the Internet again, the spammers could jump back in. Most probably, without other antispam measures, the percentage of spam email would be cyclic, and periodically the Internet would be unusable.

Although these considerations make the "do nothing" approach very unlikely to be a practical or even attempted solution, this solution is still of theoretical interest to our ethical analysis. First, it illustrates that spam on the Internet is a good example of the classic problem of "spoiling of the commons." The Internet is based on a machine architecture and protocols that emphasize scalability, ease of entry, and relatively low cost for senders (as compared to, for example, television broadcast). This design is optimal for users who act in good faith, but it is wide open for abuse. Spammers, particularly F-UCBE spammers, are bad actors who exploit the Internet and its good-faith users. If the commons (the Internet) becomes unusable because of spam, then it is spoiled for all, including the spammers. Thus the literature on the spoiling of the commons is relevant to any discussion of this problem.

It is also useful to examine the ethical significance of the "do nothing" approach. The stakeholders involved in the email spam issue are, roughly speaking, spam senders, spam receivers, those involved in email delivery, and email senders who do not send spam. (Those categories are dependent on your definition of spam.) Receivers who do nothing and eventually abandon Internet email will be losing the advantages of the Internet, but also losing their frustrations over spam. Email delivery service providers will be losing business, and in some cases their jobs, as fewer people use Internet email. Spam senders and the developers and vendors of antispam solutions will lose their customers. People who once depended on Internet email for communication will have to use alternative methods. This stakeholder analysis dramatically illustrates the common interests of spam-producing and spam-blocking professionals.

21.5.2 The Ethics of Reducing the Number of Spam Emails Read After They Are Sent

Many "antispam" measures seek to remove from spam recipients the burden of seeing and manually removing spam emails. Three such measures are blacklists, whitelists, and filters. We will refer to all three as "spam blockers." When successful, spam blockers reduce the negative effects on individual users, but they do nothing to reduce the bandwidth required to carry the spam from the senders to the receivers. Indeed, unless spam blockers are completely successful (and no one seriously claims that they are), they will probably increase the overall system spam load. The argument goes as follows: if spam blockers result in 90% of spam being blocked, and if the incremental cost for more spamming is sufficiently low, then spammers will increase the number of spams sent by a factor of 10 to counteract the spam blockers.

When a spam blocker allows an email that the recipient thinks is spam to reach the recipient, that constitutes a "false negative." A false positive occurs when, in attempting to block spam, the blocker additionally blocks an email that the recipient would not have labeled spam. As we have seen above, there is no standard definition of "spam," so it is likely that both false positives and false negatives will be common. False negatives reduce the positive effects of spam blockers, and false negatives mean that recipients are blocked from seeing emails they might have some interest in seeing.

A spam "blacklist" blocks emails on the basis of the sender's email address. The blacklist can be maintained by a third party; in such case the recipient automatically adopts the third party's preferences, false positives, and false negatives. Users can add to the blacklist, meaning that the first email from a spammer gets through, but subsequent emails from that source should be blocked. Unfortunately, bulk emailers can easily change their email address; moreover, deceptive emailers can either spoof a return address or send emails from computers they have "hijacked," turning unwitting Internet users into email drones.

Whitelists allow a user to specify senders whose messages are allowed into the user's email queue. If email addresses are spoofed, or if email sources on the whitelist have been hijacked, the whitelist is ineffective. False negatives are a particular risk with whitelists since anyone not on your whitelist is assumed to be a spammer. This precludes, for example, an email from a long-lost friend.

Spam filters are a blocker that attempts to identify spam based on the content or addresses. Some of these filters are adaptive and seek to infer a user's preferences for what email to block and what email to allow through. The false positives and false negatives of a filter can be surprising to a user, especially early in the process of fine-tuning the filter preferences. Again, the user may sacrifice the possibility of pleasant surprises in emails when a filter determines what emails are blocked. One advantage of blacklists and whitelists is that users have some sense of control over what is blocked or not blocked; since spam annoys people, there is a certain amount of satisfaction, for example, when a user can identify a sender as a spammer to a blacklist system. Filters do not allow this admittedly mild form of revenge.

One ethical consideration with spam blockers is the location of responsibility for dealing with unwanted emails. When the spam blockers are deployed by Internet Service Providers (ISPs) and system administrators on behalf of their customers and users at the receiving end, the organizations who deliver the email are taking responsibility for trying to reduce spam. When the spam blockers are bought and implemented by users, they are taking responsibility for trying to reduce spam. One group that has not been particularly active in trying to reduce spam are the ISPs and system administrators of spam *senders*. When the ISPs and users of spam *recipients* block spam, the spam has already added to the bandwidth problems of the Internet; if the ISPs of spam *senders* would block spam, that would reduce the harmful effects of spam for the entire Internet, not just for the recipients. Spammers pay their ISPs for sending spam, and this is likely a factor in reducing the ISPs' enthusiasm for blocking the spam.

If spam blockers were developed and tuned so that the ISPs servicing spam senders could detect and block spam at its source, it would seem a more efficient (for the Internet system as a whole) and (consequentially speaking) more ethical solution than requiring users to buy spam blockers. Perhaps this is a direction that will eventually be explored. These kinds of spam blockers would, however, have a far more limited market than spam blockers for individual users.

A broader analysis may suggest that a more comprehensive, system-wide responsibility is critical. If the current system encourages spammers, perhaps the system as a whole should be modified. Although this would require considerable initial investments, such an approach might be less costly to most stakeholders after a few years. Some system-wide solutions to email spam are discussed in the following section.

21.5.3 The Ethics of Suggestions to Reduce the Number of Emails Sent

If the number of spam emails *sent* is reduced at the source, this helps the Internet as a whole, and it helps individuals who are less victimized by spam emails. Unfortunately, efforts to reduce the sending of spam emails have so far not resulted in anything like the elimination of spam emails.

21.5.3.1 Changing the Economics of Email This would likely change the behavior of ISPs. The current model has ISPs who email for spammers getting paid extra for large volumes of email, while ISPs who deliver email to individuals are not getting paid extra for delivering normal quantities. Currently, bandwidth is usually charged. An alternative would be in some way to charge a sender for each email sent. Other economic schemes include senders offering micropayments to recipients who open or respond to emails. All these economic schemes would have limited effectiveness unless all ISPs adopted them; otherwise, spammers would merely become customers of ISPs who didn't penalize bulk emails. However, ISPs who change their policies to discourage spammers do achieve an improved ethical posture for themselves, if not for the entire Internet.

21.5.3.2 Legislate Against the Sending of Spam This technique has been tried, and some prominent domestic spammers have been arrested. However, this has not eliminated the problem of spam email. One problem is that state or national laws, unlike the Internet, do not reach across political boundaries. Spammers can effectively avoid antispam laws. Another problem is the definitional problem we have discussed previously. Furthermore, when spammers spoof return addresses or use drones, this complicates prosecution of the true source of the spam emails. These problems are not insurmountable, though all three are difficult and would require worldwide cooperation. One positive aspect of antispam laws is that it forces consideration of important definitional issues, and it allows for political bodies to formalize a societal disapproval of spammers. This public "shaming" can have useful effects.

21.5.3.3 Require Authentication Before Email Is Delivered This would not address all of the characteristics that bother people about spam emails. However, it would address the problem of accountability. If effective, it could help legislation against some types of email spam to be enforced. This would require significant financial investment in revising Internet protocols and infrastructure, and there is no guarantee that the resulting system would be immune to technical subterfuge. The costs and benefits, therefore, will be difficult to predict accurately in any consequential analysis.

21.6 CONCLUSIONS

Email spam, as well as other types of spam, vexes electronic communications. An analogous problem, unsolicited commercial phone calls, have been significantly reduced in the United States through legislation, though not eliminated. But email spam has so far proven resistant to legislative and technological fixes. Part of this problem is definitional, and part of the problem is the fundamental open nature of Internet protocols and architectures.

Email spam is an example of a spoiling of the commons; email spammers exploit the economic model of Internet email and the relative openness of email protocols. NF-UCBE shifts the cost of advertising to email recipients, who must clear out advertisements for products they have no interest in. F-UCBE uses the Internet to defraud victims. The ethical case against F-UCBE is straightforward, and the people who are responsible for F-UCBE are condemned. An ethical analysis of NF-UCBE is more nuanced, since the economic incentives of current email arrangements make NF-UCBE attractive.

The struggle against unwanted emails will continue. Ethical analysis can be useful in analyzing emerging strategies of email senders and email recipients. In these analyses, ethicists must be careful to look at individual stakeholders as well as systematic stakeholders; both micro- and macroissues are important. All such analyses should start with a clear exposition of the characteristics of the emails that will be considered "spam." Only then can we discuss with precision the ethics of spam.

REFERENCES

Bitpipe.com (2007). Spim definition. Retrieved (June 20, 2007) from http://www.bitpipe.com/tlist/Spim.html

Detritus.og (2001). The Monty Python spam skit! Retrieved (June 20, 2007) from http://detritus.org/spam/skit.html

Everett-Church, R. (1999). The spam that started it all. *Wired*, April 13, 1999. Retrieved (June 20, 2007) from http://www.wired.com/news/politics/0,1283,19098,00.html

FTC (2007). Federal Trade Commission. The CAN-SPAM Act: requirements for commercial emailers. June 8, 2007. Retrieved (June 20, 2007) from http://www.ftc.gov/bcp/conline/pubs/buspubs/canspam.html

Hayes, D. (2006). Free-speecher quits net-abuse, labels Spamhaus internet terrorists. Retrieved (June 20, 2007) from http://www.emailbattles.com/2006/08/22/free-speecher-quits-net-abuse-labels-spamhaus-internet-terrorists/

Hoanca, B. (2006). How good are our weapons in the spam war? *IEEE Technology & Society*, 25(1), 22–30.

Internet2 (2007). Building tomorrow's Internet. Retrieved (June 20, 2007) from http://www.internet2.edu/

Kulawiec, R. (2004). 10 reasons why involving government in spam control is a bad idea. *Circle ID*. July 19, 2004. Retrieved (June 20, 2007) from http://www.circleid.com/posts/10_reasons_why_involving_government_in_spam_control_is_a_bad_idea/

Moor, J. (1999). Just consequentialism and computing. *Ethics and Information Technology*, 1, 65–69.

Northcutt, S. (2007). Spam and flooding. Retrieved (June 20, 2007) from http://www.sans.edu/resources/securitylab/spam_flooding.php

Scanlan, E. (2005). The fight to save America's inbox: state legislation and litigation in the wake of CAN-SPAM. *Shidler Journal of Law, Commerce & Technology*. December 16, 2005.

Retrieved (June 20, 2007) from http://www.lctjournal.washington.edu/Vol2/a012Scanlan.html#_Toc122503853

Spam Defined (2007). Retrieved (June 20, 2007) from http://www.monkeys.com/spam-defined/definition.shtml

Wikipedia (2007). Spam (electronic). Retrieved (June 20, 2007) from http://en.wikipedia.org/wiki/Spam_%28electronic%29

CHAPTER 22

The Matter of Plagiarism: What, Why, and If

JOHN SNAPPER

22.1 THE CONCEPT OF PLAGIARISM

As with most ethical concepts, there is plenty of room for debate over the definition of "plagiarism."[1] Plagiarism will be treated here very broadly as expression that *improperly* incorporates existing work either *without authorization* or *without documentation*, or both. The emphasis on impropriety is important. There are a wide variety of situations where it seems acceptable to repeat prior expressions while ignoring a possible attribution and making no attempt to seek permission from a putative source. We commonly repeat jokes and report established dates for historical events without citing sources, and we do so without qualms about plagiarism. An expression is only plagiarism if it is unacceptable on some established value. But we should be careful to avoid the error of being overly narrow in identifying any particular value or standard as the basis for condemning an expression as plagiarism. Among the reasons for finding an expression to be plagiarism, we may note that it is sometimes condemned as theft of intellectual property, sometimes as a failure to live up to a standard of originality, sometimes as a violation of the moral rights of a prior author, sometimes as fraudulent misrepresentation of authorship. A debate over whether an expression is plagiarism is, therefore, a debate over the standards for and values inherent in its condemnation. The present study is an overview of the variety of standards and values that underlie accusations of plagiarism, with an emphasis on how computer technology has changed the focus for those accusations. It should come as no surprise that accusations of

[1]The present paper profits from excellent philosophical commentary from Kenneth Himma. Prof. Himma's contributions are particularly valuable because we do indeed agree to disagree on many significant points.

The Handbook of Information and Computer Ethics, Edited by Kenneth Einar Himma and Herman T. Tavani

plagiarism are often based in a complex heap of intertwined and poorly understood values.

In the present discussion, we will further assume that plagiarism involves some form of deception, broadly construed. So we may modify the above definition to say that it improperly and *deceptively* incorporates the existing work. This approach is fairly common in the literature on plagiarism. Richard Posner writes, for instance, that "concealment is at the heart of plagiarism" (p. 17 of *The Little Book of Plagiarism*). This modification of the definition is, however, not obvious. There are at least two ways to explain the fact that accusations of plagiarism commonly involve deception. On one analysis, plagiarism is always to be condemned as the passing off of unoriginal work as one's own. This analysis draws attention to a particular set of values or standards that typically underlie a condemnation of plagiarism. Alternatively, however, the common appearance of deception in plagiarism may be viewed as a piece of the ordinary human desire to conceal all our improprieties, and not intrinsic to the concept of plagiarism. In any event, we should recognize that there are situations (for example, cases of conscientious objection) where generally accepted values are pointedly and publicly flouted. Not surprisingly, we can find cases where plagiarist-like expressions are openly declared. To put too much emphasis on concealment will put our use of the term at odds with its popular use to refer to openly declared "piracy," including openly declared copyright infringement. As an initial example, consider the disputes over Jeff Koons' 1988 "Banality Show." Koons, a star in today's art world, became embroiled in a still controversial legal dispute over his translations of kitschy postcard images, most famously the "String of Puppies," into whimsical three-dimensional sculptural works. Koons' sculptures succeed in part because they are recognized as "copies" of culturally accepted clichés. Koons made no attempt to conceal the source of his image. Even so, we may note the judge's comment that Koons' work does not "escape being sullied by the accusation of plagiarism." Since the "plagiarism" is not a legal term, the judge's usage here reflects his understanding of the ethical term in general use. He provides us with an example that at first look seems like a use of the term that is contrary to our definitional inclusion of an element of deception.

As an example more relevant to the digital controversies, consider the unauthorized web distribution of copyrighted films where both the distributor and the receiver understand that the item is pirated. There is no deception. When distributed from pirate centers in localities that by policy do not enforce international copyrights, there is moreover no attempt to conceal the free distribution from the copyright holder. Even though it is easy to find examples in the press where this sort of piracy is called "plagiarism," it seems only marginally to be plagiarism on a definition that includes an element of deception. Or at least the attempt to incorporate this sort of piracy into our discussion of deceptive plagiarism will demand explanation. Regardless of how we view open piracy, the practice is important for our discussion of the values that underlie the usual condemnation of plagiarism. Open piracy is often a frontal attack on some of those values and draws attention to the controversial nature of those values.

There can be questions about whether an expression is or is not deceptive. Let us contrast, for instance, a politician's presentation of a speech written by a "ghost" speech writer and a student's presentation to a teacher of an essay written by a "ghost" essayist. The former practice is so commonplace in many countries that most citizens just assume that their politicians always read texts prepared by ghostwriters. This is not deceptive and is not seen as plagiarism. And yet in the same countries, schools punish students who hire essayists to prepare "their" works. This is deceptive and is plagiarism. The difference is not immediately obvious. I have personally confronted students at my university who seem sincerely puzzled over why they are seen as plagiarists just because they hire ghostwriters to prepare their essays. Why, they ask, are they not permitted to partake in this common practice? Indeed, both the politician and the student may be the legal owners of "their" works since, under copyright law, a work for hire belongs to the employer, not to the author. The example shows that an accusation of deception can based on the local conventions. It has become, for reasons that are not at all clear, a tradition in academia that named authors have at some level crafted the works that bear their names, although not so in politics. What is a standard nondeceptive practice among politicians is deceptive plagiarism among academics. A student who makes use of ghost essayists may be accused of plagiarism even if that student freely admits the practice because the academic context defines ghostwriting as a deception. In my university at least, students are expected to know this definition, and failure to appreciate the definition is generally not accepted as an excuse. The definition of deception is relative to context or to professional role in these sorts of cases. (We might note that even within academia, a certain level of ghosting is common practice and not condemned as plagiarism. One obvious example is the common assumption that law professors have unacknowledged assistants to prepare the legal notes for their works, even in cases where the legal notes are by far the greater portion of a manuscript.)

It is important to distinguish between issues of authorization and documentation, for these are two sorts of wrongs. Failure to have authorization is typically theft of intellectual property, most commonly a copyright infringement that deprives a copyright owner of income. This wrong is the focus, for instance, of lawsuits over the unauthorized distribution of music over the web. Although we must keep in mind that the present study of plagiarism is a study of an ethical, not a legal concept, a focus on authorization naturally draws on discussions of the values that underlie the legal criteria for copyright infringement. In contrast, failure to document is the focus of scholars who complain about passages downloaded from the web and incorporated into papers without citations. When there is no copyright infringement, this wrong is rarely the focus of a law case. A study of the failure to acknowledge sources pulls us away from the law into a study of cultural mores. The present chapter addresses the ethical bases for both wrongs, and how our sense of these wrongs has been affected by the new forms of expression that are inherent in computer technologies.

Some examples should make clear the distinction between failure to receive authorization and failure to document. As an example of a failure to document that is not a failure to receive authorization, consider my attempt to pass off an obscure and anonymous nineteenth century poem as a new work of my own. The point here is that there can be no failure to receive authorization because there is no one with a right to authorize. Since in most cases the right to authorize is based on a copyright and the copyright on a nineteenth century poem would have expired, not even a publisher could claim a right to authorize in this case. All the same, the example illustrates a case of plagiarism. A more likely modern example bearing on web usage involves the distribution of illicit pornographic images. If I distribute such images claiming them to be my own work, then I am plagiarizing. Since illicit work cannot be copyrighted, there can be no question of copyright infringement in this form of plagiarism. There is no one with a legal right to authorize distribution, and thus there can be no failure to seek authorization.

As an example of a failure to receive authorization that is not a failure to document, let us return to the question of pirated films. Pirated films typically include the credits, right down to the copyright notices. All the documentation that is expected in scholarly work is present, but the copy is nonetheless a case of copyright infringement. If the pirate intentionally conceals the lack of authorization, presenting the work as a legally authorized copy, then the work is plagiarized by our definition. Popular press discussions of "plagiarism" usually address this sort of case, which is deception over authorization, not over documentation. For present purposes, we may build up the case with an additional element of deception. Consider the common practice of providing unauthorized subtitles on pirated foreign language films. Since subtitling is a right of the copyright holder, this is an additional element of copyright infringement. The subtitles might easily also be a case of both deceptive authorization and deceptive documentation, on the assumption that a purchaser reasonably assumes that the subtitles are the work of the legal distributor, even if both the purchaser and seller understand that the item is otherwise an unauthorized, pirated copy.

We should also note that plagiarism can be unintentional, both when there is a failure to authorize and when there is a failure to document. As an example of an unintentional failure to authorize, let us consider a source of streaming popular music, where each piece is fully documented. If the distributor generally meets all royalty demands, but by oversight forgets royalties on one tune, we have a fairly clear case of unintentional plagiarism. Even if unintentional, there is deception since most parties to the distribution will be under the impression that the distribution is authorized.

The two sorts of plagiarism are based on two sorts of ethical and cultural concerns. Discussions of authorization are generally (though not always) focused on economic issues. In that case, the wrong is theft or piracy of intellectual property that has monetary value, and the redress is generally monetary compensation. The persons most worried about how the computer environment has aggravated plagiarism viewed as theft are corporations or individuals with a serious economic stake in material that is more or less easily digitally reproduced and distributed. In contrast, discussions of documentation usually appear in the context of scholarly or journalistic work. The wrong is failure to inform the user, reader, or audience of the source of the work. Here the injured party is not the owner, but the receiver of the plagiarized material who cannot identify the source. The harms are less a matter of economic loss than of scholarly deprivation of important information. The redress, if any, is usually dismissal from a scholarly or professional position. Needless to say, there are plenty of examples of plagiarism that involve both unauthorized theft and undocumented presentation.

22.2 LACK OF AUTHORIZATION—ECONOMIC FOUNDATIONS

Since most legal disputes over plagiarism concern cases of unauthorized copying that infringe a copyright, it is no surprise that this form of plagiarism has received the most attention in books, articles, and the press. We will refer to plagiarism that includes a copyright issue as "infringing plagiarism," even while we attempt a more general interest in the ethical ideals that underlie the claim that certain forms of expression demand permission from some recognized authority. The present discussion distinguishes between two theoretical bases for treating a failure to receive authorization as wrongdoing. The theoretical foundation for copyrights in the Anglo-American tradition is consequentialist, typically today in the context of an economic analysis of the business environment created by copyright law. In contrast, continental European theory, especially in the French tradition, includes recognition of the natural, moral rights of the copyright holder. We begin with the Anglo-American tradition.

Copyright law (and intellectual property law in general, including patents and trade secrets) is justified in the Anglo-American tradition as a legal tool for encouraging progress in science and technology (and the arts). American copyright law, for instance, is based on Article 1, Section 8 of the U.S. Constitution according to which copyrights are to "promote the progress of science and the useful arts." That this aim is noted in the foundation of U.S. copyright law is not to say, of course, that U.S. copyright law is restricted to that aim. For instance, the language of the Constitution tacitly appeals to a traditional distinction between the useful arts (such as the art of computer programming) and the fine arts (such as the art of music composition), and the Constitution identifies copyright policy with an interest in the applied arts. But that historical reading of the Constitution is inconsistent with the legal tradition. Copyright law is used to promote both the useful arts and the fine arts, and sometimes seems rather far removed from the attempt to encourage either art. The recent extension of the term for copyrights from 75 to 100 years was largely at the request of the Disney Corporation, which seeks to keep control of its classic films. This extension is entirely in keeping with congressional practice on copyright legislation, although it seemingly shows little consideration for the encouragement of any artistic practice. In practice, those working on legal actions relating to copyright law should consult a good lawyer or a good lobbyist, and not study the theoretical basis for the notion of plagiarism. The present discussion, however, emphasizes social and ethical theory of plagiarism, which insofar as it involves copyright issues is now seen in the United States as largely an economic study of how the law encourages or discourages some valued practices. On this theory, infringing plagiarism is to be condemned for how it harms the environment for the publication or distribution of copyrighted material.

On a standard economic approach, each feature of copyright policy is justified insofar as it maximizes wealth under some measure. In the simplest circumstance, this may be seen as maximizing economic growth. The present author prefers to focus more specifically on the extent to which the law encourages the publication of new expression, whether that expression is commercial software or popular music. The starting point for an understanding of this consequentialist approach to plagiarism must be a rejection of the naive (and surprisingly common) preoccupation with a supposed trade-off between incentives to produce new expressions and access to existing expressions. On the naive analysis, a fair use policy that gives open access to patches for security flaws in existing software would be assessed in terms of how that policy encourages user access to software patches but diminishes the incentive to write patches. On the naive analysis, infringing plagiarism of software patches is condemned to the degree that it interferes with the economic incentive to produce software patches. The problem with a purely access/incentive analysis of plagiarism is that it ignores the full range of economic issues and theories for how a copyright standard might affect the production of expressions for the general good. Certainly access/incentive issues are worthy of consideration, but they are only one piece of the picture. Even the traditional justification of copyright protections for printed material does not focus on the incentives to authors to express, so much as on the health of the industry that publishes those expressions. In the popular mind, publishers may have been seen as evil profiteers claiming profits that should by right go to the creative novelists or song writers who are forbidden to even copy their own works. But on an economic approach, the primary harm of infringing plagiarism is to the publishing industry that provides the business background for creative arts, and only very indirectly shows any concern for novelists or song writers. Similarly today, most software production is "work for hire" where the copyright is held by corporations that employ the programmers (or authors), and the real issue is how alternative standards of copyright infringement affect that economic system.

There is a range of economic theories that seek to understand the consequences of a broader or narrower definition of copyright infringement, and narrower or broader definitions of fair use, and tighter or looser plagiarism policies. It is hardly within the scope of the present chapter to review those theories, other than to warn against a simple focus on incentives. More sophisticated analyses may note the advantages of enriching corporations with a past history of producing popular software, the economic effects of higher or lower transaction costs for seeking copyright permissions, the advantage of the legal documentation system that has sprung up to support the copyright system, etc. For our purposes here, it is important to understand plagiarism as a practice that is condemned because it interferes with the economic goods that justify the copyright system. And, therefore, we may justify a practice of making unauthorized copies by showing that the economic consequences of unauthorized copying are actually good, rather than harmful, for the industries in question. One or another version of this argument has started to gain credence in the recent literature. In particular, we may note the achievements of the "open source" software community, and we may note the viability of pirate organizations that ignore copyright policies. In both cases, we see an explicit acceptance of practices that might be branded as plagiarism in traditional markets. In both cases, we see an argument that the concept of infringing plagiarism is based on an economic misunderstanding.

The proponents of open source software promote the distribution of software (including the popular Fire Fox web browser and the GNU components of the LINUX operating system) without any demand for formal authorization from the distributors. The highly charged rhetoric of the "open source movement" somewhat misleadingly says that it distributes open source software under "nonproprietary" contracts that (with some variations in details) make the software available for modification and further free distribution. The rhetoric is misleading because the so-called "nonproprietary" contracts depend on the validity of the copyrights that attach to almost all software creations by default on present intellectual property policy. It is more accurate to say that the authors of open source software renounce a portion of their proprietary rights, including any demand for formal authorization of copies, while preserving other proprietary rights such as the right to limit some commercial uses. Since the open source movement permits exact copying without authorization, there can be no such thing as that sort of infringing plagiarism that is characterized here as a failure to seek authorization. We will see later that the open source movement will complain about the sort of plagiarism that is characterized here as a failure to acknowledge.

The primary proponents of open source software argue that the elimination of the transaction costs that are inherent in demands for authorization will lead to the creation of more high-quality software than is produced under a system that treats a failure to receive authorization as plagiarism. The open source movement notes that the present copyright system often deprives competent software writers of the opportunity to personalize or to make improvements to commercial software. The obstacles are inherent, they argue, in a copyright tradition that intentionally hides its algorithms under precompiled object code and that moreover encourages software corporations to refuse to even listen to suggestions for code modifications (under the very realistic fear that any acceptance of any such suggestions would compromise their copyright claims). The result is a huge transaction cost to the implementation of even small modifications or incremental improvements to commercial software. A familiar anecdote tells us that Richard Stallman, perhaps the best-known proponent of the open source movement, entered the movement over frustration with clumsy software drivers for his office printer that, under proprietary conditions, could not be amended without huge transaction costs. Although the actual programming fix would have been trivial, the software drivers were provided in unreadable object code that could only be amended through the vendor. The argument is, in effect, that transaction costs are a more significant interference with the production of software than any encouragement for software production that may be gained from the copyright system that encourages the distribution of closed source software. The best evidence for the claim is basically the success of the open source software movement in producing and distributing high-quality software. The open source web browser is, for instance, very popular and (in the minds of many literate computer users) superior to the well-known closed source alternatives. There are two important features to be noted about this argument. In the first place, the focus is on transaction costs, not on incentive or access issues as imagined in a simplistic economic analysis. Second, the evidence is based on direct observation of the market reactions to open source and closed source software, without a deeper analysis of how copyright standards may or may not encourage the publication of closed source web browsers.

As a second example of an economic picture that is at odds with the standard economic condemnation of infringing plagiarism, let us consider the pirate market for films and for music that is presently centered in Asia, notably in India and China. Although these countries officially recognize international copyrights, there is an industry (largely based in Mumbai) that flourishes through the distribution of pirated works. Since, as noted above, there is often no deception about these practices, they are best viewed as open nonplagiarizing copyright infringement on the present definition. Our interest in the practices is, however, the insight they provide into the theoretical justification for condemning plagiarism as a wrong. In a long tradition of pirate romanticism, these communities see themselves as promoting freedom, as fighting for free expression in the face of rules designed merely to enrich corporate profiteers. They would claim that their practice of ignoring authorization leads to a lively market in which there is more expression than that available through the markets governed by plagiarism standards. And they, therefore, provide a test case for the claim that we promote expression through the recognition of the wrong of plagiarism. Interestingly, the evidence is ambiguous. It appears that there are presently works (including the film classics that appeal to a small market of film buffs) available through the pirate centers that are not available through proprietary publishers. On the contrary, there are few examples of high-cost, high-quality film production in those areas. The new productions tend to be low cost.

As a second example of a consequentialist analysis of plagiarism, let us consider the much-debated standards for sharing music held in digital files. From a consumer point of view, a digital work has greater value if you can do more with it. Within limits, file sharing obviously enhances the value of a copyrighted music file. To take a very elementary example, I am more likely to buy a music disk at a high cost if I can, for no additional cost, copy its contents to a portable mp3 player and to my home system, while retaining the original disk to play on my car's disk player. I expect to pay a lower price if I need to buy duplicate copies for each use. From the consumer point of view, it appears that an overly restrictive plagiarism standard can diminish, rather than enhance, the value of digital music recordings. If we were to judge plagiarism standards with an eye to encouraging the sale of music on disks, this argument from the consumer viewpoint suggests that we should permit more copying and lower the plagiarism barrier. This argument is, of course, an oversimplification. We may get a different conclusion if we focus on the need to meet the initial costs of symphonic music that is relatively expensive to produce. The point is that it is hard to find the right level of plagiarism standards to determine the limits of copying music recordings for multiple uses. In this context, we may note that the explosion of creative hip-hop music of the 1990s was largely dependent on the free use of digitally reproduced samples in ways that might now be seen as plagiarism, but which at that time were accepted without any stigma of wrongdoing. The hip-hop music scene made maximum use of the newly available technology, incorporating samples taken (plagiarized?) from existing recordings. On a consequentialist theory, the creativity boom in the wide-open hip-hop music scene is one tiny piece of evidence that may be used for assessing standards for plagiarism and for fair use copying in the newly digitized music industry. But in general, we should be very hesitant to draw conclusions from this sort of evidence. In software, both the open source community and the traditional commercial producers can point with pride to their software achievements. In film, we may note the high-level productions from the European and American film industries that demand authorization for any sort of copies, while at the same time noting that classic films unavailable in the United States are on the shelves in the pirate centers of Mumbai.

The present examples draw attention to how computer technology leads to reassessments of the concept of plagiarism. Intrinsic to the music example is the possibility of a technology for at-home, simple copying of digital music files, and the suggestion that some amount of copying at that level should be permitted as a means to encourage the production of high-quality music. If that at-home technology is not available, as for instance, it was not available for traditional hard copy books or for music manuscripts a century ago, then there is no value added in a loose plagiarism standard. When almost all copying is commercially based piracy by competing producers, an economic analysis provides good reasons for a stringent plagiarism standard. We may also note that, even in the present market, the technology of digital production suggests that there might be different standards for different forms of digital expression. Thus, the movie industry that produces works at extremely high costs might demand a different order of protection than the music industry that produces works at relatively very low costs. The difficulty, however, is that the legal cost of attempting to write industry-specific (or "sui generis") plagiarism standards is itself prohibitive. Just consider, for instance, the legal expenses involved in extending a sui generis standard to the international context which has no prior tradition with that standard. Since an honest economic analysis must include the cost of legal resolution of disputes, there is every reason to continue with a single standard for hard copy books and soft copy software, even though the result is not ideal in either context.

22.3 LACK OF AUTHORIZATION—NATURAL OR MORAL RIGHTS

The title of the present section refers, perhaps misleadingly, to "natural" rights, a concept usually associated with Locke's theory that a laborer has a claim to his or her labor, and consequently to items with which that labor is mixed. Although the Lockean version of natural rights theory is not popular among jurists, there is a natural rights approach to intellectual property taken from a European (largely French) tradition that is very important for current practice. In contrast to Anglo-American copyright tradition that evolves from consequentialist concerns, the French copyright tradition is associated with a Hegelian claim that an author's personhood is inherently dependent upon the author's creations. A much-quoted passage in Hegel informs us that a person has "the right of placing its will in any and every thing, which thing is thereby mine." Mankind, he tells us, has the "absolute right of appropriation" of those items whose souls "take on my will." In intellectual property jargon, the technical term "moral rights" refers to a particular list of rights justified in French copyright tradition on a romantic, Hegelian picture of an artist whose identity (and by somewhat dubious extension, even personality, personhood, or essence) are inherent in expressive works. On a moral rights theory, copyright protects the individual's natural right to his own person, a right enforced through control over personal expressions. The harm of plagiarism discussed above is harm to the publishing industry that provides the environment for publication and expression. In contrast, a moral rights theory will see harm done on a very personal level to the very identity of the victim. As a first approximation, we may see the harm as similar to the harm of slander that injures a person's reputation and sense of self, even when there is no economic loss. (Traditional Stoic philosophy would teach us that one's sense of one's self is not impaired by the publication of a lie, but I regret that the picture of a personal sense of one's self harmed by plagiarism and misrepresentation may be more true to the world.)

We may illustrate the difference between personality harms and economic harms with a glance at a couple of well-known legal disputes. The family of John Huston, speaking on behalf of the deceased film director, objected to the television presentation of a colorized version of *The Maltese Falcon.* The television broadcaster owned through purchase the U.S.-based copyright to the film. There could, therefore, be no question of any further economic compensation to the Huston estate. In fact, the option of colorization adds to the purchase value of the film's copyright, and so from a purely economic point of view a protest over colorization is counterproductive. (The point is not much different from the observation above that a music recording gains value to a purchaser when a purchaser is allowed to use it in more ways.) Under U.S. law, with its focus on economic harms, Huston had no grounds for protest. But under French law, there is the additional question of whether the publication enhances or harms the author's sense of his works, and his sense of himself as encapsulated in his works. The French courts decided for Huston, recognizing his right to oversee the "integrity" of his work. A second, well-known example is Shostakovich's objection to the use of one of his uncopyrighted music scores as background music in a film that portrayed his home country in an unpleasant light. Lacking any copyright basis for his complaint, Shostakovich's protest failed in the United States but succeeded in France, where his right to control his creation was recognized independently of the economic issues.

It is important to note that although there is a significant shift in the nature of the argument as we move from the U.S. debate over the economic structure of the publishing industry to a French debate over the personal rights of the individual victim, the issue remains unauthorized expression. Huston and Shostakovich claim authority to determine how their works may be presented. They seek redress for the wrong of making copies contrary to that authority. Huston objects to unauthorized revisions of the work. Shostakovich objects to the context in which the copies appear. In the jargon of moral rights theory, these are both attempts to preserve the "integrity" of the work. In addition to this right to integrity, moral right theory is generally said to promote the following three rights: the right to demand attribution (which we will discuss below as a right to be acknowledged), the right to withhold a work entirely from disclosure, and the right to retrieve a work (with compensation) from its owner. To this list, we may add that there is some discussion of a right of access to a work. These rights are not identical, and we will see that they are separable in both theory and practice. We may philosophically debate the conception of authorial control that underlies the French recognition of plagiarism in these cases, but we must recognize that the debate is over the scope of what counts as plagiarism and the philosophical basis of the protest against plagiarism.

Intuitively, the romantic picture of the author whose personality lies in his works applies more to protections for the creative arts than to protections for works of applied arts, such as computer software. This is certainly the view taken by the U.S. policy that is now challenged by the need to bring U.S. copyright policy into agreement with international copyright policy that recognizes moral rights. As a step toward reaching that accord, the United States enacted the 1990 Visual Artists Rights Act (VARA), which recognizes a modified version of moral rights in works of creative, fine art. Given this focus on fine art, we should expect that the first place to look for moral rights controversies over plagiarism in computer technology is the use of that technology in our new arts programs. Consider, for instance, the recent debate over the unauthorized use of music and film clips in an art form that now goes under the popular name "mash-up." This is a digital form of collage that typically superimposes film clips and music clips in surprisingly enjoyable ways. A fine example that the present writer finds particularly enjoyable is a humorous use of clips from the popular 2005 movie *March of the Penguins* to create a trailer for a Hitchcock-like thriller.

A mash-up that has recently inspired considerable debate over the values inherent in the demand for authorization is *Grey Album*, in which Danger Mouse superimposes words from Jay-Z's *Black Album* on the notes from the Beatles *White Album*. The owners of the *White Album* copyright called an end to the distribution of *Grey Album* by simply drawing attention to their legal right to authorize such use of the music, without offering any philosophical justification for their decision to block the use of the music. The philosophically interesting issue of the justification for this authority, however, appeared in the public debate over whether a copyright owner should have this authority. *Wired Magazine*, for instance, quotes Danger Mouse himself as saying, "I'm just worried . . . whether Paul and Ringo will like it. If they say that they hate it, and that I messed up their music, I think I'll put my tail between my legs and go." *Wired* then goes on to quote a leading alternative music advocate: "all kinds of artists have always borrowed and built on each others' work . . . these corporations have outlawed an art form." In a nutshell, *Wired* has captured two sides of the debate over a moral right to control artistic works through the authorization of copies.

On the one hand, Danger Mouse recognizes an artist's right to integrity, such as pressed by Shostakovich. What is distinctive about this right in the French copyright tradition (in contrast to earlier U.S. copyright traditions) is that the right is nontransferable. Whereas copyrights in the U.S. tradition are bought and sold, only the original artist (or his descendents speaking on his behalf) can enforce that right. This attitude is intrinsically recognized by Danger Mouse's respect for Ringo's likes and his disdain for any others who may happen to own the *White Album* copyright. On the other side of the debate, we see the claim that creative work in the arts depends on the opportunity for artists to borrow and build on each other's work. On this view, any recognition of a right to integrity is simply a mistake that interferes with the creative arts. In the digital context, to build on another's works is often the incorporation of bits taken from existing works. The argument has been made that the arts are stifled when such action is condemned as plagiarism, and the solution is to abandon the notion of plagiarism for the creative arts. The new electronic arts are replete with examples of works that incorporate digital copies of prior works and of cases where the electronic arts have seemingly been impeded by stringent plagiarism standards.

The attack on plagiarism is the focus of recent work by Sherrie Levine, a central figure in avant-garde photography. Levine is mostly associated with her photographs of photographs made by other photographers, and her presentation of these copies as her own work in the 1980s. Her *Meltdown* series from the 1990s is more relevant to the digital arts. A meltdown is a digital manipulation of a prior image from an established artist. In a meltdown, Levine averages the color tones of the "original" work within a crude grid. The darker grid elements correspond to darker areas in the original. Although the result is not immediately recognizable as based on any particular work, the trace to that prior work is essential to appreciation of the meltdown. Since, as with the Jeff Koons' work considered above, there is no deception about the source, these works may not be plagiarism in our present definition. Still, Sherrie Levine's problematic claim to have made new work in this way is essential to a study of plagiarism. By attacking the values inherent in the concept of plagiarism, she draws attention to the contestable nature of those values. The artistic content of her works is inseparable from a theoretic attack on the romantic Hegelian notion of creativity that underlies the moral rights justification of plagiarism standards. The consequence is a very broad, theoretical rejection of the traditional conception of plagiarism in the fine arts. Her meltdown of Marcel Duchamp's *LHOOQ* is perhaps the classic example, since Duchamp's work itself is just a slight alteration to a reproduction of a prior painting. If nothing else, her work argues that copying is central to the visual arts and that artistic claims to originality are deeply contestable.

Turning away from the fine arts, we may note that some version of moral rights has also been debated within the open source software community. Open source software is generally distributed under an open source copyright contract (that many proponents of the open source community prefer to call a "copyleft" contract, since the contract disclaims the authorization rights that are seen as central to the traditional copyright). A particularly popular version is the GNU General Public License promoted by the Free Software Foundation (see the literature review below). As part of this contract, there are no constraints on the use of the software. This portion of the contract bears on the notion of a right to integrity, such as claimed by Shostakovich. Since in U.S. law, integrity rights under VARA only apply to certain fine arts, the holder of a software copyright would only have a right to restrict the use of software if that right were written into the copyleft contract. The issue for the open source community is whether a copyleft contract may specify, for instance, that the software may not be used in the production of deadly weapons, the creation of sadistic pornography, or some such things. This issue has been a serious debate within the open source community. Interestingly, the debate in this context has less to do with Hegelian theory of personality than with a general conception of free expression. Software programmers do not generally think of "their" software as an extension of their persons. The issue has focused on a general conception of freedom, often with explicit analogy to the conception of "free speech." The fear is that a programmer's right to limit a program's uses would lead to a restrictive society where freedom would be at risk. In March 2007, the standard version of a GNU General Public License disclaims any integrity rights.

(As a personal anecdote, let me report that during a recent conference on the validity of the integrity clause in the open source contracts, I was surprised to hear some European programmers say that they did indeed think that their work was an extension of their personality. The U.S. programmers did not agree. Without any real evidence of a survey, I suspect that the Europeans at this meeting were atypical.)

With respect to a moral right to integrity, we find a complex situation. In the fine arts, Danger Mouse will agree that it is plagiarism to digitally incorporate musical samples against the artistic wishes of the original musician. In contrast, Sherrie Levine refuses to recognize any issue of plagiarism when there is a creative use of prior material. In the software industry, the open source community has decided with Levine that it is not plagiarism to use software in ways that meet with disapproval from the original programmer. There are striking differences in the underlying philosophical approach to the issue. At issue in the debate over digital sampling in the arts is a romantic notion of originality. The question is whether originality merits recognition, and the presumed harm of plagiarism is harm to the originator. If Sherrie Levine refuses to admit that there is plagiarism, it is because she refuses to recognize the romance of originality. In the debate within the open source community, the issue is the balance between the opportunity to prevent "evil" uses of software and the desire to promote free choice in an open society. If the open source community does not recognize a programmer's moral right to integrity, it is because the recognition of that right seems incompatible with its conception of social freedom.

22.4 LACK OF ACCREDITATION—NONINFRINGING PLAGIARISM

Even where there is no question of authorization to make copies, academics tend to view unaccredited copies as a scholarly wrong. Although plagiarizing failure to accredit is most often combined with copyright infringement, there are also many ways to commit the wrong of failing to accredit where there is no infringement. I might copy without accreditation from works in the public domain, such as a Supreme Court decision. I might copy without accreditation from my own earlier publications on which I hold copyrights. I might copy without accreditation from a classical source where the term of copyright protection has long expired. I might present works for hire as works that I personally crafted. This form of plagiarism has received less attention than copyright infringement in the legal literature, since there is rarely any person who has standing to initiate legal procedures or to claim compensation for either economic or personality harm. An economic analysis might suggest that there is no reason to forbid plagiarism in this sort of case, since the originals (for example, Supreme Court decisions) are likely to be produced and distributed regardless of protections against such plagiarism. A personality analysis will see no unauthorized attack on my person if I am myself copying from myself. Yet this form of plagiarism is often seen as the paradigm of the wrong of plagiarism. It is a case, most typically, of an individual (deceptively) claiming originality for expressions that are not original.

When a failure to accredit is also infringement, the accusation of plagiarism will most likely focus on the lack of proper authorization, since this focus lets us easily identify the harm as harm to the copyright holder. All the same, we can try to distinguish the harm of failure to authorize and the harm of failure to accredit, even when these are combined in a single act. The latter wrong is related to the wrong of fraudulent misrepresentation. The wrong in this form of plagiarism seems to be harm to the receiver or reader of the fraudulently represented work. There are several ways to conceptualize this harm to the reader of unaccredited work. A reader may be unable to validate a claim through a back-trace to its authoritative or nonauthoritative source. A reader may depend on the inferior expertise of the plagiarizer rather than the greater expertise of the originator. A reader may be left unaware of significant materials for further scholarly study. A reader may be put at risk for infringing plagiarism by the unintentional further publication of a work without the authorization of the true holder of a copyright. It is interesting that the opportunity for these sort of harms is aggravated by the present focus on electronic sources for research, particularly by the ever-popular use of the web as a handy reference tool.

The clear separation between the values of authorization and the values of acknowledgment may be seen by a return, once again, to the values of the open source software community. The open source proponents typically insist on acknowledgment of a software source as they at the same time make no demand for authorization. Thus, the Free Software Foundation Web site (on March 1, 2007) discussion of revisions to the GPL copyleft contract suggests that contracts may "require that the origin of the material they cover should not be misrepresented or that modified versions of that material be marked in specific reasonable ways as different from the original version." The Free Software Foundation is adamant that there is no demand for authorization in software use. But a failure to acknowledge the original source with care may be condemned. The word "plagiarize" might be left out of the legalistic discussion, for good legal reasons, but the sense of violation over a failure to identify sources is clear.

The open source position makes sense when we note the harms that are caused by a plagiarizing failure to identify software sources. In part, the harms are intrinsic to the software technology itself. Under present conditions, software tends to be fluid, going through frequent modifications, updates, and versions. As noted above, the open source community makes a point of the opportunity for easy modification without transaction costs. As a result, some well-known open source products go through faster, more fluid upgrading modifications than the software versions that are released as closed source software. This opportunity for fluid modification is the prime bragging point for the open source community. But frequent modification is both an advantage to those who like to see bugs removed and programs improved and a disadvantage to users who constantly find their applications outdated. The very technology thus demands support services that must be able to identify the canonical form of each user's software. Solid acknowledgment of modifications and citation to sources is thus essential to the industry. These practical considerations entail a condemnation of failure to site sources and identify modifications. The technology itself creates a heightened concern for the values inherent in the condemnation of plagiarism. The very advantages that the open source community sees to its elimination of the concept of nonauthorizing plagiarism aggravate the possibility of harm caused by nonacknowledging plagiarism. They entail a condemnation of plagiarism.

We may note that the use of the web as a tool for scholarly research makes plagiarism a particularly sensitive issue in scholarship, even to the point of creating new standards for citation. The problem is that web-based sources of information tend to disappear, leaving the researcher with no way to justify a claim. For instance, not long ago, I had through an honest mistake, posted a wrong birth date on my web-based professional vita. Anyone who based a comment on that now-nonexistent source might find himself accused of poor scholarship, since there is no way to confirm that there ever was an authoritative source for that wrong date. We may see this concern reflected in a number of new standards for proper scholarly acknowledgment. We now set it as a standard that any web reference includes the date at which the reference was confirmed. If inadequate citation is plagiarism, then it is plagiarism to fail to include that date. I hereby claim that the web references below were all confirmed on March 1, 2007. If I am being deceptive in this claim, then I am not living up to a newly established standard. It is not entirely clear that the word "plagiarism" is not expanding to include this new sort of falsified citation. We may have to wait and see how the concept of "plagiarism" evolves with the new standards for web citation, but in any event we can see in these sorts of concerns how the new technology bears on our conception of citation and plagiarism. We may also note that the same concerns have led to serious web reference sources to institute a policy for archiving Web sites. The popular web-based encyclopedia Wikipedia provides links from each article to earlier dated versions of the article. Given this archival standard, we might argue that it is not infringing plagiarism when a scholar referring to a web-based source makes an archival copy of that source without authorization to copy. This is to say that an evolving standard for proper accreditation can influence a policy on fair copying. We should never be surprised to learn that our conception of plagiarism is changing in interesting ways.

A particularly important form of noninfringing plagiarism concerns the use of ideas when there is no copy of the form of expression. Intrinsic to the theory of copyright law is the claim that copyrights may not impede the free expression of ideas that, in the United States anyway, are seen as central to an extremely important right to "free speech." To take a very recent example, the author of the popular novel *The Da Vinci Code* was sued under copyright law for taking without authorization the plot idea that Jesus Christ fathered children by Mary Magdalene. But to claim authority to authorize use of this "idea" would restrict that free discussion of the idea that is seen as a central value of free society. Discussion of this idea should not be a copyright infringement. But it might be scholarly plagiarism all the same. You, the reader of this passage, might wish to study the concepts of free speech raised by this example. You may wish to know where I found the example and to verify the example. Good scholarship thus suggests that I refer you to my sources, and failure to do so might be seen as plagiarism, particularly if I leave you with the impression that this discussion is my own original contribution to plagiarism scholarship, concealing an original source. It is a tricky question to decide when an idea of this sort is so distinctive that a reference to the source is demanded. In fact, I doubt that is demanded here. The basic point is commonplace, discussed in almost any book on copyrights or on plagiarism. (If you insist, you can find a prior discussion in Richard Posner's *The Little Book of Plagiarism*, p. 13. But Posner agrees that he too need not give any citation to the source of this commonplace example. You will find no further citations, no aids to further scholarship, in his book.) On the contrary, I do believe that my remarks above on the viability of pirate communities and pirate markets do deserve further reference to the source of this important new research. You will find references below in the literature review.

Again, we must distinguish between the harms of failure to receive authorization to express an idea in a distinctive form and the harms of failure to cite the source for an idea. We could imagine a legally recognized tort of failing to cite, even when there is no failure to seek authorization. We could imagine that the author of *The Da Vinci Code* were to sue under this new tort. Such a tort would not be seen as impeding open discussion because there would be no demand for prior restraint (that is, prior authorization) that could impede discussion of ideas. But we do not have a legal tort of plagiarism of this sort. Obviously, the harms are not seen at present as sufficiently serious to merit legal recognition of the plagiarism concept of nonaccreditation. And yet journalists are fired for taking stories from rivals, and students are dismissed from universities for paraphrasing without citing. This intermediate level of condemnation only makes sense if we postulate a low level of harm, enough harm to justify losing a job, but not enough to justify legal penalty. As such, nonacknowledging plagiarism in the scholarly context is an intrinsically puzzling notion.

It is tempting to see the wrong of idea plagiarism as a form of cheating in a competitive context. The plagiarizing journalist who takes a competitor's story might gather undeserved honors in a competitive profession. The plagiarizing student who paraphrases an unacknowledged text has taken an undeserved grading advantage over other students competing for good grades. The harm is then to the competitors. An interesting example concerns a recent submission of a proposal to a federal funding agency. On present practice, these submissions are sent for review to academics in relevant fields. In the case in question, the reviewer first rejected the proposal and then submitted a rewritten version of the proposal (that took the ideas without the form of expression) to an alternative government funding agency. There is certainly an element of plagiarism in this story, and it is plagiarism that takes an unfair advantage. The harm of unfair advantage may be a component of plagiarism, but it cannot be adequate in itself to justify the condemnation of plagiarism. Cheating as such is seen as a distinct form of academic dishonesty. On an informal survey, I have confirmed that students who sneak texts into closed-book exams are typically accused of "cheating," but not of "plagiarism." This is certainly the use of unaccredited sources without citation. But, at least at my university, the wrong here is seen as taking an unfair advantage, distinct from the wrong of plagiarism per se. And our indignation over the dishonest reviewer who stole a funding proposal is only partly a reaction to his plagiarism. The case seems particularly reprehensible because it also involves dishonest interference with the original proposal. The reviewer's cheating was probably worse than his plagiarism.

It seems more likely that the wrong of idea plagiarism is the wrong of hiding material information. We noted above two forms of this wrong: the reader may depend on the inferior expertise of the plagiarizer rather than the greater expertise of the originator. (In legalese, this is a form of detrimental reliance. It is detrimental to the extent that the reader relies on a copy that is inferior to the orginal.) Alternatively, the reader may be left unaware of significant materials for further scholarly study. These are closely related, but not identical wrongs. It is not unusual that the plagiarized text is a superior source to the original. In this case, the reader is still deprived of the opportunity to make that comparison and perhaps to trace a scholarly idea. Note that neither wrong applies to the student who cheats by using a text during a closed-book test, since the test reader is presumably fully aware of the existence of the text. A focus on the wrongs related to the loss of information helps us to understand the distinct wrongs of idea plagiarism.

The standards for idea plagiarism have a history. It is a common observation that Shakespeare took many of his plots from sources without citation, creating a lovely area of research for today's Shakespeare scholars who seek to fill in the missing traces. Today's standards would condemn these borrowings as not merely idea plagiarism, but moreover as infringing plagiarism. Shakespeare's practice, however, seems to have been perfectly acceptable in Elizabethan England. The standards have changed. We can now speculate on how the digital context for expression might engender further changes to the standards for idea plagiarism. We may note, for instance, that the scientific community is much enlarged and that the pace of research is much accelerated by digital publication. Whereas a mere lifetime ago, scientific communities tended to be concentrated in a few major centers, with researchers who were directly acquainted with each other, we now see research centers spread across the world, informing each other through web-based communication and publication. Digital databases and citation indexes have become central to professional work. It is reasonable to speculate that this digital community needs a higher standard of plagiarism. We can no longer depend on word of mouth among acquaintances to find the expert sources. If we moreover depend on a digital system that is prey to falsification, then we must view plagiarizing deception as particularly dangerous, and we might demand more scholarly citation than was standard in the last generation.

22.5 A PERSONAL VIEW OF THE MATTER

The present author believes that, in some contexts, new technology has tended to increase the importance we should place on acknowledgment and to lessen the importance that we should place on authorization. I further believe that, in some contexts, our present conception of plagiarism is too stringent, and we should loosen the conception to accept certain sorts of expression that are presently condemned as plagiarism. At the same time, I believe that there are contexts where we would be justified in promulgating a fiercer plagiarism standard.

As an example of a context where we can lessen authorization concerns while increasing concern for acknowledgment, consider academic scholarship. For academic publication, the digital technology has shifted the major costs of typesetting away from the publisher to the authors and editors who produce print-ready digital manuscripts. In these circumstances, the publishing industry needs fewer protections to ensure profitability. That is to say that we can loosen the conception of plagiarism to permit a greater degree of copying without authorization.

In contrast, we have noted how the fluidity of web-based information and of software production has increased the need for careful archiving of sources. Insofar as these needs are addressed through the promulgation of a plagiarism standard, the fluidity of web information suggests a need for a more stringent demand for acknowledgment and a greater awareness of when these demands are not met. That is to say, we should promote an increased emphasis on plagiarism in these contexts.

The situation, however, is really very complicated. We noted above that today's technology has created problems for a film industry that seeks to recoup high production costs when piracy is cheap and easy. In contrast, technology has lessened both the cost of production and of distribution in the music industry. In the extreme, the excitement of hip-hop music was to some extent dependent on the ability of musicians to use digital sampling to produce disks at home in the morning and hear them broadcast on the radio in the afternoon. On an economic analysis, we can argue from these changes for a heightened need for fierce plagiarism standards in the movie industry, and a lowered acceptance of loose plagiarism standards in the music industry. We should not, however, conclude that we need institute industry-specific plagiarism standards. Industry specific standards can create an ethical and legal morass with huge transaction costs. Moreover, we must remember that the present situation is the consequence of rapidly changing technology, and any attempt at a reasonable solution for today's technology may turn out to be a mistake tomorrow. So we would be ill advised to go to the effort to create a complicated set of standards just to address today's problems. Rapid advances in digital technology have created an almost intractable dilemma for the conception of plagiarism in the arts.

The standards for infringing plagiarism have created a special set of problems for the software industry. Copyrights have, from the start, been a clumsy form of protection for executable software. But the debate over the proper form of protection for software was settled in the 1970s. Copyrights, for any number of reasons, must be the preferred protection for most software, and it is simply absurd to suggest a reversal of this policy at this late date. Still, the copyright standard for software has generated problems, and the open source community has addressed these issues through attacks on some elements of infringing plagiarism, while at the same time respecting complaints about the plagiarist who does not acknowledge sources. Regardless of what one thinks of the open source project, we can appreciate the basic lesson: There is much to be learned from questioning what counts as plagiarism and what counts as nonplagiarizing use.

22.6 LITERATURE REVIEW

The Office of the General Council at the University of Texas posts (March 2007) a very practical guide to the legal constraints on the presentation of copyrighted material at http://www.utsystem.edu/ogc/IntellectualProperty/cprtindx.htm. This is an excellent place to seek answers to questions about conditions on what may or may not be distributed or posted, with or without authorization. Needless to say, everyone is well advised to consult a practicing intellectual property lawyer before performing any action about which there is a real question, risk, or ambiguity.

The In-A-Nutshell books in the West Nutshell Series published by Thomson West provide an excellent, readable introduction to the basic legal background for infringing plagiarism. Of special relevance are *Intellectual Property* by R. Miller and M. Davis, *Entertainment Law* by S. Burr, and *Intellectual Property and Unfair Competition* by C. McManis, These are uniformly up-to-date, concise, authoritative, and very readable, even for those who have no prior familiarity with the law. There is no better starting point for a theoretical interest in the legal issues, especially concerning infringing plagiarism.

Richard Posner's recent book, *The Little Book of Plagiarism* (Pantheon Books, 2007), is a lovely, short, and very readable study of noninfringing plagiarism, treated as the attempt at passing off unoriginal material as one's own. Richard Posner is certainly an outstanding scholar in the field of the economics of law. He and William Landes, who is an outstanding scholar in intellectual property law, have recently written a more scholarly study, *The Economic Foundations of Intellectual Property Law* (Belknap Press of Harvard University, 2003). This book is an excellent starting point for a scholarly study of those issues discussed above under the subtitle of "economic foundations," including a deep study of transactional costs. Given that "infringement" is a term of property law and that "plagiarism" is not a legal term, it is hard to find law articles that focus on noninfringing plagiarism. Of interest in this area is Lisa Lerman's 2001 article, "Misattribution in Legal Scholarship: Plagiarism, Ghostwriting, and Authorship," *South Texas Law Review* (42.2 467–492).

The present author has been deeply impressed and influenced by an unlikely source for the study of plagiarism in the electronic arts. Martha Buskirk's study of *The Contingent Object of Contemporary Art* (MIT Press, 2003) includes an in-depth study of copying in today's somewhat wild world of fine art, and the enforcement of those claims through demands for authorization of copies. The issues raised above over the work of Jeff Koons and Sherrie Levine are explored in depth in her book. The book is also simply fun to read and immensely informative on the concept of authorship and originality in the arts. The exchange on Danger Mouse's mash-up is found in the *Wired Magazine* (February 14, 2002) article "Copyright Enters a Gray Area" by Noah Shachtman, also on the Web (in March 2007) at www.wired.com/news/digiwood/0,1412,62276,00.html. In March 2007, the "revenge of the penguins" is viewable on YouTube at http://www.youtube.com/watch?v=fmd59Pbym2U. Nate Harrison's 2004 "Can I Get an Amen" is a useful and informative discussion of the heavily repeated use of a six-second sample of a 1960s drum "break" throughout the music industry. It is to be found (in March 2007) on the Web at http://nkhstudio.com/pages/popup_amen.html. I highly recommend the twenty-minute audio to anyone interested in the way in which free copying has encouraged creativity in music. He investigates the popular use of sampling without authorization to lower the cost of production and distribution in the popular music industry.

Moral rights theory within intellectual property law has become a very hot topic in legal scholarship. From among the many law review articles, I recommend Patrick Masiyakurima's 2005 article "The Trouble with Moral Rights" in *The Modern Law Review* (63(3) MLR 411-434. Charles Bietz's 2005 article "The Moral Rights of Creators of Artistic and Literary Work" in *The Journal of Political Philosophy* (13.3 330–358) places the dispute over moral rights in the context of political theory. A now outdated, but still useful introduction, is Russell J. DaSilva 1980 article "Artist's Rights in France and the U.S." in *the Bulletin of the Copyright Society of the U.S.A.* (28 BULL COPR SOCY 1-58). The remarks on the use of Shostakovitch's music are discussed by both Bietz and DaSilva. The Hegelian concept of personhood in such a law is further explored by Margaret Jane Radin in "Property and Personhood," *Stanford Law Review* 34 : 967. Alternative views of the nature of natural rights to intellectual property are presented in Kenneth Himma's paper "The Justification of Intellectual Property: Contemporary Philosophical Disputes" forthcoming in *The Journal of the American Society for Information Sciences and Technology.*

There is a new scholarly interest in the pirate communities that eschew the wrong of plagiarism. Adrian Johns' forthcoming book *Piracy*, University of Chicago Press (expected 2008) will put literary piracy in its historical context. Ravi Sundaram's forthcoming book (announced 2007) *Pirate Culture and Urban Life in Delhi: After Media*, Taylor and Francis, Routledge Studies in Asia's Transformations, includes an analysis of today's pirate communities that relish expression free of copyright restrictions.

The open source movement in software remains a major area of controversy, generating volumes of writings and discussions. Of special note is Christopher Kelty's *Two Bits: The Cultural Significance of Free Software and the Internet*, which at this writing is still forthcoming (expected 2008) from Duke University Press. A good collection of articles is *Perspectives on Free and Open Source Software*, ed. Joseph Feller, Brian Fitzgerald, Scott Hissam, Karim Lakhani, MIT Press, 2005. More generally, Lawrence Lessig's books on free software have acquired almost cult status among proponents of free software. These books include *The Future of Ideas*, Random House 2001, and *Free Culture*, Penguin Press 2004. (Meeting with proponents of open source software, you might get that failure to have studied *The Future of Ideas* in depth is a mark of illiterate ignorance.) Primary sources for arguments in favor of open source software are the home page of the Free Software Foundation, on the web (March 2007) at www.fsf.org and the home page for the Creative Commons, on the web at http://creativecommons.org. The debate over a right to restrict uses of open source software is to be found at these pages. The Creative Commons is also a good source for debates over the open use of digital sampling in mash-ups.

CHAPTER 23

Intellectual Property: Legal and Moral Challenges of Online File Sharing

RICHARD A. SPINELLO

23.1 INTRODUCTION

The recording industry in the United States has filed multiple lawsuits against purveyors of file sharing software. It has even initiated lawsuits against individuals who make substantial use of this technology (*RIAA v. Verizon,* 2003). The industry contends that the unauthorized "sharing" of copyrighted files is actually tantamount to the theft of intellectual property. The industry also contends that those companies that provide this software, such as Grokster and StreamCast, are culpable of secondary or indirect liability for such theft. This assertion is contentious, but recent court cases have tended to support the recording industry's claims about secondary liability, especially when there is evidence of inducement.

Lost in the thicket of lawsuits and policy challenges are the ethical issues associated with the use and distribution of file sharing software. Is the downloading or "sharing" of copyrighted music morally reprehensible? Quite simply, are we talking about sharing or pilfering? Is social welfare enhanced by a legal regime of indirect liability? And should we hold companies like Napster, Grokster, or BitTorrent morally accountable for the direct infringement of their users, particularly if they intentionally design the code to enable the avoidance of copyright liability? Or does such accountability stretch the apposite moral notion of cooperation too far? In this overview, we will present the conflicting arguments on both sides of this provocative debate. Although our primary focus will be on the ethical dimension of this controversy, we cannot neglect the complex and intertwined legal issues. We will take as a main axis of discussion the recent *MGM v. Grokster* (2005) case, in which all

The Handbook of Information and Computer Ethics, Edited by Kenneth Einar Himma and Herman T. Tavani

of these issues have surfaced. We begin, however, with a brief summary of this technology's functionality.

23.2 PEER-TO-PEER NETWORKS

The technology at the center of these copyright disputes is a software that enables computer users to share digital files over a peer-to-peer (P2P) network. Although the P2P architecture is evolving, a genuine P2P network is still defined as the one in which "two or more computers share [files] without requiring a separate server computer or server software" (Cope, 2002). Unlike the traditional client/server model, where data are only available from a single server (or group of servers), data can be accessed and distributed from any node in a P2P network. Each computer in the network can function as a server when it is serving or distributing information to others. Or it can assume the role of a client when it is accessing information from another system. As a result of this decentralization, the P2P network has the potential to be a more reliable information distribution system. For example, in the client/server model, if the server that hosts the data crashes, no data are available. But if a computer in a P2P network crashes, data are still available from other nodes in the network. Files can also be shared more expeditiously and more widely than ever before.

Thus, the most distinctive feature of this architecture is that each node in the system is a "peer" or an equal. There is no need for a central authority to mediate and control the exchange of information. The "purest" P2P architecture is flat and nonhierarchical. However, the diminished control associated with such a completely decentralized network leads to obvious scalability problems. As Wu (2003) observes, as the network grows "the loss of control makes it difficult to ensure performance on a mass scale, to establish network trust, and even to perform simple tasks like keeping statistics."

P2P software programs are usually free and easy to install. Once they are installed, a user can prompt his or her personal computer to ask other PCs in a peer-to-peer network if they have a certain digital file. That request is passed along from computer to computer within the network until the file is located and a copy is sent along to the requester's system. Each time a P2P user makes a copy of a digital file, by default that copy becomes available on the user's computer so that it can be copied by other P2P users. This process, which is known as "uploading," results in "an exponentially multiplying redistribution of perfect digital copies" (Petition for Writ of Certiorari, 2004).

Peer-to-peer networks require some method of indexing the information about the digital files available across the network so that user queries can be handled efficiently. There are three different methods of indexing: a centralized index system, in which the index is located on a central server; a decentralized indexing system; and a supernode system, in which a special group of computers act as indexing servers. The first method, which was adopted by Napster, relies on central servers to maintain an index of all the files available on the network; users search that index, and they are then referred to peers with a copy of the desired file. This method was abandoned by Napster's successors after Napster lost the court case defending its technology *(A&M Records, Inc. v. Napster, 2001)*. The supernode system, developed as part of KaZaA,

BV's FastTrack technology, relies on selected computers within the network with large memory capacity; these index servers perform the searches and return the search results to the user. The supernodes change periodically, and a given computer may never realize that it is serving in this capacity. The supernode has been the preferred solution in recent years, since it combines the advantages of the first two methods. While Grokster has depended on the supernode approach, some versions of the Gnutella protocol have relied on the decentralized method with no supernodes.

The P2P architecture represents a powerful communications technology with obvious social benefits. These include properties such as anonymity and resistance to censorship. The problem with P2P software, however, is that it has facilitated the unauthorized reproduction and distribution of copyrighted works in violation of the Copyright Act. Approximately 2.6 billion copyrighted music files are downloaded each month (Grossman, 2003), and about 500,000 infringing movie files are downloaded each day (MPAA Press Release, 2004). Companies supplying this software are obviously aware that their users are downloading copyrighted files, but they do not know which specific files are being copied or when this copying is occurring.

23.3 SHARING OR THEFT?

The Web has emboldened free riders and engendered a new ethic on copying proprietary material based on the belief that cultural products such as movies and music should be freely available online to anyone who wants them. Whatever enters the realm of cyberspace as a digital file is fair game. According to this ethic, there is nothing wrong with the use of P2P networks for sharing copyrighted material. Freenet's project leader, for example, has described copyright as "economic censorship," since it retards the free flow of information for purely economic reasons (Roblimo, 2000). He and others support an anticopyright model, which calls for the repudiation of exclusive property rights in cyberspace. Echoes of this viewpoint can also be found in the writings of other information libertarians such as Barlow (1994):

> ... all the goods of the Information Age—all of the expressions once contained in books or film strips or newsletters—will exist as thought or something very much like thought: voltage conditions darting around the Net at the speed of light, in conditions that one might behold in effect, as glowing pixels or transmitted sounds, but never touch or claim to "own" in the old sense of the word.

One could infer that Barlow would not be troubled by the "darting around" of copyrighted works on P2P networks, since these digital networks help the Net to realize its true potential and thereby accelerate the abandonment of archaic notions of intellectual property "ownership."

Nonetheless, not everyone agrees with Barlow's radical vision or the anticopyright approach. For those who recognize the value of P2P networks for sharing digital content and also respect the beneficial dynamic effects of intellectual property rights, some important questions need consideration. Are those who copy copyrighted files by means of a P2P system legally responsible for breaking the law? Does their action constitute direct infringement of a copyright? According to the U.S. Copyright Act (2004), an infringer is "anyone who violates any of the exclusive rights of the copyright owner . . .," including the right to make copies.

Defenders of the unfettered use of P2P networks come in many different stripes. But they typically concur that the personal copying of protected files on P2P networks is legally acceptable. Some legal scholars in this camp maintain that users who download files are not making copies of those files but simply "sharing" digital information over a conduit.[1] They also argue that even if sharing digital copies over a network is equivalent to making an unauthorized reproduction of a copyrighted work, that action comes under the fair use exception. Litman (2002), for example, contends that it is far from evident under current law whether individual users are liable for copyright infringement if they engage in "personal copying." On the contrary, Zittrain admits that "it is generally an infringement to download large amounts of copyrighted material without permission; even if you already own the corresponding CD, the case could be made that a network-derived copy is infringing" (Gantz and Rochester, 2005).

What about the moral propriety of sharing copyrighted files without permission? While David Lange (2003) does not argue that such file sharing is morally acceptable, he claims that there is considerable "softness" on the side of those who make the opposite claim. He maintains that those who argue that file sharing is morally wrong do so along the following lines: "Taking property without permission is wrong. Recorded music is property. Taking recorded music without permission is therefore wrong as well." But the problem in this line of reasoning lies in the minor premise. Many do not accept that music is property and, in Lange's view, there is some merit to this claim. Therefore, the issue of a legitimate property right in such intellectual objects "is still very much unsettled . . . [and] it may yet be that the idea of property and exclusivity will prove unable to withstand the popular will" (Lange, 2003).

Lange seems to assume an asymmetry between physical, tangible property, and intellectual property. He does not question a right to physical property (ownership of a house or an automobile), but intellectual property rights are more ambiguous, given the peculiar characteristics of that property. Unlike its physical counterpart, an intellectual object is nonrivalrous and nonexcludable: its consumption doesn't reduce the supply available to others, and it's difficult to "exclude" or fence out those who haven't paid.[2]

---

[1]Peter Menell et al. point out that "file sharing" is a misnomer: "a more accurate characterization of what such technology accomplishes is 'file search, reproduction, and distribution' . . . following a peer-to-peer transaction, one copy of the file remains on the host computer and another identical copy resides on the recipient's computer." See Amici Curiae Brief of Law Professors (Mennell, Nimmer, Merges, and Hughes) (2005).

[2]It should be pointed out that even if intellectual objects are not strictly speaking property, this does not negate the creator's right to limited control over his or her created products. See Himma (2007a) for more elaboration on this issue.

We cannot resolve this complex issue here, but let it suffice to say that a potent case can be made for a right to intellectual property based on the classic labor-desert argument first proposed by Locke. I have argued elsewhere that the application of Locke's theory to intellectual property seems plausible enough (Spinello, 2003). Locke's core argument is that each person has a property right in herself and in the labor she performs. Therefore, it follows that a property right should also extend to the final product of that labor. Surely one is entitled to that right by virtue of creating value that would not otherwise exist except for one's creative efforts. Who else would have a valid claim to the fruits of this person's efforts? This property right must satisfy one condition summarized in the Lockean proviso: a person has such a property right "at least where there is enough, and as good left in common for others" (Locke, 1952). The Lockean-inspired argument for an intellectual property right is that one's intellectual labor, which builds on the ideas, algorithms, generic plots, and other material in the intellectual commons, should entitle one to have a natural property right in the finished product of that work such as a novel or a musical composition. The ideas remain in the commons and only the final expression is protected, so the common domain is undiminished and the proviso satisfied.

This labor-based approach gives intellectual property rights a stronger normative foundation than consequentialist arguments, which, in my opinion, are ultimately indeterminate (Spinello, 2003). A creator or author should have a basic right to exclude others from her work because she created that work through her own labor.[3] Even the U.S. Supreme Court has recognized the general suitability of this argument: "sacrificial days devoted to. . . creative activities deserve rewards commensurate with the services rendered" (*Mazer v. Stein*, 1954). We should not focus on the nature and qualities of the product (tangible or intangible, excludable or nonexcludable), but on the value inherent in that product that is the result of labor and initiative. Also, contrary to Lange's comments, given the normative underpinnings of intellectual property rights, they should not be contingent on the support of the majority. What's of primary importance is the creator's interests—she has expended time and energy in the creative process. At the same time, while the consumers' interests cannot be completely discounted, their desire for "content" should not give rise to some sort of morally or legally protected interest.[4]

If we grant the premise that "recorded music is property," then it seems clearer that there might be something wrong with copying this music without permission. But in order to classify copyright infringement as a form of theft we need to understand more precisely what is entailed by a property right, particularly when that right is viewed from a distinctly moral perspective. The essence of such a right is the liberty to use a physical or intellectual object at one's discretion, that is, the right to determine what is

---

[3]The recognition of this right is only a starting point for policy makers who may choose to qualify it in certain ways for the sake of justice and the common good.

[4]Himma (2007b) develops this line of reasoning quite cogently. He argues that "the interests of content creators in controlling the disposition of the content they create outweigh[s] the interests of other persons in using that content in most, but not all, cases."

to be done with this created object. It includes a right to exclude others from appropriating or using that object without my permission. According to Nozick (1974)

> The central core of the notion of a property right in X, relative to which other parts of the notion are to be explained, is the right to determine what shall be done with X; the right to choose which of the constrained set of options concerning X shall be realized or attempted.

Theft, therefore, should be understood as a misuse, an "unfair taking," or a misappropriation of another's property contrary to the owner's reasonable will.[5] In the case of intellectual property (such as digital movies and music), unless the copyright holder's consent can be reasonably presumed, using that copyright holder's creative expression in a way that exceeds her permission is using it contrary to her will. We can be quite certain that downloading a copyrighted Disney movie and then uploading it for millions of other users to copy is an action that is contrary to the will of Disney. When one uses something contrary to the creator's (or owner's) will, this kind of act is equivalent to *taking something* from that owner without her volition and consent. The use of a piece of intellectual property without the copyright holder's permission (and therefore against his will) is unfair to that copyright holder since it violates his right to determine what is to be done with that property. And this unfair use of another's intellectual property constitutes a form of theft (Grisez, 1997).

This presumes that the copyright holder's will is reasonable and that his or her consent would not be given. In the case of digital music files, one can also safely presume that the copyright holder will not want his or her music uploaded on peer-to-peer networks for everyone else to copy, since this action will erode legitimate sales of this product. On the contrary, one might presume that making another copy of a purchased online song or MP3 file for one's own personal use would be reasonable and acceptable. An additional CD burn of music I already own, so I can have an extra copy to keep in my office, seems perfectly legitimate (unless the owner indicates otherwise).

When seen in this light, the moral argument against downloading and sharing music files with other P2P users seems more persuasive. But what about the objection that "sharing" within the online community is a noble act and that the sharing of digital information serves the common good through a de facto expansion of the public domain? After all, we are taught at an early age that sharing with others is a good thing to do. The Internet and P2P software facilitate sharing, so why should there be constraints that hold back the full potential of this technology?

Grodzinsky and Tavani (2005) make some noteworthy arguments along these lines as they underscore the fundamental importance of sharing. In their view, it is important to "defend the principle 'information wants to be shared,' which presumes against the 'fencing off' or enclosing of information in favor of a view of information that should be

[5]It should be presumed that the owner's will is reasonable unless a strong case can be established to the contrary. This qualification leaves room for exigent circumstances in which an owner's failure to share or license his property may yield dire consequences, and hence the appropriation contrary to his will is arguably not unfair.

communicated and shared." Of course, information has no "wants," so the argument being proposed here is the normative claim that the distribution of information as widely as possible should be promoted and encouraged.[6]

We must be careful, however, not to overestimate the value of sharing. Sharing is not a core value or a basic human good, even though it often contributes to the harmony of community and the furtherance of knowledge, which are basic human goods. The basic human goods (including life and health, knowledge, the harmony of friendship and community, and so forth) are basic not because we need them to survive but because we cannot flourish as human beings without them. These goods, which are intrinsic aspects of human well-being and fulfillment, are the primary ends or reasons for action and the ultimate source of normativity. Also, goods intrinsic to the human person are greater than instrumental goods. For example, life is a higher good than property. Hence, the basic human goods are of primary significance for justifying ethical policies.

But information sharing does not fall in this category. The obvious problem is that such sharing does not always promote the basic human goods. Sharing is an instrumental good, a means to an end, but that end may not always be morally acceptable. For example, the act of "sharing" pornographic material with young children is certainly immoral, since it contributes to the corruption of those children by hindering the proper integration of sexuality into their lives. Exposure of immature children to pornography puts their relationships at risk and often yields social disharmony. Similarly, child pornography is being "shared" with increasing frequency over P2P networks, but no reasonable person would be in favor of this form of sharing (Amici Curiae Brief of Kids First Coalition, 2005). Therefore, we cannot assume that sharing information, either as a means or as an end, always contributes to the common good. We must consider the quality and the nature of the information to be shared.

Information may "want to be shared," but for the sake of the common good we must sometimes put restrictions on the sharing of information. Some information (such as child pornography) shouldn't be shared at all, and other types of information should be shared according to the wishes of its creator, assuming a valid property right has been established. Since an intellectual property right or copyright embodies the labor-desert principle, it is difficult to dispute this right from a moral perspective. As noted earlier, "sharing" is a misnomer since what is really going on is the search for a digital file followed by its reproduction and distribution. The real question is whether or not an intellectual work, such as a movie created by Disney at considerable expense, should be shared with impunity against the will of the creator and rightful owner of this intellectual property.

But what about compulsory sharing, that is, a scheme whereby noncommercial file sharing would become lawful, and copyright owners would be compensated through a tax or levy on Internet services and equipment? Grodzinsky and Tavani (2005) favor this approach because they recognize that creators should be compensated for their efforts. This model, which has been carefully developed by Fisher (2004), would

[6]There are some scholars such as John Perry Barlow who do maintain that information has the quality of being a life-form. See Himma (2005) for a helpful discussion of this topic.

essentially displace exclusive property rights with mandatory compensation. According to Lessig (2001), it is "compensation without control." In legal terms, liability rules would take the place of property rules.

Compulsory licensing certainly has potential, and its benefits should not be discounted by policy makers. At the same time, advocates of compulsory licensing often gloss over the practical implementation problems. How and by whom will a fair compensation plan be determined? What about the high costs of administering such a cumbersome regulatory system, which will undoubtedly be subject to the vagaries of the political process? The potential for economic waste cannot be casually dismissed. Finally, what will be the effects of compulsory licensing for digital networks on the whole copyright regime—will there be a perception that copyrighted works in any format or venue are "up for grabs?" Above all, we must not naively assume that nonvoluntary licensing is a panacea. Epstein (2004) clearly summarizes the Hobson's choice confronted by policy makers: "any system of private property imposes heavy costs of exclusion; however, these costs can only be eliminated by adopting some system of collective ownership that for its part imposes heavy costs of governance—the only choice that we have is to pick the lesser of two evils."

23.4 SECONDARY LIABILITY

Now that we have considered the issue of direct copyright infringement, we can focus attention on the question of contributory or secondary infringement. If we accept, however cautiously, the reasonable assumption that downloading and uploading copyrighted files is direct infringement, what about the liability of those companies providing the software for this purpose? Do they "cooperate" in the wrongdoing in any morally significant way?

Purveyors of P2P systems fall in the category of "technological gatekeepers." Kraakman (1986) elaborated upon the limits of "primary enforcement" of certain laws and the need for strict "gatekeeper liability." According to Zittrain (2006), such secondary liability "asks intermediaries who provide some form of support to wrongdoing to withhold it, and penalizes them if they do not." To some extent, the emergence of the decentralized P2P architecture, which eliminates intermediaries, represents the undermining of the gatekeeper regime. According to Wu (2003), "the closer a network comes to a pure P2P design, the more disparate the targets for copyright infringement. . . ." For this reason, Gnutella has described itself as a protocol instead of an application. The problem is that efficiency requires some centralization. As we saw above, the FastTrack system used by Grokster relies on supernodes with generous bandwidth for the storage of an index, and it uses a central server to maintain user registrations and to help locate other peers in the network. Thus, there is some conflict between developing a P2P system that will optimize avoidance of legal liability and the technical goals of efficiency and scalability.

The debate over imposing secondary liability has been intense in recent years. Proponents argue that it's a major deterrent of infringement. According to the Supreme Court, "when a widely shared service or product is used to commit infringement, it may be impossible to enforce rights in the protected work effectively against all direct infringers, the only practical alternative being to go against the distributor of the copying device for secondary liability . . ."(*MGM v. Grokster*, 2005). However, as Lemley and Reese (2004) point out, "going after makers of technology for the uses to which their technologies may be put threatens to stifle innovation." Hence the tension in enforcing secondary liability or pressing these cases too vigorously: how to enforce the rights of copyright holders in a cost-effective way without stifling innovation. The issue has received extraordinary attention in the legal literature (see Dogan, 2001; Fagin et al., 2002; Kraakman, 1986; Lichtman and Landes, 2003), but it has not yet generated much interest among ethicists. Before we review the ethical dimension of this debate, however, it is instructive to say more about the legal framework.

There are two doctrines of secondary liability in current copyright law. First, contributory infringement pertains to "one who, with knowledge of the infringing activity, induces, causes, or materially contributes to the infringing conduct" (*Gershwin Publishing Corp. v. Columbia Artists Management, Inc.*, 1971). Second, "one may be vicariously liable if he has the right and ability to supervise the infringing activity and also has a direct financial interest in such activities" (*Gershwin Publishing Corp. v. Columbia Artists Management, Inc.*, 1971). There have been a number of secondary liability cases involving file-sharing software companies such as Aimster and Napster, but the most notable is *MGM v. Grokster*. The Grokster case exposes the legal complexity of secondary liability claims, and it can serve as a springboard for a moral analysis of this issue.

23.5 *MGM V. GROKSTER*: A BRIEF HISTORY

The plaintiffs in this case included song writers, music publishers, and motion picture studios (such as MGM, Universal City, and Disney Enterprises). They filed suit against Grokster and StreamCast for contributory and vicarious copyright infringement. Both of these companies initially used their own OpenNap software (a version of Napster that had been reverse engineered). But in 2001, they licensed the FastTrack peer-to-peer distribution technology from the Dutch company KaZaA, BV. Subsequently, KaZaA, BV sold its assets to an Australian company known as Sharman Networks, which is also named in the suit. Once the licensing agreement was in place, both Grokster and StreamCast transferred their users from their OpenNap systems to KaZaA's FastTrack software system, named "Grokster" and "Morpheus," respectively. StreamCast has since revoked its licensing arrangement with Sharman, and now it uses a variation of the open source peer-to-peer network known as Gnutella. Gnutella is a radically decentralized network, unowned and unmanaged by its designers.

The plaintiffs contended that both of these networks were employed for the purpose of swapping copyrighted music and movie files and that their business models depended on copyright infringement. The magnitude of file sharing of copyrighted works is beyond dispute: "90% of the works available on the FastTrack network demonstrably were infringing, and over 70% belonged to Plaintiffs" (Plaintiffs' Joint Excerpts of Record, 2003). On the contrary, Grokster and StreamCast have argued that "potential noninfringing uses of their software are significant in kind, even if infrequent in practice" (*MGM v. Grokster*, 2005). One example of a noninfringing use is the distribution of public domain literary works made available through Project Gutenberg.

The plaintiffs lost the first round of this case when a California District Court granted the defendants' motion for summary judgement. The case was promptly appealed to the Ninth Circuit Court of Appeals. In August, 2004 the appeals court affirmed the judgement of the lower court. The Ninth Circuit found no basis for contributory liability because Grokster had knowledge of direct infringement only after the fact, when it was not possible to take action. The Ninth Circuit agreed with the lower court, which explained that "liability for contributory infringement accrues where a defendant has actual—not merely constructive—knowledge of the infringement at a time during which the defendant materially contributes to that infringement . . . ." (*MGM v. Grokster*, 2004). Also, there was no vicarious liability because Grokster could not exercise control over users given the decentralized nature of the network. Thus, while both courts found that the defendants were deriving material benefits from the illicit activities of their users, they held that those defendants had no power to deal with the infringing conduct as it was occurring.

The Ninth Circuit judges also relied on the standard set forth in the *Sony v. Universal* (1984) case. In that case, Universal Studios had sued VCR manufacturer Sony for copyright infringement, but the Supreme Court ruled in favor of the defendant, reasoning that a VCR is capable of substantial noninfringing uses and that its manufacturers should therefore be immune from secondary liability when infringement does occur. One such noninfringing use would be "time-shifting," that is, taping a movie or television show so that it could be viewed at a different time. Thus, manufacturers of "staple articles of commerce" (like VCRs) that have "commercially significant noninfringing uses" cannot be held liable for contributory infringement just because they have general knowledge that some of the purchasers of that equipment might use it to make unauthorized copies of copyrighted material. The Supreme Court's legitimate concern was that copyright owners should not be allowed to interfere with the manufacture and distribution of new technologies just because their copyrighted works were involved. The *Sony* safe harbor standard, that is, "capable of substantial noninfringing uses," is invoked to protect innovative technologies such as P2P software, and it was a decisive precedent for the resolution of the Grokster case. The "capability" criterion acknowledges that technology is malleable and evolves over time. The *Sony* standard insists that the important interests of copyright holders must be judiciously balanced with the interests of technology innovators.

In the fall of 2004, the Grokster case was appealed to the U.S. Supreme Court, which granted certiorari. In a surprising turn of events, that court ruled unanimously in June 2005 that file sharing companies may be liable for contributory infringement. In its deliberations, the justices sought to determine "under what circumstances the distributor of a product capable of both lawful and unlawful use is liable for the acts of copyright infringement by third parties using the product" (*MGM v. Grokster*, 2005). According to the Court, the Ninth Circuit erred by interpreting the *Sony* precedent too broadly. According to the Ninth Circuit's interpretation, the distribution of a technology or commercial product capable of substantial noninfringing uses could give rise to contributory liability for direct infringement *if and only if* the distributor of that technology had actual and timely knowledge of specific instances of infringement and failed to take action. In the Ninth Circuit's view, "because contributory copyright infringement requires knowledge *and* material contribution, the Copyright Owners were required to establish that the Software Distributors had 'specific knowledge of infringement at a time at which they contributed to the infringement, and failed to act upon that information'" (*MGM v. Grokster,* 2004, citing *A&M Records, Inc. v. Napster,* 2001). Grokster and StreamCast had no such actual and timely knowledge and therefore could not be held liable.

But this broad interpretation of *Sony* is off the mark. According to the Supreme Court, "nothing in *Sony* requires courts to ignore evidence of intent . . . and where [such] evidence goes beyond a product's characteristics or the knowledge that it may be put to infringing uses and shows statements or actions directed to promoting infringement, *Sony's* staple article rule will not preclude liability" (*MGM v. Grokster,* 2005).

Thus, where there is inducement of infringement, defined as "active steps . . . taken to encourage direct infringement" (*Oak Industries, Inc. v. Zenith Electronics Corp.* 1971), the safe harbor provided by the *Sony* standard is nullified. Accordingly, the Court of Appeals judgment was vacated and the case was remanded for further proceedings. In the closing remarks of the majority opinion written by Justice Souter the Court noted the "substantial evidence in MGM's favor on all elements of inducement" (*MGM v. Grokster,* 2005). This is a portentous statement about the future prospects for Grokster and StreamCast given the court's inducement standard.

The Court's decision was sharply criticized by some legal scholars. Lessig, for example, claimed that this unfortunate decision would most likely have a "chilling effect" on innovation (Samuelson, 2005). Zittrain (2006) also alluded to such a chilling effect and suggested that the Court would have been better off affirming the Ninth Circuit's opinion "without adding an inducement counterpart to *Sony.*" Others disagreed. Samuelson (2005) opined that "as long as the courts apply high standards for inducement liability—requiring proof of overt acts of infringement . . . and a specific intent to induce infringement—there should be ample room for innovative technologies to continue to thrive." It is also worth pointing out that Lessig's argument is purely speculative. He and other critics have offered no empirical data to support the hypothesis that this ruling will deter innovation in any materially significant way.

23.6 MORAL CONSIDERATIONS

As we have noted, ethicists have not subjected the issue of secondary liability, especially as it pertains to P2P networks, to much moral scrutiny. But we can identify two salient moral issues, one at the "macro" level and the other at a more "micro" level of the individual moral agent. First, can secondary liability law itself be normatively justified in social welfare terms? Second, how can we understand indirect copyright liability from a strictly moral viewpoint?

The utilitarian arguments for maintaining a strong legal tradition of secondary liability seem especially persuasive. Given the enormous difficulties of enforcing copyright protection by pursuing direct infringers and the threats posed to content providers by dynamic technologies such as P2P software, the need for this liability seems indisputable. As Zimmerman (2006) indicates, bringing infringement cases against "private copyists" is difficult, since "private copying often takes place out of public view." Pursuing these private copyists would also be expensive since it would require frequent and repeated litigation. Moreover, it stands to reason that copyright holders will not pay to enforce their rights where the costs of doing so exceed the expected returns.

On the contrary, intermediaries are "highly visible," and a single lawsuit can deter the actions of many egregious infringers, provided that the intermediary has contributed "in some palpable way to the creation of unlicensed private copies" (Zimmerman, 2006). Thus, bringing suits against these intermediaries overcomes the disutility of pursuing private copyists, and so a compelling case can be advanced that indirect liability is a much more efficient mechanism for achieving justice. As Mennell and his coauthors point out in their Amici Curiae Brief (2005), "The social and systemic benefits of being able to protect copyrights at the indirect infringement level rather than at the end user level, are substantial. Suing thousands of end users who waste both private and public resources is not nearly as effective as confronting enterprises whose business model is based on distributing software that is used predominantly for infringing uses." In addition, third parties such as software providers often have a reasonable opportunity to deter copyright infringement by means of monitoring user activities or designing code in a way that impedes infringement.

On the contrary, there must be reasonable restrictions on the scope of secondary liability claims so that they do not stifle innovation. Those restrictions are embodied in the *Sony* precedent—dual-use technologies, capable of substantial noninfringing use, should be immune from secondary liability so long as there is no evidence of inducement. But the bottom line is that indirect liability, carefully implemented, promotes efficiency in the enforcement of copyright law, which is necessary to maximize the production of creative expression in the first place. Arguably, although we must bear in mind the difficulty of measuring welfare effects, this policy enhances social welfare, since it appears to be so strongly justified in utilitarian terms.

The second question is how we assess the moral propriety of actions that appear to facilitate the wrongdoing of others. The moral case for indirect liability centers on the question of cooperation. Under normal circumstances, cooperation associated with communal activities certainly poses no moral problems. But what about the "community" of online file sharers, which is made possible by software providers such as Grokster? Is there something problematic about the online file-sharing community, given that the primary function of the software is for sharing copyrighted music and movie files?

The basic moral imperative at stake here can be stated succinctly: a moral agent should not cooperate in or contribute to the wrongdoing of another. This simple principle seems self-evident and axiomatic. If someone intentionally helps another individual carry out an objectively wrong choice, that person shares in the wrong intention and bad will of the person who is executing such a choice. In the domain of criminal law, if person X helps his friend commit or conceal a crime, person X can be charged as an accessory. Hence, it is common moral sense that a person who willingly helps or cooperates with a wrongdoer deserves part of the blame for the evil that is perpetrated. But this principle needs some qualification since under certain circumstances cooperation at some level is unavoidable and not morally reprehensible.

First, a distinction must be made between formal and material cooperation. Second, while all forms of formal cooperation are considered immoral, we must differentiate between material cooperation that is justifiable and material cooperation that is morally unacceptable. The most concise articulation of the first distinction is found in the writings of the eighteenth century moral theologian and philosopher St. Alphonsus Liguori (1905):

> But a better formulation can be expressed that [cooperation] is *formal* which concurs in the bad will of the other, and it cannot be without fault; that cooperation is *material* which concurs only in the bad action of the other, apart from the cooperator's intention.[7]

The question of material cooperation is quite complex, so we will confine our analysis to formal cooperation, which is more straightforward and apposite in this context. Is there a case to be made that Grokster and StreamCast are culpable of such formal cooperation which, in Liguori's words, cannot be "without fault" (*sine peccato*)? Formal cooperation means that one intentionally shares in another person's or group's wrongdoing. In other words, what one chooses to do coincides with or includes what is objectively wrong in the other's choice. For example, a scientist provides his laboratory and thereby willingly assists in harmful medical experiments conducted on human beings by a group of medical doctors because he is interested in the results for his research. This scientist shares in the wrongful intentions and actions of the doctors.

Is there evidence of such "formal cooperation" among P2P software companies? Has code been designed and implemented by commercial enterprises like StreamCast and Grokster in order to avoid copyright law? Are these companies deliberately seeking to help users to evade the law, perhaps for their own material gain? If so, one can justifiably press the claim that there is formal cooperation and hence moral irresponsibility.

Although it is often difficult to assess intentionality, the assertions of both companies in the Grokster case seemed to betray their true motives. Both companies "aggressively sought to attract Napster's infringing users," referring to themselves as "the next Napster" (Plaintiffs' Joint Excerpts of Record, 2003). StreamCast's CEO boldly proclaimed that "we are the logical choice to pick up the bulk of the 74 million users that are about to 'turn Napster off'" (Plaintiffs' Joint Excerpts of Record, 2003). Moreover, StreamCast executives monitored the number of songs by famous commercial artists available on their network because "they aimed to have a larger number

[7]"Sed melius cum aliis dicendum, illam esse *formalem*, quae concurrit ad malam voluntatem alterius, et nequit esse sine peccato; *materialem* vero illam, quae concurrit tantum ad malam actionem alterius, praeter intentionem cooperantis."

of copyrighted songs available on their networks than other file-sharing networks" (*MGM v. Grokster,* 2005). Mindful of Napster's legal problems, they were careful to avoid the centralized index design employed by Napster in order to circumvent legal liability. Some products were even designed without key functional advantages (such as specialized servers to monitor performance) in order to evade legal liability. Given this evidence, even the District Court conceded that the "defendants may have intentionally structured their businesses to avoid secondary liability for copyright infringement, while benefiting financially from the illicit draw of their wares" (*MGM v. Grokster,* 2003). At the same time, both companies derived substantial advertising revenues from their users who downloaded a massive volume of copyrighted music and movie files.

It surely appears that these companies deliberately designed their software to help users share copyrighted files and evade the law, and to profit from this collusive activity. Even if the *Sony* standard is blind to design issues (since it only requires that technologies be capable of substantial noninfringing use), we cannot ignore the moral problem of designing code as a mechanism for avoiding the law. On the contrary, there is a moral requirement to design products that will support and respect valid laws. In this case, the code should have included filtering and monitoring tools that would have minimized the software's potential for misuse, unless it would have been "disproportionately costly" to do so (*In re Aimster Copyright Litigation*, 2003).

Therefore, the evidence is strong that Grokster and StreamCast have acted in "bad faith." Given the rhetoric and behavior of both companies, this case is a classic example of formal cooperation where a moral agent's will coincides with the moral wrongdoing (evasion of copyright law) of another, and the moral agent helps to bring about that wrongdoing. In this context, the illicit cooperation takes the form of providing the mechanism to download and upload copyrighted files. These companies deliberately blinded themselves to this type of content being distributed over their network, and failed to take any sincere affirmative steps to deter the exchange of these protected files. On the contrary, these companies took positive steps to facilitate an illegal and immoral activity and to materially benefit from that activity. They made no secret of their true purpose by repeatedly stating their desire to be the heirs of Napster's notoriety.

In general, companies cannot develop code with the intention of minimizing the burden of just law, including copyright law. This type of "antiregulatory code" (Wu, 2003) that has been deliberately designed to undermine the legal system cannot be morally justified. But if code, including permutations of the P2P architecture, has been designed for the purpose of a legitimate functionality (sharing of information), and its misuse as a mechanism of legal evasion is accepted as an unwanted side effect, the code designer cannot be held morally accountable. Moral prudence also dictates that reasonable efforts must be made to anticipate such misuse. One such effort would be the inclusion of tools such as filters that might mitigate or curtail the misuse of one's product. Under these circumstances, it is safe to claim that there would be no formal cooperation. What is morally decisive, therefore, is the intentions and purpose of the code designer. When code is designed as a deliberate mechanism for the evasion of a legitimate law, there is complicity and moral liability. But software developers cannot be held morally accountable when prudently designed code, created for a valid purpose, is exploited by some users for copyright infringement or some other mischief. In the case of Grokster, intention and purpose seem pretty clear; but in other situations involving potential gatekeepers, the question of moral liability will be much more ambiguous.

23.7 CONCLUSIONS

In the first part of this essay we explained the likelihood that the unauthorized downloading of copyrighted files constituted direct infringement. We also delineated the limits and problems associated with primary enforcement. This explains why secondary liability has become such a salient issue.

The Internet has many "gatekeepers," from Internet Service Providers (ISPs) and search engines to purveyors of certain types of network software. These gatekeepers are sometimes in a position to impede or curtail various online harms such as defamation or copyright infringement. Thanks to the Digital Millennium Copyright Act (1998), ISPs have been immunized from strict gatekeeper liability since they are rightly regarded as passive conduits for the free flow of information. But other companies such as Napster and Grokster have had a more difficult time navigating toward a safe harbor. We have concentrated on these popular gatekeepers, purveyors of P2P network software such as Grokster, and we have discussed the scope of their legal and moral liability.

The Supreme Court has introduced an inducement standard while upholding the basic doctrine of *Sony*. "Purposeful, culpable expression and conduct" must be evident in order to impose legal liability under this sensible standard (*MGM v. Grokster,* 2005). In this way, the Court has judiciously sought to balance the protection of copyright and the need to protect manufacturers. If P2P developers succeed in developing a pure decentralized system that does not sacrifice efficiencies they may succeed in undermining the gateway regime, which has been so vital for preserving the rights of copyright holders.

Of course, as we have intimated, inducement is also problematic from a moral point of view. It is not morally permissible to encourage or facilitate the immoral acts of others, especially when one profits by doing so through advertising revenues. If we assume that direct infringement is morally wrong, inducement and the correlation of profits to the volume of infringement represents formal cooperation in another individual's wrongdoing.

Finally, as Lessig (1999) has reminded us, "code is law," and given the great power of software code as a logical constraint, software providers have a moral obligation to eschew the temptations of writing antiregulatory code. This type of code includes some P2P programs that facilitate and promote copyright infringement. Instead, developers must design their code responsibly and embed within that code ethical values in the form of tools that will discourage and minimize misuse. This assumes, of course, that such modifications would be feasible and cost effective.

REFERENCES

A&M Records, Inc. v. Napster. (2001). 239 F 3d 1004 [9th Cir.].

Amici Curiae Brief of Kids First Coalition et al. (2005). On Petition for Writ of Certiorari, to the Supreme Court of the United States, in review of *Metro-Goldwyn-Mayer Studios v. Grokster, Ltd* (2004). 380 F.3d 1154 [9th Cir.].

Amici Curiae Brief of Law Professors (Mennell, Nimmer, Merges, and Hughes) (2005). On Petition for Writ of Certiorari, to the Supreme Court of the United States, in review of *Metro-Goldwyn-Mayer Studios v. Grokster, Ltd* (2004). 380 F.3d 1154 [9th Cir.].

Barlow, J.P. (1994). The Economy of Ideas. *Wired*, 2(3), 84–88.

Cope, J. (2002). Peer-to-Peer Network. Available at: http://www.computerworld.com/networkingtopics/networking/story

Digital Millennium Copyright Act (1998), Section 512 (c).

Dogan, S. (2001). Is Napster a VCR? The implications of Sony for Napster and other Internet-Technologies. *Hastings Law Journal* 52, 939.

Epstein, R.A. (2004). Liberty vs. Property? Cracks in the Foundation of Copyright Law, U *Olin Working Paper*, April, 2004. Chicago Law and Economics.

Fagin, M., Pasquale, F., and Weatherall, K. (2002). Beyond Napster: using antitrust law to advance and enhance online music distribution. *Boston University Journal of Science and Technology Law*, 8, 451.

Fisher, W. (2004). *Promises to Keep*. Harvard University Press, Cambridge.

Gantz, J., and Rochester, J.B. (2005). *Pirates of the Digital Millennium*. Financial Times Prentice-Hall, Upper Saddle River, NJ.

Gershwin Publishing Corp. v. Columbia Artists Management, Inc. (1971). 443 F.2d 1159 [2nd Cir.].

Grisez, G. (1997). *Difficult Moral Questions*, Franciscan Herald Press, Quincy, IL, pp. 589–598.

Grodzinsky, F. and Tavani, H. (2005). P2P Networks and the *Verizon v. RIAA* case: implications for personal privacy and intellectual property. *Ethics and Information Technology*, 7(4), 243–250.

Grossman, L. (2003). It's All Free, *Time*, 166, 88.

In re Aimster Copyright Litigation (2003). 334 F. 3d 643 [7th Cir.], cert. denied sub nom., 124 S. Ct. 1069 (2004).

Himma, K.E. (2007a). The justification of intellectual property rights: contemporary philosophical disputes. *Journal of the American Society for Information Science and Technology*, 58, 2–5.

Himma, K.E. (2007b). Justifying property protection: why the interests of content creators usually win over everyone else's. In: Rooksby, E. and Weckert, J. (Eds.), *Information Technology and Social Justice*. Idea Group, Hershey, PA, pp. 47–68.

Himma, K.E. (2005). Information and intellectual property protection: evaluating the claim that information should be free. In: Spinello, R. (Ed.), *Newsletter of Law and Philosophy*. The American Philosophical Association, pp. 2–9.

Kraakman, R. (1986). Gatekeepers: the anatomy of a third party enforcement strategy. *Journal of Law, Economics, and Organizations*, 2, 53.

Lange, D. (2003). Students, music and the net: a comment on peer-to-peer file sharing, *Duke Law and Technology Review*, 23, 21.

Lemley M., and Reese R. (2004). Reducing copyright infringement without restricting innovation. *Stanford Law Review* 56, 1345.

Lessig, L. (2001). *The Future of Ideas: The Fate of the Commons in a Connected World.* Random House, New York.

Lessig, L. (1999). *Code and Other Laws of Cyberspace.* Basic Books, New York.

Lichtman, D., and Landes, W. (2003). Indirect liability for copyright infringement: an economic perspective. *Harvard Journal of Law & Technology,* 16, 395.

Liguori, St. Alphonsus. (1905–1912). In: Gaude, L.(Ed.), *Theologia Moralis.* 4 volumes. Ex Typographia Vaticana, Rome, pp. 357.

Litman, J. (2002). War stories. *Cardozo Arts & Entertainment Law Journal,* 20, 337.

Locke, J. (1952). *The Second Treatise of Government.* Bobbs-Merrill, Indianapolis, IN.

Mazer v. Stein (1954) 347 U.S. 201.

MPAA Press Release. (2004). MPAA Launches New Phase of Aggressive Education Campaign against Movie Piracy. Available at: http://mpaa.org/MPAAPress/

Metro-Goldwyn-Mayer Studios v. Grokster, Ltd. (2005) 125 U.S. 2764.

Metro-Goldwyn-Mayer Studios v. Grokster, Ltd. (2004) 380 F.3d 1154 [9th Cir.].

Metro-Goldwyn-Mayer Studios v. Grokster, Ltd. (2003) 259 F. Supp. 2d 1029 [C.D. Cal.].

Nozick, R. (1974). *Anarchy, State, and Utopia.* Basil Blackwell, Oxford.

Oak Industries, Inc. v. Zenith Electronics Corp. (1988). 697 F. Supp. 988 [ND Ill.].

Petition for Writ of *Certiorari* (2004). To the Supreme Court of the United States, in review of *Metro-Goldwyn-Mayer Studios v. Grokster, Ltd.* 380 F.3d 1154 [9th Cir.].

Plaintiffs' Joint Excerpts of Record (2003). *Metro-Goldwyn-Mayer Studios, Inc. v. Grokster, Ltd.* 259 F. Supp. 2d [C.D. Cal].

RIAA v. Verizon (2003). No. 03-7015[D.C. Cir].

Roblimo, L. (2000). Posting to Slashdot, Available at http://slashdot.org/article.pl

Samuelson, P. (2005). Did MGM really win the Grokster case? *Communications of the ACM,* 19–24.

Sony Corp of America v. Universal City Studios, Inc. (1984) 464 U.S. 417.

Spinello, R.A. (2003). The future of intellectual property. *Ethics and Information Technology,* 5(1), 1–16.

U.S. Copyright Act, 17 U.S.C. (2004). Section 501(a).

Wu, T. (2003). When code isn't law. *Virginia Law Review,* 89, 679.

Zimmerman, D. (2006). Daddy are we there yet? Lost in Grokster-Land. *New York University Journal of Legislation and Public Policy,* 9, 75.

Zittrain, J. (2006). A history of online gatekeeping. *Harvard Journal of Law & Technology,* 19, 253.

PART VI

ACCESS AND EQUITY ISSUES

CHAPTER 24

Censorship and Access to Expression[1]

KAY MATHIESEN

24.1 INTRODUCTION

No one wants to be a censor. Or, more precisely, no one wants to be called a "censor." To describe a person as a censor, or an act as one of "censorship," is to condemn the person or the action. Although there are numerous calls to arms to resist censorship and compilations of instances of censorship across the globe, little work has been done to help us understand the concept itself. Without getting clear on what we mean by censorship, it is difficult to get a grip on exactly what is wrong with it, and indeed, on whether it is always wrong. Those who use the term simply assume that we know what it is, that it is wrong, and do not look much further.

One might hope that work in law or philosophy might provide the answer. However, the concept of "censorship" is commonly used in ways that go much beyond the strict confines of First Amendment law.[2] Nor have philosophers, even those who have written much of "freedom of expression," tried to provide a conceptual analysis of censorship itself.[3] In this chapter, I try to fill in this gap by providing a definition of censorship. With this definition in hand, I consider the sorts of justifications given for censorship and canvass the arguments against censorship.

[1] Ken Himma deserves many thanks for his extensive critical comments, which pushed me to clarify and defend my definition of censorship. I also owe thanks to Don Fallis and to my students in the School of Information Resources and Library Science at University of Arizona for many helpful discussions.

[2] I will argue in Section 24.3 that we ought not to equate censorship with infringement of first amendment rights. In other words, I argue that censorship is not by definition an act of the government.

[3] See, for example, Brink, 2001; Cohen, 1993; Dworkin, 1981; Scanlon, 1972; Scoccia, 1996; and van Mill, 2002.

The Handbook of Information and Computer Ethics, Edited by Kenneth Einar Himma and Herman T. Tavani

24.2 THE INTEREST IN ACCESS TO EXPRESSION

Censorship limits access to an expression, either by deterring the speaker from speaking or the hearer from receiving such speech.[4] By an "expression" I mean anything that may be composed by one person and communicated to another. This includes such things as speeches, personal communications, books, articles, compilations of data, art works, photographs, and music. Given that censorship limits access to expression, it is important to have clearly before us why access to expressions is valuable. Cohen (1993) provides an admirably clear and convincing account of the fundamental human interests that can only be satisfied if there is access to the expressions of others. Cohen links our concern with freedom of speech to three fundamental interests: (1) the interest in expression, (2) the interest in deliberation, and (3) the interest in information (pp. 223–230).

Cohen defines the interest in expression as "a direct interest in articulating thoughts, attitudes, and feelings on matters of personal or broader human concern and perhaps through that articulation influencing the thought and conduct of others" (Cohen, 1993, p. 224). Note that although Cohen's emphasis is on acts of expression directed to others with the goal of "influencing thought and conduct," this should not be understood as limited to those works that are clearly propositional in character. Works of art, such as novels, music, photographs, and paintings also "articulate" "thoughts, attitudes, and feelings." There are a number of ways in which *access* to expression supports this interest in expressing. First, as Cohen notes, most acts of expression are acts of communication to others.[5] By promoting access to information, we are enabling the success of such expressive acts by connecting, for instance, the writer with the reader. Second, to engage in acts of expression, people need a rich information culture that will allow them to develop their ideas and learn how to communicate them effectively.

There is, however, more to the interest in expression than Cohen's account covers. Cohen does point out how access to expression satisfies the receiver's deliberative and informational interests, but he does not recognize an independent interest in accessing expression. It would be a mistake, however, to think that the interest in accessing other's expressive acts is merely derived from the more fundamental interest that others have in expressing themselves to us. Human beings have an independent interest in accessing the expressions of others. Just as we have a need to express ourselves, we have a need to hear other's expressions. We have, in other words, "a direct interest in accessing the thoughts, attitudes, and feelings of others on matters of personal or broader human concern."

In addition to our direct interest in expressing ourselves and hearing what others have to say, access to expressions is necessary to satisfy what Cohen calls our "deliberative interests." The deliberative interest concerns our ability to revise and gain a deeper understanding of our individual and collectively held beliefs and

---

[4]This characterization will be made more precise and defended in the following section.

[5]Not all expressions are acts of communication to another person; think, for example, of a private diary (though one might argue that it is an act of communication with one's future self).

commitments. This requires access to expressions of others, because of "the familiar fact that reflection on matters of human concern cannot be pursued in isolation. As John Stuart Mill emphasized, reflection characteristically proceeds against the background of an articulation of alternative views by other people" (Cohen, 1993, p. 229). It is only in the context of free access to the full range of "alternative views," according to Cohen, that we can engage in deliberation on what to believe, value, and do.[6]

Finally, access to others' expressions allows us to leverage our collective epistemic labor and satisfy what Cohen calls our "informational interests." The "informational interest" is the "fundamental interest in securing reliable information about the conditions required for pursuing one's aims and aspirations" (p. 229). Without access to such information, individuals and groups will be unable to effectively carry out their aims. In a free society, we assume that the individual and the collective good is promoted by persons having the ability to determine for themselves what they value and having the freedom to pursue those goals effectively (as long as they do not interfere with a similar pursuit by others). Access to the information and knowledge contained in the expressions of others allows us to do this. Furthermore, the well-being of both individuals and groups requires that we base our actions on the best information available, information we are unlikely to be able to get all on our own.

To summarize, we have an interest in access to expressions based on our fundamental interests in communicating with others, both as speakers and as hearers. In addition, we have a fundamental interest in accessing expressions based on individual and collective deliberative and informational interests. Notice that the focus here is on the primary importance of the capacity of persons to communicate with each other, rather than on the interest in the freedom of the speaker to speak. Standard discussions of free expression focus almost exclusively on the importance of person's freedom to engage in acts of expression. However, it is less often noted that such expressions are fundamentally acts of *communication*. If persons do not have the right to receive information so expressed, the act of communication is prohibited.

The view advocated here is that the goal of freedom of expression is not that a speaker gets to speak, but that people are able to communicate with each other. The goal of speech is to reach a willing hearer. Indeed, often we are concerned with protecting freedom of speech for the sake of the hearer, rather than for the sake of the speaker. For example, the U.S. Constitution's First Amendment protection of freedom of the press is not there to protect the members of the press's ability to write stories, but to protect the right of the public to be informed by such stories. As the U.S. Supreme Court put it in *Griswold v. Connecticut*, "The right of freedom of speech and press includes not only the right to utter or to print, but the right to distribute, the right to receive, the right to read (*Martin v. Struthers*, 319 U.S. 141, 143) and freedom of

[6]Note again that such alternative views are not only communicated through works of opinion, but also through works of art, novels, and other creative works.

inquiry, freedom of thought, and freedom to teach (see *Wieman v. Updegraff*, 344 U.S. 183, 195)."[7]

Freedom of speech, thus, includes the liberty to express one's point of view to others and the liberty to receive any such expressions. To censor is to interfere with this liberty by either suppressing such expressions or preventing the reception of such expressions. In short, a censor wishes to prevent a willing speaker from speaking to a willing hearer. To censor is to interfere in an act of communication between consenting individuals.

24.3 DEFINING CENSORSHIP

Above I characterized censorship as limiting access to content, either by deterring the speaker from speaking or the hearer from receiving the speech. Or, more informally, it is to interfere with acts of communication between consenting adults. In this section, I develop and defend the following more precise definition of censorship: To censor is to *restrict or limit access to an expression, portion of an expression, or category of expression, which has been made public by its author, based on the belief that it will be a bad thing if people access the content of that expression.*

Before considering some possible objections, it is worth explaining some key features of this definition. First, to leave it an open question whether censorship can be justified, I avoid defining censorship in a way that would beg the question as to whether it is always wrong. In other words, this definition tries to capture only the descriptive, and not the normative element of censorship. Some might object that this fails to recognize the fact I noted above; we typically use the term "censorship" to pick out wrongful attempts to suppress expressions. Also, to the extent that suppressing an expression is not wrong, we are often loath to call it censorship. So, for example, one might think that it is not censorship to try to restrict people's access to child pornography. My view is that significant progress can be made by simply focusing on the descriptive element of what makes something a candidate for "censorship." Recently, Carson (2006) defended a similar approach to the definition of lying. Carson argues that "There are good pragmatic reasons for us to use the concept of lying to help point out and distinguish between salient features of actions and thereby assist us in making moral judgments. In order to serve this purpose, the concept of lying must be defined independently of controversial moral assumptions" (p. 288). The same reasons for avoiding normative elements apply to the definition of censorship. What we want to know is what makes the sort of limiting of expression that we call

---

[7]In a concurring opinion to *Lamont v. Postmaster General*, Brennan expanded on this theme, writing that. "It is true that the First Amendment contains no specific guarantee of access to publications. However, the Protection of the Bill of Rights goes beyond the specific guarantees to protect from congressional abridgement those equally fundamental personal rights necessary to make the express guarantees fully meaningful.... I think the right to receive publications is such a fundamental right. The dissemination of ideas can accomplish nothing if otherwise willing addressees are not free to receive and consider them. It would be a barren marketplace of ideas that had only sellers and no buyers." (U.S. at 308)

"censorship" (often or always) wrong. To do this it is not helpful to first define it as wrong at the outset.[8]

Second, this definition makes the motivation of the censor part of what makes an act count as "censorship." This is pretty standard in the definitions of censorship; however, it is worth noting a way in which this definition differs from many others that are offered. The above definition specifies that the motivation must be to avoid something "bad," but does not say it is based on disapproval or moral judgment on the content itself. Many definitions of censorship define censorship as motivated by moral disapproval of the material in question. According to the American Library Association, for example, censorship is "based on a *disapproval of the ideas expressed* and desire to keep those ideas away from public access" (ALA, *Intellectual Freedom and Censorship Q & A*) [emphasis added]. Similarly, Heins (2004, p. 230) defines censorship as "suppressing or restricting speech based on *disapproval of its content*" [emphasis added]. And, according to Boaz (2003, p. 469), censors want to prevent access to content that is "dangerous to government or *harmful to public morality*" [emphasis added].

Although the motivation for censorship is often disapproval of the content or worry about its effects on "public morality," this is not always the case. I may not morally disapprove of some content, but still think it would be bad if people had access to it. For example, I do not morally disapprove of the information about how to make a bomb out of household chemicals. I do not think having such information available will hurt "public morality." The information itself is interesting and perfectly benign. Nevertheless, I may think that it is appropriate to limit access to such material. In such a case, I do not "morally disapprove" of the content itself; I morally disapprove of what persons might *do* with such information. And, this disapproval would not be some particular perspectives of mine. It is likely that almost everyone would agree that it would be a bad thing if someone used this information to blow up a power plant. Thus, I think this definition captures the broader range of motivations that might underlie attempts to censor some expression.

Third, this definition limits censorship to those efforts to restrict access to an expression "made public by its author." I have included this proviso in the definition to exclude cases where access to private or secret expressions is restricted. To make an expression public is to communicate it with the intent that anyone may access it.[9] Above I described censorship as interfering with an act of communication between consenting persons. If I have not willingly made some information public, then you are not censoring when you prohibit others from accessing it. Thus, for example, if my doctor limits your access to my private medical records, even if he does so specifically because he thinks it would be a bad thing if you accessed the content contained in that

[8]An alternate approach would be to take my definition and add "wrongfully." One could, for example, specify that censorship only occurs when a person has a right to expression or when the harms of such restriction in fact outweigh the harm of access.

[9]This would be a case of a completely public expression. Expressions may also be public relative to a particular audience. For example, I may wish to communicate something only to my students or to my friends, in which case it would have been made public to them only.

record, he is not engaging in censorship. Nor does the government censor when it does not give anyone who asks access to information about its current investigations into a drug smuggling ring.

Some may object that by adding the "publicity" requirement, I have rendered the definition too narrow, because it would not include things that are inappropriately kept away from the public. Indeed, there may be cases where someone has the right to know details of my medical history or where the public has a right to know information that the government would prefer to keep secret. However, failing to tell people what they have the right to know is different from censorship. My sexual partner may have the right to know my HIV status, but to fail to tell him my HIV status or to lie about it is not to engage in censorship. To censor is to be a third party interfering in a communicative exchange. A censor keeps one party from telling another party something. The censor does this either by shutting up the speaker in some way or by interfering with the recipient's access to that speech. If the speaker does not wish to speak to the recipient or the potential recipient does not wish to receive that speech, there is no censorship.

This does not mean that there are not issues of intellectual freedom that concern secret or private works, but these issues are about whether such works *should* be made public. Refusing to make some information public may be detrimental to the deliberative, informational, and expressive interests discussed by Cohen. And, thus, respecting intellectual freedom may in some cases require that such information is made public. This means that censorship does not capture all types of inappropriate limiting of people's access to information. Rather than being overly narrow, however, this definition allows us to make important distinctions between different ways in which one's intellectual freedom and right to access may be abridged.

Some might also object that this definition is too narrow, because it is overly focused on restricting *access*, rather than speech. And, thus, it does not clearly classify cases where a speaker is prevented from even speaking, for example, by fear of punishment or by limitation of opportunities, as cases of censorship. As I argued in the previous section, there are good reasons to focus on access to expression. As Brennen (*Lamont v. Postmaster General*, 1965) said, "The dissemination of ideas can accomplish nothing if otherwise willing addressees are not free to receive and consider them" (p. 308). However, while this definition focuses on the restriction of a hearer's access to some expression, it would still classify punishing or in other ways restricting the speaker as censorship. If the goal of such restrictions or punishments is to prevent others from accessing the potential speech, then it satisfies the above definition. In such a case, access to an expression is restricted by muzzling the speaker.

The more serious objection to this definition, however, is that it is overly broad. Many discussions of censorship focus exclusively on actions by the government. LaRue (2007, p. 3), for example, defines censorship as "the action by *government officials* to prohibit or suppress publications or services on the basis of their content". Some would object that by not limiting censorship to governmental bans on expression, I have overexpanded the concept of censorship. In the rest of this section, I argue to the contrary; this government-focused conception of censorship is arbitrarily narrow.

It is true that what we may first think of when we hear the word "censorship" is government putting in place a law that bans an expression from being published or sold. And, it is also true that the state's "monopoly on power" makes it a particularly effective censor. If the government forbids the public display of a work, its effort at censorship will likely be very effective. By contrast, if I as a private individual refuse to have some work in my house because I don't want my visitors reading it, I would be a very ineffective censor. However, while the state may have maximal power within a society, such maximal power is not necessary for successful suppression of expression. Suppose, for example, that some entity has control over an entire mode of communication and can restrict certain categories of expression from being included in the media under its control. I fail to see what difference it would make whether it was the government or some other entity exercising this power. It most certainly would not make a difference to the ability of the public to satisfy their interests in access to expression.

For the same reason it is not clear that to censor, one must absolutely *prohibit* persons from accessing speech by banning it. There are many ways in which one could significantly limit a person's access to some expression without prohibiting it entirely. One could make the work prohibitively expensive. One could make it socially humiliating to access the expression. One could make it a laborious and time-consuming effort to access it. One could make it very difficult to access the expression without particular skills or technological devices. Furthermore, using all of the above methods, one could limit access without punishing transgressors. True, to censor one needs to have some power, but this does not need to be the power to punish. If one has the financial wherewithal to control some access points to information, one would likely have the power to do at least some of the things listed above. And, any of these acts would sufficiently inhibit the ability of willing speakers to get to willing hearers that it should count as censorship.

24.4 TYPES OF HARM AND ARGUMENTS AGAINST CENSORSHIP

I claim that someone censors when they think that it will be a bad thing if other persons access certain expressions. The idea of access to some content being "bad" is very open to interpretation—what sorts of badness do I have in mind? It is important to distinguish two ways in which one might think that access to some content is a bad thing. One may think it is *inherently bad* for persons to access an expression or that it is merely *instrumentally bad* for persons to access an expression. In what follows, I discuss how access to content might be thought to be bad in each of these ways and I canvass the arguments that censorship is not justified on these grounds.

24.4.1 Inherently Harmful Access

Often persons would like to censor content that they find offensive and objectionable. They are not primarily concerned to stop the possible bad consequences that might result from such access; they are concerned to stop the access itself. This view holds that accessing some content is simply *inherently* bad. Examples of content that people have held it would be inherently bad for people to access are explicit sexual content, racist language, violent content, blasphemous works, treasonous works, politically "subversive" works, and the like. People frequently find such content offensive on its face. Part of the harm of someone accessing such expressions is thought to lie in the act of reading or hearing itself, independently of any further consequences that may arise from such access. There are three notable ways in which one might think that access to information is inherently bad: (1) the material is offensive or insulting to the recipient, (2) the material is degrading or corrupting to the character of the recipient (and perhaps the speaker as well), or (3) accessing the material exploits the human beings who are the subjects of the expression.

Given that I characterize censorship as interfering with the communicative acts of consenting adults, I am going to put aside cases of type (1), where persons are the targets of communications that they find offensive or insulting and reasonably do not wish to receive. In particular, I will not deal here with cases such as Stanford University's policy forbidding hate speech that is "addressed directly to the individual or individuals whom it insults or stigmatizes."[10] Even most liberal theorists, who tend to promote wide access to information, make a distinction between choosing to access some content and "inadvertent" exposure to it.[11] I focus here on cases where some third party wishes to prevent some act of communication between others. So, I will only consider cases where, for example, some racist expression is offered to those wishing to listen to it.[12] In such cases, the person who wishes to engage in the censorship is not directly offended or insulted by the speech she accesses, but thinks it is a bad thing that *other* people are able to access such content.

If one objects to others' accessing "offensive" material, it may be because one thinks that it is damaging or corrupting to a person's character to access such material. For example, some believe that it is degrading or sinful to look at certain images—an argument often made with regard to viewing pornography. However, it is not the case that inherent badness of access is always merely "self-concerning" in this way. One might also argue that accessing some material is inherently bad with regard to others. In particular, it may be argued that by merely reading or looking at some material we are failing to treat others with respect. Suppose, for instance, some content is created via the exploitation or victimization of persons. If others then access this content to get some benefit from it, it may be argued that these persons are participating in and continuing this exploitation. This argument has been made with regard to using or publishing data gathered by Nazi experiments on Jewish people.[13]

---

[10]Cited in Brink (2001), p. 134.

[11]See, for example, Brink (2001); Dworkin (1981); Feinberg (1985); and Wendell (1983).

[12]The question of whether speakers may have a right in some cases to speak to unwilling hearers is an important issue that I do not explore here.

[13]See, for example, Cohen's (1990, p. 126) description of Howard Spiro's view that by using the data from experiments carried out on Jews, "we make the Nazis our retroactive partners in the victims' torture and death."

Using such material may be seen as contrary to the Kantian (Kant, 1998 [1785]) principle of respect; it treats the humanity of the victims of the Nazis as merely a "means" to get information.

According to a number of liberal theorists, the fact that some people believe that it is degrading or offensive to access some content does not show that such content should be unavailable to those who do not share this view. Most commonly Mill's "harm principle" is cited in support of the claim that such restrictions would be unjustifiable. According to Mill (1975 [1859], p. 15): "The only principle for which power can be rightfully exercised over any member of a civilized community, against his will, is to prevent harm to others. His own good, either physical or moral, is not a sufficient warrant." Similarly, Ronald Dworkin has suggested that the principle of "moral independence" gives us a reason to reject such censorship. On this principle, human beings "have the right not to suffer disadvantage in the distribution of social goods and opportunities... just on the ground that their officials or fellow-citizens think that their opinions about the right way for them to lead their own lives are ignoble or wrong" (Dworkin, 1981, p. 194). By this view, the fact that some persons do not like the idea of other people accessing this content is a very attenuated sort of "harm" that would unlikely to be enough to justify restricting the access of those who do wish to access it.

However, the harm argument does not deal with the sorts of cases I have called "participation in exploitation." In these sorts of cases the badness is not merely self-regarding—the concern is not merely that I am degrading my character. The idea here is that human beings deserve to be treated in certain ways, independently of whether they are causally affected by that treatment. Perhaps a thought experiment will help to illustrate what I mean. Suppose we are in a world where there are no children and no more children will be born (as was depicted in the film "Children of Men"). Thus, there is no possibility of any child being abused in the future. Would there be any other-regarding badness in looking at child pornography? Clearly my looking at or even buying such material can neither lead to a market in it nor to anyone actually abusing children. However, one might argue that by looking at this material I am failing to show respect to the children who were photographed—I am exploiting them. Some may object to the idea of this rather ethereal sort of harm to others. Nevertheless, it is worth noting that there may be other-regarding concerns about access to expressions, even if one does not think that such access will causally produce harm.

24.4.2 Instrumentally Harmful Access

Not all persons who wish to censor some material wish to do so because they think such access itself is bad. Rather they may be concerned about the bad consequences that might result from such access, for example, that it will cause harm to someone else. In short, one might think that it is *instrumentally bad* for persons to access the content in certain expressions. There are a number of ways in which one may think access to some content can result in bad consequences. Below I categorize four sorts of bad consequences one might want to avoid by censoring:

(1) creating a market, (2) creating a hostile atmosphere, (3) influence, and (4) implementation.[14]

(1) **Creating a Market** Given that persons may be harmed in the creation of certain sorts of content, to provide access to such content is to create a market that will lead to the creation of more content, and thus to more harm to those used in its creation. Most notably, this argument has been made in relation to child pornography.[15]

(2) **Hostile Atmosphere** The accessibility of certain sorts of content may create an attitudinal environment that undermines the equality and agency of some person or group of persons. For example, some argue that pornography should be censored, because it creates a social atmosphere that perpetuates sexism and the objectification of women.[16]

(3) **Influence** The exposure to certain sorts of content may tend to create harmful or antisocial attitudes and behaviors. For example, the most frequent argument for restricting the amount of violence in the media is that it influences children to be more violent.[17]

(4) **Implementation** Information may provide instructions or information that can be used to do something that causes harm. This argument is used to support the censorship of works that describe, for example, how to create a bomb, how to commit suicide, how to make drugs, and so on.[18]

Is censorship ever appropriate to avoid these bad consequences?

Before moving on to look at answers to this question, it is worth putting aside at this point a particularly popular, but specious, argument against censorship. According to this argument, no matter how bad allowing people access to some expression might be, we should never engage in censorship because it will take us down a slippery slope. The American Library Association voices a common sentiment when it says on its Web site that "What censors often don't consider is that, if they succeed in suppressing the ideas they don't like today, others may use that precedent to suppress the ideas they do like tomorrow" (ALA, *Intellectual Freedom and Censorship Q & A*). According to this argument, there is no way to distinguish appropriate from inappropriate censorship. But, what reason is there to think we cannot make necessary distinctions? One could

---

[14]Note that these are not mutually exclusive; one may think that access causes all of these harms at once. Nor are inherent and instrumental harms mutually exclusive—one may think that access to some content will be bad in all these ways.

[15]As was held by the Supreme Court in *New York v. Ferber*, "the advertising and selling of child pornography provide an economic motive for, and are thus an integral part of, the production of such materials. . . ." A similar argument has also been made with regard to publishing the results of unethical research; see, for example, Luna, 1997.

[16]See, for example, Langton, 1990, "pornography is seen as a practice that contributes to the subordinate status of women."

[17]See, for example, Etzioni (2004). See Heins (2004) for a critique of this view.

[18]For a discussion of restricting information about bomb making, see Doyle (2004).

just as well say about the law that, "What legislators don't consider is if they succeed in prohibiting actions they don't like today, others may use that precedent to prohibit actions they do like tomorrow." As van Mill (2002) puts it, "we are *necessarily* on the slope whether we like it or not, and the task is always to decide how far up or down we choose to go, not whether we should step off altogether." But, the slippery slope is not the only argument against censorship based on instrumental harm.

There are three standard arguments that hold that it is not appropriate to censor expressions to avoid possible negative consequences: (a) in fact there are no such negative consequences that can be tied to accessing such expressions, (b) rights to information cannot be overridden based on such consequentialist reasoning, and (c) the harms created by denying access will in almost every case outweigh any harm created by allowing access.

(a) is what Cohen (1993) calls the "minimalist" defense of freedom of speech: negative consequences can never be tied to the simple fact that someone has accessed some information. According to the minimalist, the censor fails to recognize the role of people's agency in accessing and *assessing* information. People are not passive recipients of information; they bring their own values, knowledge, and perspectives to the information. Simply because someone reads a book that says that the Holocaust did not occur or a book that describes how to build a bomb, it does not follow that they will come to doubt the existence of the Holocaust or go out and build the bomb. If I am, for example, accessing child pornography to figure out how to prevent it, then the information is not harmful. One might think of the minimalist slogan as a riff on the National Rifle Association's "Guns don't kill people" slogan—"Information doesn't harm people, people harm people." However, the fact that any information can be harmless or even beneficial in the right hands does not show that general access or access by the wrong persons to such information may not lead to great harm. Only a very strong libertarian argument would hold that it is never appropriate to limit access to those things that have some benign uses, but that also enable a person to cause grave harm (e.g., guns, explosives, powerful painkillers).

(b) is what Cohen calls the "maximalist" defense of freedom of speech. According to maximalism, even if access to information may lead to harm, it ought never be restricted. This trumping is different from the claim (that I will consider below) that as a matter of fact the benefits of access to expression will outweigh its costs. On the maximalist view, access to information is not the sort of thing that can be weighed against other goods. The maximalist typically bases the value of speech on Kantian concepts such as autonomy. Strauss (1991, p. 354), for example, argues that to restrict access to persuasive speech involves the "denial of autonomy" because it interferes "with a person's control over her own reasoning process". One limitation of the rational autonomy argument, however, is that it is not clear that it holds for expressions that cannot be understood as rational or persuasive speech.[19]

A second more serious limitation of the autonomy argument is that it relies on an overly individualistic epistemology, wherein everyone must evaluate information for

[19]Scoccia (1996) argues, for example, that the "persuasion principle" articulated by Strauss does not cover material that engages in nonrational, nonpersuasive speech.

herself. To see this, we need to clarify an ambiguity in how access to information may lead to harm because of implementation. First, consider the typical sort of case we might worry about. Suppose that Joe wants to check out *The Anarchists Cookbook*, because he wants to build a bomb to blow up his neighbor's mailbox. If he gets access to this text, then, given that the directions in it actually give him accurate information about how to build a bomb, he is able to effectively carry out his plans. The information, owing to its accuracy, allowed the receiver to attain the knowledge necessary to achieve his goals.

Things may be however, quite different. Suppose that Susan wants to check out a particular popular book about health, because she wants to learn how to control her diabetes. However, suppose that the information in the popular book is inaccurate. In that case, the information, owing to its inaccuracy, may lead the receiver into error and make it harder for her to achieve her goals.

In the first case, some may want to limit access to the information because they disapprove of Fred's goal and want to prevent him from getting the knowledge necessary to reach it. In the second case, some may want to limit access to the information because they approve of Susan's goal and want to prevent her from getting inaccurate information that will inhibit her ability to reach it. It is less clear that the second sort of "censorship" is interfering with Susan's autonomy. Rather, it seems that we are promoting her autonomy by helping her to achieve her goals by keeping her away from inaccurate information. In Cohen's terms, by censoring we would be supporting her "informational interest"; her interest is an *accurate* information that will help her carry out her plans.

According to consequentialist supporters of (c), "restricting speech is more likely to have bad consequences than protecting it" (Brison, 1998, p. 321). Mill (1975, 1859), for example, famously argues that by allowing a free flow of ideas and information, we will be more likely to get to the truth and to keep hold of the truth that we have achieved. More recently Doyle (2001) has argued that in the context of libraries, even though access to expressions may lead to harm, in almost every case the value of access will outweigh any harms created by it. By this view, whether or not to access would be a decision to make, like any other, based on whether or not it will produce the best consequences. And, such decisions will depend on the facts of the cases in question. The defender of access to expression may worry that such an approach would too easily allow censorship.[20] They worry that by allowing a balancing of access against other goods, we put intellectual freedom at risk. Woodward (1990), for example, rejects such consequentialist defenses of access, because it does not "defend intellectual freedom in principle." "The consequentialist defense of intellectual freedom then, depends upon first establishing whether or not people are better off when they are exposed to all intellectual efforts. The typical result is that one starts dividing intellectual efforts into those that are good for people and those that are not" (p. 15).[21]

---

[20]For an argument, in contrast to Doyle (2001) that utilitarianism cannot provide an absolute defense against censorship, see Fallis and Mathiesen (2001).

[21]For a critical discussion of the claim that there is a fundamental right to information, see Himma (2004).

However, it will not be quite so easy to show that censorship will lead to less harm overall than allowing access. First, it is worth reminding ourselves of the central interests in expression, deliberation, and information, which can only be satisfied when there is a general free flow of expression. The benefits we receive from having these interests satisfied (and the harms from not having them satisfied) will not be easily overridden. Second, we have to ask ourselves not what *in principle* it might be good to censor. We have to ask ourselves what *in actual practice* would be the consequences of having policies in place that restrict access. It is at this point that "slippery slope" and "chilling effect" arguments might have some force. The slippery slope may be an actual and not just a conceptual possibility, if human beings in fact tend not to be so good at distinguishing material they personally dislike from that which is harmful.[22] Also, there may be a genuine chilling effect on expression if people tend to be overly cautious in avoiding social disapprobation. As Doyle (2001, p. 69) argues, "For intellectual freedom to be genuine, people must have the confidence that they will not be harassed for what they publish or seek out." This is not to say that, even taking all of this into account, using consequentialist reasoning will never lead you to think that censoring is the right thing to do; nevertheless, the hurdle is quite a bit higher than one might have thought.

24.5 CONCLUSION

In this chapter, I have proposed the following definition of censorship: restricting or limiting access to an expression, portion of an expression, or category of expression, which has been made public by its author, based on the belief that it will be a bad thing if people access the content of that expression. A virtue of this definition is that it does not make describing an act as "censorship" a discussion stopper. In other words, this definition allows that there may be cases of censorship that are morally permissible or even obligatory. People on both sides of a debate can agree that some action fulfills the above definition and they can then go on to have the real conversation about whether such an action is justified. This conversation requires that we look carefully at both why access to expression is important and what the harms related to access might be. Then we can think through the justifications for and against censorship in a clear and systematic way. My hope is that this chapter has gotten us started on this more fruitful approach to the issue of censorship. Ultimately, I believe that given our strong interests in access to expression and the reasonable concerns about human implementation of policies that restrict access, cases of justifiable censorship will likely be relatively rare.[23]

[22]Doyle (2001) also argues that more harm than good may be produced if we let people censor, because whoever the "censor" is will be tempted to use this power in ways not justified by utilitarian reasoning.

[23]At least with regard to adult's access to expressions. There are likely more cases where it will be justifiable to restrict children's access. See the chapter by Mathiesen and Fallis in this volume for a discussion of children's rights to access information.

REFERENCES

Association, American Library (2007). *Intellectual Freedom and Censorship Q & A*. Available at http://www.ala.org/ala/oif/basics/intellectual.htm

Boaz, M. (2003). Censorship. Bates, M.J.M., Niles, M., and Drake, M. (Eds.), *Encyclopedia of Library and Information Science*. Marcel Dekker, New York.

Brink, D.O. (2001). Millian principles, freedom of expression, and hate speech. *Legal Theory*, 7, 119–57.

Brison, S. (1998). The autonomy defense of free speech. *Ethics*, 108, 312–339.

Carson, T. (2006). The definition of lying. *Nous*, 40(2), 284–306.

Cohen, B. (1990). The ethics of using medical data from Nazi experiments. *Journal of Halacha and Contemporary Society*, 19, 103–126.

Cohen, J. (1993). Freedom of expression. *Philosophy and Public Affairs*, 22(3), 207–63.

Doyle, T. (2001). A utilitarian case for intellectual freedom in libraries. *Library Quarterly*, 71 (1), 44–71.

Doyle, T. (2004). Should web sites for bomb-making be legal? *Journal of Information Ethics*, 13 (1), 34–37.

Dworkin, R. (1981). Is there a right to pornography? *Oxford Journal of Legal Studies*, 1(2), 177–212.

Etzioni, A. (2004). On protecting children from speech. *Chicago-Kent Law Review*, 79(3), 3–53.

Fallis, D. and Mathiesen, K. (2001). Response to Doyle. *Library Quarterly*, 71(3), 437.

Feinberg, J. (1985). *Offense to Others: The Moral Limits of the Criminal Law*. Oxford University Press, Oxford.

Griswold v. Connecticut. 381. U.S. 479 (1965).

Heins, M. (2004). On protecting children—from censorship: a reply to Amitai Etzioni. *Chicago-Kent Law Review*, 79, pp. 229–255.

Himma, K. (2004). The moral significance of the interest in information: reflections on a fundamental right to information. *Journal of Information, Communication, and Ethics in Society*, 2(4), 191–201.

Kant, I. (1998, 1785). *Groundwork of the Metaphysics of Morals*. Cambridge University Press, New York. (Translated by Gregor, M.J.).

Lamont v. Postmaster General. 381. U.S. 301 (1965).

Langton, R. (1990). Whose right? Ronald Dworkin, women, and pornographers. *Philosophy and Public Affairs*, 19(4), 311–359.

LaRue, J. (2007). *The New Inquisition: Understanding and Managing Intellectual Freedom Challenges*. Libraries Unlimited, Westport, CT.

Luna, F. (1997). Vulnerable populations and morally tainted experiments. *Bioethics*, 2(3 and 4), 256–264.

Mill, J.S. (1859, 1975). On Liberty. *Three Essays*. Oxford University Press, Oxford.

New York v. Ferber. 458. U.S. 747 (1982).

Scanlon, T. (1972). A theory of freedom of expression. *Philosophy and Public Affairs*, 1(2), 204–226.

Scoccia, D. (1996). Can liberals support a ban on violent pornography? *Ethics*, 106, 776–799.

Strauss, D.A. (1991). Persuasion, autonomy, and freedom of expression. *Columbia Law Review*, 91(2), 334–371.

van Mill, D. (2002). Freedom of speech. In: Zalta, E.N. (Ed.), *The Stanford Encyclopedia of Philosophy*, Winter 2002 edition. Available at <http://plato.stanford.edu/archives/win2002/entries/freedom-speech/>

Wendell, S. (1983). Pornography and freedom of expression. In: Copp, D. and Wendell, S. (Eds.), *Pornography and Censorship*. Prometheus, Buffalo, pp. 167–183.

Woodward, D. (1990). A framework for deciding issues in ethics. *Library Trends*, 39(1 and 2), 8–17.

CHAPTER 25

The Gender Agenda in Computer Ethics

ALISON ADAM

25.1 INTRODUCTION

The idea that gender is a major (possibly even *the* major) way of classifying and ordering our world has been propounded by a number of authors, mainly, although not exclusively, writing from a feminist position.[1] The recognition of continuing differences between men's and women's lots, at home, in the workplace, and in education, even in societies that have seen considerable opening up of opportunities for women in the space of a generation or so, has been a major force in developing contemporary feminist writing.[2] Coupled with this, we continue to be fascinated by differences *between* men and women, putative differences in behavior, differences in interests, differences in management style, differences in parenting approach, and so on. This suggests that, in the process of ordering and classifying our world into that which is feminine, that is, belonging to or pertaining to women, and that which is msasculine, that is, pertaining to men, we may polarize this binary in a search for, and perhaps even a maintenance of, difference.

How do the above issues relate to moral thinking? A large body of writing on feminist ethics has sprung up in the last 30 or so years, developing from earlier grass roots work on women's rights and from theoretical developments in feminist philosophy.[3] The job of feminist ethics is twofold. First of all, it forms a substantial critique of traditional ethical theories, which, it argues, can be seen as masculine in conception. Second, it seeks to develop new feminist forms of ethics, derived, at least in part, from the challenge to mainstream ethics but focusing on women's moral experiences in

[1]See Adam (1998).

[2]Wajcman, J. (1991). *Feminism Confronts Technology.* Polity Press, Cambridge.

[3]Tong (1993).

The Handbook of Information and Computer Ethics, Edited by Kenneth Einar Himma and Herman T. Tavani

order to make normative judgments on a wider range of issue, particularly those areas where women have traditionally assumed a subordinate role or where they have had negative experiences because of their gender.

As one of the leading writers on feminist ethics Rosemarie Tong [4] puts it, the job of feminist ethics is, "... to create a gender-equal ethics, a moral theory that generates nonsexist moral principles, policies and practices."

To date, the focus of feminist ethics has tended to be women's caring roles, especially mothering.[5] There are some theoretical problems with this focus, particularly in terms of the emphasis on "ethics of care" that can be seen as problematic as it reinforces women's traditional self-sacrificing role while, at the same time, emphasizing a level of control over those who are cared for. There have been few attempts to apply feminist ethics to science and technology.[6] However, given the arguments above, it is important to extend the reach of feminist ethics to domains beyond and wider than traditional caring roles because differences in men's and women's experiences and potential inequalities between them have far-reaching consequences beyond our roles as carers. This involves broadening the scope of feminist ethics in theoretical terms that will extend its range, ideally by making many of its many good ideas more widely applicable. This includes building a bank of case studies in wider areas such as technology and the workplace where a range of human and technological relations are evident, and demonstrating that there are gender issues that are important across a wide spectrum of ethical situations.

Computer ethics is a new area of applied ethics with a rapidly burgeoning portfolio of ethical case studies and problems. In this chapter, I frame the question: "What gender issues are involved in computer ethics and what contribution may feminist ethics offer computer ethics?" In the following section, I briefly introduce the topic of feminist ethics. The next section reviews existing research on gender and computer ethics. This falls into two main categories: empirical comparisons of computer ethics decision making by men and women and other aspects of gender and computing that have been considered in ethical terms in the literature—the latter usually involves a consideration of the low numbers of women in computing. In forming a critical analysis of these areas, I identify a number of gaps where extended discussion from a gender perspective would benefit several current problem areas within the purview of contemporary computer ethics. These include topics such as cyberstalking and hacking. Finally, and more speculatively, I suggest what might be offered from these ideas about gender analysis of computer ethics back to the theoretical development of feminist ethics, framing the discussion on "cyberfeminism" as a possible locus for a feminist computer ethics.

---

[4] Tong (1999), available at http://plato.stanford.edu/archives/fall1999/entries/feminism-ethics/ (accessed 24th November 1999).

[5] Puka (1993), Ruddick (1989).

[6] Adam (2005).

25.2 FEMINIST ETHICS

Feminist ethics grew out of long-running debates about the special nature of women's morality that date from at least the time of Wollstonecraft's[7] *A Vindication of the Rights of Women* in the eighteenth century and Mill's[8] concerns about the virtue of women in, *The subjection of women* in the middle of the nineteenth century. Arising from an interest on grass root issues such as sexualities and domestic labor, juxtaposed with more theoretical concerns, the topic of feminist ethics was firmly put on the feminist agenda by Gilligan's[9] *In a Different Voice*, which is often regarded as the canonical work of feminist ethics.

Care ethics is a cornerstone of most approaches toward feminist ethics. Gilligan's *In a Different Voice* was written to counter Freud's notion, echoed in Kohlberg's work, that while men have a well-developed moral sense, women do not. Gilligan argued instead that women often construct moral dilemmas as conflicts of responsibilities rather than rights and that, in resolving such conflicts, they seek to repair and strengthen networks of relationships. It has been immensely influential in spawning the tradition of feminist ethics with a focus on the ethics of care, mothering, and relationships as an approach particularly attached to women's values. This signals a commitment to responsibility rather than rights, the collective social group rather than the individual, and an ethic based on caring rather than the supposedly impartial individual reason of the Kantian moral agent.

This is the basis of an "ethic of care." Indeed, the concept of an ethic of care has emerged as a strong theme, if not the strongest theme, in feminist ethics. Jaggar[10] has termed it "a minor academic industry."

Considerable debate continues to surround Gilligan's work. Although she was criticized and subsequently revised her position, her work has made an enormous impact in the academy beyond the disciplines of ethics and psychology. When it was first published its ideas appeared very radical. On the one hand she does claim that women's moral development is different from men's, but on the other she argues that traditional scholarship on ethical development is not neutral but is designed to favor a masculine, individualistic, rationalistic, justice, and rights-based approach to ethics over a feminine, communitarian, care-based approach. In other words, she argued that the standard of morality that is valorized as the gold standard is based on a masculine model of ethical reasoning.

In her original study, Gilligan claimed that her empirical research demonstrated that women tend to value an ethic of care that emphasizes relationships and responsibilities, while men value an ethics of justice that emphasizes rules and rights. The evidence from a later study[11] was not so clear cut, as the women in that study focused equally on justice and care while only one man espoused a care ethic. Gilligan

---

[7]Wollstonecraft (1988).

[8]Mill (1970).

[9]Gilligan (1982).

[10]Jaggar (1991).

[11]Gilligan (1987).

saw her work as a refutation of Freud's argument that women are somehow morally inferior to men. As a development of Freud's work, Kohlberg[12] elaborated a six-stage theory of moral development from Stage One, as a punishment and obedience orientation, up to Stage Six, as an orientation toward universal moral principles as described by Kant. Kohlberg found that women rarely get past Stage Three (interpersonal concordance), while men usually ascend to Stage Five (social contract legalistic orientation). Such findings could be used to argue that women are less morally developed than men. Gilligan questioned these findings to argue that Kohlberg was really describing male rather than human moral development. In any case, much hinges on the interpretations of Kohlberg's study.

Although it is beyond the scope of this chapter to elaborate a detailed critique of Gilligan's work, Koehn[13] notes that care ethics may not be unequivocally better than the purportedly masculine justices and right-based ethics. Indeed, care- and justice-based approaches may be complementary and we may be in danger of polarizing men's and women's experiences by emphasizing the differences between the two approaches.

25.3 GENDER AND COMPUTER ETHICS—A MALE–FEMALE BINARY?

Turning to computer ethics, it is interesting that the topic of gender has received relatively little attention to date. However, one major strand of research concentrates on looking for differences in men's and women's ethical decision making with respect to computer ethics problems. It should be noted that such research does not appear in the philosophically inclined computer ethics journals such as *Ethics and Information Technology*; instead it belongs within a business and management research paradigm that favors an approach based on statistical surveys and where rigor is highly prized.[14] In forming a critique of this work, I am conscious that the criticisms I make could be leveled at most statistical studies that treat gender in the way I describe below, and therefore are more widely applicable than in computer ethics.[15] Nevertheless, these papers provide an approach toward ethical behavior in relation to gender that has been explicitly applied to computer problems, and a number of them appear in information systems journals, traditionally a place where some research on computer ethics appears.[16] I should make it clear that some of the argument that I make is not specifically a "gendered" critique. For example, criticizing excessive use of student populations and certain types of numerical survey are not, of themselves, gender issues, and they are criticisms that might be leveled more generally at some types of survey approaches to business and management research. My reason for alluding to

[12]Kohlberg (1981).

[13]Koehn (1998).

[14]Oakley (2000).

[15]Adam et al. (2001).

[16]Bissett and Shipton (1999), Escribano et al. (1999), Khazanchi (1995), Kreie and Cronan (1998), Mason and Mudrack (1996), McDonald and Pak (1996).

them here is partly because they are part of the rounded story of this type of research on computer ethics and partly because they reinforce some of the critical elements that *are* clearly gendered in nature. Specifically, I argue that they serve to reinforce gender stereotypes that are prevalent in this research.

Broadly speaking, the research methodology in such studies can be characterized as follows. A population of subjects (in these studies almost always a student population) is surveyed by questionnaire and is asked to rate responses in relation to either a set of questions or a set of artificial scenarios. Responses are usually Yes/No or rated against a Likert scale (a scale with a number of points (3,4,5, or more), where one end indicates the most positive response and the other end indicates the most negative response). The results are then analyzed quantitatively (some using little more than percentages, but mostly using more sophisticated statistical methods). This may involve splitting out various ethical variables and rating subjects' responses against them. The analysis is then turned back from quantitative measures into qualitative conclusions, which are, in some cases, that women are more ethical than men in relation to computer ethics problems, in other cases that there is no discernable difference. Interestingly, none of the studies I cite found that men were more ethical than women. Sometimes, these results are related, theoretically to Gilligan's *In a Different Voice*, which is still the best-known work in feminist ethics, but other prominent studies make no use of feminist or gender-based ethics in terms of explanation.[17] Interestingly, Gilligan's work was the only substantive reference to writing on feminist ethics that I discovered in any of these studies.

The above description is very broad brush, and one could argue that it is too sweeping in its generalization of detailed studies from a researcher who clearly favors qualitative approaches. Nevertheless, I hope to capture the predominant style of a set of studies that reflects its roots in the dominant North American business research paradigm and to argue why such an approach is not appropriate for gender analysis nor for computer ethics, without wishing to mount a broadside attack on an approach that may be perfectly appropriate for other areas.

The studies of Khazanchi and Kreie and Cronan,[18] respectively, mirror several other studies more focused on gender and business ethics[19] in that they focus on student audiences, using a questionnaire to elicit responses and analyzing these by standard statistical techniques in order to find support for gender differences in ethical decision making. Khazanchi's aim was to understand whether gender differences influence the degree to which individuals recognize unethical conduct in the use and development of information technology. To this end, a sample of undergraduate and graduate business students was surveyed against a set of seven ethical scenarios and were asked to rate these as to degree of "unethicalness." These scenarios reflected categories comprising the ethical responsibilities of IS professionals regarding disclosure, social responsibility, integrity, conflict of interest, accountability, protection of privacy, and personal conduct and were derived from earlier research. Subjects were asked to rate the

---

[17]Gilligan (1982), Bissett and Shipton (1999), Igbaria and Chidambaram (1997), Kreie and Cronan (1998), Mason and Mudrack (1996), McDonald and Pak (1996).

[18]Khazanchi (1995), Kreie and Cronan (1998).

[19]Mason and Mudrack (1996), McDonald and Pak (1996), Reiss and Mitra (1998).

unethical acts in each scenario against a 7-point Likert scale where 1="absolutely not unethical" and 7="absolutely unethical," with no labels for the intermediate range. Khanzanchi then derived an aggregate score of "unethicalness" and correlated this against gender. Despite concerns as to the external validity of using students in the survey, he found that the women of his survey consistently outperformed the men in identifying unethical actions across all his scenarios. "The present study shows the ability to recognize (and ultimately resist) unethical actions involving IS dilemmas rests in part on the nature of the ethical dilemma and differences in gender of the adjudicator. The findings provide an insight into gender differences in the ethical judgement of future leaders and managers in the management information systems discipline."[20]

Bissett and Shipton's[21] questionnaire survey of IT professionals studying part time used a set of scenarios with respondents rating whether they would undertake similar behavior on a scale of "always" to "never." They found a small positive correlation between female gender and a tendency to consider the feelings of others. By contrast, Escribano, Peña, and Extremora's[22] survey of university students involved Yes/No responses to a number of questions. They found the women in their survey far more interested in the ethical aspects of information technologies than the men, despite the fact that they used such technologies much less than the male respondents.

Probably the most prominent of empirical studies of gender and computer ethics in the last decade is Kreie and Cronan's research, which is published in the high-profile periodical, *Communications of the ACM*, read by practitioners and academics.[23] Although supplemented by later research on ethical decision making where gender features as a variable,[24] the earlier paper remains the most widely cited possibly because of its position in a widely circulated periodical. In addition, it is very typical of the genre I describe, and therefore worthy of close inspection.

These researchers explored men's and women's moral decision making in relation to a set of computer ethics cases. The examples were, by and large, not blatantly criminal but were designed to reflect the situations we are often presented with within the workplace where extensive computer systems and networks are pervasive, for example, viewing sensitive data, making an electronic copy. The main research method in the study involved asking respondents to rate their responses against a set of influential environmental factors such as societal, individual, professional, and legal belief systems. In addition, there are so-called "personal values." The authors proposed these factors to be those that influence ethical decision making. As is typically found in similar studies, a student population was surveyed and asked to rate whether the behavior described in a given scenario was acceptable or unacceptable.

Following the survey, it appears that some discussions with students helped explain judgments about the various scenarios. Respondents were also asked about their moral

[20]Khazanchi (1995), p. 744.

[21]Bissett and Shipton (1999).

[22]Escribano et al. (1999).

[23]Kreie and Cronan (1998).

[24]Kreie and Cronan (1999).

obligation to take corrective action and whether knowledge of negative consequences, for example, a fine or reprimand would affect what a person should or should not do. For each scenario, the respondents were asked which set of values, for example, personal values, societal, environmental, influenced their decision most. The authors' conclusion was that most people were strongly influenced by their personal values. Kreie and Cronan[25] conclude that "men and women were distinctly different in their assessment of what is ethical and unethical behavior. For all scenarios, men were less likely to consider a behavior as unethical. Moreover, their judgement was most often influenced by their personal values and one environmental cue—whether the action was legal. Women were more conservative in their judgements and considered more environmental cues, as well as their own personal values."[26] The authors make suggestions as to the policy implications of these results: "From the manager's viewpoint, men may be influenced more effectively through statements of what is legal (or not). Women might be effectively influenced by passive deterrents (policy statements and awareness training of unacceptable ethical behavior)."

Some of the findings of this study are reinforced by Leonard's and Cronan's later work,[27] which is based on Kreie and Cronan's[28] attitudinal model. This model postulates that attitudes toward ethical behavior can be broken down along dimensions of belief systems, personal values, personal, professional, social, legal, and business environments, moral obligation, and consequences. Although this study focuses on attitudes toward ethical decision making rather than on ethical decision making *per se*, it is very typical of the genre of study described here in that a student population's attitudes are subject to statistical sampling, where gender is regarded as one of the variables explaining attitude difference.

25.4 GENDER AND COMPUTER ETHICS STUDIES

These studies belong to a particular genre of research strongly representative of the business and management literature, a style of research that perpetuates its approach without substantial inroads from other types of research.

While statistical sampling of large populations remains an a important approach toward understanding some sorts of research questions, there are problematic aspects in all the studies reported under this banner. In the detailed critique below, I outline the critique in the form of a set of critical elements. The first two, viz. the question of how appropriate it is to survey a student audience and the perennial concern of quantitative versus qualitative research, are research issues that must be addressed by any researcher of business and management topics. They are not "gendered" issues as such, although I do argue that they reinforce problematic gender stereotypes. The next element, namely, what one is surveying in ethical terms, that is, general ethical

---

[25]Kreie and Cronan (1998), p. 76.

[26]Kreie and Cronan (1998), p. 76.

[27]Leonard and Cronan (2005).

[28]Kreie and Cronan (1999).

behavior, ethical decisions, or something else, is relevant to empirical research on ethics. Finally the question of how gender is dealt with theoretically is relevant to all social studies of gender. Hence, it is the intersection of these four vectors that forms a critique of gender and computer ethics empirical research.

25.4.1 Student Population

In every one of the studies detailed above, including the latest study by Leonard and Cronan,[29] a student population was surveyed. Although it is clear that in some of the studies the students also worked or had work experience, the tendency to utilize and then generalize from a student population is still problematic.

This is a methodological issue because, as Mason and Mudrack note, this may give a certain homogeneity to the results obtained as the sample is clearly not representative of a general population.[30] But there is a wider issue that can be cast as an ethical question. If we use our students in this way, we may take advantage of them because of the power differential between student and teacher. I am not arguing that this is a specifically gender question, rather that this is a problem that can be found in all the gender and computer ethics studies I surveyed and it appears to be prevalent in much business and management literature, not just in the research that forms the focus of this chapter. Interestingly, I have not found the question addressed in any empirical studies of the type described here. A student and a teacher do not stand in the same relationship as a researcher and a member of the public. The teacher grades the student's work and gives testimonials for future employment or education. A student may feel unable to opt out of the research, to make their view felt by consciously not taking part as a member of the public might do. Additionally, given that students become very adept at telling their teachers what they think they want to hear under the guise of "examination technique," there is the question of whether a student will apply the same process to a teacher's research survey, consciously or otherwise.

The *raison d'etre* of the critical wing of management and information systems research rests on exposing differences in power.[31] However, at the other end of the spectrum, it is important that more traditional approaches do not ignore questions of power. There must be ways of being confident about the use of statistical approaches while still acknowledging such issues. This also raises the question of how far one can expect results to be reproducible in populations drawn from outside the student body. For instance, would the same questions applied to a sample drawn from the general public, or from a professional group, produce a similar result? If not, there is some "tacit knowledge" in relation to being a student and how that affects the results, which is not being made explicit in this approach.

---

[29]Leonard and Cronan (2005).

[30]Mason and Mudrack (1996).

[31]Howcroft and Trauth (2005).

25.4.2 Quantitative Versus Qualitative Research Methodologies

All the studies detailed above were similar in approach, in that they all employed questionnaire surveys, either with a binary "Yes/No" or 5-point or 7-point Likert scale, which could then be analyzed quantitatively for statistical significance. Such an approach reflects the dominant quantitative paradigm of management and information systems research. Nevertheless, it is interesting that this style of research never questions research methodology (the hegemonic approach that does not demand explanation in the research literature), whereas it is quite common to see research papers reporting interpretive and critical research in business and management (not the dominant paradigm, therefore having to make more effort to justify itself) with extended discussions of research methodology.

While acknowledging the value of statistical studies, nevertheless there are clear problems with such an approach in the studies I consider. Only the Bissett and Shipton[32] paper points to the problem of whether what people say they do is the same as what they do in a real-life situation. This may be even more of a problem than usual in the present set of studies, as respondents are explicitly asked whether they would behave in some potentially immoral or even illegal way. In other words, respondents are not being asked to choose between categories that are anything like neutral. It is naturally tempting to cast oneself as more moral in the questionnaire than one might be in real life.

This is clearly a very well-trodden path in all social research, where the quantitative/qualitative or positivist/interpretivist battle continues to rage.[33] Within the social sciences, and in business and management research, there continues to be debate over the validity of quantitative methods—surveys and statistics—versus qualitative methods such as interviews and observation. Broadly speaking, quantitative approaches map onto what might be termed the "positivist" paradigm, which assumes an objective world that is amenable to measurement. This is often regarded as the method of the sciences. Although qualitative approaches are used under the positivist banner, they are more often associated with an interpretivist or constructivist approach, which looks to the research subjects' interpretations of their world and which sees knowledge as being socially constructed.

We should not ignore the power of the academy, especially in the United States, in shaping such matters into a certain type of mold from which it is difficult to escape, at least if one wants a permanent, tenured position. Within business and management research, at least, the positivist tradition dominates and many academics feel obliged to follow it. Despite the preponderance of quantitative approaches to gender and computer ethics decision making, the issue of the appropriateness of statistical methods applied to ethics cannot be ignored and points to the need for a wider consideration of the appropriateness of other research methods.

There is the question of what responses on a numerical scale actually mean and whether subjects can reliably attach meaning to the individual intervals in a 7-point

---

[32]Bissett and Shipton (1999).

[33]Oakley (2000).

scale. Is 1="absolutely not unethical" the same as "absolutely ethical" or not and does it differ from 2="not quite so absolutely unethical?" Shades of ethicality would appear to be conceptually clumsy, very hard to define, and not necessarily very meaningful. I am not suggesting that a qualitative approach could be used here where the numbers are just replaced by words. Rather, I argue that a qualitative paradigm would approach the whole question quite differently.

Although from my comments above one would not expect the dominant quantitative research paradigm to be challenged, nevertheless it is interesting that none of the authors in these studies considers the possibility of interviewing or using ethnographic techniques such as participant observation.[34] Such research is expensive. Participant observation requires an often lengthy period immersed in the culture under study. The observer becomes part of and participates in the culture. But, at the same time, the observer must retain a degree of strangeness from the culture under study, otherwise he/she will begin to take for granted aspects of that culture that need to be analyzed and made explicit. For computer ethics, the promise of participant observation lies in the potential to witness ethical reasoning and behavior as it happens in the field. This may reveal it to be a process with a much more complex and less clear-cut structure and which may not even result in a decision at all, when compared with the instant Yes/No decisions prompted by questionnaires. This is a point that will be reconsidered below where ethical behavior is reconsidered.

One cannot help but note not only that interviewing and participant observation are much more time-consuming techniques but also that their results are much less amenable to rendering into numerical form. Questionnaires can be made to yield numbers that can then be fed into the statistical mill no matter what the validity of the original qualitative assumptions on which they were based. The research is then distanced from its foundations that are not subject to scrutiny.

In performing a quantitative analysis of qualitative elements, the studies described above appear to be falling prey to the common assumption prevalent in business studies, computing, and information systems, which I have criticized elsewhere, namely that objective factors are available and that these can be easily and unequivocally identified.[35]

Indeed, in the Kreie and Cronan study and the later paper by Leonard and Cronan,[36] there is the additional assumption that, even if such factors do have some reality as discrete factors, we can reliably separate out our beliefs and rate them against things such as social, psychological, or religious beliefs. Can we do this in such a way that each belief system can be identified in an individual's response and can be treated separately? Apart from questioning the validity of such a factoring process, I argue that it allows authors to hide behind the apparent authority of their statistics, obviating the will to develop a more thoroughgoing conceptual, theoretical analysis. In other words, numbers cannot replace theoretical, conceptual explanations.

[34]See Forsythe (1993).

[35]Adam (1998).

[36]Kreie and Cronan (1998), Leonard and Cronan (2005).

The qualitative/quantitative conundrum, to which the above discussion suggests gender and computer ethics empirical studies continue to fall prey, is part of a larger debate between qualitative and quantitative research methodologies. This discussion applies not just to work in gender and computer ethics, although it is starkly visible in the studies I outline above, but is more generally a part of research in information systems and business. Oakley[37] points out that this debate has been a long-running issue in the social sciences. She argues that it is not nearly as clear-cut as it appears as it is impossible to be completely qualitative, for example, we talk of "some," "more," "less." Similarly, it is impossible to be completely quantitative as our quantities are quantities of some quality. Despite this, the debate has assumed an unwelcome polarity, a kind of "paradigm war." Inevitably, one side tends to dominate, and in many parts of the social sciences, good research is thought of in terms of quantitative research.

Although the debate has rumbled around the social sciences for some time, somewhat belatedly the qualitative/quantitative debate has filtered into business and management, computing, and information systems research, where the two camps are seen as "hard" and "soft," roughly translating into quantitative and qualitative and where the "hard," or quantitative, enjoys a hegemony with quantitative techniques favored in North America, and qualitative approaches more prevalent in Scandinavia and Western Europe.[38]

A brief example of a qualitative study on gender and computing can be found in a research project on women in the IT workplace.[39] The researchers undertook a number of in-depth interviews with women working in IT in the UK. They found that their respondents reported much discomfort with being a women in what they perceived as a masculine workplace. They adopted a number of coping strategies, one of which involved conscious attempts to "neutralize" their gender, for example, by wearing androgynous clothes such as jeans and tee shirts, adopting a "tough" style of talking and acting.

25.4.3 What is Ethical Behavior?

Taken together, the above considerations imply that empirical research in this area focuses on discrete ethical acts that are assumed to be readily identifiable and where it is assumed that an unequivocal rating of how ethical the act is can be made. The use of questionnaires with numerical scales forces us to look at the world in this way, so this is a problem not just for characterizing ethical behavior but for characterizing any type of behavior where we use such questionnaires. The particular question for studying ethical behavior is how far it is meaningful to frame ethical behavior in this way.

If we were to focus on the ethical process, rather than the end point of the ethical act, this would make the result of the ethical act less important *per se*, as quite different approaches can arrive at the same result through different routes.

---

[37]Oakley (2000), p. 31.

[38]Fitzgerald and Howcroft (1998).

[39]Adam et al. (2006).

The studies I survey above rarely make their underlying ethical position explicit. So while, as I argue below, gender is undertheorized, so, too, is ethics. Some ethical behavior may be in the form of a well-defined judgment as to whether a discrete act is ethical or not. This kind of ethical behavior may therefore be more amenable to mathematical manipulation. But it is doubtful whether much ethical behavior is amenable to neat formulation.

Looking more squarely at a broad range of ethical behavior suggests that we would have to be much more sophisticated about our theorizing. My comments about this are not purely directed at research on gender. However, I want to argue that this style of research brings with it a view of the world that packages up complex concepts in an overly simple way.

In this research, gender is treated as a unitary, unanalyzed variable. Apart from any other reason this tends toward essentialism, that is, the assumption that men and women have essential, fixed, natural, and even possibly biological, characteristics. The question of essentialism is complex, and space does not permit an extended discussion. Nevertheless, I do want to claim that essentialism can be dangerous, for men and women, as it can let in, through the back door, all sorts of unchallenged stereotypes, for example, that women are less adept at using technology than men, men are more focused on careers than are women. One hopes that the audience for such research is sufficiently sophisticated to recognize and reject simple stereotypes; nevertheless, it is interesting to find gender stereotypes abounding in these studies as the following section discusses.

The fact that a proposition tends toward essentialism does not necessarily make it objectionable, of itself. My argument is rather "where do we draw the line?" We would not be able to agree. For instance, we may assume that there are certain fixed biological characteristics that men and women have. However, how do we classify individuals with "gender dysphoria," that is, the experience of knowing one is a woman trapped in a man's body or, conversely, knowing one is a man trapped in a woman's body? Apparently clear biological characteristics are sometimes not so clear.

Additionally, we could look at seemingly innocuous statements such as "women can become pregnant; men cannot become pregnant." Clearly, there are many steps in between noting that (only) women can become pregnant and possible discrimination because of that fact. Yet, as I write, a new report on the UK workplace[40] notes that women with young children face more discrimination in the workplace than disabled people or those from ethnic minorities. The report cites a survey of 122 recruitment agencies that revealed that more than 70% of them had been asked by clients to avoid hiring pregnant women or those of childbearing age. My claim is that apparently innocuous facts that are regarded as essential qualities of men or women may ultimately lead to behavior that is repressive.

None of the studies related above is substantially reflective on the adequacies of their data gathering methods. Yet my arguments imply that, in the longer term, if we wish to gather data about real moral behavior in the field we must turn to more anthropologically inspired methods, in particular, forms of ethnography and

[40]Equalities Review Panel (2007).

participant observation where the observer participates and becomes part of the culture. Such an approach is likely to yield much richer accounts of ethical behavior than can be gained solely by questionnaire-type surveys. Additionally, the use of observational techniques is likely to move the focus of ethical behavior away from discretely identifiable moral acts as the primary activity in acting morally. The emphasis on discrete ethical acts can be seen as part of a mechanistic, Tayloristic view of management that regards a goal, achieved through a set of rational steps, as the primary activity of the manager. This puts the position rather starkly; nevertheless, the rationalist view of management has proved extraordinarily tenacious and difficult to challenge.

25.4.4 The Undertheorizing of Gender and Ethics

The arguments above suggest that existing work on empirical research on gender and computer ethics is substantially undertheorized both in terms of gender and in terms of moral behavior in regard to computer ethics issues. Part of the problem is that the field is far more fragmented than I have presented it in this review. By and large, the studies I discuss here do not appear to "know" about one another, and they refer to neither philosophically inclined ethics research, critical management research that focuses on power, nor feminist ethics. There is little sense of a tradition where one study builds on another; wheels are continually reinvented. A second aspect of the weak theoretical base of this research is displayed in the way that, for some of the papers reviewed, as I suggest above, the authors end up making often unwarranted stereotypical generalizations that do not appear to follow from their research, by way of conclusion. For instance, Kreie and Cronan[41] conclude from their study that women are more conservative in their ethical judgment than men, and that they might be best served by passive deterrents toward unethical behavior, while men might require more substantive ethical deterrents.

It is hard to see why women's apparent tendency toward more ethical behavior should make them more conservative. There appears to be a jump in reasoning here with several steps skipped over. "Conservative" in this context means less likely to take risks and more likely to conform to established societal norms. This does not follow from the research issues involved in these studies but starts to look like a stereotypical judgment about an expectation of men's more "laddish" behavior against a "well-behaved" female stereotype where women are seen as guardians of society's morals. In this case, such "bad behavior" might be seen as more acceptable for men than for women. It is against just such a stereotypical judgment that feminist ethics seeks to argue. Similarly Khazanchi's[42] conclusion is that women are better able to recognize "and ultimately resist" unethical behavior. However, it is not clear why the ability to resist unethical behavior should necessarily go along with the ability to recognize it. Once again this conclusion smacks of gender stereotypes of women's "good" behavior or moral virtues. Of course, one could argue that it hardly matters that women are seen

[41]Kreie and Cronan (1998).

[42]Khazanchi (1995).

as more morally virtuous than men. However, what happens if women are penalized for not living up to saintly expectations? A man stepping out of line might be let off the hook, if it is felt that he could not help his masculine instincts. Yet there may be harsher sanctions for women who will not stay within the confines of traditional feminine behavior.

A clear example of the problem of an approach that carries stereotypical, unanalyzed attitudes to gender with a factoring approach is evident in Leonard and Cronan's[43] study, where they report that women's attitudes to the ethical behavior depends not only on moral obligation in all the cases in the study but also on various other influencers depending on the case.

By contrast, men's responses were more uniform, depending primarily on moral obligation, awareness of consequences, and personal values. The conclusion is that "males have a given set of parameters when assessing a situation, whereas females adjust their parameters depending on the situation."[44] It is worth looking at this conclusion in some detail as it highlights the problematic aspect of this research. First, the conclusion is drawn that the data implies that women *adjust* their parameters while men do not. But the researchers have not captured any data on women making adjustments to their moral outlook; rather, they assume that this is what the data implies. It may well be that their assumed "factors" model of ethical belief does not fit very well for women as it is based on a masculine approach to ethics. Various inferences can be drawn from these findings. Nevertheless, as a minimum the model does make women appear more changeable, while men appear to be more constant in their approach.

Clearly, this could be interpreted in different ways. It could be seen as casting women in a positive light, perhaps being more sensitive to subtle ethical nuances. However, if one returns to Gilligan's[45] critique of Kohlberg's[46] model of ethical maturity, which purported to find men more morally mature than women, then women's supposed tendency to change their minds could be cast as a negative aspect of their moral reasoning. Gilligan's argument was rather that women tended to have a different approach to ethics based on caring and relationships rather than a justice and right-based approach (traditionally seen as the highest form of ethical reasoning and more associated with men). So in Leonard and Cronan's study, a different set of factors based more on feminine rather than masculine values might have made the women in the study appear more constant and less changeable. Indeed, an alternative research approach would have been to use grounded theory,[47] which would have let relevant concepts emerge from the data rather than starting off with an assumed model of ethical reasoning where respondents were required to rate their responses against a set of factors.

[43]Leonard and Cronan (2005).

[44]Leonard and Cronan (2005), p. 1167

[45]Gilligan (1982).

[46]Kohlberg (1981).

[47]Strauss and Corbin (1997).

But the most significant aspect of the undertheorizing problem relates to the way that this research makes so little reference to the, by now, quite substantial body of research on feminist ethics that could be used to help explain results. We can take citation of Gilligan's *In A Different Voice* as a kind of minimum level of reference to feminist ethics. Of the research reviewed above, only McDonald and Pak, Mason and Mudrack, and Bissett and Shipton[48] refer to it, and, indeed, it is the *only* work of feminist ethics referred in any of the studies. Although academic disciplines so often run on parallel tracks with little intersection, it is surprising how far these studies avoid feminist research.

Surprisingly, neither Kreie and Cronan nor the later paper of Leonard and Cronan[49] makes reference either to Gilligan or to any other part of the large body of writing in feminist ethics, which might have helped them explain their results. Indeed, they make no attempt to explain *why* their research apparently reveals differences between men and women. This is all the more surprising given that Gilligan's work is very widely known over a number of domains, unlike other work in feminist ethics. Importantly, had these authors engaged with the debate surrounding Gilligan's work, which also centered round an empirical study, they would have been able to apply not only her arguments but also the criticism of her arguments to good effect on their own study. On the latter point, Larrabee[50] notes that one of the criticisms of Gilligan's research was that she asked her respondents to work through a number of artificial case studies rather than observing them making real, live ethical decisions. As I have argued above, this is difficult research to undertake, it requires a time-consuming observational approach rather than a survey, and it raises unsettling questions as to the focus of so much management research on decision making or reported attitudes.

A similar concern applies with Kreie and Cronan's study.[51] Asking respondents to approve or disapprove of a scenario where software is copied illegally may well invoke disapproval in subjects. Although some people regard intellectual property rights as illegitimate, many will want to be seen as good software citizens.

However, just like driving slightly above the speed limit, small-scale software copying is rife, and this study just does not get at subjects' moral decision making and the processes behind that decision making in real scenarios where they may be faced with the decision of whether or not to copy some desirable and readily available piece of software, where, on the face of it, no harm is done if the copy is made. To some extent, this is recognized in the later Leonard and Cronan study, where respondents appeared less concerned with legal climate than in earlier work, although it is not clear why this should be the case. Questionnaires and interviews are problematic. Researchers can never be sure if people will respond to a "live" situation in the same way as they have detailed in the questionnaire. Indeed, as individual respondents, none of us can be sure that we will behave the way we thought we would and the way we may have described, in all good faith, in a questionnaire.

[48]McDonald and Pak (1996), Mason and Mudrack (1996), Bissett and Shipton (1999).

[49]Kreie and Cronan (1998), Leonard and Cronan (2005).

[50]Larrabee (1993).

[51]Kreie and Cronan (1998).

Although these questions always dog social science data gathering, there are special reasons why there are particular problems with gathering ethical data. This relates to the gap between "is" and "ought." We may well recognize good ethical behavior and therefore respond accordingly in a questionnaire, but we may not have the moral fiber to stick to our good intentions when faced with a real-life situation. This is likely to be more apparent at the "petty crime" end of the scale. For computer ethics, small-scale software copying provides a good example of something that is not legal yet is endemic and causes perpetrators little loss of sleep.

This is much like the argument in Nissenbaum's[52] "Should I Copy My Neighbor's Software?" On the face of it, taking the viewpoint of standard ethical positions, the answer appears to be "no." But following Nissenbaum's detailed arguments shows that the answer is not nearly so clear-cut when one probes the reasons for copying or not copying in more detail. The binary approval/disapproval in Kreie and Cronan[53] or scales of approval and disapproval invoked by Likert scale studies evoke too sharp a Yes/No response. Indeed, there are hints that the researchers found the responses too clear-cut in this study where the authors decided to go back and interview groups of students as to how they arrived at decisions. In other words, these authors find themselves obliged to go back in order to probe the processes behind the decisions.

25.5 WOMEN IN COMPUTING—AN ALTERNATIVE APPROACH

A substantial literature of empirical studies on the low numbers of women in IT and computing has grown up in the last 30 or so years. To some extent this can be viewed as a spillover from information systems and computing research on barriers and "shrinking pipelines,"[54] which tends to see the gender and ICT problem as one of women's access to ICTs and their continuing low representation in computing all the way through the educational process to the world of work. Indeed, if one talks of women and computing or gender and computing, it is the question of low numbers of women that many people regard as the issue. This is undoubtedly a problem (although not the only problem) and worthy of serious consideration. Until recently, such research found voice more substantially in the research areas of work, education, psychology[55] and on the fringes of computing disciplines.[56] However, papers in this general mold are beginning to appear in ethics journals and computer ethics conferences, suggesting that authors are starting to cast the women and computing access/exclusion problem as an explicitly ethical problem, although this is not how the area has been traditionally seen in the past.[57]

[52]Nissenbaum (1995).

[53]Kreie and Cronan (1998).

[54]Camp (1997).

[55]Brosnan (1998).

[56]See Lovegrove and Segal (1991), Grundy (1996).

[57]See Panteli and Stack (1998), Panteli et al. (1999), Turner (1998), Turner (1999).

Studies that discuss the low numbers of women in computing have been criticized, in the past, for adopting a traditional, liberal position that characterizes the gender and computing problem in terms of educating, socializing, and persuading women rather than challenging the subject matter and deeper structures of the subject.[58] A liberal position assumes the neutrality of the subject—here computing—and further assumes that it is just a question of attracting more women into the discipline for equality between men and women to prevail. This is a classic "add more women and stir" argument, but it does nothing to tackle the underlying structural reasons why women are absent from the discipline.

Apart from anything else, a liberal argument, by leaving the organization of the computing profession unchallenged, does little to offer a means of alleviating women's position in relation to computing education and work. Furthermore, campaigns to attract women based on such a position do not appear to work. I do not want to imply that all the gender and computing research I cite above suffers from it. Interestingly, because such work is beginning to view itself as ethics research it sidesteps some of the criticisms of liberalism because there is a growing realization that deeper, structural issues are involved in the question of women's inequality.

Although the woman and computing problem is not new, it is still there. Numbers of women though all levels of computing remain low, meaning that women are still absent from employment in well-paid and interesting careers for whatever reason, and so it is a problem yet to be adequately addressed.

Casting this more as an ethical problem than an access problem starts to make the issue look less like a question of why women are not, apparently, taking up the opportunities being offered to them; in other words, it is not the relatively simple "liberal" problem of running publicity campaigns to attract women into the discipline. It is more of an ethical and political problem of potential exclusion even if exclusion is not intended. In other words, it moves the onus for change away from women, and their apparent failure to take up challenges, toward the computer industry's failure to examine and change its potentially exclusionary practices. For instance, in research undertaken with my colleagues,[59] we found an example of a woman physically being barred from the laboratory that she managed. Furthermore, we found a working environment in computing that was often hostile to women, exacerbating the tendency of experienced women to leave an IT job mid-career. This serves to act as an important reminder of how little has changed for women in the computing industry in the last 20 or so years. Hence, casting the problem of women and computing explicitly as an ethical problem rather than purely as a failure on the part of women to take up opportunities has the positive effect of potentially strengthening the political dimension.

Another, again barely explored aspect of computer ethics in relation to gender relates to education. Turner[60] describes her battle to produce a gender-inclusive computer ethics curriculum, an effort that appears to be unique. It may only be when

[58]Henwood (1993), Faulkner (2000).

[59]Adam et al. (2006).

[60]Turner (2006).

we see gender integrated into the curriculum rather than as a kind of "bolt on" that we may have a better basis for future work.

Slender though this body of research may be, it raises many important theoretical concerns for a computer ethics informed by feminist theory to explore further—professionalization and professionalism, the masculinity of expertise, experiences of democracy and freedom, accountability and responsibility, how ethics education may be augmented to include gender issues—and thereby points the way toward a more thoroughgoing theorizing of gender and computer ethics issues.

25.6 GENDER AND COMPUTER ETHICS—CYBERSTALKING AND HACKING

With the additional consideration of a few more recent papers, the above discussion represents the conclusions I had arrived at as to the current state of research on gender and computer ethics while completing my book on the subject, *Gender, Ethics and Information Technology*. In that work, I issued a plea for more substantive empirical work on computer ethics topics informed by a gender analysis, in the hope that this could strengthen theoretical dimensions of the topic. I also hoped to see the debate move on beyond attempts to search for differences between men's and women's ethical decision making. Furthermore, it is clearly desirable to develop gender and computing research, in new directions, beyond a concentration on low numbers.

There are some clear candidates for a more extended gender analysis among computer ethics topics. Although it is beyond the scope of this chapter to examine this in detail, in *Gender, Ethics and Information Technology*, I argue that[61] theoretical tools could include not just feminist ethics but also feminist political and legal theory, where the intention is to see computer ethics problems not just in moral terms but also in social, political, and legal ways. One effect would be to give a better theoretical armory with which to uncover power relations, an important part of any analysis along gender dimensions, to be used alongside ethical theory.

Cyberstalking, or the stalking of an individual by means of information and communications technologies, represents an important computer ethics topic for feminist analysis. Although the majority of perpetrators are male and the majority of victims are female, this point is rarely adduced as an important element of the phenomenon requiring explanation. Notably, the policy document produced by Attorney General, Janet Reno, under the Clinton administration, notes the statistics without explanation, preferring to see the phenomenon more in terms of "bad for business," in other words, a potential deterrent to the spread of electronic commerce.[62] Later research echoes this surprising avoidance of the centrality of the gender dimension of cyberstalking.[63] A feminist analysis of the phenomenon draws on

[61]Adam (2005).

[62]Reno (1999).

[63]Bocij and Sutton (2004).

writing from feminist politics[64] to argue that we should look to the well-worn split between public and private, noting that historically the concept of privacy has related differently to women and to men. For instance, note the way that married women only won the right to property, person, and children well into the nineteenth century, as the state was reluctant to interfere in the private world of the home, traditionally women's domain. DeCew[65] notes that, as women have traditionally had few rights to privacy, it is not always easy to see when their rights are being transgressed. This may contain the seeds of an explanation as to why official bodies are reluctant to see cyberstalking as an issue that is especially problematic for women and that may need special measures to counteract it because of this.

Cyberstalking is a fairly obvious candidate for a gender analysis as there are clear ways in which women and men have different experiences in relation to it. However, I argue that a thoroughgoing gender analysis, informed by feminist ethics, may offer us an enhanced understanding of other computer ethics topics. One such topic is hacking. Hacking is well known to be largely a masculine phenomenon. Indeed, the interest that has been shown in finding elusive women hackers could lead us to believe that the low number of women hackers, paralleling the low number of women in the computing industry, was all there is to a gender analysis of hacking.[66]

However, a feminist analysis of the hacker ethic reveals how masculine in inspiration the hacker ethic appears to be. The "hacker ethic" is a complex phenomenon that can be explored along several dimensions. In terms of a gender analysis, possibly the most importance element is the hacker ethic's pretensions to egalitarianism.

25.6.1 Are Hacker Communities Egalitarian?

By the term "hackers" I include both the original meaning of those who apparently spend their waking hours engaged in computer-based activity and also the later meaning of the term that includes illegal breaches of security.[67] The lack of women in hacking appears to be a perennial source of interest.[68] Levy's *Hackers* is the earliest substantial work on the hacker phenomenon. It is an eloquent description of the masculinity of the early hacker world and has been immensely important in raising awareness of the history of hacking. Levy was one of the first commentators to question the paucity of women hackers. He notes, "Even the substantial cultural bias against women getting into serious computing does not explain the utter lack of female hackers."[69] Yet, apparently without irony, this quote is juxtaposed with the observation that "hackers talked strangely, they had bizarre hours, they ate weird food, and they spent all their time thinking about computers" and a long, and unfortunately detailed, description of a male hacker's personal hygiene problems. More importantly, even

[64] DeCew (1997).

[65] DeCew (1997).

[66] Taylor (1999).

[67] Levy (1984).

[68] Taylor (1999).

[69] Levy (1984), p. 84.

taking into account women's actual physical presence in hacking activities, Levy is unaware of the way in which his text renders women invisible.The Who's Who at the beginning of the book includes 3 women, 52 men, and 10 computers. Some of the computes are described in terms of their relationship to men (e.g., Apple II and Steve Wozniak), but all of the women are someone's wives. Hence women, if they do appear, are defined in terms of their relationship to men, even when they have expert hacking credentials. Women are, therefore, largely invisible in this important early work.

An early anecdote involving a woman and programming demonstrates how women may be excluded from the hacker world. Despite the fact that this particular woman, Margaret Hamilton, was an expert programmer (she went on to manage the computers on board the spacecraft of the Apollo moon shot), she is designated by the pejorative term, "Officially Sanctioned User."[70] At MIT, she used the official assembly level language supplied on the DEC machine, not the hacker-written language that had been developed by the MIT hacking group. So, when the hackers of the "Midnight Computer Wiring Society" altered the hardware of the DEC machine one night to run their own programs, hers failed to work the next day. Naturally, she complained, and hardware alterations were officially banned, although in practice, eventually tolerated. The episode was designated the "Great Margaret Hamilton Program Clobber."[71] Although she clearly had considerable programming expertise, and therefore could be expected to achieve some measure of respect, she was an official user, thereby belonging to the official world that was the antithesis of hacking. She used the despised official programming language. She kept daytime hours, rather than joining the nocturnal world of the hackers. Her complaint and the subsequent banning of nocturnal hardware alterations, although initially upheld, gradually reverted to the *status quo*. The hackers were described as "playful," and the incident was cast as a prank.

Levy[72] details the central tenet of the hacker ethic. "Hackers should be judged by their hacking, not bogus criteria such as degrees, age, race, or position." Interestingly the categories "gender" and "able-bodiedness" are absent, though it would be reasonable to conclude that "such as" could include these categories. Levy continues: "This meritocratic trait was not necessarily rooted in the inherent goodness of hacker hearts—it was mainly that hackers cared less about someone's superficial characteristics than they did about his (sic) potential to advance the general state of hacking."

We may challenge the egalitarian rhetoric of hacking where it is assumed that hacker communities are largely meritocracies where race, gender, and so on, are unimportant. Indeed, we might question how such a rhetoric of equality arose in the first place. Part of the answer may lie in the way that hackers cast themselves as a new, alternative political movement challenging the existing order with all its baggage of racism, sexism, and repression. However, just because a political movement challenges a nonegalitarian orthodoxy, this does not mean that it is free from the instruments of oppression itself. And this would be especially problematic for hackers,

[70]Levy (1984), pp. 96–97.

[71]Levy (1984), p. 97.

[72]Levy (1984), p. 43.

where many are not involved politically and others are only loosely organized along the political dimension.

The related notion that democracy is a spontaneous property of new Internet communities has been criticized by a number of authors.[73] The point here is that there is little evidence that equality and democracy should be treated as spontaneously emergent properties of new communities whether they be political or occupational groups.

This is a typical liberal expression of equality, rather in the vein of the employment adverts that claim equal opportunities regardless of ethnicity, disability, gender, religion, and so on. A statement of equality is seen to be enough to achieve it. This view also tends to dismiss as superficial (here the term used is even stronger: "bogus") characteristics of people, characteristics that are far from superficial but are deeply felt, often regarded as positive, aspects of their identities, and which should be respected rather than diminished or dismissed. Under this "differences as bogus" view, equal opportunities are then passive instruments, difference and diversity are not identified, and one need do nothing to achieve equality except state a belief in it.

For the hacker ethic, the salient features of the debate relate to the freedom of information ethic and how it relates to freedom of speech/censorship debates in hacker terms. There is little doubt that organizations such as the Electronic Frontier Foundation and prominent exponents of the hacker ethic and freedom in cyberspace have a political influence. Jordan and Taylor[74] associate a technolibertarian view with the hacker community where strong libertarian and free market principles are closely allied to the hacker ethic. In other words, hacker communities regard it as important that individual liberties are preserved at all costs.

Additionally, there are specific instances of hacker involvement in legislative processes, which do have implications for the freedom of speech/censorship debates particularly insofar as they relate to feminist concerns over pornography. Hackers have been involved in political action, to considerable effect, in relation to Internet legislation. As Raymond[75] notes:

> "The mainstreaming of the Internet even brought the hacker culture the beginnings of respectability and political clout. In 1994 and 1995, hacker activism scuppered the Clipper proposal which would have put strong encryption under government control. In 1996, hackers mobilized a broad coalition to defeat the misnamed 'Communications Decency Act'(CDA) and prevent censorship of the Internet."

In fact the "broad coalition" designating themselves the Citizens Internet Empowerment Coalition[76] consisted of various Internet service providers, broadcast and media associations, civil liberties groups, and over 56,000 individual Internet users. Although designed to protect minors from indecent and offensive material on the

---

[73]Ess (1996), Rheingold (1993), Winner (1997).

[74]Jordan and Taylor (2004).

[75]Raymond (2001), p. 17.

[76]CIEC (1997).

Internet, the successful challenge was made on the grounds that the CDA failed to understand the unique nature of the Internet and that it was so broad as to violate the freedom of speech protections of the First Amendment. We should also note the subsequent dismissal of the Child Pornography Prevention Act,[77] which means that virtual pictures of children are not illegal in the United States. The hacker argument is that such a law does not protect children in the way that is intended.

However, freedom of speech is promoted as the highest ideal. This resonates with Gilligan's[78] research where she argues that care and relatedness are more feminine in inspiration rather than the rights-based approach, here the right to free speech, which she attributed to masculine ethical approaches.

Levy's[79] work on early hackers does not explicitly describe the hacker ethic as a work ethic. Nevertheless, from the importance of the adoption of very particular working habits, there is a clearly implied hacker work ethic and this has implications for egalitarianism. This theme has been explored in more detail by Himanen,[80] who argues that the hacker ethic is a new work ethic that challenges the Protestant work ethic of Weber's classic text, *The Protestant Ethic and the Spirit of Capitalism*,[81] which has dominated Western capitalist societies for so long. Here the hacker ethic is a passionate, joyful, playful approach to work where making money is not the driving force and where access and freedom of expression are explicitly enshrined in the ethic. This contrasts with the Protestant work ethic's emphasis on work as moral and, originally, religious duty, where play and leisure are clearly contrasted with work. Hackers' relationship to time is important and different from the "time is money" ethic of capitalism that has intensified and speeded up in the new information society.[82] Even leisure becomes work as we "work out" in the gym and marshal our home lives in tight schedules, as if the whole of our lives have become Taylorized. The use of new technology can blur the boundaries of work and leisure, but not always to our advantage. Yet Himanen argues that the hacker ethic advocates organizing for playfulness, a freedom to self-organize time, and resists work-time supervision. Some hackers make money through traditional capitalist routes and then, having achieved financial independence, are free to pursue their hacker passions.

The hacker work ethic paints an ideal of freedom from financial worry and the ability to organize one's own work that few, whether men or women, can hope to achieve even in affluent Western democracies. However, from the point of view of the present argument, we need to question whether there is a gender dimension, both in terms of desirability and feasibility, lurking within the hacker work ethic. Studies of women, work, and leisure [83] suggest that women's access to leisure is different from that of men's; they have less of it. So far it seems that patterns of leisure use of information and

---

[77]Levy (2002).

[78]Gilligan (1982).

[79]Levy (1984).

[80]Himanen (2001).

[81]Weber (1930).

[82]Castells (2000).

[83]Green et al. (1990).

communications technologies and the Internet reproduce standard gendered patterns of leisure time availability rather than breaking them down.[84] In addition, many women work a "double shift," that is, they have primary responsibility for home and child care as well as their paid employment. Commentators on the economy of the household can be surprisingly blind to the contribution that women make by their largely invisible labor of looking after homes, bodies, and children.[85] Those whose lives revolve around the computer may not notice the real-life bodily care that goes into keeping them fed, cleaned, and organized for their virtual lives within the machine.[86]

25.7 WHAT MIGHT "FEMINIST COMPUTER ETHICS" OFFER FEMINIST ETHICS?

Writing on "mainstream" feminist ethics relates tenuously to social and philosophical studies of technology. Feminist ethics has concentrated on women's role as carers, with the result that the "ethics of care" has emerged as a largely feminist ethical theory. It is not easy to see how care ethics could be brought to bear on technology. However, there may be alternative ethical frameworks that can be developed on feminist themes and that can fruitfully inform computer ethics. In particular, I argue that a feminist computer ethics based on "cyberfeminism" may offer some alternatives to care ethics for feminist ethics.

The concept of the "cyborg" or cybernetic organism, a hybrid of human and machine, has been immensely popular in "cyberpunk" novels such as Gibson's[87] *Neuromancer* and in science fiction films such as *Terminator* and *Robocop*, yet it has been appropriated as a feminist icon, most famously in Haraway's *A Cyborg Manifesto*.[88] It is difficult to overestimate the influence of her essay, which Christie[89] describes as having "attained a status as near canonical as anything gets for the left/feminist academy."

In Haraway's hands, the cyborg works as an ironic political myth initially for the 1980s but stretching into and finding its full force in the next decade and beyond, a blurring, transgression, and deliberate confusion of boundaries of the self, a concern with what makes us human and how we define humanity. In our reliance on spectacles, hearing aids, heart pacemakers, dentures, dental crowns, artificial joints, not to mention, computers, faxes, modems, and networks, we are all cyborgs, "fabricated hybrids of machine and organism."[90]

For Haraway, the cyborg is to be a creature of a postgendered world. The boundary between human and machine has been thoroughly breached. The transgression of

[84] Adam and Green (1998).

[85] Cudd (2001), pp. 92–94).

[86] Helmreich (1994).

[87] Gibson (1984).

[88] Haraway (1991).

[89] Christie (1993), p. 172.

[90] Haraway (1991), p. 150.

boundaries and shifting of perspective signals a lessening of the masculine/feminine dualisms that have troubled feminist writers, and this means that we do not necessarily have to seek domination of the technology. Her cyborg imagery contains two fundamental messages:

> "...first, the production of universal, totalizing theory is a major mistake that misses most of the reality...; and second, taking responsibility for the social relations of science and technology means refusing an anti-science metaphysics, a demonology of technology, and so means embracing the skilful task of reconstructing the boundaries of daily life....It is not just that science and technology are possible means of great human satisfaction, as well as a matrix of complex dominations. Cyborg imagery can suggest a way out of the maze of dualisms in which we have explained our bodies and our tools to ourselves." [91]

Haraway's approach can be termed "cyborg feminism." In the mid-1990s, "cyborg feminism" was further developed by Stabile. Stabile[92] notes a positive shift in the feminist canon toward a technophilia heralded by cyborg feminism and allied writings. Cyborg feminism can assume an uncritical enthusiasm for technology and science and women's relationships with them. The problem with this is that unbridled technophilia may undermine the feminist political project that has traditionally been critical of science and technology. Wilson[93] puts this more strongly in her concern that cyborg feminism signals this as a complete break with the traditional political activism of the feminist project. "...the cyborg feminist need not do anything in order to be political...the fact that the cyborg signifies is enough to guarantee her politics."

"Cyberfeminism" is a later term than "cyborg feminism." Cyberfeminism is a feminist movement that sees women as subverting traditional masculine control of technology, particularly new information technologies such as the Internet, and that sees women achieving equality with men through their use of such technologies. The term was possibly coined by Sadie Plant[94] in the mid-1990s. Cyberfeminism and cyborg feminism are essentially the same movement, although cyberfeminism is the better-known term. Plant's writing on cyberfeminism centered around the idea of blurring nature/machine boundaries via the concept of the cyborg, coupled with a utopian view of women as naturally in control of their own destinies on the Internet and, indeed, in connection with other technologies. Indeed, Plant argues that women naturally control the Internet. Although the "women in control" concept may be appealing to feminism, it is contrary to the experience of many women in relation to their historical experiences of technology.[95] A number of authors have criticized cyberfeminism as being too utopian, not sufficiently historical, political, or critical,

---

[91]Haraway (1991), p. 182.

[92]Stabile (1994).

[93]Wilson (1993).

[94]Plant (1997).

[95]Adam (1998), Squires (1996).

and not sufficiently aware of women's mixed experiences of cyberspace.[96] Yet, at the same time, there are many positive aspects to the idea of cyberfeminism. Additionally, as I shall argue below, of late there has been a movement to attach political activism to cyberfeminism, which may make more recent manifestations of cyberfeminism more attractive than earlier approaches that appeared to be consciously apolitical.

For all its difficulties in connection to the political project of feminism, cyberfeminism does have a slightly subversive quality as women are seen as actively controlling the technology for their own uses. More recently, commentators have suggested that cyberfeminism is attractive in that it counters much of the pessimism of earlier work on gender and technology that saw technology as subject to relentless male domination.[97] Additionally, cyberfeminism is seen as a way of subverting traditional masculine control of technology. Although many activities that are claimed as cyberfeminist are very serious in scope, there is also an element of play.[98]

Although, there is still a dearth of empirical material on cyberfeminism, it is worth exploring as a way of opening up a more public role for feminist ethics, in contrast to what appears, on the face of it, to be a more inward-looking care ethics. As noted below, in cyberfeminism's favor we now see the seeds of political engagement that were not so clear a decade or more ago when Wilson's[99] critique of cyborg feminism "being" rather than "doing" was appropriate.

Feminist ethics is practical in its intent; therefore, all the difficulties apart, the best way of making at least a beginning to an alternative feminist technoscience or "cyberfeminist" ethics is by constructing an example where a masculine approach to hacking may be revisited and reshaped by an ethic based, at least partly, on the new politically inspired feminism.

Despite the above critique, there are positive aspects to hacker ethics that might be appropriated into a cyberfeminist ethics—notably the notion of the "hack" as a "neat" technological trick and the idea of playfulness. In constructing the outline of a feminist alternative, I look to the experiences and political stance of some women hackers who explicitly name themselves as cyberfeminists and other cyberfeminist activists.

There is some evidence that a female hacker ethic may differ from the male hacker ethic in respect of aspects of freedom of speech and pornography. For instance, Sollfrank[100] calls for a consciously political cyberfeminism for women hackers. In 1997, in protest against the practices of the art world in canonizing male artists and ignoring women artists, she "spammed" a German net art jury with submissions from 127 fictitious female net artists.[101] She wanted this to be read as a deliberate

[96]Adam (1998).

[97]Wajcman (2004).

[98]Schleiner (2002).

[99]Wilson (1993).

[100]Sollfrank (2002).

[101]Schleiner (2002).

transgression of boundaries, typical of cyberfeminism, in that the boundary between aesthetics and activism was to be seen as blurred.

Sollfrank wanted to make a feminist point by direct action and, at the same time, to create an artistic fiction through her virtual women artists. Guertin[102] makes the link between what she terms postfeminist virtual disobedience and artistic and aesthetic forms. Yet at the same time there is an element of playfulness about Sollfrank's actions as part of a "clever hack." It is clear that as a definition of "playfulness" this is, of course, entirely questionable. She did indeed make her political point, but at the expense of the inconvenience of the net art jury. This is a feature of "hacktivism," the blending of hacker techniques with political activism.[103] On the plus side, a clever technical trick can make a political point. On the negative side, the "hack" can seriously cause inconvenience to many people and can lead to loss of business, privacy and so on.

Other groups, or individuals, identifying themselves as cyberfeminists, are engaged in political activity, although not necessarily in hacker terms, which is subversive yet champions aspects of human rights. The subRosa cyberfeminist art collective looks to feminist activity across categories of embodiment as a means of change. "We favor affirming tactics of antidisciple over strategies for coping with inequities, whenever possible."[104] They argue for strategies of resistance in fighting against human rights violations based on race and disability.

Interestingly, some prominent women hackers have explicitly sought to crusade *against* Internet pornography; in other words, they are procensorship and do not elevate freedom of information over other ideals and it is here where we may find the seeds of an alternative feminist hacker ethic. I have noted above that the traditional hacker ethic, with its arguments for freedom of speech and information, does not tend to ally itself with procensorship movements. For instance, Natasha Grigori started out in the 1990s running a bulletin board for software pirates, but has now founded and runs antichildporn.org, where hackers' skills are used to track down child pornography and pass the information to law enforcement authorities. Similarly, a women hacker who called herself "Blueberry" set up another antichild pornography organization, condemned.org.[105] Jude Milhon, a hacker who initiated the alternative magazine *Mondo 2000*, argues that women hackers are more likely to be involved in hacktivism with a political or ethical end than in other areas of illegal hacking.[106] Guertin[107] consciously links cyberfeminism with the political activism of hacktivism, but Remtulla[108] questions whether there are any differences between hacktivism in general and supposed cyberfeminist versions of it. This is still very

---

[102]Guertin (2005).

[103]Jordan and Taylor (2004).

[104]SubRosa (2005), p. 44.

[105]Segan (2000a).

[106]Segan (2000b).

[107]Guertin (2005).

[108]Remtulla (2006).

much an open question, and the issue of whether women hackers are more likely to be involved in hacktivism than their male counterparts awaits further empirical evidence.

Nevertheless, there is some, albeit still slender, evidence of the emergence of a different ethic among female hackers, one where political activism is to the fore and where explicit stances and actions are taken on topics such as child pornography rather than holding to an ideal of freedom of speech at all cost. This ethic combines an ethic of care from feminist ethics of care—children must be actively protected hence the positive attempts to counter pornography on the Internet. We can also see where cyberfeminism comes in as these women are just as technically competent as male hackers, so cyberfeminism's earlier manifestation of women in control of networked technologies may not seem quite so far-fetched, after all. There may be few of them, but they have the knowledge and control of technology that is central to the appeal of cyberfeminism. They are subversive, in that they use hacking for political end through "hacktivism," and at the same time they are unwilling to accept at face value the libertarian approaches of traditional hacker ethics toward equality and freedom of information. Art and aesthetics, even playfulness are combined with political activism, Care ethics is present but brought into the public sphere rather than remaining in the private sphere of mothering. Hence, we see some signs of an emerging women hackers' ethic where hacking skills can be put to use for more conscious political ends in hacktivism and anti child pornography Web sites and enforcement. In tracking down women hackers, we must attend to the latter point especially as there is hope of sowing the seeds of a more inclusive and politically rounded hacker ethic based on cyberfeminism.

25.8 CONCLUSION

In conclusion, this chapter seeks to make a case for gender to receive a more thoroughgoing treatment within computer ethics by considering gender issues that are involved in computer ethics, and also by thinking of the contribution that a feminist version of computer ethics might offer back to the development of feminist ethics as a discipline. In reviewing current research on gender and computer ethics, I am fairly critical of current approaches to men's and women's ethical decision making that appear to be stuck in a traditional business and management research paradigm where differences between men and women are overemphasized and where the concept of gender appears undertheorized. Other current approaches to gender and computer ethics include work on women's underrepresentation in the computing profession. In looking for computer ethics examples amenable to a gender analysis using feminist ethics, cyberstalking and hacker ethics may be analyzed from a feminist position. On a final, more speculative note, I suggest that newer forms of cyberfeminism that emphasize a political intent coupled with elements of subversion and playfulness might offer a new dimension to feminist ethics and the ethics of care that can be pressed into service to offer a new theoretical dimension for a feminist computer ethics.

REFERENCES

Adam, A. (1998). *Artificial Knowing: Gender and the Thinking Machine*. Routledge, London and New York.

Adam, A. (2005). *Gender, Ethics and Information Technology*. Palgrave, Basingstoke and New York.

Adam, A. and Green, E. (1998). On-line leisure: gender and ICTs in the home. *Information, Communication and Society*, 1(3), 291–312.

Adam, A., Howcroft, D., et al. (2001). Absent friends? The gender dimension in IS research. In: Russo, N.L., Fitzgerald, B., and DeGross, J.I. (Eds.), *Realigning Research and Practice in Information Systems Development: The Social and Organizational Perspective*. Kluwer, Norwell MA and Dordrecht, pp. 333–352.

Adam, A., Griffiths, M., Keogh, C., Moore, K., Richardson, H., and Tattersall, A. (2006). Being an it in IT—Gendered identities in the IT workplace. *European Journal of Information Systems*, 15(4), 368–378.

Bissett, A. and Shipton, G. (1999). An investigation into gender differences in the ethical attitudes of IT professionals. *ETHICOMP99*, Rome.

Bocij, P. and Sutton, M. (2004). Victims of cyberstalking: piloting a web-based survey method and examining tentative findings. *Journal of Society and Information*, 1(2). Available at http://josi.spaceless.com/article.php?story=2000214050558297. (accessed August 30, 2004).

Brosnan, M.J. (1998). *Technophobia: The Psychological Impact of Information Technology*. Routledge, London and New York.

Camp, T. (1997). The incredible shrinking pipeline. *Communications of the ACM*, 40(10), 103–110.

Castells, M. (2000). *The Rise of the Network Society*, 2nd edition. Blackwell, Oxford and Malden, MA.

Christie, J.R.R. (1993). A tragedy for cyborgs. *Configurations*, 1, 171–196.

CIEC. (1997). The Internet is not a Television, available at http://www.ciec.org (accessed January 20, 2003).

Cudd, A. (2001). Objectivity and ethno-feminist critiques of science. In: Ashman, K. and Baringer, P. (Eds.), *After the Science Wars*. Routledge, New York and London, pp. 80–97.

DeCew, J. (1997). *In Pursuit of Privacy: Law, Ethics, and the Rise of Technology*. Cornell University Press, Ithaca, NY and London.

Equalities Review Panel (2007). *Fairness and Freedom: The Final Report of the Equalities Review*. HMSO, London.

Escribano, J.J., Peña, R., Extremora, J. (1999). Differences between men and women in terms of usage and assessment of information technologies. *ETHICOMP99*, Rome.

Ess, C. (Ed.) (1996). *Philosophical Perspectives on Computer-Mediated Communication*. State University of New York Press, Albany, NY.

Faulkner, W. (2000). The power and the pleasure? A research agenda for 'making gender stick.' *Science, Technology & Human Values*, 25(1), 87–119.

Fitzgerald, B. and Howcroft, D. (1998). Towards dissolution of the IS research debate: from polarisation to polarity. *Journal of Information Technology*, 13, 313–326.

Forsythe, D. (1993). Engineering knowledge: the construction of knowledge in artificial intelligence. *Social Studies of Science*, 23, 445–477.

Gibson, W. (1984). *Neuromancer.* Vanguard, NY.

Gilligan, C. (1982). *In a Different Voice: Psychological Theory and Women's Development.* Harvard University Press, Cambridge, MA.

Gilligan, C. (1987). Moral orientation and moral development. In: Kittay, E.F. and Meyers, D.T. *Women and Moral Theory.* Rowman & Littlefield, Totowa, NJ, pp. 19–33.

Green, E., Hebron, S., et al. (1990). *Women's Leisure, What Leisure?.* Macmillan Education, Basingstoke.

Grundy, F. (1996). *Women and Computers.* Intellect, Exeter, UK.

Guertin, C. (2005). From cyborgs to hacktivists: postfeminist disobedience and virtual communities. *Electronic Book Review.* Available at: www.electronicbookreview.com/thread/writingpostfeminism/hackpacifist (accessed January 20, 2006).

Haraway, D. (1991). *Simians, Cyborgs and Women: The Reinvention of Nature.* Free Association Books, London.

Helmreich, S. (1994). Anthropology inside and outside the looking-glass worlds of artificial life, unpublished paper, Department of Anthropology, Stanford University, Stanford, CA.

Henwood, F. (1993). Gender perspectives on information technology: problems, issues and opportunities. In: Green, E., Owen, J., and Pain, D. (Eds.), *Gendered by Design? Information Technology and Office Systems.* Taylor & Francis, London, pp. 31–49.

Himanen, P. (2001). *The Hacker Ethic and the Spirit of the Information Age.* MIT Press, Cambridge MA and London.

Howcroft, D. and Trauth , E. (Eds.), (2005). Handbook of Critical Information Systems Research: Theory and Application, Edward Elgar Publishing, Cheltenham, UK.

Igbaria, M. and Chidambaram, M. (1997). The impact of gender on career success of information systems professionals. *Information Technology and People*, 10(1), 63–86.

Jaggar (1991). Feminist ethics: projects, problems, prospects. *Feminist Ethics.* C. Card. University Press of Kansas, Lawrence, Kansas, pp. 78–104.

Jordan, T. and Taylor, P.A. (2004). *Hacktivism and Cyberwars: Rebels With a Cause?.* Routledge, London and New York.

Khazanchi, D. (1995). Unethical behavior in information systems: the gender factor. *Journal of Business Ethics*, 15, 741–749.

Koehn, D. (1998). *Rethinking Feminist Ethics: Care, Trust and Empathy.* Routledge, London and New York.

Kohlberg, L. (1981). *The Philosophy of Moral Development.* Harper and Row, San Francisco.

Kreie, J. and Cronan, T. (1998). How men and women view ethics. *Communications of the ACM*, 41(9), 70–76.

Kreie, J. and Cronan, T.P. (1999). Copyright, piracy, privacy, and security issues: acceptable or unacceptable actions for end users? *Journal of End User Computing*, 11(2), 13–20.

Larrabee, M.J. (Ed.), (1993). *An Ethic of Care.* Routledge, New York and London.

Leonard, L.N.K. and Cronan, T.P. (2005). Attitude toward ethical behavior in computer use: a shifting model. *Industrial Management & Data Systems*, 105(9), 1150–1171.

Levy, N. (2002). Virtual child pornography: the eroticization of inequality. *Ethics and Information Technology*, 4(4), 319–323.

Levy, S. (1984). *Hackers. Heroes of the Computer Revolution*. Penguin, Harmondsworth UK.

Lovegrove, G. and Segal, B. (Eds.), (1991). *Women into Computing; Selected Papers, 1988–1990*. Springer-Verlag, London and Berlin.

Mason, E.S. and Mudrack, P.E. (1996). Gender and ethical orientation: a test of gender and occupational socialization theories. *Journal of Business Ethics*, 15, 599–604.

McDonald, G. and Pak, P.C. (1996). It's all fair in love, war and business: cognitive philosophies in ethical decision making. *Journal of Business Ethics*, 15, 973–996.

Mill, J.S. (1970). The subjection of women. In: Rossi, A.S. (Ed.), *Essays on Sex Equality*. University of Chicago Press, Chicago, IL, pp. 125–156.

Nissenbaum, H. (1995). Should I copy my neighbor's software? In: Johnson, D.G. and Nissenbaum, H. (Eds.), *Computer Ethics and Social Values*. Prentice Hall, Upper Saddle River, NJ, pp. 200–213.

Oakley, A. (2000). *Experiments in Knowing: Gender and Method in the Social Sciences Polity*. Cambridge, UK.

Panteli, N. and Stack, J. (1998). Women and computing: the ethical responsibility of the IT industry. *ETHICOMP98*, Rotterdam.

Panteli, A., Stack, J., Ramsay, H. (1999). Gender and professional ethics in the IT industry. *Journal of Business Ethics*, 22(1), 51–61.

Pateman, C. (1988). *The Sexual Contract*. Polity. Cambridge, UK and Oxford.

Plant, S. (1997). *Zeros + Ones: Digital Women + the New Technoculture*. Fourth Estate, London.

Puka, B. (1993). The liberation of caring: a different voice for Gilligan's "different voice," In: Larrabee M.J. (Eds.), *An Ethic of Care: Feminist and Interdisciplinary Perspectives*. Routledge, New York and London, pp. 215–239.

Raymond, E. (2001). *The Cathedral and the Bazaar: Musings on Linux and Open Source by an Accidental Revolutionary*. O'Reilly, Sebastopol, CA.

Remtulla, K.A. (2006). Then isn't it all just 'hacktivism'? *Electronic Book Review*. Available at www.electronicbookreview.com/thread/writingpostfeminism/concurrent (accessed February 20, 2007).

Reiss, M.C. and Mitra, K. (1998). The effects of individual difference factors on the acceptability of ethical and unethical workplace behaviors. *Journal of Business Ethics*, 17, 1581–1593.

Reno, J. (1999). Cyberstalking: a new challenge for law enforcement and industry. A Report from the Attorney General to the Vice President. Available at http://www.usdoj.gov/ag/cyberstalkingreport.html (accessed November 30, 1999).

Rheingold, H. (1993). *The Virtual Community: Homesteading on the Virtual Frontier*. Addison-Wesley, Reading, MA.

Ruddick, S. (1989). *Maternal Thinking: Toward a Politics of Peace*. Beacon, Boston, MA.

Schleiner, A.-M. (2002). Countdown to collective insurgence: cyberfeminism and hacker strategies. *Technics of CyberFeminism Conference*. Available at www.opensorcery.net/countdown.html (accessed January 20, 2006).

Segan, S. (2000a). Female of the species; hacker women are few but strong. Available at http://more.abcnews.go.com/sections/tech/dailynews/hackerwomen000602.html (accessed January 20, 2003).

Segan, S. (2000b). Facing a man's world: female hackers battle sexism to get ahead, Available at http://more.abcnews.go.com/sections/tech/dailynews/hackerwomen000609 (accessed January 20, 2003).

Sollfrank, C. (2002). Not every hacker is a woman. In: Reiche, C. and Sick, A. (Eds.), *Technics of Cyberfeminism*. Available at http:www.obn.org/reading_room/writings/html/notevery.html (accessed August 30, 2004).

Squires, J. (1996). Fabulous feminist futures and the lure of cyberculture. In: Dovey, J. (Ed.), *Fractal Dreams: New Media in Social Context*. Lawrence and Wishart, London, pp. 194–216.

Stabile, C. (1994). *Feminism and the Technological Fix*. Manchester University Press, Manchester and New York.

Strauss, A. and Corbin, J. (Eds.). (1997). *Grounded Theory in Practice*. Sage, Thousand Oaks, CA.

SubRosa (2005). Human Rights and Cyberfeminism. pp. 43–47. Available at www.refugia.net/yes/yes_05cyberfem.pdf (accessed January 20, 2006).

Taylor, P. (1999). *Hackers: Crime in the Digital Sublime*. Routledge, London and New York.

Tong, R. (1993). *Feminine and Feminist Ethics*. Wadsworth, Belmont, CA.

Tong, R. (1999). Feminist ethics, In: Zalta, E. (Ed.), *The Stanford Encyclopedia of Philosophy (Fall* 1999 *Edition*). Available at http://plato.stanford.edu/archives/fall1999/entries/feminism-ethics/ (accessed November 24, 1999).

Turner, E. (1998). The case for responsibility of the computing industry to promote equal presentation of women and men in advertising campaigns. *ETHICOMP98*, Rotterdam.

Turner, E. (1999). Gender and ethnicity of computing, perceptions of the future generation. *ETHICOMP99*, Rome.

Turner, E. (2006). Teaching gender inclusive computer ethics. Available at. http://ict.open.ac.uk/gender/papers/turner.doc (accessed September 10, 2006).

Wajcman, J. (2004). *TechnoFeminism*. Polity, Cambridge, UK and Malden, MA.

Weber, M. (1930). *The Protestant Ethic and the Spirit of Capitalism*. Routledge, London.

Wilson, E. (1993). Is transgression transgressive? In: Bristow, J. and Wilson, A. (Eds.), *Activating Theory: Lesbian, Gay and Bisexual Politics*. Lawrence & Wishart, London.

Winner, L. (1997). Cyberlibertarian myths and the prospect for community. *ACM Computers and Society*, 27(3), 14–19.

Wollstonecraft, M. (1988). In: Brody, M. (Ed.), *A Vindication of the Rights of Women*. Penguin, London.

CHAPTER 26

The Digital Divide: A Perspective for the Future

MARIA CANELLOPOULOU-BOTTIS and KENNETH EINAR HIMMA

26.1 INTRODUCTION

The global distribution of material resources should bother any conscientious person. One billion of the world's six billion people live on less than $1 per day, whereas two billion live on less than $3 per day. Poverty in the affluent world is largely relative in the sense that someone who is "poor" simply means he has significantly less than what others around him have. But because wealth is, unfortunately, frequently associated in the West with moral worth, it is important to realize that relative poverty is a genuinely painful condition. People who live in conditions of relative poverty are generally treated with less respect, and hence are denied something that is essential to human well-being.[1]

In the developing world, poverty and the suffering it causes is considerably worse. Here poverty is characteristically "absolute" in the sense that people do not have enough to consistently meet their basic needs. People in absolute poverty lack consistent access to adequate nutrition, clean water, and health care, as well as face death from a variety of diseases that are easily cured in affluent nations. Indeed, 15 million children die every year of malnutrition in a world where the food that is disposed of as *garbage* by affluent persons is enough to save most, if not all, of these lives.

[1]Nevertheless, poverty is becoming more serious in countries like the United States, where a recent study shows an increase in the percentage of the population in "severe poverty," which is defined as having an income less than half of that defined by the federal poverty line. The number of people living in severe poverty increased by 26% from 2000 to 2005. See, for example, Pugh (2007). Moreover, there is some absolute poverty in the United States, as there are now more than 750,000 persons who are homeless. See Olemacher (2007).

The Handbook of Information and Computer Ethics, Edited by Kenneth Einar Himma and Herman T. Tavani

The digital divide is not any one particular gap between rich and poor, local and global, but rather includes a variety of gaps believed to bear on the world's inequitable distribution of resources. There is, of course, a comparative lack of meaningful access to information communication technologies (ICTs); a gap in having the skills needed to use these resources; a gap between rich and poor in their ability to access information needed to compete in a global economy; and a gap in education that translates into a gap in abilities to process and absorb information. There are, of course, nondigital gaps that contribute to the distribution of resources: poor nations have less highly developed infrastructure at every level needed to contribute to productive economic activity. There has also been the unfortunate result of pressure by organizations, such as the IMF, World Bank, and USAID, on poor nations to privatize their most economically prosperous resources, which typically get sold to a wealthy Western nation that profits from the privatization of recipient poor nations while protecting vulnerable markets such as agriculture against the competition of poor nations with subsidies that are (arguably) illegal under the World Trade Agreement.

The point here is not that global and local poverty are problems of many dimensions that are extremely difficult to solve, but rather that the moral importance of the digital divide as a problem that needs to be addressed is linked to inequalities between the rich and the poor—and especially wealthy nations and nations in absolute poverty. There may be a case for thinking that such divides are inherently unjust, but that seems somewhat implausible: economic injustices are viewed as problematic more because of the suffering they cause and less because there is some sort of deontological egalitarian principle that requires absolute equality of justice.

26.2 THE BIDIRECTIONAL RELATIONSHIP BETWEEN ABSOLUTE POVERTY AND THE DIGITAL AND INFORMATION DIVIDES

There are gaps in access to information and information communication technologies within nations and between nations. Within the United States, for example, there are such gaps between rich and poor citizens, whites and blacks, and urban dwellers and rural dwellers. According to the U.S. Department of Commerce (1999),

> the 1998 data reveal significant disparities, including the following: Urban households with incomes of $75,000 and higher are more than twenty times more likely to have access to the Internet than rural households at the lowest income levels, and more than nine times as likely to have a computer at home. Whites are more likely to have access to the Internet from home than Blacks or Hispanics have from any location. Black and Hispanic households are approximately one-third as likely to have home Internet access as households of Asian/Pacific Islander descent, and roughly two-fifths as likely as White households. Regardless of income level, Americans living in rural areas are lagging behind in Internet access. Indeed, at the lowest income levels, those in urban areas are more than twice as likely to have Internet access than those earning the same income in rural areas.

Other things being equal, poor people in the United States are less likely to have access to online information and the ICTs that makes access possible than affluent people.

Similar gaps exist between the affluent developed world and the impoverished developing world. Although Internet access is increasing across the world, it is still the case that a comparatively small percentage of the developing world's poor has Internet access. A 2005 UNESCO report indicated that only 11% of the world's population has access to the Internet, but 90% of these persons live in the affluent industrialized developed world.[2]

Although these differences in access to ICTs and information correlate with differences in wealth, there is a causal relation between them. Obviously, people who are too poor to fully meet their immediate survival needs cannot afford either ICT access and the training that prepares one to take advantage of such access. But not being able to afford such training and access is likely to perpetuate poverty in a global economy increasingly requiring the ability to access, process, and evaluate information. Lack of access owing to poverty is a vicious circle that helps to ensure continuing poverty.

26.3 THE MORAL BASIS FOR THE IDEA THAT THE VARIOUS DIGITAL DIVIDES SHOULD BE ELIMINATED

The moral basis for the case for affluent nations to eliminate the digital divide is grounded in the idea that nations and people with far more than they need to satisfy basic needs have a moral obligation to redistribute some of their wealth, at the very least, to nations and people in life-threatening or absolute poverty. If the digital divide is both a reflection of a gap between rich and absolute poor and perpetuates that gap, it follows that wealthy nations are obligated to close the divide.

Some clarification on moral terms might be helpful here. To say that X is good is to not to say that X is obligatory. Failure to do something morally good is not necessarily morally wrong and does not necessarily merit blame, censure, or punishment. It would be good if I were to run into a burning building to try to rescue someone, but it is not morally wrong for me to refrain from doing so; risking my life to save another is *supererogatory*, that is to say, morally good but beyond the call of obligation. Failure to do something morally obligatory, in contrast, is necessarily morally wrong and merits blame, censure, or punishment. We praise supererogatory acts, but not obligatory acts. We blame nonperformance of obligatory acts, but not nonperformance of supererogatory acts.

It is noncontroversial that it is *morally good* for affluent persons or nations to help impoverished persons or nations, but there is considerable disagreement about whether affluent persons and nations are *morally obligated* to help alleviate the effects of absolute poverty. As noted above, many persons in the United States take the position that the only moral obligations we have are *negative* in the sense that they

---

[2]Ponce (2005).

require us only to abstain from certain acts; we are obligated, for example, to refrain from killing, stealing, lying, and so on. On this view, we have no moral obligations that are *positive* in the sense that they require some positive affirmative act of some kind. It follows, on this view, that we have no moral obligation to help the poor; helping the poor is good, but beyond the demands of obligation.

Himma (2007) argues that this view is both mistaken and pernicious. In particular, he argues that this view is inconsistent with the ethics of every classically theistic religion, ordinary intuitions about certain cases, and each of the two main approaches to normative ethical theory, consequentialism and deontological ethical theory. Taken together, these arguments provide a compelling case for thinking the affluent are morally obligated to help alleviate the conditions of absolute poverty wherever they are found. If the various digital divides perpetuate these conditions, then, among the other inequalities that the wealthy are obliged to address (e.g., absence of schools, infrastructure) are included those that comprise the digital divide taken as a whole.

26.4 EMPIRICAL SKEPTICISM ABOUT THE RELATIONSHIP BETWEEN DIGITAL DIVIDES AND ABSOLUTE POVERTY

Some have argued that whatever the "digital divide" may mean, it does not deserve a special place either in our terminology or even in our scientific and political agenda. It has been with us for quite some time now. The bibliography of empirical studies relating to it is so vast, that "digital divide skeptics" would face real trouble trying to persuade us that the whole matter should not attract this kind of attention; it is only a "topic *du jour*,"[3] a delusion,[4] a myth,[5] a costly mistake,[6] or (worse) it constitutes a plain sham.[7] These studies attempt to cast doubt on the thesis that bridging the digital divide can make a significant dent in bridging the economic inequalities between rich and poor.

So, no matter how skeptical one is, one is bound to reflect very seriously upon the question whether all this literature, all this research, and all these programs and efforts to "bridge the digital divide," at national and global levels, have to mean that the phenomenon of the digital divide is new and different from the "divides" we have seen when other means of communication and publication (the printing press, TV, radio) first emerged in the past. Also, we have to think whether it is true, as widely supported, that this divide, this gap, has to close as a sort of first priority and perhaps, as the

[3]Thierer (2000).

[4]Oppenheimer (1997). Available at http://www.tnellen.com/ted/tc/computer.htm, last access 2007, May 1, referring especially to the policy of heavy federal funding of the goal "computer in every classroom."

[5]Compaine (2001).

[6]See Kenny (2002), referring especially to the digital divide and developing countries.

[7]The "digital divide", is a sham—an excuse for Big Government to court Silicon Valley with fistfuls of corporate-welfare dollars in exchange for campaign contributions . . . , Thierer (2000).

fulfillment of a moral obligation, born within the battle against what has been called "information poverty."[8]

26.4.1 Meanings

The "digital divide," which is a new term on its own "right," occupying a central component of the global *lingua franca* for research on the Internet,[9] comes with a whole set of other new terms, more or less accepted as deserving a place in our language, such as "netocracy,"[10] the "digerati" (the intellectual elite of technology advocates),[11] "information apartheid,"[12] "technological apartheid,"[13] "information-haves" and "information have-nots," and "digital democracy."[14] Apart from all these new words, it is the digital divide that seems to embody in two words the whole philosophy of "digital is different."

It is true that there are many ways to see a phenomenon, which, yet, has come to signify essentially one thing: the fundamental disparity between information haves and have-nots. Depending upon who is on the sides of this comparison, we refer to a national digital divide (digital information inequalities within a state) and a global digital divide (digital information inequalities among nations). Researchers have also "broken" the large questions into small ones, looking into, for example,[15] *intra*national digital divides *within* rich and poor in India,[16] China,[17] Africa,[18] Australia,[19] Asia

[8]Britz (2004). For Britz, information poverty is the situation in which individuals and communities, within a given context, do not have the requisite skills, abilities, or material means to obtain efficient access to information, interpret it, and apply it appropriately. He argues that information poverty is a serious moral concern and a matter of social justice and as such should be on the world's moral agenda of social responsibility. For a further analysis of the moral issues associated with information poverty see Himma K. (2007), who explains why affluent nations have a moral obligation to help developing nations overcome poverty in general but also, the information and the digital divide (p. 6).

[9]Linchuan Qiu (2002), p. 157.

[10]Bard and Soderqvist (2002), hold that "netocracy" means a new order, a new ethic, with technophiles and cosmopolitans having made the net their own country and viewing information as a more valuable good than tangibles.

[11]Edge, Who Are the Digerati? Available at http://www.edge.org/digerati. Accessed 2007, May 1.

[12]Davis (2001). The term is also described as a reason for combat; for example, see http://findarticles.com/p/articles/mi_qa3628/is_199401/ai_n8722216, last access 2007 May 5, "...this lack of African-American librarians and information professionals reflects what AALISA's president, Itibari M. Zulu of UCLA's Center for Afro-American Studies calls "information apartheid," which articulates a practice of differentiating information according to social status, and access to current technology...."

[13]Castells (1998).

[14]See Norris (2001), p. 95. See also (generally) Yu (2002).

[15]I will cite one study or article per continent, or state, or case, as an example, but of course, many more exist.

[16]Subba Rao (2005).

[17]Cartier et al. (2005). For an even more special approach on China, digital divide and disability, see Guo et al. (2005).

[18]Mutume (2003).

[19]Willis and Tranter (2006). The authors examined the social barriers to Internet use in Australia over a five-year period, finding that, although Internet diffusion should narrow the "digital divide," "democratization" of access was not firmly supported.

Pacific,[20] or states such as the United States,[21] Canada,[22] Italy, New Zealand,[23] regional or even tribal territories such as the Mississippi Delta region[24] or the New Mexico tribes,[25] and the "doctor–patient digital divide."[26] Relevant research has been published in journals, books, etc., belonging to a number of scientific fields, such as law, psychology,[27] sociology,[28] economics, management, political science,[29] librarianship,[30] and others.

The list goes on when other factors than territoriality are used. Other questions include the digital divide and disability (information inequalities between able and disabled), the digital divide and age, and the digital divide and gender etc. The list of "markers" to measure the digital divide is long and in essence, usually[31] no different from the known list of possible discrimination factors in other settings: age, gender, economic and social status, education, ethnicity, type of household (urban/rural), and so on.

Yet, as we have seen, other classifications present us with more (and different) kinds of divides: the access divide (whether people have or not meaningful access to a computer and the Internet—what was initially the whole question of the digital divide), the capital divide,[32] the treatment divide,[33]the global divide, the domestic divide, the political divide, and others.

[20]Saik-Yoon-Chin (2005).

[21]National Telecommunications and Information Administration (2002). Accessed 2007 May 9. The NTIA has produced earlier reports on the digital divide in the United States, starting from 1995. *We have included these reports in our references.*

[22]Rideout (2003).

[23]Howell (2001).

[24]Lentz and Oden (2001).

[25]Dorr and Akeroyd (2001).

[26]Malone et al. (2005). The researchers examined discrete geographical districts with differing patterns of health information seeking, identifying two groups, "information-hungry"/online health seekers and those who were offline information seekers.

[27]See, for example, Montero and Stokols (2003) (the authors describe, among others, a whole new area of psychological research, "digital psychology" or "cyber psychology"). See also Kalichman et al. (2006).

[28]Katz et al. (2001).

[29]Norris (2001).

[30]See Dutch and Muddiman (2001).

[31]But see van Dijk (2005), who, in a way, replaces these usual inequality "markers" with categorical differences between groups of people, such as black/white, male/female, and citizen/foreigner, fully adopting a more general argument about inequality framed by Tilly (1998).

[32]Celli and Dreifach (2002). The authors define the capital divide as the divide between those who succeeded in capitalizing Internet businesses during the "go-go" 1990s and those who did not.

[33]Celli and Dreifach (2002), defining the treatment divide as the use of Internet-related data, such as click-rates, lingering patterns, and purchasing habits, by retailers to target different users for different treatment online.

26.4.2 Expectations

If bridging the digital divide had to become a top priority in our allocation of resources, benefits occurring from this endeavor, when successful, must be (more than) reasonably anticipated. But doubts about this are not easily detected in the digital divide literature. In a way, it was almost a waste of time, a retreat to common knowledge, or plain truism, of the sort no court would ask evidence for, that the more people are connected, the more benefited they are—who would want to be deprived of such a medium such as the Internet, especially once they had more than just glanced at one of its pages?

But the problem was not just framed in terms of learning to use tools, like the Internet, so that new workers will gain the skills and familiarity with new technologies that will allow them to find jobs in a new economy,[34] or in terms of "enriching our world, facilitating our work lives and providing a skill set needed for a growing economy."[35] No, this was too narrow a vision: the digital revolution certainly brought with it the image of an entirely new world, of a wonderful new, wired global village, where kids from Africa or Indonesia, for example, would be able to acquire as much knowledge as a typical middle-class American kid. The network had to expand, by all means, perhaps at all costs: the new "good," the new interest in access to unlimited digital information meant equally unlimited opportunity, entertainment, personal growth, unlocked working potential,[36] even spirituality,[37] easing the path toward democracy and freedom of speech for so many countries,[38] to name a few.[39] Therefore, it is a real loss, both to individuals and to the states, a certain way to poverty,[40] it is almost a tragedy, a return to the dark ages,[41] to "fall through the Net"—as eloquently

---

[34]This is, as it seems, the main goal in the report NTIA (2002), p. 91 (where other goals, such as helping the disabled, are also mentioned and analyzed).

[35]NTIA report (2002) p. 91.

[36]Schwartz (1999)."If we want to unlock the potential of our workers we have to close that gap. . ." (meaning the digital divide).

[37]See Celli and Dreifach (2002), p. 54.

[38]On this, see generally Norris (2001). See also generally Budge (1996) for the proposition that digital technologies are, perhaps, the most important development in our lifetime that could fuel the process of allowing more opportunities for citizen deliberation and direct decision-making.

[39]Celli and Dreifach (2002),". . . our society" success in living up to the American ideal of freedom and equality will be measured in part by how well we fare in achieving the Internet ideal. It will be measured by how effectively we bridge the Digital Divide" p. 71.

[40]See Alexander (1996). *Unraveling the Global Apartheid: An Overview of World Politics.* Polity Press, Cambridge. He sustains that in a world governed by information, exclusion from information is as devastating as exclusion from land in an agricultural age, p. 195.

[41]See Rifkin (2000). "The future may become a wonderland of opportunity . . . it may . . . become a digital dark age for the majority of citizens, the poor, the non-college educated, and the so-called unnecessary . . ." (if they do not have access to the net), p. 228.

described in the first U.S. studies on the matter of digital divide,[42] the words in 2000 still written within brackets (digital divide[43]).

26.4.3 Empirical Studies Illustrating Perceived Failures

It is easy to detect that these expectations were really high and to become, immediately, skeptical. To support them with valid argument, we need evidence that funding projects aiming at closing the digital divide will lead to positive results. I will present a number of instances where this was not the case.

Logically, we support technology for the end of a better social, among other aspects, life. So what effects does Internet use really have in our social life? It has been reported in a major study about Internet use in the United States[44] that as people become more and more wired, they alienate themselves from traditional societal functions and activities. In particular, the more people use the Internet, the more they lose contact with their social environment (spending less time with family and friends and attending fewer social events), they turn their back to the traditional media (reading fewer newspapers etc.), the more time they spend working at home and at the office and the less time they spend commuting in traffic and shopping in stores (i.e., being outside). Surely, this was not part of the digital revolution dream. There is, however, evidence supporting opposing conclusions as well.[45]

Of course, whereas this is a cost of the so-called Information Age, this study shows only that the effects of the Information Age are not all beneficial—something that is probably true of nearly every technology, including the television and the development of antibiotics, which by curing diseases among the top ten killers in the early 1900s have enabled us to extend average life span to a point where people are so old that they have to deal with a host of conditions that severely diminish the quality of their lives.

Such studies, however, have nothing to do with showing that the empirical assumptions underlying the correlation between poverty and the digital divide are incorrect. The relevant effects have nothing to do with economic progress, but rather with the breakdown of community—a cost that is surely important, but it pales in

[42]See U.S. Department of Commerce (1999), Accessed 2000, October 16, The 2000 study was the fourth report in the Commerce Department series of studies, all called *Falling Through the Net.* This fourth one added to the theme "digital divide" the theme "toward digital inclusion," "moving in a new phase of information- gathering and policy-making by recognizing the phenomenal growth that has taken place in the availability of computing and information technology tools, tempered by the realization that there is still much more to be done to make certain that everyone is included in the digital economy," p. xiii.

[43]Same as above.

[44]See Norman and Erbring. Accessed 2007, May 7.

[45]Hampton and Wellman (2003). This was a small study of a suburb near Toronto, comparing "wired" and "nonwired" residents, using survey and ethnographic data. The conclusion was that more "wired" neighbors are known and chatted with, and they are more geographically dispersed around the suburb, in comparison to the "nonwired" ones. So, it seemed that the Internet support neighboring issues, but also facilitated discussion and mobilization around local issues.

importance to being in conditions of life-threatening poverty. It seems irrational to prefer life-threatening hunger and death to some partial breakdown in the social bonds constituting a community.

More to the point, research[46] into the digital divide and education in the United States showed that even as more and more students get connected, schools will face a deeper challenge, figuring out why so many students graduate *without* basic skills such as reading. Under the researchers' view, the answer may not be in the Internet at all: "...for students who can read, who can figure, and who have "learned how to learn," lack of exposure to digital equipment in education will not be much of a handicap ...."[47] It has also been proposed that there is no good evidence that most uses of computers significantly improve teaching and learning.[48] As proposed, billions of dollars (in 1997) that were directed toward technology should be freed and made available for impoverished fundamentals, such as teaching skills in reading, thinking, organizing inventive field trips and other rich hands-on experiences, and building up inspiring teachers.[49]

Another story, coming from LaGrange, Georgia, is also indicative of failures in the attempt to bridge the digital divide. In 2000, the city became the first one in the world to implement a program called "The Free Internet Initiative," meaning that the city would offer broadband access to the Internet for every citizen.[50] Internet access was provided through a digital cable set-top box that was distributed free of charge and every citizen could also receive free training. Access to the net was possible through people's televisions under a system called WorldGate. As reported, the project had a very limited success, especially in relation to the target group of people of lower social and economic status.[51]

It is, however, important to note that such studies tell us little about the global divides as they involve the United States, a nation with a unique history of tension between blacks and whites and rich and poor. One problem that hinders efforts to raise the performance of impoverished inner-city black youths (especially males) in schools is that they do not believe that a good education will enable them to overcome racism and get a good job, and so do not invest as much effort as affluent white and black students. Indeed, there are studies that show that educational success among blacks is disparaged by other blacks as "selling out" and "acting white"—an act of race traitorship of sorts.

---

[46]Singleton and Mast (2000).

[47]Singleton and Mast (2000), p. 33.

[48]See Oppenheimer (1997), p. 45.

[49]Oppenheimer (1997), p. 62. See (supporting this view) analytically Stoll, C. (2000).

[50]See Keil M., Meader, G. and Kvasny, L. (2003). Bridging the digital divide: the story of the free internet initiative. 36th-Hawaii-International Conference on Systems Sciences; p. 10. Available at csdl.computer.org/comp/proceedings/hicss/2003/1874/05/187450140b.pdf. Accessed 2007 May 5.

[51]Kiel et al. (2003), "...it appears that (the city officials) had an unrealistic and in some ways naive view that providing free access to technology would, by itself, be enough to bridge the digital divide...," Lack of motivation to use the Internet, intimidation by technology, (perhaps) the very low cost of $8 per month (the program was not absolutely free), the lack of the ability to print text through the program, illiteracy and other reasons are mentioned as inhibiting the goals of the project (Kiel et al., 2003).

But these studies fail to realize that the digital divides are not likely to be solved overnight; the problems that cause absolute poverty in the global South and relative poverty in affluent nations are much too complex to be solved by a one-time, short-term investment of information capital. Attitudes may have to be changed, while educational systems will also have to be improved. But, even in the long-term, there are many contingent cultural difficulties, at least in the United States and presumably in other countries with a history of institutionalized systemic racism, that will have to be overcome. Solving the problems associated with the digital divide is a long-term commitment.

Reminding us of the failure to attract the target "people of low social and economic status" in the WorldGate program of LaGrange, other research dealing with information poverty and homeless people concluded[52] that the homeless may lack needed financial resources, but this did not translate into a lack of access to their more frequently articulated information needs (mainly, how to find permanent housing, how to help children, how to find a job, how to deal with finances, how to cope with substance abuse and domestic violence). Homeless people found information mostly by person-to-person contact. Moreover, and more important, this information was not available on the Internet and, as a conclusion, it may not be true that the economically disadvantaged groups are the most vulnerable to negative effects from the inability to find needed information in electronic formats.[53] Besides, there is a question whether homeless people, along with the poor generally, are perhaps more susceptible to advertising and deceptive commercial practices, and so perhaps it is true that "any bridge across the digital divide will just lead poor people into consumerist quicksand."[54]

Although this study reminds us of the fact that information is not necessarily valuable,[55] it is of somewhat greater relevance in efforts to alleviate poverty by bridging the digital divide. Different people, at this point in time, have needs for different content: you cannot take a homeless person and turn her into a stockbroker simply by providing her with the information a stockbroker has; what she needs is information about social services. But, again, this tells us no more than that the problems comprising the digital divide and its relation to poverty are enormously complex and require a long-term, multifaceted approach to solving them.

There are a number of other studies calling attention to a different obstacles faced in bridging the digital divide as a means of addressing poverty. Another relevant question, whether, as a matter of government policy, the use of digital technology should be subsidized, has been answered, in many instances, in the negative. In Australia, many government departments do sponsor electronic marketplaces, but

---

[52]Hersberger (2003).

[53]Hersberger (2003), ". . .as more computer literate children grow up and become homeless, there could be a higher demand for access to digital information technologies. For now, the lack of access to digital information does not seem to negatively affect the everyday life of homeless parents. Having access to the Internet would be a luxury, but it is not perceived as a need at this time . . .," p. 248.

[54]Thierer (2000).

[55]See Himma (2004).

research offers evidence that this causes a number of complications, not only in terms of stifling free trade[56]; researchers recommend as saner, a "wait and see" attitude, as public intervention toward efforts to bridge the digital divide is a waste, as the market will offer more and more cheaper and simpler computer products.[57]

In Costa Rica, the Little Intelligence Communities project (LINCOS), founded by MIT, Microsoft, Alcatel, and the Costa Rican government, aiming at helping through telecenters the poor rural Costa Rican communities seems to have failed. It was not the poor, but the rich coffee farmers who tried to take advantage of the project; local residents either did not care at all, or were interested in accessing virtual pornography and vice.[58]

These studies show that efforts to bridge the digital divide will not succeed unless people are properly educated about what these technologies can accomplish economically, and people must also want to produce those various results. It is important not to dramatically change the quality and content of cultural attitudes, but at the same time people in absolute poverty should expect to change their attitudes if they are to fully alleviate the conditions that perpetuate their misery.

Some problems are simply technological in character and require more time to resolve. In the study discussed above, one difference was that what its producers were hoping for, for the developing countries' digital divide, was not realized for a new simpler computer, the simputer.[59] Severe failures of programs were reported for ICT programs, for example, in Tanzania (Africa) and Andhra Pradesh (India).[60]

Similarly, in Greece, a very expensive software program, funded by the European Community, as a telemedicine program aiming at connecting sick people and their primary care doctors with the most specialized physicians in the biggest Athens hospital for trauma (KAT), failed in its entirety when the physicians realized that all their orders, based upon digitally sent exam results, scans, etc., from the remote islands, would be stored and that the question of medical liability (who could be responsible?) was not safely resolved.[61] The attempt to use telemedicine to close the digital gap between people in remote islands (who do not have access to digital medical diagnostic technology) and people in the center (who do), in this case, was a total failure. It seems that the money would have been better spent by funding the salaries of specialized physicians who would work at these remote islands and by financing some medical equipment there.

---

[56]Standing et al. (2003).

[57]Thierer (2000), and Compaine (2001). But see also van Dijk (2005), p. 185, who hotly disagrees with these views, citing them as "...the worst advice one can give at the stage of the introduction of the new technology."

[58]Amighetti and Reader (2003), Accessed 2007, May 9.

[59]Malakooty (2007), p. 6, where she describes various low-cost computer projects worldwide (Brazil, India, Mexico etc). These projects do not appear to be successful, because for example, sales of cheap computers like the "simputers" have not met expectations. See generally www.simputer.org.

[60]Wade (2002).

[61]Comment by Kanellopoulos, N., Professor of Computer Science, Ionian University, Greece, responsible for this program, at a lecture he gave in the Department of Informatics, Corfu, Greece, May 2, 2007.

In this case, the program was never used, so (at least) no medical accidents occurred; not so for another widely publicized program, the Computer Aided Dispatch of the London Ambulance Service (LASCAD), which was a notorious failure in 1992. When the electronic system first replaced the older manual system for receiving emergency calls, dispatching ambulances, and monitoring progress of the response to the calls, the call traffic load increased, the same messages were sent again and again confusing the staff who did not know which ambulance units were or were not available (because the crew were frustrated and reported incorrectly). Ambulances arrived after people had died or 11 hours after a call for a stroke; the estimates of the total number of fatalities vary from 10 to 30, within a few days. The system crashed and it was withdrawn; compounding, though, this comedy of errors[62] is the quite unbelievable fact that more than a whole decade later, in 2006, the upgrade of a similar software system of computer-controlled call-taking crashed *again* in London and LAS had to return (once again) to the safer method of simply using *pen and paper*.[63]

But these, again, do not justify the sort of digital skepticism described at the beginning of this section. All technologies that resolve morally important problems take time to develop. Although it is widely believed that gene therapy will make possible cures for diseases that are currently incurable, the research has progressed very slowly. The same should be expected of ICTs and related measures intended to alleviate the conditions of poverty. There is simply nothing one can do to make the economic injustices of the world disappear tomorrow.

26.4.4 Bridges and Questions

What is the situation with world poverty today—generally and in connection to technology? In 2001, 33 million people in the developing world were on the registered waiting lists for telephone connections, the average waiting periods being over 10 years in some countries.[64]

The focus should change more significantly toward battling the content divide, that is, the great disparity between content in English, responding mainly to the needs of English-speaking developed countries, and content in other languages, responding to the needs of citizens of the developing countries as well.[65] Additionally, it is more important to take care so that intellectual property laws, locking content and allowing only for a pay-per-view meaningful access to works, are amended, for the benefit of both developing and developed countries.[66] In this sense, the fight against infogopolies,[67] which of course does not aim to deny their true rights, deserves a high place in our "bridging" agenda.

---

[62]As described by Finkelstein and Dowell (1996), pp. 2–4.

[63]See BBC News. Accessed 2007, May 5.

[64]Dholakia and Kshetri (2007).

[65]Guadamuz (2005). Especially on unreasonable and harmful intellectual property (*sui generis*) protection of databases, which are one of the most valuable information products internationally; see Canellopoulou-Bottis. Accessed 2007, May 9.

[66]On this, see extensively, Lessig (2000).

[67]A term from Drahos and Braithwaite (2002), p. 169.

What is important in terms of "bridging" the information gap is perhaps a reframing of a series of questions. It would not help very much to calculate and use statistics; they are abundant, relative to the digital divide, and there is no meaning in exposing a series of numbers showing disparities in access etc., to information. We should rather return and seek answers to some more fundamental questions. I will endeavor to propose some of the following:

(1) What is the relationship between the inequality produced by the digital divide and the inequalities we have known for centuries (male–female inequality, income inequality, black/white inequality, etc.)?

(2) Is it true that information and communication technologies must be seen as *not* possessing some inherent quality that enables them to leapfrog institutional obstacles and skill and resource deficiencies on the ground,[68] but they are, as seems eminently plausible, simple *tools* that make possible greater economic prosperity?

(3) And if we agree that information and communication technologies are tools, should we demand that the research on projects supporting the bridging of the digital divide contain concrete data on issues of costs (especially in terms of alternative projects left out *instead*), returns of investments and sustainability of projects, in lieu of mere "plans," "intentions," and "opportunities"?

(4) Is it true that efforts to bridge the digital divide may have the effect of locking developing countries into a new form of dependency? Is it true that the constant upgrading of Microsoft's programs places developing countries at a big disadvantage?[69]

(5) What is the relevance, if any, of the digital divide discourse (e.g., Internet access to all) with the fact that data is not information, information is not marketable knowledge, and marketable knowledge is not wisdom?[70] The gaps between these various notions must be identified to call better attention to how our efforts to bridge the various gaps should succeed. For example, we must provide education that enables people to convert data to information, and information to marketable knowledge. To ensure full human flourishing, we want to ensure that bridging the digital divide leads not only to ending life-threatening poverty, but also to full flourishing of human beings, which requires wisdom, aesthetic experience, philosophical self-reflection, and so on.

(6) If bridging the digital divide means mainly a better economy, then we must take into account the relationship between material resources and happiness.[71] For example, it is reasonable to hypothesize that people in the United States are not happier than they were, say in 1950, even though they have more wealth.

---

[68]Wade (2002).

[69]As suggested by Wade (2002).

[70]See generally Stoll (2000) p. 143.

[71]See *The Economist* (2006).

> But this much also seems right: a better economy, *which means the difference between being in life-threatening poverty and being able to satisfy one's basic needs,* does make a huge difference in both subjective (e.g., happiness) and objective measures of well-being. And this is the biggest concern about the digital divides. It also makes some difference with respect to relative poverty, such as exists within an affluent society like the United States, where worth is all too frequently equated with social status and wealth. In any event, what is generally true about wealth is this: going from absolute poverty to being able to reliably satisfy one's need produces a huge increase in measures of personal utility, but subsequent increases result in progressively diminishing increases in personal utility. That is, once you have the ability to reliably satisfy basic needs, wealth has diminishing marginal utility. A $5000 raise produces less additional utility to someone making $150,000 than to someone making $100,000 and so on.

Only if we have some persuasive answers to these kinds of questions, will we be able to arrive at equally persuasive conclusions about exactly how to approach the problems of poverty to which the digital divide contributes.

REFERENCES

Amighetti, A. and Reader, N. (2003). Internet Project for Poor Attracts Rich. *The Christian Science Monitor.* Available at http://www.csmonitor.com/2003/0724/p16s01-stin.html.

Bard, A. and Soderqvist, J. (2002). *Netocracy.* Pearson Education, London.

BBC News (2007). *Computer Problem Hit 999 Calls.* Available at http://news.bbc.co.uk/go/or/fr/-/1/hi/england/london/5279706.stm.

Britz, J. (2004). To know or not to know: a moral reflection on information poverty. *Journal of Information Science*, 30(3), 192–204.

Budge, I. (1996). *The New Challenge for Direct Democracy.* Polity Press, Oxford.

Canellopoulou-Bottis, M. (2004). A different kind of war: internet databases and legal protection or how the strict intellectual property laws of the West threaten the developing countries' information commons. *International Review of Information Ethics.* Available at http://papers.ssrn.com/sol3/papers.cfm?abstract_id=952882.

Cartier, C., Castells, M., and Qui, J.L.C. (2005). The information have-less: inequality, mobility and translocal networks in Chinese cities. *Studies in Comparative International Development*, 40(2), 9–34.

Castells, M. (1998). *The Information Age: Economy, Society and Culture, Part 3: the End of Millenium.* Blackwell, Oxford.

Celli, E. and Dreifach, K. (2002). Postcards from the edge: surveying the digital divide. *Cardozo Arts and Entertainment Law Journal*, 20, 53–72.

Compaine, B. (Ed.) (2001). *The Digital Divide: Facing a Crisis or Creating a Myth?* MIT Press, Cambridge, MA.

Davis, C.M. (2001). Information apartheid: an examination of the digital divide and information literacy in the United States. *PNLA Quarterly*, 65(4), 25–27.

Dholakia, N. and Kshetri, N. (2007). Digital divide to digital dividend. Available at SSRN: http://ssrn.com/abstract=847186. Last access 2007, May 8.

Dorr, J. and Akeroyd, R. (2001). New Mexico tribal libraries bridging the digital divide. *Computers-in-Libraries*, 21(9), 36–42.

Drahos, P. and Braithwaite, J. (2002). *Information Feudalism: Who Owns Knowledge Economy?* Earthscan Publications, London.

Dutch, M. and Muddiman, D. (2001). The public library, social exclusion and the information society in the United Kingdom. *LIBRI*, 51(4), 183–194.

Finkelstein, A. and Dowell, A. (1996). A comedy of errors: the London ambulance service case study. *In Proceedings of 8th International Workshop on Software Specification & Design IWSSD-8*. IEEE CS Press.

Guadamuz, A.L. (2007). The digital divide: it's the content stupid! *Computer and Telecommunications Law Review*, 304, 73–77, 113–118, 2005. Available at SSRN: http://ssrn.com/abstract=766624. Last access 2007, May 9.

Guo, B.P., Bricout, J.C., and Huang, J. (2005). A common space or a digital divide? A social model perspective on the online disability community in China. *Disability and Society*, 20(1), 49–66.

Hampton, K. and Wellman, B. (2003). Neighboring in Netville: how the internet supports community and social capital in a wired suburb. *City & Community*, 2(4), 277–311.

Hersberger, J. (2003). Are the economically poor information poor? Does the digital divide affect the homeless and access to information? *Canadian Journal of Information and Library Science*, 27(3), 45–63.

Himma, K. (2004). The moral significance of the interest in information: reflections on a fundamental right to information. *Journal of Information, Communication, and Ethics in Society*, 2(4), 191–202.

Himma, K. (2007). The information gap, the digital divide and the obligations of affluent nations. *International Review of Information Ethics*. Vol. 7 (09/2007).

Howell, B. (2001). The rural-urban digital divide in New Zealand: fact or fable? *Prometheus*, 19(3), 231–51.

Kalichman, S.C., Benotsch, E.G, Weinhardt, L., Austin, J., Luke, W., and Cherry, C. (2006). Health related internet use, coping, social support and health indicators in people living with HIV/AIDS: preliminary results from a community survey. *Health Psychology*, 22(1), 111–116.

Katz, J.E., Rice, R.E, and Aspden, P. (2001). The Internet, 1995–2000-access, civic involvement and social interaction. *American Behavioral Scientist*, 45(3): m 405–419.

Kenny, C. (2002). Should we try to bridge the global digital divide? *Info*, 4(3): 4–10.

Lentz, R.G. and Oden, M.D. (2001). Digital divide or digital opportunity in the Mississippi Delta region of the US. *Telecommunications Policy*, 25(5), 291–313.

Lessig, L. (2000). *Code and Other Laws of Cyberspace*. Basic Books.

Linchuan Qiu, J. (2002). Coming to terms with informational stratification in the People's Republic of China. *Cardozo Arts & Entertainment Law Journal*, 20, 157.

Malakooty, N. (2007). Closing the Digital Divide? The 100$ PC and other projects for developing countries, available at http://pcic.merage.uci.edu/papers/2007/100PC.pdf. Accessed 2008, February 1.

Malone, M., Mathes, L., Dooley, J., and White, A.E. (2005). Health information seeking and the doctor-patient digital divide. *Journal of Telemedicine and Tele-Care*, 11, 25–28.

Montero, M. and Stokols, D. (2003). Psychology and the internet: a social ecological analysis. *Cyberpsychology & Behavior*, 6(1), 59–72.

Mutume G. (2003). Africa takes on the digital divide. *Africa Recovery*, 17(3), available at www.un.org/ecosocdev/geninfo/afrec/vol17no3/173tech.htm. Accessed 2008, February 1.

National Telecommunications and Information Administration (1993). *National Information Infrastructure: Agenda for Action*. Available at http://metalab.unc.edu/nii.toc.html.

National Telecommunications and Information Administration (1995a). *Connecting the Nation: Classrooms, Libraries, and Health Care Organizations in the Information Age*. Available at http://www.ntia.doc.gov/connect.html.

National Telecommunications and Information Administration (1995b). *Falling Through the Net: A Survey of Have-Nots in Rural and Urban America*. Available at http://www.ntia.doc.gov/ntiahome/digitaldivide/.

National Telecommunications and Information Administration (1998). *Falling Through the Net II: New Data on the Digital Divide*. Available at http://www.ntia.doc.gov/ntiahome/digitaldivide/.

National Telecommunications and Information Administration. (1999). *Falling Through the Net: Defining the Digital Divide*. Available at http://www.ntia.doc.gov/ntiahome/digitaldivide/.

National Telecommunications and Information Administration (NTIA) (2002). *A Nation Online: How Americans are Expanding their Use of the Internet*. Available at http://www.ntia.doc.gov/ntiahome/dn/nationonline_020502.htm.

National Telecommunications and Information Administration (NTIA) (2002). *A Nation Online: How Americans are Expanding Their Use of the Internet*. Available at http://www.ntia.doc.gov/ntiahome/dn/index.html.

Norman, N. and Erbring, L. *Internet and Society: A Preliminary Report*. Study by the Stanford University Institute for the Quantitative Study of Society, Stanford University. Available at www.stanford.edu/group/siqss/Press_Release/Preliminary_Report-4-21.pdf.

Norris, P. (2001). *The Digital Divide: Civic Engagement, Information Poverty, and the Internet Worldwide*. CUP, New York.

Olemacher, S. (2007). Official Count: 754,000 People Believed Homeless in U.S. *Seattle Times*, February 28. Available at http://seattletimes.nwsource.com/html/nationworld/2003592874_homeless28.html.

Oppenheimer, T. (1997). The computer delusion. *The Atlantic Monthly*, 280(1), 45–62.

Ponce, M. (2005). *UNESCO Report Highlights Digital Divide*, November 4. Available at http://english.ohmynews.com/articleview/article_view.asp?article_class=4&no=256818&rel_no=1.

Pugh, T. (2007). More Americans Falling Deeper into Depths of Poverty. *Seattle Times*, February 26. Available at http://archives.seattletimes.nwsource.com/cgi-bin/texis.cgi/web/vortex/display?slug=poverty26&date=20070226&query=poverty.

Rideout, V. (2003). Digital inequalities in Eastern Canada. *Canadian Journal of Information and Library Science—Revue Canadienne des Sciences de l' Information et de Bibliotheconomie*, 29(2), 3–31.

Rifkin, J. (2000). *The Age of Access: The New Culture of Cybercapitalism Where All of Life is a Paid-for Experience*. Tarcher-Putman Books, New York, p. 228.

Saik-Yoon-Chin. (2005). Diverging information societies of the Asia Pacific. *Telematics and Informatics*, 22(4), 291–308.

Schwartz, J. (1999). US Cities Race Gap in Use of Internet: Clinton Bemoans "Digital Divide". *The Washington Post*, July 9, A1.

Singleton, S. and Mast, L. (2000). How does the empty glass fill? *A Modern Philosophy of the Digital Divide, EDUCASE-Review*, 35(6), 30–6.

Standing, C., Sims, I., Stockdale, R., and Wassenaar, A. (2003). Can e-marketplaces bridge the digital divide? In: Korpela, M., Montealegre, R., and Poulymenakou, A. (Eds.) *Working Conference on Information Systems Perspectives and Challenges in the Context of Globalization*. Kluwer Academic Publishers, Dordrecht, Netherlands, pp. 339–353.

Stoll, C. (2000). *High-Tech Heretic: Why Computers Do Not Belong in the Classroom and Other Reflections of a Computer Contrarian*. Anchor Books.

Subba Rao, S. (2005). Bridging digital divide: efforts in India. *Telematics and Informatics*, 22(4), 361–75.

The Economist, Happiness (and How to Measure It), December 23, 2006.

Thierer, A. (2000). *Divided over the Digital Divide*. Available at http://www.heritage.org/Press/Commentary/ED030100.cfm.

Tilly, C. (1998). *Durable Inequality*. University of California Press.

U.S. Department of Commerce (1999). Falling through the net: toward digital inclusion. Available at http://ntiahome/digitaldivide.

van Dijk, J. (2005). *The Deepening Divide—Inequality in the Information Society*. Sage Publications.

Wade, R. (2002). Bridging the digital divide: new route to development or new form of dependency? *Global Governance*, 8(4), 443–466.

Willis, S. and Tranter, B. (2006). Beyond the "digital divide": internet diffusion and inequality in Australia. *Journal of Sociology*, 42(1), 42–59.

Yu, P. (2002). Bridging the digital divide: Equality in the information age. Working paper series no. 44, Cardozo Law School, Jacob Burns Institute for Advanced Legal Studies. Available at ssrn.com.abstract_ID=309841.

CHAPTER 27

Intercultural Information Ethics

RAFAEL CAPURRO

27.1 INTRODUCTION

Intercultural Information Ethics (IIE) can be defined in a narrow or in a broad sense. In a narrow sense it focuses on the impact of information and communication technology (ICT) on different cultures as well as on how specific issues are understood from different cultural traditions. In a broad sense IIE deals with intercultural issues raised not only by ICT, but also by other media as well, allowing a large historical comparative view. IIE explores these issues under descriptive and normative perspectives. Such comparative studies can be done either at a concrete or ontic level or at the level of ontological or structural presuppositions.

The present IIE debate follows the *international* debate on information ethics that started with the "First International Congress on Ethical, Legal, and Societal Aspects of Digital Information" organized by UNESCO in 1997 in the Principality of Monaco and subsequent meetings, culminating in the World Summit on the Information Society (WSIS) (Tunisia, 2003, Geneva, 2005). These conferences were aimed particularly at reaching a consensus on ethical principles to be implemented through practical policy, as in the case of the "Declaration of Principles" of the WSIS.

The academic debate on *intercultural* issues of ICT takes place in biennial conferences on "Cultural attitudes towards technology and communication" (CATaC) organized by Charles Ess and Fay Sudweeks since 1998. But intercultural issues are also raised in the ETHICOMP conferences organized by Simon Rogerson since 1995, the conferences on "Ethics of Electronic Information in the Twentyfirst Century" (EEI21) at the University of Memphis since 1997, and the CEPE conferences (Computer Ethics: Philosophical Enquiry) since 1997.

The first international symposium dealing explicitly with *intercultural information ethics* was organized by the International Center for Information Ethics (ICIE) and was entitled "Localizing the Internet. Ethical Issues in Intercultural Perspective." It took

The Handbook of Information and Computer Ethics, Edited by Kenneth Einar Himma and Herman T. Tavani

place in Karlsruhe (Germany) in 2004. As far as I know, my introductory paper to this symposium was the first paper addressing the question of IIE in its title (Capurro, 2007a). The proceedings were published online in the "International Review of Information Ethics" (IRIE, 2004). A selection of papers was published as a book in 2007 (Capurro et al., 2007; Capurro and Scheule, 2007). The journal *Ethics and Information Technology* has dedicated a special issue edited by Charles Ess on privacy and data protection in Asia (Ess, 2005). The Oxford Uehiro Centre for Practical Ethics together with the Uehiro Foundation on Ethics and Education and the Carnegie Council on Ethics and International Affairs, organized an international conference, entitled "Information Ethics: Agents, Artefacts and New Cultural Perspectives" that took place in 2005 at St Cross College, Oxford. The conference addressed issues beyond the moral questions related to "agents" and "artifacts," considering also cultural questions of the globalization of information processes and flows, particularly "whether information ethics in this ontological or global sense may be biased in favor of Western values and interests and whether far-eastern cultures may provide new perspectives and heuristics for a successful development of the information society." (Floridi and Savulescu, 2006; Floridi, 2006). Soraj Hongladarom and Charles Ess have edited a book with the title *Information Technology Ethics: Cultural Perspectives* (Hongladarom and Ess, 2007a, 2007b; Weckert, 2007). The book puts together a selection of contributions on what Western and non-Western intellectual traditions have to say on various issues in information ethics (Froehlich, 2004), as well as theoretical debates offering proposals for new synthesis between Western and Eastern traditions.

In the following, an overview of IIE as discussed in these sources is given. The first part deals with the foundational debate of morality in general as well as in IIE in particular, starting with the question of the relation of reason and emotions. This question is addressed within the background of continental European philosophy with hints to Eastern traditions. It follows a review of the foundational perspectives on IIE as developed by Charles Ess, Toru Nishigaki, Terrell Ward Bynum, Bernd Frohmann, Lorenzo Magnani, Thomas Herdin, Wolfgang Hofkirchner, Ursula Maier-Rabler, Barbara Paterson, Thomas Hausmanninger, and myself. The second part presents some ethical questions about the impact of ICT on different cultures in Asia and the Pacific, Latin America and the Caribbean, Africa, Australia, and Turkey. The third part addresses succinctly special issues such as privacy, intellectual property, online communities, governmentality, gender issues, mobile phones, health care, and the digital divide as addressed in the already-mentioned IIE sources.

27.2 THE FOUNDATIONAL DEBATE

27.2.1 On the Sources of Morality

There is a classic debate in moral philosophy between cognitivism and noncognitivism with regard to the truth-value of moral claims, namely:

(1) Moral claims lack truth-value and are merely expressive of human emotions of approval or disapproval (moral noncognitivism).

(2) Moral claims have truth-value (moral cognitivism).

Moral cognitivism concerns the following alternatives:

(1) Morality is objective in the sense of being true or false in virtue of mind-independent facts about the world—and not in virtue of what cultures or individuals think about them (i.e., moral objectivism);
(2) Normative moral relativism (or conventionalism or intersubjectivism) that claims morality is manufactured by the beliefs and practices of cultures (i.e., moral claims are true in a culture only if accepted, believed, or practiced by some sufficiently large majority of the culture); and
(3) Normative moral subjectivism that claims morality is manufactured by the beliefs and practices of individuals (i.e., moral claims are true for a person only if accepted by that person).

The distinction between cognitivism and noncognitivism presupposes that human emotions have no cognitive value and, *vice versa*, that human cognition has a truth-value if and only if it is free of emotions. This is, in my view, a wrong alternative since there is no emotion-free cognition and emotions have a cognitive value as demonstrated by neurobiologist Antonio Damasio (1994). This empirical approach to the relation between reason and emotion converges in some regards with Martin Heidegger's phenomenological approach to moods and understanding (Heidegger, 1987, 172ff). According to Heidegger, moods are not primarily private feelings, but they disclose a public experience, that is, they concern the way(s) we are in a given situation with others in a common world (Capurro, 2005a). Being originally social our feelings do not separate us from each other, but even in the case in which we speak of mood as a subjective state, this belongs already to the situation in which I am embedded implicitly or explicitly together with others. The psychologist Eugene Gendlin remarks that Heidegger's conception of moods is "interactional" instead of "intrapsychic" (Gendlin, 1978). Gendlin underlines another important difference with regard to the traditional subjectivist view, namely, the relation of mood and understanding or, more precisely, the conception of moods as a specific way of understanding. Moods are not just affections coloring a situation, but an active although mostly implicit way of understanding a situation independently of what we actually say or not with explicit words. There is then, according to Heidegger, a difference as well as an intimate relation between mood, understanding, and speech as basic parameter of human existence.

Within this background, my position concerning the truth-value of moral claims is neither subjectivist, nor objectivist or simply relativistic. They have a common ground to which they implicitly or explicitly *relate*. One classical answer to the question of the foundation of morality is that moral claims relate to the basic moral principle *Neminem laede, imo omnes, quantum potes, juva* (do no harm, help where you can). I believe that even if we can give good reasons for such a fundamental moral principle the knowledge of such reasons is not enough to move the will in order to do (or not) the good.

Is there a foundation for this principle? Nietzsche questioned the ambitious theories aiming at a religious and/or metaphysical foundation of morality such as

Schopenhauer's volitional metaphysics or intellectualistic theories (Nietzsche, 1999, Vol. 11, p. 171). His plea was for a more modest and patient practice, namely the comparison of the rich variety of human moralities and their theories. We live, according to Nietzsche, in the "epoch of comparison" (*Zeitalter der Vergleichung*) (Nietzsche, 1999, Vol. 2, pp. 44–45). One example of this task of comparison between, for instance, Western moral theories and classical Chinese philosophy is the work of the French philosopher and sinologist Jullien (1995).

According to Karl Baier, basic moods, through which the uniqueness of the world and the finitude of our existence become manifest, are a transcultural experience common to all human beings. They concern our awareness of the common world (Baier, 2006). It is on the basis of the mood of anxiety (*Angst*), for instance, that we are aware of death (*Sterblichkeit*) and finitude or in the mood of "being born" (*Gebürtlichkeit*) in which we feel ourselves open for new possibilities of being. In *Being and Time,* Heidegger gives a famous analysis of two moods, namely fear (*Furcht*) and anxiety (*Angst*), borrowing basic insights from Kierkegaard's *Concept of Anxiety* (Heidegger, 1987, 228ff). The key difference between these moods is that while fear is a mood in which one is afraid about something fearsome, anxiety, in contrast, faces us with our being-in-the-world itself in such a way that no intraworldly entity is at its origin. But we are confronted with the very fact of the being there, with our existence in the world, and of the being of the world itself, without the possibility of giving an intrinsic reason for them. Hubert Dreyfus remarks: "In anxiety Dasein discovers that it has no meaning or content of its own; nothing individualizes it but its empty thrownness." (Dreyfus, 1991, p. 180) Such an experience is not necessarily accompanied by sweating and crying, but it is rather more near to what we could call today a "cool" experience of the gratuity of existence.

Ludwig Wittgenstein describes his "key experience" (*mein Erlebnis par excellence*) in the "Lecture on Ethics" with the following words: "This experience, in case I have it, can be described most properly, I believe, with the words *I am amazed about the existence of the world.* Then I tend to use formulations like these ones: 'How strange that something exists at all' or 'How strange that the world exists'" (Wittgenstein, 1989, p. 14, *my translation*). According to Wittgenstein we have really no appropriate expression for this experience—other than the existence of language itself. On December 30, 1929 Wittgenstein remarked: "I can imagine what Heidegger means with being and anxiety. Human beings have the tendency to run against the boundaries of language. Think, for instance, about the astonishment that something at all exists. (. . .) *Ethics* is this run against the boundaries of language." (Wittgenstein, 1984, 68, *my translation*). The *primum movens* of our actions lies in the very facticity or "thrownness" (*Geworfenheit*) (Heidegger) and finitude of human existence that is disclosed through moods.

In terms of Heidegger's "Being and Time," we are ontologically "indebted" or "guilty" toward the "calling" of the other, in the various senses of the word "guilty" such as "having debts" to someone or "being responsible for" (Heidegger, 1987, p. 325ff). We are primordially "guilty" in the sense that we are indebted to the "there" of our existence, between birth and death. Our existence is basically "care" (*Sorge*) of our factual and limited possibilities that manifest themselves within the framework of the uniqueness and "nullity" of our existence as well as of the fact of the world itself. Our moods, or more specifically, our "basic moods" (*Grundstimmungen*), play a key role in what we could call a holistic ethics that makes theoretically explicit the mechanism of our well-being thus only a necessary but not a sufficient condition for our (moral) actions. The moral imperative is precisely this call for care and our capacity to give a finite or "guilty" answer. It is a categorical imperative (take care of yourself) insofar as we cannot not take care of our lives, but it allows at the same time, due to its indeterminate form, multiple options of life interpretation and design that arise from the open possibilities that "call" our attention and challenge our practical reason. This basic human experience gives rise to different interpretations and their corresponding cultural articulations. As historical beings, humans accumulate, as individuals as well as societies, unique existential experiences that constitute what we could call their dynamic cultural *a priori* laid down in their cultural memory. The uniqueness of the facticity of the world and human existence can therefore be understood as a common abyssal ground for morality and for moral theory, both being subject to different cogno-emotional interpretations.

This is not a plea for a kind of naturalistic fallacy of deriving "ought" from "is," but the awareness that we cannot not take care of our lives and the given world we live in. In *saying* this I am not even providing a sufficient reason for doing the good just because such linguistic utterance would be insufficient without the experience of the "call" itself to which a theory can only point without being able to give a foundation, in which case the phenomenon of the "call" and the facticity of the world would be negated as originating such utterance.

The enigmatic "fact" of our being-in-the-world, our facticity, is the "first call" or *primum movens* of our will. This provides, I believe, an experimental and theoretical frame of reference for different ethical theories and practices, which is not a metaphysical ground. Both experiences are contingent, but at the same time prescriptive or normative in the sense that they urge or "call" for situative, that is, historical responsible thought and action by letting us become conscious of our ontological "guilt." This kind of responsibility does not therefore aim necessarily at identical shared moral norms as answers to such a call, although such a search is theoretically reasonable and pragmatically necessary in a given situation or with regard to a global phenomenon as in the case of ICT that is no less situated or "localized" (Capurro et al., 2007).

Buddhism, for instance, experiences the world in all its transitoriness in a mood of sadness and happiness, being also deeply moved by suffering. This mood "opens" the world in a specific way. According to Baier, there is something common to all human beings in the basic or deep moods, but at the same time there are specific moods at the beginning of human cultures, such as astonishment (*thaumazein*) in the Greek experience of the world. Baier is also well aware of the danger of building stereotypes, particularly when dealing with the differences between East and West, considering, for instance, the search for harmony as an apparently typical and unique mood of Asian cultures or the opposition between collectivity and individuality. As there are no absolute differences between cultures there are also no exclusive moods. Experiences such as nausea, pangs of moral conscience, or the "great doubt" are common to

Japanese Buddhism and modern Western nihilism. For a sound future intercultural methodology Baier suggests that we look for the textual basis from literature, art, religion, and everyday culture to pay attention to complex phenomena and to the interaction between moods and world understanding. I would also like to add the role of legal and political institutions as well as the historical and geographical settings in which these experiences are located. If there is a danger of building stereotypes, there is also one of overlooking not only concrete or ontic but also structural or ontological differences by claiming a world culture that mostly reflects the interests and global life style of a small portion of humanity.

From this perspective, moral cognitivism and noncognitivism are partial views of human existence that is grounded on moods *and* understanding. Normative moral subjectivism takes for granted that individuals can be conceived of as separated from their being-in-the-world with others, that is, of the social and historical network of practices and beliefs, without critically asking about the origin of this conception of an isolated individual itself. Morality is not founded on independent facts about the world, but arises spontaneously (*sponte sua*) from (Greek: *hothen*) the awareness and respect for the abyssal facticity and uniqueness of the world itself and human existence that are the invaluable and theoretically nonprovable truth-values on which all moral claims rest (Lévinas,1968). Beliefs, institutions, and practices of cultures give a long-term stability to such claims and make them obvious. Cultural frameworks are not conceived as closed worlds but as grounded in common affective human experiences of sharing a finite existence in a common world. In other words, the ontic differences between human cultures are refractions of the common world awareness. Every effort to determine the nature of this awareness gives rise to different experiences and interpretations. We speak of multicultural ethics in case we juxtapose such interpretations instead of comparing them. The opposite is a monocultural view that conceives itself as the only valid one. Human reason is genuinely plural with regard to common tentative transcultural expressions of this common ground such as the "Universal Declaration of Human Rights," whose principles are subject to permanent scrutiny and intercultural interpretation and, being linguistic utterances, build the necessary but not sufficient condition for moving the will of, say, the member states of the United Nations to put it into practice (Ladd, 1985).

27.2.2 On the Foundation of IIE

27.2.2.1 Charles Ess Charles Ess' "global information ethics" seeks to avoid imperialistic homogenization while simultaneously preserving the irreducible differences between cultures and peoples (Ess, 2006). He analyzes the connections of such an ethical pluralism between contemporary Western ethics and Confucian thought. Both traditions invoke notions of *resonance* and *harmony* to articulate pluralistic structures of connection alongside irreducible differences. Ess explores such a *pros hen* pluralism in Eastern and Western conceptions of privacy and data privacy protection. This kind of pluralism is the opposite of a purely *modus vivendi* pluralism that leaves tensions and conflicts unresolved, thus giving rise to a cycle of violence. Another more robust form of pluralism presupposes a *shared* set of ethical norms and standards but without overcoming deep contradictions. An even stronger form of pluralism does not search identity but only some kind of coherence or, as Ess suggests, complementarity between two irreducible different entities. The problem with this position is that it still asks for some kind of unity between irreducible positions. In order to make this goal plausible and somehow rational one must show where the possible focus that allows complementarity lies. Otherwise I see a contradiction between irreducibility and complementarity. This is a similar problem as the one raised by Thomas Kuhn concerning the question of the incommensurability of scientific theories arising from a paradigm change through scientific revolutions (Kuhn, 1962). Ess' concepts of resonance, or complementarity, raise the Aristotelian question of equivocity, analogy, and univocity. I think that irreducible positions cannot be logically reduced to some kind of complementarity, but it may be a deeper experimental source of unity such as the one I suggested at the beginning that is beyond the sphere of ontic or, to put it in Kantian terms, categorial oppositions. Kant's solution was the presupposition of a *noumenal* world that manifests itself practically through the categorical imperative. I believe that the facticity and uniqueness of the world and human life offers an empirical *hothen* dimension if not for overcoming categorial differences at least for a dialogue on cognitive-emotive fundamental experiences of our common being in the world (Eldred, 2006).

There are pitfalls of *prima facie* convergences, analogies, and family resemblances that may be oversimplified by a *pros hen* strategy. In many cases we should try to dig into deeper layers in order to understand where these claims originate or simply accept the limits of human theoretical reason by celebrating the richness of human experience. In his critical response to Charles Ess, Kei Hiruta questions the necessity and desirability of *pros hen* pluralism. As he rightly stresses, it is not clear what the points of shared ethical agreements are and how this call for unity fits with a call for diversity concerning the judgments of such "ethical perspectives" (Hiruta, 2006, p. 228). It looks as if the advocates of ethical pluralism would like to avoid the untolerable, such as child pornography in the Internet, working on the basis of a (pragmatic) problem-solving strategy leading to "points of agreement" or "responses" on the basis of Socratic dialogue. The problem with Socratic dialogue is that it is based on the spirit of *parrhesia,* which is a key feature of Western philosophy. I will discuss this issue.

27.2.2.2 Toru Nishigaki In his contribution on information ethics in Japan, Toru Nishigaki makes a difference between the search of ethical norms in the context of new information technologies (IT) on the one hand, and the changes "on our views of human beings and society" becoming "necessary to accompany the emergence of the information society" on the other hand (Nishigaki, 2006, p. 237). Such changes concern, for instance, the Western idea of a "coherent self" being questioned by information processing in robots. Although this change may lead from a Western perspective to nihilism, Buddhist philosophy teaches that there is no such a thing as a "coherent self" ethics having to do with compassion as well as with the relationship between the individual and the community, instead, as with the preservation of a "coherent self," the key ethical question being how our communities are changing instead of how far the "self" is endangered. As Nishigaki remarks: "It is possible to say, therefore, that in a sense the West now stands in need of Eastern ethics, while the East stands in need of Western ethics." Nishigaki stresses at the same time that there is no "easy bridge" between IT and Eastern philosophy. IT as looked from a cultural standpoint "has a strong affinity with the Judeo-Christian pursuit for a universal interpretation of sacred texts." Although we in the West look for some kind of unchanged meaning of terms, such as in Charles Ess' *pros hen* search for shared values and a tolerant or benevolent view of judgment diversity, the Zen master is eager to exercise himself in his disciple "by doing away with universal or conventional interpretations of the meanings of words" (Nishigaki, 2006, p. 238). In other words, the Buddhist stance teaches us Westerners another strategy beyond the controversy between monism and pluralism, by way of a kind of practice different from the Socratic dialogue. Nishigaki points to the controversy in the West between cognitive science and its view of cognition as a "representation" of the "outer world" and the view shared by our everyday experience as well as, for instance, phenomenology. Biologist Francisco Varela's theory of autopoiesis offers an alternative based on the Buddhist view of cognition as "a history of actions performed by a subject in the world" being then not representation of a pregiven world by a pregiven mind but "enactment" of such a history in the world (Nishigaki, 2006, p. 239). Nishigaki calls "ethical norms" the code or "behavior pattern" as perceived by a social system's observer. I would prefer to speak here of "moral norms" and reserve the concept of ethics for the reflection of such an observer on the factual norm. This is no less than the Aristotelian distinction between "*ethos*" and "*techne ethike*" or between morality and ethics. This terminological and conceptual difference has been proposed, for instance, by sociologist Niklas Luhmann (1990), being also broadly used in Western ethics. The undifferentiated use of these terms, as is mostly the case in everyday life, might lead to an uncritical approach of the role of ethics as observer-dependent reflection, which is the standpoint addressed by Nishigaki's "fundamental informatics." From this perspective, the conflict raised by globalization does not consist in the universal application of Western ethics but of Western morality. The universal application of Western ethics means that the discussion on morality would take place only on the basis of Western conceptual schemes. This is exactly what intercultural information ethics questions, understood as a permanent process of reflection and "translation," intend to avoid. For a comprehensive view of this East/West dialogue see Nishigaki and Takenouchi (2007).

27.2.2.3 Terrell Ward Bynum The information society is (and has always been) culturally fragmented into different information societies. Consequently, what is (morally) good for one information society may be considered as less appropriate in another one. Terrell Ward Bynum advocates, borrowing insights from Aristotle, Norbert Wiener, and James Moor, for a "flourishing ethics" (FE), which means that "the overall purpose of a human life is to flourish as a person" according to the basic principles of freedom, equality, and benevolence and the principle of minimum infringement of freedom (Bynum, 2006, p. 163). If the goal is to maximize the opportunities of all humans to exercise their autonomy—a conception of human existence that is culturally grounded in Western social philosophy—Bynum rightly follows that "many different cultures, with a wide diversity of customs, religions, languages, and practices, can provide a conductive context for human flourishing." In other words, Wiener's principles provide a foundation for a nonrelativistic global ethics that is friendly to cultural diversity. Bynum widens the scope of this human-centered ethics into a "general theory of Flourishing Ethics" (General FE) that includes the question of delegation of responsibility to "artificial agents" and the consequent need for ethical rules for such agents. Although Bynum welcomes different ethical traditions, he is well aware that some of them would not be compatible with General FE.

27.2.2.4 Bernd Frohmann Following the ethical thought of Michel Foucault and Gilles Deleuze, Bernd Frohmann proposes a philosophical interrogation of the local effects of the Internet through three main concepts: effect, locality, and ethics (Frohmann, 2007). He discusses the relationships between the global and the local or, more specifically, between the flows of capital, information, technology, and organizational interaction by pointing to the similarities and difference of today's "space of flow" (Manuel Castells) with some of its predecessors, for instance, in England's global empire. According to Frohmann, who follows Foucault, "ethical action consists in a 'mode of subjectivation' not eclipsed by the will to truth's drive to knowledge, transcendence, and universality. A philosophical *ethos* seeks contingencies and singularities rather than universal determinants, which block the aim of getting 'free of oneself'"(Frohmann, 2007, pp. 64–65). This is a plea for a kind of IIE that focuses on a careful situational analysis starting with the local *hothen* conditions that does not mean monocultural chauvinism, but critical appraisal of the way(s) computers control societies and the strategies people can develop in order to become "digitally imperceptible." Frohmann asks for strategies of "escaping" the Internet rather than "localizing" it as far as it can become a local instrument of oppression.

27.2.2.5 Lorenzo Magnani Lorenzo Magnani analyzes the rise of human hybridization with ICT and the building of what he calls, following Karl Roth, "material cultures" (Roth, 2001). Material cultures refer to people's material environments consisting of food, dwellings, and furniture in contrast to immaterial interactions dealing with language as well as the actors' perceptions, attitudes, and values. Magnani writes: "In our era of increasing globalization, ICT artifacts, such as the Internet, databases, wireless networks, become crucial mediators of cross-cultural relationships between human beings and communities" (Magnani, 2007, p. 39). If new artifacts become, ready-to-hand, the question is "at what ethical and cultural cost?" (Magnani, 2007, p. 40). According to Magnani, there is evidence that technical instruments such as cell phones and laptops vary significantly in their use according to their cultural differences. Local cultures are thus used as countercultures to globalization, such as the case of the role played by cell phones in ensuring the success of the people's revolution in the Republic of the Philippines (Magnani, 2007, p. 45). Magnani introduces the concept of "moral mediator" to indicate "a cultural mediator in which ethical aspects are crucial and the importance in potential intercultural relationships is central." A "moral mediator" consists of objects or structures that carry ethical or unethical consequences beyond human beings' intentionalities. An Internet Web site used to sell online not only realizes an economic transaction "but also carries ethical effects insofar as it implies certain customer's behaviors related to some policies and constraints." ICTs can enhance but also jeopardize local cultures. Magnani advocates in favor of the "principle of isolation" as a means to protect the self-identity of cultures that has to be equilibrated with the need to promote cyberdemocracy counterpoisoning the negative effects of globalization.

27.2.2.6 Thomas Herdin, Wolfgang Hofkirchner, and Ursula Maier-Rabler Thomas Herdin, Wolfgang Hofkirchner, and Ursula Maier-Rabler discuss the mutual influence between culture and technology on broad inter- and transcultural levels. They write: "The cultural-social framework of a society is formed mainly by the political-social system, by the legislative system, and particularly by the predominating ethic and religious values. As a result of these diverse dimensions, a continuum between the poles of information-friendly versus information-restrictive cultures emerges" (Herdin et al., 2007, p. 57). Following the concept of "transculturality" coined by Welsch (1999) the authors claim that cultures cannot be perceived as homogenous units anymore. They suggest that this concept should be enhanced with regard to the permeability between global and local cultures, allowing individuals to switch between different identities. The concept of "digital culture" is used to describe the model of mutual influence between cultural and ICT technology. Digital culture allows vast numbers of people with different cultural backgrounds to share knowledge, but it also gives rise to what has been called the "digital divide" based on low economic levels, as well as the "cultural divide" based on low educational levels. They discuss the dialectic of shaping, diffusion, and usage of ICTs along the following dimensions: digital content culture, digital distribution culture, and digital context culture. A main challenge concerns the creation of one global culture on the basis of the "reductionist way of thinking in intercultural discourse that is called *universalism*. Cultural universalism reduces the variety of different cultural identities to what they have in common. Identities are homogeneized by a sort of melting pot that was named McWorld (Barber, 2001)" (Herdin et al., 2007, p. 65). According to the authors, cultural thinking that reconciles the one and the many is achievable only on the basis of a way of thinking that allows integration and differentiation for which such terms as "transculturalism" (Welsch, 1999), "glocalization" (Robertson, 1992) and "new mestizaje" (John Francis Burke in Wieviorka, 2003) have been proposed.

27.2.2.7 Barbara Paterson According to Barbara Paterson, not only does the computer revolution threaten to marginalize non-Western cultural traditions, but the Western way of life also has caused large-scale environmental damage (Paterson, 2007), the task of computer ethics being to critically analyze such holistic effects. She proposes that the Earth Charter can function as a framework for such holistic research as it addresses, unlike the WSIS declaration, a broader public. In sum, "computer ethics needs to acknowledge the linkages between computing, development, and environmental conduct" (Paterson, 2007, p. 164).

27.2.2.8 Thomas Hausmanninger According to Thomas Hausmanninger, the right to differ that can be observed in the realm of religious belief (Martin Luther) gains today, since the "turn to contingency" in the epistemological debate of the twentieth century, something like the quality of a human right (Hausmanninger, 2007). The ethical obligation to respect the difference and plurality of belief systems is grounded, according to Hausmanninger, in picturing human beings as persons or subjectivities owing to each other the right to free self-realization. What has to be respected in order to respect human dignity may differ between cultures. Hausmanninger intents to regain the concept of subjectivity as a basis of "our" intercultural information ethics. The task of encompassing it with other endeavors remains open.

27.2.2.9 Rafael Capurro In today's information society we form ourselves and our selves through digitally mediated perceptions of all kind. The power of networks does not lead necessarily to slavery and oppression but also to reciprocity and mutual obligation. Globalization gives rise to the question of what does locally matter. Cyberspace vanishes into the diversity of complex real/virtual space-time connections of all kinds that are not any more separable from everyday life and its materiality. The boundaries of language against which we are driven appear now as the boundaries of digital networks that not only pervade but also accelerate all relationships between humans as well as between all kinds of natural phenomena and artificial things. For a more detailed analysis of the relation between moods and understanding with explicit relation to the information society see Capurro (2005a) and Wurman (2001). There are no neutral natural and/or artificial things within the realm of human cognitive-emotional existence. Every appropriation of, say, the "same" ICT creates cultural and moral differences. The task of IIE, understood as a reflection on morality, is not only to bridge these differences creating common moral codes but also to try to articulate and understand them as well. In my introductory paper to the ICIE symposium, I situate IIE within the framework of intercultural philosophy and analyze the question of universality with special regard to the WSIS discussions, particularly to the question of the human right to communicate and the right to cultural diversity. I point to society's responsibility to enable cultural appropriation. Following Walzer (1994) and Hongladarom (2001), I conceive moral arguments as "thick" or "thin" regarding whether they are contextualized or not, but I question the view that there is no third alternative (*tertium non datur*) between mono- and metacultural ethical claims. A purely metacultural information ethics remains abstract if it is not interculturally reflected. The task of IIE is to intertwine "thick" and "thin" ethical arguments in the information field.

In a contribution on the ontological foundation of information ethics, I point, following the analysis by Michel Foucault, to the Western tradition of *parrhesia* or "direct speech" as a special trait of Western moral behavior and democratic practice in contrast to the importance of "indirect speech" in Eastern traditions. I have developed this difference with regard to Confucian and Daoist thought and their relevance for the development of the Chinese information society (Capurro, 2006b). I point to the fact that the debate on an ontological foundation of information ethics, its questions, terminology, and aim, is deeply rooted in Western philosophy so far (Capurro, 2006a, p. 184).

In *resonance* to Charles Ess' Aristotelian concept of an ethical *pros hen* (toward one) that looks for the pluralist interpretation and application of *shared* ethical norms (Ess, 2006), I argue in favor of a *hothen* approach that turns the attention to the question of the source(s) of ethical norms including the multiple cognitive-emotional experience of such source(s). The task of IIE is not only to describe and criticize different kinds of cognitive-emotional interpretations of the common origin (*arché*) of moral experiences, but also to open the endless task of ethical comparison or *translation* between such interpretations. As Susan Sontag suggests, the task of the translator can be seen as an ethical task if we conceive it as the experience of the otherness of other languages that moves us to transform our mother tongue—including the terminologies used by different philosophic schools—instead of just preserving it from foreign or, as I would say, *heretic* influences (Sontag, 2004).

27.3 THE IMPACT OF ICT ON LOCAL CULTURES FROM AN IIE PERSPECTIVE

The ICIE symposium addressed the question of how embodied human life is possible within local cultural traditions and the horizon of a global digital environment. This topic with its normative and formative dimensions was discussed in three different perspectives, namely: Internet for social and political development, Internet for cultural development, and Internet for economic development. The symposium dealt with questions such as: How far does the Internet affect, for better or worse, local community building? How far does it allow democratic consultation? How do people construct their lives within this medium and how does it affect their customs, languages, and everyday problems? It also dealt with the impact of the Internet in traditional media, on cultural and economic development, as well as on the environment.

Charles Ess reviews examples of the cultural conflicts that occur when computer-mediated communication (CMC) technologies are deployed "outside" the boundaries of the Western cultural values and communicative preferences that shape their initial design and implementation, leading to the danger of "computer-mediated colonization" (Ess, 2007). Ess argues that ethnocentrism and its attendant colonization on the part of those who design and implement CMC technologies ought to be resisted through the use of "culturally aware" approaches to implementation and design.

27.3.1 Asia and the Pacific

Frances Grodzinsky and Herman Tavani analyze the question of whether the Internet has benefited life overall (Grodzinsky and Tavani, 2007a). They point to the cultural and linguistic diversity of countries in Asia and the Pacific where governments have established either a monopolistic model of development under their strict control or one that opens ICT infrastructure to private and international organizations. The global network involves a tension between cultural homogenization and heterogenization that can lead to increased fragmentation as well as increased homogenization. They believe that there are good reasons why cyberspace should not be homogenized. Even if cultural sovereignty may disappear along with national borders, the particulars of cultural autonomy should be preserved. They see no contradiction between the cultural richness of the cyberspace incorporating some aspects of universalism in ways that do not erode such diversification by ending in an "e-McDonalds."

Makoto Nakada examines the relationships between people's attitudes toward their society and culture and the meaning of the Internet in Japan (Nakada, 2007). According to the empirical evaluation, it seems as though the Japanese live in a world consisting of old Japan (*Seken*) and new Japan (*Shakai*) (see the discussion below on privacy in Japan).

Tadashi Takenouchi questions the prevailing materialism and individualism in today's Japanese society as well as what he calls "digital reductionism," according to which humans and other living beings are "nothing but" digital processing machines (Takenouchi, 2006, p. 188). As a remedy he advocates for an "informatic turn" understood as an "unrestricted capability of interpretation" that comes near to Western philosophical hermeneutics as well as to Eastern Buddhist concepts of "nothingness" (*mu*) and "self-understanding through relationships with others." He analyzes some correspondences between this turn in "fundamental informatics" with some ideas by Viktor Frankl, particularly the relation between *homo patiens*, who fulfills his/her life by trying to give an interpretative answer to the sufferings of others by taking care of them, and *homo sapiens* who manages things rationally and effectively through high-tech information processing skills. Frankl's *homo patiens* closely resembles the Buddhist idea of compassion "as it is produced by applying imagination to patience with regard to other fellow beings." (Takenouchi, 2006, p. 191). According to Takenouchi the axis of *homo patiens*, with the tension between despair and fulfillment, is more essential to humans than the axis of *homo sapiens,* with the poles of low information and high information skilled. The Japanese present information society debates on how far mastering IT skills will allow social participation of the handicapped or create a gap.

Along with this line of reasoning and experience, Vikas Nath reports on the diversity of digital governance models in countries such as India, Brazil, South Africa, Bangladesh, and the Philippines (Nath, 2007).

According to Lü Yao-Huai, a basic universal set of ethical standards is needed; otherwise global information interaction will be thrown into chaos (Lü, 2007). This minimum set includes three basic principles, namely information justice, information equality, and information reciprocity. He points to the concept of "Shen Du" (be watchful of oneself when one is alone) as having a special value in raising the moral consciousness of the individual beyond legal frameworks.

27.3.2 Latin America and the Caribbean

Daniel Pimienta reports on his experience with a Latin American/Caribbean virtual community leading to discussions about the intersection and boundaries of ethics and cultures in the new social movements based on the Internet (Pimienta, 2007).

According to Pimienta these models of communication are not the same in different parts of the world. Another example of such difference is given by Hugo Alberto Figueroa Alcántara in his report on collective construction of identity on an Internet basis in Mexico (Figueroa and Hugo, 2007) and by Susana Finquelievich with regard to Latin America (Finquelievich, 2007).

27.3.3 Africa

The first African Information Ethics Conference was held in Pretoria/Tshwane, South Africa, February 5–7, 2007 (African Information Ethics Conference, 2007). It was organized by the University of Pretoria, the University of Wisconsin-Milwaukee, and the International Center for Information Ethics. Not much has been published on the challenges to African philosophy arising from the impact of ICT on African societies and cultures(Jackson and Mandé, 2007). In my keynote address I explore some relationships between information ethics and the concept of *ubuntu* (Capurro, 2007b). One of the few detailed analyses of the relationship between *ubuntu* and privacy was presented by H. N. Olinger, Johannes Britz, and M.S. Olivier at the Sixth International Conference of Computer Ethics: Philosophical Enquiry (CEPE, 2005). According to Barbara Paterson "African thought emphasizes the close links among knowledge of space, of self, and one's position in the community," participation being the "keystone of traditional African society" (Paterson, 2007, p. 157). As Mogobe Ramose remarks, *ubuntu* is "the central concept of social and political organization in African philosophy, particularly among the Bantu-speaking peoples. It consists of the principles of sharing and caring for one another" (Ramose, 2002, p. 643). Ramose discusses two aphorisms "to be found in almost all indigenous African languages," namely: *Motho ke motho ka batho* and *Feta kgomo tschware motho*. The first aphorism means that "to be human is to affirm one's humanity by recognizing the humanity of others and, on that basis, establish humane respectful relations with them." Accordingly, it is *ubuntu* that constitutes the core meaning of the aphorism." The second aphorism means "that if and when one is faced with a decisive choice between wealth and the preservation of life of another human being, then one should opt for the preservation of life" (Ramose, 2002, p. 644). Following this analysis we can ask: what is the role of *ubuntu* in African information ethics? How is the intertwining of information and communication technology with the principles of communalism and humanity expressed in aphorisms such as *Motho ke motho ka batho* that can be translated as "people are other people through other people?" What is the relation between community and privacy in African information society? What kind of questions do African people ask about the effects of information and communication technology in their everyday lives? The proceedings of the Pretoria conference as well as an *Africa Reader on Information Ethics* will provide substantial contributions to IIE in and from Africa. The already started dialogue on the basis of the ANIE platform promises a fruitful intercultural exchange in the mood expressed in the motto of this conference: *The joy of sharing knowledge*.

This conference was unique in several respects. First, it dealt with information ethics in Africa from an African perspective. Second, it encouraged African scholars to articulate the challenges of a genuine African information society. Third, it was devoted to fundamental ethical challenges such as the foundations of African information ethics, the issue of cultural diversity and globalization dealing particularly with the protection and promotion of indigenous knowledge, and the question of the impact of ICT on development and poverty, as well as on socio-political and economic inclusion/exclusion, North-South flow of information in terms of information imperialism, and the flight of intellectual expertise from Africa. One important outcome of the conference was the *Tshwane Declaration on Information Ethics in Africa* (Tshwane Declaration, 2007) as well as the creation of the African Network for Information Ethics (ANIE, 2007). A *Reader on Africa Information Ethics* is in preparation.

27.3.4 Australia

Maja van der Velden analyzes how far the preoccupation with content and connectivity obscures the role of IT by making invisible different ways of knowing and other logics and experiences (van der Velden, 2007). She presents an aboriginal database in Northern Australia useful for people with little or no literacy skills. According to van der Velden, "the technological design of an information system controls, to a large extent, how information is produced, categorized, archived, and shared in the system. This design reflects the politics, culture, and even race, gender, class, and ethnicity of the people involved" (van der Velden, 2007, p. 85).

27.3.5 Turkey

Gonca Telli Yamamoto and Faruk Karaman deal with IIE in Turkey, a country in which Western, Islamic, and Turkish cultures compete. They write: "With its Westernization efforts, Turkey presents a very special case for analyzing IT ethics. In spite of the great efforts to become part of the Western civilization, Turkey is still struggling to decide to which civilization it wants to belong—Western civilization or Islamic or Eastern civilizations" (Yamamoto and Karaman, 2007, p. 190). Even the Western-oriented population do not see, for instance, an ethical issue in copying intellectual property. The Internet revolution is felt in a delayed fashion in Turkey, which means that the digital divide has become a serious problem.

27.4 SPECIAL ISSUES

27.4.1 Privacy

Privacy is a key question as it deals with basic conceptions of the human person. Intercultural dialogue is the antidote to the danger of getting "lost in translation" that arises from a monological discourse aiming at reducing all differences to its own language. In his introduction to the special issue of the journal *Ethics and Information Technology* on privacy and data privacy protection in Asia, Charles Ess underlines the importance of "an informed and respectful *global* dialogue in information ethics" (Ess, 2005, p. 1).

27.4.1.1 China Lü Yao-Huai analyzes the privacy experience of today's ordinary Chinese people (Lü, 2005). According to Lü there is a transformation of contemporary Chinese consciousness of privacy starting with economic and political reforms since 1980. He writes: "Before 1978, if someone publicly expressed the intention of pursuing individual interests, he or she would have certainly been called an egoist." (Lü, 2005, p. 12). This cultural and moral change concerns mainly three aspects: (1) Individual freedom is not any more a taboo topic. A conversation partner can refuse to answer a question on the plea that "this is my privacy." (2) There is also a tendency not to interfere with what one perceives to be the privacy of the other. (3) The common Chinese concept of privacy, *Yinsi* (shameful secret), has been expanded to include all personal information whether shameful or not that people do not want others to know.

With the rise of the Internet in the 1990s the question of data privacy emerged in China. Lü provides an overview of the legal framework of Chinese data protection and discusses the following ethical principles: (1) the principle of respect, (2) the principle of informed consent, (3) the principle of equilibrium (between the safety of personal privacy and the safety of society), and (4) the principle of social rectification. The last two principles take society as the higher value. Lü questions the claim that privacy remains a foreign concept for many Chinese people. He argues that in rural areas, following the tradition of collectivism, people are more interested in other people's lives than in the cities. But also all Chinese papers on information ethics dealing with privacy interpret it as an instrumental instead as an intrinsic good. Chinese researchers argue that privacy protection has a function with regard to social order. Although many Chinese still think that there is no right to privacy within the family, a survey among the young generation shows the opposite interest. Lü foresees a strong influence of Western views on privacy without Chinese traditional culture becoming fully Westernized in this regard (Lü, 2007).

27.4.1.2 Thailand Krisana Kitiyadisai explores the changes of the concept of privacy in Thai culture, based on collectivism and nonconfrontation (Kitiyadisai, 2005). "Being private" applies in traditional Thailand to the space shared by family members. The lack of a Thai word for privacy is due, according to Kitiyadisai, to the feudal heritage of Thai society with a system of hierarchical ranking, politeness protocols, and patronage. Strong relationships are based on the principle of nonconfrontation avoiding the disastrous results of "losing-face" (*siar-na*) instead of face-saving (*koo-na*). According to Kitiyadisai, "the combination of privacy as 'private affairs' (*rueng-suan-tua*) and the right of 'noninterference' works in support of 'saving face'" (Kitiyadisai, 2005, p. 18). These values are similar to Confucian values of "ancestor reverence, respect for 'face,' responsibility, loyalty, modesty, and humility" (Kitiyadisai, 2005, p. 24). According to Buddhism, human rights are not intrinsic to human individuals but they are necessary for conducting a virtuous human existence. Kitiyadisai provides an overview of the data protection legislation in Thailand. She stresses the ongoing tensions between "imported liberal democratic values" and "traditional Thai values."

Soraj Hongladarom describes a grave challenge to the privacy of Thai citizens. The Thai government plans to introduce a digital national identification card in a country with no specific law protecting personal information. The threat of political misuse raises the question of the nature of privacy and its justification. Hongladarom explores this question from the perspective of two famous Budhhist sages, namely Nagarjuna (c. 150–250 AD), founder of the Mahahāyāna Buddhism, and Nagasena (c. 150 BC). He writes: "The reason I believe the Buddhist perspective is important in this area is that Buddhism has a very interesting claim to make about the self and the individual on whose concept the whole idea of privacy depends." (Hongladarom, 2007, p. 109).

The fact that Buddhism rejects the individual self does not mean that it rejects privacy. In order to understand this counterintuitive argument, Hongladarom distinguishes between the absolute and conventional levels of assertion. From an absolute point there is no distinction between subject and object. If there is no inherently existing self then privacy is grounded on the conventional idea that it is necessary for democracy, which means that privacy has an instrumental instead of an intrinsic or core value. But, according to Hongladarom, the distinction between intrinsic and instrumental values has an insecure foundation as all values rest on our attachment to them (Hongladarom, 2007, p. 116). According to Nagasena, the conventional self exists in conventional reality and is shown to be a mere illusion after analysis in terms of the "ultimate truth." Hongladarom parallels Nagarjuna's distinction between "conventional truth" and "ultimate truth" with Kant's distinction between a "phenomenal" and a "noumenal" realm. But in contrast to Kant, there is no "I" providing a transcendental unity of apperception. Privacy, as used in everyday life, is not denied in Buddhism. It is in fact justified as an instrument for the purpose of living harmoniously according to democratic ideals. But "from the ultimate perspective of a Buddha, privacy just makes no sense whatsoever" (Hongladarom, 2007, p. 120). Violations of privacy are based on the three "mental defilements" (*kleshas*), namely greed, anger, and delusion, the antidote being to cultivate love and compassion. He writes: "Compassion naturally arises from this realization when one realizes that other beings are no different from oneself. All want to get rid of suffering, and all do want happiness. The benefit of this realization for information ethics is that compassion is the key that determines the value of an action"(Hongladarom, 2007, p. 120). Compassion is, I would say, the "basic mood" of Buddhist experience of the uniqueness of the world and our existence of which we have *nolens volens* to take care. Pirongrong Ramasoota Rananand examines information privacy in Thai society. Classical Buddhist teaching may not necessarily reflect the behavior of relatively secularized Buddhists in contemporary Thai society (Ramasoota, 2007, p. 125). She presents an overview of privacy and data protection in Thai legislation. The Thai public is aware of the importance of control over the circulation of one's personal information, particularly in the Internet, in order to limit state surveillance. Pattarasinee Bhattarakosol indicates that there are various factors related to the development of IT ethics in Thailand, one main factor being family background. She writes: "When ICT was implemented as a necessary tool, people became independent, self-centered, object-oriented, and careless. Therefore, most of the time, people spend time to serve their own needs more than sharing time with family members" (Bhattarakosol, 2007, p. 149). Given the fact that Thai culture is a part of religion, the author concludes that religion is the only antidote to the present ethical challenges.

27.4.1.3 Japan and the West Similar tensions can be found in Japan as Makoto Nakada and Takanori Tamura analyze in a paper that was originally conceived as a dialogue with them and myself published also in this issue of *Ethics and Information Technology* (Nakada and Tamura, 2005; Capurro, 2005b). It is a pity that the constraints of Western monologic academic culture did not allow the publication of this original dialogical essay. According to Nakada and Tamura, Japanese people live in a threefold world, namely *Shakai* or the world influenced by Western values, *Seken* or the traditional and indigenous worldview, and *Ikai*, which is a world from evils, disasters, and crime seem to emerge. On the basis of the analysis of the way an homicide was portrayed in the quality newspaper *Asahi Shimbum*, Nakada and Tamura show the ambiguities of the concept of privacy in modern Japan. They write: "while the standpoint of *Shakai* would consider publishing personal information about the Tutiura murder victims to be an invasion of privacy and violation of human rights, from the standpoint of the traditional values and beliefs of *Seken*, this publication at the same time functions as a warning against the breakdown of moral and ethics—an breakdown, finally, that is rooted in *Ikai* as the domain of such betrayal" (Nakada and Tamura, 2005, p. 28).

Living in three worlds creates a kind of discontinuous identity that is very different from Western metaphysical dichotomies, as I show in my contribution to this intercultural dialogue (Capurro, 2005b). One main difference concerns the question of "denial of self" (*Musi*), which seems to be one of the most important Buddhist values for the majority of Japanese people. This view is the opposite of the idea of Western subjectivity from which we, Westerners, derive the concepts and "intrinsic values" of autonomy and privacy. It follows from this that for Japanese people private things are less worthy than public ones. But our modern dichotomy between the public and the private sphere offers only loose parallels to the Japanese distinction between *Ohyake* (public) and *Watakusi* (private). *Ohyake* means the "big house" and refers to the imperial court and the government, while *Watakusi* means "not *Ohyake*," that is, what is partial, secret, and selfish. Japanese imported the notion of "privacy," taking it in the form of *puraibashii* as a loan word that means data privacy in the sense of "personal information" used in the West. In other words, there are two axes and they are intermixed. This is, I believe, another outstanding example of cultural hybridization that gives rise to a complex intercultural ethical analysis.

In Western societies I perceive a no less dramatic transformation of the concepts of autonomy and privacy toward what I call "networked individualities" (Capurro, 2005b, p. 40). Our being-in-the-networked-world is based not only on the principle of autonomy but also on the principle of solidarity. As an example, I present the discussion of data privacy in Germany since 1983 that led to the principle of informational privacy. Without becoming Easternized, we now speak of privacy in reference to communities, not just to isolated subjects. Behind a conceptual analysis there is the history of societies with its correspondent cognitive-emotional perceptions of the world, that is, of the web of human relations and contingent experiences, laid down in language and shared through oral and/or written traditions as the primordial medium of social cultural memory. Even if we agree on the surface of our intercultural dialogues that one concept in one culture "in some way" corresponds to the other, their factical or historical resonance is different and leads to different options about to what is considered morally good or bad.

27.4.2 Intellectual Property

Dan Burk examines the question of intellectual property from the perspective of utilitarian and deontological traditions in the United States and Europe in contrast to some non-Western approaches (Burk, 2007). In the United States "intellectual property rights are justified only to the extent that they benefit the public in general," which means that they could be eliminated "if a convincing case against public benefit could be shown" (Burk, 2007, p. 96). The industries supporting copyright usually make the case for public benefit arising from the incentives offered by such constraints. The European tradition regards creative work as reflecting the author's personality. According to Burk, two similar models of privacy regulation have emerged. The United States has adopted a *sectoral* approach, "eschewing comprehensive data protection laws in favor of piecemeal treatment of the issue," while the European Union has adopted an approach "based on comprehensive legislation, and grounded in strong, even inalienable individual rights" (Burk, 2007, pp. 97–98).

In China, the Confucian tradition largely denied the value of novel creative contribution by instead promoting the respect for the classical work that should be emulated. Under this perspective, copying becomes a cardinal virtue. For New Zealand Maori, creative works belong to the tribe or group, not to a single author. Similarly, among some sub-Saharan communities as well as in the case of many Native American tribes the control of cultural property may be restricted to certain families. In all these cases the goal of ownership is "to maintain such control, rather than to generate new works" (Burk, 2007, p. 102).In line with arguments by Lawrence Lessig, Wolfgang Coy explores the question of sharing intellectual properties in global communities from a historical point of view. Although there is a growing interest in commercially useful intellectual artifacts, there are still vast unregulated areas, for instance, native cultural practices, including regional cooking, natural healing, and use of herbal remedies (Coy, 2007).

Similar alternatives to Western individualist conceptions and practices of privacy can be found in non-Western cultures, such as the indigenous African norms based on the concept of *ubuntu* that emphasizes communal values or in Japanese norms of information access as defined by "situated community."

27.4.3 Online Communities

Wolfgang Sützl compares different conceptions of locality in the Internet on the one hand and in the emerging localized "free networks" on the other, investigating the ethical and intercultural status of both conceptions (Sützl, 2007). Free networks are guided by the idea of the commons and the principle of sharing and participating in contrast to a closed conception of location as the negation of freedom. Following Martin Heidegger and Emmanuel Lévinas, Lucas Introna argues that communities are communities because their members already share concerns or a meaning horizon of ongoing being, that is, a world. According to Introna the boundary between the insiders and the outsiders must continually remain unsettled. Virtual strangers raise the possibility of "crossing" and questioning these boundaries. But virtuality may also function to confirm them (Introna, 2007).

Frances Grodzinsky and Herman Tavani examine some pros and cons of online communities particularly with regard to the digital divide and its effects at the local level, that is, in the United States as well as in other nations such as Malawi (Grodzinsky and Tavani, 2007a).

27.4.4 Governmentality

Fernando Elitchirigoity discusses various facets of the Internet in the context of Michel Foucault's notions of "governmentality" and "technologies of the self" (Elichirigoity, 2007). He argues that the emergence of new forms of informational empowerment do not function independently from the informational practices that make them possible and, thus, need to be understood less as an absolute gain of freedom and more as the way freedom and power are continually produced and reproduced as processes of governmentality. He analyzes the significance of these tools in connection with significant changes in retirement and pension programs in the United States and other Western countries.

27.4.5 Gender Issues

Britta Schinzel criticizes common attitudes within the computer professions and the working cultures in which they develop. Alternative perspectives for responsible technological action may be derived from (feminist) situational, welfare-based close-range ethics or micro-ethics (Schinzel, 2007).

According to Johny Søraker, it is possible to broaden the moral status of digital entities in case they have become "an irreplaceable and constitutive part of someone's identity" (Søraker, 2007, p. 17). The author draws insights from Western as well as from East Asian classical philosophy.

27.4.6 Mobile Phones

Theptawee Chokvasin shows how the condition of self-government arising from hi-tech mobilization affects Thai culture (Chokvasin, 2007). Buddhism encourages us to detach ourselves from our selves, the self having no existence of its own. The Buddhist teachings of "self-adjustment" and "self-government" should not be misunderstood as if there is a "persistent person who acts as their bearer" (Chokvasin, 2007, p. 78). Autonomy means to adjust oneself to the right course of living. According to Chokvasin this Buddhist concept of autonomy can only be conceived by those who know the Buddhist teachings (*dhamma*). There is a kind of freedom in the Buddhist concept of autonomy that is related to impermanence (*Anitya*), suffering (*Duhkha*), and not-self (*Anatta*). Not clinging to our individual selves is the condition of possibility for moral behavior, that is, for "human nobility." Chokvasin claims that the mobility made possible by the mobile phone makes possible a new view on individuality as an instrumental value at the cost of disregarding the morally good.

Richard Spinello argues that all regulators, but especially those in developing countries, should refrain from imposing any regulations on IP telephony intended to protect a state-sponsored telecom and its legacy systems (Spinello, 2007).

27.4.7 Health Care

In their analysis of cross-cultural ethical issues of the current and future state of ICT deployment and utilization in healthcare, Bernd Stahl, Simon Rogerson, and Amin Kashmeery argue that the ethical implications of such applications are multifaceted and have diverse degrees of sensitivity from culture to culture (Stahl et al., 2007). They use the term "informatics" instead of information systems or computer science because it is more inclusive and socially oriented. For the purpose of this study, culture is being defined as the totality of shared meanings and interpretations. They write: "An important aspect of culture is that it has a normative function. This means that cultures contain an idea of how things should be and how its members are expected to behave. This means that they are inherently utopian and imply a good state of the world" (Stahl et al., 2007, p. 171). The normative character of culture is transmitted through morality, values as well as tenets and creeds that are called by the authors "metaethics." Cultures are deeply linked to the question of identity. The authors see a close link between culture and technology starting with agricultural cultures and, nowadays, with the importance of ICT for our culture(s). Applications of ICT in health care raise not only a policy but also an ethics vacuum that becomes manifest in the debate on values-based practice (VBP) versus evidence-based practice (EBP) of decision making. The authors analyze cases of Western and non-Western cultures in order to show the complexity of the issues they deal with. British culture is an example of Western liberalism, utilitarianism, and modernism that is fundamentally appreciative of new technologies. This modernist view overlooks the pitfalls of health care as a complex system with conflicting actors and interests. In Islamic cultures, governed by the Shari'a code of conduct, the question of, for instance, "a male healthcare provider to examine a female patient (or vice versa) are hot debate topics" (Stahl et al., 2007, p. 178). The authors present six scenarios in order to give an idea of such ethical conflicts when dealing with ICT in health care.

27.4.8 Digital Divide

Lynettee Kvasny explores the existential significance of the digital divide for America's historically underserved populations (Kvasny, 2007). According to Kvasny, the increased physical access to ICT does not signal the closure of the digital divide in the United States. She writes: "For me, the digital divide is fundamentally about evil—it is a painful discourse softened through statistics and dehumanized by numbers. [. . .] Instead of understanding the everyday practices of people who historically have been excluded from the eWorld and developing technology services and information sources to serve their unique needs, the more common response is to convert and educate the backward masses. We produce discourses that discount their values and cultures and show them why they need to catch up." (Kvasny, 2007, p. 205). In other words, she refuses the instrumental depiction of the digital divide (Britz, 2007; Himma, 2007).

27.5 CONCLUSION

IIE is an emerging discipline. The present debate shows a variety of foundational perspectives as well as a preference for the narrow view that focuses IIE on ICT. Consequently comparative studies with other media and epochs have mostly not been considered so far. With regard to IIE issues in today's information societies, there are a lot of cultures that have not been analyzed, such as Eastern Europe and the Arabic world. Asia and the Pacific is represented by Japan, China, and Thailand. Latin America and Africa are still underrepresented. I plead for the enlargement of the historical scope of our field beyond the limited horizon of the present digital info-spheres even if such a view is not an easy task for research. IIE is in this regard no less complex than, say, comparative literature.

IIE not only deals with the question of the impact of ICT on local cultures but explores also how specific ICT issues or, more generally, media issues, can be analyzed from different IIE perspectives. The present debate emphasizes the question of privacy, but other issues such as online communities, governmentality, gender issues, mobile phones, health care, and, last but not least, the digital divide are on the agenda. New issues such as blogs and wikis are arising within what is being called Web 2.0.

We have to deepen the foundational debate on the sources of morality from a IIE perspective. According to Michel Foucault, ethics can be understood not just as the theory but as the "problematization" of morality (Foucault, 1983). IIE has a critical task to achieve when it compares information moralities. This concerns the ontological or structural as well as the ontic or empirical levels of analysis. One important issue in this regard is the question of the universality of values versus the locality of cultures and *vice versa* that is related to the problem of their homogeneization or hybridization as well as the question of the relation between cognition and moods and the corresponding (un-) successful interplay between information cultures.

REFERENCES

African Information Ethics Conference (2007). Proceedings. *International Review of Information Ethics*, 7. Online: http://www.i-r-i-e.net.

ANIE (African Network for Information Ethics) (2007). Online: http://www.africainfoethics.org

Baier, K. (2006). Welterschliessung durch Grundstimmungen als Problem interkultureller Phaenomenologie. *Daseinsanalyse* 22, 99–109.

Barber, G. (2001). *Jihad vs. McWorld*. Ballentine, New York.

Bhattarakosol, P. (2007). Interactions among thai culture, ICT, and IT ethics. In: Hongladarom, S. and Ess, C. (Eds.), *Information Technology Ethics: Cultural Perspectives*. Idea Group, Hershey, PA, pp. 138–152.

Britz, J. (2007). The Internet: the missing link between the information rich and the information poor? In: Capurro, R. Frühbauer, J. and Hausmanninger, T. (Eds.), *Localizing the Internet. Ethical Aspects in Intercultural Perspective*. Fink, Munich, pp. 265–277.

Burk, D.L. (2007). In: Hongladarom, S. and Ess, C. (Eds.), *Information Technology Ethics: Cultural Perspectives*. Idea Group, Hershey, PA, pp. 94–107.

Bynum, T.W. (2006). Flourishing ethics. *Ethics and Information Technology*, 8, 157–173.

Capurro, R. (2005a). Between trust and anxiety: on the moods of information society. In: Richard, K. (Ed.), *Communication Ethics Today*. Troubadour Publishing Ltd., Leicester, pp. 187–196. Online: http://www.capurro.de/lincoln.html

Capurro, R. (2005b). Privacy: an intercultural perspective. *Ethics and Information Technology*, 7, 37–47.

Capurro, R. (2006a). Toward an Ontological Foundation of Information. *Ethics and Information Technology*, 8, 157–186.

Capurro, R. (2006b). Ethik der Informationsgesellschaft. *Ein interkultureller Versuch*. Online: http://www.capurro.de/parrhesia.html

Capurro, R. (2007a). Intercultural information ethics. In: Capurro, R. Frühbauer, J. and Hausmanninger, T. (Eds.), *Localizing the Internet. Ethical Aspects in Intercultural Perspective*. Fink, Munich, pp. 21–38.

Capurro, R. (2007b). Information Ethics for and from Africa. *International Review of Information Ethics*, 7 Online: http://www.capurro.de/africa.html

Capurro, R., Scheule, R.M. (2007). Fruit, water, and philosophy. Intercultural perspectives on the Web. In: Capurro, R. Frühbauer, J. and Hausmanninger, T. (Eds.), *Localizing the Internet. Ethical Aspects in Intercultural Perspective*. Fink, Munich, pp. 323–332.

Capurro, R., Frühbauer, J. and Hausmanninger, T. (Eds.) (2007). Localizing the Internet. *Ethical Aspects in Intercultural Perspective*. Fink, Munich.

Chokvasin, T. (2007). In Hongladarom, S. and Ess, C. (Eds.) *Information Technology Ethics: Cultural Perspectives*. Idea Group, Hershey, PA, pp. 68–80.

Coy, W. (2007). On sharing ideas and expressions in global communities. In: Capurro, R. Frühbauer, J., and Hausmanninger, T. (Eds.), *Localizing the Internet. Ethical Aspects in Intercultural Perspective*. Fink, Munich, pp. 279–288.

Damasio, A. (1994). *Descartes' Error: Emotion, Reason, and the Human Brain*. Putnam Publishing.

Dreyfus, H.L. (1991). *Being-in-the-World. A Commentary on Heidegger's Being and Time*, Division I. The MIT Press.

Eldred, M. (2006). *Technology, Technique, Interplay: Questioning Die Frage nach der Technik*. Online: http://www.webcom.com/artefact/untpltcl/tchniply.html

Elichirigoity, F. (2007). The internet, information machines, and the technologies of the self. In: Capurro, R., Frühbauer, J. and Hausmanninger, T. (Eds.), *Localizing the Internet. Ethical Aspects in Intercultural Perspective*. Fink, Munich, pp. 289–300.

Ess, C. (2005). Lost in translation?: Intercultural dialogues on privacy and information ethics (Introduction to special issue on privacy and data privacy in Asia). *Ethics and Information Technology*, 7, 1–6.

Ess, C. (2006). Ethical pluralism and global informaion ethics. *Ethics and Information Technology*, 8, 215–226.

Ess, C. (2007). Can the local reshape the global? ethical imperatives for humane intercultural communication online. In: Capurro, R., Frühbauer, J., and Hausmanninger, T. (Eds.), *Localizing the Internet. Ethical Aspects in Intercultural Perspective*. Fink, Munich, pp. 153–169.

Figueroa, A. and Hugo, A. (2007). Collective construction of identity in the internet: ethical dimension and intercultural perspective. In: Capurro, R., Frühbauer, J., and Hausmanninger, T. (Eds), *Localizing the Internet. Ethical Aspects in Intercultural Perspective*. Fink, Munich, pp. 229–241.

Finquelievich, S. (2007). A toolkit to empower communities in Latin America. In: Capurro, R., Frühbauer, J. and Hausmanninger, T. (Eds.), *Localizing the Internet. Ethical Aspects in Intercultural Perspective*. Fink, Munich, pp. 301–319.

Floridi, L. (2006). Information technologies and the tragedy of the Good Will. *Ethics and Information Technology*, 8, 253–262.

Floridi, L. and Savulescu, J. (2006). Information ethics: agents, artefacts and new cultural perspectives. *Ethics and Information Technology*, 8, 155–156.

Foucault, M. (1983). *Discourse and Truth: The Problematization of Parrhesia*. University of California at Berkeley, October–November 1983. Online: http://foucault.info/documents/parrhesia/ (visited on May 30, 2007).

Froehlich, T. (2004). *A brief history of information ethics*. In Textos universitaris de biblioteconomia i documentacio. Nr. 13. Online: http://www.ubes/bid/13froel2.htm (visited on May 30, 2007).

Frohmann, B. (2007). Foucault, Deleuze, and the ethics of digital networks. In: Capurro, R., Frühbauer, J. and Hausmanninger, T. (Eds.), *Localizing the Internet. Ethical Aspects in Intercultural Perspective*. Fink, Munich, pp. 57–68.

Gendlin, E.T. (1978). Befindlichkeit: (1) Heidegger and the Philosophy of Psychology. *Review of Existential Psychology & Psychiatry: Heidegger and Psychology*. Vol. XVI, pp. 1–3. Online: http://www.focusing.org/gendlin_befindlichkeit.html (visited on May 30, 2007).

Grodzinsky, F.S. and Tavani, H.T. (2007a). The internet and community building at the local and global levels: some implications and challenges. In: Capurro, R., Frühbauer, J. and Hausmanninger, T. (Eds.), *Localizing the Internet. Ethical Aspects in Intercultural Perspective*. Fink, Munich, pp. 135–149.

Grodzinsky, F.S. and Tavani, H.T. (2007b). Online communities, democratic ideals, and the digital divide. In: Hongladarom, S. and Ess, C. (Eds.), *Information Technology Ethics: Cultural Perspectives*, Idea Group, Hershey, PA, pp. 20–30.

Hausmanninger, T. (2007). Allowing for difference. some preliminary remarks concerning intercultural information ethics. In: Capurro, R., Frühbauer, J., and Hausmanninger, T. (Eds.), *Localizing the Internet. Ethical Aspects in Intercultural Perspective*. Fink, Munich, pp. 39–56.

Heidegger, M. (1987). *Being and Time*. Translated by John Macquarrie & Edward Robinson, Basil Blackwell, Oxford.

Herdin, T., Hofkirchner, W., and Maier-Rabler, U. (2007). In: Hongladarom, S. and Ess, C. (Eds.), *Information Technology Ethics: Cultural Perspectives*. Idea Group, Hershey, PA, pp. 54–67.

Himma, K.E. (2007). The information gap, the digital divide, and the obligations of affluent nations. *International Review of Information Ethics*, 7.

Hiruta, K. (2006). What pluralism, why pluralism, and how? A response to Charles Ess. *Ethics and Information Technology*, 8, 227–236.

Hongladarom, S. (2001). *Cultures and Global Justice*. Online: http://www.polylog.org/them/3/fcshs-en-htm (visited on May 30, 2007).

Hongladarom, S. (2007). Analysis and justification of privacy from a Buddhist perspective. In: Hongladarom, S., and Ess, C. (Eds.), *Information Technology Ethics: Cultural Perspectives*. Idea Group, Hershey, PA, pp. 108–122.

Hongladarom, S. and Ess, C. (Eds.), (2007a). *Information Technology Ethics: Cultural Perspectives*. Idea Group, Hershey, PA.

Hongladarom, S. and Ess, C. (Eds.) (2007b). Preface. *Information Technology Ethics: Cultural Perspectives*. Idea Group, Hershey, PA, pp. xi–xxxiii.

International Review of Information Ethics (IRIE) (2004). *International ICIE Symposium. Online*: http://www.i-r-i-e.net.

Introna, L.D. (2007). Virtual Strangers—on the social and ethical conditions of virtual communities. In: Capurro, R., Frühbauer, J., and Hausmanninger, T. (Eds.), *Localizing the Internet. Ethical Aspects in Intercultural Perspective*. Fink, Munich, pp. 95–108.

Jackson, W. and Mandé, I. (2007). New Technologies and Ancient Africa: the impact of new information and communication technologies in sub-Saharian Africa. In: Capurro, R., Frühbauer, J., and Hausmanninger, T. (Eds.), *Localizing the Internet. Ethical Aspects in Intercultural Perspective*. Fink, Munich, pp. 171–176.

Jullien, F. (1995). *Fonder la Morale*. Grasset, Paris.

Kitiyadisai, K. (2005). Privacy rights and protection: foreign values in modern Thai context. *Ethics and Information Technology*, 7, 17–26.

Kuhn, T.S. (1962). *The Structure of Scientific Revolutions*. The University of Chicago Press.

Kvasny, L. (2007). The Existential significance of the digital divide for America's historically underserved populations. In: Hongladarom, S. and Ess, C. (Eds.), *Information Technology Ethics: Cultural Perspectives*. Idea Group, Hershey, PA. pp. 200–212.

Ladd, J. (1985). The quest for a code of professional ethics: an intellectual and moral confusion. In: Johnson, D.G. and Snapper, J.W. (Eds.), *Ethical Issues in the Use of Computers*. Belmont, CA, pp. 8–13.

Lévinas, E. (1968). Totalité et Infini. *Essai sur l'extériorité*. Nijhoff, The Hague.

Lü, Y. (2005). Privacy and data privacy in contemporary China. *Ethics and Information Technology*, 7, 7–15.

Lü, Y. (2007). Globalization and informatioan ethics. In: Capurro, R., Frühbauer, J., and Hausmanninger, T. (Eds.), *Localizing the Internet. Ethical Aspects in Intercultural Perspective*. Fink, Munich, pp. 69–73.

Luhmann, N. (1990). Paradigm lost. *Über die ethische Reflexion der Moral*. Suhrkamp, Frankfurt am Main.

Magnani, L. (2007). The Mediating Effect of Material Cultures as Human Hybridization. Online Communities, Democratic Ideals, and the Digital Divide.

Nakada, M. (2007). The Internet within *Seken* as an old and indigenous world of meanings in Japan. In: Capurro, R., Frühbauer, J., and Hausmanninger, T. (Eds.), *Localizing the Internet. Ethical Aspects in Intercultural Perspective*. Fink, Munich, pp. 177–203.

Nakada, M. and Tamura, T. (2005). Japanese conceptions of privacy: an intercultural perspective. *Ethics and Information Technology*, 7, 27–36.

Nath, V. (2007). Digital governance models: towards empowerment and good governance in developing countries. In: Capurro, R., Frühbauer, J., and Hausmanninger, T. (Eds.), *Localizing the Internet. Ethical Aspects in Intercultural Perspective*. Fink, Munich, pp. 77–94.

Nietzsche, F. (1999). Menschliches, Allzumenschliches. In: Giorgio, C. and Mazzino, M. (Eds.), *Kritische Studienasugabe*. Vol. 2. dtv, Munich.

Nietzsche, F. (1999). Nachgelass 1884–1885. In: Giorgio, C. and Mazzino, M. (Eds.), *Kritische Studienausgabe*, Vol. 11. dtv, Munich.

Nishigaki, T. (2006). The ethics in Japanese information society: consideration on Fracisco Varela's *The Embodied Mind* from the perspective of fundamental informatics. *Ethics and Information Technology*, 8, 237–242.

Nishigaki, T. and Takenouchi, T. (Eds.) (2007). *The Thought of Information Ethics*. Communis 05 (in Japanese) Tokyo.

Olinger, H.N., Britz, J.J., and Olivier, M.S. (2005). Western privacy and ubuntu: influences in the forthcoming data privacy bill. In: Philip, B., Frances, G., and Lucas, I. (Eds.), *Ethics and New Information Technology*, CEPE 2005, Enschede, The Netherlands, pp. 291–306.

Paterson, B. (2007). We cannot eat data: the need for computer ethics to address the cultural and ecological impacts of computing. In: Hongladarom, S. and Ess, C. (Eds.), *Infor-mation Technology Ethics: Cultural Perspectives*. Idea Group, Hershey, PA, pp. 153–168.

Pimienta, D. (2007). At the boundaries of ethics and cultures. Virtual communities as an open ended process carrying the will for social change (the "MISTICA" experience"). In: Capurro, R., Frühbauer, J., and Hausmanninger, T. (Eds.), *Localizing the Internet. Ethical Aspects in Intercultural Perspective*. Fink, Munich, pp. 205–228.

Ramasoota Rananad, P. (2007). Information privacy in a surveillance state: a perspective from Thailand. In: Hongladarom, S. and Ess, C. (Eds.), *Information Technology Ethics: Cultural Perspectives*. Idea Group, Hershey, PA, pp. 124–137.

Ramose, M.B. (2002). Globalization and *ubuntu*. In: Coetzee, P. and Roux, A. *Philosophy from Africa. A Text with Readings*. 2nd edition. Oxford University Press, pp. 626–650.

Robertson, R. (1992). *Globalization*. Sage, London.

Roth, K. (2001). Material culture and intercultural communication. *International Journal of Intercultural Communication*, 25, 563–580.

Schinzel, B. (2007). Gendered views on the ethics of computer professionals. In: Capurro, R., Frühbauer, J., and Hausmanninger, T. (Eds.), *Localizing the Internet. Ethical Aspects in Intercultural Perspective*. Fink, Munich, pp. 121–134.

Sontag, S. (2004). Die Welt als Indien. In Lettre International, September, Vol. 65, 82–86.

Søraker, J.H. (2007). The moral status of information and information technologies: a relational theory of moral status. In: Hongladarom, S. and Ess, C. (Eds.), *Information Technology Ethics: Cultural Perspectives*. Idea Group, Hershey, PA, pp. 1–19.

Spinello, R.A. (2007). Laissaz-Faire or regulation? Social and policy implications for IP telephony. In: Capurro, R. Frühbauer, J., and Hausmanninger, T. (Eds.), *Localizing the Internet. Ethical Aspects in Intercultural Perspective*. Fink, Munich, pp. 109–119.

Stahl, B.C., Rogerson, S., and Kashmeery, A. (2007). Current and future state of ICT development and utilization in healthcare: an anaylsis of cross-cultural ethical issues. In: Hongladarom, S. and Ess, C. (Eds.), *Information Technology Ethics: Cultural Perspectives*. Idea Group, Hershey, PA, pp. 169–183.

Sützl, W. (2007). Internet and free networks. From world-networking to place-networking. In: Capurro, R., Frühbauer, J. and Hausmanninger, T. (Eds.), *Localizing the Internet. Ethical Aspects in Intercultural Perspective*. Fink, Munich, pp. 243–262.

Takenouchi, T. (2006). Information ethics as information ecology: connecting Frankl's thought and fundamental informatics. *Ethics and Information Technology*, 8, 187–193.

Tshwane Declaration on Information Ethics in Africa (2007). Online: http://www.africain-foethics.org/tshwanedeclaration.html

van der Velden, M. (2007) In: Hongladarom, S. and Ess, C. *Information Technology Ethics: Cultural Perspectives*. Idea Group, Hershey, PA, pp. 81–93.

Walzer, M. (1994). *Thick and Thin: Moral Arguments at Home and Abroad*. University of Notre Dame Press, Notre Dame.

Weckert, J. (2007). Foreword. In: Hongladarom, S. and Ess, C. (Eds.), *Information Technology Ethics: Cultural Perspectives*. Idea Group, Hershey, PA, p. x.

Welsch, W. (1999). Transculturality—The puzzling form of cultures today. In: Featherstone, M. and Lash, S. (Eds.), *Spaces of Culture: City, Nation, World*. Sage, London, pp. 194–213. Online: http://www2.uni-jena.de/welsch/Papers/transcultSociety.html (visited on May 30, 2007).

Wieviorka, M. (2003). *Kulturelle Differenzen und kollektive Identitäten*. Hamburger Edition, Hamburg.

Wittgenstein, L. (1984). Zu Heidegger. In: McGuiness, B.F. (Ed.), *Ludwig Wittgenstein und der Wiener Kreis. Gespräche, aufgezeichnet von Friedrich Waismann*. Suhrkamp, Frankfurt am Main.

Wittgenstein, L. (1989). *Vortrag über Ethik*. Suhrkamp, Frankfurt am Main.

Wurman, R. (2001). *Information Anxiety 2*. Indianapolis, IN.

Yamamoto, G.T. and Karaman, F. (2007). Business Ethics and Technology in Turkey: An emerging country at the crossroad of civilizations. In: Hongladarom, S. and Ess, C. (Eds.), *Information Technology Ethics: Cultural Perspectives*. Idea Group, Hershey, PA, pp. 184–199.

INDEX

The Handbook of Information and Computer Ethics Edited by Kenneth Einar Himma and Herman T. Tavani